THE LORE OF
FLIGHT

THE LORE OF FLIGHT

World copyright © 1971,
fully revised and updated 1990,
AB Nordbok, Box 7095
402 32 Gothenburg, Sweden

Mallard Press and its accompanying design
and logo are trademarks of BDD
Promotional Book Company, Inc.

ISBN 0-792-45413-8

THE LORE OF
FLIGHT

MALLARD
PRESS

BDD Promotional Books Company, Inc.
666 Fifth Avenue
New York, N. Y. 10103

THE LORE OF FLIGHT is the result of international co-operation between aviation, technical and other specialists living and working thousands of miles apart, as shown on the map above where each spot marks the location of one or more of the contributors. Such international co-operation would not have been possible without the rapid communication which flying has made available, continuously shortening travel times and bringing people closer together. This revised edition is supervised by Peter Billing, a producer of scientific and technical programs for Swedish Television, and a devoted amateur aviator, who has published several technical articles in the aviation field.

CHIEF ADVISER AND SUPERVISING EDITOR

JOHN W. R. TAYLOR

participated closely with the publishers from the planning stage to completion of The Lore of Flight, *and guided the team of writers, artists and contributors. He wrote the historical section and has been responsible for the general editing of the book.*
Mr. Taylor's technical background includes seven years as a member of Sir Sydney Camm's design team at Hawker Aircraft Limited. He is editor and chief compiler of the international yearbook Jane's All the World's Aircraft, *consultant editor of Air BP, and author of more than 215 books on aviation and space flight. He is a Fellow of the Royal Aeronautical Society, Royal Historical Society, a Member of the Society of Licensed Aircraft Engineers and Technologists, and an Associate Fellow of the Royal Aeronautical Society.*

MEMBERS OF THE MAIN EDITORIAL TEAM:

MAURICE ALLWARD

of the Technical Publications Department of British Aerospace, started his career in the Design Office at Hawker Aircraft Limited. He shares responsibility for the production of technical servicing handbooks for such aircraft as the BAE 146 airliner, the BAES 125 business jet and the A300 series of European Airbuses. He is a member of the British Interplanetary Society, and several of his published books have been on the subject of rockets and space travel.
He wrote the sections on aircraft structures and on flying lightplanes and gliders, and collaborated in the parts dealing with airports and space.

BILL GUNSTON

is Technology Editor of *Science Journal*. He was a pilot in the RAF and has since built a reputation as a technical aviation journalist and writer, with over 100 major books to his credit including the latest edition of *Flight Handbook*. He wrote the section on aero-engines and equipment and collaborated in the production of the index, and he worked closely with the artists to ensure complete accuracy of the technical illustrations throughout the book, for which he wrote many of the captions.

KENNETH MUNSON

has specialized for many years in the study of manned aircraft and is the author of some three dozen books on the subject. He is the assistant editor of *Jane's All the World's Aircraft* and is author and editor of the international series of aviation handbooks known as the *Pocket Encyclopaedia of World Aircraft in Color*.
He wrote nearly half of, and edited the complete text of, the illustrated Encyclopaedic Index.

JOHN D. R. RAWLINGS

served in the RAF as a pilot during and after World War Two, and later entered Holy Orders in the Church of England. He has made a particular study of British military aviation, past and present, and writes on this and other aviation topics. He has been awarded the C. P. Robertson Memorial Trophy for 1983 for his work in this field.
He did a great deal of the preparatory and checking work for, and wrote part of the text of, the Index.

The publisher wishes to give special thanks to the following experts from many countries, who have contributed information and illustration material to this book:

H. G. ANDERSSON
Sweden, Public Relations Manager, Saab-Scania AB.

GUNNAR ATTERHOLM
Sweden. Air traffic control officer.
Captain, Royal Swedish Air Force (Reserve).

ARTHUR BLAKE
South Africa. Historian.

PETER M. BOWERS
USA. Engineer with Boeing. Aviation historian.

STANLEY BROGDEN
Australia. Former Director of Public Relations for The Royal Australian Air Force. Author of several books including the standard History of Australian Aviation. He is associate editor of the magazine *Aircraft*.

B. BRUNNERFUP
Sweden. Former test pilot and flight instructor.

J. B. CYNK
UK. Historian specializing in Polish Aviation.

WALTER DOLLFUS
Switzerland. Former aviation historian.

BART van der KLAAUW
Netherlands. Aviation journalist.

JEAN LIRON
France. Aviation historian.

HARRY McDOUGALL
Canada. Aviation journalist.

DAVID MONDEY
UK. Aviation historian.

VACLAV NEMECEK
Czechoslovakia. Aviation historian and journalist.

VICO ROSASPINA
Italy. Aviation consultant. Former senior officer in the Italian Air Force.

EIICHIRO SEKIGAWA
Japan. Aviation commentator.

PIERRE SPARACO
Belgium. Editor in chief of *Aviation et Astronautique*.

DARIO VECINO
Spain. Aviation journalist.

MANO ZIEGLER
Germany. Wartime pilot of Me 163 rocket fighter, former editor of *Flugwell*, and former Public Relations Manager of *Deutsche Airbus* GmbH. Author of several major works.

JOHN W. WOOD

ILLUSTRATIONS

The technical drawings have been produced under the leadership of John Wood, who worked closely with the supervising editor and the editorial team.
John Wood was earlier a member of the advanced Projects Group, which was responsible for all new designs for the entire Hawker Siddeley Group. Today he and his team of artists have built up a considerable reputation in the field of Technical Illustration, and have produced more than 150 books, which have been sold in many different countries including America, France, Spain, Germany, Italy, Sweden and Japan. The illustrations in this book pay particular attention to the authentic detail and technical accuracy of the periods.

ARTISTS

Apart from John W. Wood the following aviation artists have been responsible for the technical drawings:

NORMAN DINNAGE
BRIAN HILEY
WILLIAM HOBSON
TONY MITCHELL
JACK PELLING

The other illustrations were drawn by
AKE GUSTAVSSON
and other members of the Tre Tryckare studio.
Test and illustrations were supervised and coordinated by **WERA FAHLSTROM**

Acknowledgement is also made to the following for advice and material:

Aermacchi, Italy
Avions Marcel Dassault, France
Janusz Babiejczuk, Instytut Lotnictwa, Poland
Beagle Aircraft, UK
Bell Aerosystems, USA
BOAC, UK
Boeing Company, USA
British Aircraft Corporation, UK
British Airports Authority, UK
British United Airways, UK
Canadair, Canada
Decca Navigator, UK
Dornier, Germany
H. C. Ellehammer, Denmark
Henri Fabre, France
Fiat, Italy
Flight International, UK
Fokker-VFW, Netherlands
General Dynamics, USA
Hawker Siddeley Aviation, UK
Library of Congress, USA
Lockheed Aircraft Corporation, USA
G. E. Lowdell, UK
McDonnell Douglas Corp., USA
Messerschmitt-Bölkow-Blohm, Germany
Ministry of Defence, UK
Musée de l'Air, France
National Aeronautics and Space Administration, USA
Nord-Aviation, France
North American Rockwell, USA
Novosti Press Agency, USSR
Rolls-Royce, UK
Saab, Sweden
SAS, Sweden
Short Bros & Harland, UK
Smiths Industries, UK
Society of British Aerospace Companies, UK
John Stroud, UK
Sud-Aviation, France
Tass Agency, USSR
Union Syndicale des Industries Aéronautiques et Spatiales, France
United Aircraft Corporation, USA
US Information Services, UK
The Lord Ventry, UK
VFW-Fokker, Germany

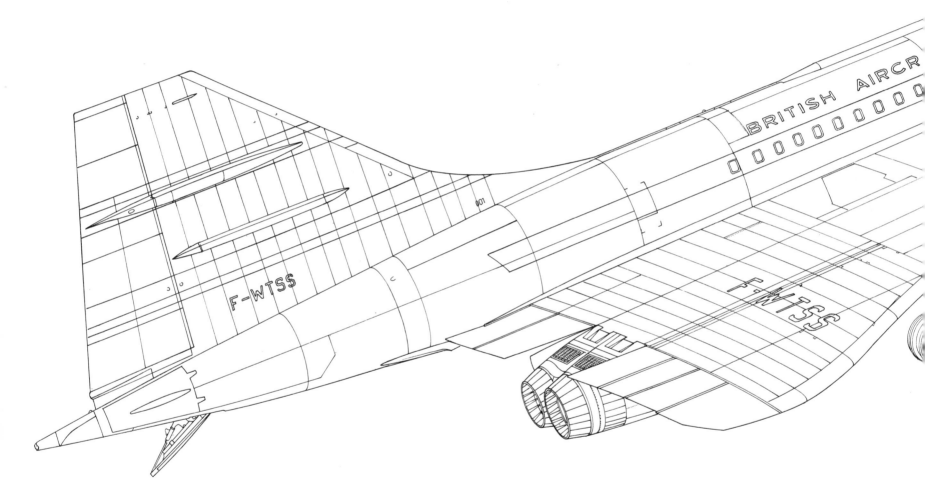

Concorde, a supersonic
airliner, will enter airline
service during 1975 and be
capable of carrying more
than 100 passengers at
twice the speed of present
day long-haul aircraft.
Constructed jointly by
Aerospatiale and the
British Aircraft
Corporation, it first flew in
November 1969. It has
many of the new
aerodynamic features
described in this book,
which resulted from an
exacting research and
development programme,
with trials spanning 12
years.
The Concorde was
ordered only by Air
France and British
Airways, and a limited
number of 16 operational
planes was built.
Though the Concorde
was a commercial
miscalculation, this
remarkable aircraft has
earned a significant place
in the history of air travel.

Some sixty-five years ago on the sand dunes at Kitty Hawk, North Carolina, man realized the fulfilment of a dream that he had coveted for centuries. That event was destined to change the course of history.

From this first faltering flight by Wilbur and Orville Wright to the jet-powered giants of today which hurdle oceans and continents in a few short hours, the story of flight is a fascinating one.

It is a story of failure, disappointment, and even tragedy. But it is also a story of courage and determination and eventual triumph. It has all the drama, suspense and excitement of the best fiction.

Having experienced the trials of a pioneer in aviation, designing, building and flying multi-engine airplanes, flying-boats and helicopters, I have lived this story personally and am proud and grateful for having been a part of it.

Naturally, I am always delighted to see a book on any phase of my chosen field. I am particularly stimulated by this one, and I am privileged to introduce it with these words.

It has the distinction of being a comprehensive treatment, not only of the history of aviation, but of virtually all facets of the structure and flight characteristics of airplanes. Each section of the book has been written by a specialist in his field, and beautifully illustrated.

IGOR SIKORSKY

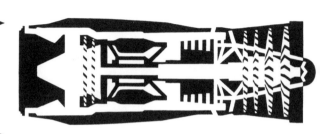

HISTORICAL

In this section of *Lore of Flight* we see how, from the earliest times, man longed to fly like the birds. One of the most magical faculties with which he could invest his pagan gods was the ability to fly. A few highly-favoured men of legend and mythology, such as Daedalus and Icarus, were credited with sharing the secret of the gods; but real-life "bird-men" who tried to copy the gods and supermen were lucky indeed if they emerged with nothing worse than broken limbs and dampened ardour.

At last, in 1783, a means of becoming airborne was found. For a century after that balloons, drifting at the mercy of the wind, represented the limit of human achievement in the air. Powered, navigable airships followed, while more ambitious, daring or scientifically-minded men experimented with primitive powered aeroplanes and gliders.

On December 17, 1903, the Wright brothers made their first brief flights in a powered aeroplane, little realising that within the lifetime of many then born the birds would be left far behind and below as men outpaced the Sun and reached for the Moon.

STRUCTURES

Today aeroplanes can fly faster than the speed of sound, carry hundreds of passengers in armchair comfort over the widest oceans and land automatically under electronic control in bad weather. Some take off and land vertically, others change the angle of their wings in flight.

The simple wood and canvas aeroplanes of the pioneers have given way to the most complex metal structures ever produced—packed with equipment and miles of pipelines and cables reminiscent of the heart, brain, arteries and nerves of a living body.

In this section of *Lore of Flight*, we look beneath the smooth skin of modern aeroplanes and learn how they work. In doing so, we begin to realise not only why airliners are so safe and efficient but why they often cost many millions of pounds or dollars to buy.

ENGINES AND EQUIPMENT

Had the petrol-engine been invented earlier, the Wright brothers might not have been first to fly a powered aeroplane. Generations of pioneers before them had been frustrated by the lack of a powerful, lightweight engine.

In this section we see how the progress of powered flight has been dictated by advances in power plant design. Not until more reliable piston-engines like the Rolls-Royce Eagle became available could airmen face the unknown perils of flying across great oceans. Supercharging took them over the highest mountains. The coming of the jet-engine brought new speed and comfort to the airlines and, together with the rocket-engine, enabled pilots to overcome the "sound barrier" so that aeroplanes can now fly faster than the speed of sound as everyday routine.

Here the history of aero-engine development is given new meaning by cutaway drawings which show how different types of aero-engine are built and how they operate. No pages of *Lore of Flight* reflect better the skill of both the engineers who created the engines and the artists who illustrated our book.

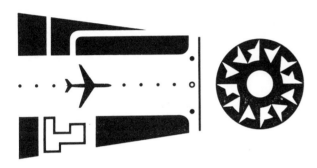

FLYING

There is a world of difference between flying and air travel. The men and women who pilot light aeroplanes and gliders come nearest to achieving the ambition of the pioneers who dreamed of flying like the birds, wheeling and soaring gracefully and effortlessly for the sheer enjoyment of flight itself.

In this part of *Lore of Flight* we are introduced first to the lighter, simpler, more personal types of flying machine. We learn how they are made and operated, and a little of what they are capable of in terms of aerobatics.

In sharp contrast, we can follow the story of a typical transatlantic flight by a large jet airliner, from the moment when its crew is briefed at Prestwick until the aircraft approaches the John F. Kennedy International Airport for its landing in New York, We see how intricate navigation aids, weather ships, and a host of other equipment and services, all play their part in making transatlantic flying one of the safest forms of modern transport, carrying millions of people annually between the Old and New Worlds.

ENCYCLOPAEDIC INDEX

This Index does more than tell us what is where in the earlier sections of *The Lore of Flight*. Its more than 1,550 entries supplement the main text and illustrations with detailed explanations and added facts which make the Index itself a unique and valuable work of reference. Important military aircraft are shown in colour, as are the symbols of the world's leading civil airlines. Other entries detail the major airports of the world and show the pattern of their runways. Chronologies outline milestones in the flying history of many nations, and pay tribute to the men and women who pioneered aviation in every corner of the Earth. Many of the photographs are unique; they come from the personal files of leading air historians of many countries, and some have never before been published.

PICTURE INDEX

The large drawing on these pages has been prepared to enable a reader totally unfamiliar with aircraft to learn the names of some of the main components of a modern aeroplane. These items are numbered, and their identity is given in the correspondingly numbered list on the facing page. In addition, selected items in this list also have references (the numbers in brackets) to columns of text in the main part of this book where the subject is dealt with in greater detail. For example, the equipment in the extreme nose of the Short Belfast is a search radar, and such radar equipment is described in columns 707, 1391 and 1516-18. But this list of key numbers is by no means the only index to this book. The main index begins on page 299 and is of an encyclopaedic nature. Having found from this drawing what an item is called it is generally possible to look it up in the alphabetically arranged index and learn much more about it, before plunging into the main body of the book at all.

Subject of the drawing is the Short Belfast, for the carriage of heavy freight, including the largest types of guns, vehicles, guided missiles and other equipment used by the British Army and Royal Air Force.

As a troop transport it could carry up to 250 men; as a missile carrier its payloads include a wide range of weapons, from strategic offensive missiles to light tactical rockets. Small missiles, such as the Thunderbird and Bloodhound, can be crated and stowed in quantity.

Large "beaver-tail" doors at the rear extend the full width and height of the hold, for easy loading and to permit the air dropping

of heavy loads. Loads which cannot be driven into the hold can be taken on board by winch and a system of roller conveyors. The distinctive strakes on the underside of the rear fuselage were added as part of a drag-reduction programme to increase the aircraft's range.

Power is provided by four 5,730-ehp Rolls-Royce Tyne turboprops, each driving a fully-feathering propeller with four aluminium blades of 16 ft diameter.

The military career of The Short Belfast in the RAF was ended at the beginning of the 1980s but a couple were converted to civil transporters for heavy-lift cargo airlines.

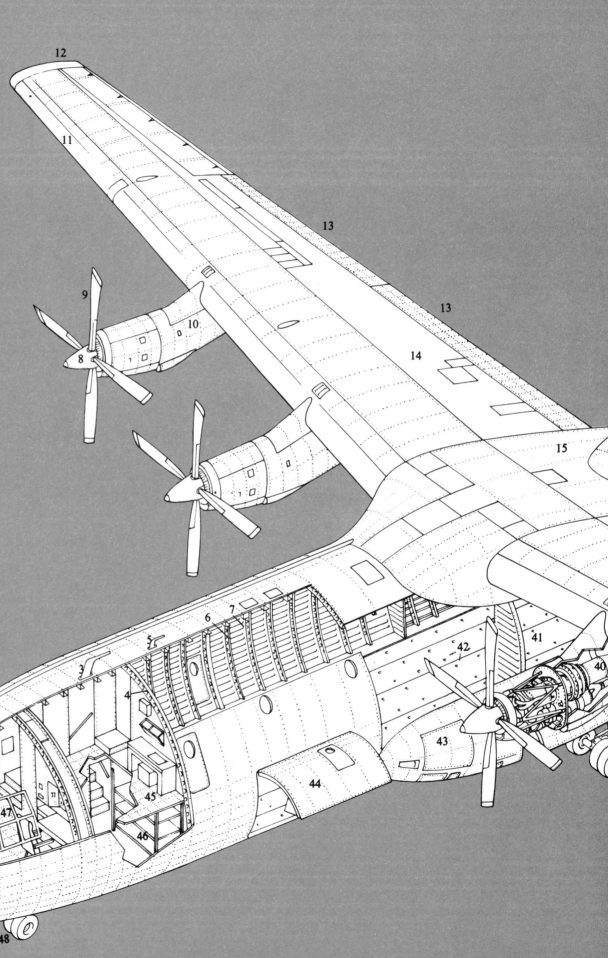

1 Search radar (707, 1391, 1516-18)
2 In-flight refuelling probe
3 VHF aerial
4 U-section frame (397)
5 UHF aerial
6 U-section frame (397)
7 Z-section stringer (397)
8 Spinner (1343)
9 16 ft dia. propeller (1321 et seq)
10 Nacelle (817)
11 Detachable leading-edge
12 Detachable wingtip
13 Flaps (four sections) (613-18)
14 Wing (mainplane)
15 Centre-section wing/fuselage fairing
16 Fuselage
17 HF stand-by
18 Tailplane (627)
19 Elevator (609 et seq)
20 VOR aerial (1373)
21 Detachable leading-edge
22 Fin (627)
23 Beacon
24 Rudder (609 et seq)
25 Trim tabs (627)
26 Door (769 et seq)
27 Emergency exit
28 Box-section frame
29 Ramp jack (781 et seq)
30 Ramp (781 et seq)
31 Wing rear spar
32 Aileron (linked to rudder) (625)
33 Trim tabs (627)
34 Wing front spar
35 Petal cowling
36 Engine mounting
37 Rolls-Royce Tyne (1285)
38 Main landing gear of bogie type (571)
39 Disc brakes
40 Jet pipe (1171)
41 Main spar frame
42 Lashing points
43 Blister for retracted landing gear
44 Front freight hold hatch
45 Aircraft galley and crew rest area
46 Off-duty crew sleeping quarters
47 Flight deck (661)
48 Nosewheels (571)

INTRODUCTION

Man has always sought to see what lies beyond the mountains. The hope of finding better hunting grounds, the need to flee from natural catastrophes, pressure of population, not to mention curiosity, all have helped to mould our natural urge to travel. So it is that the history of mankind is closely bound up with the development of various means of transport. All major inventions, all forms of production, are dependent for their exploitation and growth on effective communications. The Industrial Revolution, for instance, would have been inconceivable without a corresponding advance in transport techniques. But the historic road was long and the journey slow.

At first man had only his legs. Then there came the dugout canoe for use on rivers and lakes or along the coasts. Thousands of years were to pass before animals were domesticated for riding and before the wheel was invented. The use of wheeled vehicles drawn by draught animals permitted slightly faster travel over the rude tracks connecting settlements, but their chief importance lay in providing a means of transporting goods with less expenditure of effort.

Development of travelling facilities was a very slow process and nearly every advance worth the name has happened during the past 150 years. Take Marco Polo, that famous Venetian traveller of the Middle Ages. In the spring of 1275 he reached Changtu where the Court of the Chinese Emperor Kubla Khan was established. A fast modern aircraft can do this journey in five hours—but it took Marco Polo five whole years. And his journey home shows what such slow travel can lead to. He was commissioned to escort the Kubla Khan's daughter to Persia where she was to marry the Regent Argun Khan. They travelled by sea, following the coast all the way, but even then it took four years to reach their destination. And by this time the Argun Khan was dead.

Marco Polo was in a quandary. He could hardly leave the princess to fend for herself in Persia, nor did he feel it proper to bring her back with him to his own home in Venice. Fortunately, the Gordian knot was cut by events, for the Argun Khan's son, having reached a marriageable age, found no objection to wedding the Chinese princess who ought to have been his stepmother.

As recently as 250 years ago things were not all that much better. The average speed of a passenger coach rarely exceeded five miles an hour, and in such conditions it is scarcely surprising that the word "communications" was not yet to be found in the dictionary.

The great migrations of peoples in prehistoric and early historical times, that formed the pattern for Europe's present-day ethnic distribution, had their corollary during the 19th century. There was a new continent to be peopled. Poverty and want led some 200,000 persons from England, Scotland and Ireland, more than 100,000 from Scandinavia, and corresponding quotas from Germany and most other European countries, to emigrate during that century and seek a new life in Canada and the USA.

The conditions under which they travelled would have had little appeal for us today. First, a jolting journey by horse and cart over appalling roads to reach their port of departure, then embarkation in a sailing ship.

Such ships were very small by modern standards and, with hundreds of passengers and their humble baggage on board, space was insufferably cramped. Sanitary conditions were indescribable. The longer the voyage continued, the worse became the food; disease was rampant and deaths a commonplace. Many thoughtful Ship's Masters used to bring a small store of earth—it meant so much to these poor emigrants to see their nearest and dearest committed to the deep with a handful of soil from their own home country.

With favourable winds the voyage could be accomplished in ten weeks and it is interesting to remark that it took Columbus, more than 300 years earlier, no longer than that to reach the Bahamas from Spain. So much for progress!

To travel for pleasure was quite unthinkable, at least for ordinary people. But those who did travel had at any rate plenty of time to admire the countryside, always provided they were tough enough to overcome the hazards and difficulties that travel in those days involved.

Yet birds travelled easily enough. Why couldn't humans fly? To say that flight was only meant for birds, and maybe for gods, was all very well, but then . . . if a humble little creature like a sparrow could fly, surely the powerful arms and shoulders of man could also manage to work a pair of wings?

It was a fatal deduction. For centuries, men leapt to their deaths from cliffs and towers, vainly flapping wings made to resemble those of a bird. Only the mythological Daedalus achieved success, because he was a myth. Even royalty dabbled

continued on page 18

Many hundreds of years ago hardly anybody travelled further than a mile or two from their homes; many were born, lived and died without ever visiting the next village. Only a compelling pressure, such as religious persecution, a religious crusade or sheer starvation, caused families to tear up their roots and undertake a major journey. There was no question of doing such a thing lightly; a long journey was a formidable prospect, likely to occupy months or years. Here a family has at last decided that economic pressure forces a move to a distant place. The grandparents waving goodbye expect never to see their children and grandchildren again. And who knows what difficulties lie ahead for the travellers? How will they find their way? How will they avoid being robbed of all their possessions? How can they be sure of finding prosperity at the end of it all? Truly, in those days the world seemed very large, and usually very hostile.

About 130 years ago the number of people undertaking long journeys began to increase in a dramatic fashion. The main reason for this was that the bulk of the population of Europe suffered a life of poverty and sometimes stark starvation. There seemed no hope for any improvement in the Old World, so the more adventurous or desperate members of the community sought a better world across the Atlantic. This illustration depicts a scene rather less than a century ago. Not all the voyagers here are poverty stricken; but for all of them the journey ahead is a major undertaking. In many cases they have to decide what to sell, for the best price they can get, and what to take with them. They will need food for weeks or months; but they may be charged a few extra precious shillings if they have much luggage. What will happen if they fall sick? What storms lie ahead during the long weeks on the mighty ocean? The world was still very big and hostile.

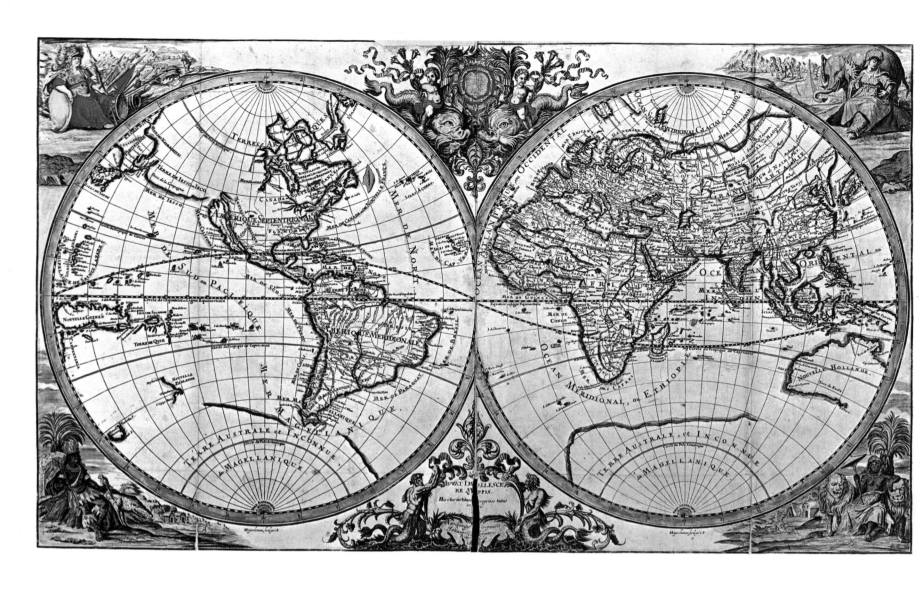

and lost when King Bladud of Britain crash-landed on the Temple of Apollo in what is now London, and was killed.

How could the bird-men, with their home-made feathered wings, know anything of the intricate mechanism that enables a bird to fly—mechanism in which muscle, sinew, heart, breathing system and devices not unlike wing flaps, variable-camber and spoilers of a modern aeroplane, all combine to make sheer brute strength unnecessary.

By the time Borelli was able to prove scientifically that man could not fly without mechanical aid, in 1680, the genius of Leonardo da Vinci had already produced drawings of an aircraft that used the power of both arms and legs to work wings through a system of pulleys and levers. This is often claimed to represent the beginning of aircraft design, but like so many "firsts" it is disputable.

Back in the year 1250, a scientifically-minded English monk named Roger Bacon recorded that he was "exceedingly acquainted with a very prudent man" who had invented a flapping-wing aircraft that was worked by the pilot turning a handle. Bacon was quite a prophet and some authorities claim that he fore-

The map of the world above was completed in the year 1635. Most of the continents look quite familiar but large regions, especially in high latitudes, are a total blank. Drawn like this the world seems small and compact; but what really counts is how long a journey takes in time rather than its length in miles. On the facing page are representatives of the vehicles man has devised to carry him across the Atlantic. If one re-drew a world map so that its scale varied *according to the speed of human transport the result would be startling. Suppose the scale were one foot for each day needed to cross the Atlantic. The early seafaring map would be some 150 feet across. The map for the Concorde passenger would measure just 1½ inches. And the map for Cosmonaut Titov would be one-sixth of an inch across, the size of the capital letters in the text of this book. This is a measure of how flight has shrunk the world.*

TIME TAKEN FOR A TRANSATLANTIC JOURNEY

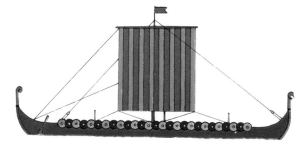

Viking longboat
AD 1001
Several months

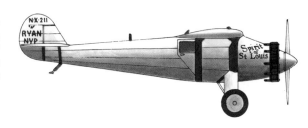

Spirit of St Louis
(Lindbergh)
AD 1927
33 hours 30 minutes
(New York-Paris)

Hindenburg airship
AD 1936
52 hours
(New York-Frankfurt)

Mayflower
AD 1620
66 days

Boeing 314 Clipper
AD 1939
27 hours 35 minutes
(New York-Southampton,
via Newfoundland)

Clipper *Dreadnought*
AD 1853
16 days

Lockheed Constellation
AD 1947
16 hours 30 minutes
(New York-London,
two intermediate stops)

Boeing 707
AD 1968
6 hours 40 minutes
(New York-London)

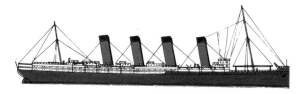

Mauretania
AD 1909
4 days 10 hours 41 minutes

BAC/Sud-Aviation
Concorde
AD 1974
2 hours 56 min. 35 sec.
(London-New York)

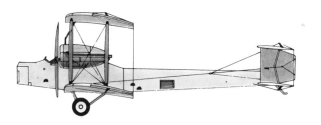

Vimy (Alcock and Brown)
AD 1919
16 hours 27 minutes
(Newfoundland-Ireland)

shadowed the lighter-than-air balloon and airship by suggesting that "a large hollow globe . . . filled with ethereal air . . . would float on the atmosphere as a ship on water".

Unfortunately, nobody knew a ready source of "ethereal air". French writer Cyrano de Bergerac noted that the morning dew rose when the Sun's rays fell on it, and told a fictitious tale of a gentleman who went flying in a large "dew-powered" vertical take-off glass ball. What happened when the Sun disappeared behind a cloud he omitted to explain.

"Ethereal air" was brought a stage nearer by Otto von Guericke's invention of the air-pump in 1650. A Jesuit priest named Francesco de Lana-Terzi realised, quite correctly, that if an object could be made lighter than air it would float in the air: the air-pump seemed to make this practicable. De Lana designed a flying boat that could be raised into the air by four thin copper globes from which the air had been pumped, making them lighter-than-air. It was soon clear that if the globes were made thin and light enough to become lighter-than-air they would collapse under atmospheric pressure. Conversely, if they were made stronger, they became too heavy. De Lana consoled himself with the explanation that this was the will of God, to prevent less scrupulous people from using his flying boat for war.

The year in which man finally got off the ground was 1783. Two French paper-makers, named Joseph and Etienne Montgolfier, noting the way that smoke from a fire whisked upwards pieces of charred paper, began experimenting with paper bags. They held large bags, open end downwards, over a fire for a while and then released them. The bags promptly rose to the ceiling. Here was the lifting force for which would-be fliers had been waiting. They did not realise that their "Montgolfier gas" was no more than hot air which, being rarefied, was lighter than the colder surrounding air and consequently rose. Not that this mattered. On 15 October, 1783, Pilâtre de Rozier became the first man to leave the ground in a tethered Montgolfier balloon of paper-lined linen. On 21 November, the same man accompanied by the Marquis d'Arlandes flew $5\frac{1}{2}$ miles over Paris in another Montgolfier hot-air balloon—and flying became practicable.

The hot-air balloon was a somewhat dangerous contraption, as the brazier which maintained the supply of hot air in flight tended to set light to the fabric. Far better was the completely enclosed spherical balloon invented in the same year by physicist J. A. C. Charles, which used as its lifting force "inflammable air", or hydrogen, discovered by Henry Cavendish in 1766. The balloon he built was so advanced that modern free balloons are similar in all their essentials.

Although man was now airborne, he was far from being a bird. His balloon could drift only where the wind carried it and return journeys were made on the ground, with the balloon folded and packed. After another century of experiment, the balloon had become elongated and had been fitted with propulsion and steering gear. Lift was thus combined with the other two essentials of power and control. Round trips were possible, as the little

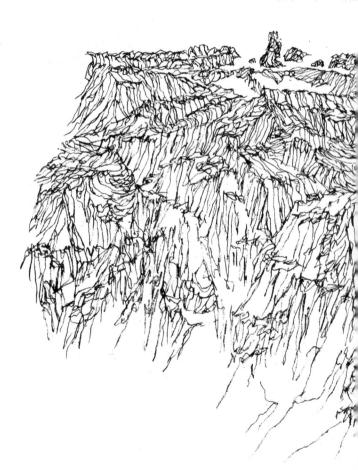

If an astronaut could have orbited the Earth 10,000 years ago he would have found it looking almost exactly the same as it does today. But throughout the past 10,000 years the Earth's natural features— oceans, mountains, parched deserts and frozen wastes—have combined with man's own barriers formed by hostile tribes and nations to render long journeys both slow and hazardous. But in the early years of the present century the pioneer aeroplanes of flimsy wood, canvas and wire heralded a new age in which virtually all the barriers have disappeared. Now the sleek jetliner flies direct from city to city, ignoring the icy crevasse, the coastline, the stormy ocean, the impenetrable jungle and the political frontier.

A typical scene from an airport terminal some decades ago. The aircraft types seen on the ramp, Boeing 707, Sud Caravelle and BAC VC-10, pioneered jet air travel. Flying was beginning to be a means of mass transportation and had taken over practically all ocean-crossing travellers from the ships. A generation ago it used to be thought that air travel was only for the exotic and the rich; and some of the travellers we see here appear to qualify on both counts. But it is not only the high—caste Indian, the film star, the Highland chieftain and the company president who fill departure lounges in every land. In many countries almost every traveller has a black skin, or a yellow or a deep red one. Some are school teachers going to a conference; others are families going on holiday, children going to school, engineers, soldiers and salesmen, people bent on business and pleasure of every kind. They add up to the whole of humanity. How many stop to consider how suddenly man has shrunk the Earth?

Brazilian, Alberto Santos-Dumont, demonstrated to Parisians by flying around the Eiffel Tower in one of his airships and by using these craft as town runabouts which could be tethered outside his club while he was lunching inside.

Ballooning had, meanwhile, become a fashionable sport, and a circus stunt for people who jumped from balloons by parachute, let off fireworks in the air and even ascended on horseback. They had a military use, too, as aerial observation posts, from which specially-trained officers could keep an eye on enemy troop movements over a large area. Yet none of this was flying in the way that birds fly—fast and exciting, wheeling and soaring and diving with no more than a deft flick of their wings.

Another form of lift was needed, that would break free from the limitations of lighter-than-air craft. An English baronet, Sir George Cayley, was first to realise that people had held the answer in their hands for centuries, whenever they flew a kite. By mounting a kite on a stick and attaching a movable tail, he produced the first heavier-than-air craft—a model glider—which worked. Five years later, in 1809, he built a full-size glider on the same lines. By 1853, he had refined his design sufficiently to float his coachman across a valley on his estate—an exploit which prompted the reluctant "pilot" to resign on the spot.

Cayley laid down most of the basic principles on which the modern science of aeronautics is founded. Henson followed in his footsteps with a remarkable design for a full-scale powered aeroplane. It could not be built, as the power to fly was still missing. Du Temple in France and Mozhaisky in Russia advanced one step further by building piloted aeroplanes that would hop short distances after gaining speed down an inclined ramp. Ader, still lacking an entirely suitable engine, succeeded in making similar hops without the use of a ramp.

The invention of the internal combustion engine at last provided the power plant for which everyone had been waiting. The Wright brothers solved the other problem of devising a practical control system and, by combining for the first time adequate lift, power and control in a single aircraft, became the first men to fly properly on December 17, 1903.

Since then aircraft, and with them travel, have undergone revolutionary development. Mention was made earlier of the European emigrants who sailed to North America in about 10 weeks. Nowadays, a business-man can breakfast at his home near Paris or Stockholm or London, get into his car and drive to the airport from where his plane takes off at say 0800. By 1445 European time that same day he will land at John F. Kennedy Airport, New York, where the airport clock will show him that the local time is 0845. That is how things are now; in the near future he will be able to get there in even less time. During the trip he sits in a comfortable armchair seat, is served with drinks and meals by an attractive stewardess and, because the noise level in the cabin is so low, he can chat with his fellow passengers, read, work, or adjust the back of his seat to the reclining position and doze.

The whole globe is covered by a network of air-routes. You can be in London one minute and a couple of hours or so later in Rome. Moscow and Tokyo are almost as close—the world has shrunk indeed! There is now nowhere on Earth which cannot be reached within 24 hours by air.

Every advance in transport method and speed has presented "ordinary folk" with the opportunity to travel to places that were previously beyond their reach. With the continuing development of charter flying and "packaged holidays" millions of people, particularly those living in the cold and austere northern climes, are enabled to follow the sun every year like the birds do and to enjoy, if only for a brief spell, the pleasure of lazing in the hot sunshine and swimming in the warm seas.

Certain scientists concerned with the question of world food supplies have expressed the opinion that famine and scarcity are linked above all with the problem of transport. Whilst in one country tons of food may have to be destroyed owing to over-production, in another country people may be starving to death. Admittedly, aircraft have not yet reached the stage of being used as a major weapon for combatting what we may term "normal" famine conditions, but as a means for providing relief in emergencies their impact is truly enormous. In cases of earthquake or other catastrophe, food, clothes, medical aid, tools and equipment can be rushed to the stricken area wherever it be, and this is of immeasurable importance in saving life, preventing epidemics, and giving fresh hope to the unfortunate victims.

The rapid development of air transport has greatly changed the pattern of commerce and trade. For one thing, thanks to the great reduction in transit time—hours instead of days—fresh goods can be sold still fresh hundreds or even thousands of miles from their place of origin—Mediterranean strawberries eaten fresh in Northern Europe, flowers from Holland, Canadian prawns, the list is almost endless. One of the most important factors is that air transport enables manufacturers to cope with problems of supply and demand more easily than ever before. Rapid transport means that smaller stocks need be kept, which in turn means less capital frozen and less risk of being landed with unsaleable stock. Should there be a sudden demand for shoes in Japan, the very next day an Italian plane can leave Milan with five tons or more of shoes on board.

Air transport is of essential importance to industry. An expensive plant breaks down, specially qualified repair staff arrive by air, followed on a later flight by the necessary spare parts from some other country. The production stoppage, which without the assistance of aircraft would have lasted several weeks, is actually reduced to a few days.

Yes, flying has surely changed travel out of all recognition. This you can see at any international airport, for nowhere is more cosmopolitan. There you will hear every conceivable language, there you will see people from all corners of the Earth. And if some are wearing exotic national dress no one will bat an eyelid, for an international airport is the world in microcosm.

The world has shrunk and is shrinking still. Modern aircraft are becoming so fast that the actual flying time is small compared with the time required for getting to and from the airport on the ground.

But we still have a long way to go before we have to compete with the effect of speed as postulated by Einstein's special theory of relativity. The extensibility of time certainly gives food for thought. Still, one cannot but feel—or at any rate hope—that the Limerick, evidently composed by a physicist, about the young lady who flew so fast that she got back the day before she started is something of an exaggeration.

But let us now return to history, to the 17th of December, 1903, when the Wright brothers made the first flight in a heavier-than-air machine. This was but a beginning. Their aircraft had no wheels, or seat, or instruments beyond a length of wool which streamed in the airflow and showed if they were flying straight or side-slipping. In this book, we can trace not only the history of manned flight, but can study in intricate detail how every part of the modern aeroplane has been evolved from simple beginnings to a component worthy of the aircraft of today. It is a book that tells in an unprecedented way how one of man's earliest dreams has been fulfilled in the form of wings to lift a world.

Scenes such as this have appeared in innumerable works of science fiction. Today, however, the creators of such books find travel through space a less inviting subject because it is difficult to keep pace with reality. All the craft illustrated here are real designs; many have already been launched into space. Can this really be said to be flight? If so, surely the Earth and Moon themselves fly. And why not, for does flight have to be restricted to being borne on wings through our Earth's atmosphere?

85 86

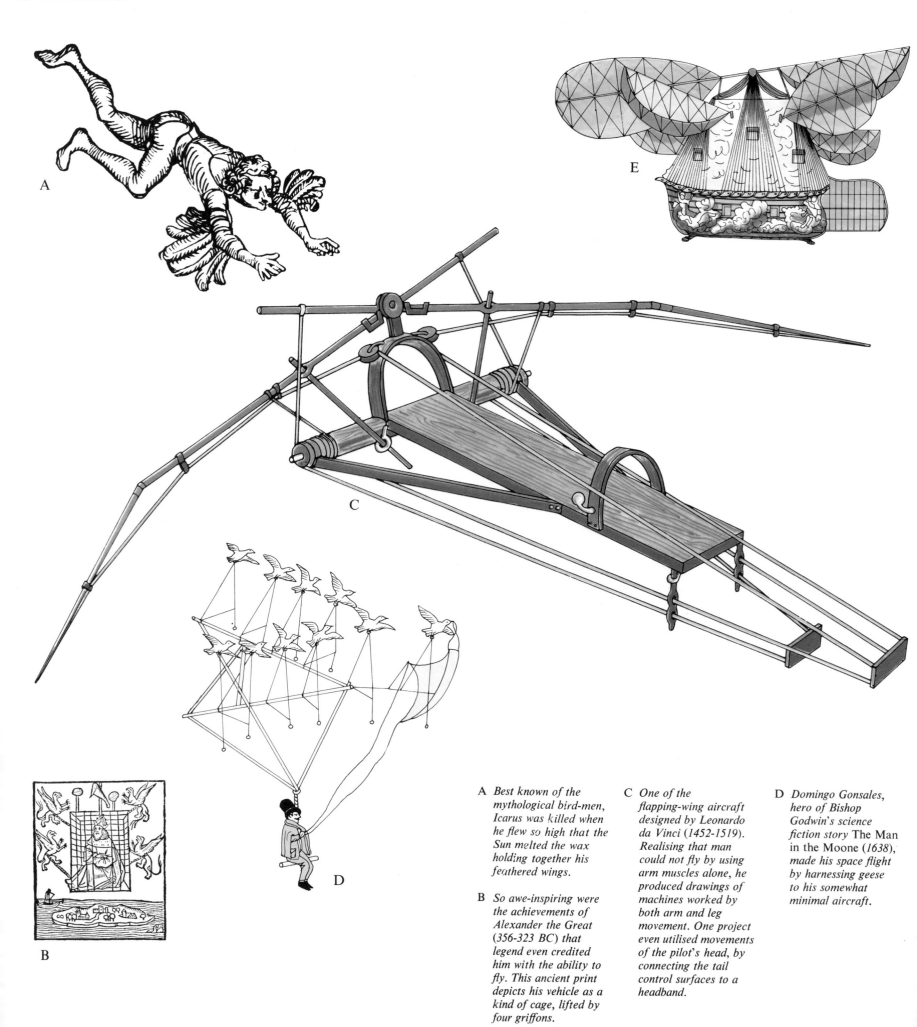

A *Best known of the mythological bird-men, Icarus was killed when he flew so high that the Sun melted the wax holding together his feathered wings.*

B *So awe-inspiring were the achievements of Alexander the Great (356-323 BC) that legend even credited him with the ability to fly. This ancient print depicts his vehicle as a kind of cage, lifted by four griffons.*

C *One of the flapping-wing aircraft designed by Leonardo da Vinci (1452-1519). Realising that man could not fly by using arm muscles alone, he produced drawings of machines worked by both arm and leg movement. One project even utilised movements of the pilot's head, by connecting the tail control surfaces to a headband.*

D *Domingo Gonsales, hero of Bishop Godwin's science fiction story* The Man in the Moone (1638), *made his space flight by harnessing geese to his somewhat minimal aircraft.*

THE BIRD-MEN

From prehistoric times, men had the urge to fly but lacked the power to do so. Thousands of years and countless lives were wasted in trying to copy the birds. Such attempts were logical enough. If the birds' tiny muscles could lift them into the air and sustain them in flight, it was reasonable to assume that the greater strength of a man would do at least as well.

The generations of bird-men who tried, and died, knew nothing of the intricate mechanism of a bird's wing. No-one had discovered that a sparrow's heart needs to work at the rate of 800 beats a minute during flight, or that a pigeon respires 400 times a minute in the air. Such "high-revving" engines are something that the human body can never match.

Leonardo da Vinci suggested augmenting muscle-power with the levers and pulleys that represented the limit of technological progress and mechanical power in his age. The results were clearly not good enough, and there is no evidence to show that anyone tried to build and fly a Leonardo ornithopter.

Today, all the technological skills and new light-weight materials of the 20th century are being utilised in an attempt to prove, at last, that man has, in his own body, the strength to fly.

HOT AIR AND HYDROGEN

After all the dreams and brave experiments by centuries of bird-men, when men did finally leave the ground it was in a craft almost as frail and fickle as a bubble. Some historians claim that a 13th century English scientist-monk named Roger Bacon first pointed the way to the lighter-than-air balloon and airship and even drew up the basic specifications for such a craft. A translation of his work *Secrets of Art and Nature*, produced in Paris in 1542, nearly 250 years after Bacon's death, credits him with writing that:

"Such a machine must be a large hollow globe of copper or other suitable metal, wrought extremely thin in order to have it as light as possible. It must then be filled with ethereal air or liquid fire and launched from some elevated point into the atmosphere where it will float like a vessel on the water."

Apparently, Bacon omitted to point out from where one obtained the necessary "ethereal air". If he believed that such an element existed in the sky

—perhaps in clouds—it led to a state of affairs similar to the controversy over which came first, the chicken or the egg. An aircraft of some kind would have been needed to collect the ethereal air, without which it could not leave the ground. . . .

For nearly five centuries, the idea of flotation in the air vied in popularity with flapping wings as a potential method of getting airborne. Monasteries were the centres of science and learning in this period, so it is hardly surprising that most of those whose theories survive, except for Leonardo da Vinci, were monks and priests. Francisco di Mendoza, who died in 1626, suggested that a wooden vessel would float through the air if filled with "elementary fire". Shortly afterwards, John Wilkins, Bishop of Chester, put Bacon's ideas into more scientific terms by stating that since the air of the upper atmosphere was known to be of lower density than the lower air, a container filled with air of the upper atmosphere would rise.

Empty egg-shells were advanced by some people as being the kind of thin, light weight bodies that would make suitable containers for upper-atmosphere air. The most promising idea seemed to be to fill them with water vapour so that, on exposure to the Sun's rays, they would rise as dew rises from the grass in the early morning. One's mind boggles at the thought of how many egg-shells would be needed to raise a "jumbo-jet" off the ground—on sunny days only, of course.

Francesco de Lana-Terzi, a Jesuit, had the good fortune to be still a teenager when Otto von Guericke invented the air-pump in AD 1650. Shunning such impractical ideas as trapping ethereal air or elementary fire, he decided that the real answer was to have nothing at all inside the egg-shell-thin lifting spheres. A near-vacuum, made practicable by the air-pump, was clearly lighter than any kind of air or other lifting agent. So, all one had to do was to make some thin copper spheres, extract the air, attach the spheres to a boat-shaped carriage and take off—strangely, nobody at this period seems to have given any thought to how they would get down again!

Unfortunately, de Lana's "flying boat" was soon caught up in a vicious circle. Careful experiments and calculations revealed that if the globes were made thin and light enough to become lighter-than-air they would collapse under atmospheric pressure as

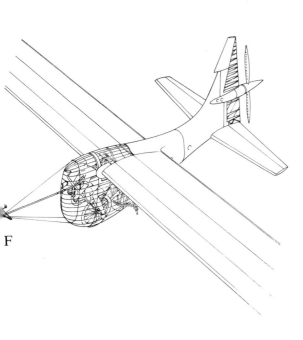

F

E *This boat-hulled craft was designed in 1781 by Jean-Pierre Blanchard, who became famous later as a balloonist. The six hinged paddles were intended to provide both propulsion and steering. They were operated by the pilot, who was inspired to greater efforts by a back-seat musician playing a horn.*

F *If men ever fly solely by muscle-power, it will probably be in a craft like the Puffin, built largely of balsa wood and covered with plastic film. The 118-lb Puffin was flown 993 yards on May 2, 1962, by its pedalling pilot.*

the air was extracted. If, on the other hand, they were made stronger, they became too heavy. Nonetheless, to de Lana goes the credit of being the first person to design a lighter-than-air craft based on definite scientific principles.

Who was it who finally discovered that the lifting force which everyone sought was in their own homes, and had been since some ingenious cave-man learned to make fire? We may never know, but there is increasing evidence to believe that the answer may be different from that which appears in most aviation histories.

The Brazilian priest Bartholomeu Lourenço de Gusmão is seldom rated very highly as a pioneer of flight, because the usually-published drawings of his *Passarola* (Great Bird) show a ridiculous contraption with every conceivable kind of lifting agent— parachute, flapping wings, rarefied air, magnets, rockets—the lot!

In fact, the drawings were probably produced by some imaginative artist who learned second-hand of Gusmão's experiments and put down his own interpretation of the *Passarola's* design.

There still exists in a university in Portugal a contemporary account of one of the experiments, made before the King of Portugal in the Hall of the Ambassadors at the royal palace. It tells how Gusmão lit a fire in a model which promptly took off and travelled through the air. Its journey was brought to an abrupt halt when it hit some curtains. This caused the craft to tumble and, in falling to the ground, it set fire to the curtains and various other objects in the room. The chronicler related that the King took it all in good part—as well he might, for he had undoubtedly witnessed the first successful lighter-than-air flying machine in history.

This historic event remained unconfirmed until recently, when the contemporary account was translated from the original Portuguese. Moreover, when the drawings of the *Passarola* are stripped of the less convincing—and probably imagined— features, it is easy to see how this might have been conceived merely as a de luxe carriage to be raised aloft beneath a larger hot-air balloon.

Gusmão is said to have built a full-scale man-carrying version in 1709, but there would have been little chance of making this light enough to fly.

It was, therefore, the brothers Montgolfier, in France, who earned recognition as the manufacturers of the first successful man-carrying lighter-than-air craft, in 1783—as explained earlier in this book. They believed that by burning wool and straw under the open neck of their balloons they produced a special "Montgolfier gas" that was lighter than air and so caused the inflated balloons to rise. In fact, the gas was no more than hot air, which became rarefied as its temperature increased and began to rise, taking the balloon with it.

Hot air gave way quickly to hydrogen—a much more efficient lifting medium, *and* safer as the balloonists no longer had to keep a fire stoked up under their inflammable vehicle in order to remain

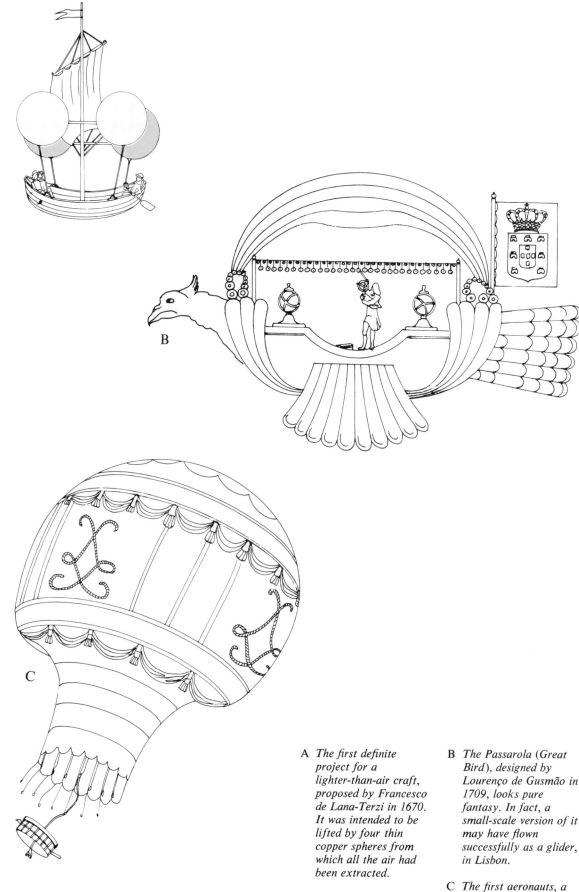

A *The first definite project for a lighter-than-air craft, proposed by Francesco de Lana-Terzi in 1670. It was intended to be lifted by four thin copper spheres from which all the air had been extracted.*

B *The Passarola (Great Bird), designed by Lourenço de Gusmão in 1709, looks pure fantasy. In fact, a small-scale version of it may have flown successfully as a glider, in Lisbon.*

C *The first aeronauts, a sheep, a cock and a duck, travelled two miles in a wicker basket under a Montgolfier hot-air balloon on September 19, 1783.*

D First man to ascend in a balloon, Jean-François Pilâtre de Rozier was also on board the Montgolfière which made the first aerial journey, across Paris, on November 21, 1783. His companion was the Marquis d'Arlandes, who was kept busy stoking the brazier which produced the hot air and sponging the envelope of the balloon when sparks made it smoulder.

E In this hydrogen balloon, Professor J. A. C. Charles and M. N. Robert made a two-hour flight from the Tuileries in Paris, in December 1783. This first ascent in a hydrogen balloon was watched by some 200,000 Parisians.

F A modern hydrogen balloon. It differs little in its essentials from the very first balloon of this type invented by Prof. J. A. C. Charles.

airborne. But even the hydrogen balloon was a mere plaything of the elements, capable of no more than a one-way, down-wind journey. Limited usefulness, coupled with the high cost of inflation—even when hydrogen was replaced by cheaper coal-gas in the 19th century—prevented the gas-filled free balloon from ever becoming much more than a sporting craft for the wealthy folk in search of thrills.

ADDING POWER AND CONTROL

The balloon offered only lift, without power or control. What was needed was a self-propelled, steerable balloon—a true air-ship. As early as 1784, in France, General J.-B.-M. Meusnier designed an elongated balloon 260 ft in length, to be powered by three hand-turned propellers. It was, however, the great English pioneer of flight, Sir George Cayley who, in 1837, drew up a design for a really practical airship, with a streamlined envelope and steam-driven propellers for propulsion and steering.

Theories were translated into reality by another Frenchman, Pierre Jullien, who built and flew a clockwork-powered streamlined model airship named *Le Précurseur* in 1850. It paved the way for the first full-size airship, in which Henri Giffard flew from Paris to Trappes at the breakneck speed of 6 mph two years later.

Giffard's airship was only partially controllable and was powered by a 3-hp steam-engine. Lack of a really suitable lightweight power plant was to continue to hamper progress for the remainder of the 19th century, but designers did not allow this to ground them. When Paul Haenlein of Germany raised the unofficial speed record to 10 mph in 1872, his airship was driven by an engine fed with gas from the envelope. In America, Professor Richell built a tiny, virtually-uncontrolled airship with a pedal-driven propeller.

The first completely successful airship, able to be steered in flight back to its starting point, irrespective of the wind, was *La France*, built by Renard and Krebs in 1884. It used yet another kind of power plant, being electrically-driven. By then, however, the internal combustion engine had been perfected. The little Brazilian pioneer, Alberto Santos-Dumont, proved its capabilities by fitting a petrol engine to one of his airships in 1898 and, later, by circling the Eiffel Tower in Paris—a feat which earned him 100,000 francs.

It was left to the Germans to find a practical use for the first really satisfactory aircraft. Graf Ferdinand von Zeppelin began to plan a series of giant military airships as early as 1874. They were not merely elongated balloons but real flying ships, built around a rigid metal framework. The first aircraft of this type was constructed by David Schwarz in 1897, but it was Zeppelin's airships that were destined to become in time among the most feared and, later, most respected aircraft ever flown, as bombers in war and airliners in peacetime.

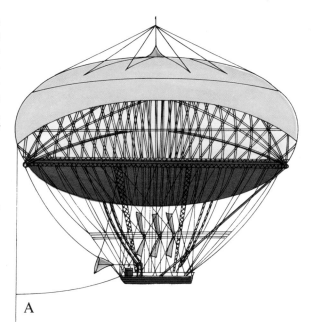

A

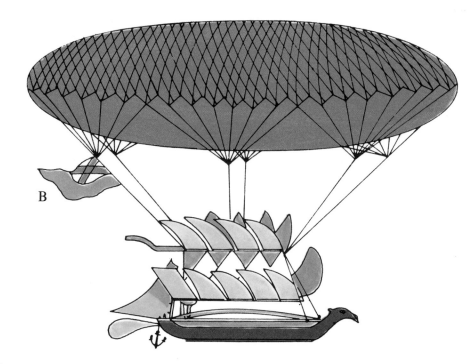

B

A *This advanced design for an airship was produced by Gen. J. B. M. Meusnier, a French military engineer. Absence of funds prevented its construction and the idea was almost forgotten after Meusnier had been killed in battle in 1793.*

B *Greatest early pioneer of airship and aeroplane flight was Sir George Cayley (1773-1857). Realising that the balloon would have only limited use until it could be propelled and steered, he suggested the adoption of elongated streamlined envelopes and rudders, as on the design illustrated, and large propellers driven by steam engines.*

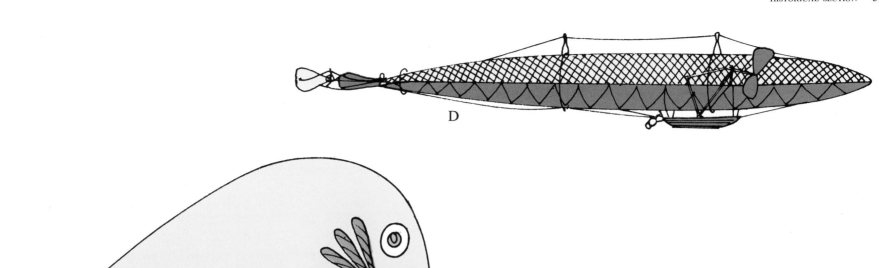

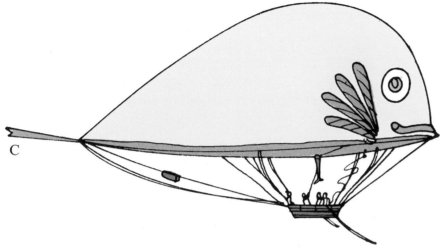

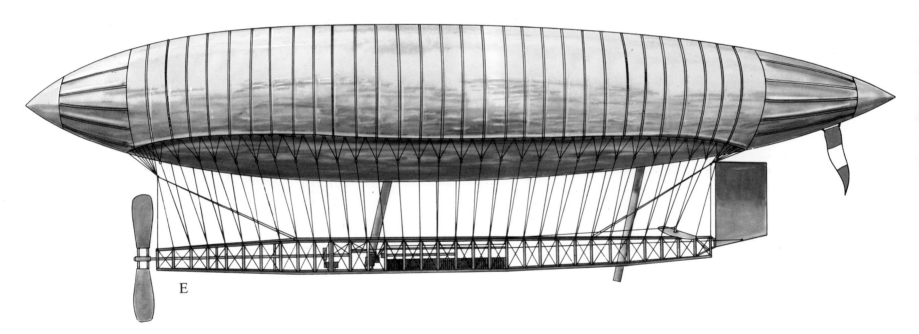

C *More fanciful and less practical was the dolphin-shaped balloon designed in 1816 by S. J. Pauly and Durs Egg, two Swiss armourers who lived in London. The oars and sliding box of ballast helped to justify its nickname of "Egg's Folly"*

D *First major step towards the construction of a practical airship was made by Pierre Jullien who, in 1850-51, flew successfully two clockwork-driven, gas-filled airship models, respectively 23 and 49 ft long. Unfortunately, the full-size version, known as* Le Précurseur, *was never tested.*

E La France, *built by Renard and Krebs in 1884, was the first airship that could be steered in flight in any direction, irrespective of wind, but still lacked a really efficient power plant. It was electrically-driven and had a speed of 14½ mph.*

THE FATHER OF AERIAL NAVIGATION

Sir George Cayley was born in Yorkshire in 1773, ten years before the Montgolfier brothers made their first hot-air balloons. Before he died, at the age of 84, he had discovered the basic principles on which the modern science of aeronautics is founded, built what is recognised as the first successful flying model aeroplane, foreshadowed the airship and the modern convertiplane, and tested the first full-size man-carrying aeroplane.

He began by making a small toy helicopter in which two rotors, consisting of feathers stuck in corks, were driven by a bow-string. It was not original, as two Frenchmen named Launoy and Bienvenu had demonstrated a similar device before the Academy of Sciences in Paris on April 28, 1784; but from then on he was on his own, formulating an entirely new science.

By 1799 he understood the problems well enough to be able to engrave a small silver disc with a diagram showing the forces of lift, thrust and drag acting on a wing. On the other side of the same disc, he engraved a design for an aeroplane with a fixed wing, dart-like tail surfaces and two propulsive paddles. This basic configuration, with its boat-like hull, was to be used in various forms in all his later full-size aircraft and was conceived when he was only 26 years old.

Having worked at the forces affecting flight, it was not long before he realised that one of the oldest of all playthings—the kite—held the key to heavier-than-air flying. He promptly mounted a kite-wing on a five-foot long stick "fuselage", with the leading-edge of the kite raised to give a 6-degree angle of incidence, and attached a cruciform tail unit to the rear end, through a universal joint. This meant that he could control the direction of flight of the model, and make it climb or dive, by adjusting the position of the tail unit. He had, in fact, resurrected an idea first suggested by Leonardo da Vinci (although he knew nothing of the Italian's drawings) and in so doing had devised control surfaces not unlike the modern rudder and elevators. He also fitted the model with a moving weight by which he could adjust the centre of gravity.

Cayley's kite-model flew successfully in 1804, and the event marked the true beginning of the fixed-wing aeroplane. By 1809, he had scaled up the design into a glider big enough to carry a boy and was sufficiently convinced of its potential to write that: "Aerial navigation will form a most prominent feature in the progress of civilisation."

To measure the efficiency of various lifting surfaces, he built a whirling-arm device of the kind that had been used to study air resistance on windmill sails. He quickly appreciated the importance of having the wing at the correct angle in relation to the airflow (angle of attack). He discovered that curved surfaces produced more lift than flat ones, creating a lower pressure above the wing. He suggested that superimposed wings (i.e. biplane and

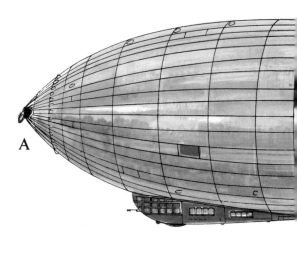

A

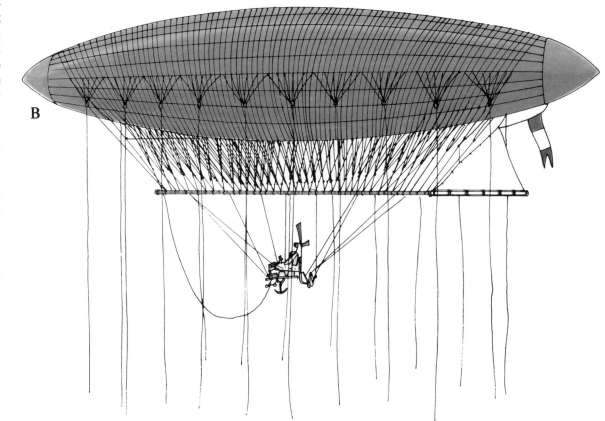

B

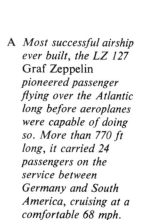

A *Most successful airship ever built, the LZ 127 Graf Zeppelin pioneered passenger flying over the Atlantic long before aeroplanes were capable of doing so. More than 770 ft long, it carried 24 passengers on the service between Germany and South America, cruising at a comfortable 68 mph.*

B *Benefiting from the work of Jullien, Henri Giffard built this steam-powered airship and, on September 24, 1852, flew it 17 miles from Paris to Trappes at 6 mph. Further progress was delayed for 32 years by lack of a better, lightweight power plant.*

C *The completely practical airship was born when Alberto Santos-Dumont (1873-1932) fitted a petrol engine for the first time in 1898. He built 14 airships, of which that illustrated is typical, and gained worldwide fame on October 19, 1901, by flying the sixth one around the Eiffel Tower in Paris.*

D *Among the pioneers of practical airship design were the brothers Paul and Pierre Lebaudy, who achieved a speed of 30 mph in this 1902 model.*

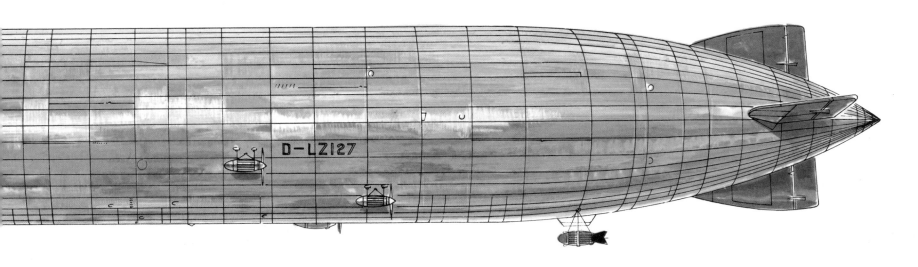

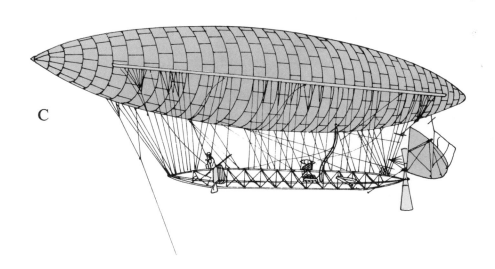

C

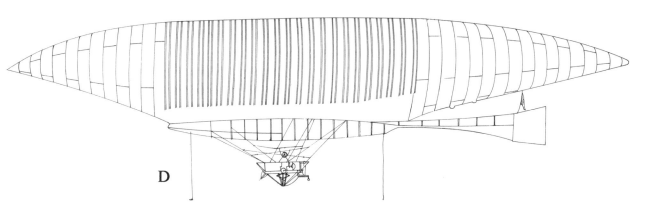

D

triplane designs) would combine maximum lift with minimum structure weight. And he proved that stability could be improved by setting the wings at a dihedral angle (i.e. in the form of a flat Vee in front elevation).

Anyone taking advantage of all of Cayley's theories might have built and flown a practical glider: to do more was not possible in the absence of a suitable lightweight power plant. In the event, Cayley himself was the only person to translate the theories into workable ironmongery in his lifetime. With his eightieth birthday approaching, he put the finishing touches to a triplane man-carrying glider, had it carried to a hillside on his estate and ordered his coachman on board. The name of that first reluctant conscript pilot has not been recorded, which is a pity, as there seems little doubt that he made a brief, comparatively uncontrolled glide across the valley—although he resigned immediately afterwards to avoid having to repeat the experience.

THE FIRST POWERED HOPS

It was William Samuel Henson who first referred to Sir George Cayley as the "Father of Aerial Navigation", in 1846. Nobody in the 19th century tried harder than Henson to prove that the title was justified. After careful study of Cayley's theories and a series of experiments with gliders, he produced in 1842 one of the most remarkable designs in aviation history. Called simply the "Aerial Steam Carriage", it was to be a huge monoplane, spanning no less than 150 ft, with cambered double-surface wings, tail control surfaces, a tricycle landing gear and an enclosed cabin for the passengers. Two six-bladed pusher propellers were to be driven by a steam-engine housed inside the fuselage.

To our modern eyes, the Aerial Steam Carriage looks far more workmanlike and practical than many of the stick-and-string contraptions that were built and flown more than sixty years later; but to the sceptical public of the 1840's it was a big joke. Henson did nothing to improve matters when he issued imaginative drawings showing the aircraft in flight over London, Paris and even the Pyramids. Instead of persuading Parliament and the public to finance the formation of an Aerial Steam Carriage Transit Company to operate world-wide services with Steam Carriages, he became an object of ridicule.

Together with his friend, John Stringfellow, he built a 20-ft-span model of the Aerial Steam Carriage in an effort to prove his claims for the imminence of international air travel. When tested in 1847, it could achieve no more than a "descending glide", weighed down by its steam-engine. So died a project which, given a lightweight engine, might have saved a further half-century of wasted effort. The original model still remains among the most treasured exhibits of London's Science Museum.

To offset the shortcomings of the power plants then available, the pioneers who followed Henson

employed an assisted take-off device in the form of a downward-sloping launching ramp. The first was Félix du Temple de la Croix, a French naval officer. His achievement was little known until a few years ago, when indisputable evidence was discovered proving that he was—so far as we know—the first person to fly a powered aeroplane, in about 1857. It was only a small model with a clockwork engine, replaced after a time by a tiny steam engine; but it flew. What is more, by 1874 du Temple had built a full-scale aeroplane on similar lines, powered by a hot-air engine.

We know very little about this aeroplane. The drawing on the facing page is based on plans left by du Temple, but he may have made changes as construction and testing progressed. All that we know for certain is that the aircraft made a short hop after gaining speed down a ramp, at Brest, in about 1874, piloted by an anonymous young sailor.

This was the first known hop by a man-carrying powered aeroplane. About eight years later, similar hops were made in Russia by a pilot named I. N. Golubev, in a large steam-powered monoplane designed and built by Alexander Mozhaisky.

By comparison with the designs of Henson and Mozhaisky, the enormous biplane built by Sir Hiram Maxim in 1894 looks like a step backward, but it was by no means as impractical as it might appear. Maxim, inventor of the machine-gun, took his aviation work seriously. Wings were tested on a whirling arm 200 ft long. Countless propellers were tested in a wind-tunnel. A steam-engine of unprecedentedly good power-to-weight ratio was designed and built. Large size was no product of grandiose ideas but a carefully-considered choice as "it is much easier to manoeuvre a machine of great length than one which is very short, because it gives more time to think and act."

Maxim, proceeding cautiously, built a system of guard rails to prevent his aircraft from rising into the air during its early tests. It developed so much lift that it broke free of the rails and became the first powered aircraft to take off with a man on board, although its brief flight was not controlled. No further tests were possible, as Maxim's flying ground was bought by London County Council for a mental home and he had to leave.

FLIGHT WITHOUT POWER

Almost without exception, the bird-men had tried to fly by flapping their artificial wings as they fell. Even Leonardo da Vinci drew little but ornithopters. There is, however, one design in his notebooks which, if translated into a full-size flying machine today, now we know how a bird is propelled, might be made to work. In this particular drawing, only the outer panels of the wings are hinged for flapping, the inboard portions being fixed.

It is surprising that none of the early would-be fliers appear to have thought of using easier-to-make

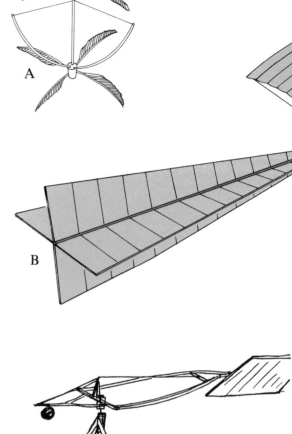

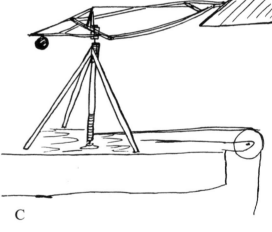

A *Sir George Cayley's aeronautical interests included rotating-wing studies. In this helicopter, made in 1796, the opposite-rotating rotors were made of feathers stuck into corks and rotated by a bowstring. Somewhat similar models had been tested 12 years earlier by Launoy and Bienvenu in Paris.*

B *Engraved on a silver disc by Cayley, in 1799, this is regarded as the first design for an aeroplane with wing, fuselage, tail unit and a means of propulsion (paddles) divorced from the lifting system.*

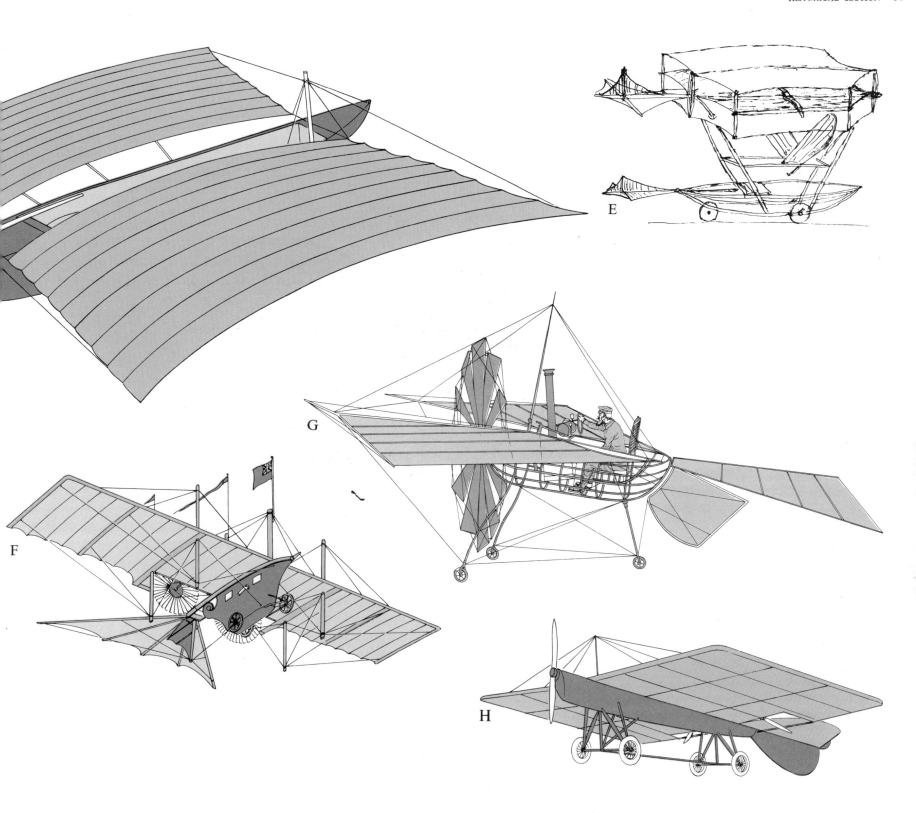

C *True heavier-than-air research dates from 1804, when Cayley built this whirling-arm device, spun by a weighted cord, to measure the lifting power of a wing surface at varying angles of attack. This illustration, together with D and E, is reproduced from Cayley's original sketch books.*

D *In the same year, Cayley flew the first successful model aeroplane, consisting of a kite-shape wing mounted on a pole and with a universally-jointed tail unit. The modern aeroplane has a basically similar configuration.*

E *Cayley's experiments reached a triumphant climax in 1849, when a boy skimmed down a hillside in this triplane glider, and in 1853 when his coachman was sent on a gliding flight across a small valley in Yorkshire.*

F *One of the most inspired designs in early aviation history was the Aerial Steam Carriage of W. S. Henson (1805-88). A model of it, spanning 20 ft, was built but did not fly. Nevertheless, this was the first-ever design for a complete mechanically-powered aeroplane.*

G *In 1874, Félix du Temple's monoplane, powered by a hot-air engine, became the first powered aeroplane to leave the ground with a pilot on board. It did so after gaining speed down an inclined ramp.*

H *Another aeroplane which made a short flight after taking off down a slope was the huge steam-powered monoplane of Alexander Mozhaisky, in Russia in 1884. This was creditable, as the aircraft weighed nearly a ton and was powered by two small British-built engines giving a total of only 30 hp.*

fixed wings. They studied the flight of birds before making their wings yet do not seem to have been impressed by the fact that many birds, such as gulls and albatross, can wheel and soar for long periods without needing to flap their wings.

Perhaps the birds' capability of flight without apparent effort was too much of a mystery to appeal to them; but it was no mystery to Sir Hiram Maxim, who commented "There is no magic in a bird soaring. Constant interchange of air is taking place, the cold air descending, spreading itself out over the surface of the Earth, becoming warm and ascending in other places. Soaring may be accounted for on the hypothesis that the bird seeks out an ascending column of air, and while sustaining itself at the same height in the air, without any muscular exertion, it is in reality falling at considerable velocity through the air that surrounds it."

A French sea captain named Jean-Marie Le Bris appears to have been one of the first to attempt a fixed-wing soaring flight in 1857. His glider was based on the shape of an albatross and was controlled by twisting the flexible wings. Towed by a horse, it made at least one successful gliding flight, thereby earning Le Bris a place among the true pioneers of flying.

It was, however, the German Otto Lilienthal who became the first great exponent of gliding. From an artificial hill, he made more than 2,000 successful flights in his graceful bird-like craft of peeled willow wands covered with waxed cotton cloth. Carefully recording the results, he improved his designs gradually, until he could cover distances of around a quarter of a mile at heights up to 75 ft above the ground. Unfortunately, he relied on movements of his body in the air—by swaying fore and aft and to each side—to control the aircraft's flight. On August 9, 1896, he lost control, crashed and died soon afterwards.

Lilienthal's last words were "Sacrifices must be made". His particular sacrifice was a tragedy, as he was about to fit an engine to one of his gliders and might well have become the first to fly a powered aeroplane. His "disciples", Percy Pilcher in England and Octave Chanute in America, carried on his work, with considerable success. By a tragic coincidence Pilcher also died as the result of a crash just as he was preparing to fly a powered machine.

WHO WAS FIRST?

The great achievement of Lilienthal was that he proved beyond any shadow of doubt that flight in a heavier-than-air machine was entirely practicable. With the petrol-engine already becoming available as a power plant, it was clearly only a matter of time before somebody made a proper powered, controlled and sustained flight.

Bitter controversy raged for years on the subject of who was really first to do so. Until quite recently, the Russians insisted that it was Golubev in Moz-

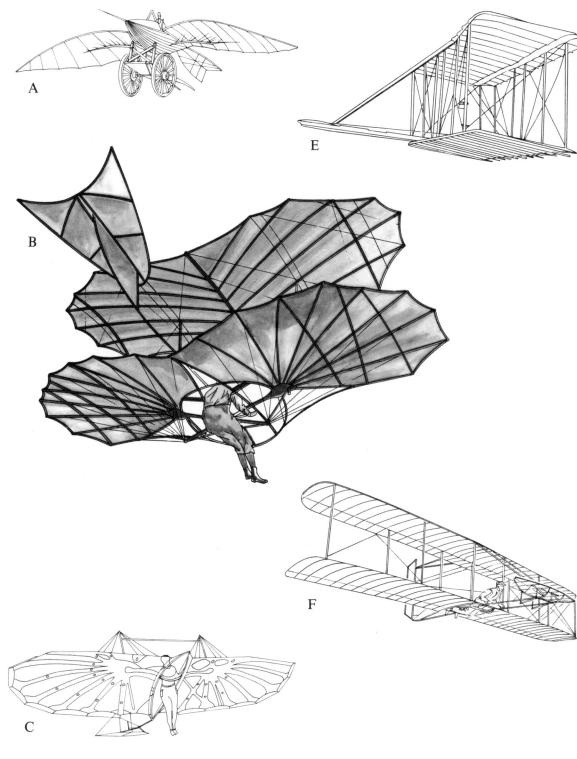

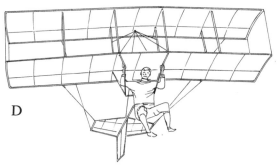

A *This glider was built by a French mariner named Jean-Marie Le Bris. Inspired by the albatross, it made a short glide over a quarry after being launched from a horse-drawn cart.*

B *Otto Lilienthal of Germany, more than any other person in the 19th Century, proved that human flight in heavier-than-air craft was entirely practicable. He flew more than 2,000 times in his elegant gliders before crashing fatally in 1896.*

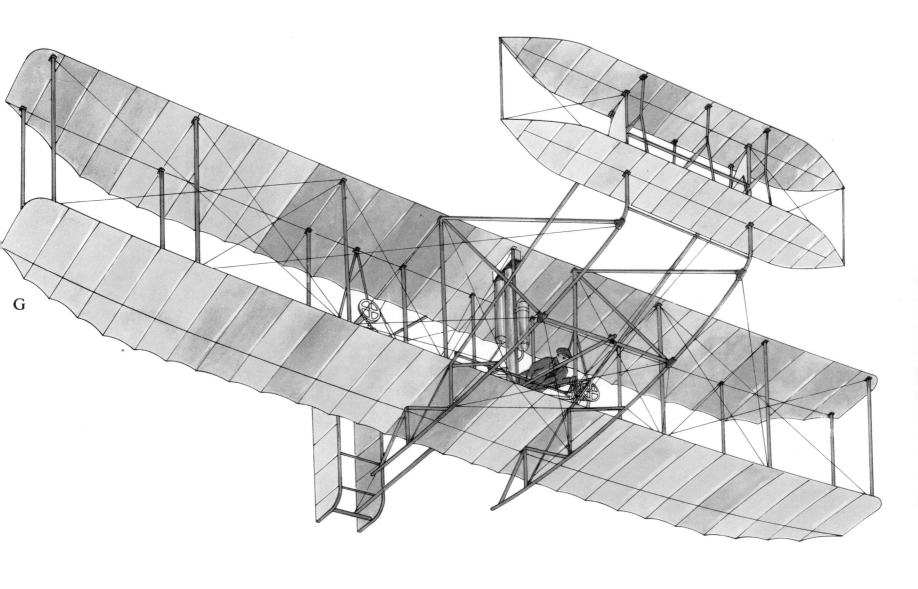

G

C *The Hawk glider of Percy Pilcher of England. Had he not been killed, Pilcher might well have pre-dated the Wright brothers, as he intended to fit an engine to a developed version of this successful design.*

D *Among the disciples of Lilienthal was Octave Chanute of America, who built this much improved glider for display at the World's Fair at St Louis in 1904. The wings were rigid with a cambered shape, as suggested by Cayley.*

E *Realising the short-comings of Lilienthal's technique of controlling his gliders by body movement, Orville and Wilbur Wright devised the warping-wing control system. Although less advanced than the now-standard aileron (invented by Matthew Boulton in England in 1868), it played a big part in enabling them to make the first sustained, powered and controlled aeroplane flights in December 1903. First, however, they tested the warping-wing system on a 5-ft biplane kite and on this 17-ft span No. 1 glider, which was flown mainly as an unpiloted kite.*

F *Satisfied that the warping-wing control system worked, the Wrights added a front elevator control surface on their second glider and then fitted a rear rudder on No. 3, shown here. In this form it made hundreds of flights in 1902-3.*

G *Not until 1903 were the Wrights ready to progress to a powered aeroplane. Then, four flights totalling 97 seconds in the air were sufficient to make the Flyer the most honoured aeroplane in history.*

haisky's monoplane, until they were prepared to admit that his flight was no more than a ramp-assisted hop.

Even today, some historians still credit Clément Ader of France with the distinction of being first. But if he did, as claimed, cover 150 ft in his bat-like steam-powered *Eole* monoplane, on October 9, 1890, it could not have been a controlled flight. His later Avion III twin-propeller aircraft achieved even less.

The Germans are on slightly firmer ground with their claims for Karl Jatho. There is little doubt that he made a number of hops and short flights of up to 200 ft, at heights of up to 12 ft, between August and November 1903. He then gave up, with the comment: "In spite of many efforts, cannot make longer or higher flights. Motor weak."

In the end, therefore, it was left to three Americans to seek success where so many had failed. The first of them was Dr Samuel Pierpont Langley, eminent astronomer and Secretary of the Smithsonian Institution, who designed a series of tandem-wing aeroplanes which he named "Aerodromes". The first of them was a steam-powered model, spanning 14 ft, which flew well over three-quarters of a mile at 25 mph in 1896.

Impressed that so eminent a man should foresee a future for the aeroplane, the US War Department offered him $50,000 if he would build a full-size version for military use. After further model tests, in which a petrol engine was used in an aeroplane for the first time, Langley had his man-carrying "Aerodrome" ready for testing by October 7, 1903. Piloted by Charles Manly—who had built its very advanced 52-hp radial petrol-engine—the aircraft was catapult-launched from a house-boat on the Potomac River. Unfortunately, it struck a post on the launch-gear and plunged into the water. The same thing happened in a second attempt, on December 8, causing one newspaperman to write sarcastically: "If Professor Langley had only thought to launch his air-ship bottom up, it would have gone into the air instead of down into the water."

Heartbroken and with no money available for further experiments, Langley gave up. There was no longer any effective competition for Wilbur and Orville Wright, the bicycle-manufacturing brothers of Dayton, Ohio, who were already preparing their tail-first biplane for flight on the sand-dunes of Kitty Hawk, North Carolina.

They deserved success. For more than four years they had been experimenting, first with a 50 ft kite to prove the efficiency of their wing-warping control system, then with full-size piloted gliders. They had evolved their own wing sections with the aid of a home-made wind-tunnel, and had built their own 12-hp petrol-engine to power their biplane, which they optimistically named *The Flyer*.

The optimism of the Wrights was justified. On December 17, 1903, they made four flights. The air age had come at last.

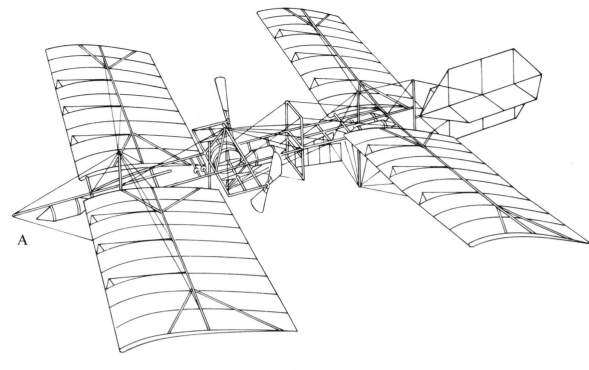

A

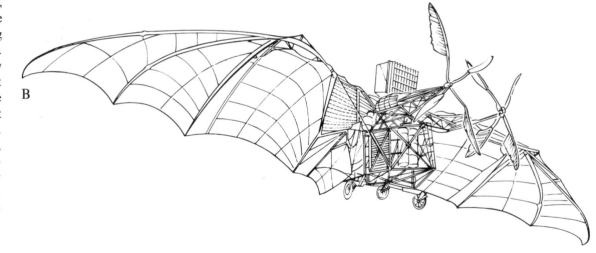

B

A *Controversy over who was the first to fly in the world, in Europe and in Britain has raged for more than 60 years.*
Dr S. P. Langley flew this 14-ft span, steam-powered model aeroplane well over three-quarters of a mile, at 25 mph, in 1896. Backed by the US War Department, he then built a full-size piloted version, but this came to grief when tested in October and December 1903.

B *Clément Ader of France made brief hops in his first bat-wing steam-powered aeroplane, the Eole, in 1890. The later Avion III (illustrated) was less successful. Neither type had a practical control system.*

C *Most successful of the pre-Wright powered flyers was Karl Jatho of Germany. He covered up to 200 ft in his 9-hp kite-like aeroplane in November 1903, but it is not accepted that his flights were controlled and sustained.*

D *J. C. H. Ellehammer of Denmark was airborne for about 140 ft in this kite-like machine, on September 12, 1906. It was not a free flight, as the aircraft was tethered to a central mast on a circular track.*

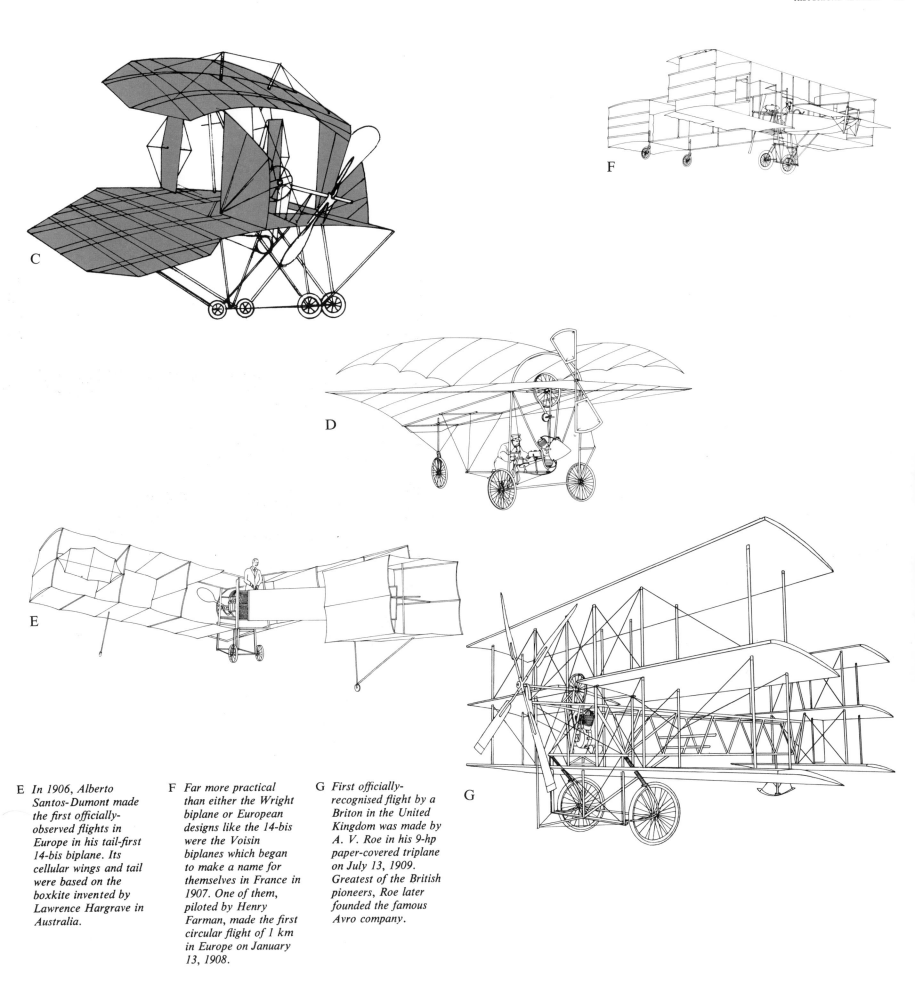

E *In 1906, Alberto Santos-Dumont made the first officially-observed flights in Europe in his tail-first 14-bis biplane. Its cellular wings and tail were based on the boxkite invented by Lawrence Hargrave in Australia.*

F *Far more practical than either the Wright biplane or European designs like the 14-bis were the Voisin biplanes which began to make a name for themselves in France in 1907. One of them, piloted by Henry Farman, made the first circular flight of 1 km in Europe on January 13, 1908.*

G *First officially-recognised flight by a Briton in the United Kingdom was made by A. V. Roe in his 9-hp paper-covered triplane on July 13, 1909. Greatest of the British pioneers, Roe later founded the famous Avro company.*

MAGNIFICENT MEN AND THEIR FLYING MACHINES

How great was the Wright brothers' achievement? The first flight, by Orville, covered only 120 ft—less than the wing span of many modern airliners—and consisted of a series of dives and climbs until one slightly steeper dive brought the 12-second hop to an abrupt end. Even the last flight, by Wilbur, covering 852 ft, was not terribly impressive.

If the Wrights had decided to rest on their laurels at that point, they would hardly be remembered today. But they were not that kind of person. Instead, they went home to Dayton and began work on a better *Flyer*. A falling-weight catapult was designed to speed take-off. A more powerful engine was fitted. Stability and control were improved. At last, the Wrights felt that their aeroplane was good enough to offer to the US Army. The reply they received on October 24, 1905, was that "the Board of Ordnance and Fortification does not care to formulate any requirements for the performance of a flying machine or take any further action on the subject until a machine is produced which by actual operation is shown to be able to produce horizontal flight and to carry an operator."

It was all rather frustrating as, 19 days earlier, they had covered 24½ miles in 38 minutes, 3 seconds, in their best flight of the year.

Even in 1908, the Wrights were still unchallenged masters of the air, flying with far greater confidence than anyone else; but already the limitations of their aircraft were becoming clear. Its lack of wheels, and dependence on a launching catapult, limited its versatility. The front-elevator design was much inferior to the tail-at-the-rear layout suggested by Cayley and Henson; and the very fact that others were taking so long to match their capabilities made them over-confident.

In retrospect, the great achievements of the Wrights were that they proved the value of a scientific, rather than a "build-it-and-see" approach to flying, and that by their example they inspired others who eventually built better "go-anywhere" aeroplanes.

Those better aeroplanes did not emerge overnight. The "14*bis*" biplane in which Santos-Dumont made the first officially-recognised flight in Europe, three years after the Wrights' early success, was less advanced than even their original *Flyer*. The aircraft in which Horatio Phillips covered some 500 ft in Britain in 1907 looked like four venetian blinds with a propeller; but a hint of future possibilities was given by the neat little triplane which A. V. Roe flew in July 1909 and the businesslike monoplane of Louis Blériot.

Aviators were not encouraged in Britain, and Roe was about to be prosecuted for disturbing the peace when Blériot flew across the English Channel in his Type XI monoplane. It would have seemed strange to punish Britain's greatest pioneer for the kind of exploit that had won world renown

A *The flight which first focused world attention on the future potential of the aeroplane was that made by Louis Blériot across the English Channel on July 25, 1909. He might have failed had not a shower of rain cooled his overheating 25-hp Anzani engine en route. Within two days after his achievement, Blériot received orders for more than 100 similar aircraft.*

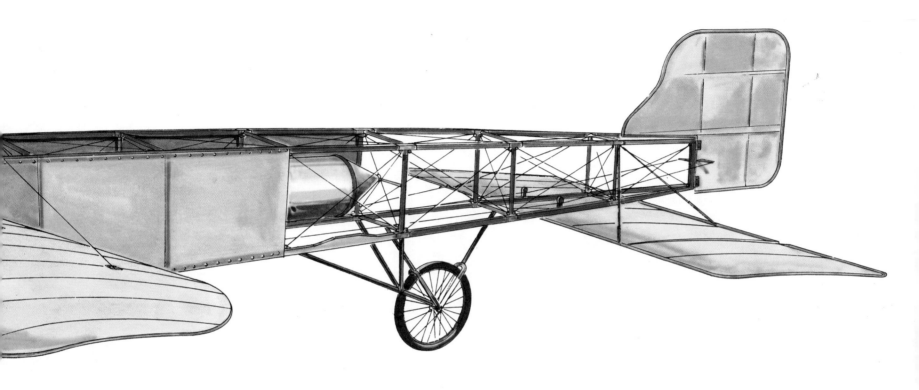

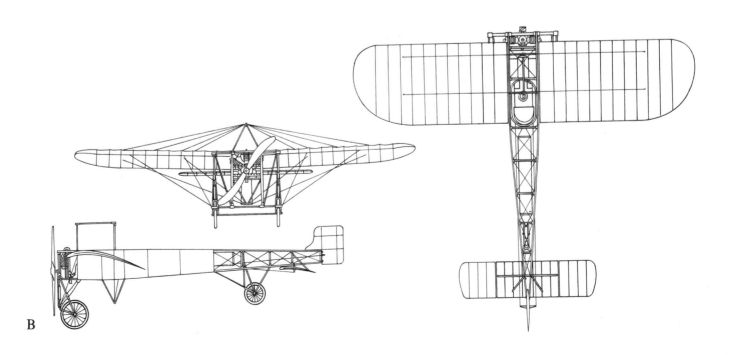

B   *Three-view drawing*
of the Blériot Type X
monoplane. A replica
of this historic machine
was flown at the
Paris air show in May
and June 1969

B

A One of the most
publicised balloon
flights of the 19th
Century was made by
Charles Green and two
passengers in the
Royal Vauxhall
Balloon on November
7-8, 1836. They flew
480 miles from London
to Weilburg in the
German Duchy of
Nassau in 18 hours.
The balloon, later
renamed the Great
Balloon of Nassau,
was made of 2,000
yards of white and
crimson silk and held
70,000 cu ft of coal gas.

B The early history of
flying in Europe is well
illustrated by the maps
on this page, showing
the routes followed by
five outstanding pioneer
flights. This map
illustrates the first
aerial voyage in
history, made by
Pilâtre de Rozier and
the Marquis d'Arlandes
over Paris on
November 21, 1783.

C The 27-mile flight by
Professor J. A. C.
Charles and M. N.
Robert across Paris,
on December 1, 1783,
was the first in a
hydrogen balloon.

D On August 9, 1884, the
airship La France
made a fully-controlled
circular flight of nearly
five miles, at a speed
of 12-14 mph, after
taking off from
Chalais-Meudon.
Although several more
successful flights
followed, the airship's
heavy electric motor
was not really suitable
for use in aircraft.

E The first circular flight
of one kilometre in
Europe was made
by Henry Farman, on
an Antoinette-powered
Voisin biplane, on
January 13, 1908. On
October 30 of the same
year he made the first
cross-country flight in
an aeroplane from
Châlons to Reims, a
distance of 16½ miles.

F Louis Blériot was not
able to follow a
straight course over
the English Channel
on July 25, 1909.
Blown northward by
the wind, he eventually
found Dover by going
in the same direction
as ships bound for
the port.

G One of the greatest
over-water flights of
the pre-1914 era was
the first crossing of
the Mediterranean, by
Roland Garros of
France on September
23, 1913. His Morane-
Saulnier monoplane
was fitted with a 60-hp
Gnome engine.

B (blue)

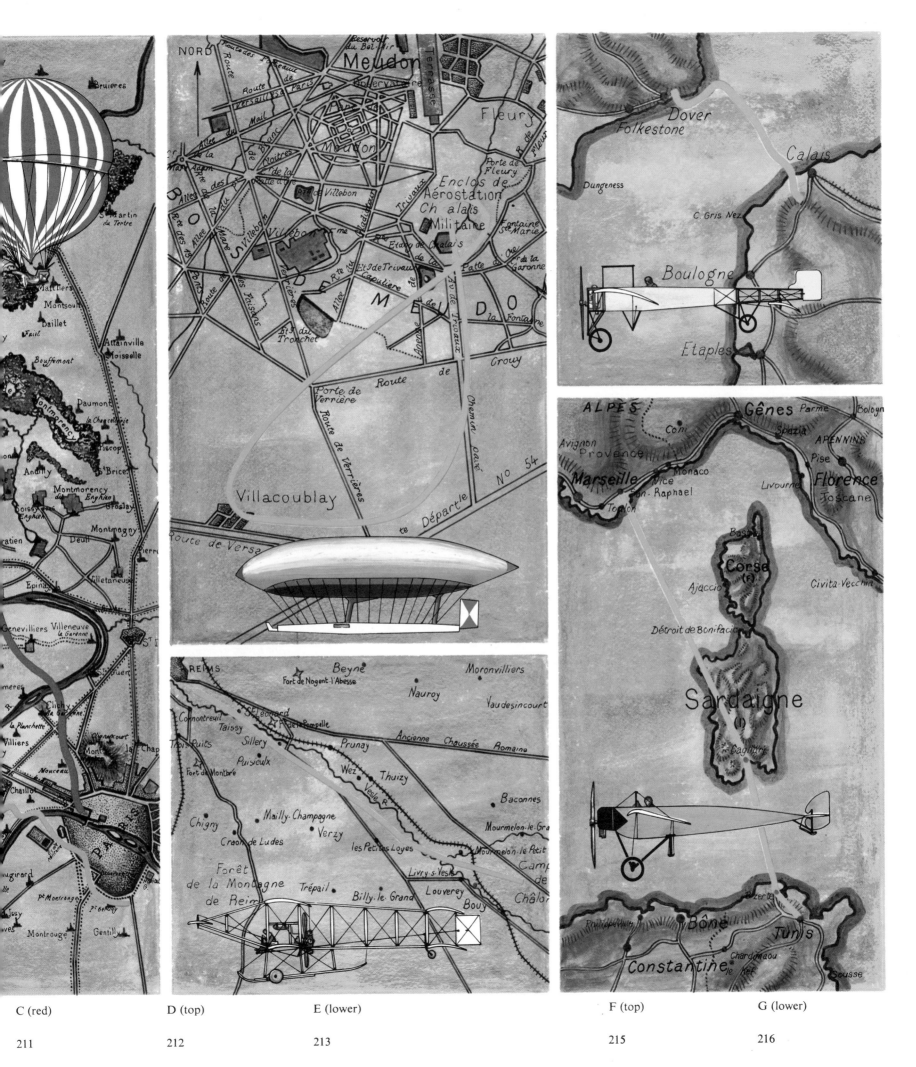

C (red)

D (top)

E (lower)

F (top)

G (lower)

211

212

213

215

216

for a Frenchman, and the case was dropped.

A more significant result of Blériot's great flight was that the aeroplane had, for the first time, shown its ability to overcome natural and political barriers. Britain no longer felt quite so secure behind her "Channel moat" and the Royal Navy.

SEAPLANES AND SHIP-PLANES

During Louis Blériot's historic flight from France to England, the 25-hp Anzani engine of his monoplane began to overheat. Just as he was about to brace himself for a ditching in the sea, a providential shower of rain cooled the cylinders and enabled him to complete his trip.

It would be wrong to suggest that this incident was sufficient to create a major interest in floatplanes able to land on, and take off from water. Blériot himself had experimented with floatplanes in 1906, in collaboration with Gabriel Voisin. Even the Wrights had tested a set of hydrofoils on the Miami River at Dayton in 1907, when they began to realise the limitations of their wheel-less *Flyers*. Nevertheless, it is a fact that designers in several countries began to look more closely at the possibilities of flying from the water in 1909-10.

The first powered aeroplane to take off successfully on floats, on March 28, 1910, was the incredible contraption illustrated above. Built in France by Henri Fabre, it utilised his special lattice type of wing spars, which had the same depth and strength as a solid spar, but did not create nearly so much drag. Perhaps the best part of the aircraft was the float landing gear, for long after the seaplane itself had ceased flying—to become eventually an exhibit in the French *Musée de l'Air*—Fabre was still turning out very similar floats for some of the more efficient seaplanes of the 1912-14 era.

The great pioneer of marine flying was Glenn Curtiss of America. During the winter of 1910-11 he fitted floats to one of his sturdy pusher biplanes and flew this off water for the first time on January 26, 1911. It was the first really practical seaplane, and Curtiss followed it with the first flying-boat on January 10, 1912. This latter aircraft differed little from the seaplane, except that the central float was enlarged sufficiently to accommodate the pilot and controls; but from it was evolved the whole long line of great and gracious flying-boats that played such a big part in the progress of civil and military flying.

Curtiss made a further contribution to military aviation at this period by building the aeroplane that made the first take-off and landing on a ship. The pilot on each occasion was Eugene Ely, and the success of his exploits proved the practicability of the aircraft carrier. In doing so—although this could not be foreseen at the time—it also knocked the first nail in the coffin of the great capital ships that were the pride and joy of the maritime powers.

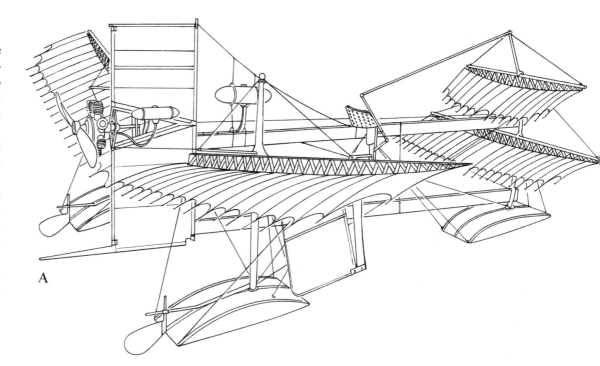

A

A *The first successful flight on a powered seaplane was made by Henri Fabre, at Martigues, France, on March 28, 1910. His strange-looking tail-first aircraft had a 50-hp Gnome engine.*

B *Glenn Curtiss of America was the great pioneer of water-based aircraft, as his seaplanes and flying-boats were far more practical than that of Fabre. The first Curtiss flying-boat (illustrated) flew on January 10, 1912.*

C *One of Louis Blériot's less-successful early designs was this ellipsoidal-wing floatplane, built in 1906. Even after it had been fitted with conventional wings and a wheel landing gear it did not fly.*

D *The Blériot floatplane of 1906 in an intermediate form with conventional wings and floats. In the following year, Blériot switched to monoplane layouts, with immediate success.*

E *The first seaplanes to fly in Britain were built by A. V. Roe and Company. A pusher seaplane of the Curtiss type, produced for Capt E. W. Wakefield, was followed by this 45-hp seaplane of Avro's own design.*

F *First seaplane able to out-perform most contemporary land-planes was the Sopwith Schneider. Powered by a 100-hp Gnome, it won the contest for the important Schneider Trophy in 1914, at a speed of 86.78 mph.*

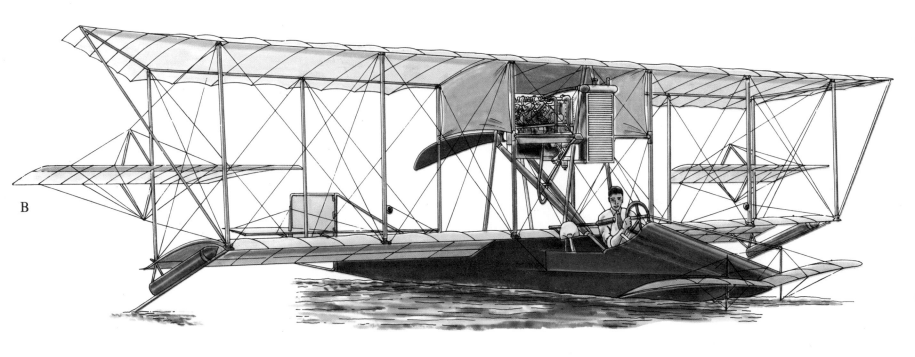

B

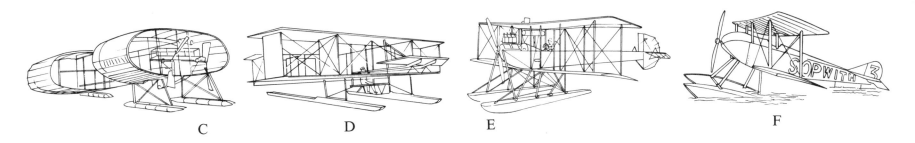

C

D

E

F

BIGGER, AND SOMETIMES BETTER, AEROPLANES

From the start, the power plant held the key to progress in flight. Little progress at all had been possible until the advent of the lightweight, efficient petrol-engine. Now that the art of flying had been learned, designers became insatiable in their demand for more powerful and more reliable engines.

France, the acknowledged centre of world aviation from about 1907 onward, was inevitably the first country to establish any kind of aeroplane industry. The Voisin brothers set the lead in laying down an assembly line of boxkite biplanes for anyone with sufficient money and courage to buy them. Antoinette and Blériot monoplanes were also bought by sporting flyers. Best of the early engines were probably the fan-type three-cylinder Anzani of 25 hp and the 50 hp and 100 hp Vee-type Antoinettes, with 8 and 16 cylinders respectively.

Then, in 1909, there appeared on the aviation scene a revolutionary little engine named the Gnome, with a crank-shaft that had to be bolted to the aircraft structure, so that the seven cylinders and propeller rotated round it. Scepticism quickly gave way to admiration when the Gnome's inventors, the brothers Louis and Laurent Seguin, showed that it would develop an honest 50 hp for an engine weight of only 165 lb. There were a few penalties, the main one being that it had to be lubricated with liberal quantities of castor oil, which was flung out in a fine spray as the engine whirled round at 1,200 revolutions per minute. But this was a small price to pay for such a power plant and the Gnome made possible most of the great flights of pre-war years. With its imitations, it also powered many of the most famous aircraft of the 1914-18 War.

To meet demands for ever-increasing power, the Seguins mounted two Gnomes together in 1913, to produce a 14-cylinder two-row engine giving 160 hp. But reliability suffered and designers began to prefer two small engines to one big one. The problem was where to put them. The Short brothers, in England, had been among the first to face up to the problem with their Triple Twin biplane, in which the front engine drove two tractor propellers by means of long chains, while the rear engine powered a single pusher propeller. This was followed by the Tandem Twin of 1911. As the pilot sat between the two 50-hp Gnomes, it was known usually as the "Gnome Sandwich".

A new idea was introduced in the Radley-England Waterplane, which had three separate engines, mounted one behind the other, driving through chains a single pusher propeller. But the person with the best idea, because it was the simplest, was Igor Sikorsky. When, in 1913, he completed the world's first four-engined aeroplane, named appropriately *Le Grand*, he simply mounted the engines in a row on the lower wings, driving tractor propellers, and set the fashion that has lasted to the present day.

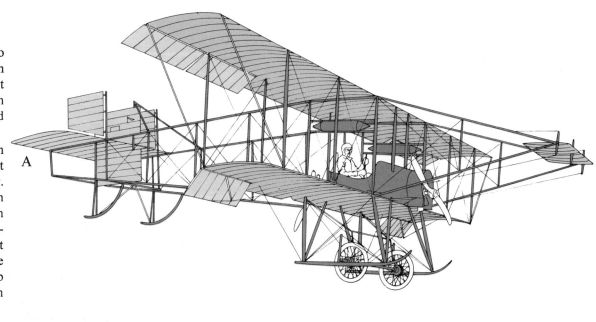

A

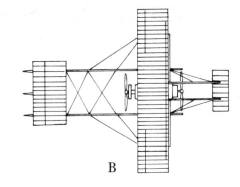

B

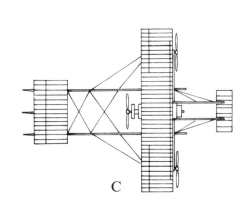

C

A *First successful twin-engined aeroplane was a Sommer biplane of 1910. It was followed by the Short S.39 Triple Twin, in which two 50-hp Gnomes drove one pusher propeller and a pair of wing-mounted tractor propellers; it was the first aeroplane able to fly safely after failure of an engine in the air. Then came the Short S.27 (illustrated), with a single tractor propeller instead of the twin forward propellers. It was known as the Tandem Twin or Gnome Sandwich.*

B *Variations on a twin-*
& *engined theme: the*
C *Short Tandem Twin and Triple Twin.*

D *Some designers "thought big" from the start—none more so than Capt Arlington Batson. His 12-winged Flying House of 1913 contained a sitting-room and sleeping apartments and was intended to fly the Atlantic. It never took off.*

E *Equally ambitious for its time, and outstandingly successful, was the Russian Knight biplane (sometimes called Le Grand) designed, built and flown by Igor Sikorsky in Russia, in 1913. It was the first four-engined aeroplane to fly.*

F *One of the four 100-hp Argus engines of the Sikorsky Russian Knight.*

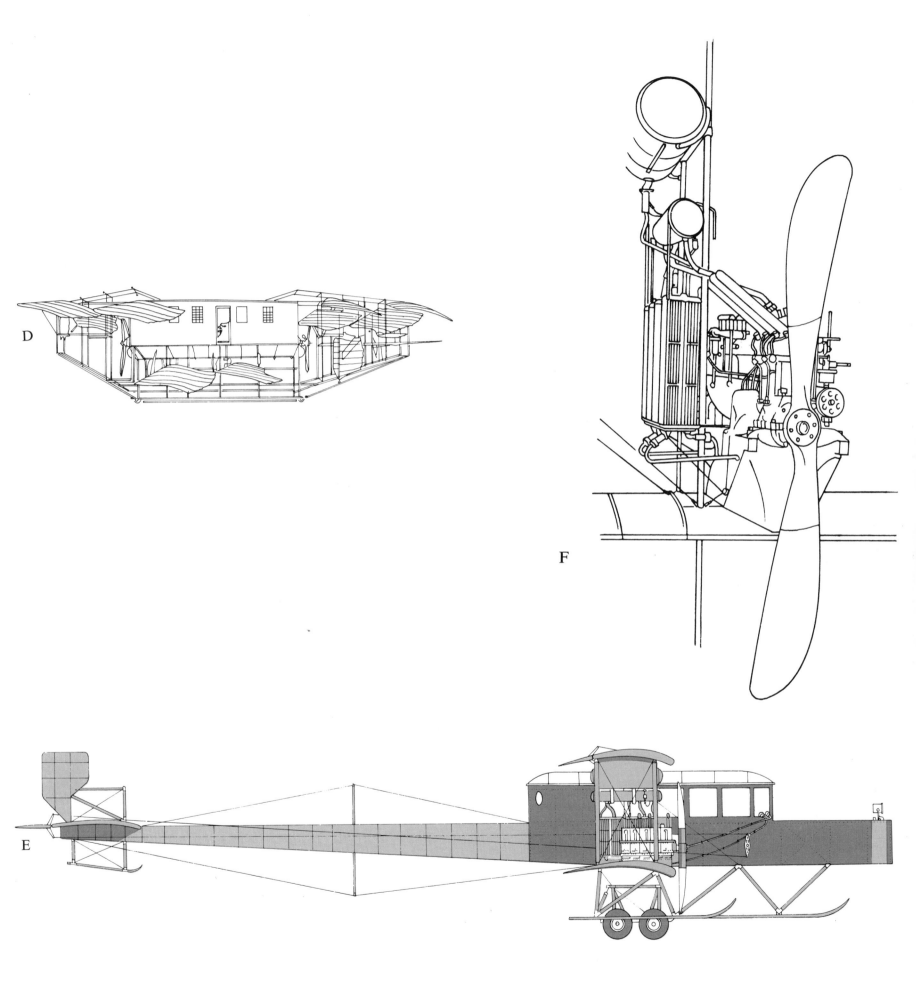

D

F

E

Even with four engines, *Le Grand* had a total of only 400 hp with which to lift its 4½ tons into the air. That it did so safely, and repeatedly, is a tribute to Sikorsky, who became recognised in due course as one of the greatest designers of all time.

THE FIRST AIR WAR

Before 1914, few people believed that the aeroplane had any military use except for reconnaissance. Experiments in firing guns from aeroplanes, and dropping bombs and torpedoes from them, were not taken very seriously by senior officers of the world's armies and navies. As a result, when the Royal Flying Corps crossed the Channel to support the British Expeditionary Force in France, its pilots flew unarmed aeroplanes. Their only operational order was to ram any enemy Zeppelin airship they might encounter on the way—a far-from-heartening prospect as they had no parachutes. The best that could be provided in the way of survival equipment was a motor-car-tyre inner tube which, worn around the waist, formed a makeshift lifebelt in case the wearer had to ditch in the Channel.

It did not take the military leaders in France long to appreciate the value of aerial reconnaissance. While requesting more aircraft with which to keep a constant watch on enemy movements, they began to press for more positive action to stop similar activities by the enemy. Eventually, aircraft designers devised ways of firing machine-guns from aircraft without shooting off the propellers in the process, and reconnaissance aircraft began to suffer heavy casualties at the hands of the new fighter-planes. Friendly fighters began to accompany the reconnaissance machines to protect them from enemy attack. So, the evolution of the combat aircraft began.

By the end of the 1914-18 War, pilots in swift, heavily-armed fighters found themselves involved daily in great swirling dog-fights as whole squadrons engaged in deadly combat. Reconnaissance machines reported and photographed every movement of the unhappy armies on the ground and directed artillery fire so that it did the greatest possible execution. Bombers showered high-explosive and incendiary bombs on battlefield and town alike. The first dive-bombers had been built to harass further the infantryman in his muddy trench. Torpedo-planes had scored the first victories against ships at sea. Frail seaplanes had, to a large extent, given way to fighters and bombers based on aircraft carriers. And long-range flying-boats ranged far out to sea to help check the menace of the submarine.

Vast industries were built up to supply aircraft in the numbers needed for all these tasks, and others such as training. The British aircraft industry alone grew in four years from a few small, scattered workshops, peopled with visionaries and craftsmen, to a giant network of manufacturers employing 350,000 men and women and turning out aeroplanes at the rate of 30,000 a year—all for war.

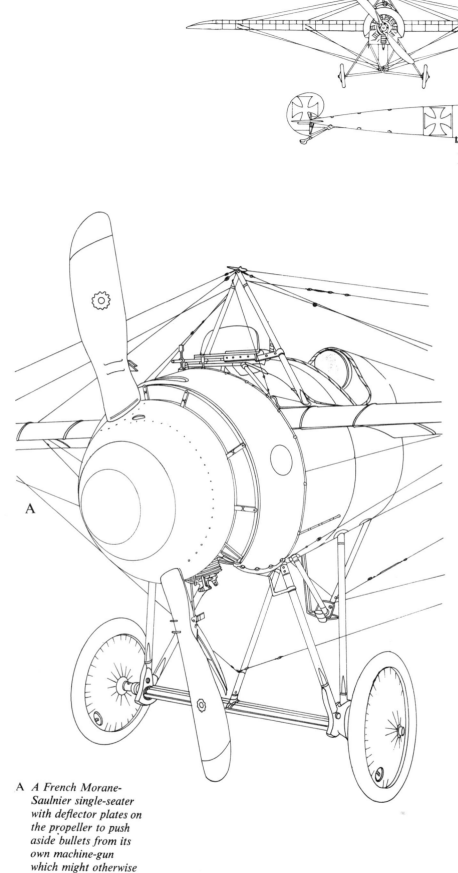

A *A French Morane-Saulnier single-seater with deflector plates on the propeller to push aside bullets from its own machine-gun which might otherwise have struck and damaged the blades. The interrupter gear fitted to the Fokker Monoplane was inspired by this more primitive idea.*

B *Three-view drawing of the Fokker Monoplane, the first true fighter aircraft.*

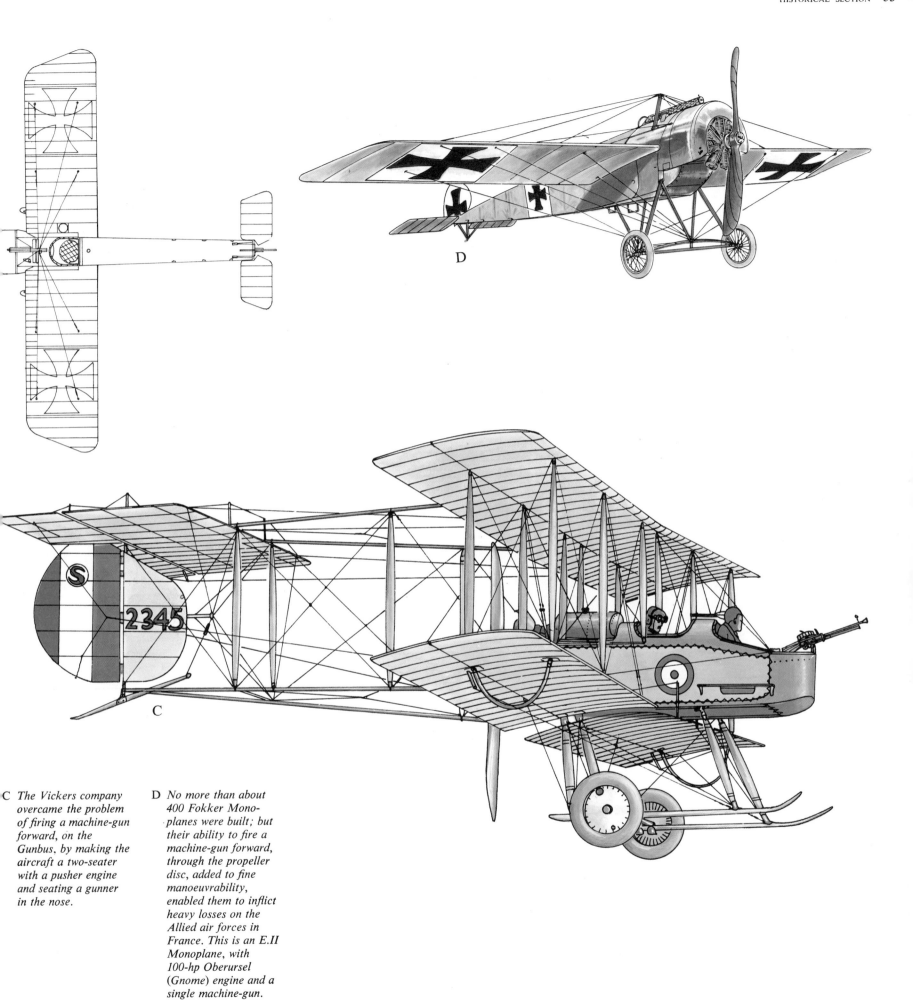

C The Vickers company overcame the problem of firing a machine-gun forward, on the Gunbus, by making the aircraft a two-seater with a pusher engine and seating a gunner in the nose.

D No more than about 400 Fokker Mono-planes were built; but their ability to fire a machine-gun forward, through the propeller disc, added to fine manoeuvrability, enabled them to inflict heavy losses on the Allied air forces in France. This is an E.II Monoplane, with 100-hp Oberursel (Gnome) engine and a single machine-gun. The more powerful E.III had two guns.

AGE OF GREAT FLIGHTS

It is often claimed that the 1914-18 War advanced aviation progress to an unprecedented degree, but such an assessment depends on the terms by which we measure progress. The best fighter aircraft of 1918 were not so fast as the experimental S.E.4 scout built at the Royal Aircraft Factory, Farnborough, in 1914. As for the prospects facing the giant new aircraft industry, the victorious Allied governments, having just won the "war to end wars", with vast stocks of combat aircraft in service and in store, made it quite clear that they would order no new warplanes for years. Many famous companies went out of business. Others worked and waited for the predicted boom in civil flying which never came.

Only in the field of aero-engine design and manufacture had the war brought significant progress. Even an attempt to fly the Atlantic seemed less fearful now that aircraft could be fitted with engines like the superb 360-hp Rolls-Royce Eagle, and several British crews made their way to Newfoundland with aircraft which they considered adequate for such a flight.

They were forestalled by airmen of the US Navy, who left Newfoundland in three Curtiss flying-boats on May 16, 1919, to attempt the first transatlantic flight, via the Azores. Each aircraft was powered by four 400-hp engines and every possible safety precaution was taken, to the extent that strings of warships were spaced out along the route to assist the airmen in an emergency. Despite all this, of the three big flying-boats only NC-4 succeeded in reaching Lisbon, thirteen days later.

Within little more than a fortnight, NC-4's feat was overshadowed completely by the non-stop transatlantic flight of Alcock and Brown in a converted Vickers Vimy bomber with two Eagle engines.

The Atlantic continued to challenge pilots throughout the nineteen-twenties and thirties. Charles Lindbergh became a world hero in 1927 by making the first solo crossing, all the way from New York to Paris, in the Ryan monoplane *Spirit of St Louis* with only a single 220-hp engine. General Italo Balbo, Italy's Air Minister, stirred the imagination of his countrymen by leading two mass flights of Savoia-Marchetti flying-boats over both the South and North Atlantic.

Less publicised, but more significant, were the crossings made by men like Jean Mermoz of France. By opening a mail service across the great oceans, they paved the way for the passenger airliners that were to follow.

OPENING UP THE AIRWAYS

To Germany goes the distinction of having operated the world's first airline services. The year was 1910, the aircraft a huge Zeppelin airship named the *Deutschland II*. With her sisters, she carried 35,000 passengers a total of 170,000 miles between Lake Constance, Berlin and other cities, before the out-

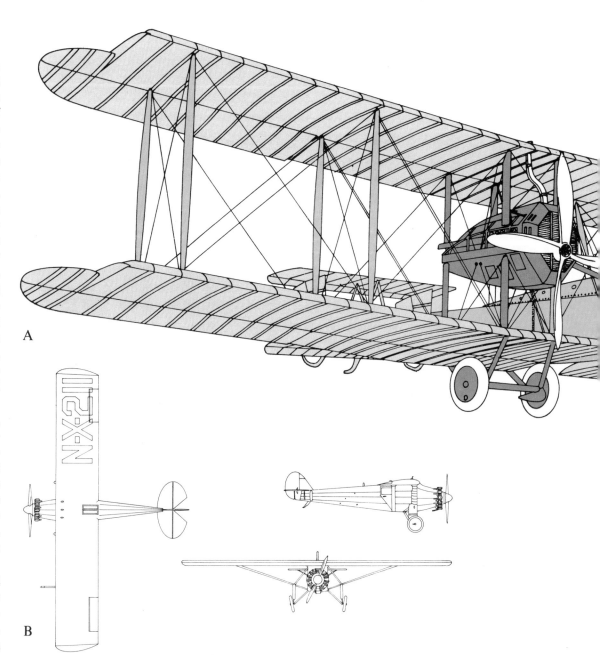

A

B

A *One of the greatest trail-blazing flights in history took place on June 14-15, 1919, when Capt John Alcock and Lt Arthur Whitten-Brown flew non-stop in this converted Vimy bomber from St John's, Newfoundland, to Clifden in Ireland in about 16 hours.*

B *Three-view drawing of the Ryan monoplane* Spirit of St Louis *in which Charles Lindbergh made the first solo transatlantic flight on May 20-21, 1927. Its Wright Whirlwind engine developed only 220 hp, and the 3,600-mile flight took 33½ hours.*

C *The growing practicability of transatlantic flying was demonstrated dramatically by the Italian Air Minister, General Italo Balbo. In 1931 he led a mass flight of 12 Savoia-Marchetti flying-boats across the South Atlantic from Rome to Brazil. Two years later he led an even more impressive armada of 24 flying-boats across the North Atlantic and back.*

D *Commercial aeroplane flying over the Atlantic was pioneered by Jean Mermoz of France. On May 11, 1930, he took off from Dakar, West Africa, in a Latécoère 28 floatplane and flew through the night to land at Natal, Brazil, 19½ hours later. The air mail he carried reached Buenos Aires four days after leaving France, compared with the usual eight days. The aircraft which set out on the return trip was lost and it was not until four years later that a regular transatlantic mail service to South America was inaugurated by the German Lufthansa company.*

E *First fixed-wing "airliners" in history were two Benoist flying-boats of the St Petersburg-Tampa Airboat Line. Carrying a pilot and one passenger, they made five return flights daily between the two Florida cities for three months from January 1, 1914.*

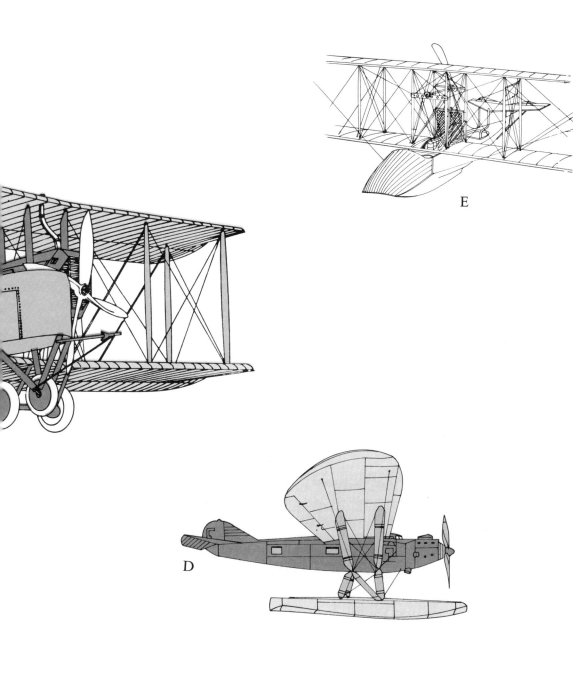

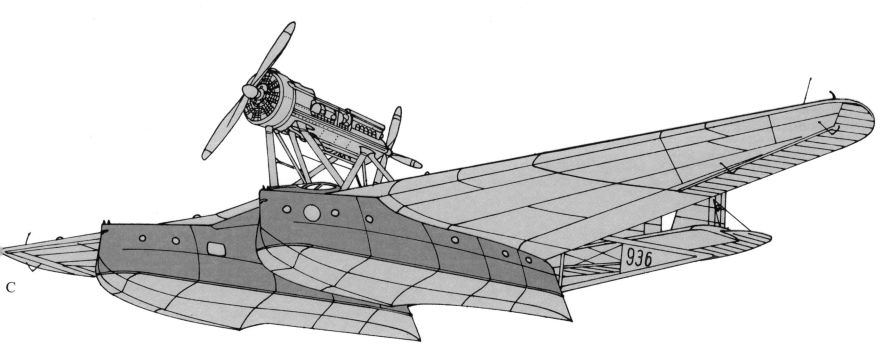

break of war gave the Zeppelins more sinister tasks to perform. This was not the end of the Zeppelin as an airliner. In the late nineteen-twenties and thirties, the *Graf Zeppelin* and *Hindenburg* operated the first passenger services across the Atlantic; but by then the airship was beginning to be outclassed by the faster aeroplane, and when the *Hindenburg* was lost, with many passengers and crew members, while landing at Lakehurst in America, on May 6, 1937, the whole concept of lighter-than-air commercial airliners died with her.

A flying-boat service was operated briefly by the Saint Petersburg and Tampa Airboat Line, in America, in 1914. One passenger at a time could be carried across the Bay of Tampa, the 23-minute flight costing five dollars. Then war halted progress for five years. When international airline flights did begin in August 1919, the best that could be offered passengers was a wicker chair inside the draughty cabin of a converted bomber. Forced landings were frequent and, in the days before radio was fitted, pilots had to rely on familiar features such as straight roads and railways to guide them to their destination.

Serious accidents were surprisingly rare. One of the worst occurred when the pilot of a Paris-bound airliner, following a main road in France, failed to spot a London-bound airliner following the same road at the same height but in the opposite direction. They met head-on.

Gradually, new standards of comfort and safety were achieved. By the late nineteen-twenties, machines like the Fokker F.VII/3m were offering a combination of all-metal construction, multi-engined

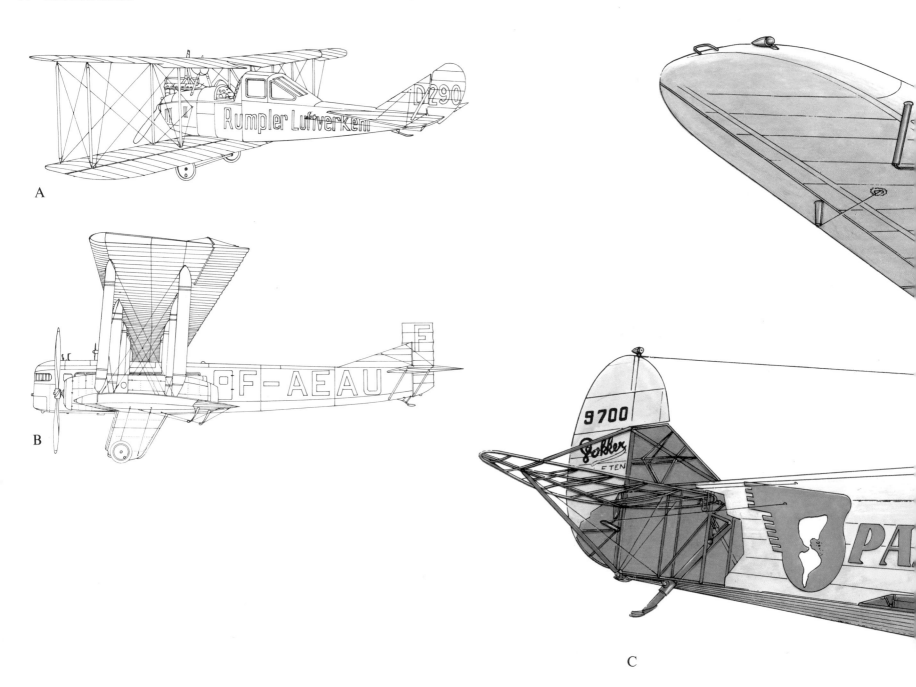

A

B

C

A Surprisingly, perhaps,
the first post-war
passenger air services
were operated by the
recently-defeated
Germans in 1919. The
aircraft used were
converted Rumpler C.I
two-seat reconnaissance
biplanes, only one of
which had an enclosed
cockpit for the two
passengers, as
illustrated.

B The aircraft with which
French airlines opened
up their first inter-
national services in
1919 were Farman
Goliaths. Developed
from a wartime bomber
design, they carried 12
passengers in two
cabins. Cruising speed
was 75 mph.

C Typical of the fine
Fokker monoplane
airliners of the late
1920s and early '30s
was this F.X Trimotor
of Pan American
Airways. Powered by
three 300-hp Wright
engines, it cost £16,000
to buy and carried
eight passengers
at 103 mph.

A First of the modern generation of all-metal streamlined monoplane airliners was the Boeing 247 of 1933. It was able to cross the United States in under 20 hours, carrying ten passengers, and was the first twin-engined monoplane able to climb with a full load after failure of either engine in flight. In addition to a retractable undercarriage, it embodied an automatic pilot, control surface trim-tabs and de-icing equipment.

B Another aircraft which had a profound effect on aviation development was the Supermarine S.6B seaplane which gained the Schneider Trophy outright for Great Britain in 1931. First aeroplane to set up a world speed record of over 400 mph, it provided the Supermarine and Rolls-Royce companies with experience which enabled them to produce a few years later the famous Spitfire fighter and its Merlin engine. Illustrated is the similar S.6 which won the 1929 contest.

E

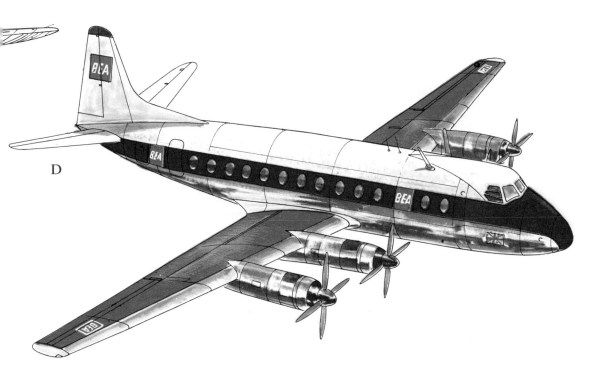

D

reliability and the luxury of an enclosed cabin. On some services, passengers even received full-course meals in flight, served by a smartly-uniformed steward.

RETURN OF THE MONOPLANE

The object of international race meetings, such as the contests for the Schneider Trophy, was to foster the design and development of advanced new types of aircraft. In the case of the Supermarine seaplanes with which Britain won the Schneider Trophy outright, with victories in 1927, 1929 and 1931, the aim was certainly achieved.

From the S.6B seaplane which won the final contest—and set up the first over-400-mph speed record afterwards—Supermarine's brilliant chief designer, R. J. Mitchell, evolved the Spitfire fighter. Its tiny, graceful shape prompted a German Air Attaché to describe it as a toy: how wrong he was became clear in the Battle of Britain, when Spitfires and Hurricanes shot the far-larger German air force from the sky over Southern England in 1940.

The same switch to all-metal low-wing monoplane design, with refinements such as a retractable landing gear, control surface trim-tabs, variable-pitch propellers, an automatic pilot and de-icing equipment, made the Boeing 247 airliner of 1933 quite revolutionary in a largely-biplane age. Its maximum speed of 182 mph made it as fast as many single-seat fighters of its day, and it reduced the travelling time on US transcontinental airline services to under 20 hours, carrying ten passengers.

Such aircraft brought to an end the long reign of the biplane. Its traditional advantage, of combining great strength with small span and weight, was less ·significant now that designers could build their aeroplanes of high-strength metals. In any case, the cleanest possible aerodynamic shape was essential if advantage was to be taken of new features like retractable landing gears and the greater power of new aero-engines.

Good as the Boeing 247 was, it was completely eclipsed by the new low-wing monoplane airliners introduced by the Douglas company soon afterwards. First of these was the DC-1 prototype, from which were evolved the DC-2 and DC-3. The latter was already gaining worldwide acceptance when World War II brought a demand for huge numbers of military transport aircraft. No type met the requirement as well as the DC-3 and more than 10,000 were built, serving under designations such as C-47 in the US Services and as the Dakota with the Royal Air Force. In Burma, a complete army was kept alive and in action by supplies delivered by and dropped from these aircraft.

When the war ended, the C-47s and Dakotas were sold to civilian airlines, to rebuild war-shattered networks of passenger services. They did the job so well, and were so unrivalled as money-makers in their class, that 30 years after the first of them flew,

C *For a long time passenger carrying aircraft had been three-engined. Certain aircraft like this Dewoitine with its retractable undercarriage and variable speed propellers were still in use after 1944.*

D *Another aeroplane which revolutionised air transport was the Vickers Viscount, which entered scheduled service with British European Airways on April 17, 1953. First turboprop airliner in the world, it not only offered new standards of performance but*

carried its passengers in vibrationless comfort above most of the "bumpy" weather that had previously induced air-sickness.

E *With its contemporary, the Hawker Hurricane, the Spitfire fighter won the Battle of Britain for the RAF in 1940,*

against great odds. In its earliest form, with a 1,030-hp Merlin engine, it had a top speed of 355 mph and carried the unprecedented armament of eight machine-guns. Later Spitfires, with Griffon engines, flew at up to 460 mph.

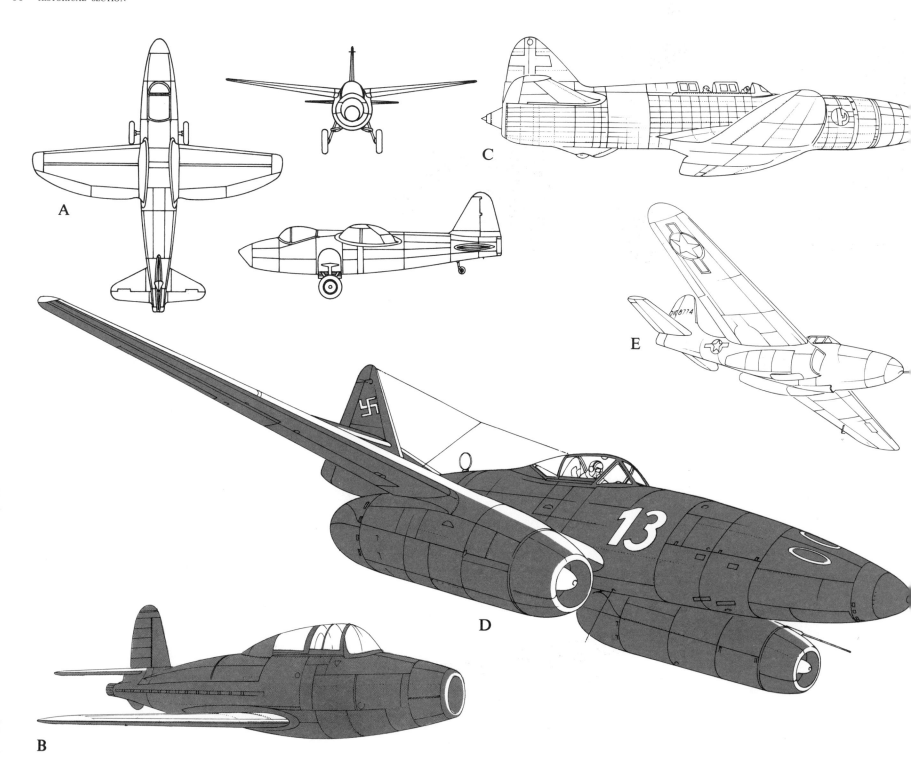

A   *The greatest single advance in aviation development since the Wright brothers' first flights was the introduction of jet propulsion. The earliest practical turbojet engine was perfected by Frank Whittle, in Britain, and ran successfully on April 12, 1937. The first jet-plane was, however, the German Heinkel He 178, flown on August 27, 1939. Three-view drawing of the Heinkel He 178.*

B   *First British jet to fly, on May 15, 1941, was the Gloster E.28/39, powered by an 860-lb thrust Whittle W.I turbojet. With a more powerful engine, the second prototype reached 466 mph and proved convincingly the capabilities of jet-propulsion.*

C   *Italy's first jet aeroplane, the Caproni-Campini CC 2, used an ordinary piston-engine to drive its compressor and was not very efficient. Top speed was only 233 mph. However, the CC 2 was the first jet aircraft to make a proper cross-country flight, from Milan to Rome, on November 30, 1941.*

D   *In August 1944, the Luftwaffe's K.G.51 fighter-bomber group became operational with Messerschmitt Me 262 twin-jet fighters. Although fast and heavily armed, the Me 262 suffered heavily at the hands of Allied piston-engined fighters like the Hawker Tempest.*

E   *America's first jet-plane, the Bell XP-59A Airacomet, flew on October 1, 1942. Powered by two I-16 turbojets, based on Whittle designs, it had a top speed of only 413 mph and was used mainly for training.*

F   *The RAF was first to use jet aircraft in action, on July 27, 1944, when Gloster Meteor fighters of No. 616 Squadron engaged German V-1 flying bombs over Southern England.*

in 1935, DC-3s still outnumbered any other type of airliner in worldwide service.

DROPPING THE PROPELLER

Back in 1928, a young cadet named Frank Whittle, at the Royal Air Force College, Cranwell, wrote a thesis on *Future Developments in Aircraft Design*. At a time when RAF fighters were flying at around 150 mph, he looked forward to the time when speeds of 500 mph would be attainable, at heights where the air is far "thinner" than at sea level. Propellers and pistons would be no good, and he suggested the use of rockets or gas-turbine engines.

There was nothing new in the basic idea. Gas-turbines, or jet-engines, for aircraft had been proposed in the old pioneer days, but serious work had to await a time when somebody produced metals capable of withstanding the intense heats and stresses involved. Believing that the time was approaching, Whittle patented his ideas for a jet-engine in 1930 and began translating theories into ironmongery, with the grudging approval of the Air Ministry. After overcoming every kind of difficulty and discouragement, he completed the first practical aircraft jet-engine and tested it successfully on April 12, 1937.

Unknown to Whittle, another young man named Pabst von Ohain was working on similar types of powerplant in Germany. Having rather better financial support, von Ohain was in fact the first to install a jet-engine in an airframe—the Heinkel He 178. It was not a very inspiring or very successful aircraft, but it was the first jet-plane to fly, on August 27, 1939.

The second, Italy's Caproni-Campini monoplane, was rather a side-track design. The compressor of its engine was driven by an ordinary piston-engine and top speed was a mere 233 mph.

When Britain's first jet aircraft, the Gloster E.28/39, took off for the first time on May 15, 1941, powered by a Whittle W.1 turbojet, it was far more impressive. Top speed was about 300 mph, although the engine's output was equivalent to only 688 hp at that speed. In other words, the E.28/39 flew nearly as fast as some of the first-line piston-engined fighter aircraft of its time on only half their power. When fitted with a more powerful turbojet, it achieved 466 mph.

Both Britain and Germany pushed ahead as quickly as possible with design and production of jet-fighters, unbeknown to each other. German air force chiefs hoped that the twin-jet Messerschmitt Me 262 taking shape in their country would be able to blunt the great Allied bomber offensive that was burning the heart out of German cities and industrial areas. But Hitler was obsessed with ideas of attack rather than defence and wasted precious months having the Me 262s modified into fighter-bombers. Thus, the first Squadron to go into action with jet-fighters, on July 27, 1944, was No. 616 of the Royal Air Force, equipped with Gloster

Meteors, each powered by two Rolls-Royce engines.

FASTER THAN SOUND

Higher speeds brought their problems. Even the faster piston-engined fighters sometimes ran into trouble during high-speed dives, losing their wings or tails for no apparent reason. Designers knew that their enemy was the invisible, seemingly harmless air, which became so compressed by the speeding aircraft that it formed almost solid shock-waves that hammered the structure until it broke up.

The shock-waves form when airflow over any part of the aircraft's structure reaches the speed of sound, which is 760 mph at sea level, dropping to 660 mph above 36,000 ft. By sweeping back the wings of their aircraft and making them thinner, designers managed to delay the shock-wave effects and gain a few extra precious miles per hour; but many experts doubted that aeroplanes would ever be able to fly above the speed of sound and newspapers began to write about the "sound barrier".

To discover if a specially-designed aircraft with a very powerful engine would be able to penetrate the barrier, the British government ordered a bullet-shaped research monoplane from the Miles company. It then got cold feet and cancelled the project.

In the USA, work continued on a small rocket-powered research aircraft known as the Bell XS-1 (later X-1), intended for a similar purpose. The pilot chosen for the attempt to smash through the sound barrier to supersonic speed was Captain Charles Yeager, who had no illusions about the hazards confronting him. Everything about the programme was strange and rather like science-fiction. To conserve fuel, it was even necessary to drop the X-1 from a converted bomber at a height of around 30,000 ft, instead of taking off normally.

Each time he flew, Yeager approached a little nearer to the speed of sound. Eventually he reached a speed of Mach 0.94 (94 per cent of the speed of sound), and felt the aircraft bucking under the hammer blows of shock waves that would have smashed anything else in the air at that time. But he had complete confidence in the X-1's structure and finally, on October 14, 1947, he opened up the four-chamber rocket engine to full power for an all-out attempt to reach supersonic speed. Fighting to keep control, he watched the needle on the Mach-meter swing past Mach .94, on to .96, .98 . . . suddenly, instead of getting worse, the hammering stopped. Yeager had become the first man to fly into the calmer conditions that lie beyond the "sound barrier". In doing so, he had proved that the barrier does not really exist.

By 1956, Britain's delta-wing Fairey Delta 2 research aircraft, powered by an ordinary jet-engine, was able to demonstrate that a properly-designed aeroplane can approach and pass the speed of sound with no more noticeable effect than a flicker of needles on the cockpit instruments as it does so. From that moment, it became only a matter of time

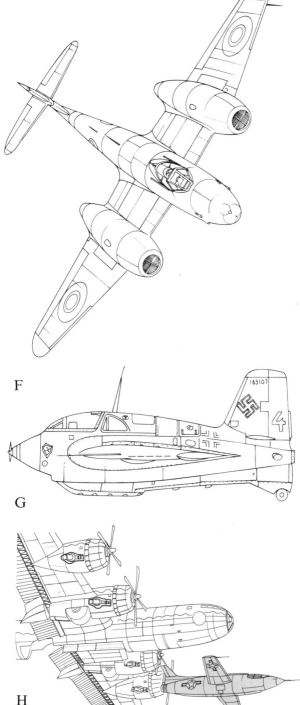

F

G

H

G *Fastest aircraft of its day, the 596-mph Messerschmitt Me 163 rocket-powered fighter went into action against American B-17 bombers on August 16, 1944. It was armed with two 30-mm cannon and 24 rockets, and had an endurance of only 12 minutes at full power, but this could be extended by switching off the engine and gliding.*

H *On October 14, 1947, Capt Charles Yeager proved that aeroplanes could fly beyond the speed of sound by accelerating the Bell X-1 rocket-plane to supersonic speed. This illustration shows the X-1 leaving the B-29 'mother-plane' from which it was launched at a height of 30,000 ft to save the fuel that would otherwise have been expended during take-off and climb.*

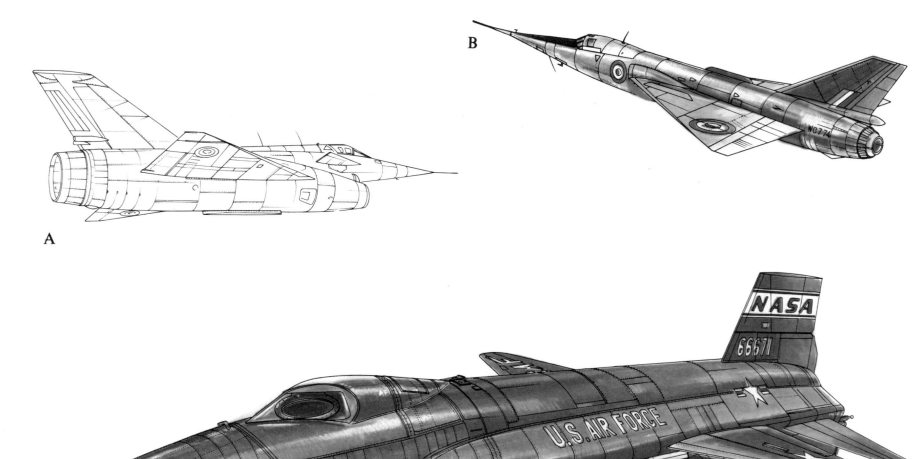

A

B

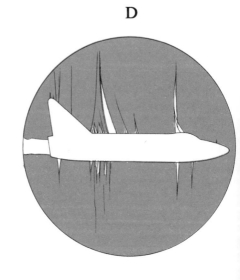

C

D

A In their quest for ever-higher performance, designers are testing a wide variety of different power plants. The French Nord Griffon had a turbojet engine mounted in the centre of a huge ramjet. Take-off and climb power was provided by the turbojet, which also started the ramjet at a predetermined speed and height.

B The world speed record was raised above 1,000 mph for the first time on March 10, 1956, when test pilot Peter Twiss achieved 1,132 mph in the Fairey Delta 2 research aircraft. More important, this aircraft showed that a carefully-designed machine could accelerate to super-sonic speed without any of the unpleasant buffeting that had been experienced with earlier high-speed aircraft.

C Fastest piloted aeroplane yet flown, the North American X-15A-2 was powered by a rocket-engine which thrust it to speeds of up to 4,534 mph (6.72 times the speed of sound). Like the X-1, it was launched in mid-air from a 'mother-plane'. The large external fuel tanks were jettisoned when empty.

D This drawing, based on a wind tunnel photo-graph, shows shock-waves on a model of the Lightning fighter in supersonic airflow.

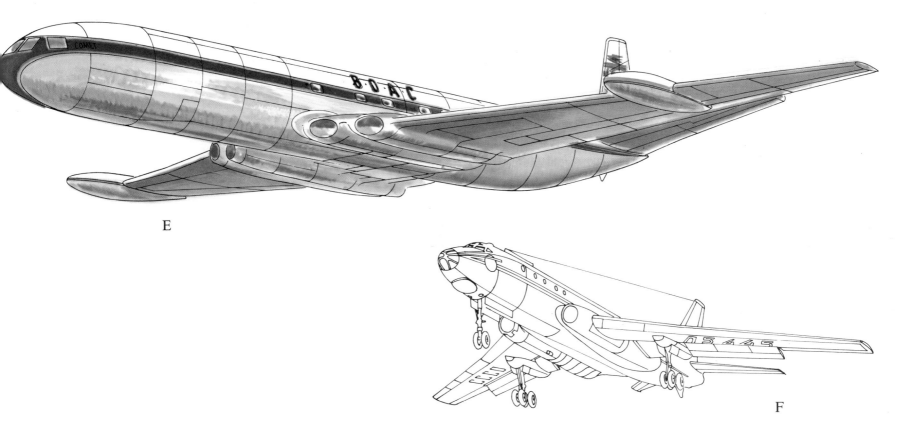

E

F

G

before airliners were designed to carry passengers at supersonic speed.

Nor has there been any slackening research that will lead one day to even higher performance for military and commercial aeroplanes. The rocket-powered North American X-15 has already carried pilots well beyond 4,000 mph and who can doubt that, one day, passengers will fly from A to B at the kind of speeds experienced so far only by globe-encircling astronauts?

SPEEDING THE PASSENGER

After World War II, Britain led the world in the new science of jet propulsion. Engines designed in the UK were being built under licence in the USA. Russia was grateful to be able to buy a handful of Rolls-Royce turbojets, which it stripped down to the tiniest component parts, copied and put into production, without licence, to power a new generation of combat aircraft that included the famous MiG-15 fighter.

Britain had no such lead in airliner design. On the contrary, a wartime agreement had designated America as the source for the large numbers of transport aircraft needed by the Allies, leaving Britain's industry free to concentrate on the manufacture of combat machines. When the war ended, therefore, Britain had no airliners to match the DC-4s, Constellations and Convairliners that were coming off the assembly lines on the other side of the Atlantic.

It was a time for bold decisions. BOAC and BEA, Britain's two state-owned airlines, were told that they would have to make do with airliners based on wartime bomber designs, supplemented by a few long-range airliners bought in America, until British factories could take advantage of their jet know-how by developing a range of revolutionary new airliners powered by gas-turbine engines.

Prototypes of the turbojet-powered de Havilland Comet and turboprop-powered Viscount were both flying less than four years after the war's end. They soon proved what their designers had hoped—that the greater speed and cruising height of turbine-powered airliners offered not only a saving in journey time but a smoother ride than passengers had ever before experienced, by climbing above most of the rough weather.

In the early nineteen-fifties, Britain's aircraft industry felt on top of the world. Nobody else could offer the airlines such fast, comfortable transports and operators queued up to buy them. Then disaster struck the Comet. Constant pressurising and de-pressurising of its cabin produced minute cracks, due to the little-understood phenomenon of metal fatigue. Two Comets fell out of the sky over the Mediterranean, with the loss of all on board, and the graceful flagship of Britain's civil air fleet was grounded, never to carry passengers again in its original form.

There followed one of the most skilful "detective stories" in aviation history. Using the latest techniques of television and salvage, the Royal Navy recovered from the floor of the Mediterranean almost the entire remains of one of the lost Comets. Scientists at the Royal Aircraft Establishment studied and analysed the shattered evidence and, to their credit, made their findings available to the aircraft industries of the world, to avoid any repetition of the

E *Britain's de Havilland Comet I was the aeroplane which introduced the speed and comfort of jet travel to the airline industry on May 2, 1952. It carried 36-44 passengers at a cruising speed of 490 mph. This drawing shows the later Comet 4.*

F *Second jet transport to enter scheduled service was the Soviet Tupolev Tu-104, evolved from the Tu-16 twin-jet bomber. Operation on Aero-*

flot's network began on September 5, 1956, initially in a 50-seat configuration.

G *A completely new fashion in jet airliner design was set by the French Caravelle, which carries its two Rolls-Royce Avon turbojets on the sides of the rear fuselage. Offering reduced cabin noise, easier maintenance and a 'clean' wing, the rear-engine layout became standard for many years.*

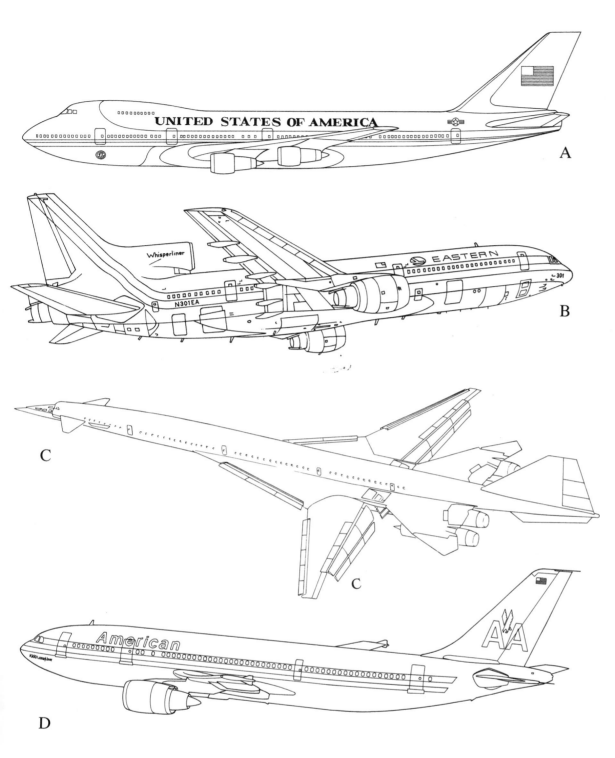

A *Another trend-setter is the huge Boeing 747, first of the 'jumbo jets'. Flown for the first time in early 1969, it is designed to carry up to 500 passengers, ten-abreast, in a 20-ft wide cabin.*
The 747 is used by the President of the USA as a personal transport called "Air Force One". It is equipped with an airborne command central.

B *Lockheed's L-1011 TriStar is the second of a new generation of wide-body airbuses to be conceived in the United States.*
The TriStar is powered by three Rolls-Royce RB.211 turbofan engines in a tri-jet layout.
It carries a total crew of 13 and has seating for up to 345 passengers in an all economy class layout.

C *First supersonic airliners to fly were the Anglo-French Concorde and Russian Tupolev Tu-144. They were to have been followed by the larger Boeing SST, designed originally with a 'swing-wing' as shown here but later intended, before the project was cancelled, to have a fixed deltashaped wing.*

D *One of the first long-range twin-jet airliners to fly the transatlantic route was The Airbus A300-600R. With a maximum load of 267 passengers, the range is 4,350 nautical miles.*

disasters. In doing so, they threw away for all time Britain's leading position as a supplier of transport aircraft to the airlines of the world.

By the time the entirely new Comet 4 was in service, and operating the first-ever jet transatlantic passenger services in October 1958, the Soviet Tupolev Tu-104—based on a jet-bomber design—was already in large-scale operation. Within weeks, the first big Boeing 707s followed the Comet across the Atlantic and, with Douglas's DC-8s, became the standard types used on long-range services in the West.

The Viscount continued to sell in large numbers, as it had no counterpart elsewhere. Similarly, when France produced its unique engine-at-the-rear Caravelle short-haul jet-liner, this also found a ready market. Sizes have continually increased, to cope with the demands of an ever-growing mass-travel market. Boeing's 747 Jumbo-jet was the first and biggest of the new "wide-body" airliners that could accommodate up to 500 passengers. The three-engined Lockheed L-1011 TriStar and McDonnell Douglas DC10 are somewhat smaller. The aircraft industries of Britain and France took a bold step forward by producing a 1,450-mph supersonic airliner, Concorde. It flew for the first time in 1969, and inaugurated air travel twice the speed of sound in 1972, enabling passengers from Europe to land in America hours before they took off by the clock.

In order to compete with the dominating manufacturers of passenger aircraft in the USA, a European tri-national consortium, Airbus Industries, was formed. The first Airbus, A-300, took its maiden flight in 1972, and since then a steady flow of technically advanced models has left the company's final-assembly line in Toulouse, France. Airbuses have found customers all over the world, even in America, thus breaking the total dominance of the domestic aircraft industry.

STRAIGHT UP, STRAIGHT DOWN

Unimpressed by the Wright biplane, Thomas Alva Edison once commented sourly: "The aeroplane won't amount to a damn until it can fly like a hummingbird, go straight up, straight down, hover like a hummingbird". History has shown that he was far from correct in his assessment of the fixed-wing aeroplane; but there has never been any doubt that aircraft able to take off and land vertically would offer many advantages compared with those that operate conventionally.

Most aircraft accidents occur during take-off and landing, when the planes are travelling at high speed close to the ground and the pilot has many things to occupy his mind and hands simultaneously. The hazards are increased when darkness and bad weather are added to his other problems. The ever-increasing size and speed of aircraft has only magnified the difficulties, and even passenger air-liners now hurtle along two-mile concrete runways, at the kind of speeds attainable by only the fastest racing cars, as they claw their way into the air.

Brilliant design, reliable power plants, skilful piloting and every conceivable kind of electronic aid combine to reduce the danger to the point where the average passenger can happily forget it; but aviation engineers have never forgotten the words of Edison.

Helicopter toys, in which a small rotor is spun up into the air by pulling on a cord, are almost as ancient as kites. The principle was first applied successfully to a full-size man-carrying aircraft in France in 1907; but more than 30 years of research were needed before the helicopter was developed into a practical vehicle by Igor Sikorsky—the same Sikorsky who built the world's first four-engined aeroplane back in 1913 but now living in America.

His little VS-300 prototype of 1939 has given birth to the thousands of helicopters that are today performing almost impossible feats all over the world.

Unfortunately, because they derive their lift and propulsion from a rotating wing, helicopters are unable to fly at speeds anything like as fast as those attained by fixed-wing aeroplanes. This has led designers to investigate all kind of different techniques for vertical take-off and landing. An entirely new form of flight emerged in 1954, when Rolls-Royce flew their fantastic "Flying Bedstead" vertical take-off and landing (VTOL) research machine, which was lifted off the ground by the thrust of two vertically-mounted jet-nozzles.

From the "Flying Bedstead" has evolved an incredible variety of different VTOL concepts, some of them already combining the ability to meet Edison's specification with supersonic performance. A few are illustrated on this page; others appear later in this book. Together, they offer exciting proof that the science of flight is still in its infancy and that future volumes of the "Lore of Flight" will have an even more fantastic and thrilling story to tell.

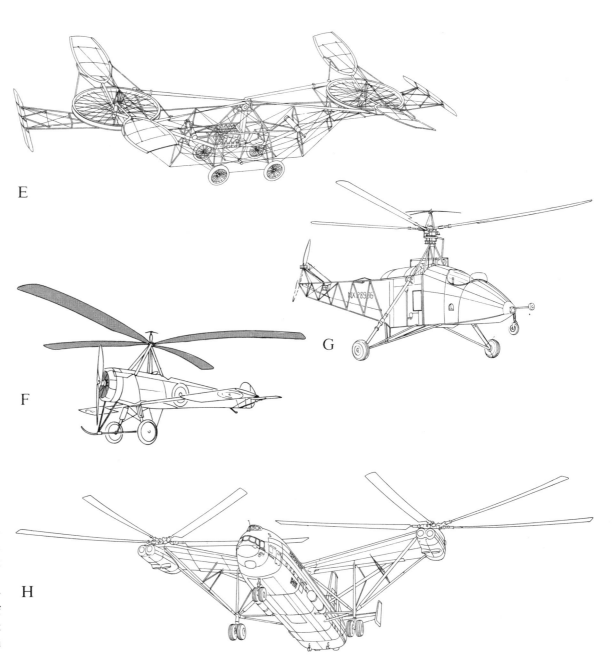

E Vertical take-off and landing (VTOL) is the key to improved safety and versatility. First heavier-than-air aircraft to take off vertically in free flight, carrying a pilot, was this twin-rotor helicopter built by Paul Cornu, on November 13, 1907.

F Greatest name in early rotating-wing design was Juan de la Cierva. The principles he evolved with his Autogiros, using unpowered rotors, made possible the modern helicopter.

G First entirely practical helicopter was Igor Sikorsky's VS-300 of 1939. It was the progenitor of all the 'single-rotor' designs of the present day and its success encouraged Sikorsky to found the helicopter industry.

H The Russian helicopter Mil Mi-12 weighs 105 metric tons: built for the transportation of civil or military loads, it can put these loads down in places where cargo planes cannot land.

I One of the world's most widely used helicopters is the Bell 206 Jetranger. First flown in 1962, it is still in production 30 years later. Powered by a 400-hp turboshaft engine, the largest version can carry up to five passengers and the pilot. The Jetranger has also found military use.

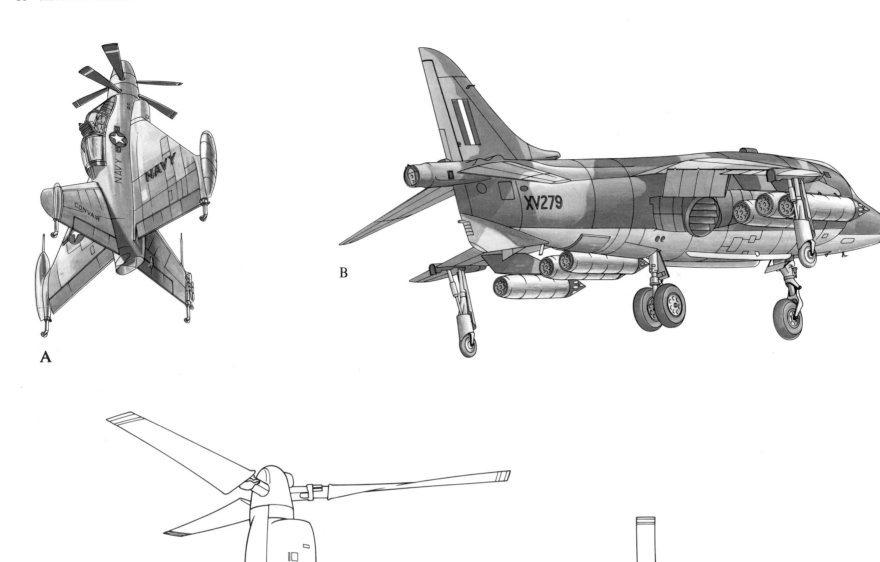

A

B

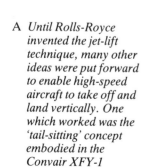

C

A *Until Rolls-Royce invented the jet-lift technique, many other ideas were put forward to enable high-speed aircraft to take off and land vertically. One which worked was the 'tail-sitting' concept embodied in the Convair XFY-1*

Pogo of 1954. The propellers, driven by a turboprop engine, acted as helicopter rotors during take-off and landing.

B *The Hawker Siddeley Harrier represents one of the truly significant combat aircraft in*

flying history. Its Pegasus turbofan engine is fitted with four rotating nozzles, by which the exhaust gases can be deflected for vertical take-off and landing. This enables the Harrier to combine the performance of a conventional jet-fighter

with unprecedented versatility.

C *The BellBoeing V-22 Osprey has all the advantages of a helicopter, and can take off and land vertically or hover with the engine nacelles in a vertical position. After lift-off,*

the engines are tilted forward and the Osprey becomes an aircraft capable of twice the possible speed of a helicopter. Designed for use by all four US armed services, it has a civil version with seating for 25 passengers.

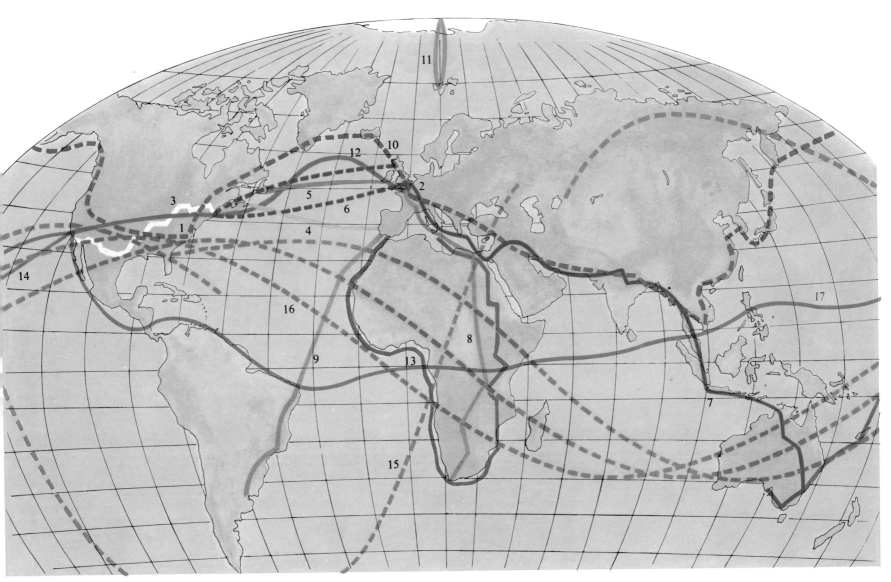

1 Site of the first flights by Orville and Wilbur Wright, Kill Devil Hills, North Carolina, USA. December 17, 1903.

2 First cross-Channel flight, from Barraques, France, to Dover, England. Louis Blériot on a Blériot XI monoplane, July 25, 1909.

3 First US coast-to-coast flight, from New York to Pasadena, California. Calbraith P. Rodgers on the Wright biplane Vin Fiz, in 69 stages, September 17-November 5, 1911

4 First transatlantic flight, from Trepassey Bay, Newfoundland, to Lisbon, Portugal, via the Azores. Curtiss flying-boat NC-4 of US Navy, carrying Lt Cdr

A. C. Read and crew of five, May 16-27, 1919 (flying time 25 hours).

5 First non-stop transatlantic flight, from St John's, Newfoundland, to Clifden, Ireland. Vickers Vimy aircraft, flown by Capt John Alcock and Lt Arthur Whitten-Brown, June 14-15, 1919.

6 First two-way transatlantic flight, from East Fortune, Scotland, to New York and back to Pulham, Norfolk. H.M. Naval Rigid Airship R.34, commanded by Major G. H. Scott, July 2-13, 1919.

7 First flight from England to Australia, from Hounslow to Port Darwin. Vickers

Vimy aricraft, piloted by Capt Ross Smith, and Lt Keith Smith, with crew of two, Novermer 12-December 10, 1919.

8 First fligth from England to South Africa, from Brooklands to Cape Town. Lt Col Pierre van Ryneveld and Major C. J. Quintin Brand, with crew of two, in two Vickers Vimys (one wrecked at Korosko, the other at Bulawayo) and a D.H.9 successively, February 4-March 20, 1920.

9 First flight across South Atlantic, from Lisbon to Rio de Janeiro, Brazil, via Las Palmas, Cape Verde, Porto Praia, Fernando Noronha. Capt Sacadura Cabral

and Capt Gago Coutinho of Portugal in a Fairey IIID Mk II Transatlantic seaplane (wrecked at St Paul's Rocks) and two Fairey IIID seaplanes successively (one wrecked at Fernando Noronha), March 30-June 17, 1922.

10 First round-the-world flight, from Seattle, USA, and back to Seattle. Two Douglas World Cruiser aircraft, commanded by Lts Nelson and Smith, April 6-September 28, 1924.

11 First flight over North Pole. Lt Cdr Richard Byrd, piloted by Floyd Bennett, in Fokker monoplane Josephine Ford. May 9, 1926.

12 First non-stop New York-Paris flight.

Charles Lindbergh in Ryan monoplane Spirit of St Louis, May 20-21, 1927. Flying time for 3,600-mile route was 33½ hours.

13 Circuit of Africa. Short Singapore flying-boat commanded by Sir Alan Cobham (23,000 miles). November 17, 1927- June 11, 1928.

14 First trans-Pacific flight, from San Francisco to Brisbane, Australia, via Honolulu and Suva, by Charles Kingsford Smith and crew of three in Fokker F. VIIB/3 m monoplane Southern Cross. May 31-June 9, 1928.

15 First manned space flight. Major Yuri Gagarin of Russia in the spacecraft Vostok. Single orbit (108

minutes), April 12, 1961.

16 First US orbital flight. Lt Col John Glenn in the Mercury spacecraft Friendship 7. Three obits (295 minutes), February 20, 1962.

17 First non-stop unrefueled round-the-world flight. Dick Rutan and Yanna Yeager in the all composite two pistonengine Voyagar. 65.000 mile in nine days. December 23,1986 – Janurary 1,1987.

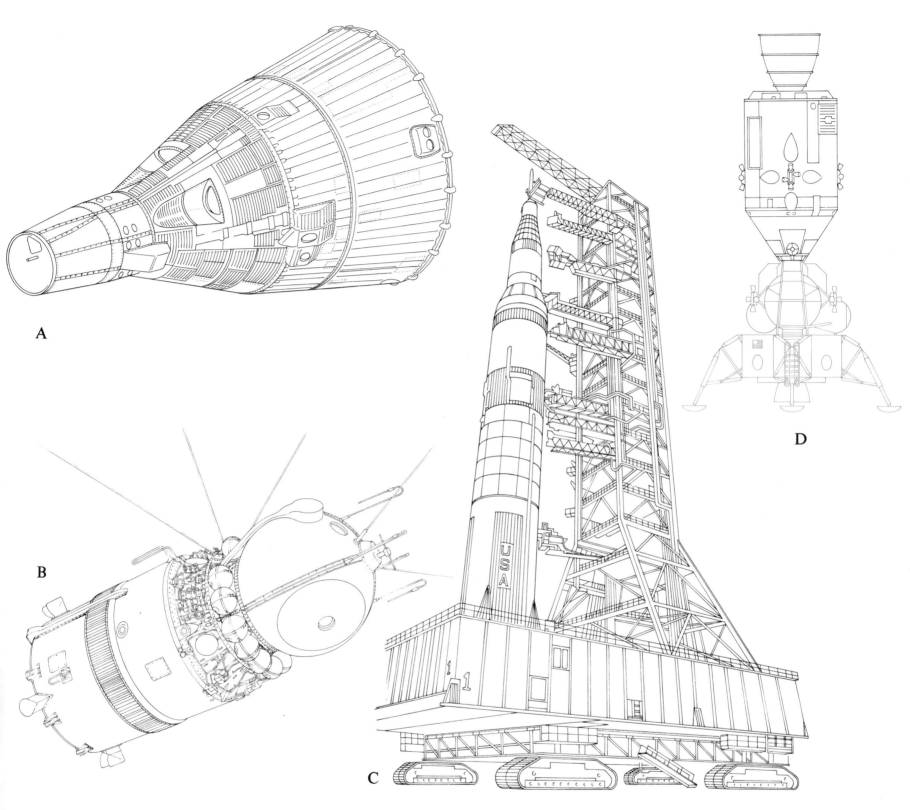

A

B

C

D

A Successor to Mercury in America's space programme was the two-man Gemini spacecraft. All ten launches, made between March 1965 and November 1966, were successful and included the first-ever rendezvous and docking experiments with target satellites in orbit.

B Since the late 1950s, the aircraft industry of the world has changed its name to "aerospace" industry. First human being to travel in space, on April 12, 1961, was Yuri Gagarin of Russia, in the Vostok I spacecraft. He made a single orbit of the Earth in 1 hr 48 min.

C Launch vehicle for the Apollo spacecraft is the huge Saturn V three-stage rocket, shown here on the crawler which carries it from the assembly building to the launch site at Cape Kennedy. Saturn V is 353 ft tall and has a launch weight of 2,723 tons, including the 42-ton spacecraft.

D The Apollo spacecraft was designed and built for the first American expeditions to the Moon. It is made up of three components. The drum-shape service module contains fuel, electrical power supply and propulsion units. The conical command module houses the three-man crew. The lunar module is used to take two of the astronauts down from the main spacecraft, in lunar orbit, to the surface of the Moon and then return them to the command module. Only the command module returns to Earth.

337   338   339   340   341   342

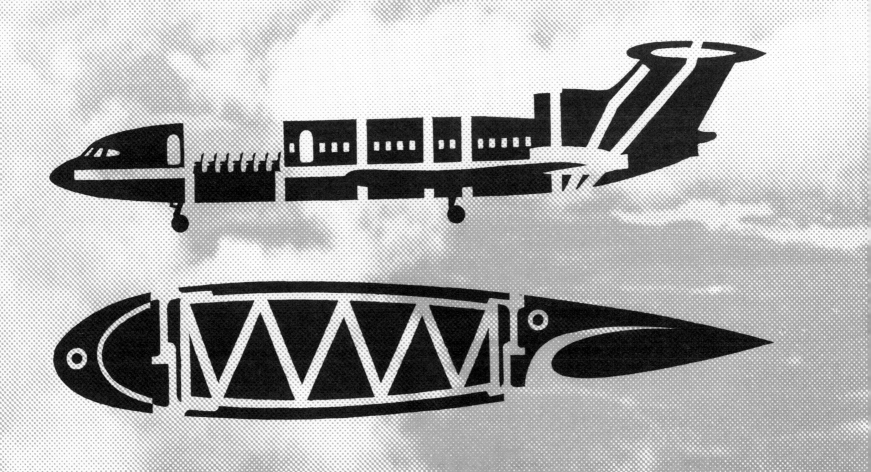

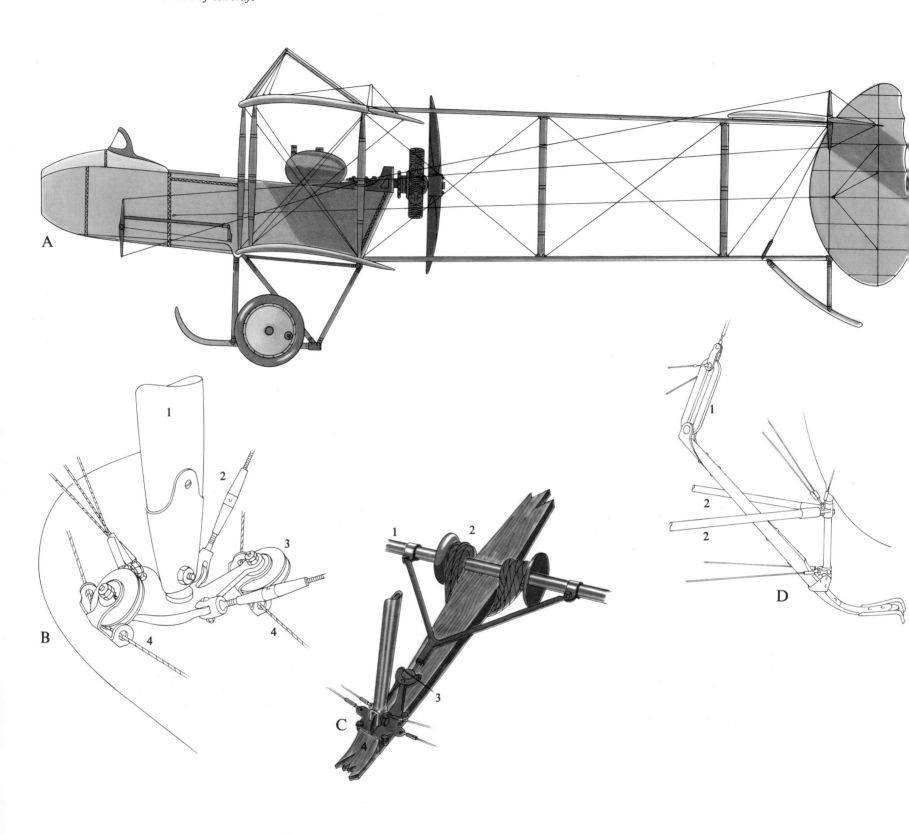

A Henry Farman 1913
biplane. The uncovered,
strutted and braced
fuselage structure is
typical of that employed
on many early
aeroplanes. The detail
drawings refer to an
earlier Henry Farman
design.

B This wing strut fitting
on the Henry Farman
also mounted the wing
warping cable pulleys
and lugs for the
interplane bracing.
1 Interplane strut
2 Turnbuckle
3 Pulley
4 Wing warping control
cables

C Henry Farman landing
gear. Rubber-cored
cables were often used
for absorbing landing
shocks.
1 Axle
2 Rubber cord
3 Ball joint
4 Skid member

D Sprung tail skid on the
Henry Farman. The
end of the skid was
turned down, to dig
into the ground on
landing and so bring
the machine quickly
to rest.
1 Rubber cord
2 Fuselage members

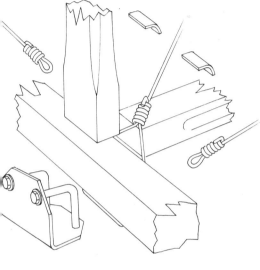

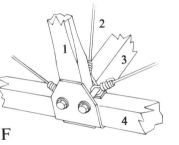

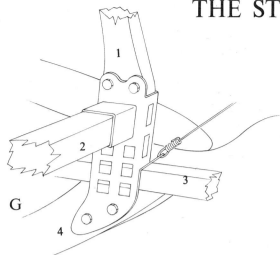

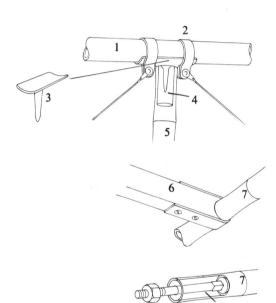

THE STRUCTURE OF AIRCRAFT

Aeronautical engineers have been defined as "men who must build for one pound of weight what any fool could do for two." This reference to weight highlights the primary difference between the science of aeronautics and that of other forms of engineering. It dominates every aspect of the structure of aircraft.

In the early days, the principal requirement of an aeroplane was that it should fly, and aerodynamic considerations made the methods of construction a secondary problem. Would-be aviators were, through lack of funds, usually the constructor as well as the designer and test pilot of their flying machines, and these early pioneers often used bamboo for the framework of wings and fuselages.

Bamboo, with its circular cross-section, was not all that could be desired from the viewpoint of reducing wind resistance, but this was not a serious disadvantage at the low speeds attained at the time. Essentially a natural tube, bamboo provided strength with lightness and had the additional attraction of not requiring any working. Once pieces of the required diameter had been selected, the constructor could use them in their natural state, by simply cutting them to the required length. An early book containing hints on the construction of flying machines advocated bamboo, adding that when using this material the only tools one needed to construct a full-size, man-carrying glider were a wood saw, pliers and a hack-saw.

When cut to the required length, the bamboo struts were connected together by simple fittings fashioned from sheet metal, usually brass because of its ease of working and resistance to corrosion. The resultant frames were then braced with wire to give them the required rigidity. The bracing was often extensive, and a joke current at the time was that builders of aircraft used to test their handiwork by releasing a bird in the centre of the structure; if the bird escaped a bracing wire was missing! In addition to providing rigidity, the bracing wires allowed the structures to be "rigged" into the correct shape by tightening or loosening the appropriate wires.

Spruce and ash ousted bamboo in due course. These light, strong woods made possible the evolution of better fuselage structures, with streamlined interplane struts, and were more suitable for the improved shapes of wing which were being

E *Strut on the Henry Farman with guides for control cables.*
1 Strut
2 Guide

F *Typical fuselage joint showing the attachment of two cross-members to a longeron.*
1 Strut
2 Bracing wire
3 Cross member
4 Longeron

G *The multi-duty steel fitting on the 1913 Sopwith Tractor biplane joined a vertical member, rear wing spar and landing gear strut to a longeron.*
1 Strut
2 Cross member
3 Longeron
4 Landing gear member

H *Bamboo, a natural tube of great strength and low weight, was used widely for the structure of many early aeroplanes. Shown here are typical methods of fitting a bolt into the end of a bamboo strut (bottom), the attachment of a rectangular member to a bamboo strut and a joint between two bamboo members.*
1 Longeron
2 Clip
3 T-piece
4 Wooden plug
5 Strut
6 Wood member
7 Bamboo
8 Wooden plug

developed. The very thin aerofoils of the early pioneers were replaced by thicker wing sections, developing more lift and permitting the use of deeper and more efficient wing spars. In order to concentrate material where it would do the most good, designers utilised spindled I-section spars, thereby taking advantage of experience gained by civil engineers with H-section girders.

A few early attempts were made to produce true "monocoque" fuselages. A monocoque structure is one in which all loads are taken by the outer shell, and not by an internal skeleton structure—the claw of a lobster is a good example of a monocoque in nature. Credit for the first is given to the British air pioneer Handley Page, who exhibited at the 1911 Olympia aero show a monoplane which had a highly polished mahogany shell fuselage.

Somewhat later came the British Deperdussin Company's monoplanes—the Thunderbug landplane and Seagull seaplane. Another early British, monocoque design was the Parnall Panther, intended for carrier duties, the fuselage of which, ingeniously, could be broken down into sections to facilitate stowage. The Sopwith Aviation Company fitted a Snail with an experimental planked monocoque fuselage early in 1918. The name, incidentally, was not a reflection on its performance, but indicated that the fuselage of the aircraft was essentially a "shell". During the 1914-18 War most German Albatros, Roland and Pfalz scouts had monocoque fuselages.

Many so-called monocoque fuselages, such as that of the Sopwith Snail, embodied certain internal stiffening members—the skin bearing most but not all of the loads—and these are better described as "stressed-skin structures". The true wooden monocoque fuselage, produced by wrapping and glueing sheets of thin plywood, came later and reached its zenith in the balsa-sandwich airframe of the de Havilland Mosquito fighter-bomber of World War II.

Wings are hardly ever of monocoque construction, because of the different nature of the loads imposed on them. A wing skin is strong in tension, but poor in compression unless supported by closely spaced stiffeners—for example, it can be dented easily by slight collisions, or by people walking on it on the ground. When an aircraft is on the ground, the skin on the undersurface of the wing is under compression due to the weight of the wing itself. In the air the situation is reversed and loads are greater. The wing carries the weight of the whole aeroplane, with the undersurface in tension and the upper surface in compression. The loads are often sufficiently high to enable passengers to see the skin wrinkling slightly between the ribs as a result of the compression loads in flight.

The world war of 1914-18 changed aircraft construction from an experiment to an industry. Up to that time the problems of aerodynamics had been more urgent than those of construction. Wood and fabric were readily available, and could be

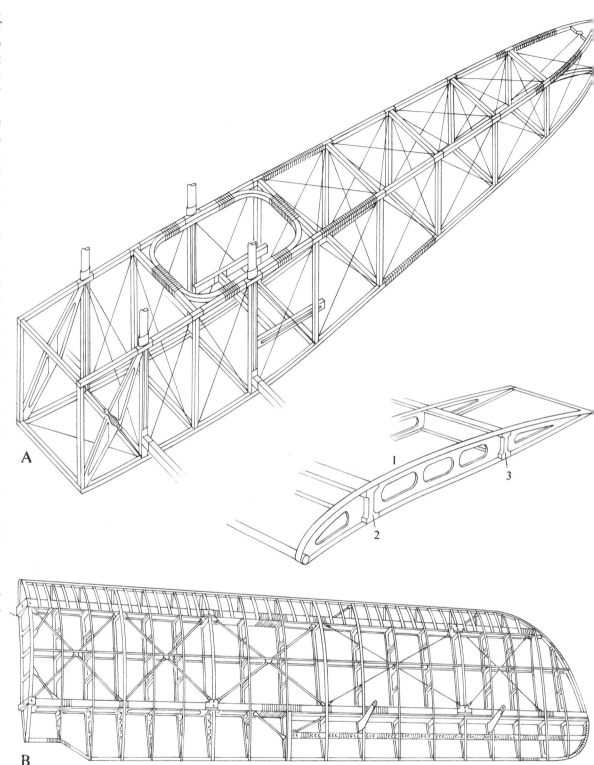

A *Wire-braced wooden fuselages of this type were used widely up to and often into the nineteen-twenties. The structure is basically an N-type girder, the four main longitudinal members, known as longerons, often being curved to provide the fuselage shape. Alternatively, light* shaped frames (known as formers) could be attached to the main fuselage members, notched to take stringers covered by either fabric or veneer.

B *Wooden wings like this were typical of the early nineteen-twenties. They were usually built around two spars, these being either spindled to an I-section out of solid wood, or fabricated as a hollow box. Ribs were built either as light girders or made of ply. The wings were braced* either by internal cables or by covering the leading-edge back to the front spar with ply. The rest of the wing was covered with fabric, tautened and made waterproof by the application of dope.
1 Rib
2 Front spar
3 Rear spar

fashioned easily by backyard inventor-aviators and their wives. Even established "manufacturers" used these materials, as quantity production was unknown and, anyway, it was far easier to find skilled wood-workers than men experienced in making lightweight metal structures.

When war broke out, the military potential of aircraft was quickly emphasised. An enormous demand was created, and from scattered sheds and motor-car workshops there grew the great aircraft manufacturing companies. Timber was soon at a premium, not only because of the demand but also because, so far as countries like Britain were con-cerned, it had to be imported, using shipping space needed urgently for food and troops. For Britain the position was made worse as she then relied on Russia for her supplies of three-ply wood and the latter's military collapse, coupled with the declining supplies of first-class spruce, led to much improvisa-tion. Searching for substitutes, manufacturers tried other woods, such as hickory, larch, cedar, cypress and poplar, but none was entirely successful.

The inevitable happened and metal began to re-place wood. Experiments were begun and towards the end of the war metal wings and fuselages were produced in considerable numbers. The most not-able developments took place in Germany. There, the first all-metal aircraft of light alloy was the Dornier RS1 three-engined flying-boat, produced in 1914. This was the forerunner of a number of large flying-boats and seaplanes used by Germany during the war.

Of greater significance were the metal aircraft produced by Hugo Junkers. In 1910 Junkers patented a thick-section cantilever wing, without external bracing. However, this epoch-making invention was not utilised successfully until 1915, in conjunction with metal construction. Junkers had considered using wood, but had concluded that his idea could be carried out efficiently only in metal. Of the relative merits of the two materials, Junkers commented: "Wood is obtainable only in fixed sizes and shapes of trunk and branch furnished by nature, whereas metal may be obtained in a nearly unlimited variety of qualities and dimensions. Furthermore, it can be shaped into any form, is more reliable, and its strength can be established more accurately and is unaffected by climate and atmospheric conditions."

To Junkers, metal was the only possible material from which to construct his revolutionary aero-planes. Metal was also more suitable for the mass production techniques which were being introduced to produce aircraft in the numbers demanded by military authorities.

The first Junkers machine, the J.1, a fully-cantilever monoplane of iron and steel, made its first flight in December 1915. Powered by a 120-hp Mercedes engine, it attained a speed of 105 mph. The authorities were sceptical of such a revolutionary design, mainly because they considered that a steel structure must be impractically heavy, and tried to

prevent further development. But Junkers went on.

In 1917 the improved J.2 appeared, still con-structed of iron and steel, followed in the same year by the J.4. This remarkable aeroplane, constructed mainly of duralumin, was a cantilever sesquiplane, intended for ground-attack duties. A two-seater, it was powered by a 200-hp Benz engine which gave it a top speed of 95 mph. The J.4 was followed by the J.7/9, the first cantilever low-wing monoplane fighter, and then by the J.10 of 1918, from which was developed the F.13, a small passenger machine.

It is interesting to note that one of the main reasons given by Junkers for adopting a low wing position was to minimize injury to the crew in the event of a crash. The idea was that the low wing, hitting the ground first, would absorb a large part of the initial landing impact. When retractable land-ing gears became feasible, the low wing position proved ideal as it allowed short, and hence relatively light, landing gear structures.

The influence of Junkers was to affect the whole course of aircraft evolution. To this pioneer goes the credit for designing and constructing the first practical cantilever wing aeroplanes, the first practical all-metal aeroplanes and the first low-wing monoplanes, all of which Junkers continued to develop during the ensuing years.

Junkers aircraft did not play an important part in the war, but the cantilever wing was used success-fully by Reinhold Platz, chief designer for Dutchman Anthony Fokker, in the outstanding Fokker fighters of 1917-18. Platz's wing employed box spars with plywood sides (webs) and heavier wooden tops and bottoms (booms). To prevent the wings from distorting, the early style of lattice ribs, which ran fore and aft across the width of the wing and gave it profile, were replaced by a plywood web type of rib, stiffened by flanges. These ribs divided the wing into a series of cells, each capable of resisting torsion, or twisting.

A further important contribution to structural evolution was made by another German designer, Dr Adolf Rohrbach. Junkers used corrugated metal for his wing and fuselage skins, which resulted in a high drag. As this type of skin was not able to carry a high proportion of the loads involved, it was also inefficient, particularly where wings were concerned. So, in 1919 Rohrbach started building metal wings embodying box-spars in conjunction with smooth-metal skinning which absorbed a relatively high proportion of the loads. This was the beginning of the modern idea of stressed-skin con-struction. The term "stressed-skin" is thought to have been used by Rohrbach for the first time in 1924. The technique was further evolved and improved by one of his colleagues, H. A. Wagner.

It was also Rohrbach who, through a brilliant lecture in the USA in 1926, helped to sow the seeds of thought from which evolved the revolutionary series of American transport aircraft of the nineteen-thirties.

All-metal structures were, however, in a minority

in the early 'twenties. At that time nearly all aero-planes were built as a skeleton, assembled on trestles and "rigged" by wires to the correct shape. The skeleton carried all the loads and was covered with fabric, tautened by brushing with cellulose dope, except where thin ply or unstressed metal was needed to form a "solid" decking around the cock-pit and the leading-edge of the wing ahead of the front spar.

Understandably, the end of the war brought a great drop in the demand for aeroplanes and a slowing up of structural development. In Britain the few hundreds of men employed on the construc-tion of aeroplanes in 1914 had grown to nearly 350,000 men and women by 1918. The United States, which entered the conflict with only 55 antiquated aircraft, had built up an industry capable of turning out aircraft at the rate of 21,000 a year, from 24 plants, at the time of the Armistice.

Civilian companies, formed to operate air trans-port services, give joy-rides or carry out air surveys, found it much cheaper to convert ex-military machines for their purposes than to support the development of new aircraft.

Technical advances were hindered therefore by the lack of money and the absence of demand for new aircraft for some six years.

In Britain the Air Ministry, remembering the shortage of timber which had tended to affect pro-duction at the end of the war and determined that it should not be a hindrance in the next, issued instructions that all future designs must be worked out in metal.

A half-way stage in the change-over at this point was that existing wooden aircraft which had been ordered for the Royal Air Force were re-designed in metal. Aerodynamic problems were shelved and designers concentrated on construction. Metal-lurgists, scientists and mathematicians undertook research and the structural engineers interpreted and applied the results. The problems of producing light alloys of lower weight and greater strength, of their treatment to facilitate production, of instability, fatigue and corrosion, were studied more closely than ever before.

Interested parties were soon offered the rare experience of being able to compare machines con-structed of different materials but having identical aerodynamic characteristics. The results were un-equivocal: the aircraft of metal construction were superior. The all-metal aeroplane had arrived in Britain, but the manner in which it came meant that it was barely possible to distinguish a metal machine from one of wood, without physically examining the internal structure. Thus, to a degree, structural development stagnated in Britain. The basic idea was bad from the structural evolution standpoint. It did not use efficiently the inherent strength of the metals, which made them ideal for the construction of cantilever wings.

A few metal fuselages were of welded construc-tion, but this technique was subjected to extensive

1 *Voisin LA5 (France):*
  Two-seat bomber. 1914
2 *Sopwith Tabloid (UK):*
  Scout/bomber: 1914
3 *Caproni Ca.46 (Italy):*
  Strategic bomber. 1918
4 *Gotha G.V (Germany):*

Strategic bomber. 1917
5 *D.H.4 (UK): Day-*
  bomber. 1917
6 *Handley Page O/400*
  (UK): Bomber. 1918
7 *Sikorsky Ilya Mouro-*
  metz (Russia): 1916

8 *Zeppelin (Staaken)*
  R.VI (Germany): 1917
9 *D.H.9A (UK): Two-*
  seat day bomber. 1918
10 *Caproni Ca.42 (Italy):*
  Triplane bomber. 1918
11 *Sopwith Cuckoo (UK):*

Torpedo-bomber. 1918
12 *Vickers Vimy (UK):*
  Strategic bomber. 1919
13 *Martin MB-2 (USA):*
  Night bomber. 1920
14 *Breguet 19 (France):*
  Two-seat bomber. 1924

15 *Fairey Fox I (UK):*
  Day-bomber. 1926
16 *Hawker Hart (UK):*
  Day-bomber. 1930
17 *Tupolev TB-3 (ANT-6)*
  (USSR): Bomber. 1932
18 *Martin B-10B (USA):*

Bomber. 1934
19 *Marcel Bloch MB 200*
  (France): Bomber. 1935
20 *Bristol Blenheim I*
  (UK): Bomber. 1937
21 *Junkers Ju87 D-5 (Ger.):*
  Dive-bomber. 1943

22 Savoia-Marchetti
S.M.79-II Sparviero
(Italy): Torpedo-
bomber. 1940
23 Boeing B-17G Fortress
(USA): Day-bomber.
1943

24 Junkers Ju 88A (Germ-
any): Bomber. 1940
25 Avro Lancaster I (UK):
Heavy bomber. 1942
26 Boeing B-29 Superfort-
ress (USA): Bomber.
1944

27 Ilyushin Il-2 Stormovik
(USSR): Attack. 1941
28 De Havilland Mosquito
B.IV (UK): Bomber.
1942
29 Convair RB-36D
(USA): Heavy

bomber. 1952
30 English Electric
Canberra B(I).8 (UK):
Bomber. 1956
31 Avro Vulcan B.2 (UK):
Strategic strike. 1960
32 Tupolev Tu-20 (USSR):

Strategic strike. 1961
33 Tupolev TU-160
(USSR): Strategic
bomber 1988.
34 Boeing B-47 A
Stratojet (USA):
Strategic bomber. 1950

35 Dassault Mirage IV-A
(France): Bomber. 1965
36 Convair B-58A Hustler
(USA): Bomber. 1960
37 Tupolev Tu-22 (USSR):
Missile-carrying
bomber. 1965

criticism in its early stages on the grounds that the joints were unreliable. This was not altogether justified, as with experience and care sound welded joints can be made; but they call for skilled labour, and repairs in the field are difficult.

A simple method of constructing metal-tube fuselages without welding was developed by Hawker Aircraft in Britain and was used on the long line of famous biplane types produced by that company, as well as on the Hurricane of Battle of Britain fame and for the front fuselages of the later Typhoon and Tempest fighters. The Hawker technique required that the tubes, where joined, should be square or rectangular in section, so that flat metal jointing plates would lie readily against them. Hence, either square tubes could be used throughout or, where round tubes were used, these could be "swaged", that is, pressed to a square section where a joint was required. In this construction the transverse members were often round tubes fitting into sockets attached by bolts to the fittings, the bracing being completed by wires.

Nowadays, many light aircraft utilise a similar steel-tube fuselage structure, but with a reversion to welded joints instead of the bolted type.

Some manufacturers, such as the Bristol Company, specialised in the construction of fuselage members from high-tensile steel strip during the between-wars period, claiming a significant reduction in weight at some increase in cost. In these fuselages some of the struts were built up from sheet strips, each strut comprising two halves riveted together.

Development was somewhat different on the Continent and in the United States, where a major innovation in aircraft construction in the mid-twenties was the advent of high-wing single-engined transport monoplanes able to carry a pilot and six to eight passengers over ranges of up to 500 miles. An outstanding example of this type was the Lockheed Vega, produced in 1927. Powered by a radial engine of 220 hp, the Vega could carry a pilot and six passengers for up to 550 miles at 110 mph. With a 425-hp engine, the aircraft could cruise at 135 mph, carrying eight passengers. It had a fully-cantilevered wooden wing and a wooden stressed-skin fuselage of an advanced streamlined form, built in two moulded sections.

Of particular importance to structural evolution was the practical demonstration that a stressed-skin structure could give the same volume of cabin space as that of a framework structure, at a saving of 35 per cent of the overall cross-sectional area, thus reducing drag and saving a great deal of weight. The success of the Vega set the fashion for this type of aircraft. It inspired world-wide development, and its influence extended to the design of larger transport aircraft.

Up to this time the monoplane formula had made little headway in the military field. Development in Britain had never recovered from an official ban on monoplanes enforced temporarily in 1912, after a series of accidents involving aircraft of this type.

Elsewhere, relatively-slow biplane fighters were preferred to monoplanes because of their excellent manoeuvrability. Biplane fighters reached their zenith with the Hawker Fury of 1929, which was the first military machine to exceed 200 mph, and the Gloster Gladiator, the last biplane fighter operated by the Royal Air Force.

Another significant event of the mid-nineteen-twenties was the appearance of the small fast monoplane. Its development was stimulated by air-racing, notably the Pulitzer Trophy races in the USA and the international Schneider Trophy races, and led directly to evolution of the modern fighter.

The 1927 Schneider Trophy race was especially significant for Britain, for it was won by the Supermarine S.5, designed by R. J. Mitchell. In 1929 the contest was won by Mitchell's S.6, and the third successive British victory was secured with the S.6B in 1931. The high-speed Schneider floatplanes —mainly British, American and Italian—pointed the way to the fighters of the future and provided ample opportunities for developing strong and light metal structures, perfecting streamlining, and evolving powerful engines. The S.6B was particularly important, as it was from this aircraft that the outstandingly successful Spitfire was developed.

The ever-growing use of steel and aluminium alloys led to the general adoption of semi-monocoque stressed-skin construction, first with a mixture of wood and metal and then with metal alone.

Fundamentally, this consists of using a number of formers or frames to give the fuselage its transverse shape and a metal skin stiffened by stringers to enable it to carry compressive loads. In wings, the place of the fuselage frames is taken by ribs, and spars are necessary to help provide the required stiffness.

An outstanding aircraft using stressed-skin construction was the Douglas DC-2, from which the even better-known DC-3, or Dakota, evolved. Not only did designer Donald Douglas (aided by a small group of engineers which included James "Dutch" Kindelberger and Arthur Raymond) evolve a first-class stressed-skin monoplane, he also—somewhat fortuitously—incorporated a durable multi-spar wing, thus pointing the way to airframes with "fail-safe" characteristics and long fatigue life.

A fail-safe structure is one in which several members carry the major loads, rather than one big member, the strength of the members being such that should one fail, the remainder can still carry the loads and enable the flight to be completed safely. Previously, few aeroplanes had flown more than 5,000 to 6,000 hours in their lifetime, but hundreds of Dakotas have flown more than 40,000 hours and one has exceeded 80,000. Before the advent of the DC-2, the importance of employing more than one load path had not been appreciated. (Fail-safe structures are described in greater detail later on).

A *Wooden tailplanes and elevators of early aircraft were structurally similar to wings of the period. Basically they consisted of spars, ribs, and leading- and trailing-edge members, with internal bracing and fabric covering.*

1 Tailplane

B *In typical early wooden fins and rudders, the rear member of the fixed fin was often made integral with the stern post or fin post of the fuselage. Rudders were either of the simple hinged type, or balanced—that is, a portion was carried forward of the hinge line so that wind pressure helped to relieve the effort required to move the rudder. The trailing-edge was usually formed from steel or aluminium tubing, bent to the required shape. Sometimes, however, wire was used, and the tight fabric covering pulled this to the scalloped shape seen on many aircraft of the 1914-18 War.*

1 Tailplane

C *Typical wing strut and bracing wire arrangements. It was usual to overbrace wing*

structures, as the duplication of members provided additional safety if one failed. This was particularly true in the case of military aircraft, where most specifications called for duplication at least of the flying wires. In addition to bracing the wing against the loads imposed during flying and landing, the wires were tautened or loosened to rig the surfaces and obtain the desired dihedral and angle of incidence.

1 Two-bay biplane.
2 Single-bay biplane.
3 Two-bay biplane with wire-braced upper wing extension.
4 Warren strut braced biplane.
5. Complex strut bracing to accommodate fuel tank in upper wing.
6 Two-bay biplane with strut-braced upper wing extension.
7 Unstaggered biplane with parallel interplane struts.
8 Staggered biplane (sesquiplane) with vee interplane struts.
9 Staggered biplane with parallel interplane struts
10 Staggered biplane with N-type interplane struts

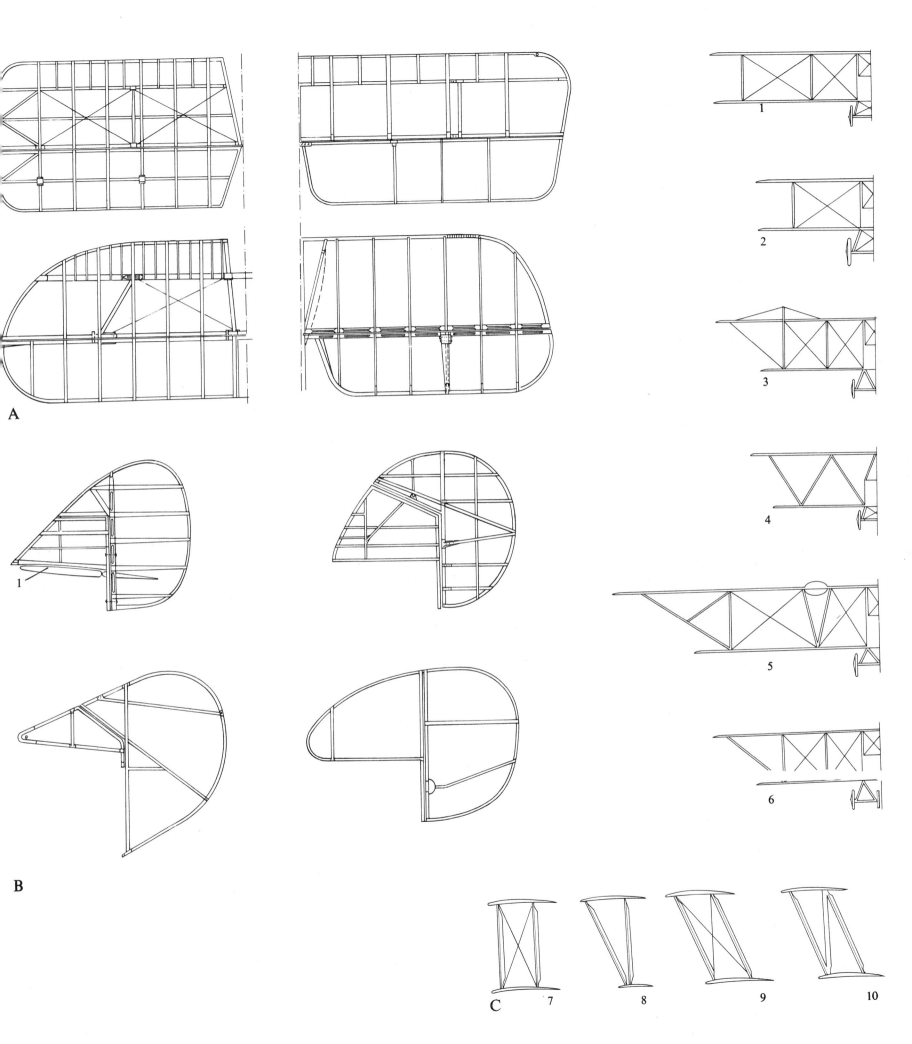

A

B

C   7   8   9   10

1   2   3   4   5   6

By the end of the nineteen-thirties, nearly all aircraft were of stressed-skin construction, made from light alloys and assembled with great precision in jigs. Jigs are structures which hold the component members of a particular assembly in their exact relationship to each other, to ensure that the finished part is accurate and interchangeable with others. Finally, a load-bearing stressed-skin is riveted in position to complete the section of fuselage or wing concerned, with the loads shared between the skin and the structure beneath.

The detail design of wing spars, fuselage frames and stringers shows great variety. Stringers of almost every conceivable section are used, individual designers and companies tending to use a particular section for their range of aircraft. For light aircraft, Z-section stringers and formers are usual. For big aircraft, lipped "hat"-section stringers are more widely favoured. Generally, the frames are cut away in the path of the stringers to allow the latter to be continuous members.

There is no general agreement as to whether the stringer should be attached to the frames at the point of intersection. In theory, in certain structures, there is no need for such a fixing, as the loads from one to the other will be transmitted by the skin. In other cases there is no doubt that a positive joint results in a more stable and stronger structure. Sometimes the two are attached to facilitate production, such an attachment holding the stringers conveniently in place prior to the attachment of the skin covering.

Aircraft using this technique included the pre-war de Havilland Flamingo, on which the stringers were located by a bent-up tab on the frame, and the Junkers Ju 88 and the "Zero", Japan's standard shipboard fighter during World War II. On this last aircraft the location was by a small angle piece riveted to both the frame and the stringer. Also of interest on this Japanese Mitsubishi aircraft was the use of exceedingly thin skin, by contemporary standards. This gave a relatively low weight, a high rate of climb and great manoeuvrability. It could not, however, withstand much combat damage and the "Zero" was outclassed by US fighters of more conventional construction.

The structure of an aircraft is a matter of careful compromise; too much emphasis on any one factor —in the case of the "Zero", ultra lightness—rarely results in a successful aeroplane.

An unusual arrangement was used on the Boeing B-17 Flying Fortress bomber. The stringers passed through holes in the web of the frames, leaving the outer flange of the frame continuous. Both stringers and frames were thus continuous across the intersection. The stringers were not joggled, but rose gently over the flange of the frame at each intersection. The same result can be obtained by attaching the stringers to the inside of the frames or—the method more widely used— the frames to the tops of the stringers without notching either member. This reduces significantly

the amount of riveting, only one or two rivets being required to secure a stringer to a frame at each intersection. This method of construction is used extensively on the Boeing 707 family of jet airliners, and was proposed for the Boeing supersonic transport aircraft, which was to have been constructed largely of titanium.

Tests on small panels indicate that tubular stringers could be up to fifteen per cent more efficient than extruded stringers, but problems of manufacture and internal corrosion have prevented their use.

To reduce the number of rivets required, some manufacturers have avoided the use of conventional stringers. The fuselage skinning is in the form of long narrow strips, with one edge turned up to provide the stiffness of a conventional stringer. The strips overlap, the flat edge of one sheet being riveted to the turned up edge of the adjacent strip.

A unique departure from the pre-war trend towards stressed-skin construction was afforded by the Vickers Wellington bomber, which utilised the principles of geodetic construction evolved by Sir Barnes Wallis. Based on experience gained during construction of the R.100 airship, it was employed first on the Wellesley bomber which set up a world long-distance record. In the Wellington, the geodetic airframe consisted of a basket-like light alloy frame-work built up from short curved sections, the whole being covered in fabric. Such a structure was immensely strong and able to withstand considerable distortion without failure. It was somewhat strange to see a test piece of geodetic construction still withstanding a heavy load in spite of severe distortion. One spectator, when shown a badly bent test fuselage, remarked that it looked like "a cockroach trying to bite its tail."

Geodetic construction enabled the designer to achieve a fine streamlined form for the fuselage, a high aspect ratio for the wings (i.e. long narrow wings) and maximum internal space. It also enabled the wings and fuselage sections and panels to be made in any convenient size, as the usual heavy transport joints between sections were eliminated. Disadvantages included the difficulty of incorporating the cut-outs necessary for doors, windows, gun positions and bomb-bays. Large-scale production also presented problems, the final solution being to construct the individual geodetic members on jigs with curvature in one plane only. The necessary twist was then imparted when the members were assembled into panels on the erection jigs.

An interesting feature of the Wellington was that the fuselage and wings were designed in subsections, the original intention being that each subsection would be finished as a complete unit before being finally assembled. All internal components, electrical cables, hydraulic pipelines and outer fabric covering were to be assembled while the sub-sections were in their respective jigs and readily

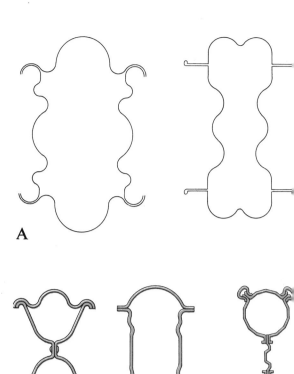

A

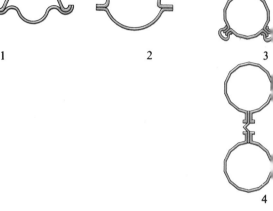

1　　　　　　　2　　　　　　　3

4

A *The curious shapes of these early metal strip spars resisted buckling under compressive loads. Generally speaking they fell into two broad types. One type was made in the form of a box from four strips of corrugated metal: the second type consisted of two tubular booms joined by a corrugated web. Spar 1 formed from* *four strips riveted together, was developed for use on biplanes and so was parallel throughout its length. Quite a lot of metal in this spar was not in the best position for taking applied loads, but in its time it was a light form of construction which pointed the way to more highly-developed types. Spars similar to 2 were used widely by*

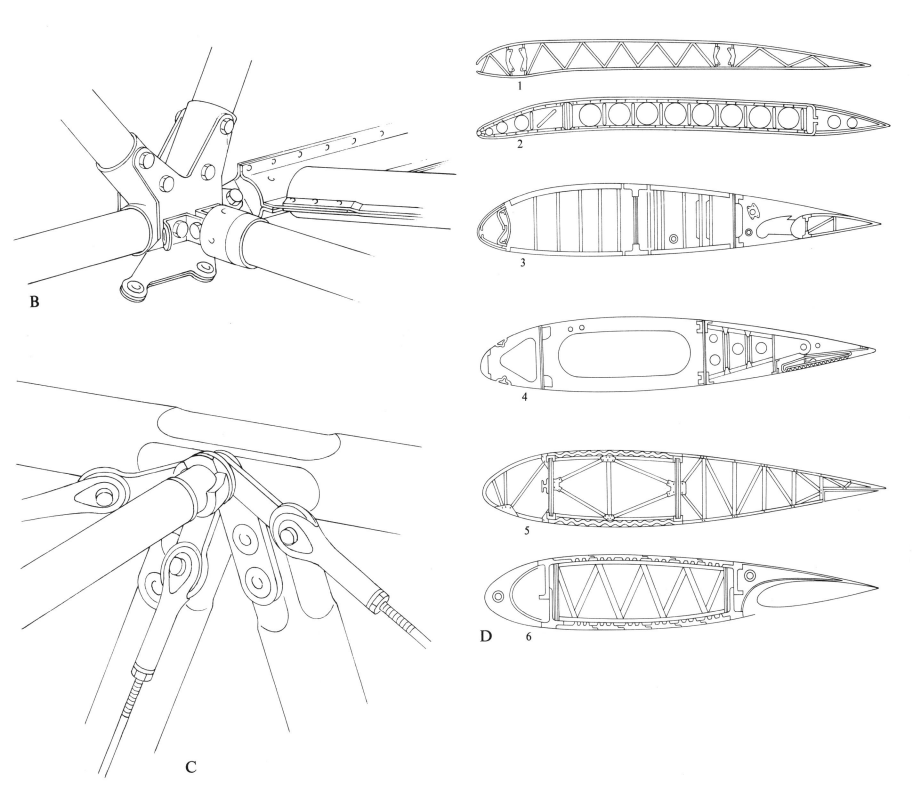

B

C

D 6

many manufacturers. The tubular-boom spar 3 consisted of no fewer than eleven strips, attached by six lines of rivets. This kind of spar was later simplified so that it could be made from only three strips, as in 4. Spars of this type were used in Hawker military biplanes in the period between the two World Wars.

B This fuselage joint was developed for use with members constructed from metal strips. Such members were more expensive but lighter than simple tubes of equal strength.

C This simple fuselage joint was developed by Hawker Aircraft. Round tubes were made square in section at the joints, so that the metal jointing plates could be bolted flat against them. Thus, either square or round section tubes could be used.

D Metal wing construction arrived in various forms. Rib 1 has channel-section flanges and ties. Rib 2 incorporates a web made from sheet metal, lightened by circular holes. Later such holes were flanged to provide stiffness. In 3, the wing is built up on a single spar fabricated from two T-section extruded booms and a sheet web. 4 shows a wing built around two spars. 5 is a box-spar wing, with vertical sides of sheet metal reinforced by vertical stiffeners and internal bracing members. The upper and lower surfaces of the box have large corrugations running the length of the wing. 6 shows a box-spar wing incorporating fail-safe features. Each spar is fabricated from two pieces to prevent a crack from extending the depth of the spar. Similarly, the wing skin is applied in narrow strips; any crack starting in one panel does not spread to the next. This technique ensures an adequate, though reduced, factor of safety until the crack is discovered.

accessible. The intention was to overcome the crowding and delay which always happen when such equipment has to be installed within the confined space of fuselage and wings.

Unfortunately, at the time these theories were worked out there appeared little prospect of orders big enough to justify the elaborate production organisation entailed. Thus, no steps were taken to put the original production scheme into practice. When big orders did come, the need for early deliveries was so urgent that the original plan was impractical under the circumstances and was never used. The sub-assembly idea has, nevertheless, since been applied successfully to many aircraft.

One of the outstanding features of the Wellington was its capacity to withstand heavy battle damage. So well did the geodetic structure spread the loads that many a "Wimpey", as they were affectionately called, returned from a bombing raid despite damage so severe that aircraft of conventional construction would have fallen apart in the air. On one occasion a Wellington returned and landed safely with its fuel tanks literally hanging down through the shattered underside of the wing, having flown several hundred miles in that condition. Repairs to damage caused by enemy cannon and anti-aircraft fire proved easier to effect than might have been expected, the small size of most of the members—coupled with the fact that they were made in only a small number of different sizes—facilitating replacement.

The geodetic principle was not adopted by any other manufacturer, but it served the Wellington well, this aircraft being the only British bomber to remain in production for the whole duration of the war.

With the advent of jet propulsion and the major increase in speeds that this brought, stressed-skin structures underwent considerable refinement. The rapid growth of air travel after the war, and the resultant increase in the number of hours flown in a year, emphasized the importance of the fatigue problem. Combat aircraft such as fighters and bombers rarely log more than a total of 3,000 hours in the air during the whole of their service life, but a civil airliner can exceed this figure in nine months. As a result fatigue failure occurred in a number of post-war airliners.

Even quite modest loads, if applied in endless reversals, can cause metal fatigue. Few people have not broken—or fatigued—a simple wire paper clip by bending it back and forth. Test pieces can be used to indicate the fatigue life of individual components, but a disturbing feature is that such tests reveal a wide variation, or scatter, in the number of reversals that can be withstood. Another difficulty is that tests and test pieces rarely reproduce exactly either the part or the conditions on an actual aircraft.

Careful design is essential in order to produce structures with good fatigue lives. Of vital importance is the prevention of stress concentrations,

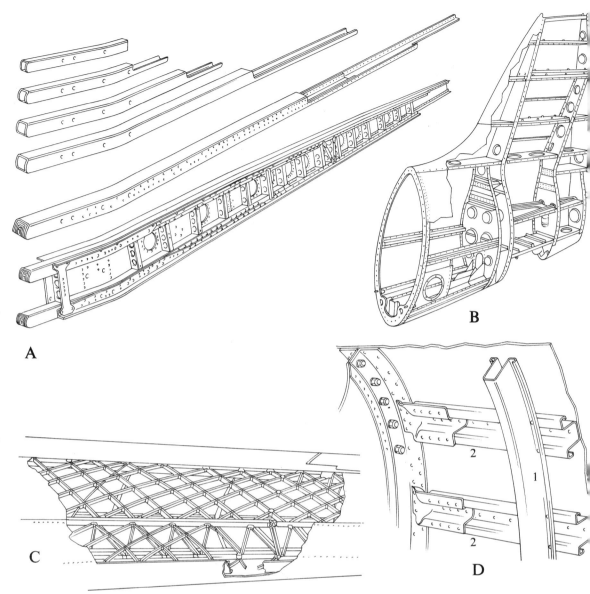

A

B

C

D

A *The wing spar of a Spitfire fighter was built up on two booms of square section, each fabricated from a number of tubes fitted inside each other. At the root end, the boom comprised five tubes and was almost solid. As the loads on the spar decreased progressively towards the wing tip, tubes were eliminated, beginning with the innermost, until only the two outer tubes remained. These were cut away on the upper side to form a doubled-up channel section. Quantity manufacture presented problems, as each tube had to be a push fit*

into and over its neighbours; thus, both outside and inside dimensions had to be accurate within very close limits.

B *The Spitfire tail unit was a typical stressed-skin structure. An advanced feature was the extension of two frames upward to serve as spars for the integral fin.*

C *Geodetic construction was used with great success on the Wellington bomber. This unique multi-load-path structure was able to absorb a great deal of battle damage without failure.*

D *Fuselage frame and stringer joints can be made in many ways. Here both members are continuous and a saving in riveting is effected, at the expense of some reduction in cabin width. This last consideration is not so important in large aircraft, so the technique is used on the Boeing 707 family of jet liners and is to be used also for the titanium structure of the same company's supersonic transport.*
1 Frame
2 Stringers

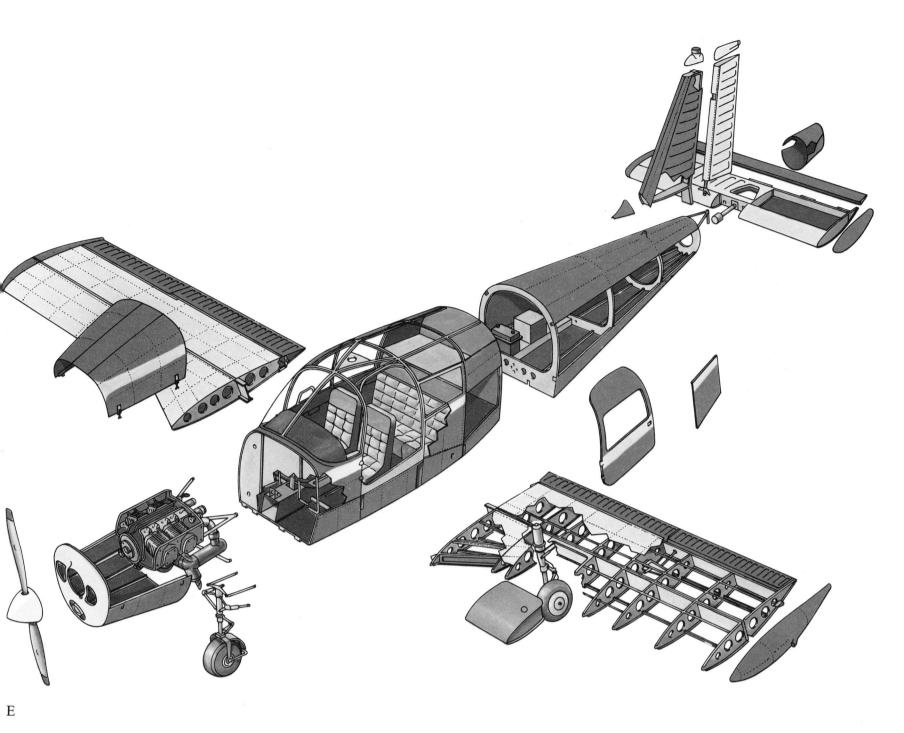

E

F *This drawing of a Piper Cherokee shows the main structural components of a typical modern small aircraft. Intended for mass production, the two-to-four seat model is made up of only 1,200 parts. The fuselage is built up on a punt-like floor assembly of press-* *formed sheet aluminium alloy components. The sub-assembly is built in one fixture, where it is riveted together using hand-held pneumatic rivet guns. The cabin super-structure, the front section incorporating the firewall, and the floor are then brought together and riveted in* *an assembly fixture, the process occupying three men for about four hours. The simple tailcone is attached in another fixture and the fuselage is then ready for the installation of controls and equipment Wing construction is conventional. The main spar is an extruded I-section light alloy* *beam. Stringers are attached to the main skin panels by an automatic machine which drills and countersinks the holes, and pops the rivets in position while the panel is held securely in a fixture which is manhandled to bring the head of the riveting machine into position.* *The main fuselage skin panels are treated similarly. Wing ribs and fuselage frames, and many small brackets, are shaped on a rubber-press which works nine hours a day and can be loaded and unloaded simul-taneously. Of interest is the use of glass-fibre* *for the cabin doors on the Cherokee Six, and for the wingtips containing integral fuel tanks. Piper has experimented with all-plastic airframes and is one of the leaders in the use of this promising new material in light aircraft.*

415   416   417   418   419   420

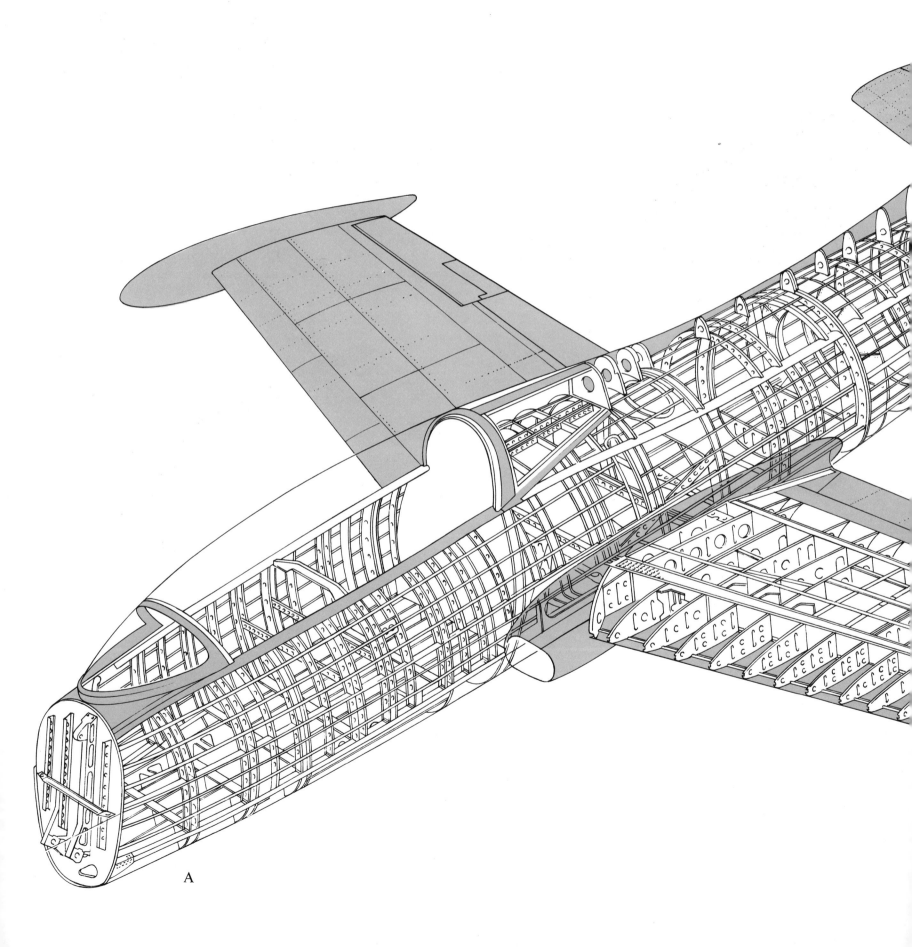

A

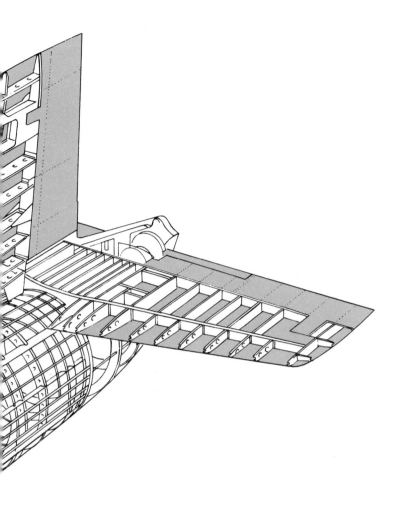

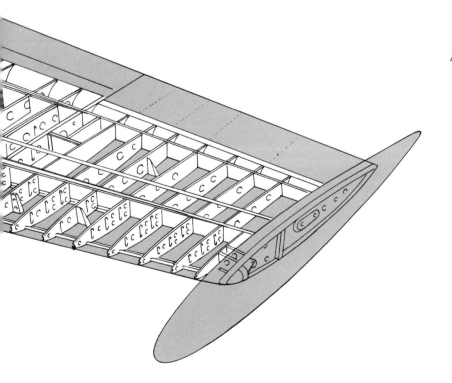

A *Aermacchi MB.326.
The structure of this
widely-used jet trainer
has to be rugged and
simple, to absorb rough
handling by pupil
pilots, but the aircraft
has to be fast enough
to prepare them for
faster, more sophisti-
cated types they will
fly next. The fuselage
is a stressed-skin, light
alloy structure, built
up on four main
longerons with formed
frames and close-
pitched stringers. It is
constructed in four
sections: the detach-
able nose section,
containing the nose
landing gear and the
electronic bay; the
main section, containing
the pressurized cabin,
the wing and engine
attachment points and
systems bays; the rear
section, attached with
four quick-disconnect
bolts, containing the
jet-pipe and carrying
the tail surfaces; and
a stainless steel tail-
cone. Two robust keel
members are incor-
porated in the lower
fuselage to provide
maximum protection
for the crew and
minimize structural
damage in a wheels-up
landing. The wing is
built around a main
front spar and an
auxiliary rear spar,
forming a two-cell
torsion box. The skin
is of flush-riveted
panels of constant
thickness. The skin
panels are stiffened by
closely-spaced ribs and
false ribs, with only
two spanwise stringers.
Attachment to the
fuselage is by two bolts
at the front spar and a
single bolt at the rear
spar. Landing gear
loads are absorbed
directly by the main
spar, the web of which
is double between the
landing gear attach-
ment and the fuselage.
Fail-safe and safe-life
design criteria have
been employed, to
obtain a structure of
great reliability and
strength.
See also illustration
on pages 84-85*

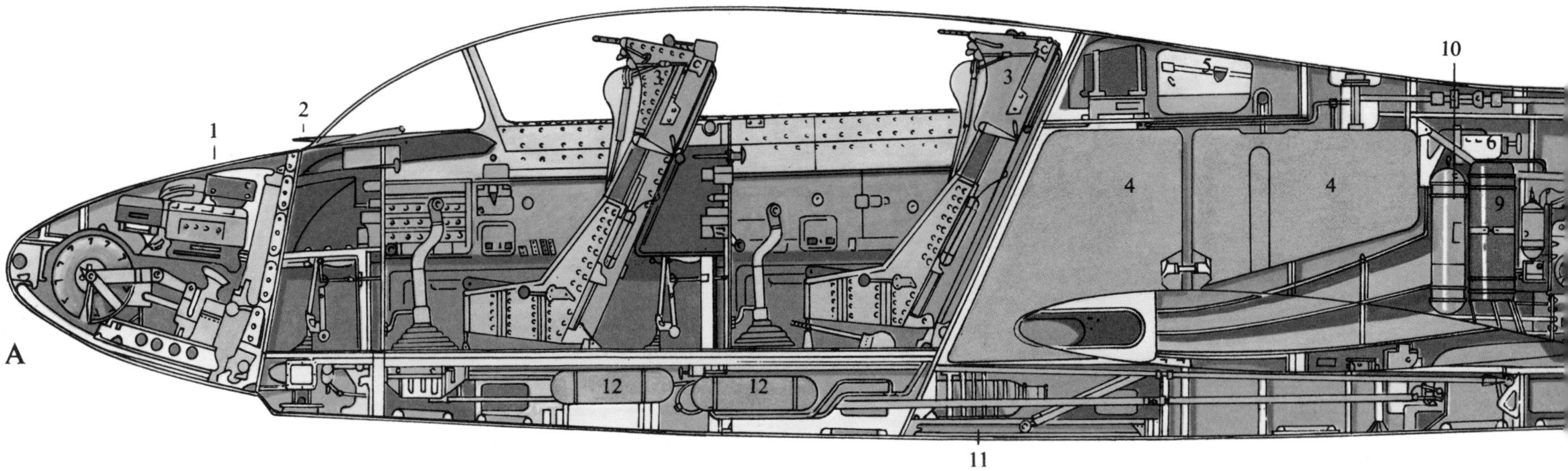

A *See also caption on
   page 83*

1 Radio equipment
2 Pitot head
3 Ejection seat
4 Fuel tank
5 Luggage compartment
6 Batteries
7 Rear fuselage junction
8 Firewall
9 Hydraulic system tank
10 Hydraulic accumulator
11 Air brake
12 Oxygen cylinder

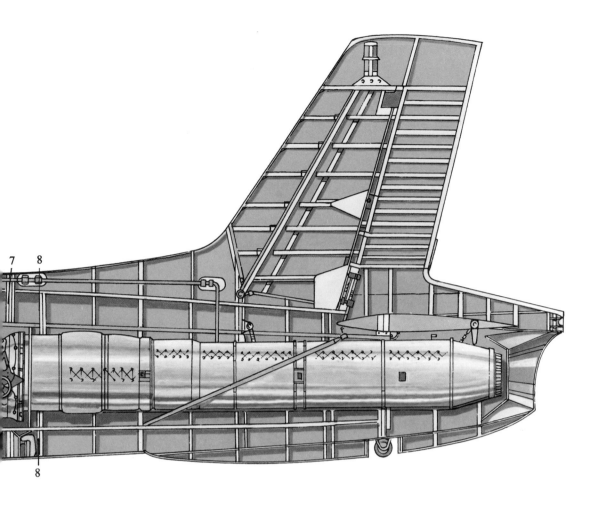

7 8

8

by avoiding abrupt changes of section and by the use of gently blending radii and good surface finish. Bolts have to be precision ground and attachment holes carefully positioned and reamed (i.e. bored to very fine limits).

Two main structural concepts have been evolved to overcome the fatigue problem. One is known as "safe life" and the other "fail safe". A safe life structure is one so designed that, according to calculations and experience gained in testing similar parts, it will last a given number of hours before a major failure occurs. Cracks may develop before the end of this period, but they will not extend catastrophically. The rate of propagation is such that any cracks will be detected during one of the regular inspections before they have significantly weakened the part. It is rarely practicable to "life" an entire airframe, but it is done frequently in the case of individual parts of the structure which are considered to be particularly susceptible to fatigue.

As indicated by the name, a "fail-safe" structure is one which, if it fails, does so safely. This is achieved by duplicating primary load-carrying members to ensure that an alternative path is available for every load. Then, if one member should fail, the remaining structure will continue to carry the loads and no catastrophic chain-reaction will result. The idea is similar to that of a man wearing both a belt and braces to make sure his trousers do not fall down. If one of his supporting structures breaks,

the remaining support will keep the trousers up.

Fatigue is also being combated by the development of better materials. The first generations of turbojet and turboprop airliners were built of high-strength "zinc-rich" aluminium alloys. Unfortunately, high strength did not automatically lead to long fatigue lives. On later airliners "copper-rich" alloys have been introduced. Although these are of lower strength, they tend to remain crack-free; if a crack does develop, its rate of propagation is much slower.

Increases in aircraft performance and size have necessitated the development of manufacturing techniques such as welding, bonding, chemical etching and the use of machined integrally-stiffened skin panels.

Although used widely before the importance of fatigue was generally appreciated, welding and bonding have gained rapidly in popularity owing to their ability to eliminate stress concentrations. Welding has rarely been used in primary structures of anything but light aircraft, but it is being used increasingly in the manufacture of advanced supersonic aircraft, as explained later.

Bonding, a process similar to glueing, is used to attach stringers to wing and fuselage skins, and to attach doubler-skins around door and window cut-outs, or wherever a local increase in skin thickness is required to reduce stresses. Bonding can produce a joint as strong as the basic metal itself, saves time and weight by eliminating riveting,

and reduces leakage from pressurized fuselages and integral wing fuel tanks. Most widely used bonding is the Redux process, in which the parts to be joined are placed on each side of a piece of adhesive cut from sheet, and bonded in an autoclave under heat and pressure.

Chemical etching—or chemical milling as it is sometimes known—consists of forming a part from a sheet thicker than required, which is then placed in a bath of weak acid until areas not protected by an acid-resistant coating have been etched away with great precision. Chemical etching is often used on wing skins, the skin being thinned down where loads are lowest, to save every possible ounce in weight, and left thick around access panel cut-outs and in those places where attachment is required. The process permits gradual changes in section, and fatigue-inducing machining lines are avoided.

The machined, integrally-stiffened skin is produced by a similar process involving conventional machining. Initiated in the United States, it involves the literal cutting of components from solid metal. During the process the various stiffeners and ribs, and any local strengtheners around access panel cut-outs, are machined integrally with the skin. The skin is often machined so that it tapers in thickness, being thinned down wherever possible to save weight.

As with bonding and chemical etching, this method of construction, by eliminating rivet holes, eases problems of sealing. More important, good fatigue properties can be obtained more readily, through the absence of regions of high localised loading which can occur at rivet positions, and also because large radii can be provided at intersection points to "smooth out" loads. A machined part can often be made smaller and lighter than its fabricated counterpart.

There are disadvantages in integral machining. The properties of the thick billets of metal required are not always as good as those of sheet. Anti-corrosive protection of sheet can be achieved by external coating of aluminium, but the protection of machined components involves surface treatment and paint processes. Large, expensive machine tools are also required, giving an accuracy to within thousandths of an inch on slabs of light alloy 30 ft or more in length.

Mistakes in machining can be extremely costly, for even a minute error, such as machining away a little too much from a stiffener, may necessitate scrapping of the complete panel. To eliminate mistakes, the machine tools are often controlled automatically by an electronic tape. There are other advantages in using automatic controls, including the saving of the time that otherwise would be spent in tedious manual positioning of the cutter and a great improvement in accuracy. In addition, cutter positions can be reproduced precisely at any time, without the need for expensive master jigs or templates.

The mechanical properties of many light alloys are at their highest when they are rolled or drawn without any subsequent machining, and this has led to the forging press technique in which large panels are squeezed to shape under a force of thousands of tons while they are in the hot plastic state. The forging presses required are immensely expensive, but the simplicity of manufacture, once the dies to shape the part have been made, and the strength of the finished product make this process increasingly attractive for sophisticated structures.

All the structural concepts so far described rely upon a degree of flexibility to allow aeroelastic deflection under load—e.g. the bending wing that "gives" slightly in turbulent conditions. However, the advent of supersonic flight, with its greatly increased air loads, the need for exceptional aerodynamic accuracy, and the phenomenon of kinetic heating—heating caused by air friction—brought new problems for structural designers.

The need for aerodynamic accuracy makes structural stiffness of vital importance. This is especially true for aircraft designed to fly fast low down, such as military ground attack machines, on which immensely strong and stiff wings are required.

Kinetic heating can induce high temperatures which have two main effects on the strength of conventional light alloys. First, the static strength begins to decrease significantly at around 100° Centigrade. Secondly, the resistance to creep, that

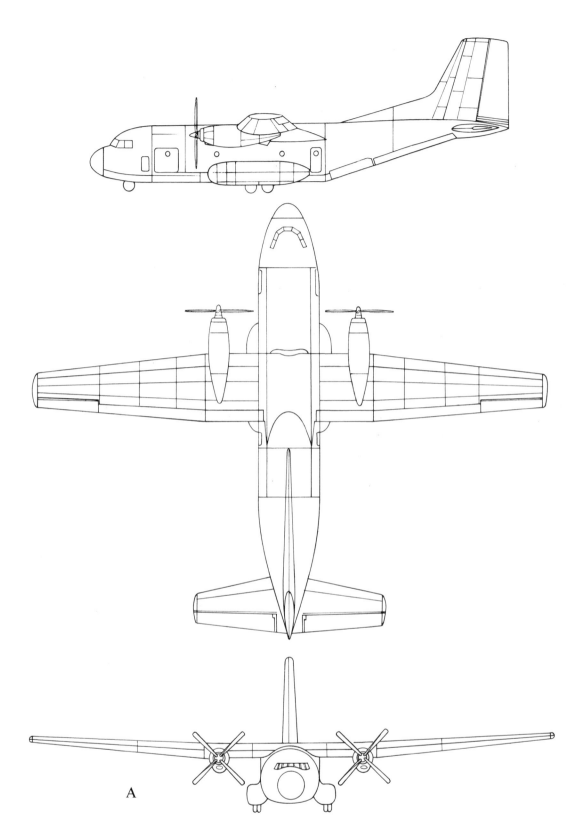

A *Transall C.160. Designed for military transport duties, the Transall C.160 is an interesting example of international collaboration. The aircraft was developed to meet the specific requirements of the Federal German and French governments and is the product of Transall (Transporter Allianz), a group of* French and German companies specially formed for its production. Vereinigte Flugtechnische Werke is overall manager of the project and is responsible for the design and manufacture of the centre fuselage and horizontal tail surfaces. Hamburger Flugzeugbau produces the front and rear *fuselage and vertical tail surfaces. The wing and powerplant assemblies are the responsibility of Nord, who assembled the first prototype. The engines are Rolls-Royce Tyne turboprops, produced by a consortium made up of Hispano-Suiza in France, MAN in Germany, and FN in*

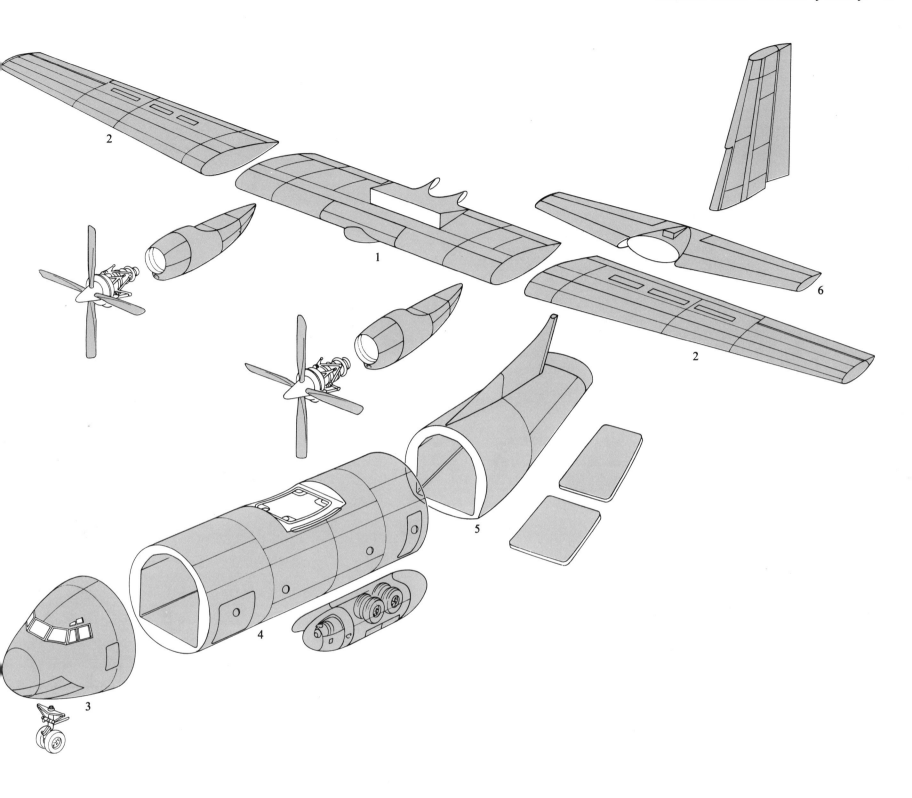

Belgium, in association with Rolls-Royce. The structure is of conventional light alloy stressed-skin construction, special attention being paid to the provision of transport joints in view of the inter-nation production programme and the need to transport individual sub-assemblies over large distances.

Thus the wing, built up around a two-spar box, comprises three major components: the centre section and the two outer wing panels. The three components are assembled by tie-bolts located in light alloy attachments mounted between two longitudinal stringers.

Similarly, the fuselage consists of three main sections: the front section, embodying the flight deck, nose gear and underfloor equipment bay; the centre section, comprising the cargo compartment, with forward loading door and two aft paratroop doors; and the rear fuselage, incorporating the loading ramp, aft cargo door and tail surfaces. The three main sections of the fuselage are joined by riveting. The tailplane is attached to the fuselage by four fittings.

Within the fuselage, the cross-section of the usable portion of the cargo compartment corresponds throughout its length to the International Railway Loading Gauge and has a length of 44 ft. The loading ramp and aft cargo door form the slanting under-surface of the rear fuselage. The aft cargo door is hinged at its rear end and opens upward and inward. This arrangement permits the loading of cargo that will fill the whole cross-section of the cargo compartment. The loading ramp and aft cargo door can be opened in flight for the aerial delivery of single loads weighing up to 17,640 lb.

1 Wing centre-section
2 Wing outer panels
3 Fuselage nose section
4 Centre fuselage
5 Rear fuselage
6 Tail unit

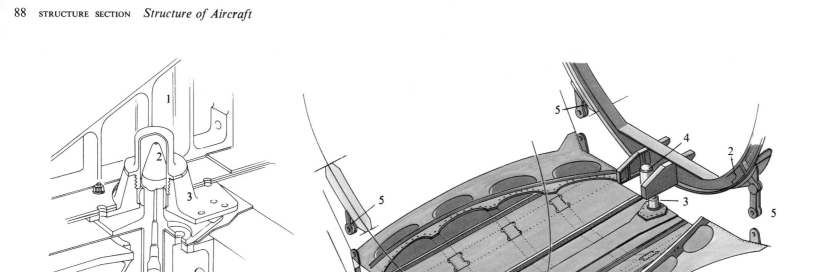

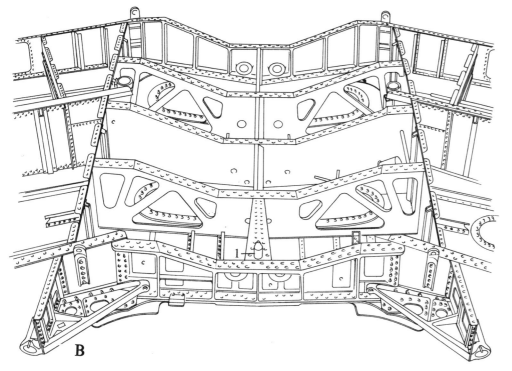

A *Wing spigot (shear pin) engaging with socket in underside of fuselage.*
  1 Fuselage keel stiffener
  2 Spigot
  3 Spigot housing
  4 Wing centre rib

B *Wing centre-section. of the H.S.125. This is dished to accommodate the fuselage and is sealed to form an integral fuel tank. Five machined members* carry the wing bending moments, the centre one extending as a third spar over the inboard section to give extra rigidity.
  1 Spigot

C *Simplified drawing showing wing-to-fuselage attachments as detailed in* A *and* E.
  1 Wing/fuselage seal
  2 Fuselage frame
  3 Spigot
  4 Spigot housing
  5 Vertical links

D *Hawker Siddeley 125. Determining the fatigue life of this twin-jet executive aircraft posed problems. An airliner is invariably designed with particular route patterns in mind, but an executive aircraft may be used in different ways by different operators or by the same operator on different occasions, which means that a wide variety of missions must be catered for.*

For example, if a jet executive aircraft is to be used mostly on medium-range work, with cruising altitudes of 20,000 ft and above, the pressure cabin needs to be capable of withstanding a lot of repeated pressurizations and depressurizations. On the other hand, if it is to be used a great deal on short-range work below the air lanes, at say 5,000 ft, pressure

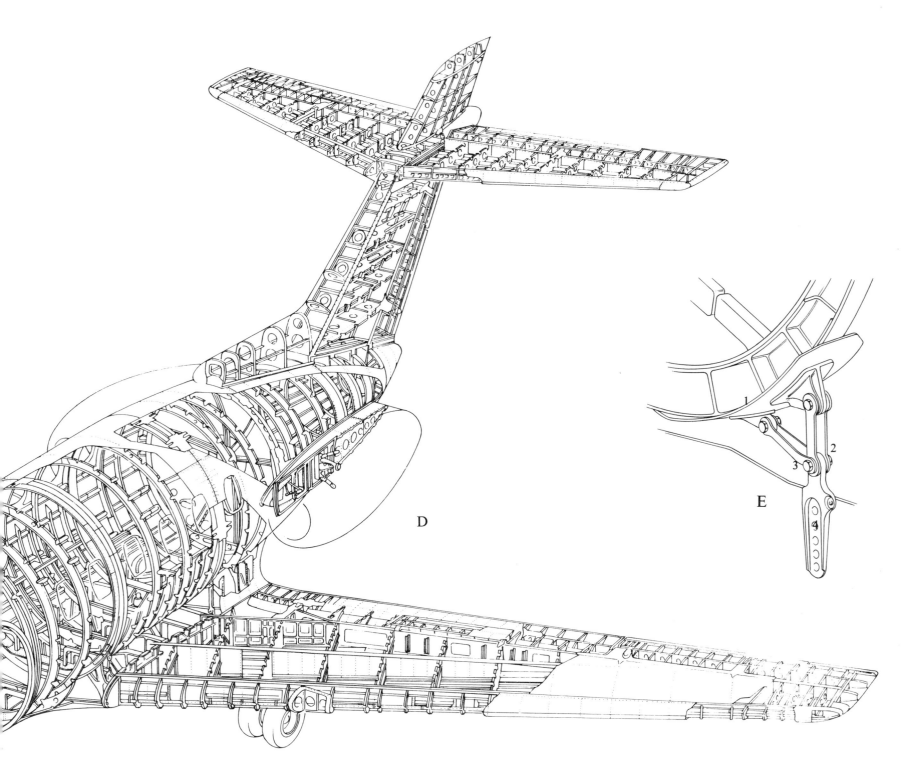

D

E

1
2
3
4

cabin fatigue becomes less vital, but the structure of the wings and tail surfaces will have to withstand a much more severe hammering from the greater atmospheric turbulence encountered at low altitudes. The final target selected for the HS. 125 was a service life of 20 years, assuming an average 500 flights a year. This is achieved by a conventional light alloy stressed-skin structure, extensive use being made of fail-safe principles. The key to the jet's rugged strength is the unique method of attaching the wing to the fuselage. The wing is an all-metal, two-spar, fail-safe structure, which is built in one piece and passes completely under the fuselage. The fuselage incorporates resin-bonded stringers and riveted frames and is of constant circular cross-section throughout most of its length: since it rests on the wing, no space is taken up by the spar which normally passes through the bottom of the cabin on aircraft of this type. Attachment to the wing is by four link-joints, drag loads being taken by a spigot on the centre-line of the wing rear spar, engaging with a fitting on the fuselage.

Not only does this give the optimum design for a pressure cabin, but the unobstructed floor provides the maximum cabin height.
In the interests of maximum safety, the main entry door, the emergency exit and the flight compartment windows which open have all been designed to fit as plugs. This method eliminates any possibility of a blow-out during pressurized flight.
The top and bottom skins of the wing are stiffened by bonded stringers. The close spacing of these contributes to the fail-safe qualities of the wing.

E *Front wing-to-fuselage link joint.*
1 Fuselage frame
2 Vertical link
3 Side link
4 Wing front spar fitting

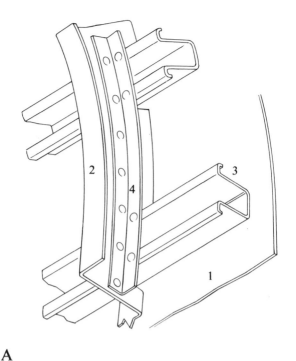

A

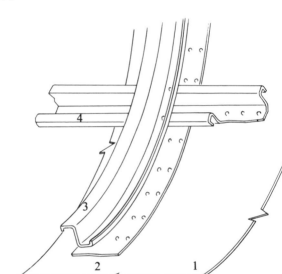

B

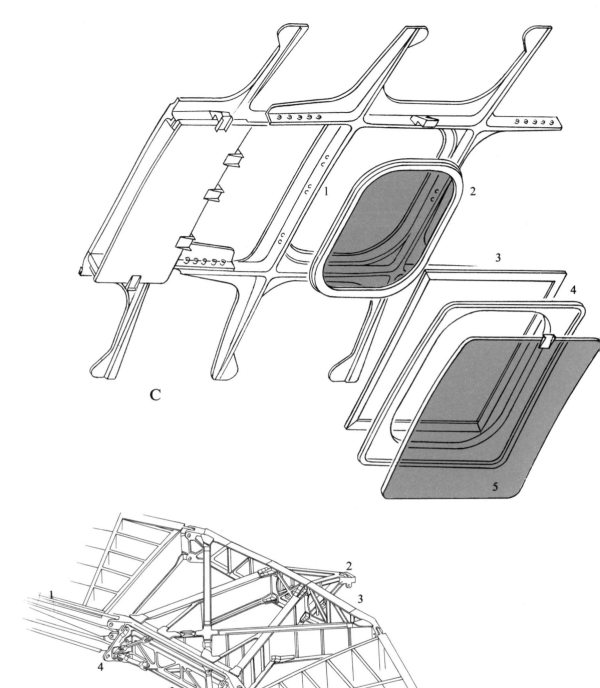

C

D

A *Typical frame shear
tie used on the Boeing
707 fuselage.*
1 Skin
2 Frame
3 Stringer
4 Reinforcing angle

B *Typical fuselage
stringer/frame joint
used on the Boeing 707
series of jet airliners.
At selected intervals a
strap is attached to the
skin, to prevent any
cracks that might*

*develop from extending
catastrophically.*
1 Skin
2 Strap
3 Frame
4 Stringer

C *Windows present
special problems in a
pressurized cabin. The
structure must diffuse
the loads smoothly
round the aperture;
this is achieved by skin
doublers or integrally
machined window*

*frame planks. Each
window is sealed by
two panes, either of
which is capable of
taking full pressuriza-
tion loads.*
1 Fuselage window
frame
2 Outer window pane
3 Seal
4 Reveal
5 Inner window pane

D *This truss-type tail-
plane centre-section,
used on the Boeing 737,
replaces the traditional
built-up box structure
with its many fasteners
and joints. The star-
shaped pattern of
beams increases
structural efficiency
and improves access
for inspection and
maintenance. The two
forgings forming the
centre-section of the
rear spar provide a*

*dual, fail-safe load
path for the tension
chords and attachment
fittings. A fail-safe
multiple-chord design
is used also for the
rear spar of the fin,
which has four chords
and attachment fittings
that fasten to a fuselage
bulkhead. This enables
the rear spars of both
fin and tailplane to
withstand normal flight
loads with any single
chord completely*

*severed. The tail
surfaces make extensive
use of structural
glass-fibre and glass-
fibre honeycomb
materials.*
1 Fail safe members
2 Jack screw support
3 Truss-type forging
4 Hinge support

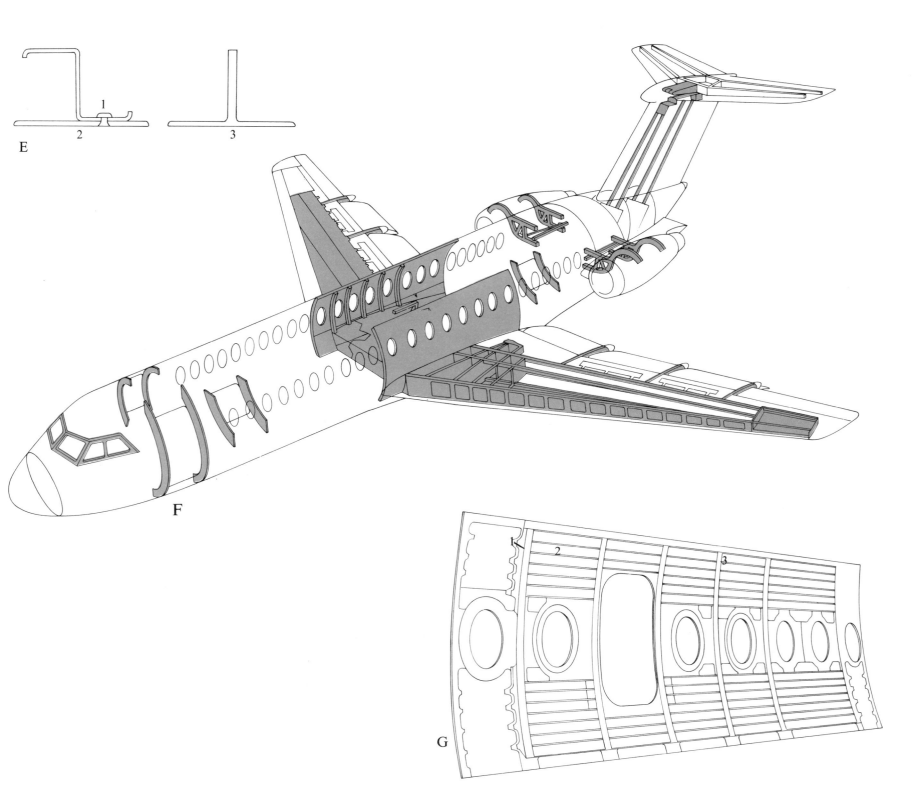

E

F

G

E *Key structural parts are often machined from solid metal, to increase structural efficiency, improve fatigue properties and, in skin panels, to ease air-pressurization and fuel tank sealing problems by eliminating rivets.*

1 Traditional method
2 Rivet
3 New method

F *Outlined here in blue are the machined structural components on the BAC One-Eleven passenger airliner. Following conventional practice, the structure is designed to fail-safe principles wherever possible. The design is such that any faults will become evident on visual inspection. A simple example of this is that a crack in a wing skin*

plank would show up because of seepage of fuel. In areas where reliance is placed on side-by-side double load paths, special inspection techniques are required to find a crack in one member, since the other holds the crack and prevents it from being seen easily. Access to the interior of the wing is provided by holes in the top surface or by detach-

able webs. These enable the structure to be inspected and the integrity maintained.

G *A typical integrally-machined fuselage skin panel, incorporating stringers and thickened areas around door and window cut-outs. This can be fastened directly to the fuselage frames.*

1 Integral stiffening
2 Integral stringers
3 Integral frames

is, deformation, falls sharply after the light alloys have been heated for 100 hours to around 120° Centigrade. Almost as important as the effect of high temperatures on the strength of materials is the problem of stresses induced by uneven heating. This can arise in subsonic aircraft undergoing rapid changes in altitude and becomes of particular importance on supersonic aircraft. Thus it requires special attention during both design and structural test programmes.

On aircraft capable of supersonic speeds for brief periods only, the structural problems are relatively minor, as there is insufficient time for the heat generated on the outer skin to "soak" into the structure inside. The situation changes when prolonged flights at supersonic speed are involved, as the temperature can rise to the point where the internal structure begins to lose its strength.

One solution for prolonged supersonic flight is to use the capacity of the aircraft itself to act as a "heat-sink". The high skin heat is absorbed or soaked away, and dangerous peaks in load-carrying members prevented. One of the first aircraft designed to soak in supersonic conditions was the Convair B-58 Hustler bomber. This aircraft can exceed the speed of sound for over two hours continuously and has a maximum speed of approximately twice the speed of sound. At such speeds, parts of its skin are raised to over 120° Centigrade by air friction heating, the proximity of the jets from the four afterburning engines, and by heat from the extensive electronic and other heat-generating systems on board.

In order to provide the high strength and rigidity required, large portions of the B-58 are fabricated from brazed stainless-steel honeycomb sandwich panels. These consist of a honeycomb core built up from strips of thin stainless-steel brazed together. The cores are then machined, and placed between pre-formed inner and outer skins. After light brazing, to locate the components in the correct position, the assembly is positioned on a large graphite block, machined to the exact contour of the finished part, and heated in an inert atmosphere until the brazing joints have been made satisfactorily. When complete the panel is relatively light, extremely rigid, and can retain its strength up to a temperature of 260° Centigrade. The cost is, however, extremely high—up to £500 ($1,200) per square foot. This is one of the reasons why the airframe of the B-58 cost more than its weight in gold.

The Concorde supersonic airliner will cruise at just over twice the speed of sound, which means that the effects of kinetic heating can just be catered for by conventional light alloys and familiar design techniques throughout most of the structure, although the state of the art is stretched to the limit.

On the Concorde, the tip of the nose reaches the highest skin temperature, around 150° Centigrade, followed by the leading-edge of the wing, which reaches 130° Centigrade. The rest of the structure is mostly below 125° Centigrade. Of interest is the fact that some relief is obtained by applying a white finish to the surface, which reduces the temperature of certain areas by more than 10° Centigrade.

To withstand the temperatures involved, the basic structure is fabricated from a special heat-resistant aluminium alloy that was developed initially for turbojet engine components. The strength of this alloy does not begin to decrease significantly until temperatures above 130° Centigrade are reached, at which temperature creep-resistance characteristics remain good. Fatigue properties are actually improved and continue to be good up to 150° Centigrade.

As already mentioned, construction of the Concorde is mostly conventional. The majority of the fuselage stringers are extruded and, wherever possible, are attached to the skin by welding. In the wing, integrally machined components are used for highly-stressed members and for the majority of the skinning. Departures from the conventional include the nacelle structure between the engine firewall and wing rear spar, and the flying control surfaces, which are of stainless-steel honeycomb construction. Aft of the rear spar, conventional steel construction is used.

A conventional structure was not feasible for the North American XB-70 research aircraft, as it is designed to cruise at three times the speed of sound. On this craft, the entire airframe is designed to soak at around 280° Centigrade. To permit this, the structure in general is of stainless steel and of a unique design. There is no conventional primary structure of ribs and frames on which a skin is laid. Instead, most of the structure is built up from pre-moulded panels, consisting of machined stainless-steel extrusions used as boundary members, with transverse struts and attachment angles inserted into stainless-steel sandwich members.

Most of the joints are welded and great difficulty was experienced in achieving the high degree of accuracy required. Some of the double-curvature panels have an area of hundreds of square feet, and these tended to grow and warp during fabrication. The single weld of the inner wing to the fuselage on each side was over 80 ft long. In all, the XB-70 contains over six miles of welds joining detail parts and over two and a half miles in major assembly. Several of these miles were the edges of fuel tanks. To obtain the correct dimensional tolerances, some of the attachment angles had to be welded in position after the completion of other operations. Final assembly of many of the parts was accomplished by precision welding, although conventional mechanical joints are employed where access is required.

The XB-70 was criticised as a "technological dud" even before the first one flew. It may be true that technology advanced and concepts changed, at least in the United States, while the aircraft was under construction; but this is largely attributable to the XB-70 programme itself. As one engineer commented at the time, "The XB-70 may be considered to embody in one aircraft an entire generation of technical advances ordinarily achieved by many aircraft in combination. The dream of ocean-spanning journeys in a comfortable aeroplane at a speed much faster than the Earth's rotation has suddenly become more attainable than anyone might have believed a few years ago. This is why the XB-70's maiden flight may prove to be one of the most important since Kitty Hawk." Later events and the experience gained with the XB-70 justified this optimism.

More recent studies have indicated that much of the structure could have been built more cheaply by utilising integral machining techniques. Even so, any transport aircraft designed to fly for extended periods at three times the speed of sound is likely to embody a great deal of stainless steel sandwich construction, and for this the vast experience gained during construction of the XB-70 will be invaluable.

Another aircraft embodying advanced constructional techniques is the Lockheed YF-12A prototype fighter. To make possible a cruising speed of around 2,000 mph, almost the whole of the primary structure is of titanium. By using titanium, the need for expensive honeycomb sandwich panels is reduced.

Many structural parts of the YF-12A are in ultra-high-tensile steel, partly for resistance to heat, but more often because these components are subject to severe stress and of very small size. One of the many unusual structural features of this aircraft is the centralised mounting of the engines in the wing. Most advanced American aircraft, and the Concorde, mount the engines under the wing. Presumably, the arrangement was adopted on the YF-12 to provide optimum engine intake conditions; but it undoubtedly posed severe structural and manufacturing problems, which were not eased by the location of the massive all-moving fins above the rear part of the engine nacelles.

The most advanced airframe structure yet built is undoubtedly that of the North American X-15. This remarkable research aircraft, which is virtually a piloted ballistic missile, has reached altitudes above 50 miles and speeds of over 4,000 mph. At these speeds and heights, its skin temperature ranges from minus 185° to over 650° Centigrade, and as a result it requires a unique structure. Almost the entire airframe, including the outer skin, the propellant tanks and the solid, heat-sink leading-edges of the wings, is fabricated in Inconel X, an alloy consisting principally of nickel and chromium, and similar to those developed for jet-engine turbine wheels. Under the Inconel skin is a primary load-bearing structure of stainless steel and titanium.

Manufacture of the airframe involved the development of completely new fabricating, heat treatment and welding techniques. Nearly three-quarters of the basic structure is welded, and for annealing at temperatures of up to 1,095° Centigrade and for

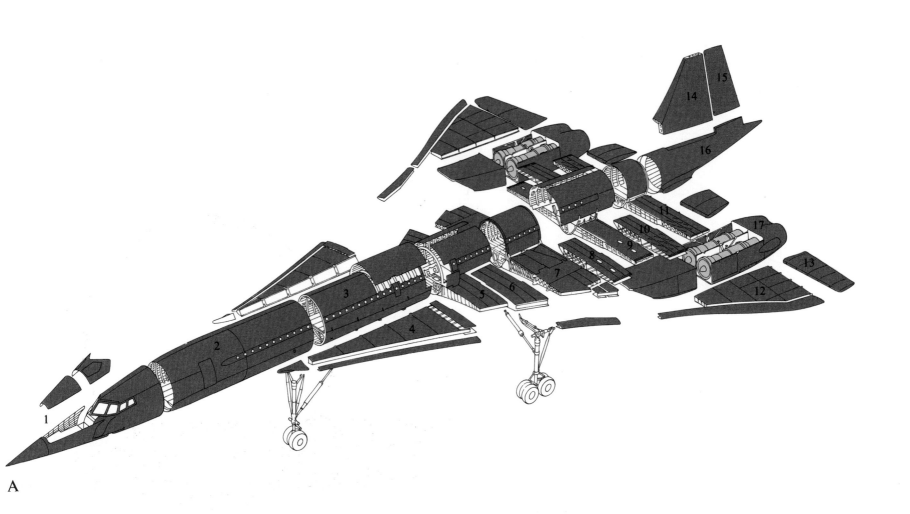

A

A THE CONCORDE:
The aerodynamic shape of the Anglo-French entry into the supersonic airliner field is the result of a careful compromise between the conflicting demands of minimum drag at supersonic speed and good handling and control at low speeds. The slender delta wing has the additional advantage of a natural "high-lift device" during the final approach to land, when the airflow over the wing forms stable vortices along the line of the leading-edge. This and the "ground cushion" effect of the wing itself are built-in safety factors.
The structure presented new problems because of the kinetic heating caused by the friction of the air passing over the skin surfaces in supersonic flight, the thermal stresses set up during acceleration and deceleration from supersonic speed—and, last but not least, the aircraft's design and production in two countries. The choice of Mach 2.2, or about 1,450 mph, as the maximum cruising speed was influenced largely by the fact that the resulting skin temperature, averaging about 120° Centigrade, is just within the tolerance of selected aluminium alloys. The structure is thus built almost entirely from conventional alloys, using advanced but well-proven manufacturing techniques. The fuselage is basically a pressurized cylinder of almost constant cross-section, with a slight "double-bubble" shape. The delta wing, a multi-spar torsion box, houses the integral fuel tanks and carries the powerplants and main landing gear. For maximum integrity, the wing and fin skin panels, fuselage window panels and other highly-loaded members are integrally machined from solid metal.

1 Hinged nose and vizor
2 Forward fuselage
3 Intermediate fuselage
4 Forward wing panel
5-11 Centre wing panels
12 Outer wing panels
13 Elevons
14 Fin
15 Rudder
16 Rear fuselage
17 Nacelles

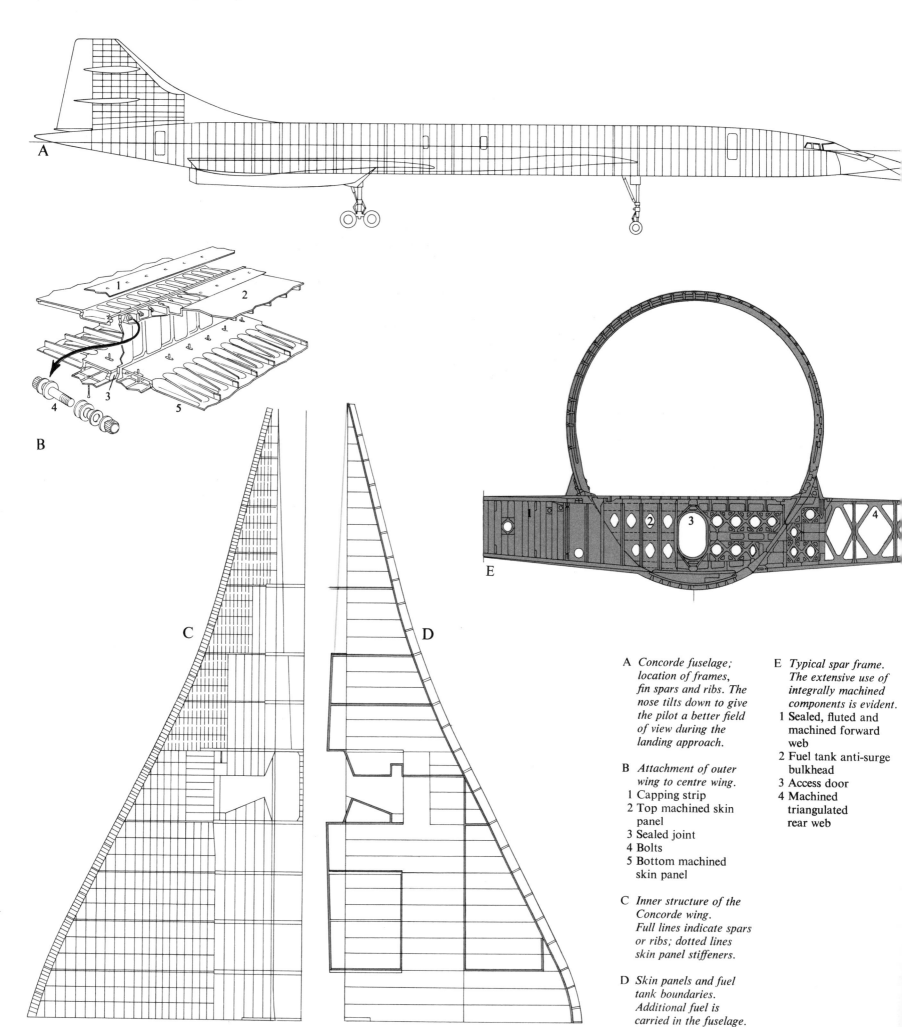

A Concorde fuselage;
  location of frames,
  fin spars and ribs. The
  nose tilts down to give
  the pilot a better field
  of view during the
  landing approach.

B Attachment of outer
  wing to centre wing.
  1 Capping strip
  2 Top machined skin
    panel
  3 Sealed joint
  4 Bolts
  5 Bottom machined
    skin panel

C Inner structure of the
  Concorde wing.
  Full lines indicate spars
  or ribs; dotted lines
  skin panel stiffeners.

D Skin panels and fuel
  tank boundaries.
  Additional fuel is
  carried in the fuselage.

E Typical spar frame.
  The extensive use of
  integrally machined
  components is evident.
  1 Sealed, fluted and
    machined forward
    web
  2 Fuel tank anti-surge
    bulkhead
  3 Access door
  4 Machined
    triangulated
    rear web

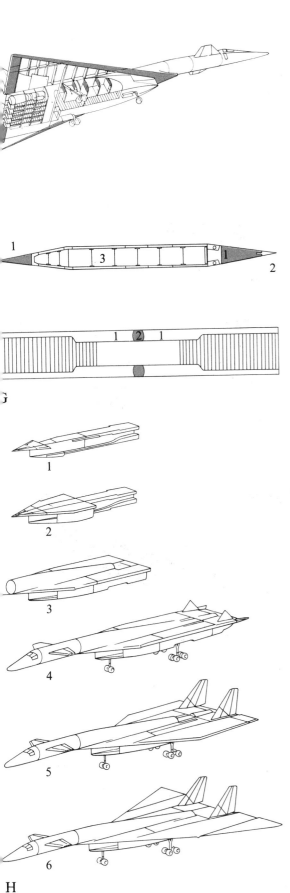

ageing at up to 705° Centigrade special furnaces were required, big enough to house virtually the entire aircraft. Nearly all internal fuel and hydraulic pipelines are welded or brazed.

With regard to future developments in metal construction, for aircraft designed to travel through the atmosphere at speeds greater than 2,000 mph, special cooling may be required to ensure that the strength of the structure is not weakened dangerously through kinetic heating.

At least six possible methods have been suggested by which cooling can be provided. One is by radiation cooling, in which the outer surface of the aircraft is allowed to reach a high temperature so that the energy is dissipated as heat. Although simple, this may result in local temperatures high enough to melt the metal in that area. Ablation cooling involves coating the outer surface of the aircraft with layers of a special material; as the layers erode away, each particle takes heat away with it. This is the method adopted to protect the heat shields of the US Mercury, Gemini and Apollo manned spacecraft, and was also used on the X-15 during tests resulting in particularly high skin temperatures.

Injection cooling involves the passage of a cooling fluid out through the skin; active cooling is similar but involves the use of a fluid circulated internally to absorb and then carry away excessive heat. A fifth method is to construct the surface to act as a heat sink and so protect the inner structure, but this is not suitable for prolonged flight, as there is a continual build-up of temperature. One method which may become practicable in the distant future is the use of magnetohydrodynamic (MHD) cooling, in which a powerful magnetic field is used to push the hot shock layer of air away.

But still in the age of titanium, stainless-steel sandwich and aluminium/lithium alloys, it should not be overlooked that wood continues to be used in some types of aircraft. Timber is still a favoured material for small aircraft, especially home-built and special acrobatic planes. Reinforced plastic has become a more commonly used material in aircraft construction. In the beginning as nonstructural components in small aircraft and gliders, but from its introduction in the early 60:s, composite material has found a steadily increased importance in the aerospace field. The pioneering development work in this field was done by sailplane manufacturers and homebuilders like Burt Rutan, one of the main figures in composite technique development and applications.

The word "composite" signifies that two or more materials are combined to provide a useful material that can be given the desired shape and characteristics. The first application of composites can be traced back to ancient history, when straw was used to reinforce mud in primitive building construction. The most significant advantage of composite material for use in aircraft is the high strength-to-weight and stiffness-to-weight ratios, but the high initial costs of the advanced composites limited their use to the military aircraft and space industries. Advanced composites are built up of continuous fibres embedded in a polymeric matrix, where the most commonly used fibres are glass, carbon and aramid (Kevlar). A variety of high-performance resins can be used as matrix materials, but the most common by far is epoxy, although there is an increasing interest and research in new thermoplastic matrix materials.

The basic material is bought in the form of prepreg, a thin film consisting of continuous parallel or, in some cases, woven fibres preimpregnated with the matrix material employed. A typical thickness of such a film, or layer, is approx. 0.125 mm. When producing a laminate, which is a stack of such layers, a major purpose is to tailor the stiffness and strength by organizing the direction of the fibres in order to obtain a laminate which matches the loads that will be applied to the structure. In this way laminated composite structures can be designed to have distinctive characteristics built in from the outset, which is impossible with conventional materials.

F *The XB-70: supersonic innovations. Designed to cruise at Mach 3, or around 2,000 mph, the XB-70 necessitated the development of new manufacturing techniques and materials. Kinetic heating raises the average skin temperature to 280° Centigrade; to accommodate this the bulk of the structure is made up of stainless steel honeycomb sandwich panels.*

Section through wing. Leading- and trailing-edges (1) have solid honeycomb cores. (2) shows a machined trailing-edge member; (3) shows honeycomb sandwich panels.

G *Honeycomb structure. Individual panels (1) were joined by special welding processes (2).*

H *The XB-70 assembly sequence involved (1) Welding the centre sections, (2) Welding the forward wing section, (3) Welding the aft wing sections and mating the upper intermediate fuselage, (4) Mating the front and aft fuselage sections, installing the landing gear and canard foreplanes, (5) Assembling the vertical tail surfaces, and (6) Mating the outer wings.*

1 Vickers F.B.5 Gun-
   Bus (UK): Fighter.
   1915
2 Morane-Saulnier L
   (Fance): Fighter. 1915
3 Fokker E.III (Germ-
   any): Fighter. 1915
4 Nieuport 11 C.1 Bébé

(France): Fighter. 1915
5 Sopwith 1 ½-Strutter
   (UK): Fighter. 1916
6 Fokker Dr. I
   (Germany): Fighter.
   1917
7 Sopwith Camel F. 1
   (UK): Fighter. 1917

8 Bristol F. 2 B Fighter
   (UK): Fighter. 1917
9 Fairey Flaycatcher
   (UK): Fighter. 1923
10 Curtiss P-6E Hawk
   (USA): Fighter. 1932
11 Boeing P-26A (USA):
   Fighter. 1933

12 Grumman F3F-1
   (USA): Naval fighter.
   1936
13 Fiat CR. 32 (Italy):
   Fighter. 1933
14 Hawker Hurricane I
   (UK): Fighter. 1938
15 Supermarine Spitfire

IIB (UK): Fighter.
   1940
16 Messerschmitt Bf 109F
   (Germany): Fighter.
   1941
17 Polikarpov I-16
   (USSR): Fighter. 1934
18 SAAB JA 37 Viggen

(Sweden): Fighter and
   attach 1977
19 F-16 Fighting Falcon
   (USA): Fighter 1979
20 Bristol Beaufighter I
   (UK): Night fighter.
   1940
21 Mitsubishi A6M3

ZeroSen (Japan):
Naval fighter. 1942
22 Yakovlev Yak-9D
(USSR): Fighter. 1943
23 Lockheed P-38J
Lightning (USA):
Escort fighter. 1944
24 Gloster Meteor III

(UK): Fighter. 1944
25 Messerschmitt Me
163B (Germany):
Fighter. 1944
26 D. H. Vampire N.F.
10 (UK): Night fighter.
1951
27 North American F-

86A Sabre (USA):
Fighter. 1949
28 Mikoyan-Gurevich
MiG-15 (USSR):
Fighter. 1948
29 SAAB J 35F Draken
(Sweden): Fighter.
1967

30 Dassault Mirage III-
CZ (France):
Interceptor. 1963
31 Lockheed F-104C
Starfighter (USA):
1958
32 BAC Lightning F6
(UK): Fighter. 1965

33 MIG-29 (USSR):
Fighter 1983
34 F-117A (USA): Stealth
fighter 1985
35 HSA Harrier GR.1
(UK): VTO tactical
strike and
reconnaissance. 1968

36 General Dynamics F-
111A (USA): Tactical
fighter. 1968
37 Spad 7 (France):
Fighter. 1916
38 Morane Saulnier 406
(France): Fighter. 1938

The biggest problem involved in building aircraft structures in composite materials has been to develop industrial production methods. The construction work has been very man-hour consuming, which has added to the high costs. But now there has been a breakthrough for the use of composites on a big scale. In the latest generation of airliners, up to 30 % of the primary structure is made of composite material, thus saving a lot of weight, which pays back in higher payload and better economy. The two-engine business plane Beechcraft Starship, a futuristic-looking aircraft, was the first plane on the market built in wholly composite material to be certificated by the civil aviation authorities. Also, the new materials made it possible to build the record-breaking Voyager that, in 1986, flew around the world without refuelling.

Continued advances in the structure of aircraft are the key to future development. So far, engineers have been able to keep pace with developments in engine power and aerodynamic evolution. There is no reason to doubt that they will continue to do so.

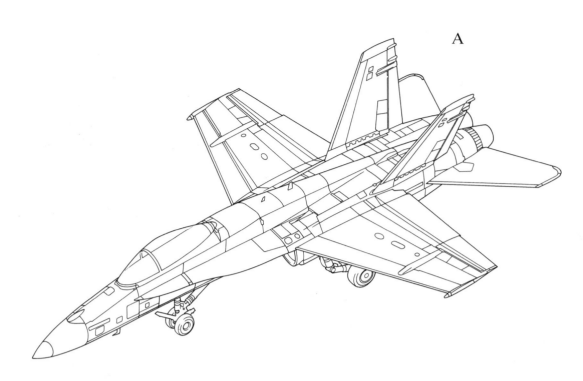

A

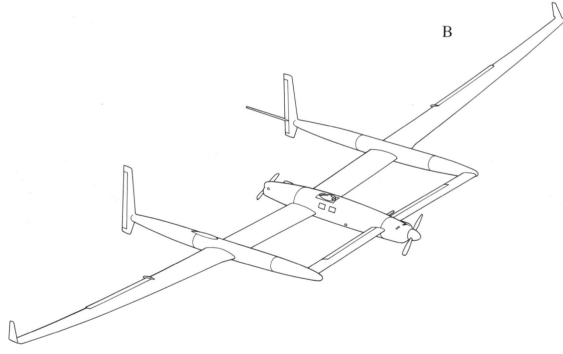

B

A The US Navy F-18 Hornet, a multimission fighter/attack aircraft, is representative of combat aircraft design of the 1980s.

Most of the structural weight is contributed by aluminium (50.4 %) Giving the airframe exceptional strength is titanium (13.2 %). The

carbon-epoxy composite material (12 %) reduces weight, retains strength and does not corrode. A specially processed carbon-epoxy material covers about 40 % of the aircraft's surface.

B Voyager, the first aircraft to circle the world at the equator non-stop without

refuelling, flew 25,000 miles in nine days. The aircraft, designed and built by Burt Rutan, is an all-composite construction. Most of the plane's structure is made of carbon-magnamite, in thin sheets bonded to paper honeycomb cores for most of the aircraft's skin.

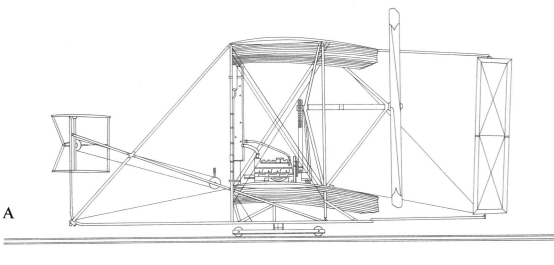

A

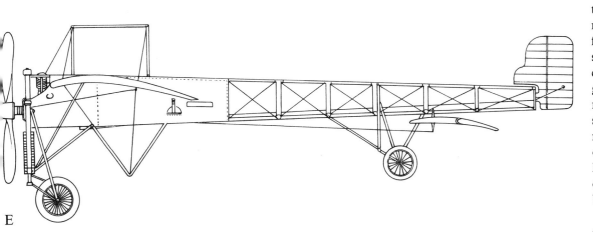

D

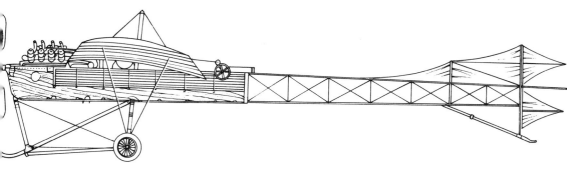

E

LANDING GEAR

No mechanism has required greater mechanical ingenuity or has resulted in more mechanical complexity than aircraft landing gear. Aircraft design is always a matter of carefully matched compromises, but as aeroplanes have evolved the landing gear seems to have called for more compromise than other parts of the structure. Occasionally, as on the vertically-launched Bachem Natter rocket-propelled interceptor, the landing gear has been compromised right off the aircraft!

In cases where their colleagues have left at least some space, landing gear designers have produced gears retracting upward, forward, sideways inward, sideways outward and backward. To effect retraction, while conforming to the design requirements for energy absorption—still the prime function—and trying to bear in mind matters such as maintenance, the landing gear designer contends with abstruse three-dimensional or skew-geometrical problems, and has had to devise mechanisms involving deformable quadrilaterals, self-breaking struts, telescopic legs, extending legs, rotating legs, hinged axles, articulated gears, ordinary bogies, articulated bogies, damper struts, locking members and other devices not readily describable. Shock-absorbers, wheels and wheel brakes necessitate specialised industries of their own.

Landing gears were less of a problem in the early days. Of the many noteworthy features of the Wright Flyer No. 1, which made the first successful powered flights in 1903, the absence of a wheeled landing gear is one of the most interesting. For its historic first take-off, the Flyer used a small trolley running along a wooden launching rail. Even if wheels were not essential for taking off and landing, they were very convenient for manoeuvring aircraft on the ground; but for some unknown reason the Wright brothers did not fit wheels to their aircraft until seven years later, in August 1910.

Yet, thought had been given to the question of wheels for aircraft long before heavier-than-air flight was achieved, In a notebook entry dated 19 March, 1808, Sir George Cayley, the famous British nineteenth century air pioneer, wrote:

"In thinking how to construct the lightest possible wheel for aerial navigation cars, an entirely new mode of manufacturing this most useful part of all locomotive machines has occurred to me—vide, to do away with wooden spokes alto-

A *The Wright Flyer of 1903 did not use conventional wheels for its historic flights. It took off on a small trolley fitted with two adapted bicycle wheel hubs, running on a wooden rail. To manoeuvre the aircraft on the ground the Wright brothers used two small wheeled trolleys placed under each wing.*

B *1909 Antoinette monoplane with projecting fore skid. This prevented the aircraft from nosing over.*

C *1909 Blériot monoplane landing gear. Carried in cycle-forks, the wheels were free to castor, this facilitating crosswind landings.*

gether, and refer the whole firmness of the wheel to the strength of the rim only, by the intervention of tight strong cording, as exhibited in the front and side elevations on the opposite page. Such a wheel would I am confident be almost everlasting for light travelling vehicles, would be very beautiful, give no shake and therefore require much slenderer axle trees, and have less friction. If the rim were of thin cast iron and likewise the nave, with some proper screw apparatus to stretch the cordage, nothing would ever require renewing but the cord, and this seldom, as it would never receive any jar like a wooden wheel. No noise would be given to the loose parts of the vehicle and with good springs travelling would be much pleasanter.''

The function of the aircraft landing gear is to dissipate the energy of descent during landing without undue shock to the aircraft or its occupants, and to provide a reasonably well-sprung carriage on which an aircraft can taxi and take off. But since the landing gear is in use for a relatively short part of an aircraft's life and, except when taxying, taking off and landing, is simply an undesirable weight carried at the expense of useful payload, it is natural that designers should endeavour to reduce its weight to a minimum. The operations of taking off and landing, however, impose heavy concentrated loads on the airframe structure and result in stringent design requirements.

Apart from a handful of later aircraft designed for special tasks, the Wright machines were almost unique in being wheeless. The ingenious machine devised by the Hungarian Trajan Vuia in 1906 was, prophetically, fitted with pneumatic tyres, the first aircraft to be so equipped. Many landing gears of the period included long, forward-protruding skids, to eliminate the possibility of nosing over. Some aircraft had small wheels fitted to the front of the skids, acting as nose wheels, although they were supported on the rear pair and a tail skid when at rest.

By 1914 a conventional landing gear configuration had evolved which was to enjoy almost universal use for more than twenty years. Aircraft were supported at the rear by a tail skid, later a wheel, and at the front of the fuselage by a pair of main wheels attached to vee-braced struts. The wheels, fitted with large-diameter, narrow-section tyres, rotated on, or with, a common cross-axle secured by lengths of rubber cord, known as bungee. In place of the rubber cord, some aircraft utilised springs. This crude springing absorbed landing shocks satisfactorily, but tended to cause excessive rebound.

Oil damping to restrict rebound was first used in a leg made by the French engineer Esnault Pelterie in 1908, and an early form of oleo (hydraulic shock-absorbing) strut was fitted to an experimental British B.E.2 at the Farnborough factory in 1912. Louis Breguet is generally credited with the successful development of the oleo strut, which he used on his aircraft from 1910 onwards.

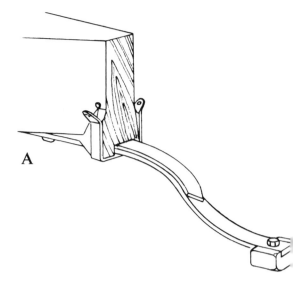

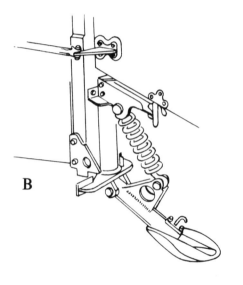

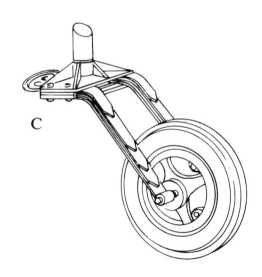

A *Simple leaf-spring tail-skid with cast iron shoe. Many materials were tried for the shoe, including hard tool steels, but none proved sufficiently superior to cast iron to warrant general adoption.*

B *Sprung, castoring tail skid. This facilitated steering.*

C *Tailwheel mounted on twin leaf-springs.*

D Multi-wheeled landing
gear of a Handley
Page airliner of the
nineteen-twenties.

E Blackburn Bluebird
divided landing gear.
Elimination of the
cross-axle, which
fouled long grass,
assisted take-off and
reduced the risk of
damage.

F The Gee Bee Super
Sportster racer had the
landing gear legs
faired and wheels
spatted to reduce drag.

G The Armstrong
Whitworth Atalanta
displayed an early
attempt to clean up
the landing gear, only
the wheels and the
outer ends of the axles
being exposed. The
track, however, was
very narrow for the
90 ft span of the
machine. The axles
consisted of
double-tapered tubes
hinged on the fuselage
centre-line. The
compression legs, sunk
in the fuselage, ran up
to the top longerons.

537   538   539   540

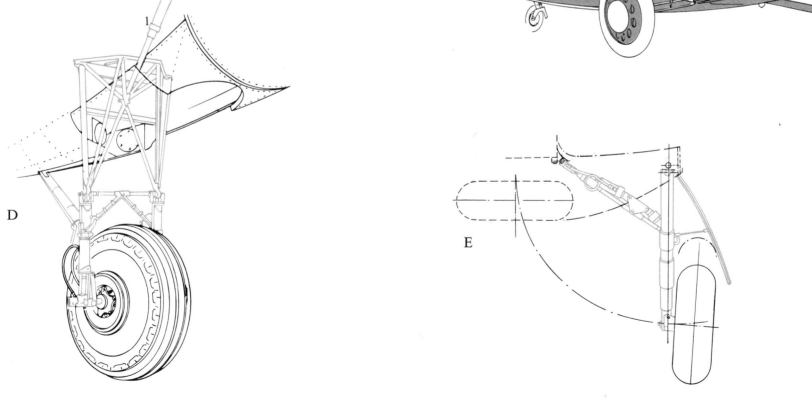

A

B

C

D

1

E

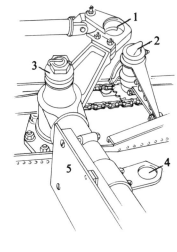

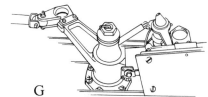

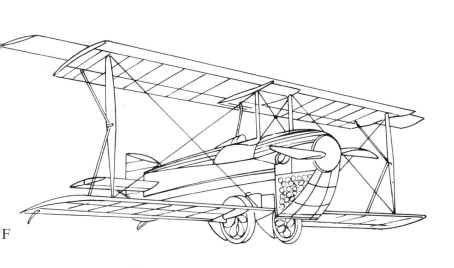

F

G

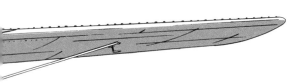

A The fine aerodynamic form of the cantilever legs of the Gloster Gladiator was made possible by the use of internally-sprung wheels.

B The ingenious internally-sprung wheel developed by the British Dowty company incorporated the shock-absorber completely within its rim, thus providing an exceptionally neat appearance. Later versions even packed a brake within the wheel confines.

C The Beardmore Inflexible was one of the first aircraft to use hydraulic brakes. Fitted to its eight-foot diameter wheels, the brakes could be operated independently to assist steering. An ingenious feature was a piston in the tailwheel, connected to the braking system, so that the main wheel brakes were applied automatically when the tailwheel contacted the ground.

D The forward-retracting landing gear of the famous Douglas DC-3.
1 Retraction jack

E The Heston Phoenix utilised a Dowty nutcracker-strut retraction mechanism. The two opposed jacks on the strut operated on an offset link at the centre of the strut which was divided by a knuckle joint. Operating simultaneously, the jacks pushed the joint upward, the consequent shortening of the strut pulling the leg inward and upward.
1 Nutcracker strut

F A semi-retractable landing gear was used on the Martin K-3 Kitten. An unusual feature was the sprung spokes fitted to the wheels. These were very like the wheels that will be utilised on some of the lunar exploration vehicles currently under development in the United States.

G A simple main leg retraction and locking mechanism was developed for the Supermarine Spitfire. Retracting outward, the legs rotated on a cantilever pivot attached to the rear spar. At the top of each leg a bellcrank extension was attached to the hydraulic operating jack. The legs were locked down by a spring-loaded locking pin engaging in a yoke. The same pin was used to lock the leg up, the pin rotating through 180° so that its chamfered face was turned towards the engaging eye on the leg.
1 Yoke
2 Locking pin
3 Pivot
4 Eye
5 Leg

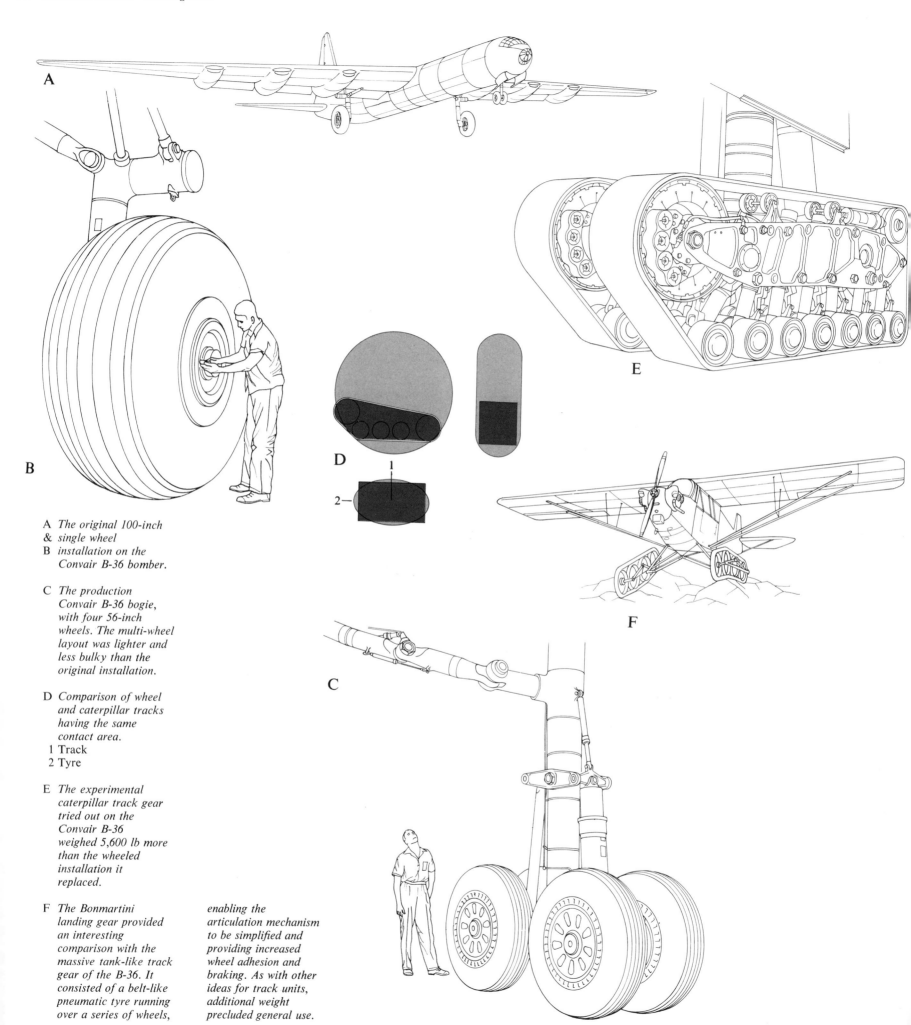

A

B

D

E

C

F

A *The original 100-inch*
& *single wheel*
B *installation on the*
Convair B-36 bomber.

C *The production*
Convair B-36 bogie,
with four 56-inch
wheels. The multi-wheel
layout was lighter and
less bulky than the
original installation.

D *Comparison of wheel*
and caterpillar tracks
having the same
contact area.
 1 Track
 2 Tyre

E *The experimental*
caterpillar track gear
tried out on the
Convair B-36
weighed 5,600 lb more
than the wheeled
installation it
replaced.

F *The Bonmartini*   *enabling the*
landing gear provided   *articulation mechanism*
an interesting   *to be simplified and*
comparison with the   *providing increased*
massive tank-like track *wheel adhesion and*
gear of the B-36. It   *braking. As with other*
consisted of a belt-like *ideas for track units,*
pneumatic tyre running *additional weight*
over a series of wheels, *precluded general use.*

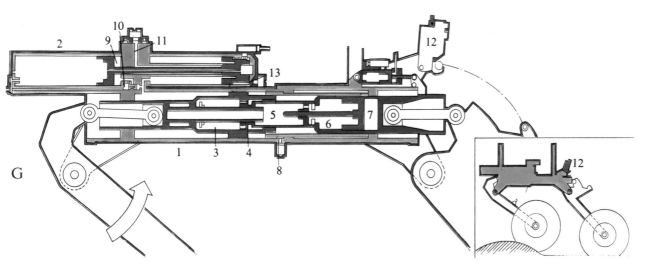

G

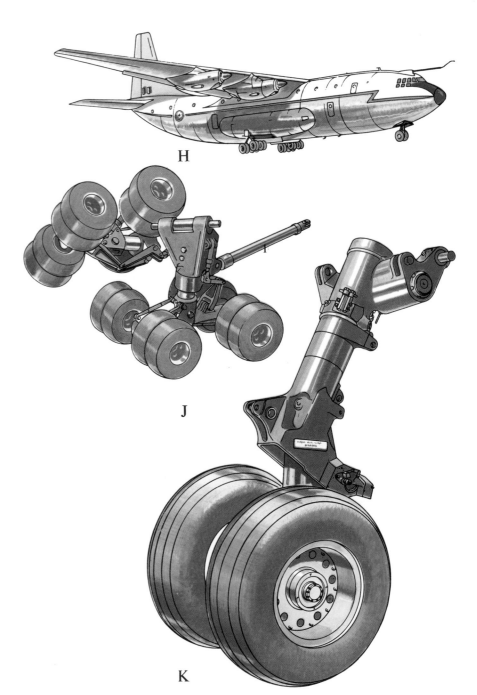

H

J

K

G *Messier "Jockey" landing gear.*
During taxying, the shock absorber-actuator unit 1 is kept isolated from the cylinder 2. The compression of one or both swinging legs causes the fluid to be transferred from either chamber or both chambers 3-4 into the shock absorber chambers 5-6. Oscillation damping of the two swinging legs is obtained by throttling the fluid passing through restrictor orifices, when the compressed nitrogen trapped in the sole gas chamber 7 expands. After takeoff, the legs are retracted by applying hydraulic pressure into the outer chamber of the shock absorber-actuator unit 8. The shock absorber fluid is directed to the front portion of the cylinder 9 after the valve 10 opens,

due to the "retraction" pressure.
During extension the hydraulic fluid is directed on to the differential piston of the cylinder 11 which returns the fluid from the front portion of the cylinder to the shock absorber, the pressure simultaneously actuating the uplock mechanisms 12, thus unlocking the legs. To raise the aircraft for alignment with the platform of a truck, the legs are made to pivot by means of the hydraulic fluid acting on to the piston, in the rear portion of the cylinder 13; this increases the oil volume within the shock absorber. The return to normal attitude is achieved by returning to the reservoir the hydraulic fluid stored in the rear of the cylinder, through the aircraft's weight.

H *The multi-wheel bogie*
& *main landing gear of*
J *the Short Belfast military transport.*
1 Retraction jack

K *The main leg of the Hawker Siddeley Andover military transport can be made to "kneel" while extended, so that the height of the floor is lowered to facilitate loading and unloading.*

However, oleos did not come into general use until the late twenties, with the development of improved designs.

A drawback of the "standard" cross-axle type of landing gear was that, in taking off from fields with long grass, speed was reduced and the take-off impeded by the axle fouling grass and scrub. Damage was frequent, particularly on rough, unprepared ground. To overcome this, the "split" landing gear was devised, with no cross-axle between the wheels.

The landing gear of early high-wing monoplanes was sometimes complicated by its integration with the lift struts used to brace the wing. In their simplest form, the lift struts were arranged to meet the top of the long landing gear leg. Landing loads were thus shared by the fuselage and the wing. Although simple and structurally economical, in that two members performed a dual rôle, this arrangement suffered from the disadvantage that if the wings were removed, the landing gear had to come off as well, a feature not appreciated by maintenance engineers.

Another arrangement, although not as clean aerodynamically, was used widely in various forms by many companies, including de Havilland, Fokker, Curtiss, Dewoitine, Ryan and Stinson. In this, a relatively short compression leg was braced to the fuselage, with a separate upper strut taking the leg loads directly to the wing root. The wing lift struts were taken from the top of the leg to the wing. Wing folding could be accomplished with this arrangement, and the machine remained supported if a wing was removed, but the drag loads were considerable.

The split landing gear, in addition to overcoming the axle damage problem, also tended to reduce drag, through the elimination of cross bracing. As speeds increased, further aerodynamic cleanliness was achieved by streamlining the leg and drag strut. The final stage was to stiffen the leg sufficiently to allow deletion of the back stay thus making the leg a full cantilever member.

Increased landing speeds led to the development of wheel braking in the late nineteen-twenties. At first simple motor-cycle-type brakes were used, the shoes being forced against the inside of a drum rotating with the wheel.

A major problem introduced with braking was the tendency of aircraft to nose over when the brakes were applied. This was due to the fact that the centre of gravity was just aft of the main wheels —an important factor in getting the aircraft into a flying attitude quickly on take-off. To overcome the nosing-over problem, Mr Cowie, a director of the British Soda Water Syphon Company, suggested fitting an extended tailwheel. The idea was to lift the tail into the flying lattitude for take-off, enabling the main wheels to be brought well in front of the centre of gravity. This was the forerunner of the tricycle landing gears of today, the development of which is covered in greater detail later.

An early innovation was the bag or expander-tube drum brake. This developed a great deal of braking power for a low weight, but was not used to maximum advantage until the introduction of nose-wheel landing gears, which solved the nosing-over problem and permitted the full use of brakes. Drum brakes were used widely until about 1950, when it became progressively more difficult to absorb the increasing kinetic energy as weights and landing speeds edged ever upward.

Disc brakes provided the solution to this problem, being capable of absorbing more energy and operating at greater friction loadings than other types. In these, a disc is keyed to the wheel, and braking power is derived from friction pads which press against it. Disc brakes are now used almost exclusively, except on very light aircraft.

On multi-wheeled landing gears, special anti-skid devices are used to enable full braking power to be utilised without fear of locking the wheels. Locking not only reduces retardation but leads to tyre bursting and skidding.

Attempts in the late nineteen-twenties to reduce drag still further led to the development of retractable landing gears. As with many other aeronautical innovations, the idea had been conceived in the early years of aviation. The first retractable gear on a flyable machine appeared in 1911, on the German Wien#ers monoplane. It was a primitive affair, each leg being fitted with a hinge so that the wheels came up rearward to lie flush beside the fuselage. Another early attempt was made by the American pilot-designer James Martin, who fitted his little Martin K-3 Kitten scout with backward-retracting gear of such narrow track that the wheels fitted snugly into the sides of the fuselage.

The first proper fully-retractable gear appeared in 1920 on the American Dayton-Wright RB high-wing racing monoplane, the wheels retracting into the fuselage. The year 1922 saw the first aircraft, the American Verville-Sperry R3 racer, in which the landing gear retracted into the wings.

At this stage of aircraft development, the extra weight and complication more than counteracted the advantages of the saving in drag. As a result, retractable landing gears did not become generally popular until the mid nineteen-thirties, when higher cruising speeds began to make the reduction in drag worthwhile despite the extra weight and complication.

The successful development of the cantilever landing leg resulted initially in some simple retraction mechanisms. A typical example was the electrically-operated gear on the Vultee transport monoplane, on which the leg retracted sideways into the wing. Manual operation was possible in the event of power failure. Another classically simple mechanism was that of the hydraulically-operated gear on the Supermarine Spitfire.

Manual operation was employed on the Grumman JF-1 amphibian. On this aircraft the wheels retracted into recesses in the sides of the hull. The gear required about 45 turns of a winding handle for retraction which, besides being tiring physically, took a long time. Similarly, the electrically-operated leg on the Vultee monoplane, mentioned previously, took about 15 seconds to retract. A rapid rate of retraction is desirable since the reduction in drag improves the rate of climb. Although with modern techniques the desired speed could be obtained with electric motors, this method of actuation has other snags, a primary one being the difficulty of providing operation in an emergency. Little progress with electrical retraction has been made since the days of the de Havilland Albatross, Short Stirling or Boeing B-17 Flying Fortress, on which emergency hand winding of the landing gear took over half an hour! For this and other reasons the majority of landing gears today are operated hydraulically.

An important innovation was the development of the Dowty "nutcracker" strut in the mid nineteen-thirties. This provided a member which could form a rigid bracing for a landing gear leg when extended, and yet could "break" at a joint to allow the strut to fold for retraction. The operating jacks were an integral part of the strut.

Development of retractable landing gears naturally concentrated on installations for low-wing monoplanes, but a few biplanes and high-wing monoplanes have had retractable gears. A particularly neat installation was devised for the Heston Phoenix light aircraft, on which the legs retracted sideways into a small stub wing, the wheels being stowed in the fuselage.

As with aviation generally, World War II accelerated landing gear techniques. One important development was the perfection of the Dowty Liquid Spring shock-absorber. Conceived initially in 1938, this was developed in great secrecy during the war. Normally liquids are regarded as incompressible, but the Liquid Spring uses a synthetic silicon-based fluid with molecules so large that they distort appreciably under pressure. The resulting shock-absorber requires no air cushion and is thus relatively compact.

Ordinary steel coil springs are the simplest and most reliable method of absorbing shocks, and, were it not for their relatively great weight compared with air or oil springs, would undoubtedly be used almost universally. As it is, their use has generally been confined to light aircraft, as an alternative to the more popular rubber discs in a telescopic strut.

A special form of shock-absorber incorporating ring springs enjoyed some success in Germany on relatively large aircraft. Basically, these springs consisted of a series of rings, chamfered alternately on the inner and outer faces so that under pressure they fitted into each other. The energy storage capacity was greater than on a normal coil spring. Perhaps the outstanding example of a ring-spring landing gear was that used on the Junkers Ju 88 bomber. However, although the Ju 88 gear was light, the rings gave trouble, due to sticking, and ater versions of the aircraft were fitted with conventional oleo units.

A major turning point in landing gear evolution was the general adoption of the tricycle or, more correctly, nose-wheel type of gear. The "text book" method of landing tailwheel aircraft is to stall the aircraft just clear of the ground, so that the main and tail wheels touch simultaneously in what is called a "three-point" landing. This gives maximum drag after touchdown to reduce the landing run, but the manoeuvre requires considerable skill, particularly at night. With a nose-wheel, by contrast, the aircraft lands in a level attitude: it can be, so to speak, "driven" on to the runway like a car.

The continuing increase in touchdown speeds demanded increased braking, to pull aircraft up within a reasonable distance, and nosewheels offered a means of preventing aircraft nosing over. Additionally, nose gears offer many other advantages over tailwheel layouts, including easier manoeuvring on the ground, improved view for the pilot on the ground, a horizontal floor—important on airliners—and reduced drag during take-off.

The final swing to nose-wheels was brought about by the arrival of the jet engine. The big increase in speed made nose-wheels essential to permit heavier braking without the aircraft nosing over. Although nose or tricycle installations are inherently heavier than those using tailwheels, the absence of propellers on the new generation of jet aircraft meant that ground clearances could be reduced. Landing gears could thus be shorter, with a consequent saving in weight, partly offsetting the inherent weight penalty of a nose-wheel layout.

With the gradual evolution of landing gears, the factors of wheel loading, wheel size and tyre pressure combined to raise the problem of permissible ground loading pressures. Generally speaking, DC-3 size aircraft with loads of 20,000 lb and above on each leg cannot be operated continuously from grass surfaces at tyre pressures of more than 35 to 45 pounds per square inch. Yet bigger aircraft need higher tyre pressures to prevent the tyres from becoming too bulky and heavy. Hence the need for paved runways and concrete aprons in front of passenger buildings at airports. To keep pace with the increasing weight and speed of aircraft, runways have grown steadily longer, although it looks now as if the limit has been reached, with some runways over 12,000-ft long at major international airports.

The problem immediately after the end of World War II was how a new class of both civil and military aircraft, with take-off weights of from 100,000 lb upward, could utilise the great number of existing runways. A temporary expedient was the use of shallow-depth, extra-wide tyres to spread the load. In Britain, development was concentrated on bogie landing gears and this provided the eventual solution, the first multi-wheel installation being made in 1947. Not only do bogie installations increase the ground contact area, thus enabling the aircraft to operate from a wider range of airfields,

but they are significantly lighter and less bulky. They are also safer, as duplication is provided in the event of a burst tyre.

Although the use of multiple wheels eases the runway loading problem, it by no means solves it completely. This is due to the basic fact that the effective "footprint" of an aircraft wheel is very small in relation to its diameter, since by far the greater portion of its periphery is not under load. One possible answer appeared to be a caterpillar track to provide a greater footprint area, and considerable effort has been expended on developing such substitutes for wheels. As in the case of retractable gears, the first tentative experiments were made long ago. The pioneer track installation was designed by the French engineer Chevreau in 1927. It consisted of an elongated belt running round a deformable frame on balls and rollers, and it had brakes. Produced by Louis Vinay, this gear was fitted to a Loire-Gourdou-Leseurre monoplane and was flown in about 1929. In those days, however, there was no apparent advantage in using tracks and interest waned for several years.

In the United States, the first machine equipped with a caterpillar landing gear was a Fairchild Cornell monoplane. This particular gear embodied an articulated framework designed specially for running over rough ground. Following successful tests, a Douglas Boston was equipped with tracks of similar construction. Many hundreds of landings were made with this gear, and obstacles up to 8 in. square were negotiated at high speed. Tests included runs over mud and sand, during which the standard nose-wheel buried itself; the tracks remained on the surface.

After the war, a Fairchild Packet was fitted with tracked gear, but this was dropped because of the excessive weight penalty, amounting to 650 lb more than the conventional gear. The most impressive tracked gear was that fitted to a Convair B-36. In theory a tracked gear for an aircraft of this size should have effected a considerable saving in weight, but in practice the B-36 track weighed 5,600 lb more than the wheeled installation it replaced!

It is mainly because of their relative lightness that multiple-wheel installations have gained ascendancy over single-wheel layouts. An interesting application of the multi-wheel principle was on the wartime Messerschmitt Me 323 heavy transport, which had no fewer than ten wheels arranged in lines of five in blisters on the sides of the fuselage. Variations of this layout have become universal on heavy military transport aircraft, the most striking example being on the Lockheed C-5A, which lands on no fewer than 28 wheels, arranged in four side groups of six wheels and a quadruple nose-wheel. The Douglas C-133 Cargomaster is unusual in having separate tandem pairs of main wheels.

An outstanding example of a heavy-load, soft-field gear is that on the Short Belfast military transport. Each main gear on this aircraft comprises a compact eight-wheel bogie unit which retracts into

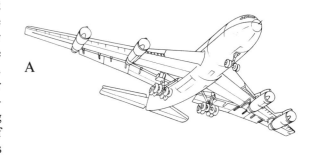

A

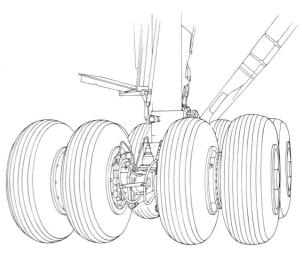

B

A *A Boeing 707 was*
& *fitted with special*
B *20-wheel landing gear
for soft-field
experiments. With
extra wheels mounted
adjacent to the normal
wheels, the aircraft
was operated from
ground barely able to
support a motor car.*

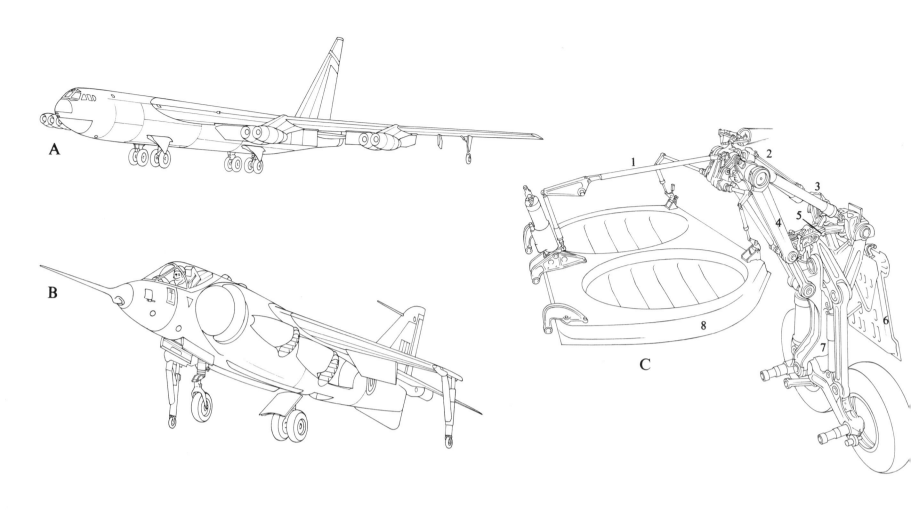

a blister on the side of the fuselage. Also of interest is the Jockey landing gear developed by the French Messier company for the Breguet 941, a transport aircraft designed especially for short take-offs and landings. The main gear comprises tandem legs on each side of the fuselage, utilising a common horizontal double-acting shock-absorber. The installation has excellent rough-ground taxying characteristics.

A wide variety of landing gear types is utilised on military aircraft. A tandem bicycle layout, with subsidiary "balancer" wheels under the outer engines to prevent toppling, was adopted for the Boeing B-47 six-engined bomber. This arrangement was continued, in developed form, on the later and much heavier B-52 Stratofortress, an eight-engined bomber. On this aircraft the main weight is taken by four twin-wheel units mounted in the fuselage, each provided with steering for use during cross-wind landings and to improve ground manoeuvrability. Lateral stability is provided by small "balancer wheels" on legs mounted outboard of the engines, these normally touching the ground only when the wings are full of fuel.

Other aircraft using tandem layouts include the Lockheed U-2 reconnaissance aircraft and the Hawker Harrier vertical take-off and landing strike aircraft. The outrigger legs on the U-2 are merely slipped under the wings and are jettisoned after take-off to reduce weight and drag. Jettisonable gears are

not new; one was used by Harry Hawker during his attempt to fly the Atlantic in 1919 and since then they have found several specialised applications.

Britain's Hawker Siddeley Vulcan and Handley Page Victor medium bombers utilise bogie main gear units, each with four twin-tyred wheels. An eight-wheeled bogie is used on the General Dynamics B-58 supersonic bomber. Bogie landing gears have also found widespread use on large civil airliners. The bogie units of the Comet, Caravelle, Convair 880, Boeing 707 and 720, and DC-8 retract sideways into the wing, but the main gears on the Soviet Tupolev series of airliners retract backward into fairings extending behind the wing trailing-edge.

One class of aircraft posing relatively few landing gear difficulties is the helicopter The major problem that may be encountered is that of "ground resonance", a violent pitching and rolling of the aircraft when on the ground caused by the rotor spinning round slightly out of balance, or through the rotor frequencies coinciding with those of the structure and landing gear. This problem of resonance is usually overcome by a careful compromise between the spring stiffness and damping characteristics of the landing gear.

Of interest is the unique landing gear on the Westland Wasp naval helicopter. Designed for operation at maximum loaded weight from restricted platforms on small ships in rough seas, the gear has to withstand exceptional vertical and horizontal

A *The Boeing B-52 tandem landing gear includes outrigger wheels, outboard of the outer engine pods, to stabilize the aircraft laterally.*

B *The Hawker Harrier VTOL strike aircraft embodies a tandem main gear, with small outrigger wheels at the wing-tips.*

C *Articulated bogie main gear of a de Havilland Comet jet airliner.*
1 Door mechanism
2 Main jack
3 Auxiliary jack

4 Side stay
5 Stabilizer
6 Fairing
7 Leg assembly
8 Landing gear bay door

D *The Convair B-58 Hustler's multi-wheel bogie main gear.*

E *In the Tupolev Tu-134 main gear, retraction is backward, the bogie assembly rotating through 180° so that the front wheels end up to the rear. This arrangement minimises the stowage space required.*

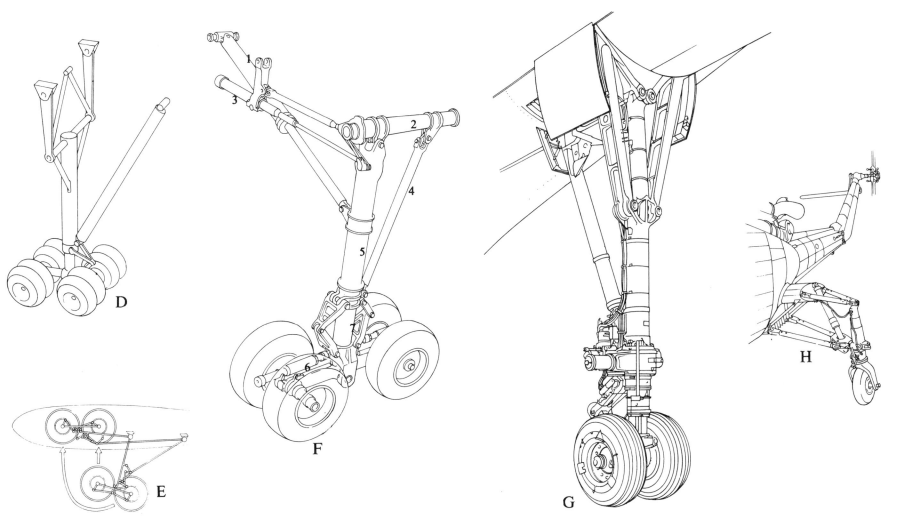

D

E

F

G

H

1 Telescopic side stay
2 Cross beam
3 Retraction jack
4 Drag strut
5 Leg
6 Pitch dampers
7 Shock absorber

decelerations. Despite violent movements of the deck, the machine must not slide or topple over and the gear thus comprises four legs spaced as far apart as practical considerations allow. The oleo struts have exceptional travel, and the four wheels can either castor freely or be locked in the fore and aft position. When they are locked to form a circle, the machine is effectively prevented from rolling forward or sideways. The Army counterpart of the Wasp, the Scout, is fitted with a simple skid landing gear.

Skids have also been fitted in modern times to a number of specialised fixed-wing aircraft, but have failed to gain general acceptance. However, although land skids have not proved very successful a close relative, the snow ski, has enjoyed persistent success where conditions have demanded its use.

An interesting application of skis was their installation on the jet-propelled Convair Sea Dart flying-boat in the mid nineteen-fifties. Twin hydro-skis were fitted, so eliminating the need for a conventional drag-inducing hull. The skis were extended for take-off and landing and retracted for slow-speed taxying, when the watertight fuselage acted as a hull.

Perhaps the most striking use of a skid landing gear was on the Messerschmitt Me 163B Komet rocket-propelled fighter of World War II. This was fitted with a single plain metal skid, hinged and provided with a shock-absorber. It was successful in that it enabled the aircraft to land, but in anything

but a precise touchdown, in still weather on smooth ground, it was brought to a stop so quickly that several pilots were injured on landing. To reduce the shock of impact, springing was introduced on the pilot's seat. For taking off, the Me 163 used a small twin-wheeled trolley, which was jettisoned once it became airborne.

A major disadvantage of this skid installation was the reduced manoeuvrability of the aircraft on the ground. In an attempt to overcome the problem, special ground handling trolleys were developed for the Me 163, but it is this factor which has probably prevented the wider use of skid landing gears on interceptor aircraft.

An entirely successful skid installation was that on the North American X-15 rocket-propelled research aircraft. This aircraft did not have to take off, being lifted for its test flights by a specially-modified B-52 bomber. After a relatively short high-speed, high-altitude test run, the pilot was preoccupied with lining up for landing. Fortunately, dry salt lakes, extending for miles, provided an almost unlimited landing area at the California test base. The touchdown was made on two skids at the rear of the aircraft, the nose being cushioned by a conventional twin-wheel nose leg as the speed fell off. More than 150 landings were made safely. However, the lack of ground manoeuvrability makes it unlikely that skids will ever enjoy widespread use.

F *With the Concorde bogie main landing gear, retraction is by hydraulic power through a single jack for each leg. A telescopic side-stay mechanically locks the leg in the extended position and also takes up side loads, but does not take any part in the gear retraction. A mechanical linkage shortens the gear automatically by drawing the shock-absorber up into the leg during retraction, so reducing the length of the leg and permitting the gear to be housed in a smaller bay. The mechanics of the gear are commendably simple, the only complication, compared with conventional gear, being the need to cool the wheel bays in flight, by cabin discharge air, so that the temperature does not exceed 80° Centigrade.*

G *The Concorde's nose gear. On all delta-winged aircraft it is necessary to keep the nose higher than usual, especially if the engines are slung under the wings. The unit is triangular in shape, which gives it greater rigidity: special alloys have been utilised to reduce the weight.*

H *One leg of the castoring long-stroke landing gear of the Westland Wasp helicopter*

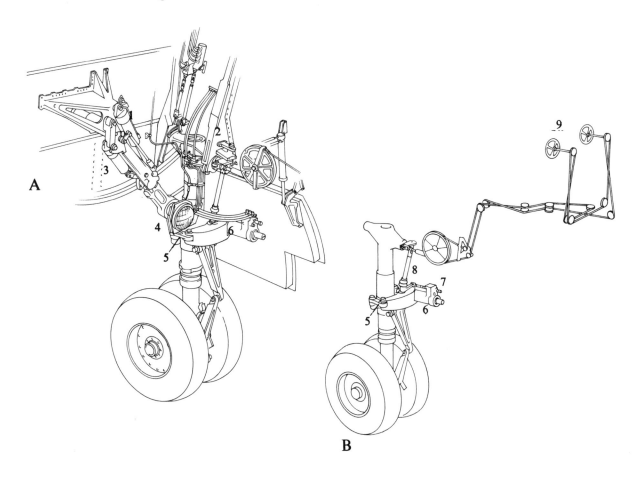

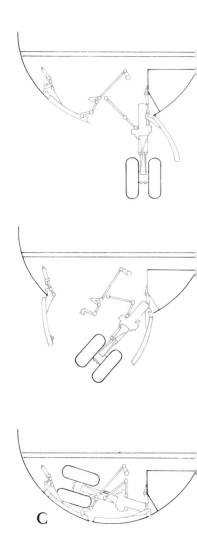

A *An unusual feature of*
B *the twin-wheeled nose*
& *gear of the Hawker*
C *Siddeley Trident airliner is that it is offset from the aircraft centre-line and retracts sideways. Advantages of this transverse stowage include a saving in weight, as the transverse retraction slot requires less reinforcing than its longitudinal counterpart, and an increase in space available below the floor for freight and equipment. On orthodox nose gears, which retract into a longitudinal slot in the fuselage, the small segments of underfloor space on either side of the slot are largely wasted because of their awkward shape and difficulty of access. Additionally, by locating the Trident gear directly beneath the front passenger door, the two frames*

which spread the loads from the nose gear into the fuselage also serve as edge members for the passenger door cut-out, giving further weight saving. The gear is hydraulically operated, but will drop freely under gravity in an emergency. If the free-fall is incomplete, the leg can be wound down by screw jack. When the nose unit is extended, the gear fairing door is closed, as are the doors of the main gear, to reduce drag. On selecting "gear up" the doors open, the legs retract and the doors close again.

1 Retraction jack
2 Leg front mounting fitting
3 Down lock spring strut
4 Landing light
5 Link
6 Steering jack
7 Control valve
8 Torque shaft
9 Steering wheels

D *Each main gear leg on*
& *the Hawker Siddeley*
E *Trident has four tyres abreast on a common lever-suspended axle. Retraction is sideways, the legs twisting through almost 90° and extending in length by six inches to make possible a neat stowage along the line of flight in the fuselage. Most unusual feature of the design is the helical cam mechanism within the top of the leg. It is movement of this cam which twists the oleo unit, and hence the wheels, and lengthens the leg. At the final stage of extension, the top of the oleo unit engages with a Curvic coupling; all vertical and torsional loads imposed on the extended leg are carried by this, and not by the cam mechanism. Advantages of the layout include a small turning circle,*

without serious tyre scrubbing, and reduced space requirement in the fuselage for stowage. When retracted, the frontal area presented by the wheels is smaller than that for a corresponding bogie, and the length of bay required is reduced by some 13 inches, providing additional stowage space for freight below the floor. A bonus arising from the mechanism used to twist the wheels is the lengthening of the leg during retraction. This allows the mounting points of each main gear unit to be six inches further outboard than would otherwise be the case, giving an increase in the track without increasing the height of the aircraft above the ground. The Trident landing gear is a good example of the mechanical ingenuity and complexity that

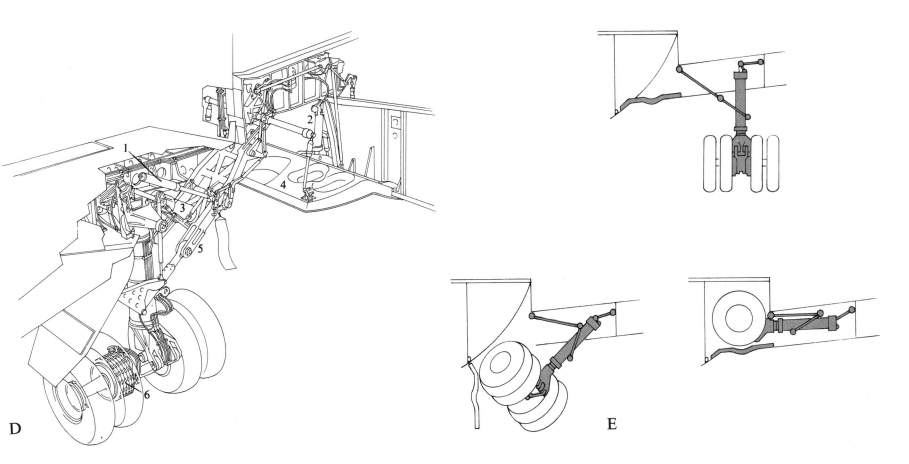

D

E

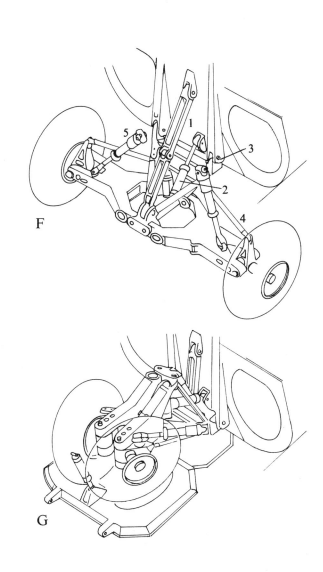

results from severe
limitations of
location, weight and
bulk placed on
designers.

1 Emergency air strut
2 Door jack
3 Retraction jack
4 Gear bay door
5 Side stay
6 Multi-disc brakes

F Ingenuity and
& complexity are evident in
G the main gear of the
General Dynamics
F-111 swing-wing
fighter. The wheels are
mounted on two sturdy
yoke beams, hinged to
a common member on
the centre-line of the
fuselage. During
retraction the forward
end of each yoke beam
is pulled up and
rearward, the two
outer ends of the yokes
folding forward. As the
yokes fold, the wheels
twist so that when
fully retracted they
are stowed almost
parallel with each
other.

1 Breaker strut
2 Retraction jack
3 Main pivot
4 Wheel parallel linkage
5 Shock strut

F

G

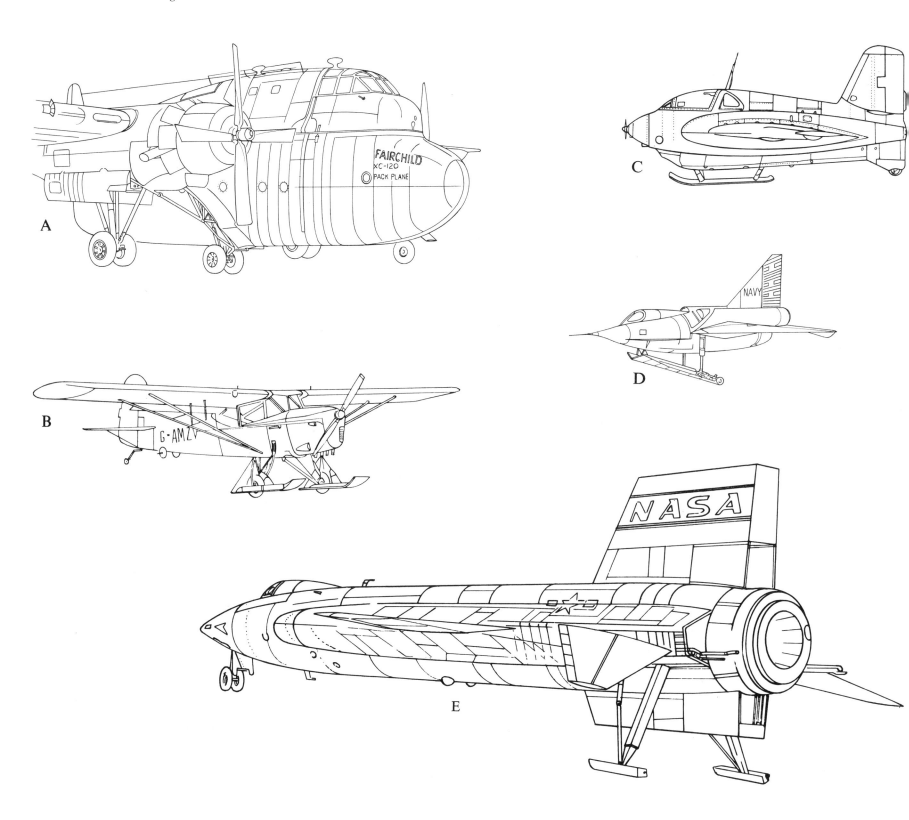

A *The Fairchild XC-120 Pack-Plane had a four-leg landing gear.*

B *Many light aircraft are fitted with snow skis. This Auster had an experimental water-ski installation.*

C *The Messerschmitt Me 163B Komet rocket-propelled interceptor on its landing skid. Considerable skill was required on landing, and several pilots were injured when touching down on rough ground.*

D *On the Convair Sea Dart the ski eliminated the need for a conventional drag-inducing flying-boat hull. Small wheels at the end of the ski enabled the aircraft to emerge from, and enter, the water under its own power.*

E *The landing gear of the North American X-15 research aircraft consisted of a twin nose-wheel and twin skids at the rear of the fuselage. In spite of a very high touchdown speed the gear proved satisfactory, more than 150 landings having been completed safely.*

A

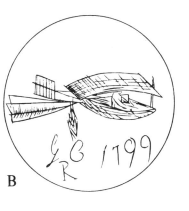

B

"When once a machine is under proper control under all conditions, the motor problem will be quickly solved." These prophetic words were written by Wilbur Wright while he and his brother Orville were building their first glider. The experience gained in testing this at Kitty Hawk in October 1900 strengthened the brothers' determination to concentrate on control in the air. In doing this the Wrights were more far-sighted than many other pioneers whose machines, even if they had been able to generate enough lift to support themselves, would not have flown properly, as insufficient attention had been paid to the problem of control.

The Wrights were not the first people to appreciate the importance of efficient controls for an aircraft; this honour goes to Leonardo da Vinci, the great medieval Italian scientist-painter, who sketched an ornithopter with a cruciform tail unit operated by a harness attached to the head of the pilot. Movement of the pilot's head could deflect the rudder-cum-elevator to either side, and up or down. This form of tail unit was based upon the rudder of a boat and the tail of a bird. A head-harness was adopted because both hands and feet would have been occupied in operating the flapping-wing mechanism.

Unfortunately, owing to a remarkable twist of fate, Leonardo's manuscripts remained generally unknown for nearly 300 years, and his sketches did little to help subsequent inventors.

The first practicable use of a rudder and elevator in the modern sense was made by the British pioneer, Sir George Cayley. On one side of a silver disc, now a treasured exhibit in the Science Museum in London, is an engraving of an aeroplane with a cruciform rudder-cum-elevator and the date it was drawn—1799. In 1804 Cayley used such a tail on his model glider, so providing the first example in history of the practical and successful aeroplane rudder and elevator. Cayley also incorporated a tail unit of this kind in the full-size glider he built and flew some time between 1804 and 1809.

The first idea for a separate rudder and elevator on an aeroplane was embodied in the famous design for an Aerial Steam Carriage patented by William Samuel Henson in 1842. This separation of the two primary flying controls has been used on virtually every aeroplane since and is standard practice today.

An elevator for a full-size machine was being constructed by Otto Lilienthal at the time of his death in 1896. It was to have been operated by a head-harness, similar to that proposed by Leonardo. The first pilot-operated tail controls on a full-sized aeroplane were those fitted to the unsuccessful Langley *Aerodrome* of 1903, which was abandoned after two crashes during launching.

To the Wright brothers goes the credit of developing the first movable elevators and rudders on successful aircraft, the former on the 1900 and 1901 gliders, and the latter on the modified No. 3 glider of 1902.

So far one vital control has not been mentioned—that required for lateral control. This is provided nowadays by ailerons, but in the early days the favoured technique was to twist or "warp" the entire wing.

The idea of twisting the wing in this manner—i.e. of increasing the angle of incidence on a wing to restore equilibrium—had occurred to several men through their observations of birds. Wilbur Wright recorded his first thoughts on the matter in 1899, as follows:

"My observations of the flight of buzzards lead me to believe that they regain their lateral balance when partly overturned by a gust of wind, by a torsion of the tips of their wings."

Although the idea had occurred to people earlier, it was the Wright brothers who first used it on a practical aeroplane. The Wrights' great contribution to the problems of control, however, was not so much their successful application of wing warping as the simultaneous use of wing warping and rudder control. Initially they had hoped to steer their aircraft entirely by wing warping; they then added a fixed rear fin. Using this combination on their 1902 glider, they experienced warp drag. This is the tendency of the down-going wing to create more drag than the up-going wing, resulting in the craft turning in the opposite direction to that intended. To overcome this, they replaced the fixed fin with a movable rudder, inter-linked with the warping control. This had never been suggested before—which is hardly surprising, as until a practical stage of flying had been reached no-one could reasonably have been expected to foresee the need for it. The movable rudder provided a powerful turning moment which overcame that of the wing drag.

A *This historic design by Leonardo da Vinci was for an ornithopter with movable tail surfaces controlled by the pilot.*

B *The silver disc, engraved by Cayley in 1799, bears the design of an aeroplane with a combined cruciform rudder-cum-elevator.*

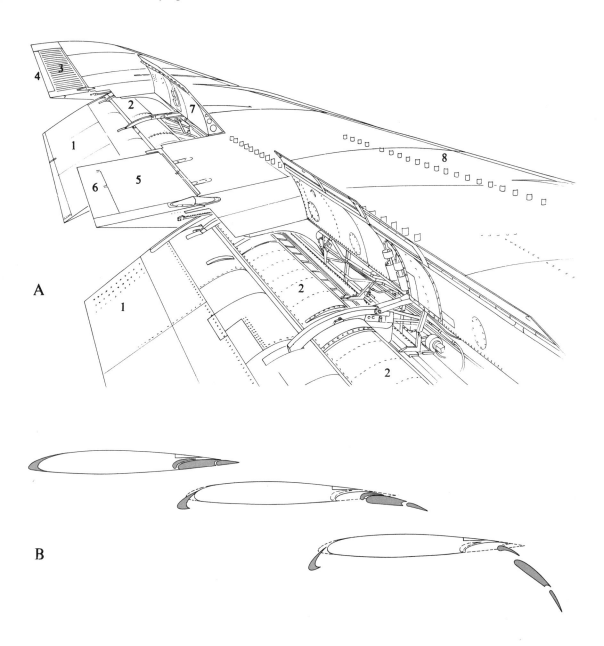

A

B

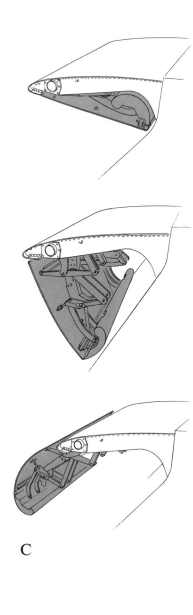

C

A *Passenger view of a wing. The swept wings of present-day airliners are designed primarily for efficient high-speed flight, with low drag, and have basically-poor lifting capabilities at low speeds. High-lift devices are therefore fitted to improve the lifting qualities and so reduce take-off and landing speeds. These lifting devices take the form of trailing-edge flaps and gadgets on the leading-edge, and work by increasing the wing area and improving the overall camber of the wing so that more air is deflected downward more rapidly. The trailing-edge flaps are quite large and, when* operated, significantly alter the shape of the wing. To inexperienced passengers it must seem as if the wing is beginning to fall apart! This illustration shows some of the things that a passenger can see on the wing of a Boeing 707. The flaps (1), embody the plates (2) to give a double-slotted effect. The outboard aileron (3), is operated by the manual servo-tab (4), for use during low-speed flight. The inboard aileron (5), is used for high-speed flight, and is also operated by a manual servo-tab (6). The spoilers (7) are operated differentially to assist the ailerons when required, and can also be operated in unison as airbrakes. The row of little metal plates on top of the wing are vortex generators (8). These improve the general airflow by swirling up the boundary layer of air close to the wing. Additional control surfaces visible on some other aircraft include lift dumpers. Similar to air-brakes, these are located in front of the flaps. After touch-down they are extended to spoil the airflow over the flaps and thus "dump lift", to make the aircraft settle firmly on its wheels and so prevent it from bouncing and permit efficient braking.*

B *The wing of the Boeing 737. The area-increasing trailing-edge flap is of the triple-slotted type. The novel leading-edge lift system consists of slats and a downward-hinging surface known as a Kruger flap. This combination represents one of the most efficient lift-increasing arrangements yet developed.*

C *The leading-edge flaps developed for the Boeing 747 "Jumbo jet" provide passengers with a sight not seen since the days of the original Wright Flyer and warping wings. The new flaps, ten of which are mounted on each wing, are said to have "variable camber" because of their ability to change shape. During normal flight the flap is tucked neatly into the wing for efficient high-speed flight. For landing and take-off the flap literally unfolds like a pocket knife. When* nearly extended the main portion of the flap is bent into a curve for maximum low-speed lift. The new flaps are made of laminated glass-fibre and glass-fibre honeycomb. This is similar in structure to the beehive honeycomb after which it is named, and was originally developed to produce light and rigid aircraft parts. The 747 is the first aircraft on which a structural honeycomb assembly is flexed intentionally.*

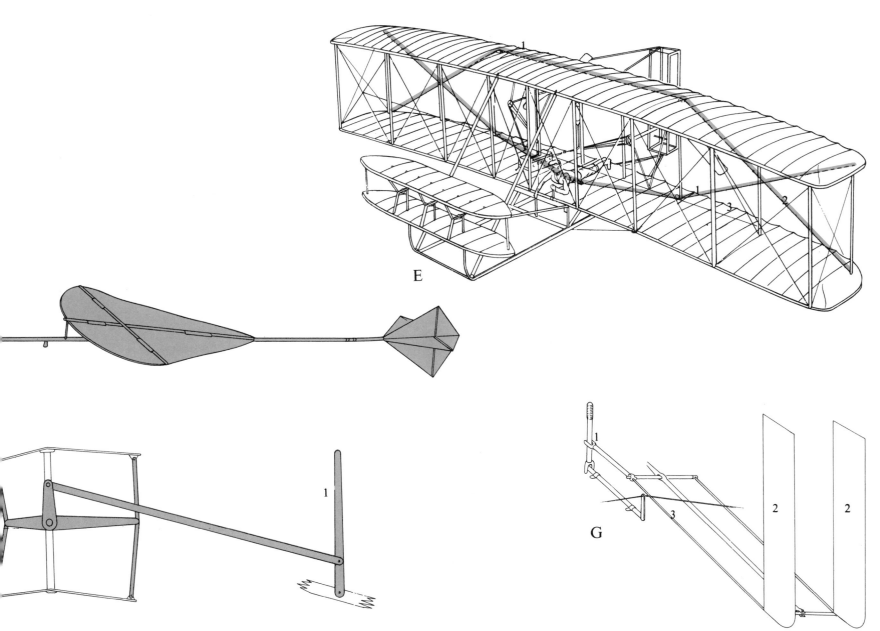

D *Cayley's model glider of 1804 incorporated the first successful rudder and elevator.*

E *In the wing warping system of the 1903 Wright Flyer, two cables ran from the pilot cradle, over pulleys, one cable leading to the top of the outer rear starboard wing strut and one to the top of the port strut. The lower ends of these outer struts were interconnected by a separate auxiliary cable which ran upward and passed over pulleys attached to the top of the two centre struts. This auxiliary cable was not attached to the control and cradle. To bank to port, the pilot-cradle was twisted to port; this pulled on the starboard cable which, acting on the top of the starboard strut, pulled down the outer part of the trailing-edge of the top starboard wing, the struts pushing down the lower wing in sympathy. Simultaneously, this downward warping of the starboard wing pulled on the auxiliary cable (connecting the lower ends of the outer struts) and automatically pulled up the outer port strut, to give negative warp on the port side and lower the port wing. On later aircraft the cradle was replaced by hand-operated control levers.*

1 Pulley
2 Auxiliary warping cable
3 Main wing warping control cables connected to pilot cradle

F *The Wright's front elevator and its method of control. The lever was held in the pilot's left hand.*

1 Lever

G *The ingenious combined rudder control and wing warping mechanism devised by the Wright brothers. Initially, the controls were interconnected, so that when the aircraft was, say, banked to the left the rudder automatically moved to the left. Later, the control was altered as shown here to permit separate, or any desired combination of control movements. Sideways movement warped the wings and fore-and-aft movement controlled the rudder.*

1 Control lever
2 Rudders
3 Wing warping cable

In 1905 the brothers realised that, although simultaneous use of warping and rudder was sometimes essential, it was undesirable at other times. In certain situations, it could be necessary to cross the controls—as in side-slipping, for example. They thus abandoned the automatic warping/rudder linkage and substituted controls working separately or together in any combination as desired, a practice which has since been adopted universally.

Although a system of control using ailerons was patented by M. P. W. Boulton, an Englishman, in 1868, the modern invention of ailerons stems directly from the Wrights' wing warping. This is due to the fact that Boulton's truly original and remarkable invention was overlooked, primarily because those parts of the patent which were published dealt mainly with ingenious but useless ideas for ornithopters. Had early pioneers been aware of Boulton's idea for lateral control, the course of aeronautical history might have been different, since lack of lateral stability and control was a major stumbling block to Langley, Curtiss, European pioneers generally and even the Wrights themselves.

As it was, no reference to an aileron occurs until January 1905, when the French pioneer Robert Esnault-Pelterie gave a lecture to the French Aero Club. Describing the construction of a Wright-type glider in 1904, he said:

"The warping of the surfaces, vaunted by the Wright brothers, which we tried, gives good enough results for the maintenance of transverse equilibrium, but we consider this system dangerous. It may, in our opinion, cause excessive strains on the wiring, and so we fear breakages in the air, which cannot occur with the ordinary rigid system. . . . We therefore felt we had to abandon warping. Nevertheless, in order to be able to control lateral balance, we then employed at the front two independent horizontal rudders, one placed at each extremity of the aeroplane. These two rudders were each connected to a little steering (operating) device within reach of the two hands of the operator. . . . This arrangement gave satisfaction, although it was not as powerful as the wing-warping."

In this passage the term "horizontal rudders" refers to what we now call ailerons. "Aileron" is the French word for the ends of a bird's wing, and was first used in its current context in 1909.

This represents the birth of the modern aileron system, although wing-warping died hard, being used on several aircraft right into the 1914-18 War including the famous Fokker Monoplane. Wing warping is still used today in certain specialised aircraft, such as a few high-performance gliders. The winner of the 1958 World Gliding Championship, a West German HK S-3, utilised wing-warping for both lateral control and flap functions.

In the early days of flying, a wide variety of mechanisms were devised for operating the three sets of control surfaces, but the advantages of hand and foot-operated controls soon became evident and have since been adopted almost universally. Feet are used to operate the rudder, with a hand-operated central control column for the elevators and ailerons. Operating movements are instinctive: fore and aft movement of the control column raises or lowers the nose of the aircraft; sideways movement of the control column (or rotation of a handwheel on top of the column) rolls the aircraft by deflecting one aileron up and the other one down; and the aircraft is yawed to right or left by pushing on the respective foot pedal.

The forces to be overcome by the control surfaces vary with differing conditions of flight. Thus the horizontal forces to be overcome by the elevators vary with alterations in speed, engine power or centre of gravity position, such as may occur when fuel is consumed. When this happens, the elevator has to be moved to maintain the aircraft in balance. It would be extremely tiring for the pilot to have to apply the necessary load indefinitely, and to remedy this a procedure known as trimming is applied to cancel out the load. This is achieved by a secondary control, known as a trimmer, or trim tab, which can be set to exert a continuous force in the required direction. The elevator is particularly sensitive to speed variations, but can be adjusted, or trimmed, to achieve a condition where the aircraft will fly itself, with the pilot's hands off the control column. A trim control may also be provided on the ailerons, to correct any tendency for the aircraft to fly one wing low, and on the rudder, to compensate for asymmetric thrust should one engine fail, or for variations in propeller torque.

On light aircraft, trimming is often achieved by small adjustable tabs (which are not controllable in flight) on the trailing-edge of the surface concerned. On advanced aircraft with powered controls, trimming can be effected by bodily displacing the control surface or, in the case of elevator and rudder trims, by movement of the normally-fixed part of the surface—i.e., of the horizontal stabilizer (tailplane) or the vertical stabilizer (fin).

Control surface movement is assisted by embodying an aerodynamic balance forward of the hinge line, and on some aircraft this is supplemented by balance tabs. These take the form of a trailing-edge tab which is arranged to move in the opposite direction to the control surface, usually through mechanical gearing so that the balancing moment is progressive with deflection of the surface.

On some aircraft the control surfaces are actuated indirectly by operation of a subsidiary or servo tab. (On early designs these tabs often took the form of a large separate surface mounted some way aft of the main surface). When a servo tab is fitted, the main surface is not connected to the pilot's controls at all, being moved solely by the loads generated by the deflected servo tab. The flying controls on the Britannia airliner are operated in this manner, the effort required from the pilot being so small that artificial "feel" has to be

A *This control column, for wing warping and elevator control, was used on the Star monoplane of 1911. Initially it was intended that lateral control on this aircraft would be effected by differential control of the cruciform elevators and rudders, a system used successfully today on the Jaguar and Vigilante attack and reconnaissance aircraft.*
1 Wing warping handwheel
2 Throttle lever
3 Cable to wing
4 Cable to elevator

B *The control column used for operating the rudder and elevator on a Deperdussin monoplane.*
1 Wing warping handwheel
2 Engine controls
3 To elevator
4 To wing
5 To rocker for wing warping
6 Rocker
7 Landing skid

C *The unusual control system of the 1911 Nieuport two-seat monoplane. Wing-warping was effected by a foot lever, elevator and rudder control by the central hand lever.*
1 Hand lever for elevator and rudder control
2 Foot lever for wing warping
3 To elevator
4 To rudder
5 Rocker shaft

D *Centralised control system on the Blackburn monoplane of 1911. Wing warping elevator and rudder were all operated by movement of the single control wheel.*
1 Universal joint
2 To elevator
3 To rudder
4 To wing
5 Rocker shaft

E *In a conventional flying control system, foot pedals operate the rudder, while a central control column operates the elevators and ailerons.*
1 Control column
2 Aileron
3 Rudder pedals
4 Rudder
5 Elevator

F *The elevator flying control system of a VC10 jet airliner. Control column movements are transmitted by rods and cables to duplicated electro-hydraulic power units, each unit acting on half an elevator. Artificial feel (p. 118) is provided by hydraulic feel units.*
1 Tailplane screw-jack
2 Electro-hydraulic power units, moving separate surfaces
3 Duplicated feel units

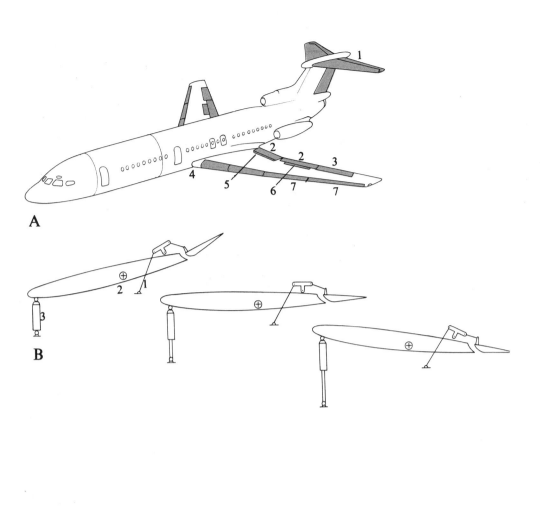

A

B

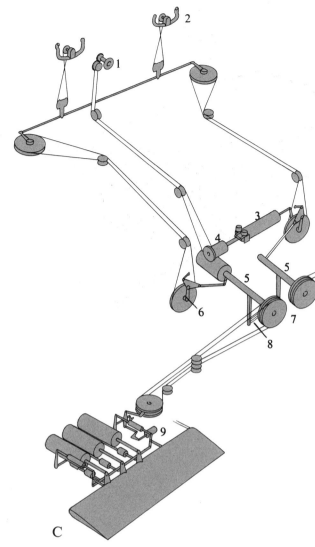

C

applied to the control circuits (see the text below).

At high speeds heavy loads are required to operate the control surfaces, and this led to the development of powered controls. Initially, power was supplied only to assist the pilot, by means of hydraulic boosters. With such a system, a proportion of the load required to move the control surfaces is still applied by the pilot. In the event of the booster failing, the systems revert to full manual control. Aircraft using such semi-manual systems include the Boeing 707 series of airliners and the B-52 Stratofortress bomber.

As speed increases, it becomes more and more difficult to keep the loads required within the capability of the average pilot, and this has led to the development of fully-powered systems. With these, the pilot does little more than operate a small hydraulic valve, requiring a load of around half a pound. The load remains constant irrespective of the speed or the deflection of the control surface, and to give the pilot some idea of the aerodynamic loads being imposed, artificial "feel" characteristics are built into powered systems. These may take the form of a simple spring, providing an increased load with increase of control column movement, or they may be provided by a complex simulator, providing forces proportional to the airspeed or Mach number.

With fully-powered control systems, a reliable supply of the necessary power is of paramount importance, because, should it fail, the aircraft would be uncontrollable. The power supply is ensured by providing duplicate operating jacks, with separate hydraulic pipelines, each supplied from an independent source. The two jacks can be arranged so that they both operate on the same control surface, or the control surface can be divided into two or more sections, each section being operated by a separate jack.

At very high speeds, around the transonic mark, ordinary elevators tend to become ineffective, and large movements are required to produce a significant force. One solution to this is the use of a variable-incidence tailplane—long used for trimming purposes—as the primary control in pitch, and many supersonic aeroplanes have a one-piece "slab" tailplane, with no elevator at all. Such a technique has also been applied to the vertical fin and rudder.

It is common for aircraft to have auto-pilots to relieve the pilot of the task of holding the aircraft in a constant attitude. The auto-pilot exercises its authority by means of servos imposed in the flying control circuits. Most auto-pilots incorporate trim or bias controls, enabling a steady turn or climb to be performed. Other information can be fed

A *Typical of the flying controls on modern high performance aircraft are those of the Hawker Siddeley Trident. Leading-edge devices and slotted flaps provide increased lift for take-off and landing.*
1 Elevator (geared flap)
2 Flaps
3 Aileron
4 Kruger flap
5 Lift dumper
6 Air brake
7 Leading edge slats

B *The "elevator" is really a geared trailing-edge flap. The tailplane (horizontal stabilizer) is connected to the control column, the flap being operated by a simple mechanical link. The gearing ensures a high negative lift force for take-off and landing, but little flap movement at high speed.*
1 Mechanical link
2 Pivot
3 Jack

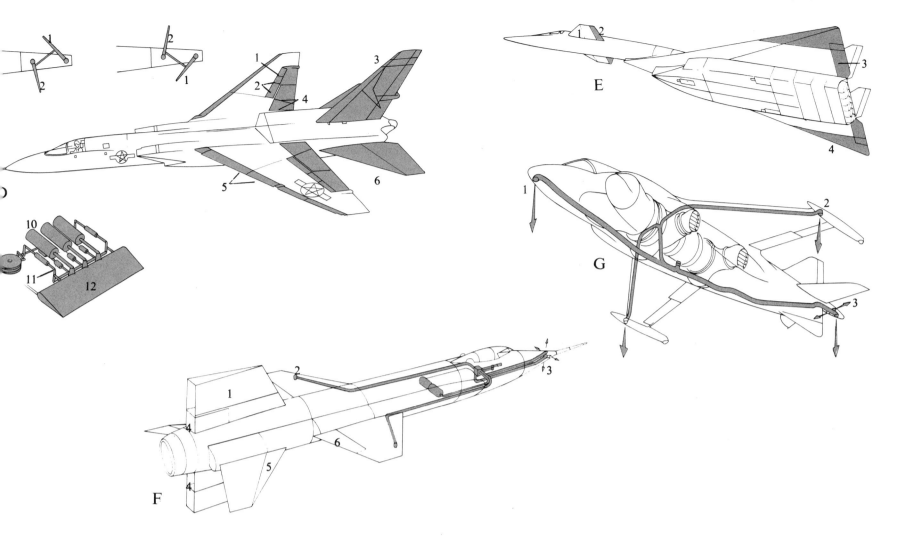

C *Trident aileron control system. This aircraft employs a fully-powered flying control system, with no recourse to a manual system. Single-piece control surfaces are used, each powered by three hydraulic jacks. The three jacks are normally working all the time and are thus considered to be "triplexed". This is quite different from those systems in which the jacks are duplicated, or triplicated; in such systems only one jack is working at a time, and, if it fails, the spare or duplicate jack is brought into operation. The advantage of a triplex installation is that, if one system fails, the remaining systems continue to work without any action on the part of the pilot,* thus eliminating the need for any hurried change-over from one system to another. Each control jack is supplied by an independent engine-mounted pump, backed up by two electrically-driven standby pumps. Should this fail, an emergency air-turbine driven pump takes over.

1 Aileron trim wheel
2 Aileron handwheel
3 Feel and centring spring strut
4 Trim screw jack
5 Duplex shafts
6 Non-linear gear unit
7 Cable compensators
8 Inter-connecting rod
9 Roll damper actuator
10 Aileron operating jacks
11 Stuck valve detector struts
12 Aileron

D *The North American Vigilante combat aircraft utilises an all-moving vertical fin,* with an all-moving horizontal tailplane, used differentially in conjunction with wing spoilers for lateral control. Spoiler down and deflector up to direct airflow through upgoing wing; movement reversed on downgoing wing.

1 Deflector
2 Spoiler
3 All-moving fin
4 Flaps
5 Drooping leading edge
6 Differential all-moving horizontal tailplane

E *Flying controls on the North American XB-70 research aircraft. The trimming foreplane is noteworthy. The elevons, combining the functions of elevators and ailerons, are used also as flaps.*

1 Trimming foreplane
2 Flaps
3 Elevons, drooped for flap effect
4 Tips, drooped at high speed

F *The North American X-15 research aircraft has a jet-reaction control system for use at high altitudes, where the near-vacuum renders the aerodynamic flying controls ineffective.*

1 All-moving fin
2 Roll control jets
3 Pitch and yaw control jets
4 Air brakes
5 Elevons
6 Flaps (no ailerons)

G *The jet-reaction control system on the Hawker Siddeley Harrier attack aircraft is used at low speeds when the aerodynamic flying controls are ineffective. Such auxiliary control systems are necessary on all direct-lift VTOL aircraft.*

1 Pitch control jet
2 Roll control jets
3 Yaw and pitch control jets

to the auto-pilot so that it can effect an airspeed hold, Mach/speed hold or height hold. Auto-pilots can also be locked on to radio or radar transmissions—e.g., to the glide-slope radio beams of the instrument landing system—or, in the case of military aircraft, to target-tracking radar. Finally, auto-pilots can be made to accept information from external references and compute the outputs to the flying controls as necessary to effect an automatic approach and landing.

The development of specialised aircraft, such as the North American X-15 and the new generation of VTOL (vertical take-off and landing) machines introduced additional problems affecting control and stability. At its operating altitude, the X-15 was in the near-vacuum of space and the ordinary aerodynamic control surfaces ceased to be effective. To provide control forces under these conditions, the X-15 incorporated a system of reaction jets, similar to those fitted to spacecraft.

VTOL aircraft also require a system of jet reaction controls to provide control forces until sufficient speed is gained for the aerodynamic controls to function. The development of a reliable reaction control system is a most complex technical problem and one which must be solved if VTOL aircraft are to experience the rich future confidently predicted for them.

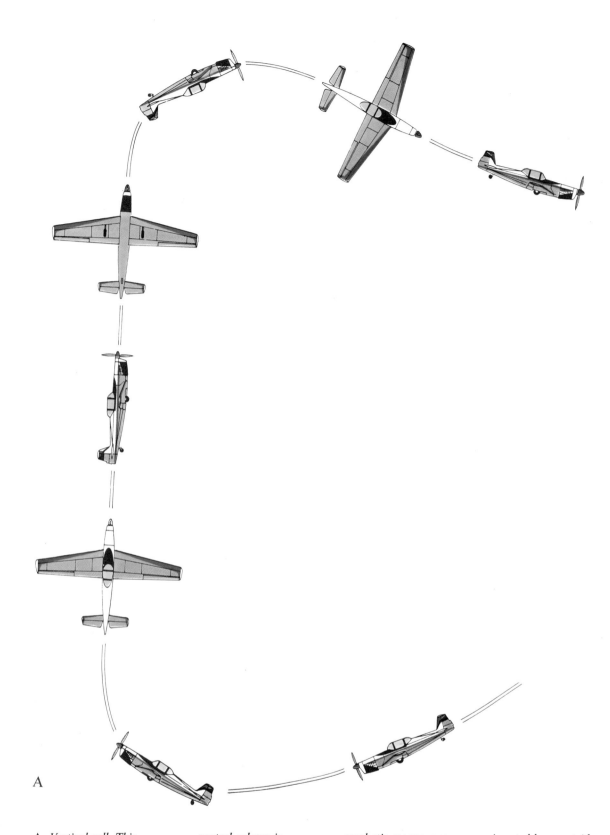

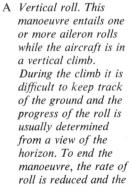

A *Vertical roll. This manoeuvre entails one or more aileron rolls while the aircraft is in a vertical climb. During the climb it is difficult to keep track of the ground and the progress of the roll is usually determined from a view of the horizon. To end the manoeuvre, the rate of roll is reduced and the* control column is pulled back to start the nose-down towards the horizon. As the aircraft reaches level flight, the rate of roll is reduced further and is stopped when the wings are level.

B *Loop. Consisting of a 360° turn in the vertical plane, the loop is the easiest of all* aerobatic manoeuvres. The elevators provide the basic control, the ailerons and rudders being used only to maintain a straight heading. In a perfect loop the exit is made at the same height as the entry, the loop path itself being circular.
Variations on the basic loop include the inverted loop, outside loop, and the inverted outside loop. In an inverted loop, the aircraft is inverted before entry and the loop is made downward, the aircraft being half rolled at completion, back to level flight.
In the outside loop the aircraft is inverted before entry, the loop being made with forward pressure on the control column; a half roll brings the aircraft back to level flight at completion. The inverted outside loop begins from level flight, the nose being put down as for a steep dive. Forward pressure on the control column is maintained until level flight is regained. On both the outside loop and the inverted outside loop the pilot is on the outside and the manoeuvres subject both aircraft and pilot to negative "g" stresses.

C *Barrel roll. This is a manoeuvre in which the aircraft rolls through 360° while the nose of the aircraft describes a circle. In a* perfectly executed manoeuvre, the circle is of constant radius and the aircraft is rolled through 90° at each quarter of the circle.

D *The Chandelle. The Chandelle is a maximum performance climbing turn, with a precise entry speed and an accurate 180° turn. The manoeuvre is entered from a shallow dive and ends with the 180° turn completed, the wings level and the aircraft near the stall. At this point the nose is lowered slowly to the level flight attitude. The manoeuvre can be executed slowly or quickly, leading to wide variations in its style.*

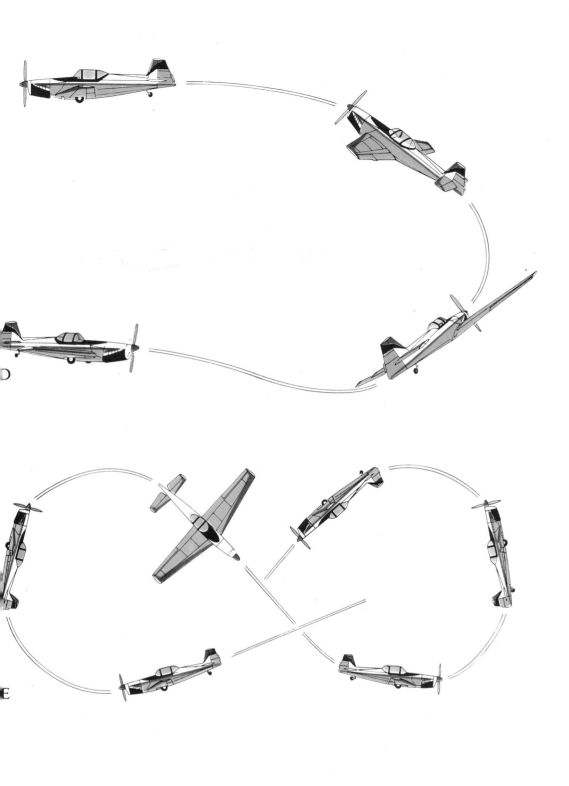

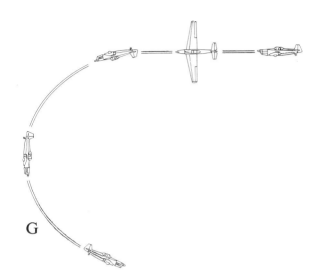

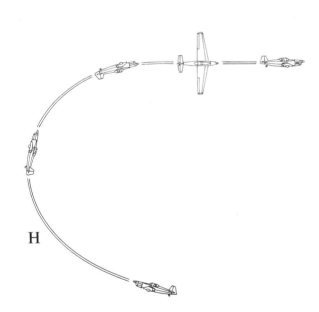

The Cuban eight. Also known as a variant of "the spectacles", this is a combination manoeuvre, combining the loop and a diving aileron roll. Each half of the "eight" is basically an Immelmann turn, completed in a 45° dive.

F *Slow roll. In this manoeuvre the nose of the aircraft is made to rotate 360° on a point, rather than around a circle as in the barrel roll. The manoeuvre is useful for demonstrating the function of each flight control in rolling attitudes, as the elevator and rudder exchange functions and movement of the controls must be exaggerated to meet the requirements of the*

manoeuvre. Modern aircraft do not slow-roll well and because of this the manoeuvre has lost its popularity.

G *Half-roll and split S. This consists of a half-roll followed by the second half of an ordinary loop. Height is lost during the manoeuvre, which, therefore, should not be entered without plenty of altitude.*

H *Immelmann turn. This is a combination manoeuvre, consisting of the first half of a loop followed by a half-roll to level flight. It is named after the German World War I ace who developed the manoeuvre to gain altitude and reverse direction 180° simultaneously.*

AIRCRAFT AEROBATICS

Aerobatics are often performed purely for fun. The thrill of completing successfully intricate and difficult manoeuvres in the air is akin to that experienced in skiing or, in the world of boats, in sailing close to the wind. Like skiing and sailing, aerobatics are a challenge to a pilot's skill and daring. However, for the majority of pilots, aerobatics are an essential part of military flight training. By performing them a pilot learns to control his aircraft accurately throughout its design range of air speeds and altitudes. This provides a sense of confidence which is as important as the flying skills attained.

FLIGHT DECK

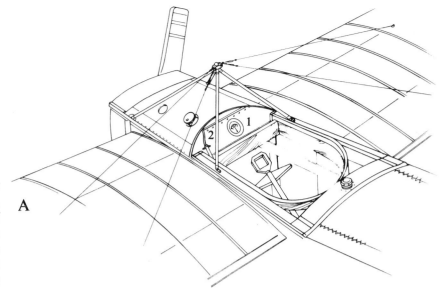

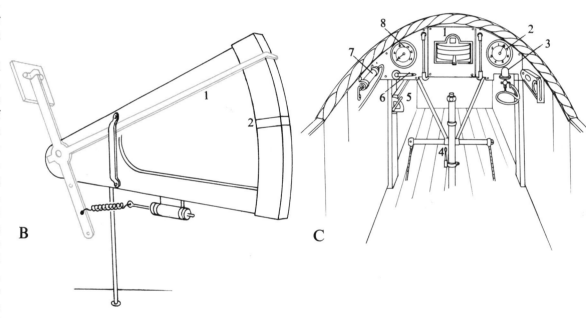

Not too long ago the flight deck of a modern airliner or a military jet looked like a glittering array of dials, switches, knobs, lights and labels. The flight decks of the aircraft leaving the manufacturers today have a much cleaner layout. A few small television displays have taken over from the earlier myriad of small watchlike indicators scattered everywhere over the panels in front of the pilots.

But it is still difficult to believe that the first aviators managed to fly without any instruments at all. This simple state of affairs did not, however, last for long, although for many years the instrumentation provided was sparse indeed by today's standards.

The first instruments that were generally fitted gave information on the performance of the engine, usually indicating revolutions per minute and the speed of the aircraft through the air. Knowledge of the speed was particularly important, as in even a quite shallow dive the speed could build up sufficiently to overstress the flimsy structures of the day. So great was the need for an instrument indicating airspeed that the British aeronautical journal *Flight*, in 1910, offered a prize of £5 for the best suggestion for a "safety speed alarm" for aeroplanes. During the ensuing weeks the correspondence pages were filled with designs for alarms which, when the speed exceeded a safe value, blew whistles, sounded sirens, rang bells, switched on lights or played tunes on the chimes from a musical box. We should not smile too unkindly at these ideas, because the high-speed aeroplanes of today incorporate "speed alarm systems" which ring bells, sound sirens and flash warning lights—or even put the aircraft into a climb—should the speed exceed a safe value!

It is thus not surprising that the first instrument developed specifically for use on aircraft was a speed indicator. Invented by the French Captain A. Etévé in 1910, and tested in January 1911, this did not indicate actual speed, but showed whether the aircraft concerned was flying faster or slower than its specified normal speed. In Britain, the Aircraft Factory at Farnborough, forerunner of the Royal Aircraft Establishment, developed an airspeed indicator which was fitted to an aircraft in 1911. Early aviators were content with this and a fuel gauge. Later, they found altimeters useful to give an indication of height and, if they were brash enough to think of a cross-country flight, a compass was carried. The first workable aircraft compass was designed in 1910 by Captain Creagh-Osborne of the British Royal Navy.

A *The sparsely instrumented cockpit of the Bristol Monoplane of 1911 shows why pilots had to fly "by the seat of their pants". A contemporary description records: "The pilot is kept well acquainted with the condition of his fuel by a petrol gauge, and a revolution indicator fitted to a small dashboard before him tells him his engine speed".*
1 Engine speed indicator
2 Fuel gauge, glass tube indicating level

B *The Etévé speed indicator, a French invention, was the first instrument developed specifically for use on aircraft. The device did not indicate the actual speed, but was designed to show whether the aircraft was flying above or below its normal airspeed. The pilot was able to maintain constant speed by keeping the pointer aligned with the red mark. The makers of the indicator advised that "a few trial flights have to be made before the instrument can be correctly adjusted".*
1 Pointer
2 Red mark

C *The cockpit of a monoplane built in 1913 to the design of G. M. Dyott, an experienced pilot, shows a marked improvement in instrumentation. Equipment includes a compass, petrol and oil gauges, altimeter, inclinometer and engine revolution indicator.*
1 Compass
2 Altimeter
3 "Tell-Tale" glass inclinometer
4 Fuel cut-off lever
5 Fuel adjustment
6 Air adjustment
7 Pressure pump
8 Engine rev. counter

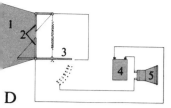

D

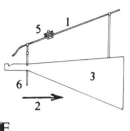

E

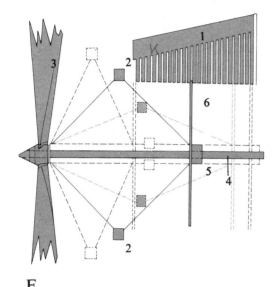

F

SUGGESTIONS SUBMITTED IN ANSWER TO THE OFFER OF A CASH PRIZE BY AN AERONAUTICAL JOURNAL IN 1910 FOR AN AIRCRAFT SPEED ALARM.

SPEED ALARMS

D *This suggestion involves the use of electrics. A funnel incorporates a pair of hinged gates, normally held closed by a spring. During flight, air pressure opens the gates, moving an arm over a series of electrical contacts, each corresponding to a particular speed. At the pre-determined limiting speed, the contact would complete the circuit to sound a siren.*
1 Funnel
2 Gates
3 Contact-arm
4 Battery
5 Siren

E *This device consists essentially of a funnel facing in the direction of flight. A hinged plate above the funnel embodies a weight, and a shutter. As the speed approaches a pre-determined level, the plate lifts, raising the shutter and allowing the whistle to sound.*
1 Hinged plate
2 Direction of flight
3 Funnel
4 Whistle
5 Weight
6 Shutter

F *This remarkable device consists of a propeller free to revolve on a shaft. Attached to the propeller and to a sliding guide are a pair of weight-type governors. Attached to and revolving with a collar is a toothed disc, engaging with the teeth of a chime from a musical box.*
At low speed the collar would be at the end of the shaft, the teeth on the disc sounding a deep note. As the speed increases, the disc slides along the shaft, making a distinctive sound at various speeds. The inventor suggests that
the teeth could be cut away over the normal speed range, so that "at danger point a musical sound would be noticeable even amongst the various whistling and humming sounds attendant to a machine in motion".
1 Chime
2 Weight governors
3 Propeller
4 Shaft
5 Sliding guide
6 Toothed disc

Instrumentation gradually increased during the 1914-18 War, and at the end of hostilities aircraft were generally fitted with an airspeed indicator, altimeter, compass, engine rpm indicator and oil pressure gauge. Communications facilities were poor. Although Britain had successfully transmitted wireless messages to a free balloon, the *Pegasus*, as early as 1908, and in America radio messages had been both sent from and received by a Curtiss biplane flying over Sheepshead Bay in New York in 1910, pilots flying over the water in 1914-18 simply carried a homing pigeon inside their aircraft. Radio was used, however, to direct gunfire from the air, and one or two RAF squadrons even used primitive transmitter-receivers in air-to-air combat.

From such crude beginnings have developed the complex flight decks of today. The instruments displayed in a modern airliner flight deck can be divided broadly into two main groups: the flight instruments and the systems instruments.

The flight instruments show the relationship between the aircraft and its environment, including the ground. They give the pilot information needed for the maintenance of controlled flight. The systems instruments provide information about the aircraft itself—about the engines and the various ancillary systems.

Important, but separate from these two groups, are the instruments and equipment required for navigation.

Of the instruments, those relating to the flight of the aircraft are the most important.

In the early days pilots flew "by the seat of their pants", trying to imitate birds by inherently sensing their attitude. Experience soon showed that normal senses are inadequate; an aviator flying into cloud, and unable to see the ground or horizon, might feel that he is flying straight and level while the aircraft is actually in a steep bank, and turning. In such circumstances, pilots frequently became confused and allowed their aircraft to slip into a spiral descent or a spin, from which they could not recover.

Tentative steps towards the development of such instruments began before the end of the 1914-18 War, but real progress was not made until the nineteen-twenties. In 1928 the Daniel Guggenheim Fund for the Promotion of Aeronautics initiated a programme for the development of blind flying. To carry out the flying, Harry Guggenheim requested—and obtained—the assistance of an outstanding Army Air Corps pilot named Lieutenant "Jimmy" Doolittle, who was later to acquire world-wide fame for his part in the bombing of Tokyo from an aircraft carrier in 1942.

Tests soon indicated the inadequacy of existing instruments. The magnetic compasses then available were no good for low-altitude flight near obstacles. Altimeters, depending upon changes in atmospheric pressure, could be as much as 100 ft in error at low levels. Doolittle also criticized the turn-and-bank indicators, which consisted of a ball that rolled from

A *McDonnell Phantom II. The cockpit of a high-performance fighter in many respects presents more of a challenge than the flight deck of an airliner. Space is more limited and, in addition to the normal flight instruments and systems indicators, accommodation has to be found for advanced navigation and communications systems, missile, gun and bomb selectors and attack systems controllers. Although for some missions there is a degree of automation in the flight control, navigation and attack functions, the pilot is kept fully occupied monitoring the array of indicators. Noteworthy in this illustration are the missile (1) and bomb (2) control panels and the missile status panel (24).*

1 Missile control panel
2 Bomb control panel
3 Control stick grip
4 Engine control panel
5 Fuel control panel
6 Auxiliary armament control panel
7 Anti "G" suit control valve
8 Emergency hydraulic pump handle
9 Flap controls
10 Emergency canopy release handle
11 Approach indexer light

12 Azimuth-elevation-range indicator
13 Standby compass
14 Manual canopy unlock handle
15 Caution light panels
16 Bomb control monitor panel
17 Temperature control panel
18 Wingfold panel
19 Stabilator trim position indicator
20 Wing trim position indicator

21 Rudder position indicator
22 True airspeed indicator
23 Radio altimeter
24 Missile status panel
25 Accelerometer
26 Angle-of-attack indicator
27 Airspeed and Mach number indicator
28 Attitude director indicator
29 Horizontal situation indicator
30 Altimeter

31 Vertical velocity indicator
32 Clock
33 Exhaust nozzle position indicators
34 Exhaust gas temperature indicators
35 Tachometers
36 Engine fuel flow indicators
37 Fuel quantity indicator
38 Pneumatic pressure indicators
39 Hydraulic pressure indicators
40 Oil pressure indicators

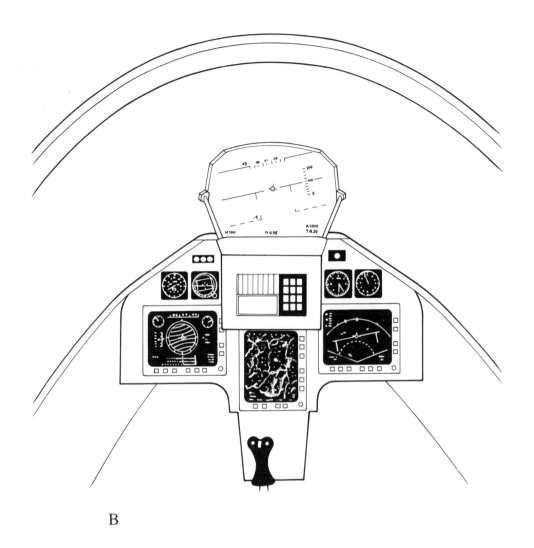

B

B *SAAB JAS 39 Gripen. To exploit the full potential of the latest generation of combat aircraft, the pilots have to be provided with a highly sophisticated working environment in which they can function efficiently while sustaining high load factors. New technologies have been developed to reduce the workload and the vast amount of information generated by the aircraft's systems and sensors.*

In the Swedish SAAB JAS 39 Gripen all conventional instruments and indicators have been replaced by three-colour CRTs where vital information is presented in easily assimilated form. Also, in front and in line with the pilot's forward field of view, there is a transparent screen on which information and essential data can be displayed. The Head Up Display allows the pilot to keep his gaze

outside the cockpit at the same time as he gets all information needed for the mission presented in front of him.

The screen to the left displays mainly data from the aircraft's systems, warnings and flight data information. In the middle, a computer—generated digital map is continually displayed, orientated in the aircraft's direction, and to the right is the tactical indicator on which

target data from radar or IR-sensors is displayed. In case of failure on any of the indicators, the information can be switched over—either manually or automatically—to any of the others. Four basic flight instruments of conventional mechanical fashion are kept as a back-up in case of total break-down of the plane's electronic systems.

side to side in a groove when the wings dipped, and which were needed to help keep the wings level during the landing manoeuvre.

Doolittle tackled the altimeter problem first. Of the barometric type, this consisted of an airtight capsule connected by a mechanical linkage and gear train to a pointer indicating height. The capsule expanded with increase in height and decrease in atmospheric pressure, with the linkage moving the at-pointer.

Doolittle found that an instrument-maker named Paul Kollsman had built an instrument which he claimed would indicate height to within 10 ft. This degree of accuracy was much greater than that of altimeters in current use and was made possible by exceedingly fine workmanship. To achieve this standard Kollsman had contacted a Swiss watch-making firm, telling them: "I want you to cut me a better gear for this altimeter than you ever cut for a watch." This the firm proceeded to do, and on test the altimeter worked perfectly.

Meanwhile the shortcomings of both the compass and the turn-and-bank indicator were being tackled by the Sperry Company. This company was already the world's leading manufacturer of the gyroscope, a device based on the principle that a rapidly spinning body opposes any change in the direction of its axis. This principle was applied to the development of an artificial horizon, to replace the current turn-and-bank indicators.

The heart of the new instrument was a tiny top, spinning at several thousand revolutions per minute, with its axis set vertically. The front of the instrument displayed a line representing the horizon and a bar representing the wings of the aircraft. The gyroscope was linked to the horizon line so that it remained level with respect to the Earth. When the aircraft climbed or dived, the bar moved above or below the horizon line. If the aircraft banked, the bar would tilt to reflect the degree of bank.

The principle of the spinning gyroscope was similarly applied to a replacement for the magnetic compass. Known as the directional gyro, this instrument could be set to a compass heading on which it would remain steady despite any change in direction of the aircraft.

The capabilites of these three instruments—altimeter, artificial horizon and directional gyro—were demonstrated dramatically on 24 September 1929. Early that day a thick fog enveloped Mitchell Field, near New York, which was being used as the base for the blind flying experiments. With a set of the new instruments installed in a Consolidated NY-2 biplane, Doolittle climbed into the cockpit and fitted a hood over his head, to cut off all the outside view. Using a radio beacon to line up with the runway, Doolittle took off, climbed to a thousand feet and flew several miles along two different headings, using the directional gyro and a radio beam for guidance. Then he flew back to Mitchell Field, homed on the radio beacon and, watching the instruments

closely, set the aircraft down in a safe, if bumpy, landing. As recorded at the time Doolittle had "taken off, flown a specific course and landed without reference to the Earth." A new era of flight deck instrumentation had arrived.

During the nineteen-thirties instrument flying became a routine accomplishment for airline and military pilots.

By 1939 Britain's Royal Air Force had standardized on a group of six flight instruments, by which pilots could control their aircraft accurately in cloud or at night. The six instruments were an altimeter, airspeed indicator, vertical speed indicator, turn and slip indicator, gyro compass and artificial horizon, the array being known as the "blind flying panel". Other countries also standardized on similar groups of "basic" instruments, although the number of instruments and their layout did not necessarily follow the pattern evolved by the RAF.

The best pressure-operated flight instruments are still not wholly accurate, however, and even gyro-operated instruments have their limitations, such as a lag in response. Such inaccuracies, which could be accepted or allowed for at moderate heights and speeds, became unacceptable with the advent of jet propulsion.

Accordingly, several new instruments have made their appearance. One of the most important is the Machmeter, which indicates the speed of an aircraft relative to the speed of sound at any height. On this type of instrument the figure "1" indicates the speed of sound at any altitude, although the actual value varies with altitude and air temperature. A Machmeter became necessary because the handling characteristics of early jet aircraft deteriorated as the speed of sound was approached. At this speed, some aircraft were dangerous to fly or became over-stressed and broke up. Modern aircraft can exceed Mach 1 or Mach 2 with virtually no change in handling characteristics, but a Machmeter is required because all aircraft have a limiting Mach-speed-number beyond which they should not be taken.

Attention was also directed to the altimeter. Conventional instruments had three pointers like the hands on a watch, denoting hundreds, thousands and tens of thousands of feet. They were adequate while aircraft climbed or descended relatively slowly and while they rarely operated much above 20,000 ft. But with the advent of jet aircraft, crashes began to occur through pilots misreading their altimeters during rapid descents from high altitudes, the most common instance being a tendency to mistake the 1,000 ft indication for 11,000 ft. At certain heights, one needle could be obscured by another, leading easily to a mistake.

To overcome this problem, the three needles were first of all redesigned to make them more clearly distinguishable. Then a special marker was added to warn the pilot when he was below 10,000 ft. The next step was to provide a number counter to indicate height, similar to the mileometer on a motor car.

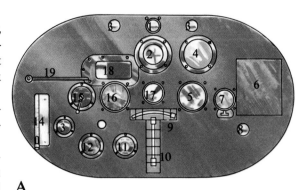

A

1 Mirror
2 Voltmeter indicator for Earth inductor compass
3 Instrument lights
4 Altimeter
5 Airspeed indicator
6 Compass correction chart
7 Eight-day clock
8 Primer operating knob
9 Lateral level
10 Longitudinal level
11 Oil-temperature gauge
12 Fuel gauge
13 Oil-pressure gauge
14 Mixture control
15 Magneto switch
16 Tachometer
17 Turn-and-bank indicator
18 Periscope
19 Periscope handle

A *"Spirit of St. Louis" instrument panel. Grouped here are the instruments used for one of the most remarkable navigational feats of all time—the 33-hour, 3,610-mile non-stop flight from New York to Paris made by Charles Lindbergh in 1927. Set close to his face, this was considered the most advanced instrument panel developed up to that time. Of interest is the periscope, used by Lindbergh to enable him to see in front of the aircraft, there being no direct view forward. Lindbergh spurned the use of radio sets, as they "cut out when you need them the most", and a sextant because it would be too heavy. He made his journey by dead reckoning, using check points wherever possible. On the 27th hour of the flight, and fifteen hours after his last check-point in Newfoundland, Lind- bergh began to worry about his position. According to his calculations he ought to have been near Ireland. Suddenly he sighted a small fishing boat— land seemed near, but what land? He cut his engine and glided down past the startled fishermen, leaning out of the window and shouting, "Which way is Ireland?" This surely must be the first and only time an airman has "asked the way" in this fashion! Unfortunately, the fishermen either did not hear Lindbergh or were too startled to answer him. He continued to fly on, and after an hour made landfall near the town of Valentia, on the south-western tip of Ireland. After 16 hours of over-water navigation he was less than five miles off course. In view of the undeveloped nature of the instruments this was a remarkable performance.*

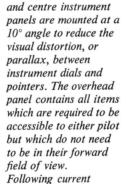

C

C *Douglas DC-9 flight deck. This is designed for a two-man crew, with all systems controlled from either seat. The overall arrangement was derived from Douglas study programmes and from suggestions by airline flight and engineering personnel, the determining factors being the space required for items, priority and frequency of use. System design was influenced in many cases by the work-load requirements of a two-man crew. Instruments and switches are arranged in patterns, to*

distinguish one system from another.
The captain's, co-pilot's and centre instrument panels are mounted at a 10° angle to reduce the visual distortion, or parallax, between instrument dials and pointers. The overhead panel contains all items which are required to be accessible to either pilot but which do not need to be in their forward field of view.
Following current practice, the random location of malfunction and warning and caution lights evident on some earlier flight decks has been replaced by a

master light and central warning panel system. Master caution and warning lights are located in the instrument coaming, directly in front of each pilot, to ensure that they are noticed immediately upon activation. All warning lights are located along the forward end of the overhead panel. Some fifty operations are monitored and the panel is divided into seven systems or areas where failures require rapid action. When a malfunction occurs, the master lights in front of each pilot and the

appropriate warning legend are illuminated. Both captain and co-pilot can quickly scan the warning panel with a minimum of head motion.
A chart holder, which does not rotate with the control wheel, is attached to each control column (not shown) to hold an approach chart over each wheel.
A variable-intensity light is built into each holder, eliminating the need for a remote source of light for each holder and reducing the overall level of light in the flight compartment.

1 Digital Mach/airspeed indicator
2 Sensitive airspeed indicator
3 Servo altimeter with digital encoder
4 Liquid contents indicator
5 Percentage tachometer indicator
6 Oil pressure indicator
7 Low range airspeed indicator
8 Mach/airspeed indicator
9 Vertical speed indicator
0 Pressure/temperature indicator

11 Mechanical counter/ pointer altimeter
12 Position indicator
13 Machmeter
14 Expanded scale jet pipe temperature indicator
15 Electrical turn and slip indicator
16 Sensitive differential pressure gauge
17 Sensitive differential pressure gauge
18 Vertical speed indicator
19 Synchro position indicator
20 Mach/airspeed indicator

1 Mach/airspeed indicator
2 Flight director indicator
3 Altimeter
4 Clock
5 Radar altimeter
6 Compass indicator
7 Course indicator

8 Vertical speed indicator
9 Turn and slip indicator
10 Autopilot trim indicator
11 Fuel indicator
12 Auxiliary fuel indicator
13 Engine pressure

ratio indicator
14 Tachometer
15 Exhaust gas temperature indicator
16 Tachometer
17 Fuel flow/fuel used indicator
18 Oil pressure indicator
19 Oil temperature indicator

20 Oil quantity indicator
21 Flap position indicator
22 Hydraulic system pressure indicator
23 Hydraulic quantity indicator
24 Fuel temperature indicator (dual)

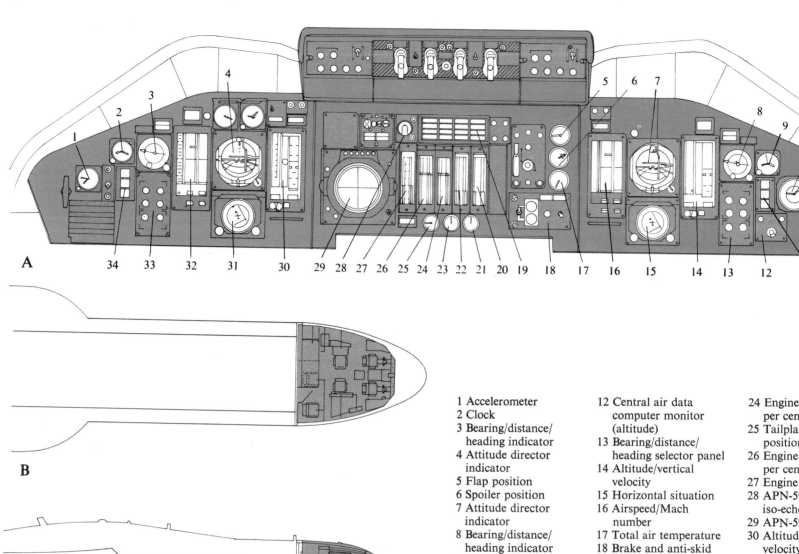

A

B

C

1 Accelerometer
2 Clock
3 Bearing/distance/
  heading indicator
4 Attitude director
  indicator
5 Flap position
6 Spoiler position
7 Attitude director
  indicator
8 Bearing/distance/
  heading indicator
9 Clock
10 Cabin pressure
  altitude
11 Marker beacon
  indicator

12 Central air data
  computer monitor
  (altitude)
13 Bearing/distance/
  heading selector panel
14 Altitude/vertical
  velocity
15 Horizontal situation
16 Airspeed/Mach
  number
17 Total air temperature
18 Brake and anti-skid
  control panel
19 Low oil pressure
20 Fuel flow
21 Aileron trim position
22 Exhaust gas
  temperature
23 Rudder trim position

24 Engine speed (N_2,
  per cent)
25 Tailplane trim
  position
26 Engine speed (N_1,
  per cent)
27 Engine pressure ratio
28 APN-59 radar
  iso-echo control
29 APN-59 indicator
30 Altitude/vertical
  velocity
31 Horizontal situation
32 Airspeed/Mach
  number
33 Bearing/distance/
  heading selector panel
34 Marker beacon
  indicator

A *Flight deck of the*
B *Lockheed StarLifter.*
& *A long-range strategic*
C *freighter, the Lockheed*
  C-141 StarLifter was
  produced to meet the
  U.S.A.F's requirement
  for a jet freighter to
  modernize the Military
  Airlift Command fleet
  of piston-engined and
  turboprop transports.
  The craft is built
  around an 81 ft long
  cargo compartment,
  10 ft by 9 ft in cross-
  section, with rear
  loading at truck-bed
  height and provision
  for air dropping.

*The fuselage is a
conventional aluminium
alloy semi-monocoque
structure, and the wing
is an all-metal two-spar,
box-beam structure.
Both structures are
built on fail-safe
principles. Wing
sweepback at quarter-
chord is 25°. Of the
tricycle type, the
landing gear comprises
a twin-wheel nose unit,
and four-wheel bogie
main units, all units
retracting forward;
the main bogies into
fairings on the sides
of the fuselage.*

*Conventional ailerons
are fitted with full
power activation by
dual hydraulic units
and emergency manual
control through geared
tabs. A variable-
incidence tailplane
provides pitch trim,
actuated normally by
an electric motor or
more rapidly by an
hydraulic motor. A
third hydraulic system
takes over elevator
control if one of the
normal systems fails.
The controls can be
operated manually in
an emergency. Rudder*

*trim is by positioning
the rudder servo-unit
control valve with
electric trim actuators.
Power is provided by
four Pratt and Whitney
JT3D turbofan engines,
each rated at 21,000 lb
s.t. Total fuel capacity
is 24,000 US gallons,
giving a range with a
32,500 lb payload of
over 7,000 miles.
The main cabin
accommodates up to
154 troops, or 123
fully-equipped para-
troopers, in fore-and-
aft seating, or 80
litters with seats for*

*up to 16 other patients
or medical attendants.
With a "comfort
pallet"—a toilet and
galley—at the forward
end of the cabin,
accommodation is
reduced to 120 seats.
Two bunks and two
seats are provided for
a relief crew. The crew
access door is on the
port side of the fuselage
nose, and two para-
troop doors are
provided near the aft
end of the cabin, one
on each side. The rear
loading ramp permits
straight-in cargo*

*loading; the opening
being equivalent to the
full cabin width and
height. The ramp can
be opened in flight for
air drops.
With minor modifica-
tions only, the
StarLifter can carry
the Minuteman ICBM
in its special container
a total weight of over
86,000 lb. A StarLifter
established the world
record for heavy cargo
drops of 70,000 lb and
was the first jet
transport from which
US Army paratroops
jumped.*

During the eighties, a completely new way of displaying information to the pilots began to appear on new aircraft as well as a retrofit on updated models of earlier origin. It is based on six TV-displays replacing most of the earlier clock- and thermometer-type indicators as well as providing new information that was not previously available to the pilots in this simple form. This new kind of electronic wizardry is sometimes referred to as a "glass cockpit". One of the pioneers to have it designed into the cockpit from the outset was the commuteliner Saab 340 that began to come off the production line in 1985.

The full six-display glass cockpit has two displays in front of each pilot showing flight information and two displays on the centre panel showing parameters from the operation of the engines and ancillary systems as well as written information, i.e. actions to be taken in case of an emergency. Some of the displays act as a safety backup for others.

The upper display in front of each pilot gives an electronic representation of the artificial horizon, providing the necessary information to manoeuvre the aircraft in areas of reduced visibility such as clouds or fog or in darkness when proper outside clues for reference are inadequate. It also includes the flight director which indicates the best rate of turn and change of altitude to intercept radio beams along an airway or approach path to the destination airport.

In the left part of the horizon display there is a thermometer—type electronic speed display. This also contains colour—coded warnings for high or low speed limits, as well as a prediction of what the speed will be in a defined period of time if power and other flight parameters are kept unchanged. This serves as a warning of approaching potentially dangerous flight conditions.

The lower display gives electronic compass information, often expanded to show a limited part of the compass arc in the upper part of the instrument to facilitate accurate flying. The main part of the instrument contains a map of the surrounding area including radio beacons, holding patterns, restricted areas, airports and recommended tracks to follow. This comprehensive information enables the pilot to determine his exact position at all times with a glance of the eye, reducing his dependence on ground—based radar for separation from other air traffic in the area, although modern cockpits are also equipped with collision avoidance indicators that flash a warning and indicate recommended avoiding action, should another aircraft get too close.

Superimposed on this map is the weather radar picture, previously shown on a separate instrument. It is designed to pick up returns or "echos" from adverse weather, mainly thunderstorms whose strong turbulence might compromise the safety of passengers and crew and in extreme circumstances even threaten to break up the aircraft. This information is also colour—coded from "cold" blue or green areas indicating little danger to "hot" red and purple areas with strong turbulence that must be avoided.

The data needed to determine the position of the aircraft can be derived from several different sources. Earlier pilots simply selected the frequency of the desired ground—based nondirectional NDB or a processed information directional VOR, a method that is still very popular in the light aviation segment. The appropriate needle in the cockpit then swung to the direction of the beacon, and the pilot simply had to follow it to get there.

The NDB or NonDirectional Beacon has been part of aviation since the two World Wars. It is a simple radio transmitter in the MW band, inexpensive and still much in use in undeveloped parts of the world. Its main drawback is a strong sensitivity to atmospheric conditions, rendering it almost useless when thunderstorms are present in the area. In the early days, crew members had to manually turn the reciever antenna to the direction of the strongest signal, but today Automatic Direction Finders (ADF) find the direction and display it with a needle on a compass dial at the selection of the desired frequency.

The VOR (Very high frequency Omnidirectional Range) is operated in much the same way by the pilot. Information from the ground beacon is transmitted in coded form for higher accuracy on the VHF band and is less prone to atmospheric disturbances; 180 different two-way signals called radials define all the 360 degrees of magnetic direction from the VOR.

The ILS (Instrument Landing System) is the most commonly used method for making accurate approaches to the runway in bad weather. It operates on the same frequency band as the VOR and uses the same displays in the cockpit. It was developed during World War II, and is now standard equipment on all major airports worldwide and most minor ones served by regular airline traffic.

ILS consists of two transmitters, the localizer providing guidance in azimuth and the glide path defining the correct angle of approach. The localizer antennas can be seen as a sometimes quite formidable setup of metal poles just beyond the far runway end, while the glide-path antenna is a single pole with two or three boxes just to the left of the first part of the runway.

Marker beacons are associated with ILS, placed under the approach path and transmitting a narrow bandshaped signal. It triggers a sound and light signal in the cockpit to show position and distance to go to the runway.

DME (Distance Measuring Equipment) is another and more complete way of displaying distance-to-go to the pilot. It is activated as soon as the pilot selects a VOR/DME or ILS/DME frequency and consists of an electronic interrogator in the aircraft. The signal from this device is "bounced back" to the aircraft from the DME function on the ground transmitter. Aircraft equipment measures the time elapsed for the signal to go from the interrogator to the beacon and back again and converts it to distance which is displayed to the pilot in plain figures.

Information from VOR, DME and sometimes also NDB is automatically received and processed in Area Navigation equipment. R-nav' as it is sometimes called, selects suitable nearby ground transmitters and calculates aircraft position from this information, requiring no switching or other enroute manual inputs from the pilot—unless he wants to change previously inserted information.

Navigating on VOR or NDB, the most common way is to select a beacon, fly there, then select another and fly on to that one. With area nav it is easy to select turning points between beacons, making flying on direct tracks possible, and giving important gains in flying time and fuel consumption.

Obviously none of these aids are suitable for long-distance navigation other than for short overland stretches. The range of VOR/DME transmissions varies with the flight level up to a maximum of 250 nautical miles under extremely favourable conditions. Powerful NDBs may reach longer, but then their accuracy has deteriorated to such a degree that it is not compatible with the separation standards required from modern airliners.

A commonly used long—distance navigation aid among airliners is the INS, Inertial Navigation System. It is completely self—contained—it requires no earth—based transmitter to support its position finding. It derives the necessary information from a gyro platform equipped with several accelerometers that sense changes in aircraft speed and direction. The pilot inserts desired flight plan and the actual position of the parking stand, and from there the INS will calculate changes in aircraft position so accurately that the difference between the INS position and the actual position after a transatlantic flight may be less than half a mile, a difference of little importance since the minimum airway width is 10 nautical miles.

Loran and Omega are two examples of modern but less expensive equipment that is used for navigation over long distances, but like the INS also suitable for medium and short—distance hops. They both derive their information from ground—based long—distance transmitters—the Omega actually uses the VLF very low frequency transmitters that were originally developed as a navigation aid for US nuclear submarines.

Omega is highly dependent on accurate time information and has a built-in atomic clock to provide this. Both systems are affected by atmospheric disturbances and need good signals from at least three stations around the world to be able to navigate accurately. If this is not possible, there is a "dead reckoning" function that takes over, basing navigation on last received information until an update is possible.

GPS, Global Positioning System, is a satellite-based navigation system for aircraft of all sizes. It is highly accurate while being reasonably cheap to acquire. A system of this kind developed in Sweden shows such accuracy it will even make it possible to find taxiways on the ground and the assigned parking spot on the apron in absolutely nil visibility.

703 705 707

Airbus A320 is the first wholly computerized civil airliner. The most visible evidence of the advanced system's technology is the flight deck. Large cathode ray tubes can provide the pilots with more information than they receive with conventional instruments. Data is displayed automatically according to flight phase or need, or manually if desired. A centralized fault display system monitors the aircraft's systems constantly by collecting diagnostic data that, in case of any trouble, automatically will be displayed in the cockpit. The data is accessible after a flight and can also be transmitted to a ground station, enabling preparations for rectification to be made before the aircraft lands.

The A320's automatic flight system is controlled by a sidestick, which is a completely new feature for a civil airliner. By manipulating the stick, the pilot gives electronic input to the flight system's computers that will manage the flight in the most effective and safe way. This technique is commonly known as "fly by wire".

The modern digital computerized systems are integrated: engine control system, flight control system and the flight management system, which provides the pilot with a full and automatic flight envelope protection. If he, for example, pulls the stick fully back, the A320 will automatically enter a climb at full thrust and maximum lift.

A320 is a short/medium range airliner with seating for 130 to 179 passengers depending on the customer's special requirements. A stretched version, A321, has increased seating capacity for an additional 20 passengers. The aircraft's structure incorporates composite materials in many parts such as the vertical and horizontal tail surfaces, ailerons, doors and panels. A total of about four tons of composite is used in the A320, which saves 800 kg compared to conventional metal structure.

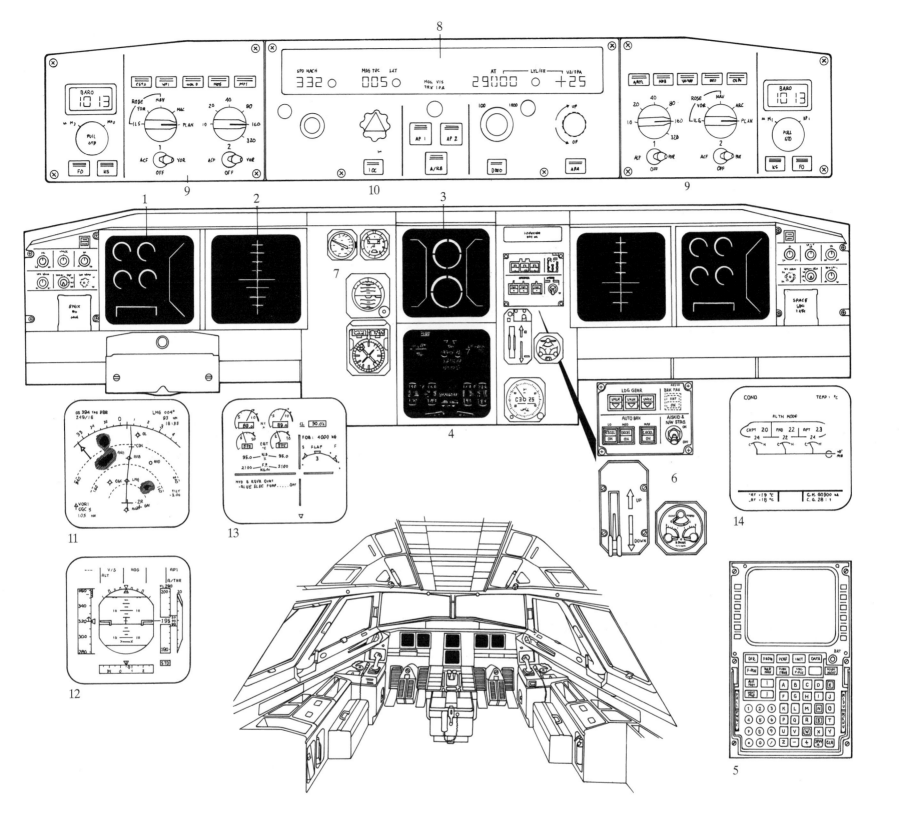

1 Primary flight display
(PFD), captain/flight
officer

2 Navigation display
(ND), captain/flight
officer

3 Engine/warning
display

4 System display

5 Multipurpose control
and display unit
(MCDU)

6 Landing gear and
brake controls and
indicators

7 Back-up mechanical
instruments

8 Digital flight data
display

9 Navigation display
control unit, captain/
flight officer

10 Flight control unit

11 ND presentation in

ARC-mode. Aircraft
symbol in mid-lower
part pointing up to the
magnetic heading
reference line.

12 PFD presentation
during climb

13 Engine/warning
presentation:
– Primary engine
parameters
– Operational status;

flap setting, fuel
volume etc
– Memo and warning
information; seat
belts, anti-ice etc

14 System display
presentation:
– Flight related system
data
– System malfunction

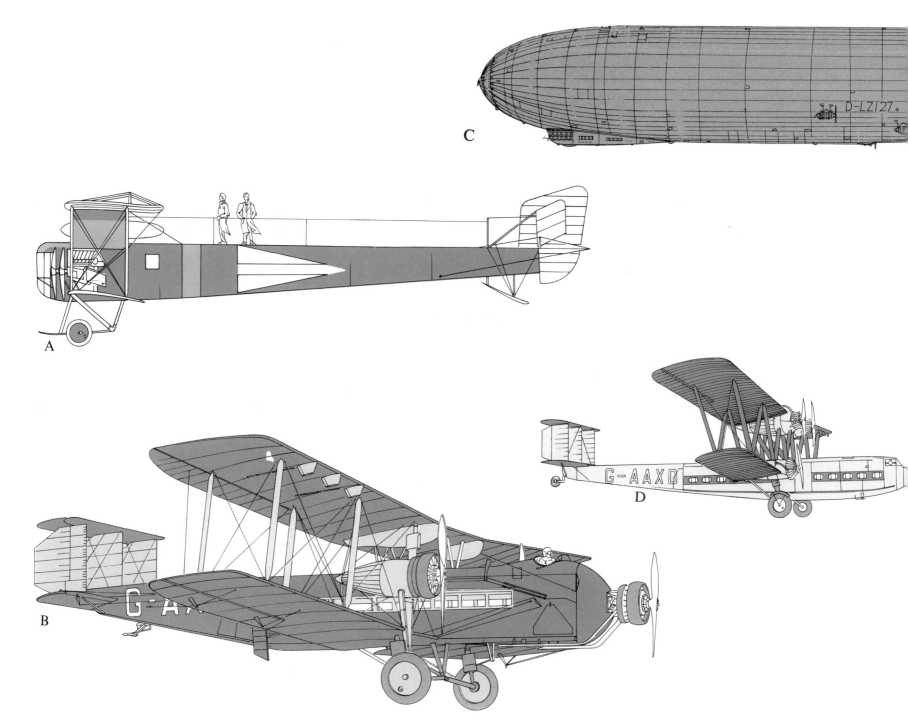

A *Amenities offered by the Sikorsky "Ilia Mourometz" of 1913 included a heated cabin, fitted with tables, chairs and couches, in which meals were served— and an open promenade deck along which passengers could walk in flight.*

B *Armstrong Whitworth Argosy. This three-engined airliner introduced the first luxury "named" service— Imperial Airways' London-Paris "Silver Wing" service of 1927. In contrast to the comfort inside the cabin, the pilot sat in an open cockpit as designers considered his view would be too restricted if he was enclosed.*

C *In the nineteen-twenties and thirties, airships offered passengers a degree of spaciousness and comfort that has not yet been equalled on a heavier-than-air craft. A voyage on board the famous "Graf Zeppelin" or "Hindenburg" combined something of the luxury and adventure of sea travel with the speed of air travel. On board were bedrooms, lounges with room to dance, a dining saloon, promenade decks— even an orchestra.*

D *Handley Page 42. Aircraft such as this offered Pullman-class comfort for the passengers and an enclosed cockpit for the crew.*

723   724   725   726

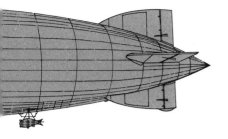

COMFORT IN THE CABIN

E

F

At first air travel offered passengers little but speed. After a time, it managed to combine a degree of reliability with speed. At a still later stage comfort began to be added to speed and reliability.

The modest first step towards comfort was to provide passengers with some form of protection from the slipstream and rain. The first "airliners" were converted bombers, and on these the protection often consisted of little more than a hinged lid containing windows, over what had been open cockpits. The draughts were still considerable, but passengers were lent a leather flying coat, gloves, helmet—and a hot-water bottle if it was very cold.

The next generation of airliners embodied bigger and less draughty cabins. On these, although the passengers were protected, the pilot continued to sit in an open cockpit, as designers considered that it was essential to give the "driver" an all-round, unobstructed view. Also, some older pilots of the time were not convinced of the desirability of enclosed cockpits, as they considered that a pilot needed the wind in his face to tell him how his aircraft was flying. So, the cockpit remained open even on such advanced aircraft as the Boeing Monomail of 1930.

The Handley Page HP.42, along with other airliners of the early thirties, introduced enclosed cockpits, whereupon the crews promptly exchanged their leather flying suits and goggles for blue serge uniforms, gold braid and peaked caps.

The size and comfort of the famous four-engined Handley Page biplanes came as a complete revelation to the air-travelling public, transforming an air journey from a primitive adventure into an experience of speed and luxury. The cabins contained tables and wall-mounted lights and were sound-proofed so that it was possible to converse easily. Heating and ventilation were provided, and passengers travelled in luxurious "Pullman" type armchairs. Five- or six-course hot meals were served by stewards.

The idea of serving refreshments in flight had been pioneered by the British Daimler Airway, which introduced new standards of passenger service and efficient maintenance. During operations in 1922, to Paris and Brussels and later to Amsterdam and Cologne, Steward Sanderson became the first person ever employed to serve refreshments during flight.

E *Corner of the 56-seat dining room on the British airship R.100.*

F *The lounge on the British airship R.101. Flowering shrubs graced the base of the vertical columns.*

G *The passenger cabin of the Bristol Pullman, 1920.*

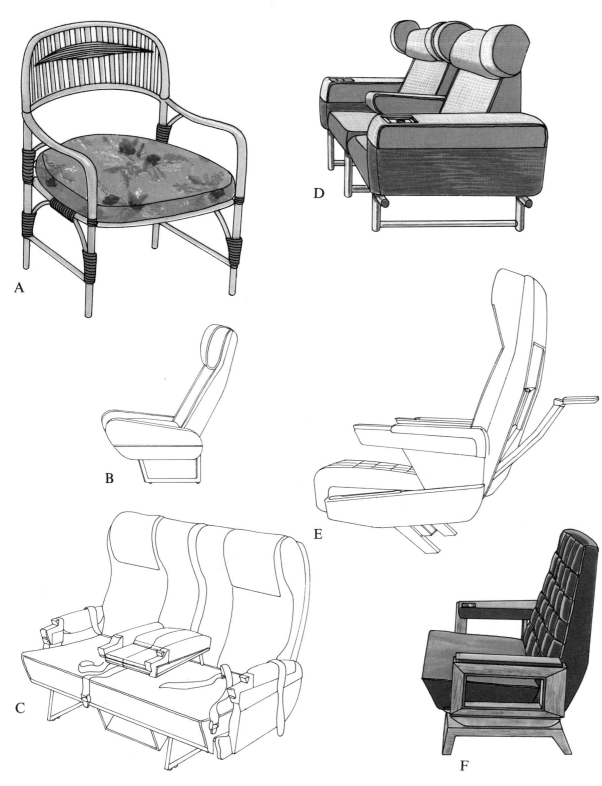

The first airline stewardesses were eight youn[g] nurses hired by Boeing Air Transport, later Unite[d] Air Lines, in May 1930, to take care of passenger[s] on their Boeing Tri-Motors on the San Francisco[-] Chicago route. Uniforms of green wool twill wer[e] provided, complete with what looked like a bathin[g] "shower cap" and Batman-like cloak.

Since then, attractive and superbly efficien[t] stewardesses have been an indispensable part of th[e] service provided inside the cabin. In addition t[o] serving meals and drinks, and answering question[s] on matters ranging from religion to aircraft techni[-] calities, stewardesses are trained to administer firs[t] aid and to assist passengers in emergencies.

For airline employment managers, the stewardes[s] "turnover" rate due to marriage is absurdly high[,] but any attempt to downgrade either looks o[r] efficiency is frowned upon. No-one has managed t[o] calculate the goodwill value of having stewardesse[s] on board, but it is generally conceded that they ar[e] the most effective "salesmen" in an airline. In th[e] early days, at least, it was even suggested that s[o] far as the predominantly masculine traffic wa[s] concerned the stewardesses represented by them[-] selves a good reason for travelling by air!

Although the newer airliners of the nineteen[-] twenties and early thirties were an improvemen[t] over the converted bomber aircraft used initially[,] travel by air was by no means really comfortable[.] With few exceptions cabins were draughty, cold an[d] noisy, and the vibration was often terrible.

This was fully appreciated by Arthur Raymond[,] a designer at Douglas Aircraft who was busil[y] engaged in negotiations with TWA to build a ne[w] 12-passenger airliner, to be known as the DC-1[.] To obtain first-hand experience, Raymond decide[d] to make a coast-to-coast flight in one of the lates[t] all-metal Ford Tri-Motors, then used by TWA[.] Known affectionately as the "Tin Goose" the For[d] was representative of the airliners of that time[.] Afterwards, Raymond said he would never forge[t] the trip. . . .

"They gave us cotton wool to stuff in our ears[;] the Tin Goose was so noisy. The thing vibrated s[o] much it shook the eye glasses right off your nose[.] In order to talk with the guy across the aisle, yo[u] had to shout at the top of your lungs.

"The higher we went, to get over the mountains[,] the colder it got inside the cabin. My feet nearl[y] froze. The leather-upholstered, wicker-back chair[s] were about as comfortable as lawn furniture.

"When the 'plane landed on a puddle-splotche[d] runway, a spray of mud, sucked in by the cabin ai[r] vents, splattered everybody".

At the first opportunity, Raymond described th[e] flight in disgust to Donald Douglas, emphasizing[,] "We've got to build comfort, and put wings on it[.] Our problem is far more than just building a satis[-] factorily-performing transport aeroplane".

Raymond's comments started everybody thinkin[g] seriously about such things as soundproofing, cabi[n] temperature control, improved plumbing, mor[e]

A *Typical wickerwork seats in a Handley Page airliner of the nineteen-twenties.*

B *A Stratocruiser chair, the design of which absorbed 100,000 man-hours and £125,000.*

C *This seat was specially developed for British European Airways. Basically a first-class double chair, it can be converted quickly into a tourist-class triple chair. This transformation is made possible by the special four-piece back rest. For first-class use, the centre backrests are fixed to*

the side backrests to form two large backs, and the armrests are moved to a wider position. A table is fitted between the inner armrests. For tourist-class use, the two inner backrests are fixed together to form three smaller backrests and the armrests are moved accordingly.

D *An upholstered seat of the nineteen-thirties.*

E *The exceptionally comfortable economy-class seat developed for the BAC VC10.*

F *A typical seat in a modern small executive jet transport.*

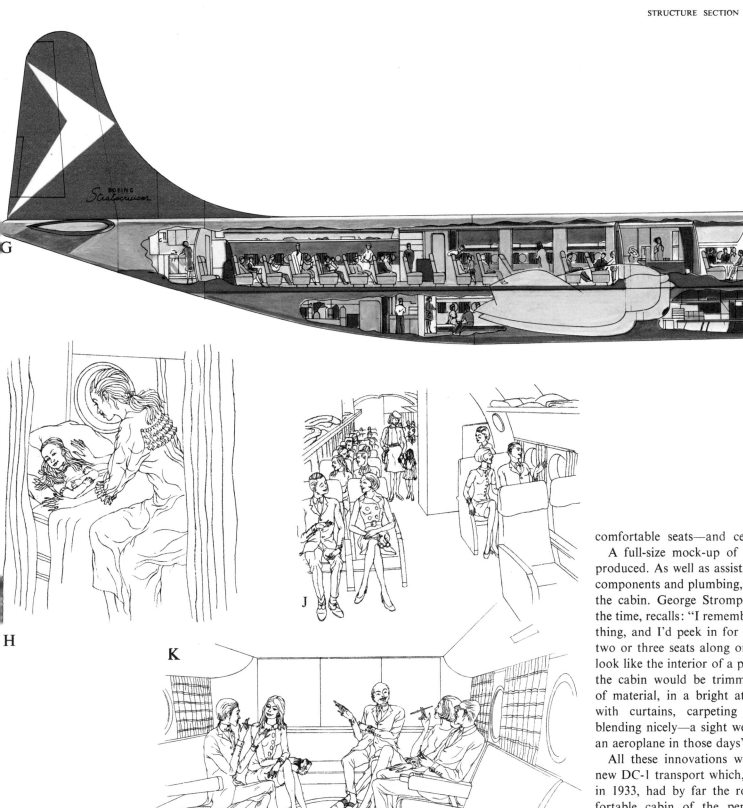

G *The unique layout of the Boeing Stratocruiser enabled passengers to relieve the monotony of long flights by getting up and walking around or visiting the lower-deck lounge which was reached by a spiral stairway from the main passenger cabin.*

H *Stratocruiser sleeping berth.*

J *The spacious Stratocruiser cabin.*

K *The luxurious below-deck lounge on the Stratocruiser.*

comfortable seats—and certainly no mud baths!

A full-size mock-up of the new aeroplane was produced. As well as assisting in the installation of components and plumbing, it was used to help plan the cabin. George Stromple, Douglas engineer at the time, recalls: "I remember, I'd be passing by the thing, and I'd peek in for a look—there would be two or three seats along one side, and it began to look like the interior of a passenger plane. One day the cabin would be trimmed with a certain kind of material, in a bright attractive colour scheme, with curtains, carpeting and chair upholstery blending nicely—a sight we didn't expect to see in an aeroplane in those days".

All these innovations were engineered into the new DC-1 transport which, when it made its debut in 1933, had by far the roomiest and most comfortable cabin of the period. Upholstered seats were fitted—the largest of their kind and pitched at 40 inches to give plenty of leg room. Each seat had an individual small reading lamp and a foot rest. There was thick carpeting on the floor.

To the rear of the cabin was a small buffet or galley, with space for Thermos flasks and boxed lunches, and an electric hotplate for keeping soup warm. Beyond the buffet was a toilet and wash basin. A sound-deadening bulkhead, two-and-a-half inches thick, separated the cockpit from the cabin, and special precautions were taken to prevent noises from the ventilators and heat intake ducts entering the cabin. Even the metal handrail on each wall was stuffed with sound-proofing material.

The net result was a cabin quieter than a Pullman car.

One engineering report summed it up this way, "For the first time in aeronautics and, perhaps, in any moving vehicle, the principle of balanced acoustics has been successfully accomplished. This aeroplane is not only the most quiet aeroplane flying, but also has a noise spectrum which seems to be less fatiguing to passengers."

From the passenger standpoint, this was one of the most important contributions to improved comfort in the cabin.

From the DC-1 was evolved the slightly bigger DC-2, one of which, operated by KLM, was entered for the 11,123-mile race from England to Australia organised in 1934. It was in competition with 19 other aircraft, some of them racing machines specially designed for the contest. This caused one London newspaper to comment that it was "an audacious assumption that such a ship could expect to compete with the fastest planes and designs on the Continent." In the event, the KLM DC-2 did not win the race—but it came in a close second, and the fresh and smartly dressed crew and passengers who disembarked at Melbourne were eloquent testimony to the degree of comfort provided on the flight deck as well as in the cabin.

From the DC-2 was developed the even more famous DC-3. The comfort and speed of this airliner really *sold* air travel and it has been described aptly as "the plane that changed the world".

In surveying briefly the development of passenger comfort, mention must be made of the peculiar comfort afforded by flying-boats. The bodies of these aircraft, designed primarily as watertight hulls, were voluminous by comparison with the fuselages of landplanes and they provided passengers with a spaciousness not obtainable on their landbased counterparts. Flying-boats, in fact, offered a degree of comfort surpassed only by that offered on the great airships and not equalled on aircraft until the advent of the giant "Jumbo-jets" thirty years later.

Most famous of the civil flying-boats were the four-engined Short Empire 'boats introduced in 1936 by Imperial Airways on the great air routes linking Britain to Africa, Australia and New Zealand, and the Boeing 314 "Clippers". These aerial ships incorporated ladies' rooms and bars, and their size enabled passengers to get up and "stretch their legs" during the long flights of which they were capable. Most comfortable flying-boat of all was undoubtedly the giant post-World War II Saunders-Roe Princess which, unfortunately, never went into service. The prototype was, however, partially fitted out with examples of the state-rooms, passenger cabins, powder rooms and lounges which would have been enjoyed by passengers had not this type of aircraft been rendered obsolete by the general switch to landplanes.

A major increase in passenger comfort came with pressurisation. For many reasons, airliners

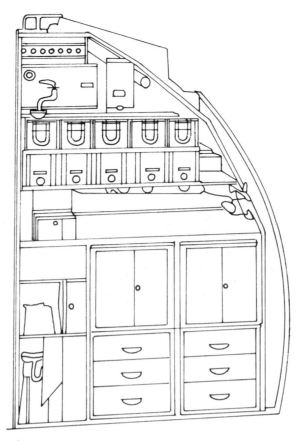

A

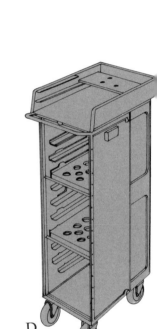

C

B

D

A *A typical aircraft galley and bar unit. Such units permit the warming and serving of traditional meals such as "roast beef and Yorkshire pudding". This unit is of the non-closed type—that is, the various containers and control switches are exposed.*

B *On the closed type of aircraft galley, up-and-over doors conceal the containers, electrical panels and the steward-warning light panels.*

C *Up to 220 passengers can be accommodated in the big Soviet Tu-114 turboprop airliner. When it is equipped with the more usual 170 seats, the central cabin, seating 48, is used as a restaurant and is not normally occupied at take-off—a feature that would be unthinkable on Western airliners for economic reasons. The kitchen is located on the lower deck and employs two or three full-time cooks. Food, once prepared, is despatched*

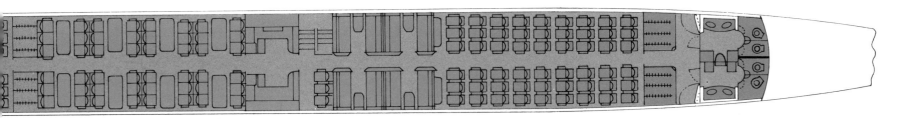

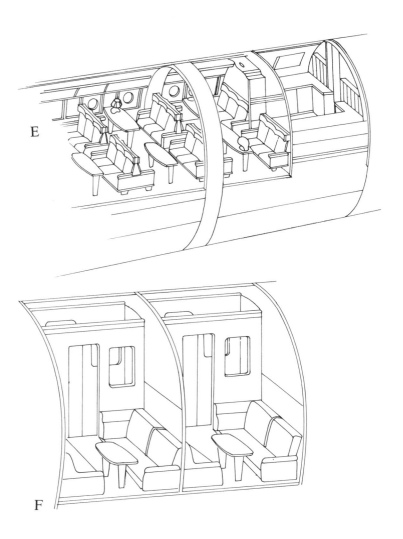

G

by two lifts to
stewardesses in a
roomy serving area
located aft of the
restaurant.

D Cabin serving trolley.

E *The 48-seat dining
room of the Tu-114.
Food is served through
the doorway at the rear
after being prepared in
a kitchen on the lower
deck.*

F *Tu-114 "roomettes"
are normally fitted with
two divans and one
folding bunk. Alter-
natively, six seats can be
installed.*

G *The cabin of the Boeing
737 airliner has the
same width as those of
the big intercontinental
jets and offers a similar
standard of comfort on
short feeder-liner-
routes. Particular
attention has been paid
to sound-proofing and
to the isolation of
equipment and system
noise from the cabin,
to increase passenger
comfort.*

were tending to fly higher and higher, in order to
get clear of congested airlanes, to cross high ground
or to overfly bad weather. With the advent of jet
propulsion, high cruising altitudes became essential, as jet engines use fuel in prodigious quantities
at low levels.

As height increases, the atmospheric pressure
decreases, making breathing difficult, as on a high
mountain. To overcome this problem, high-flying
aircraft are pressurised—i.e., the pressure in the
cabin is maintained at a level above that of the
atmosphere outside, so that passengers can breathe
normally. The first pressurised airliner was the
pre-war Boeing 307 Stratoliner, and passengers
were quick to appreciate the increased comfort it
offered.

Another innovation introduced by the Stratoliner
was the carriage of an extra crew member as a
flight engineer, to relieve the pilot of responsibility
for maintaining the engine power settings, to keep
track of the cabin pressurisation and to control
operation of the various sub-systems.

After World War II, the comfort of air travel

took a temporary step backward, in that a number of the first post-war airliners were developments of bombers and thus far from ideal for passenger carrying. Also, in order to make the aircraft economical, passengers were beginning to be packed in much more closely than they had been in some of the airliners used before the war.

A uniquely comfortable airliner of this period was the Boeing Stratocruiser, a civil counterpart of the C-97 Stratofreighter which, in turn, was based on the B-29 bomber. Its main passenger cabin consisted of a new fuselage built on top of the original bomber fuselage, the top one being the larger and the two streamlinging into each other. The lower deck contained forward and aft cargo holds, separated by a snack lounge and bar, which was connected to the upper deck by a circular staircase.

It was this lower-deck lounge and bar that made the Stratocruiser so popular with passengers—a popularity which it retained even when the aircraft was outclassed by newer, faster airliners. During a long flight, passengers could go down to the bar for a drink and get away from their fellow travellers for a while. The Stratocruiser also had a fully equipped galley, near the tail, and men's and women's washrooms separated the forward compartments from the main passenger cabin.

The introduction of turboprop and turbojet airliners revolutionized passenger comfort. But, although they brought quiet and vibration-free flight, they accelerated the trend of cramming passengers closer and closer—a trend that continues today and is not likely to be arrested until the advent of the "Jumbojets", and even then for a short period only.

The item affecting passenger comfort most intimately is undoubtedly the cabin seat. The earliest seats were of wickerwork construction, as this gave strength with light weight, if not much comfort. From these modest beginnings have evolved the complex seats of today. Perhaps the major problem is the difficulty of determining the size and shape of seat that is comfortable to young and old, short and tall, and thin and not-so-thin passengers When Boeing elected to design their own seats for the Stratocruiser, 100,000 man-hours and £125,000 were expended before the design was finalized so that, in Boeing's own words at the time, "Whether you're a petite blonde of 100 lb or a buxom bread-and-butter-and-egg man of 300 lb you'll be equally comfortable. The same seat which comfortably supports the 100-lb blonde, without springing her up into the luggage rack, is equally pleasing to the 300-pounder without sinking him to the floor."

Before the war, all air travel was mainly of one class, but post-war expansion has seen the development of different categories of travel: first class, business, and economy or tourist class. In principle, all domestic routes and charter flights only have one economy class, but on international flights the two, or three—class systems are offered to the traveling public. The classes differ mainly in the on-board meal service and seat comfort and the airline companies name the different classes in exotic, imaginative ways as Senator Class, Club Class, Euroclass etc. Since a first—class passenger takes up nearly twice the space of an economy class passenger, the ticket prices are set accordingly. The first—class traveller enjoys wider seats and more legroom and in some cases, on very long routes, the seat back can be folded to a vertical position and together with a footrest, the seat is converted to a "sleeperette". Built usually in pairs, first—class seats often embody a small console or a table between the passengers.

The design of aircraft seats is a highly specialized business. In addition to the obvious aim of comfort, there are many basic requirements and air-worthiness regulations to be met. There are high demands for strength and the need to ensure that the seat is free from any feature which would injure a passenger in case of a crash. Considerable anthropological science is applied to the shaping, material and safety provisions, and elaborate testing techniques have been devised to establish crashworthiness and energy-absorption characteristics. Also, it is now regulated by civil aviation authorities that the seats' upholstery and fabric cover must be of fire-blocking, nonflammable material. The same safety requirements go for the material in the cabin interior and there must be an illuminated ramp in the cabin floor, leading the way to the nearest exit.

All seats in all classes nowadays have reclinable backs and a folding table that enables the passenger to eat comfortably regardless of the reclining position of the person in front of him.

What airlines must never forget is that although air travel is exciting to the person making a first flight, and to people going on holiday, a more boring form of transportation has yet to be devised for those who travel regularly by air.

To help overcome this boredom some airline operators show in-flight movies. This form of entertainment was pioneered in Great Britain in 1924 and then by Pan American Airways who put on experimental film shows in 1939, on board a Boeing Clipper on the 234-mile flight between Miami and Havana, with the purser operating the projector. The experiment lasted several months and was then dropped, to be revived 16 years later, when the airline began showing sound films on its transatlantic services. In November 1950, the film *Native Son* was the subject of the first airborne world première.

Another pioneer of in-flight entertainment was TWA, which showed a picture called *Flying Hostess* on board one of the first DC-2s in 1934. Today, almost all airlines offer film shows to their passengers on long-haul routes. When it was implemented as a common feature, which coincided with the beginning of the jet age, the film was shown from a downfolding 16-mm projector that threw the picture on a large screen at the forward end of each passenger compartment. To hear the sound track, passengers use lightweight earsets that are connected to the plane's central sound and public address system via connections in each armrest. A variety of stereo music is also available to choose from individually at each seat position. The quality of the film shows has greatly improved today mainly because of the introduction of video tape systems that have supplanted the old-type 16-mm film projectors that were clumsy and subjected to frequent malfunctions.

The airline companies select the films to be shown on board with utmost carefulness not to upset or frighten any of the travelling spectators. It's a delicate task to select films that can interest all categories of passengers regardless of age, nationality, religion and social status and the first command in this difficult selection is not to show films that include any air crashes or disasters which can add to the discomfort among those numerous passengers who have a fear of flying. On long overseas flights often news programmes can be viewed on board, transmitted either from national terrestrial stations or from direct broadcasting TV-satellites. Instructional videos, describing on-board safety routines and practical tips how to manage at the destination airport are becoming a common issue that can be viewed on board today's passenger aircraft. In the future, each passenger position will be equipped with individual TV-sets placed on the back of each seat. This arrangement will allow for more flexibility if you don't want to watch the picture show as it's not necessary to draw the cabin's window-curtains in order to darken the whole cabin when the films are presented.

For the travelling businessman many kinds of services are offered which enable him to carry out much of his work even when not in physical contact with the ground. Via communications satellites, telephone calls can be made and telefax can be sent to any place on earth. Also, as we live in the age of increasing computerization, it will be made possible to transmit digital computer in formation two-way if a passenger for example wants to get in contact with a databank from his portable computer.

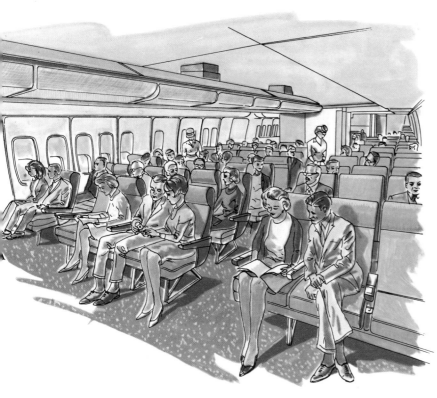

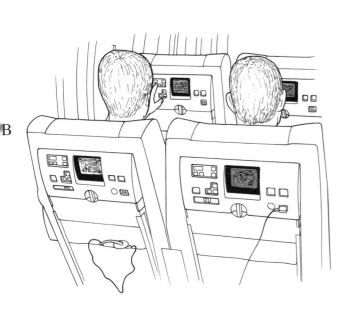

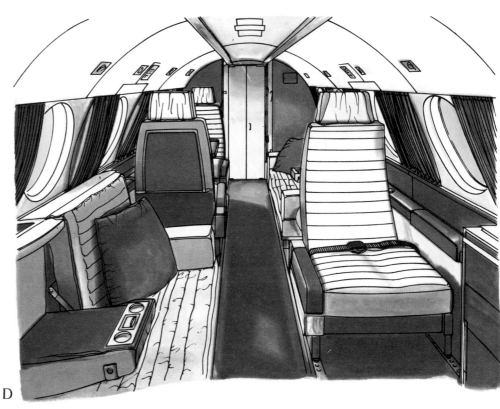

B

D

A *A typical cabin on the Boeing 747 "Jumbo-jet". Depending upon the operating airline's choice of interiors, this airliner can carry from 350 to 490 passengers. The 20-foot wide fuselage allows the use of wider seats than those previously used in economy-class configurations, giving increased comfort.*

B *The future airline passenger will have his own maneuvering panel in front of him. Individually he will be able to select from a number of films and stereo music entertainment. There will also be connections for on-board use of portable computers and telefax machines.*

C *A spiral staircase connects the main passenger cabin of the Boeing 747 with the upper cabin area behind the flight deck.*

D *The cabins of planes used by directors of large companies always have a luxurious and ultra comfortable appearance, as in this Dassault Falcon 20.*

763 764 765

DOORS

Today, big loading doors on jet airliners mean big business, but in the early days of aviation entry into an aircraft presented no problem. Often the aviator was more concerned with ensuring that he did not fall out rather than how he got in.

The steady increase in speeds made it necessary to provide the pilot with some kind of protection from the airflow. Initially this took the form of an elementary wind-deflecting transparent shield. If it rained, the aviator still got wet. The first attempt to protect the pilot more adequately occurred in 1912, with the appearance of the Avro cabin biplane and the Avro cabin monoplane, the latter being publicised as the "first totally-enclosed aeroplane to fly in the world". However, enclosing the pilot on these two aircraft was for the purpose of "streamlining" rather than pilot protection. In the same year, Blériot constructed a machine in which the passengers were in an enclosed cabin and the pilot in an open cockpit.

As in other fields of aviation development, these early experiments were made long before it became the generally-accepted practice to enclose the pilot and passengers. In fact, as we have already seen, for many years it was deliberate policy to keep the pilot out in the open, even when the passengers were being offered cabins. So there was little need for more than a single door on the passenger cabin.

This was no mere concession to class distinction. Many designers felt that pilots would lack an all-round field of vision if they sat in an enclosed cockpit. Others thought that pilots would tend to doze on a long flight without the icy slipstream on their faces. Another very practical reason for keeping them out in the open was that if they were inside a cabin they would not be able to put their heads over the side to follow the roads and railways which formed the major navigational aids of that era!

The end of the first World War in 1918 had brought disaster to the British aircraft industry, as it had to those of other countries. The "war to end wars" had been won and the Royal Air Force alone had over 22,000 warplanes in service or in store. Clearly, there was little scope for new production in the military field. So, like most aircraft companies at the time, the British west country firm of Westland decided to try its luck in the civilian market.

The aeroplane that began to take shape on the drawing boards was no multi-engined, multi-passenger airliner; but in a modest way it was even more revolutionary. Named the Limousine, it was just that—a pioneer attempt to provide passenger-car comfort for three air travellers. Today, this aeroplane would be known as an executive transport. The possibilities of business flying were demonstrated on one flight when letters were drafted and handed to a secretary who had them typed and ready for posting by the time the flight ended. The cabin contained four seats, with the pilot occupying the rear one on the port side. He was, in effect, the chauffeur, separated from the passengers by the simple expedient of raising the seat 30 inches, so that his head poked through a hole in the roof!

The prototype Limousine was operated on an experimental service between Croydon and Paris. Later models, known as Limousine IIs, were used on air services from London to both Paris and Brussels.

In 1920, the British Air Ministry held a Commercial Aeroplane Competition, with the object of finding the best possible replacements for the converted bombers being used by early British airlines. As their entry for the "Small Aeroplane" class, Westland scaled up the Limousine into a six-seater, the result being known as the Limousine III.

To reduce the fire hazard and to permit smoking in the cabin, the fuel was carried in external tanks under the wings. During the flight to Martlesham for the competition, the pilot took advantage of this, and of the machine's excellent flying qualities, by lowering himself into the cabin for a chat and a smoke. The extent to which the passengers appreciated this early demonstration of automatic flight is not recorded, but the aircraft arrived quite safely and went on to win the first prize of £7,500 in its class.

Fitting doors to these early aircraft was a relatively easy matter. The main fuselage loads were taken by the internal structure, so that a hole for the doorway could be cut in the canvas or plywood covering of the fuselage with little weight penalty.

The situation changed with the advent of monocoque and stressed-skin construction. In these, the stresses are taken by the skin, and where a cut-out occurs the loads have either to be taken through the access panel or door, or around the aperture. The difficulty of making the door a close fit in the frame usually makes it impractical for hinged doors to

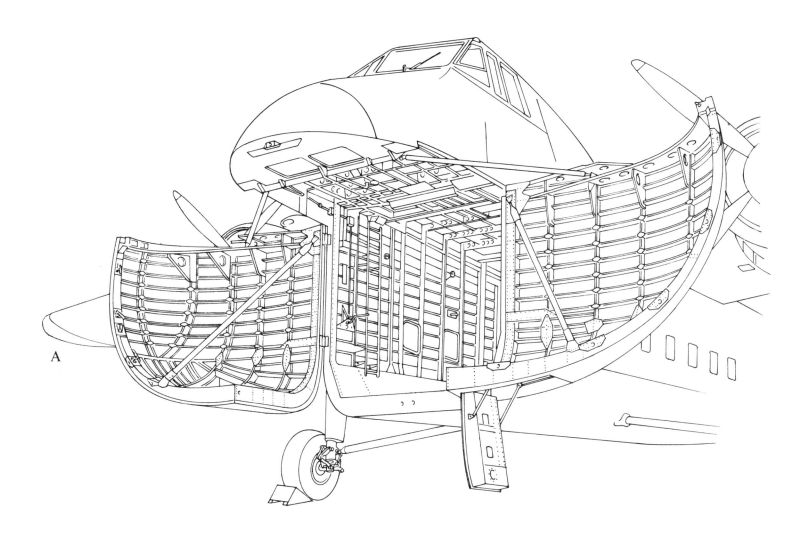

A

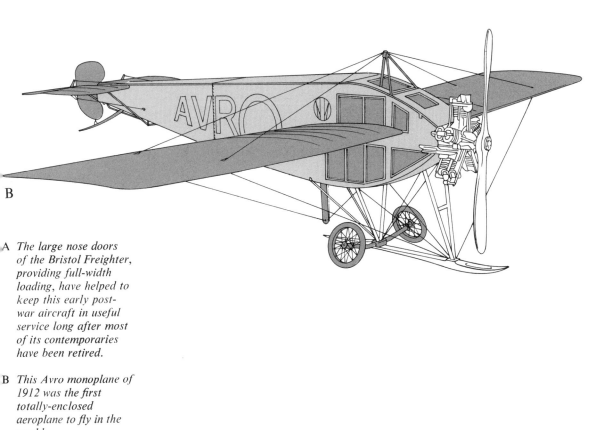

B

A The large nose doors
  of the Bristol Freighter,
  providing full-width
  loading, have helped to
  keep this early post-
  war aircraft in useful
  service long after most
  of its contemporaries
  have been retired.

B This Avro monoplane of
  1912 was the first
  totally-enclosed
  aeroplane to fly in the
  world.

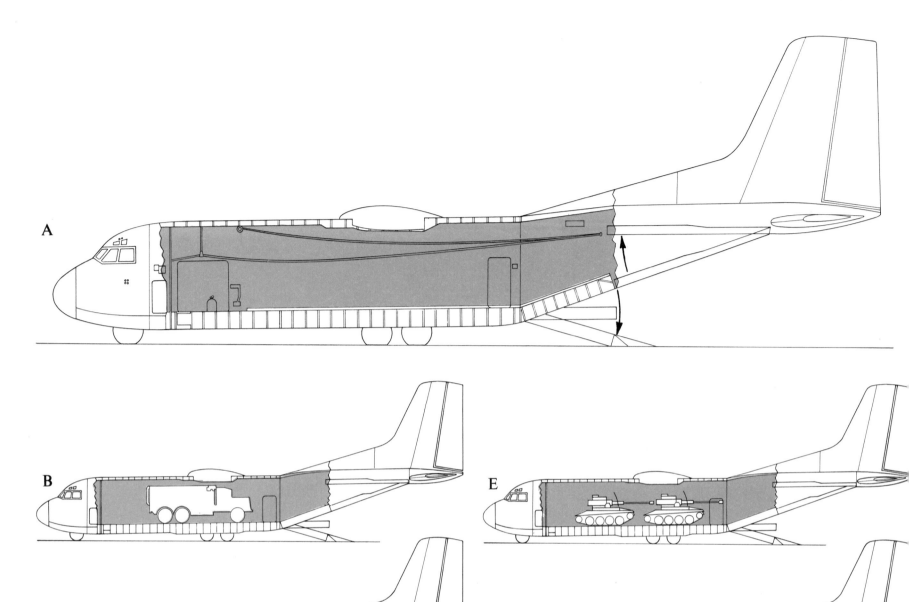

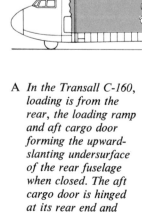

A *In the Transall C-160, loading is from the rear, the loading ramp and aft cargo door forming the upward-slanting undersurface of the rear fuselage when closed. The aft cargo door is hinged at its rear end and*

opens upward and inward. This permits the loading capacity to be fully utilised. The loading ramp and cargo door can be opened in flight for the aerial delivery of heavy loads.

Typical Transall C-160 loads.

B *One 22,822-lb. tanker truck.*

C *Two 12,000-lb vehicles.*

D *One 32,000-lb tank.*

E *Two 16,000-lb self-propelled guns.*

F *One 155-mm, 17,000-lb gun.*

G *One 14,000-lb. Starfighter.*

781   782   783   784

arry significant loads. At first, however, the doors were relatively small and the loads light.

The first indication of the true importance of doors came during World War II, when it became necessary to airlift heavy and bulky items of military equipment, requiring larger-than-normal doors through which to load them. The problem was accentuated by the development of large-scale air supply operations. Doors were required that could be opened safely and closed in flight, and the problems involved in manhandling and ejecting equipment at precisely the right moment to ensure that it would land within the target area were considerable.

The advent of jet propulsion, and the development of high-flying aircraft with pressurized cabins, raised the design of doors and cut-outs to a fine art. A door aperture in such aircraft has to be designed carefully to provide a stress-path round the hole. A strong rim by itself is not sufficient; the loads must be diffused into the surrounding structure so that there is no abrupt concentration of stress to cause fatigue. The problem, of course, is common to any aperture in a stressed-skin structure, but it is much more critical in a cabin subjected to repeated pressurization.

In flight, the pressure differential creates a heavy load on doors that tends to force them open. The load can be as high as 15 tons on a passenger door and inadvertent opening in flight would be catastrophic.

To eliminate this hazard, the de Havilland Comet, the world's first jet airliner, embodied plug-type doors. These opened inwards and were wedge-shaped so that, when closed, they fitted snugly into an aperture of the same shape. The higher the differential pressure, the tighter the doors fitted. However, doors which open inward tend to obstruct the entrance and take up valuable space. Considerable ingenuity has been expended in developing doors which overcome this last objection and yet are a plug-fit when closed. Perhaps the simplest is the pull-in and slide-along door, as used on aircraft such as the Caravelle, Britannia and Friendship, and on the Russian Il-18 and Tu-114.

Another way of avoiding the limitations on cabin layout imposed by inward-opening doors is to get them right out of the way, by making them slide "up and over". This type of door is used on some Comets and on Tridents and Electras, and has also been adapted successfully for Hawker Siddeley HS.125 executive jet aircraft. However, most machines in this latter category have simple outward-opening doors, as this facilitates the embodiment of built-in airstairs.

Many US airlines specify outward-opening doors, and two neat solutions that achieve this and yet retain the inherent safety of a plug door when closed have been developed for the Convair CV-880 and the Boeing family of jet airliners. The CV-880 door slides up outside the fuselage, freeing itself from its tapering aperture, and is then hinged outward and sideways in the normal manner. On the Boeing airliners the main doors have wedge sides and folding tops and bottoms. With the ends folded, the doors can be eased outward to lie against the side of the fuselage.

The military need to carry heavy and bulky items led to the development of specialized freight aircraft, and this in turn introduced major door problems. During World War II, small army vehicles and light weapons were manhandled through doors in the side of the fuselage, but for rapid loading and unloading, and for the maximum utilization of cargo space, end loading is essential. Loading from the front presents the fewest problems, a method used on aircraft like the Douglas Globemaster. However, nose doors preclude the air dropping of large loads, for which a rear exit is essential. Rear doors are now universal on military transports such as Lockheed's Hercules Star Lifter, Russia's Antonov An-22 and Ilyushin Il-76, and the giants Lockheed C-5A Galaxy and Antonov An-124 Condor.

The embodiment of doors providing access to the full width and height of the freight hold can give rise to drag problems on the rear fuselage—a problem accentuated still more by the need for the doors to open in flight. On some early aircraft, the latter difficulty was overcome by removing the doors completely during air-dropping missions.

In the civil aviation field, the growth of air freighting has brought with it a need for large cargo aircraft. It has not proved economically practical to utilize military freighters in the civilian rôle, but this situation may change in the future. A small number of Lockheed Hercules and part of the former RAF Short Belfast fleet are operated by civilian cargo airlines. All of Russia's "civil" freight aircraft are essentially military transports.

In the West the demand for civilian freight aircraft was met, in the main, by conversion and adaption of passenger airliners.

The most radical method of conversion is to create an opening in the fuselage so that the freight can be loaded from either the front or the rear, thus permitting bulky and long items to be taken on board. One of the most spectacular conversions is the Superguppy, used for transportation of wings and other components between the different manufacturing sites of Airbus Industries throughout Europe. Originally a Boeing Stratocruiser, the Superguppy is equipped with new turboprop engines; the fuselage has been "blown up" to three times its original volume, and the whole nose section can be swung sideways for loading. By cutting the fuselage just ahead of the tailplane surface, a number of Canadian CL-44s and DC-6s were adapted to freighters with a swing-tail to permit utilization of the full width of the aircraft for loading. Conversion för freight-carrying duties is, however, effected usually by the incorporation of a single large upward-hinged door in the side of the fuselage. Such modifications are expensive and require extensive strengthening around the door aperture and complex door locking arrangements.

A recent innovation has been the development of "quick-change" aircraft. This facilitates the use of aircraft for carrying passengers during the day and for carrying freight at night, thus increasing their utilization.

The quick-change concept requires considerably more than a large door; it involves a whole system. To permit rapid removal and installation of the seats, they are supplied in the form of eight- or twelve-seat pallets. These can be removed, complete with sections of the floor and carpeting, into specially designed seat-storage trucks. When removed, the pallets expose built-in rollers on the floor which facilitate the loading and unloading of freight pallets and, of course, the seat units themselves. Special locks secure the seat-pallets, eliminating any tendency to vibration.

Galley units used in quick-change aircraft also have to be designed as pallets and are fitted with quick-release water and electrical connections.

Other interior details such as overhead baggage holds, bulkheads and lavatories are made easily detachable, and conversion of the cabin from passenger to cargo configuration usaually does not take more than an hour.

To permit the change in direction inevitable with side-loading, a ball-transfer mat is located adjacent to the loading door. This consists of a flat plate embodying a number of closely-pitched balls rotating in sockets, over which the pallets can be pushed easily.

Modern cargo planes even have automatic, built-in conveyors which handle the containers and pelleted goods inside the aircraft without any human effort.

Many widebodied aircraft are equipped with cargo doors and built—in automatic cargo-handling equipment in the lower belly-deck, for easy loading of passenger baggage in standard containers and to take freight on regular passenger flights. There also exist combi-type aircraft where the rear part of the cabin is permanently fitted out to take cargo, while the front section is equipped for passengers. The non-flexibility of the combi layout never made the concept very popular, so the development has been towards specialized aircraft for either passengers or cargo.

The demand for air freight capacity has risen constantly as the market has become dramatically more international. Many aircraft manufacturers today offer pure freight versions of passenger aircraft, designed and built for their role from the outset. The cargo version of the Boeing 747 can take a load of 100 tonnes, and can be distinguished from the passenger aircraft by the lack of cabin windows and by a big, upward-swinging nose door.

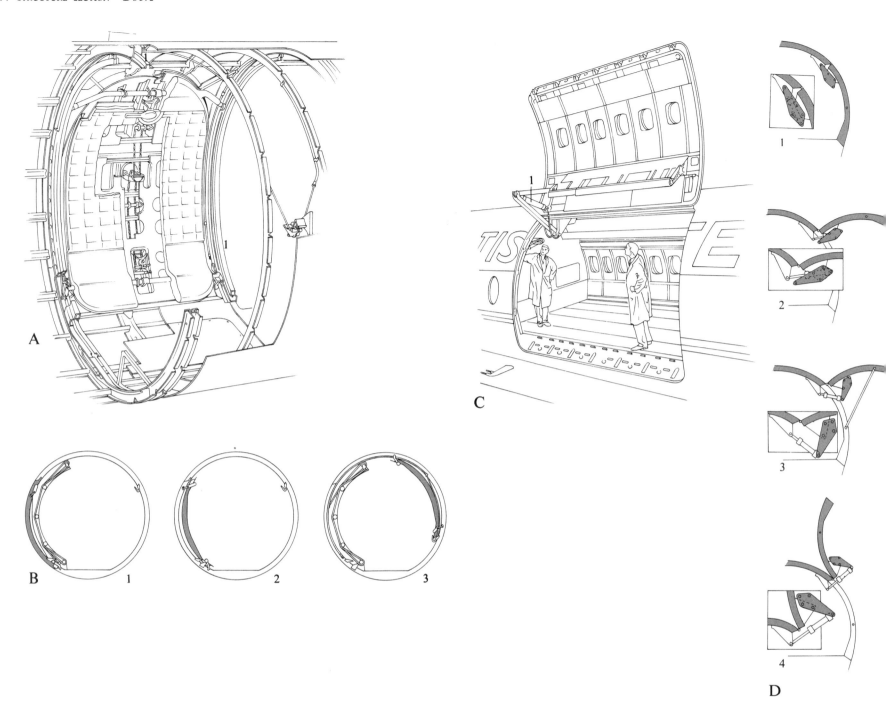

A *The Hawker Siddeley HS.125 cabin door is of the "up and over" type, sliding in rails up into the roof, to leave the entrance unobstructed. When closed, it is a plug fit in the door aperture.*

1 Door track

B *Operation of the HS.125 cabin door.*

1 Door locked
2 Door unlocked (operating mechanism omitted for clarity)
3 Door stowed

C *Measuring twelve feet long by seven feet, the Super VC10 freight door is opened and closed by a single hydraulic jack. Heavy circumferential loads due to cabin pressure are taken by the structurally-stiff door, which embodies hook-type latches and an "over-centre" mechanism for straining the door into position to accommodate the deflection due to loading. In the closed position, the bottom edge of the door is locked to the sill by seven hydraulically-operated toggle latches. When these latches are fully home, two bolt-type locks engage in the two vertical members of the door frame. If required, the door can open to a near-vertical position to permit the crane-loading of freight. The door incorporates six passenger windows and the inside is trimmed to match the rest of the cabin. Normal air conditioning, sidewall heating, lighting and passenger service panel facilities are embodied, to cater for the all-passenger role.*

D *Opening sequence of the Super VC10 freight door (1-4)*

E *The Douglas Globemaster I transport had an under-belly elevator for raising freight into the cargo-hold.*

1 Operating jack

F *The Convair Tradewind was designed in 1950 as a long-range patrol flying boat, but seven were built with an upward-opening nose, permitting vehicles and troops to be disembarked directly on to beaches. Such operations were possible only in fine weather—a restriction which severely reduced the logistic value of the aircraft.*

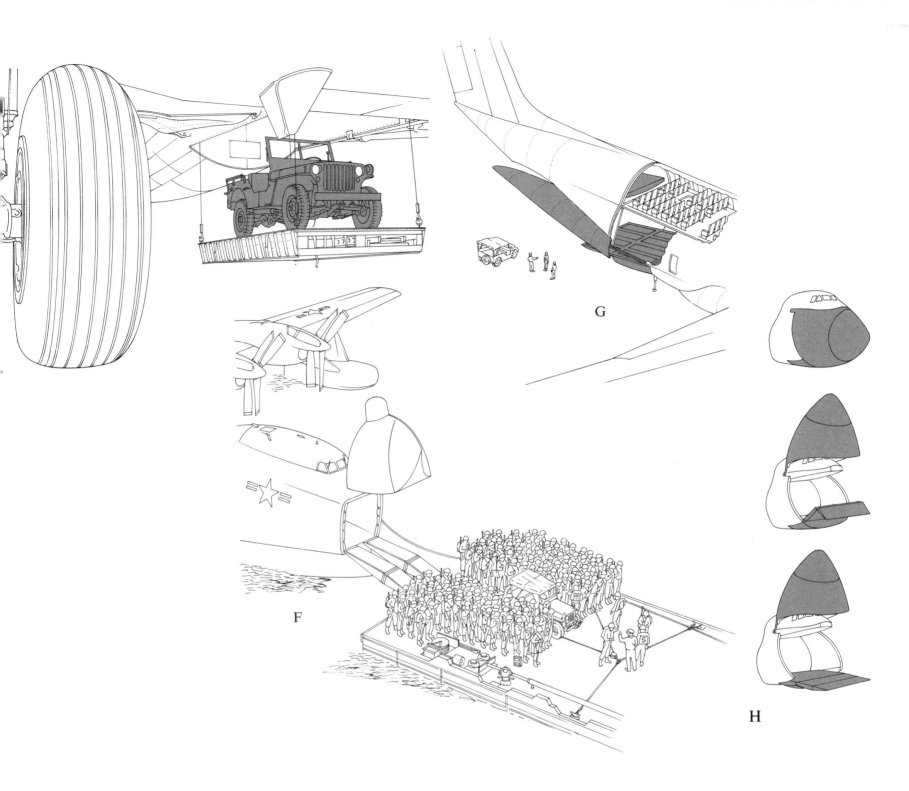

G

F

H

Loading and unloading the huge Lockheed C-5A logistics transport presented particularly difficult problems to the designers. Very large surfaces were required to open and yet had to be able to withstand heavy aerodynamic and pressurization loads in flight. Designed to carry a 125,000-lb payload 8,000 miles, the C-5A has a wing span of 222 ft, a length of 245 ft and a maximum weight of 769,000 lb. Typical loads for the cavernous hold include two M-60 tanks and sixteen ¾-ton lorries; or one M-60 tank, two Iroquois helicopters, five M-113 personnel carriers and two trucks; or 10 Pershing missiles complete with tow and launch vehicles.

To permit rapid loading and unloading, ingenious door arrangements are provided at both ends. At the rear of the C-5A, there is a three-piece door. A 13-ft wide centre panel folds upward to permit parachute dropping of cargo in flight without creating drag. The two large clamshell side doors cannot be opened in flight, but on the ground they permit unobstructed access to the hold. A ramp, which also forms part of the hold floor, can be adjusted to suit varying loading heights.

H At the front, the entire nose takes the form of a vizor which swings upward. This offers several advantages. When open, the door is well clear of the loading operations, the engines can be run with the vizor open, and flight deck visibility is such that the aircraft may be taxied in this configuration. Attached to the floor is a built-in three-section loading ramp, the outermost section being divided into various longitudinal sections to conform with uneven terrain. Loading is facilitated by the 28-wheel landing gear. The main gear consists of four similar six-wheeled bogie units, housed in blisters on the sides of the fuselage. The nose gear is a single four-wheeled unit.

Each of the four main units and the nose gear can be raised and lowered individually to vary the ground clearance from the normal 3 ft "truck-bed" height down to 1 ft. The same technique can be used for wheel-changing and to free an aircraft bogged down in soft ground.

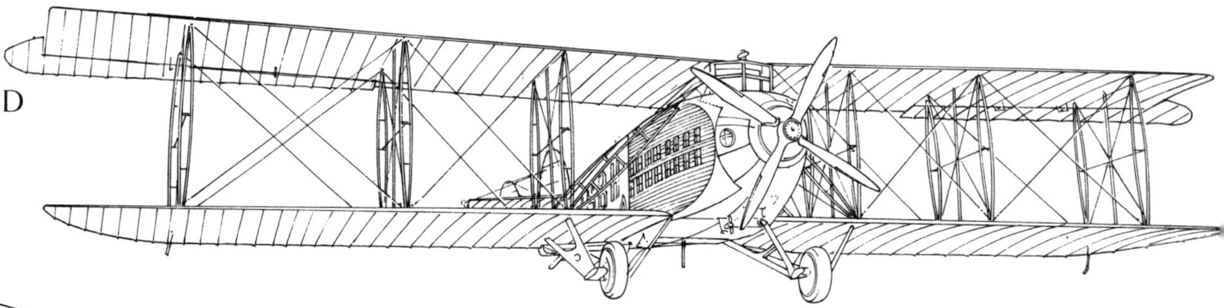

D

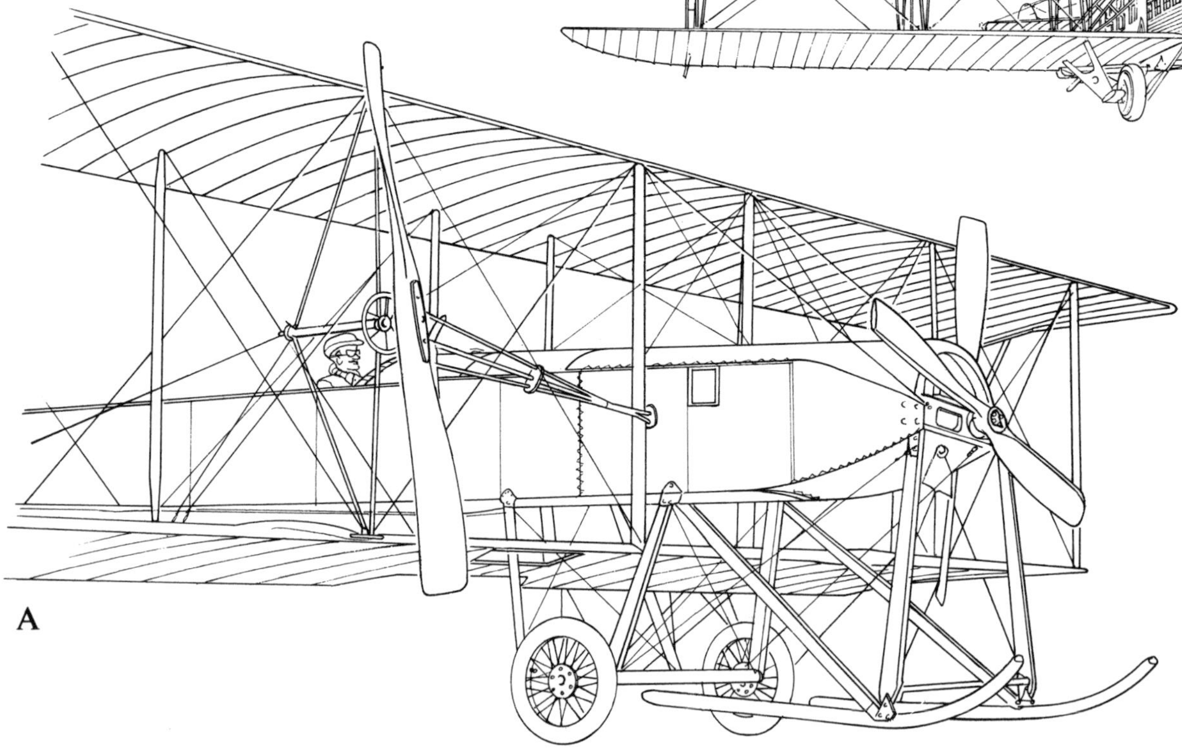

A

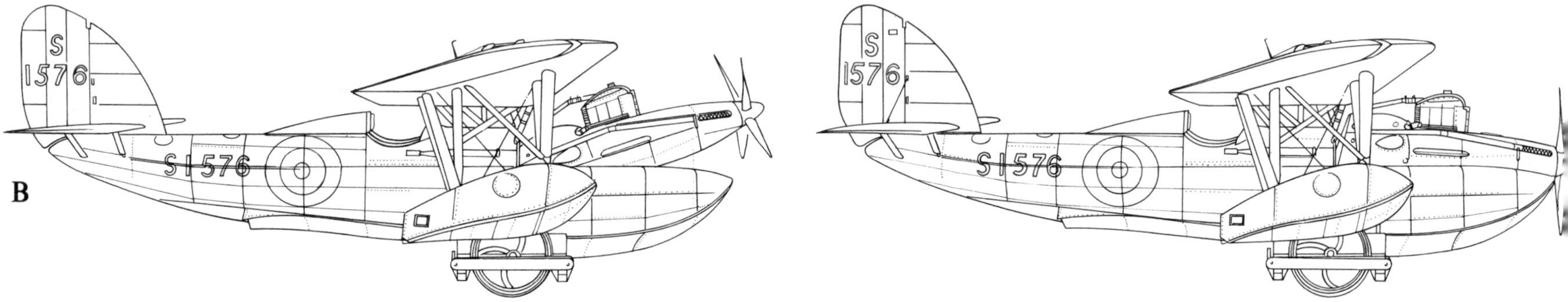

B

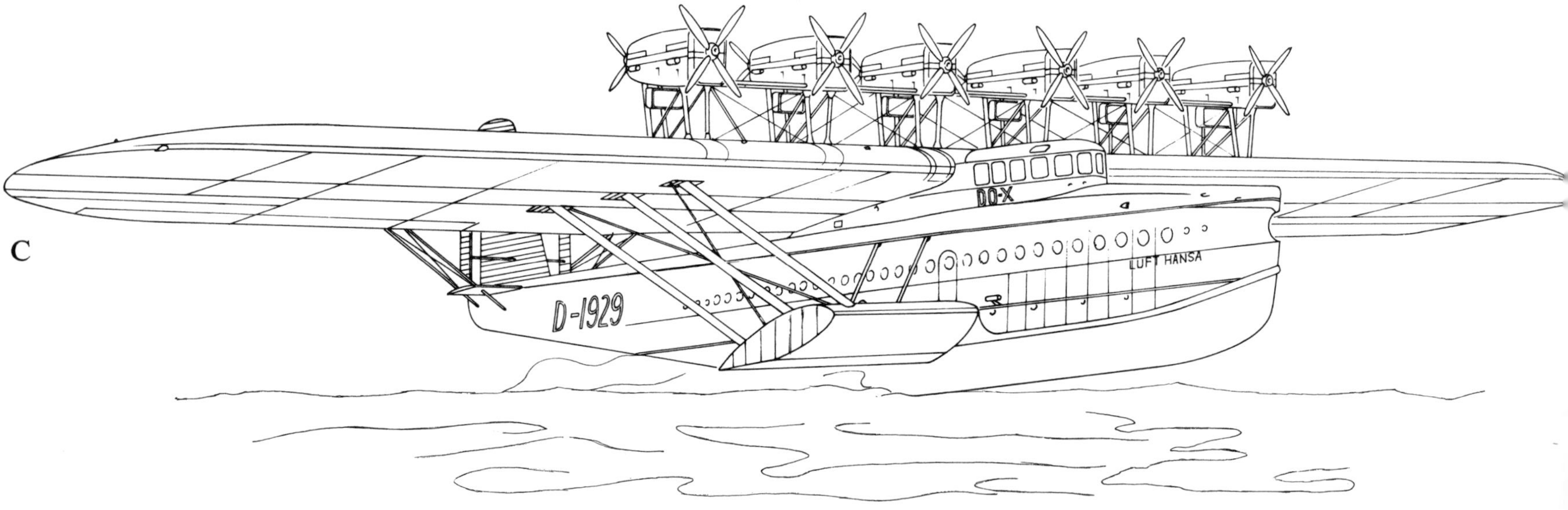

C

Mounting the engine on their aircraft presented a major structural problem to the pioneers, and even today the question of how and where to mount power plants is among the most difficult confronting designers. Somehow, the compact and heavy bulk of the engine must be placed in such a way that the highly-localized loads can be diffused safely from the pick-up points into the relatively fragile airframe structure. On jet-powered aircraft, there must also be provision for efficient intake and exhaust systems.

The loads involved are many and considerable. First of all there is gravity, as the engine is probably the heaviest single component of any aeroplane. Then there are the thrust loads, and the propeller torque and gyroscopic-couple moments. Because of the basic need to change an engine rapidly, detail design is complicated by the need to provide quick-release mounting points, and quick-release couplings on the fuel pipelines and engine controls. The Wright brothers solved the problem simply by sitting the engine on top of the centre-section of the lower wing of their 1903 Flyer and its developments. While facilitating the task of mounting the engine, this installation required a chain or bevel-gear to drive the remotely-mounted propellers. Such drives were generally unreliable and were not widely used.

As the traditional aircraft configuration evolved, with the stabilizing and control surfaces at the rear, the nose of the fuselage became accepted as the best and most logical place to mount the engine. If two engines were required, they were generally mounted on the leading-edge of the wings. On multi-engined biplanes, it was common practice to mount engines midway between the wings, on supports provided by the interplane struts.

Some of these early installations provided a degree of in-flight accessibility that would be unthinkable today. Had this not been so on the Vickers Vimy, the first non-stop flight across the Atlantic by Captain John Alcock and Lieutenant Arthur Whitten-Brown, in 1919, would probably have ended in disaster.

Misfortune dogged the flight from the start. The radio failed within the first hour. A section of exhaust pipe fell off, exposing the aviators to the harsh roar of the unsilenced gases. The airmen's electrically-heated suits failed, leaving them cold as well as tired. Half-way across, the Vimy flew into a fierce storm cloud, was tossed around like a leaf and

got into a spin from which Alcock miraculously regained control only feet above the sea. A strong headwind reduced their speed.

Then it began to snow. Alcock climbed to nearly 9,000 ft, to try and get above the snow clouds, without success. Suddenly the flyers heard the noise they had been dreading. The comforting steady beat of the engines faltered. The air intakes of the two 360-hp Rolls-Royce Eagle engines—mounted midway between the wings on each side of the fuselage—were becoming blocked with ice.

This time it was Brown who saved the day. Although severely handicapped by a war wound, he clambered out of the cockpit onto the ice-covered wing. Bracing himself against the freezing slipstream he chipped the ice away from the engine with a knife. He then moved back into the cockpit and out over the other side to deal with the second engine. All was well for a short while; but he had to repeat the performance five times in all, without a parachute, 9,000 feet up, in the snow.

The clearance required by the propeller provided obvious limitations to the location of piston-engines, and the main alternative to putting them on the fuselage nose and the wings was a "pusher" installation. Many "pusher" aircraft were developed before and during the 1914-18 war; more recent types have included the Convair B-36 bomber and the Cessna Super Skymaster, an aircraft in which twin engines and propellers both "push" and "pull".

Early "pusher" types included the French-built Farman "Longhorns" and "Shorthorns", and the British Vickers Gunbus and de Havilland D.H.2. Initially, the Gunbus and D.H.2 achieved some success through their forward-firing machine-guns, but they were both superseded by more speedy tractor biplanes as soon as an efficient interrupter gear had been developed to enable guns to be fired through the propeller "disc".

Between the wars an attempt toward originality was made with the Westland F.7/30, produced to meet an Air Ministry specification for a high-speed single-seat fighter powered by the 600-hp Rolls-Royce Goshawk steam-cooled engine. A basic feature of this aircraft was the use of a long shaft connecting the propeller and engine, the latter being mounted amidships so that the pilot could be placed well forward. The concentration of weight

A *Short Triple-Tractor biplane, 1912. The extended cowling conceals two 50-hp Gnome engines mounted in tandem. The rear engine was connected by chains to the two propellers mounted on the wing struts. The installation generated so much heat under the cowling that it became known as the "Flying Field Kitchen".*

B *The Parnall Prawn of 1930 was powered by a spare auxiliary engine from the airship R101. The engine could be pivoted upward during take-off to clear the spray. The Prawn is shown here on its beaching trolley.*

C *Dornier Do X, 1929. This large flying-boat was powered by no fewer than twelve engines, mounted in opposing pairs, along the top of the wing. This is the largest number of engines ever installed on an aeroplane.*

D *In 1921 Breguet started trials on the Leviathan, one of the first specially constructed transport aircraft. It was equipped with two Bugatti engines.*

around the centre of gravity gave the aircraft an exceptionally high degree of manoeuvrability and the forward location of the cockpit gave the pilot an excellent field of view. Upward and rearward fields of view were unimpaired, as the top wing was of gull-form, sloping down to the fuselage on each side of the canopy.

Operationally, this would have been superb, but the test pilot found it a mixed blessing. After the first slow roll in the aircraft he was treated to an unhampered rearward view of 20-foot flames burning away the fabric covering of a large portion of the fuselage and one side of the tailplane. The fuel system was altered to put things right, but on the next slow roll the same thing happened. The trouble was finally cured by modifying the exhaust system.

The ultimate "pusher" was the Convair B-36 heavy bomber of the nineteen-fifties. It was powered by six Pratt and Whitney Wasp Majors, delivering a total of 23,000 hp and mounted on the trailing-edge of the wing, driving pusher propellers. The performance of later versions of the aircraft was increased by the addition of four General Electric turbojets, mounted in pairs in two pods under the wing and developing a total thrust of 20,000 lb, equivalent to about 19,000 hp. Turbojets have been added to increase the performance of other piston-engined aircraft, such as the Avro Shackleton, on which Rolls-Royce Bristol Vipers have been unobtrusively installed in the rear of the outboard piston-engine nacelles.

The Super Skymaster is a twin-engined executive aircraft embodying both a tractor and a pusher propeller, one at each end of the fuselage—a layout reminiscent of the German Do 335 fighter of World War II. This configuration produces an aircraft combining the traditional advantages of two engines with the operating simplicity of one. With conventional wing-mounted engines, considerable training and skill is required to handle an aircraft in the event of one engine failing, or the sudden loss of power on one side produces a violent yaw. On the Super Skymaster most of these problems are avoided.

The advent of jet propulsion opened up new possibilities for the location of the engines. Such powerplants transmit little torque and, being relatively compact, can be located within the fuselage or wings, or they can easily be mounted outside the airframe in streamlined nacelles.

This new freedom was not fully appreciated at first, and many early multi-jet aircraft appeared with the turbojets mounted in the position that would otherwise have been occupied by piston-engines. Typical of these pioneering jets were Germany's Messerschmitt Me 262, Britain's Gloster Meteor and America's North American Tornado bomber. On all of these aircraft the turbojets were mounted on the wing, as on twin piston-engined aircraft.

One of the first aircraft to exploit the new freedom offered by the turbojet was the Hawker P.1040,

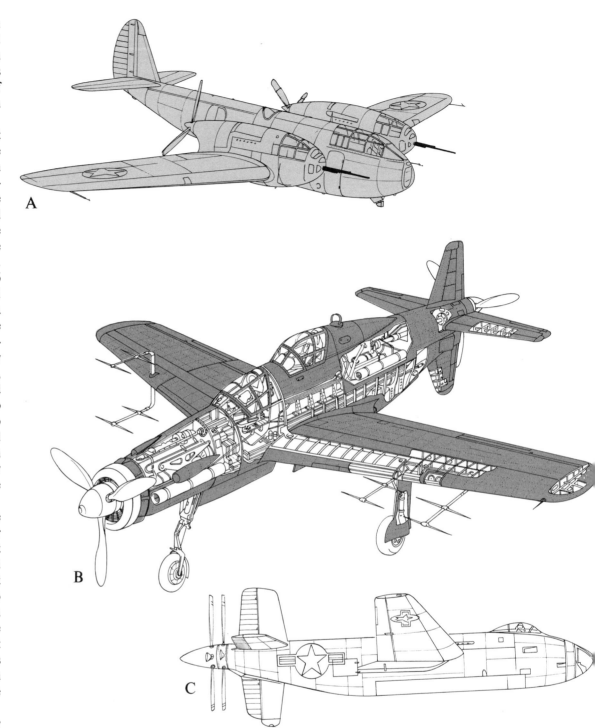

A

B

C

A **Bell Airacuda, 1937.** The two 1,000-hp Allison engines of this fighter were mounted on the wing, driving pusher propellers. This unorthodox installation permitted gun compartments to be embodied in the nose of each engine nacelle.

B **Dornier Do 335, 1943.** This unconventional German fighter-bomber had one engine in the nose, driving an orthodox tractor propeller, and a second engine aft in the fuselage, driving a pusher propeller behind the tail. Service pilots experienced difficulty in handling the aircraft, which was prone to "porpoising" and "snaking" at high speeds. Maximum speed was 413 mph at 26,000 ft. The aircraft illustrated is the two-seat night fighter version of the Do 335

C **Douglas XB-42 Mixmaster.** Produced in 1945, this unconventional bomber was powered by two 1,000-hp Allison engines buried in the fuselage and driving counter-rotating, co-axial propellers located aft of the tail surfaces. The installation resulted in high aerodynamic cleanliness. Single-engine performance was good in that the installation eliminated all tendency to yaw.

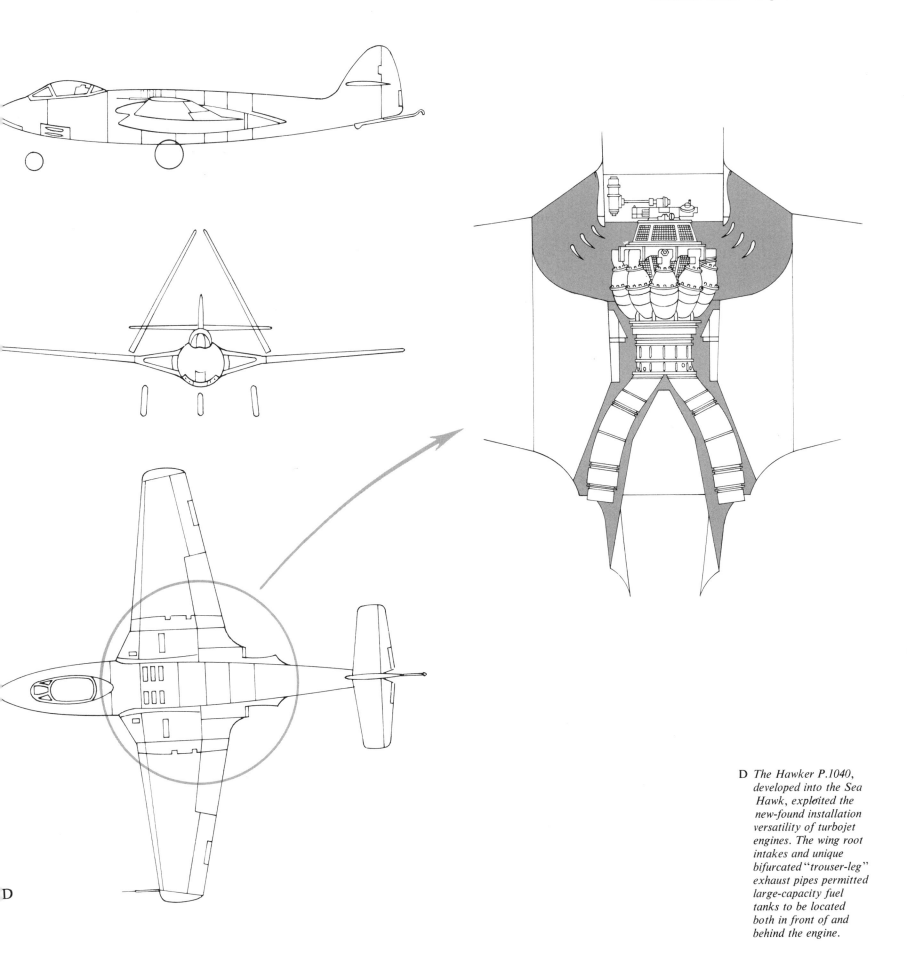

D *The Hawker P.1040, developed into the Sea Hawk, exploited the new-found installation versatility of turbojet engines. The wing root intakes and unique bifurcated "trouser-leg" exhaust pipes permitted large-capacity fuel tanks to be located both in front of and behind the engine.*

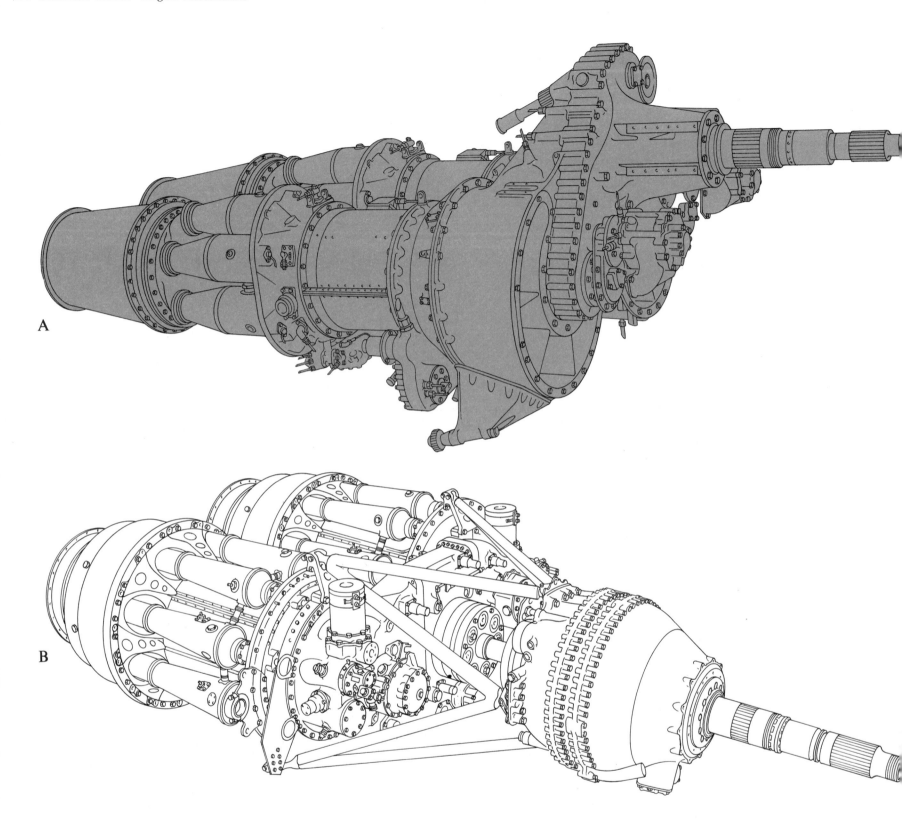

A

B

A *The unique Armstrong Siddeley Double Mamba turboprop powerplant was fitted in the Gannet anti-submarine aircraft. Each of the two engines powered one of two contra-rotating propellers, with the following advantages: the safety inherent in twin engines.*

maintenance and fuel economies, as one half of the engine could be shut down for cruising; full power of the double engine was available during take-off and emergencies.

B *Coupled Proteus. Developing 7,000 hp this coupled Bristol Siddeley turboprop powerplant was used on the Saunders-Roe Princess flying-boat and was intended for later marks of the Bristol Brabazon landplane. It drove counter-rotating propellers.*

om which the highly-effective Sea Hawk carrier-orne fighter evolved.

While providing sufficient power for very high eed, turbojets are extravagant on fuel, and the awker design team was faced with the problem of nding space for approximately twice the amount f fuel carried by the P.1040's predecessor, the entaurus piston-engined Sea Fury, in order to ve the new aircraft a comparable range.

The problem was solved by mounting the engine midships with air intakes in the leading-edge of e wing and the jet-efflux exhausting via "trouser-g" jet-pipes, at the trailing-edge of each wing root. his novel installation enabled one large fuel tank be located aft of the engine and another between e engine and the cockpit. Other advantages of this nique layout included an outstanding field of view r the pilot past the short pointed nose and an fficient, conventional, rear fuselage and tail ructure.

De Havilland designers were also quick to appre-ate the versatility of the turbojet engine. In 1943, hile war was still raging fiercely, this company roduced a scheme for a small jet transport based pon the Vampire fighter, with two engines buried the tail of the fuselage. The idea was so promising at in the following year the company designed a igger version of the jet, capable of carrying 12 assengers at 500 mph for 2,200 miles. This new esign was powered by three engines at the rear.

De Havilland returned to the engines-at-the-rear neme in the early days of the development of the omet, the world's first jet airliner. One project tudy sketch of this historic aircraft depicted a il-first configuration with three engines at the ear.

However, in its final form the Comet appeared ith four Ghost engines buried in the wings, two on ach side, close to the fuselage. This installation ave the aircraft a remarkably clean appearance erodynamically, and permitted maintenance to be one at shoulder height without the use of steps r platforms.

Operationally, the location of the Comet's ngines close to the fuselage meant that there was ttle tendency to yaw in the event of one engine ailing. Disadvantages included the difficulty of ccommodating improved versions or other types f engines and the difficulty of providing a fail-safe ing structure. These problems did not prevent uried engine installations being adopted on Britain's rio of V-bombers, the Valiant, Vulcan and Victor. imilar wing-root installations were adopted by ussia for her Tu-16 and Mya-4 bombers, and u-104 and Tu-124 jet airliners.

In America, engine installation technique deve-oped along a quite different line. The optimum ength of intake and exhaust systems is much less han that enforced by a buried powerplant, so that n engine carried outside the airframe can be nstalled with more efficient intakes and nozzles. hus, America's great fleets of Boeing B-47 and

B-52 bombers, and the Boeing 707 and Douglas DC-8 families of airliners all appeared with their jet engines mounted in pods under the wing. In addition to providing optimum intake and exhaust systems, the engines—by "holding down" the wings—offset the loads tending to bend up the wings in flight, permitting a saving in structure weight. Powerplant accessibility is excellent with a podded installation, and it is relatively easy to install alternative types of engine.

There are, however, disadvantages. Debris is more easily sucked into the engines; it is more difficult to keep the aircraft straight in the event of engine failure, and operators faced almost certain loss of most engines in the event of a wheels-up landing. Wing flaps have to be interrupted in the path of the engine efflux, and the undersurface of the wing is fully exposed to noise damage. Additionally, the plane of the engine turbine wheels is in line with fuel tanks in the wing and with the pressurized portion of the fuselage, increasing the risk of fire or explosive decompression in the event of an engine break-up.

In France, Sud-Aviation overcame many of these problems in a novel and hitherto untried fashion. On their Caravelle airliner they mounted the engines in pods on each side of the fuselage at the rear. Advantages of this location include the shielding of the engine intakes from debris thrown up by the wheels, a quieter cabin, and—because the engines are well away from the fuel in the wings—a reduced risk of fire in the event of a crash landing. With engines at the rear, the wing is left clean, unspoilt by bulges, pods or nacelles, and can thus perform its basic task of producing lift most efficiently.

The advantages of this type of installation appeared so overwhelming that a whole new generation of rear-engined aircraft appeared. The configuration seemed ideal for nearly all categories and sizes of aircraft, ranging from small executive types to large airliners. At the time of writing, all twin-jet executive aircraft, such as the Jet Commander, North American Sabreliner, Hawker Siddeley HS.125 and Dassault Fan Jet Falcon, embody rear-mounted engines.

The aft engine installation was also adopted by BAC for the One-Eleven "bus-stop" jet, by Douglas for the DC-9, by BAC on the VC10 and by Russia on the Tupolev Tu-134 and Ilyushin Il-62 airliners. A variation of the usual layout was adopted for the advanced Tupolev Tu-22 supersonic bomber, the engines being mounted unusually high, on each side of the base of the fin.

Rear installations were chosen unanimously by the first manufacturers of three-jet airliners—Boeing for the 727, Hawker Siddeley for the Trident, Tupolev for the Tu-154 and Yakovlev for the Yak-40. With three engines, installation at the rear is the logical location, since the third engine—otherwise so difficult to locate—can simply be mounted in the tail.

However, as with most aviation innovations, rear-mounted engine installations are not without disadvantages. Perhaps the most serious of these are the undesirable stalling characteristics which may result, due to the wing blanketing the tail surfaces at high angles of attack. In certain conditions of flight, this can render the elevators ineffective, with the result that the aircraft can get into what is known as a "stable stall" condition from which recovery is impossible. To prevent aircraft from running into this danger, "stick pusher" systems are often installed, which lower the nose automatically if a dangerous stall condition is being approached. But these add weight and further complexity—and are a hazard themselves if they operate inadvertently.

With rear-mounted engines the extra lift obtained from the "clean" wing is not all profit. Without engines to reduce the bending moment in flight, a clean wing has to be stronger—and heavier—offsetting some of the additional lift. More of the lift goes in overcoming the weight penalty incurred in strengthening the rear fuselage to accommodate the loads imposed by the engines.

Other disadvantages include problems involving the distribution of passengers and freight, to maintain the correct centre of gravity "balance", and the relative difficulty of "stretching" or lengthening the fuselage as part of the natural process of development. The engines may be out of the line of spray from the wheels, but they are also out of reach of engineers standing on the ground and some form of ladder is required for the most minor adjustment or blade inspection.

Finally, one of the big advantages claimed initially for rear-mounted engines—that of cabin quietness—is only significant at moderate speeds, up to say 500 mph. Above this speed, during normal cruising flight, most of the noise in the cabin is aerodynamic due to the passage of the fuselage through the air. At 600 mph, the quietness of a cabin depends more on the quality of the soundproofing than on the location of the engines!

When initiating the 737 short-range airliner programme Boeing decided that the disadvantages of a rear installation outweighed the advantages, and reverted to hanging the engines under the wings. A return to this traditional location is also evident in the second generation of small feeder-liner projects currently under development, and in the 500-seat "jumbo-jets" and 300-seat "airbuses".

Even on the second generation of three-engined aircraft, such as the Douglas DC-10 and Lockheed L1011, underwing locations were chosen for two of the engines, the centre one only being mounted at the rear.

The under-wing pod-mounted engines on the 737 posed particular problems. In order to keep the intakes as far away from the ground as possible, to reduce the debris ingestion problem, the engines are tucked closely under the wings. The effects of this combination required careful consideration and the pylons had to be specially designed to accommodate the resulting rather tricky airflow pattern. Thus, the pylon centreline is not in a straight plane, but is curved to match the local airflow.

On supersonic aircraft the problem of engine installation is significantly more difficult, as the engine has to operate under all conditions from static at sea level to, perhaps, Mach 2.2 at 60,000 ft. Probably the most important change is the dramatic increase in importance of the intake. The efficiency of a simple intake decreases drastically at Mach numbers in excess of 1.4, owing to the formation of strong shockwaves across the airflow. Instead of a simple orifice, a centre-body or side-wedge generating an inclined shockwave is required, to decelerate the supersonic air and increase its pressure.

An intake designed to be ideal for cruising is liable to starve the engine of air at take-off; if it is designed only for the sea-level static condition, its "swallowing" capacity at supersonic speeds greatly exceeds the engine's need. Excess air then spills around the intake, causing drag and buffeting. This problem of too much air can be overcome by embodying either a special sub-system to bleed it off, or an intake which can be adjusted for varying conditions.

The former method was utilised on Canada's Avro Arrow interceptor, excess air being diverted through a ring of gills around the intake duct which opened automatically when the speed reached Mach 0.5. A similar solution is utilised on the Lockheed Starfighter.

However, relatively simple "fixed-geometry" intakes are not so efficient as fully-variable systems. The first aircraft with adjustable, or variable-geometry, intakes to go into production was the General Dynamics (Convair) B-58 Hustler bomber. Its four 15,600-lb thrust turbojets are installed in individual pods, each embodying a circular, sharp-edged intake surrounding a conical centre-body. The latter is retracted at low speeds, but at supersonic speeds it extends progressively to create the desired pattern of shock waves and reduce the air-swallowing capacity of the engine. The BAC Lightning interceptor employs a similar idea in reverse fashion, its fixed-geometry intake being designed for optimum performance at a Mach number of about 2, with auxiliary intake doors to ensure the correct

flow of air at low speeds. The location of the Lightning's engines is also of interest, the two 14,400-lb thrust Avons being mounted one above the other in the fuselage, with the lower engine well forward of the upper one. Individual exhaust nozzles are provided, each with variable reheat.

The most challenging civil jet installation yet built is that of the BAC/Sud-Aviation Concorde supersonic airliner. On this, the four Rolls-Royce Bristol/SNECMA Olympus engines, each developing 35,080 lb of thrust, are mounted in two twin underwing nacelles, positioned outboard of the main landing gear units. This arrangement brings the intake into a region where it is sheltered under the wing, which tends to straighten the flow and minimise the effect on intake performance of variation in angle of attack.

As previously implied, the efficiency of a supersonic engine is largely dependent on the efficiency of its air intake and exhaust systems. On the Concorde, the air intakes embody a variable-angle ramp in the upper surface which adjusts the shock-wave pattern and operates in conjunction with spill vents in the lower intake surface to regulate the supply of air to the engine.

The exhaust system has a variable-geometry primary nozzle, a secondary nozzle, thrust reverser, and a low thrust-loss silencing system. An afterburner to provide a low-thrust boost in certain conditions is installed on early engines, but will not be required on production Concordes with fully-rated engines.

A similar underwing location for the engines was chosen for Russia's supersonic airliner, the Tu-144. Originally its four Kuznetsov 28,660-lb thrust turbofan engines were mounted side by side in the rear of a single large duct, the entrance of which was divided into two; later, however, the engines were moved further outboard and paired in a similar manner to those of the Concorde.

During the evolution of the bigger, faster Boeing SST, many different wing shapes and engine locations were investigated. Early studies showed the engines mounted aft of a sharply-swept wing; in pods under a large delta wing; and with two engines under the wing centre-section and a third at the base of the fin. The final choice was for individually-podded engines under a delta wing, with conical centre-body intakes.

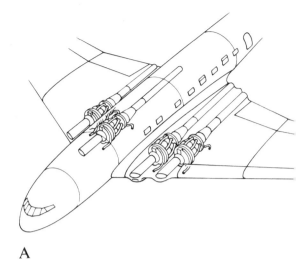

A

B

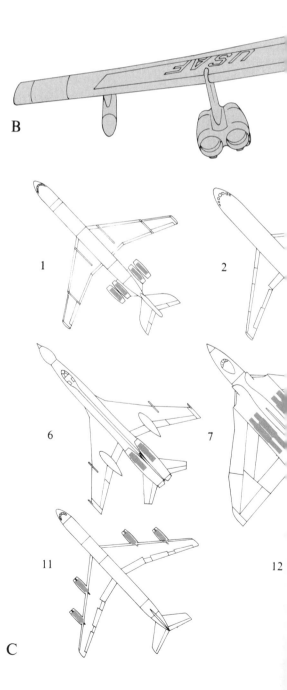

1   2

6   7

11   12

C

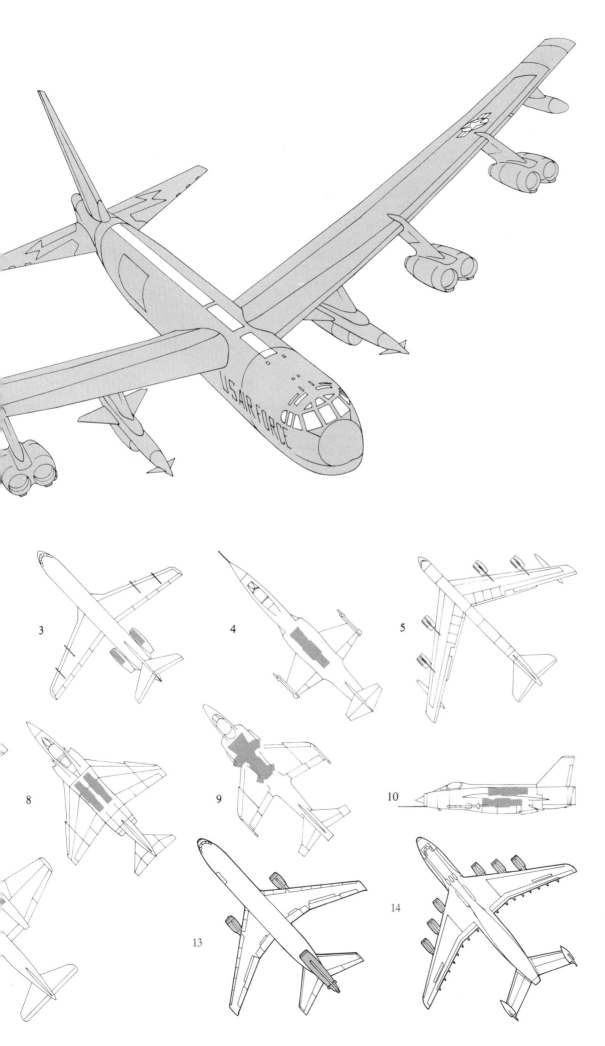

A De Havilland Comet.
  The four Ghost engines
  of the world's first jet
  airliner were buried in
  the wing, close to the
  fuselage. The
  installation resulted in
  little asymmetric thrust
  in the event of engine
  failure and easy
  maintenance, but
  presented structural
  problems.

B Boeing B-52
  Stratofortress. In its
  time the heaviest
  aircraft ever put into
  production, this
  488,000-lb long-range
  bomber is powered by
  eight 13,750-lb
  turbojets, mounted in
  pairs in four nacelles
  supported on forwardly
  inclined cantilever
  struts under the wings.

C The versatility of
  installation conferred
  by turbojet engines is
  well indicated by the
  accompanying selection
  of jet-powered aircraft:

1 BAC VC10; four
  engines in rear-
  fuselage pods.
2 Hawker Siddeley
  Trident; three engines,
  two in pods and one in
  the tail.
3 Sud-Aviation
  Caravelle; two engines
  in rear-mounted pods.

4 Lockheed F-104
  Starfighter; one
  engine, mounted
  centrally in the
  fuselage, fed by cheek
  air intakes.
5 Boeing B-52
  Stratofortress; eight
  engines, in four
  underwing pods, each
  housing two engines.
6 Tupolev Tu-22; two
  engines in pods
  adjacent to the fin.
7 Hawker Siddeley
  Vulcan; four engines,
  buried in the wing.
8 McDonnell Phantom:
  two enignes, mounted
  side-by side in the
  fuselage.
9 British Aerospace BAe
  Harrier; one engine
  mounted centrally in
  the fuselage. Vectored
  thrust exhausts through
  four nozzles in the
  sides of the fuselage.
10 BAC Lightning; two
  engines, staggered one
  above the other in the
  fuselage.
11 Boeing 707; four
  engines in underwing
  pods.
12 English Electric
  Canberra, two engines
  centred in the wing.
13 DC-10/MD-11; three
  engines, two in
  underwing pods and
  one in the tail.
14 Antonov An-225; six
  engines in underwing
  pods.

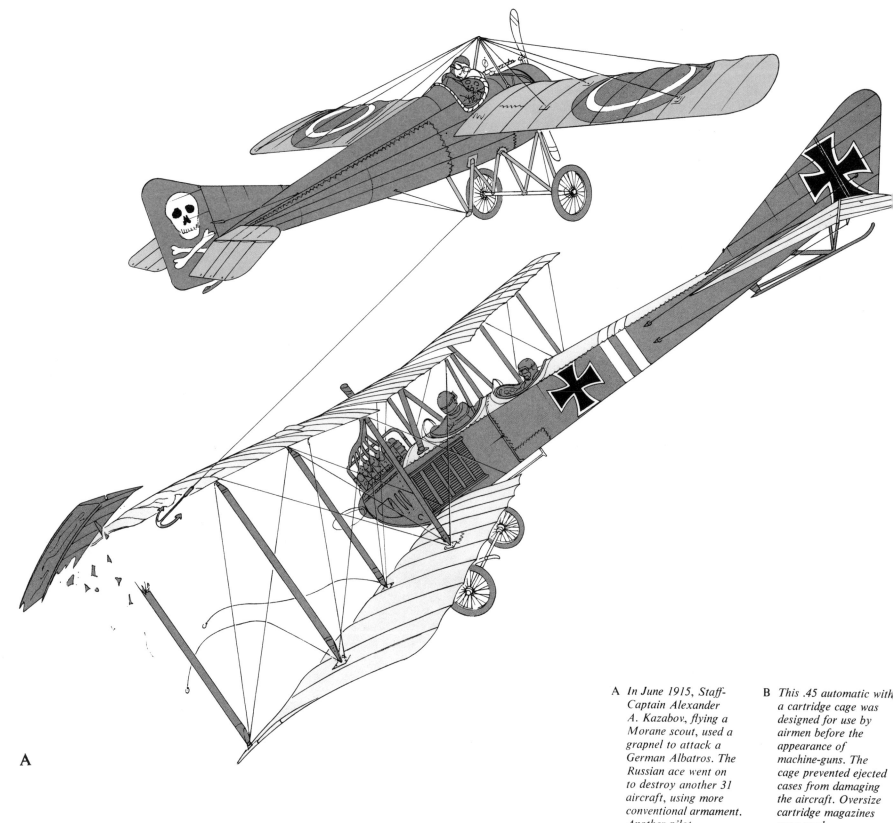

A

A *In June 1915, Staff-Captain Alexander A. Kazabov, flying a Morane scout, used a grapnel to attack a German Albatros. The Russian ace went on to destroy another 31 aircraft, using more conventional armament. Another pilot considered trailing a bomb fitted with hooks; the idea was that, having "fished" for and caught an enemy aircraft, the bomb would be exploded electrically.*

B *This .45 automatic with a cartridge cage was designed for use by airmen before the appearance of machine-guns. The cage prevented ejected cases from damaging the aircraft. Oversize cartridge magazines were used.*

1 Cage

C *The wedge-shaped deflector plates on the propeller blades of the Sopwith Tabloid deflected bullets from the unsynchronised gun that would otherwise have damaged the blades.*

1 Deflector

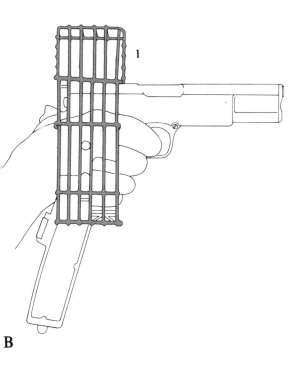

B

C

Many of the early pioneers of flying believed that they were working on an invention that would make further wars impossible. Their reasoning was logical. Military history shows frequently that a key to victory is to know in advance the plans of the enemy. In the nineteenth century captive observation balloons contributed to success in many battles by enabling commanders to keep track of the movement of opposing armies. The aeroplane, being mobile and fast, clearly exceeded the capability of the captive balloon, making possible reconnaissance sorties deep into enemy territory.

Human nature being what it is, however, aircraft were used for more aggressive purposes than reconnaissance right from the beginning. During the almost-forgotten Italo-Turkish war of 1911-12, the Italians used aircraft to drop propaganda leaflets and bombs, as well as for taking reconnaissance pictures, including cine film. During the equally-forgotten Balkan War in the following year, Bulgaria used aircraft to bomb Turkish positions.

When World War I started, Britain carried out an air raid on German territory within weeks. The first attack, on 22 September, did no damage, but the second raid, on 8 October, caused the destruction of the newly-completed Zeppelin Z.9 at Dusseldorf. During this raid bombs were also dropped on Cologne railway station.

These raids were made by Sopwith Tabloids, small single-seaters capable of dropping only 20-lb bombs. They left no doubt about the value of air attacks, and the Air Department of Britain's Admiralty decided it was time to stop playing with toy bombs. The Director issued a specification calling for a twin-engined aircraft suitable for extended patrols over the sea. The Handley Page company produced such a promising proposal that the Director requested them to design and build an even bigger "bloody paralyser of an aeroplane". The result was the twin-engined O/100, first of a long series of British heavy strategic bombers.

It is not known for certain when the first gun was fitted to an aeroplane. August Euler, a German pioneer, took out a patent on a machine-gun installation in 1910 and a French Nieuport two-seater was armed with a machine-gun during the following year.

In America, a low-recoil machine-gun, invented by Samuel Neal McClean but promoted and im-

proved by Col. Isaac Newton Lewis, was tested in June 1912. The test was made on a Wright biplane fitted with a crossbar on which the pilot and observer rested their feet during flight. The muzzle of the gun rested across this bar and Captain Chandler fired the gun at a target as they flew over it.

This first experimental firing of a new type of machine-gun from an aircraft created intense public interest, but brought no production orders. So Colonel Lewis sailed over to Europe and formed a company in Belgium. Soon there was a steady demand for the Lewis gun, which was revolutionary in its time. It was light and truly portable, and its 47- or 96-round ammunition drums were easy to store and load, and protected the bullets from dirt.

In August 1914, two British airmen took a Lewis gun aloft with them without authorization. At an altitude of 5,000 ft they emptied a drum at a German aircraft. Although they did not bring it down, this represented the first operational use of the machine-gun in aerial warfare.

The British airmen were not commended for their initiative. On the contrary, instructions were issued that aircraft were not to be so equipped—because of the danger of inciting retaliation on the part of the enemy!

On early reconnaissance sorties, German and Allied pilots sometimes waved to each other if they met in the air. However, when it became clear that air reconnaissance could affect the course of the infantry battles being fought in the trenches, efforts had to be made to bring down enemy aircraft, to prevent their reporting what they had seen.

Before there was any general recognition of the desirability of equipping aircraft with weapons, British and German observers began carrying a revolver, rifle, or even a shotgun, with which to take pot shots at each other. There is just one recorded instance of a German aircraft being downed by shotgun fire from a British observer. A number of weird devices were developed to increase the effectiveness of pistols. The most common of these was a wire cage attached to a .45 automatic pistol to catch ejected cartridge cases and prevent them from striking the pilot or damaging the aircraft. Used in conjunction with an oversize magazine, it enabled up to 20 rounds to be fired in standard pistol fashion.

The need for an automatic weapon, however, was

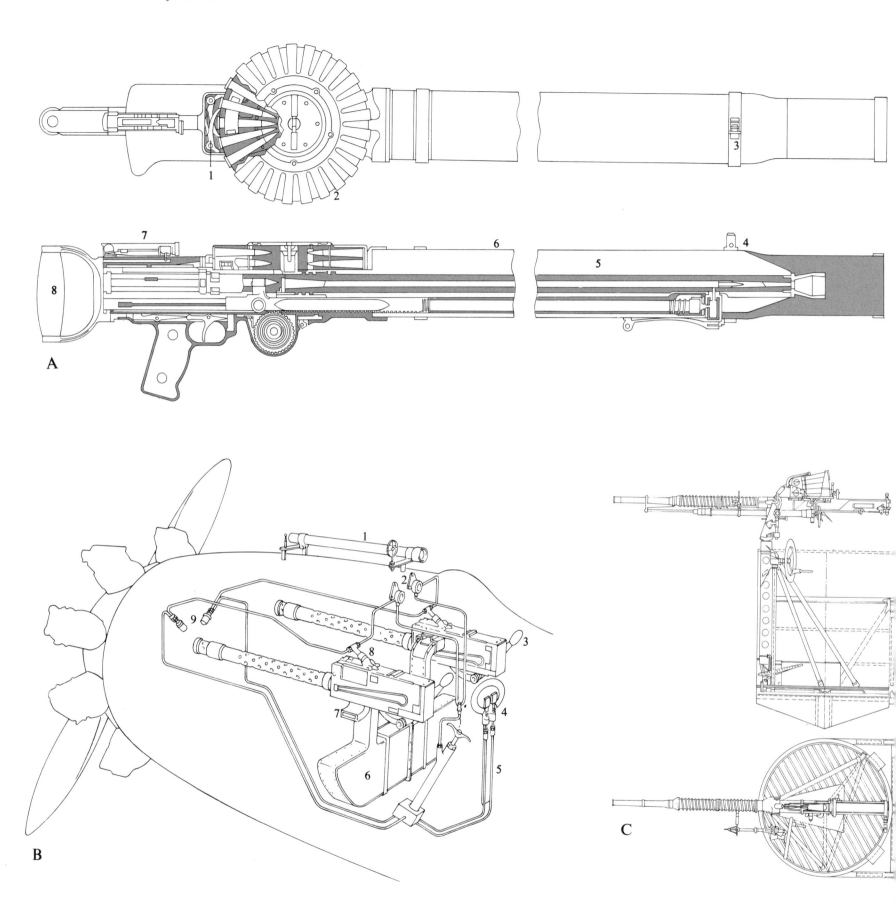

A

B

C

obvious. The Lewis gun had proved a suitable weapon for aerial use; the problem was to find a way of mounting it on types of aeroplane with a performance good enough to enable them to catch enemy aircraft.

The fastest aeroplanes were usually tractor types —that is, with the engine and propeller at the front. Unfortunately, if a machine-gun was mounted in the obvious place, on top of the fuselage in front of the cockpit, the bullets hit the propeller. As a result, machine-guns were first widely used on two-seat pusher aircraft, since this layout gave the gunner the widest field of fire with the minimum of problems. Vickers had exhibited their pusher-type Experimental Fighting Biplane No. 1 (EFB.1) named the Destroyer, at an Aero Show in London as early as 1913. This aircraft had a belt-fed Vickers Maxim machine-gun mounted in the nose, with a field of fire of 60° up and down, and to left and right. From it was developed the famous FB.5 Gunbus, in which the gunner, armed with a Lewis gun on a spigot in the nose, had an even greater field of fire.

The "pusher" Gunbus was by no means the only armed aircraft used in France in the first year of the war. Considerable ingenuity was expended in trying to devise a satisfactory method of mounting and firing machine-guns on the faster "tractor" aircraft. The most successful idea was to mount the gun on the side of the fuselage at an angle, so that it could be fired outside the propeller arc. Guns fitted in this way included even a breech-loading duck-gun firing chain-shot! In July 1915, Captain Lanoe G. Hawker, patrolling in a Bristol Scout armed with a single-shot Martini carbine mounted at an angle, forced down three German two-seaters armed with machine-guns. The feat earned him the Victoria Cross.

Aircraft armed in this manner had to be flown "crabwise" in order to line up with the target, and it is a tribute to the skill of the pilots that quite a few victories were scored with angle-mounted weapons. It was, however, only a temporary expedient. A better solution was to mount the gun above the centre-section of the top wing and fire it by pulling a cable attached to the trigger. This enabled the pilot to point his aircraft directly at the target, as the gun was aligned with a sight mounted in front of the cockpit. The main disadvantage of this method was the difficulty of reloading the guns in flight. Other, upward-firing, gun installations were devised specially for anti-Zeppelin operations.

Explosive darts were also carried for use against Zeppelins. Invented by Engineer Lt. Francis Ranken, they were released three at a time from a 24-round container. Four vanes spread from the tail of each dart, the idea being that they would jam against the envelope of an airship after the head had penetrated it, giving time for the charge to detonate inside.

Somewhat similar were the simple steel darts showered upon the enemy from aircraft. Known as "flechettes", they were about 5 in. long and $\frac{3}{8}$ in. in diameter. Carried in canisters of about 250, they were released over enemy horse and troop concentrations. The darts, of course, did no damage unless a direct hit was scored and they were not widely used. Small bombs were more effective.

Another weapon that was tried out briefly was the stick-stabilised, firework-type rocket, mounted in sets of four on the interplane struts of a fighter and fired electrically. The main targets for such weapons were highly-inflammable balloons and airships, and some successes were achieved before the missiles were rendered obsolete by the development of efficient incendiary ammunition for machine-guns.

The device which made machine-guns on fighter aircraft really effective was the interrupter gear, or gun synchronising gear, which enabled bullets to pass between the blades of a revolving propeller. The original use of interrupter gear by the Fokker Monoplane of the German Air Force put it so far in advance of its contemporaries in 1915 that it almost shot the Allied air forces from the skies over France.

The idea of a synchronising gear, however, was not new. Most of the contestants had experimented with such devices before the war. The original inventor was Franz Schneider, chief designer for the German L.V.G. company, who patented his gear in July 1913. At about the same time one Lt. Poplavko hit upon a similar idea in Russia, as did the Edwards brothers in England, but their designs were simply filed by officials and forgotten.

Schneider had a little more success, in that his gear was fitted eventually to the prototype L.V.G. E.VI two-seat monoplane; but this aircraft crashed in 1915 while on its way to the front for operational testing. Another synchronising gear had been produced pre-war by Raymond Saulnier of the Morane-Saulnier company in France. This worked quite well with the Hotchkiss machine-gun. However, some of the bullets were a little late in firing and so, to prevent the propeller from being damaged, Saulnier fitted a steel deflector plate to each blade to divert bullets that otherwise would have struck them. When the war started Saulnier had to return the gun, which he had borrowed from the government, and work on his device ceased.

The effort was not entirely wasted, for the deflector plates used in the early tests were fitted to a Morane-Saulnier single-seat monoplane by Roland Garros, a famous pre-war sporting pilot. In the spring of 1915, armed with a single machine-gun, he shot down several enemy aircraft whose crews, seeing him flying straight at them, assumed they were safe from attack. Unfortunately, on April 19 Garros suffered engine failure and force-landed behind the German lines. His aircraft was captured before he could set fire to it.

With the secret revealed, the Germans at once ordered Anthony Fokker to copy the idea on his new monoplane. Fokker's engineers suggested that he should resurrect Franz Schneider's idea of an interrupter gear instead. In a few days Fokker and his designers had devised a simple cam-and-pushrod linkage between the oil-pump drive of the Oberursel

A *The Lewis gun, the first aerial machine-gun, was tested in America in 1912 and put in service on British aircraft in September 1914. The gun was light and its 47- or 96-round detachable drums were easier to store and load than ammunition belts, and protected the bullets from dirt. The Lewis gun was a standard aircraft weapon for over a quarter of a century.*
1 Feed pawl
2 Magazine pan
3 Clamp ring
4 Front sight
5 Radiator
6 Radiator casing
7 Back sight
8 Spade handle
(for air use)

B *World War I saw the emergence of twin, fixed forward-firing machine-guns as the standard fighter armament. This typical twin-Vickers installation shows the Constantinesco synchronising gear, ammunition boxes, feed arrangements and gun-sight. Twin guns remained the conventional fighter armament for twenty years.*
1 Sight
2 Air release valves
3 Loading handle
4 Firing levers
5 Control column
6 Ammunition boxes
7 Case chute
8 Trigger motor
9 Generators

C *The Vickers-Armstrongs 37-mm quick-firing gun was developed in 1930, specifically for use on aircraft. Firing shells weighing 1.8 lb, it was accurate up to ranges of 4,500 yards. Continuous pressure on the trigger fired all the rounds in the magazine, but single shots were possible. Two types of ammunition were used—a high-explosive shell fitted with a sensitive fuse ar.d an armour-piercing type.*

871   872   873   875

rotary engine of his M.5K monoplane and the trigger of an ordinary belt-fed Parabellum machine-gun. The linkage pulled the trigger once during every revolution of the propeller, giving an average rate of fire of 400 per minute, the gun firing so that the bullets always passed between the blades.

The first victory with the interrupter gear was gained by Lieut. Max Immelmann on 1 August 1915, when he forced down a British aircraft that was bombing Douai aerodrome. His Fokker E.I monoplane was not very fast, but it was manoeuvrable and its synchronised gun gave it an immense advantage over opposing Allied aircraft. The Fokkers gained air superiority for Germany from the Autumn of 1915 to mid-1916. Immelmann himself became the leading exponent of the new art of air fighting, and went on to claim another 14 victims.

In Britain, Vickers were also experimenting with a mechanical interrupter gear, and in 1916 a few Bristol Fighters fitted with this mechanism were sent to France. Introduction of the Vickers gear led to the adoption of the Vickers machine-gun as the standard forward-firing armament of British fighters, instead of the Lewis gun. The main reason for this was that the Vickers interrupter gear was more compatible with this type of gun and, since the guns were fixed, objections to the Vickers gun on the grounds of its unwieldiness disappeared. Also, pilots preferred the belt-fed Vickers gun, as it did not involve them in the frequent changes of ammunition drum that had been necessary with the Lewis gun.

Not until 1917 did the British develop an interrupter gear superior to that fitted to German aircraft. This new gear was devised by Georges Constantinesco, a Rumanian, who had experimented with the transmission of power through fluids. With the Constantinesco gear, the guns were fired by means of impulses transmitted through a column of liquid contained under pressure in a pipe, in such a manner that no blade of the propeller was hit by bullets. Not only did it overcome the unreliability of mechanical gears; it was also easily adaptable to any type of engine.

The Vickers guns used on aircraft were slightly modified versions of the standard land infantry weapon. One important change was that the rate of fire was increased from about 500 rounds per minute to nearer 900. Later a distintegrating belt was developed for the ammunition. These belts were composed of articulated metal links connected by the rounds themselves. As the cartridges were fired, the cases and links were separately ejected down small chutes. Disintegrating belts are now a standard feature for all aircraft machine-guns.

In spite of the superiority of fixed forward-firing guns for offensive operations, a need remained for movable weapons for defensive purposes, and the Lewis gun was retained as the standard free-mounted gun until the advent of the power-operated gun turret.

The mounting of free guns on aircraft proved a

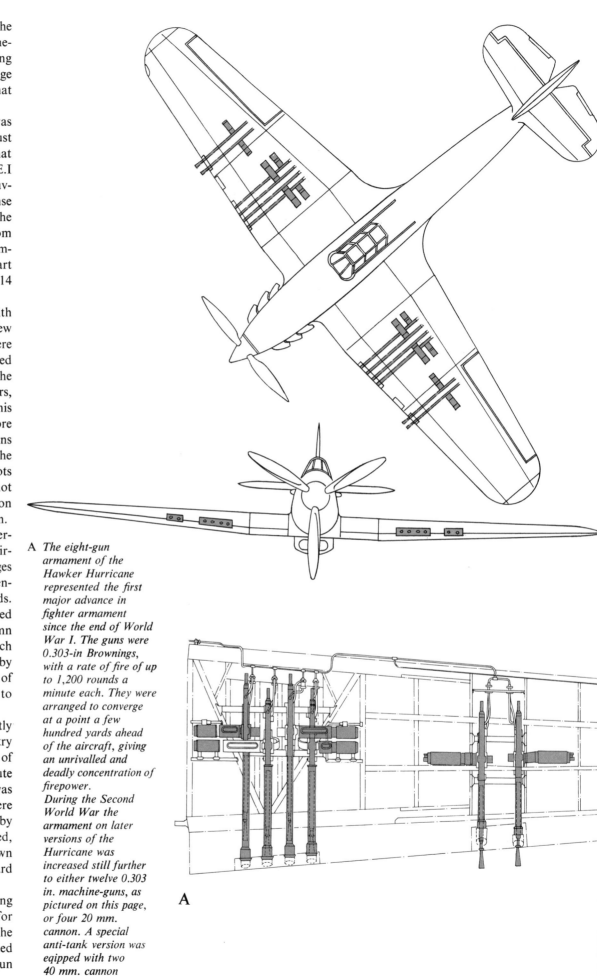

A *The eight-gun armament of the Hawker Hurricane represented the first major advance in fighter armament since the end of World War I. The guns were 0.303-in Brownings, with a rate of fire of up to 1,200 rounds a minute each. They were arranged to converge at a point a few hundred yards ahead of the aircraft, giving an unrivalled and deadly concentration of firepower.*
During the Second World War the armament on later versions of the Hurricane was increased still further to either twelve 0.303 in. machine-guns, as pictured on this page, or four 20 mm. cannon. A special anti-tank version was eqipped with two 40 mm. cannon mounted under the wings.

A

difficult problem, but Warrant Officer F. W. Scarff, of the Royal Naval Air Service developed a simple but efficient mounting which in modified form was used up to and during World War II.

When speeds approached the 200 mph mark, shields were developed to protect the gunner from the strong slipstream and to enable him to operate his weapon more freely and effectively.

The standard fighter armament of twin fixed forward-firing machine-guns persisted throughout the world until the 1930s. Many special gun installations were, however, evolved for particular duties. In addition to the upward-firing anti-Zeppelin installations referred to previously, downward-firing guns had been installed for ground strafing as early as 1917-18.

The standard gun calibre throughout this period was roughly 0.30 in, but heavier-calibre guns were also used. Heavy-calibre weapons fitted during World War I had included a one-pounder Vickers pom-pom gun, employed for night attacks on ground targets, and the French *moteur-canon*. The latter was conceived as early as 1911 by Louis Blériot, who tested a gun that fired through the hollow propeller shaft of an aero-engine, but the idea was not regarded seriously until 1917.

Then, at the suggestion of the French fighter ace Georges Guynemer, Hispano-Suiza mounted a 37-mm Puteaux shell-firing gun between the cylinder blocks of their 220-hp geared V-engine, so that it could be loaded with one hand in flight and fired through the hollow propeller shaft. The *moteur-canon* was installed in the specially-designed Spad 12 and 14 fighters and was used to shoot down several enemy aircraft.

The year 1935 brought the first major advance in fighter armament since World War I. In that year the Hawker Hurricane introduced the then-unrivalled fire-power of eight machine-guns. As they were mounted in the wings, outside the propeller disc, no interrupter gear was needed and the guns could develop their maximum rate of fire. By 1940 the Hurricane offered an ever-heavier alternative armament of four 20-mm cannon. Later Hurricanes were fitted with two 40-mm guns under the wings, which were used with great effect for destroying tanks during the North African campaigns. Bombs and air-to-surface rockets were also added to the Hurricanes' armament.

The deadly effectiveness of rockets for ground attack duties was demonstrated to the full by the later Hawker Typhoon and by the Russian Il-2 *Stormovik*. Since then, fighter armament has evolved to 30-mm cannon, rotating-barrel Vulcan guns with a rate of fire of 6,000 rounds a minute and nuclear-armed missiles.

Similar dramatic progress has taken place in the armament of bomber aircraft. Bombs have developed from small hand-held weapons dropped over the side of aircraft in World War I to the mighty 22,000-lb Grand Slam "earthquake" bombs dropped by Lancasters on Germany in World War II and the incomparably more terrible atomic bombs dropped by B-29s on Japan.

Bomber defensive armament has also increased dramatically. Initially guns were installed to protect bombers from attack from all quarters until they resembled flying porcupines. Increased speeds led to the development of power operated turrets, located in the nose and tail, and sometimes above and below the fuselage.

The advent of jet propulsion, which increased bomber speeds to nearly that of fighters, diminished the need for defensive gun armament, although the General Dynamics B-58 Hustler still carries a sting in its tail in the form of a 20-mm multi-barrel cannon. Today, however, a bomber's defence is usually entrusted to complex electronic equipment which generates signals designed to confuse and upset the intricate guidance of attacking missiles.

The former gunners have become electronic-countermeasure engineers, and the H-version of the eight-engined B-52 even carries two electronics-countermeasure officers on board, one for the handling of active and the other for passive countermeasure actions. This version was developed especially to carry cruise missiles, but the original task for the strategic, long-range B-52 was deep penetration into hostile territory for mass carpet bombing with conventional ironbombs.

The British V-bombers (Valiant, Victor and Vulcan) represented another approach to the nuclear aerial warfare at the end of the 1950s. With transonic capability, the V-bombers would fly into enemy airspace at high speed and high altitudes, armed with bombs equipped with tactical nuclear warheads.

At the same time, guided air-to-air missiles began to supplant the built-in heavy-calibre guns in the new fighter designs. Such was the overconfidence in missiles that the F-4 Phantom originally came off the drawing-board without any fixed gun armament at all. Low-level dog-fight tactics were considered a thing of the past. The new role of the fighter depended on early warning of hostile actions, quick climb and interception at high altitudes, where enemy bombers and attack planes were to be shot down by infra-red heat-seeking missiles. This scenario of future aerial warfare was drawn by desk-tacticians, but the reality proved to be totally different. In Vietnam most of the air combat was fought at extremely low levels, and the American losses of fighters were initially much higher than could be accepted. The lack of dog-fight gunnery training was disastrous. The old art of air-to-air gunfighting was hastily brought into honour again, and special gunnery training units were established, the best-known being the US Navy Weapons Training School "Top Gun" at Miramar NAS.

Two conflicts during the 1980s have had vast impact on the development of aerial-weapon technique and tactics. In the air war over the Bekaa valley the use of RPVs, remotely piloted reconnaissance aircraft, proved to be an invaluable source of information about the positions of enemy anti-aircraft installations. The next step was to jam all radar and radio communication so that the targets could be attacked without any interception from the enemy. The fighting has been described as the first electronic air battle, where a massive effort of sophisticated countermeasures had a decisive effect on the outcome.

From the Falklands conflict many conclusions were drawn; most important was the need for airborne early warning and command. The British suffered heavy losses in men and ships due to lack of information about low-flying attacking aircraft. The Argentine Super Étendards could launch their guided sea-skimming anti-ship Exocet missiles from well out of sight and return safely to their bases on the mainland. On the other hand, when Argentina's Mirages and A-4 Skyhawks were encountered by the British Harriers, which were launched from aircraft carriers, the outcome was mostly in favour of the Harrier. The VTOL fighter proved itself to be a formidable weapon, even if it was inferior to its opponent in speed, and war historians do not hesitate to give the Harrier's multi-role capability much of the credit for the British retake of the Falklands.

The airborne weapons that are being developed for the next century will be integrated in complex systems that will enable much of the fighting to be carried out at long range. Modern fighter radars can "see" aircraft at distances of many miles, pick out a succession of individual targets, track these while scanning for others, validate a hostile target, work out tracks and future positions, assign a sequence of priorities, and get everything ready for the pilot to engage all targets with a succession of air-to-air missiles. The US Navy F-14 Tomcat can have six of its Phoenix radar-guided missiles in the air at once, each heading toward a different preselected target at distances up to 100 miles. At high altitudes, the speed can reach Mach 5 and the Phoenix is guided to the target by an active radar seeker in the nose. The missile even has a built-in electronic countermeasure device which makes it quite resistant to jamming. But the active radar seeker is the Achilles' heel of the radar-guided missile, as the radar signals disclose the missile, thus making it vulnerable to counteractions. Therefore, new types of AAMs that are being developed act in two steps after firing. The American AMRAAM (Advanced Medium-Range Air-to-Air Missile) in the first phase leaves the fighter as a "dumb" missile that streaks away to a point based on steering data fed into the missile's system by the launching aircraft's radar calculations. When it nears the predicted target position, the missile goes active and switches on its own radar seeker, locks on the target and steers to it. The moment to go active is of crucial importance. Too early means that the enemy is warned and has time to surround himself with massive jamming and decoys. Too late means that the missile's system cannot process the seeker's information in time to pass within lethal radius of the target.

The infrared seeker, which is sensitive to heat and homes for the target's hot engine exhaust, is the most widely used technique for target-seeking air-to-

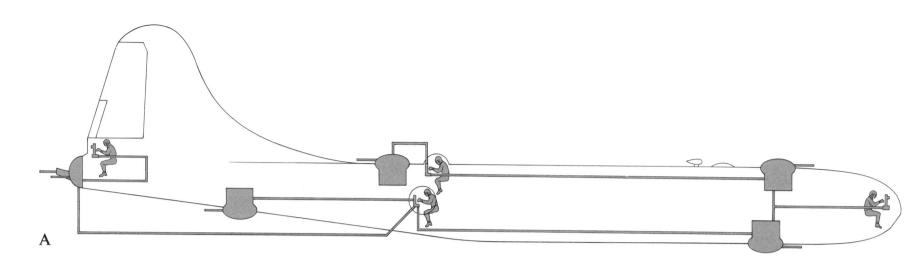

A

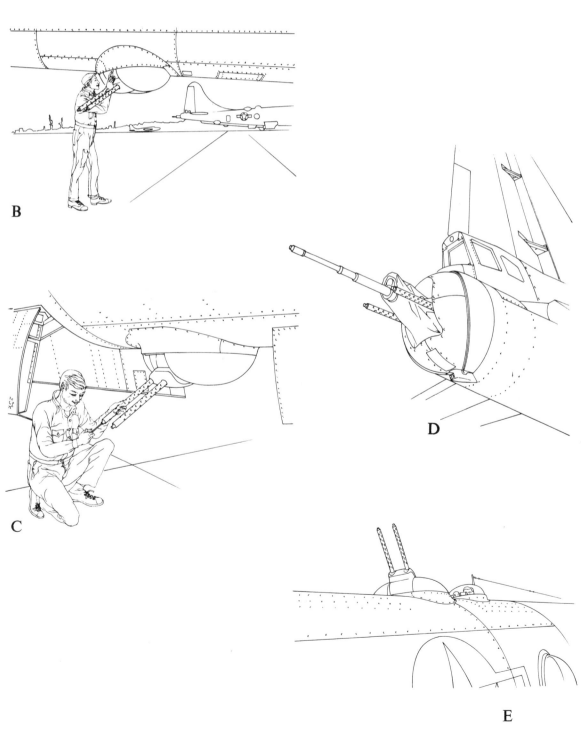

B

C

D

E

A *The Boeing B-29
Superfortress was the
best-protected bomber
of World War II. Its
defensive armament
comprised twelve .50-in
calibre machine-guns
mounted in five low-
silhouette power-
operated turrets. The
guns were fired by
remote control and
were linked to a unique
central gunfire control
system. Gunners had
primary control of
certain turrets and
secondary control of
others. The gunner with
primary control of a
turret had first call on
its services, which he
could relinquish to
gunners with secondary
control. An intricate
signal system between
the gunners permitted
this exchange of turret
control. The idea was
that if a gunner found
no target within his
range of vision in
combat, he could,
under certain
conditions, pass control
of his turret to another
gunner, who could use
it. Thus, in the event of a
concentrated attack on
the bomber from one
direction, the gunner
whose vision covered
the point of attack
could borrow a second
turret to double his
firepower. The
bombardier in the nose,
with the widest range
of vision, usually
commanded both
forward turrets, either
separately or together.*

B *Lower rear turret. The
low silhouette of the
electrically-operated
remotely-controlled
turrets is evident.*

C *The lower forward
turret was located
under the B-29's front
fuselage.*

D *The tail turret was the
post for the lonely
"Tail-end Charlie"
who, once he was in,
had no access to other
parts of the aircraft.*

E *The rear upper turret
was located forward of
the tail-fin.
Immediately in front
was the upper gunner's
aiming point. The
gunner sighted
directly on an attacking
fighter, pin-pointing it
by a tiny spot of light,
and enclosing it in a
tiny circle of light.
An automatic
computer adjusted for
wind velocity, gravity,
lead correction
(depending upon the
speed of the enemy
fighter) and parallax—
that is, correction for
the distance between
the gunner and guns.*

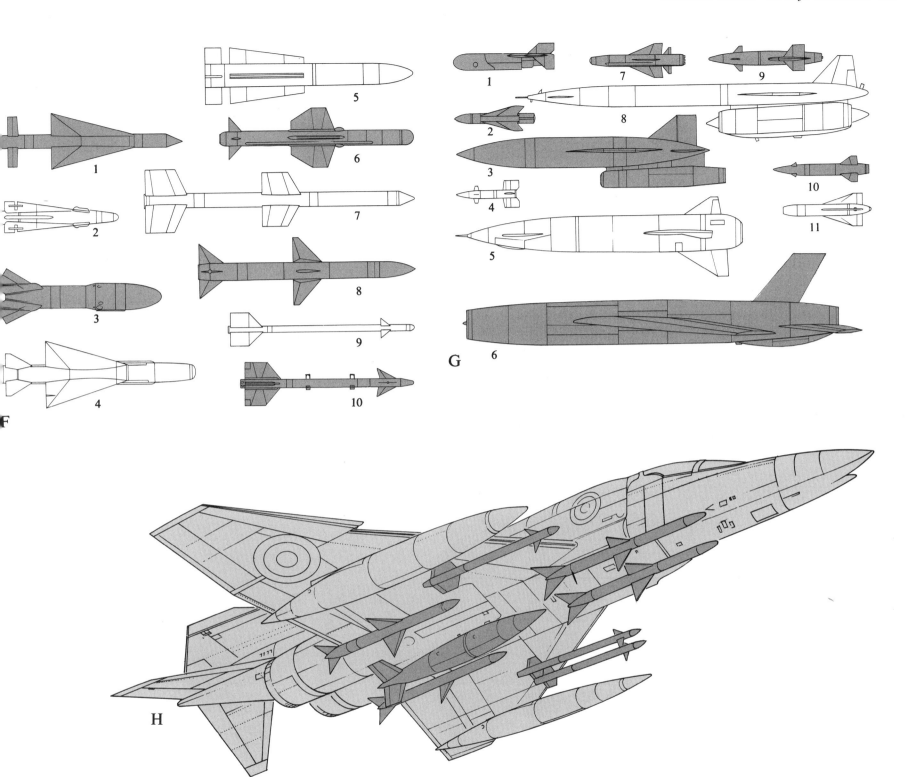

F *Air-to-air missiles.*
1 "Ash"
2 Nuclear Falcon
3 Genie
4 Matra R530
5 Phoenix
6 Red Top
7 "Awl"
8 Sparrow
9 Atoll
10 Sidewinder 1C

G *Air-to-surface missiles.*
1 Quail
2 Nord AS 20
3 "Kipper"
4 "Harp"
5 Blue Steel
6 "Kangaroo"
7 Martel
8 Hound Dog
9 Robot 04
10 Bullpup
11 Walleye

H *Warloads of up to 16,000 lb can be carried by the Phantom. Standard armament is six Sparrow III, or four Sparrow III and four Sidewinder air-to-air missiles on four under-fuselage and four under-wing mountings, plus bombs. Typical weapon*

*loads include:
18 × 750 lb bombs,
15 × 680 lb mines,
11 × 1,000 lb bombs,
7 smoke bombs,
11 × 150 gal Napalm
bombs, 4 Bullpup
air-to-surface missiles
and 15 packs of
air-to-surface rockets.*

air missiles. The American Sidewinder has been in production since 1955 and is still leading the market in its latest version, AIM-9L/M. One of the gigantic advantages of an IR missile is that it is passive: it emits no radiation of its own, so the target is unikely to be aware of its approach. If detected, the IR seeker can be fooled by high-intensive flares, launched from the target aircraft. The drawback of the IR guided missile is shorter range compared to radar AAMs, and the fact that clouds put the IR seeker out of action. So, in the future, we can probably expect air-to-air weapons to be equipped with a combination of many guiding systems.

1. **Heinkel He 111H**
   Wing span: 74 ft 1½ in
   Length: 53 ft 9½ in
   Max take-off
     weight: 27,400 lb

2. **Messerschmitt Bf 109E**
   Wing span: 32 ft 4½ in
   Length: 28 ft 6 in
   Weight: 5,523 lb
   Max speed: 354 mph
   Service
     ceiling: 36,000 ft
   Range: 412 miles

3. **Junkers Ju 88A1**
   Wing span: 60 ft 3½ in
   Length: 47 ft 1 in
   Max take-off
     weight: 27,060 lb

4. **Supermarine Spitfire I**
   Wing span: 36 ft 10 in
   Length: 29 ft 11 in
   Weight: 5,784 lb
   Max speed: 365 mph
   Service
     ceiling: 34,000 ft
   Range: 575 miles

5. **Hawker Hurricane I**
   Wing span: 40 ft 0 in
   Length: 31 ft 11 in
   Weight: 6,600 lb
   Max speed: 324 mph
   Service
     ceiling: 34,000 ft
   Range: 900 miles

901   902   903   904   905

BATTLE OF BRITAIN ADVERSARIES

In the summer of 1940 the air forces of Britain and Germany engaged in what has become known as the Battle of Britain. Although winning this vital Battle did not at once gain Britain victory, if the RAF had been beaten she would have lost the war.

At the start of the Battle, Royal Air Force Fighter Command had only 704 aircraft with which to meet the onslaught of 3,500 aircraft in three Luftwaffe fleets. Of the 704 aircraft, 620 were Hurricane and Spitfire fighters. Individually, these were not markedly superior to the opposing Me 109s, but they formed a most formidable team. They were also helped immeasurably by the newly developed radar network which, by detecting every enemy attack, avoided the need for maintaining standing patrols and enabled the British fighters to engage the enemy with the minimum consumption of fuel.

When the Battle was over, the Luftwaffe had lost 1,733 aircraft, together with their crews. Cost to the Royal Air Force was 915 aircraft, from which many pilots parachuted to safety and returned to their squadrons.

The main weight of the German attack was provided by Heinkel He 111, Dornier Do 17 and Junkers Ju 88 bombers, protected by Messerschmitt Bf 109 fighters. A few attacks were mounted by Junkers Ju 87 dive-bombers, but these aircraft were easy prey to the defending Spitfires and Hurricanes.

1. Heinkel He 111H. One of the main bombers in the Battle of Britain, the He 111H was a five-seat medium bomber, powered by two 1,350-hp Junkers Jumo 211 in-line liquid-cooled engines. Defensive armament comprised one 20-mm cannon and six machine-guns. Up to 4,400 lb of bombs could be carried. Cruising speed was 180 mph and the normal range 1,280 miles.

2. Messerschmitt Bf 109E. The prototype of this single-seat fighter, which first flew in September 1935, was powered by a 695-hp Rolls-Royce Kestrel engine. This was replaced by a 700-hp Junkers Jumo engine in the version which showed its paces with the German Condor Legion in the Spanish Civil War. In turn this engine was replaced by a 1,100-hp Daimler-Benz in-line liquid-cooled engine, with which the Bf 109E had become the standard Luftwaffe fighter by the time of the Battle of Britain. Armament consisted of one 20-mm cannon firing through the propeller spinner, two 20-mm cannon in the wings and two machine-guns above the engine.

3. Junkers Ju 88A1. One of the most effective German bombers, the Ju 88 had a crew of four grouped closely together in the portion of the fuselage ahead of the wing. Powered by two 1,410-hp Junkers Jumo in-line liquid-cooled engines, the Ju 88 could carry its maximum bomb load of 5,720 lb for 832 miles, cruising at 250 mph. Alternatively, nearly 2,000 lb of bombs could be carried for over 3,000 miles. Defensive armament consisted of either three or six machine guns.

4. Supermarine Spitfire I. When the Battle of Britain began on August 12, 1940, nineteen squadrons of Spitfire I's were available to the British defence forces. Powered by a 1,030-hp Rolls-Royce Merlin III in-line liquid-cooled engine, the Spitfire I had a top speed of 365 mph, making it slightly faster than its main adversary, the Messerschmitt Bf 109E. The Spitfire was also much more manoeuvrable, particularly in the vital turn; but the German fighter could outclimb and outdive it, enjoyed altitude superiority and had longer-ranging armament. The standard armament of the Spitfire I was eight machine-guns.

5. Hawker Hurricane I. Thirty-two squadrons of Hurricane I's formed the backbone of the British defence system during the Battle of Britain. Compared with the Bf 109E, the Hurricane was inferior in climb and speed at all altitudes, but was much more manoeuvrable, out-turning the German fighter with ease. It was also an excellent gun platform and was capable of withstanding extensive battle damage. Because of its overall inferior performance, the Hurricane was directed mainly against the Luftwaffe's bombers, while the faster Spitfire engaged the escorting fighters, and for this task the Hurricane was unsurpassed. The net result was that Hurricanes destroyed more aircraft during the Battle than all other fighters and anti-aircraft guns added together.

The Hurricane I was powered by a 1,030-hp Rolls-Royce Merlin III in-line liquid-cooled engine. Standard armament was eight machine-guns.

In addition to these aircraft, many other types were used in smaller numbers and special mention should be made of the Dornier Do 17. Popularly known as the "Flying Pencil", because of its long slim fuselage, the Do 17 was a twin-engined medium bomber which during 1937-41 formed the backbone of the German bomber fleets. It was powered usually by two 1,000-hp Bramo Fafnir radial air-cooled engines. A crew of four was carried and the defensive armament consisted of six machine-guns, or five machine-guns and one 20-mm cannon. Maximum bomb load was 2,200 lb, maximum range 745 miles and cruising speed 186 mph at 14,000 ft.

SYSTEMS

The structure, although costing a great deal of money and necessitating massive test rigs, is only a relatively small part of a complete aircraft. Much more money and much more development time are absorbed by the various aircraft systems. These to an aeroplane are what your breathing, blood-circulating, digesting and nervous systems are to you; while an aircraft's structure is the equivalent of your skeleton.

The most important system of an aircraft is the one by which it is controlled and manoeuvred—the flying control system—and this has been covered in a previous chapter. The major remaining systems are those for fuel, hydraulics, electrics, pneumatics, cabin air conditioning, oxygen and protection against icing in flight.

Common to all powered aircraft is a fuel system which stores the fuel and feeds it to engines in the required quantity. It can vary from a simple system in which the fuel is stored in a single tank above the engine, into which it feeds through an on/off valve, to a complex pressure-refuelling system storing thousands of gallons in integral tanks with sub-merged fuel pumps and sophisticated venting, jettison and contents indicating systems.

Fuel represents a high proportion of the weight of an aircraft and care must be taken to ensure that the centre of gravity and lateral balance are not affected adversely as fuel is consumed. For example, on a swept-wing aircraft if the outer tanks in the wings were used first, the aircraft would become progressively nose-heavy. The problem is often solved by using devices that draw off, or meter, fuel in the required proportions from all tanks simultaneously.

Fuel tanks must be vented to atmosphere, to prevent pressure from building up during rapid changes of height or during refuelling, which could cause the tank to rupture.

It is, of course, vital for the pilot to know just how much fuel he has, and the contents of the tanks are measured either by a simple float-operated gauge similar to those used on motor cars, or by means of what are known as capacitance gauges. These operate by detecting the change in electrical capacity between two plates as the fuel between them is replaced by air. To permit readings on the ground, simple motor-car oil type "dipsticks" are provided. To make this easier on large aircraft, inverted "dip-

sticks" have been developed which can be with-drawn from the underside of a tank without any loss or dripping of fuel.

As aircraft fuels are not all of the same density, special devices have been developed which measure the mass-flow, as distinct from the volume-flow. This information is needed for accurate flight-planning and cruise control.

* * * * *

At heights much above 15,000 ft the atmosphere is too thin and too cold for normal breathing; aircraft operating above this height require to be pressurized, and the temperature and humidity of the cabin air has to be controlled. It is normal to provide such a degree of pressurization that the cabin pressure is equivalent to an altitude of 8,000 ft when an aircraft is actually flying at 40,000 ft.

Fresh air can be provided by two basic means—from a mechanically-driven compressor, or by tapping air from the compressor of a jet engine. Air tapped from an engine is hot and has to be cooled before it can enter the cabin, and complex heat-exchangers and refrigeration systems have been developed for this purpose.

Operation at high altitudes makes necessary another system—for supplying oxygen. Surprisingly, perhaps, this was one of the earliest systems to be installed in aircraft, as the need for breathing equipment at high altitudes had been demonstrated in balloon experiments long before the first aero-plane flights. One of the most interesting of these experiments was made at Wolverhampton, England, in September 1862, when James Glaisher, a meteoro-logist, and Henry Coxwell, a professional balloonist, went up far higher than they had intended.

At about 17,000 ft Coxwell, who was controlling the balloon, became short of breath and turned purple in the face. When they reached 21,000 ft, even the slightest exertion caused Glaisher to experience severe difficulty in breathing. Soon afterwards his vision became so impaired that he was unable to observe his instruments. Then his arm and neck became paralysed and, finally, he fainted.

Meanwhile Coxwell, not having succumbed to the same extent as his companion, had been trying to get the balloon under control. The valve line seems to have got out of reach, probably due to the en-velope expanding unevenly. In an effort to reach it

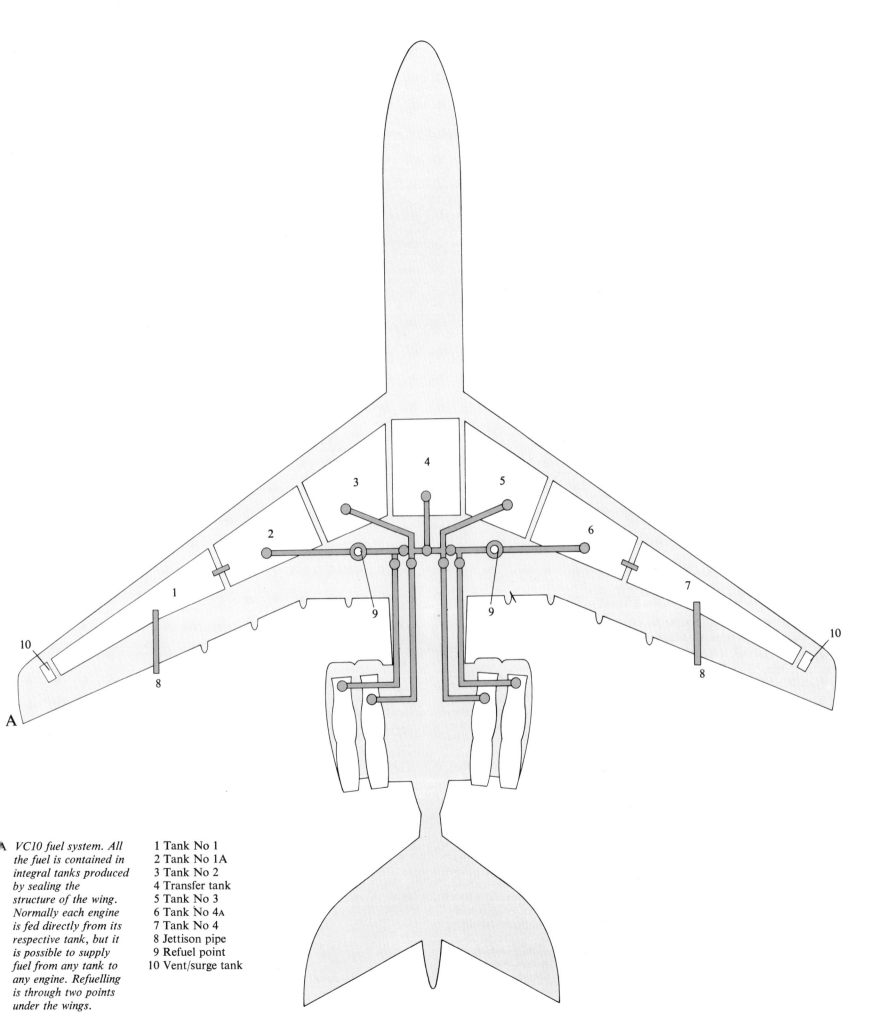

A *VC10 fuel system. All the fuel is contained in integral tanks produced by sealing the structure of the wing. Normally each engine is fed directly from its respective tank, but it is possible to supply fuel from any tank to any engine. Refuelling is through two points under the wings.*

1 Tank No 1
2 Tank No 1A
3 Tank No 2
4 Transfer tank
5 Tank No 3
6 Tank No 4A
7 Tank No 4
8 Jettison pipe
9 Refuel point
10 Vent/surge tank

and be able to release gas from the balloon, Coxwell climbed up the rigging above the basket. Unfortunately, he had removed his gloves and, grasping the ice-cold hoop, his hands became numb. So, to operate the valve he had to pull on the cord with his teeth.

His action undoubtedly saved the lives of both men. After the flight Glaisher claimed that he had become unconscious at 29,000 ft and that the balloon rose to at least 36,000 ft before Coxwell opened the valve. This was an exaggeration, for if they had reached this altitude they would not have survived to tell the tale. It is thought that the two aeronauts might have reached 24,000 ft, or a little higher.

At the time, the cause of the men's disability was attributed to fatigue, the reduced atmospheric pressure and the extreme cold; no one had sufficient understanding of the human respiratory system to suggest the real cause or to provide a remedy.

Not until 1869, when Paul Bert, a Frenchman, made a thorough investigation of the cause of "mountain sickness", was it appreciated that the ill effects of exposure to high-altitude conditions resulted from reduction in the oxygen content of the air rather than from reduction in the pressure of the atmosphere.

The 1914-18 War hastened the need for oxygen in high-flying aircraft and by the time of the Armistice no fewer than four main types of oxygen equipment had been developed, the basic elements of which are still in use today. The Germans used liquid oxygen; but although equivalent apparatus was developed in England, British aircraft utilised gaseous systems.

In the years following the war, the performance of aircraft increased steadily until, at heights of around 38,000 ft, pilots began to suffer mental and physical decline in spite of the availability of an oxygen breathing system. The need to supply pressure as well was quickly realised, leading to the development of pressure cabins. The first successful pressurized aircraft was the Lockheed XC-35, built in 1937. It was of great significance, for it paved the way for pressurized airliners as we know them today. The first pressurized airliner, the Boeing 307, went into service in 1939.

On high-flying airliners, pressurization normally provides adequate passenger comfort without need for a separate breathing system; but the possibility of decompression occurring above 25,000 ft must be considered. Exposure at such altitudes, as already described, could be serious unless oxygen was available immediately for passengers. A conventional approach to the problem would be inadequate, as it would take far too long for cabin attendants to help a hundred or more passengers to don normal oxygen masks.

The problem has been overcome by the installation of masks which, in the event of pressure cabin failure, are ejected automatically in front of each passenger. The masks are normally housed unobtrusively either in the rack or in the back of the

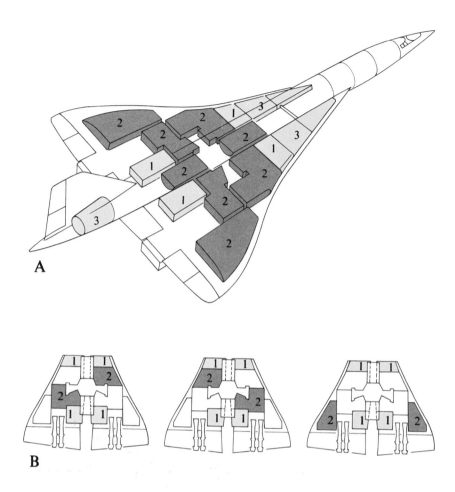

A *Concorde fuel system. This is used not only to supply fuel to the engines, but also for cooling purposes and for trimming the aircraft. Each group of main tanks has two collector tanks, kept full at all times, which feed the engines on their side of the aircraft. The fuel is used as a "heat sink" to cool the air-conditioning system, the hydraulic system, electrical equipment and engine lubricating oil.*
Special tanks at each end of the fuselage are provided for trimming purposes (i.e., to maintain the correct relationship between the centre of gravity and the aerodynamic centre of pressure). *Fuel is transferred rearward while the aircraft accelerates to supersonic flight and forward during the return to subsonic flight.*
1 Collector tanks
2 Main tanks
3 Trim tanks

B *The centre of gravity is maintained within close limits during supersonic flight by an arrangement whereby the fuel is pumped to small collector tanks from diagonally-opposite main tanks as shown above. The main outer wing tanks feed directly into the aft collector tanks on completion of the climb. They are kept full during the climb to reduce wing bending moments.*
1 Collector tanks
2 Main tanks

C *Hawker Siddeley Trident hydraulic installation. This consists of three separate systems each having its own engine-driven pump, reservoir, pipelines, valves and operating jacks. All three system power the flying controls. Normally, all three systems work in unison and no hurried change-over is necessary in the event of failure. If one system fails, the two remaining systems automatically provide full control. Should a further system have to be shut down for any reason, the one*

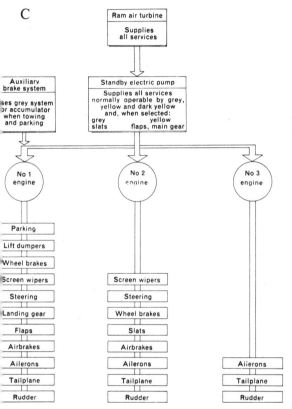

C

Ram air turbine

Supplies
all services

Auxiliary
brake system

ses grey system
or accumulator
when towing
and parking

Standby electric pump

Supplies all services
normally operable by grey,
yellow and dark yellow
and, when selected:
grey   yellow
slats   flaps, main gear

No 1
engine

No 2
engine

No 3
engine

No 1 engine	No 2 engine	No 3 engine
Parking		
Lift dumpers		
Wheel brakes		
Screen wipers	Screen wipers	
Steering	Steering	
Landing gear	Wheel brakes	
Flaps	Slats	
Airbrakes	Airbrakes	
Ailerons	Ailerons	Ailerons
Tailplane	Tailplane	Tailplane
Rudder	Rudder	Rudder

remaining system provides sufficient power for normal control operation, including landing. Two electrically-driven pumps provide a standby source of power should one of the main pumps fail. In the unlikely event of stoppage of all engines, an air-driven turbine drops out into the airstream to provide an emergency source of power.

seats and drop out in an emergency. The act of pulling the mask on to the face switches on a supply of oxygen sufficient to last until the aircraft has descended to a lower altitude.

It is particularly important that the flight crew should be able to obtain oxygen rapidly in an emergency, and the masks on the flight deck are usually of a special type which can be ejected and fitted extremely quickly. However, even this might not be good enough under all circumstances and, to ensure that control would be maintained even during this brief period, some airlines have a rule that at least one flight crew member is "on oxygen" all the time.

A system found on all but a few simple aircraft engaged on limited daytime operations is that which supplies electricity. There are many services, such as lighting, radio and weather radar, which can be operated only by electricity; and others, such as air-liner galleys and windscreen anti-icing, for which electricity is the most practicable or economical source of power. Electricity has the added advantages of providing precise control, and of being clean, flexible and versatile. Both dc (direct current) and ac (alternating current) systems are available and many considerations affect the choice.

For simple aircraft and some large aircraft, a dc system has always been regarded as the conventional choice. However, dc generators tend to wear rapidly when operated above 25,000 ft. As most modern military and civil transport aircraft operate above this height, the trend is towards ac systems with brushless alternators. Many large aircraft have more than one system; for example, a secondary system may serve the radio and radar equipment or be used for de-icing.

A generator is often mounted on each engine of a multi-engined type, to provide multiple sources of

supply. Great care is taken to ensure that in the event of malfunction the circuits "fail safe", and that no single fault affects more than one electrical channel.

Another system that is common to all but the simplest aircraft is the hydraulic system. Hydraulic power is used to operate flying controls, landing gear retraction, flaps, air-brakes, wheel-brakes, nose-wheel steering, bomb doors, air-stairs, wing folding and cargo doors. The wide-spread use of hydraulic power is due mainly to the advantages inherent in using a relatively incompressible fluid. Fluid flow can be controlled accurately and the operating jacks are locked by the fluid trapped in the operating cylinder.

On light aircraft, hydraulic power can be provided by a simple hand-pump, but usually power is supplied by engine-driven pumps, backed up, if necessary, by electrically-driven pumps. Where failure of the hydraulic system would be serious, as in the case of powered flying controls, it is usual to operate them from two, or even three, independent hydraulic systems.

In normal duplicated or triplicated systems, if one fails, a physical change-over is necessary to bring in one of the other serviceable systems. This takes time and could occur at an awkward moment. To prevent problems, duplex and triplex systems have been evolved. On such installations all the individual systems work in unison all the time. Such systems are "fail-safe" since, as all the systems are working continuously, no hurried change-over is necessary if one system fails, and power is maintained uninterrupted. Triplex systems are used only for flying controls, and are such that in the unlikely event of two systems failing the remaining system can still provide sufficient power to enable the aircraft to complete its flight and make a safe landing.

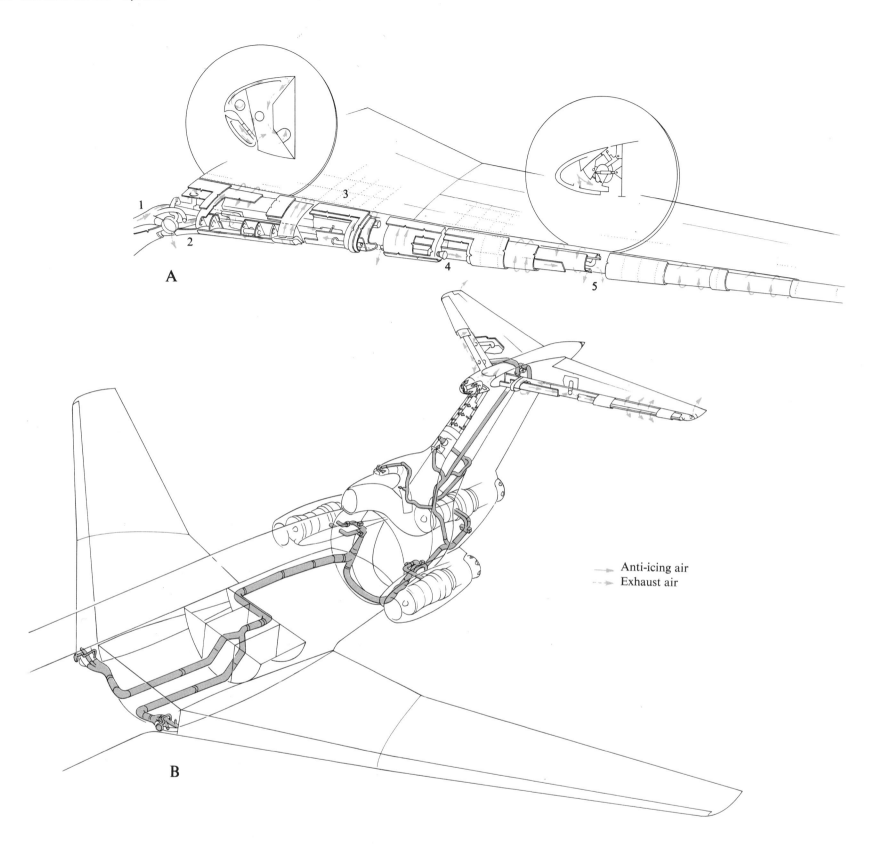

A

B

→ Anti-icing air
--→ Exhaust air

A Anti-icing system.
& On turbine-engined
B aircraft it is convenient
to use hot air tapped
from the engines
for anti-icing of
the wings, tail and
power plant. Flight
deck windows and
pitot heads are
usually warmed
electrically.

In place of warm air,
de-icing fluid systems
can be used for
protecting leading-
edges. In such
systems de-icing fluid
is fed under pressure
through porous panels
secured along the
leading edges of the
wing. The fluid gives
a powerful de-icing

capability while
preserving aerodynamic
efficiency. Another
system utilises rubber
boots, which can be
inflated and deflated
successively to break
off any ice that may
have formed on the
leading-edge to which
they are attached.

1 Supply to droop
leading edge
2 Supply to fixed
leading edge
3 Rotating joint
4 Connection betwee
inner and outer
droop leading edge
5 Exhaust from
droop leading edge

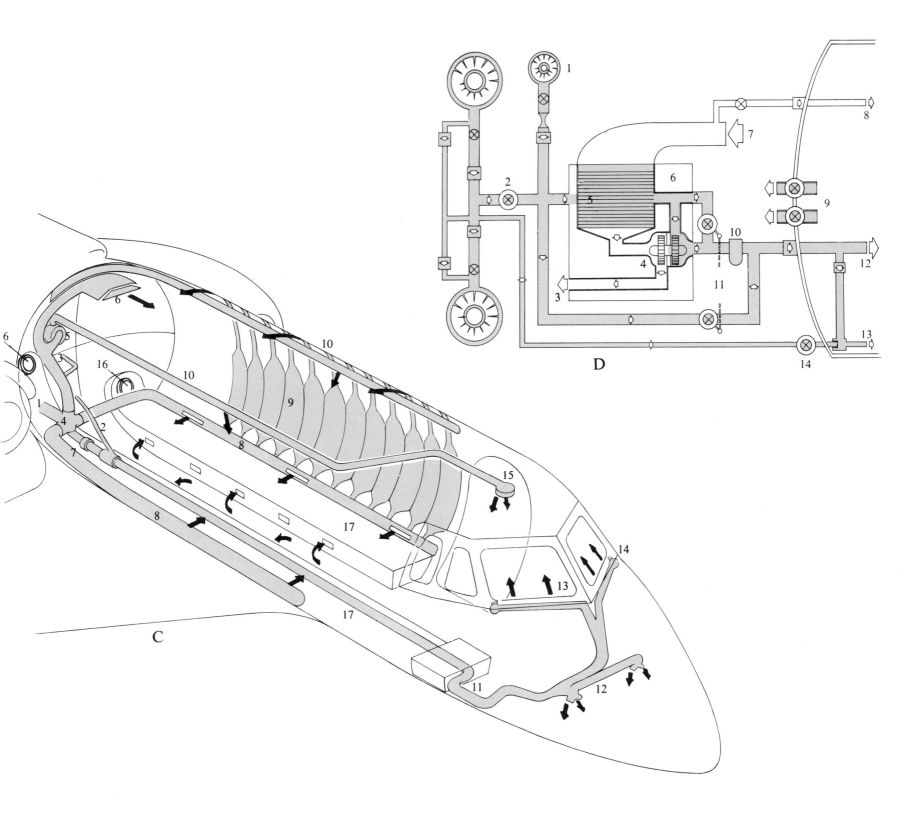

C

1 Cabin air supply
2 Flight deck auxiliary
   heat supply
3 Fresh air supply
4 Two-way valve
5 Fan
6 Flood-flow outlet
7 Check valve
8 Floor level duct
9 Wall bags (both sides)

10 Personal air ducts
11 Flight deck duct
12 Crew's foot warmers
13 Windshield demister
14 Fan-augmented air
   outlet
15 Flight deck ventilator
16 Discharge valves
17 Underfloor discharge
   ducts

D

1 A.P.U.
2 Flow control valve
3 Cooling air exhaust
4 Cooling turbine
5 Heat exchanger
6 Refrigeration unit
7 Cooling ram
   air intake
8 Fresh air supply
9 Dual discharge valves

10 Water separator
11 Interconnected cabin
   air temperature
   control valves
12 Cabin air supply
13 Flight deck
   auxiliary heat supply
14 Control valve

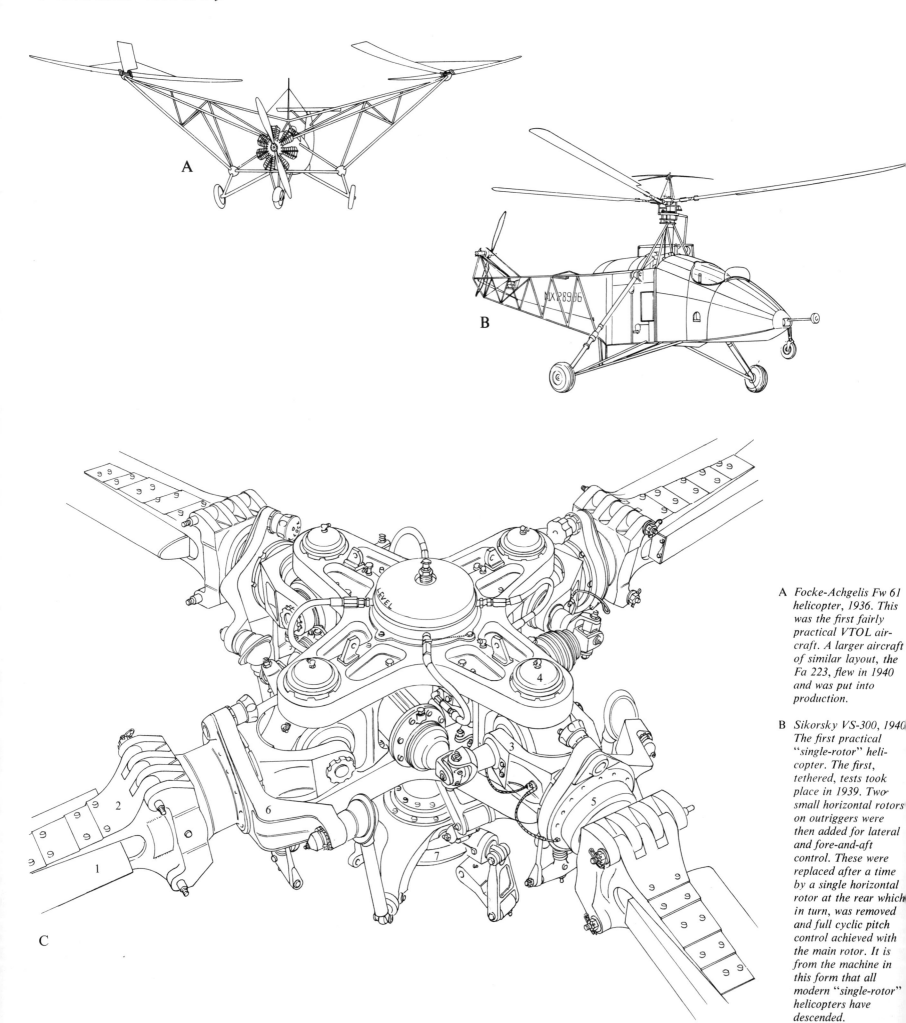

A Focke-Achgelis Fw 61
   helicopter, 1936. This
   was the first fairly
   practical VTOL air-
   craft. A larger aircraft
   of similar layout, the
   Fa 223, flew in 1940
   and was put into
   production.

B Sikorsky VS-300, 1940
   The first practical
   "single-rotor" heli-
   copter. The first,
   tethered, tests took
   place in 1939. Two
   small horizontal rotors
   on outriggers were
   then added for lateral
   and fore-and-aft
   control. These were
   replaced after a time
   by a single horizontal
   rotor at the rear which
   in turn, was removed
   and full cyclic pitch
   control achieved with
   the main rotor. It is
   from the machine in
   this form that all
   modern "single-rotor"
   helicopters have
   descended.

VTOL AIRCRAFT

Heart of the helicopter is the rotor head. Each blade is usually secured to the head by a special hinged fitting which allows the blade freedom of movement about three axes. The fitting embodies a "flapping" hinge, allowing the blade to rise and fall; a "drag" hinge on which it can move backward and forward slightly; and a "pitch change" hinge about which the blade's angle of incidence can be varied. Since the blades are rotating and the aircraft fuselage is stationary, cockpit control lever movements are transferred to the rotor head through what is called a swashplate, the upper part of which revolves with the blades while the lower part remains stationary relative to the fuselage. The swashplate is mounted on universal bearings, so that it can be tilted in any direction. The swashplate can be moved bodily up and down, changing the pitch of all the blades equally and simultaneously (i.e.

collectively), or it can be tilted so that the pitch of the blades is altered unequally as they revolve. In the latter case, as each blade approaches a certain position in its cycle of rotation, its pitch is decreased causing it to fly low. As it continues round to the other side, the swashplate tilt increases the pitch, causing the blade to fly high, and so on. This is known as cyclic-pitch change, and the effect is to tilt the whole "rotor disc" to the same angle as that of the swashplate. This causes the helicopter to move in the direction of the tilt. The rotor head is thus an exceptionally complicated mechanism and the loads imposed by the blades, being both high and of a varying nature, present severe fatigue problems.

1 Rotor blade
2 Blade root fitting
3 Flapping hinge
4 Drag hinge
5 Pitch change bearing
6 Pitch change control arm
7 Swashplate

The best known and most widely used form of vertical take-off and landing (VTOL) aircraft is the helicopter. Yet helicopters were rarely, if ever, referred to as VTOL aircraft until the need arose for faster machines able to share their ability to rise vertically. Even today, there is a tendency in some quarters to consider helicopters as a race apart, not to be classified with other VTOL vehicles. They are, however, the true pioneers of this fascinating class of aircraft.

Two primitive man-carrying helicopters, one built by Paul Cornu and one by the Breguet brothers, flew in France as long ago as 1907; but the first practical machine of this type did not appear until 1936. It was the German Focke-Achgelis Fw 61, which made flights of over an hour, reaching a height of 11,243 ft and a speed of 76 mph. Also tested in 1936 was a Breguet helicopter in France, followed in 1938 by the Weir helicopter in Britain.

As is well known, the rotation of a piece of machinery in one direction produces a torque reaction in the opposite direction, and on these machines some means had to be found to prevent the fuselage from spinning in the opposite direction to the rotor.

The method adopted was to employ two rotors of equal size, revolving in opposite directions, so that the torques balanced each other. The Breguet helicopter had the two rotors positioned co-axially, that is, one above the other; the other two machines had them placed side-by-side, on outriggers extending from the fuselage. This had the desired effect of keeping the aircraft free from torque problems, but the solutions entailed complicated mechanical systems. As a result, all three machines were relatively heavy, and could carry little fuel or useful payload.

Credit for producing the first really successful helicopter thus goes to Russian-born Igor Sikorsky. After an early flirtation with a co-axial rotor design, built in Russia in 1910, Sikorsky, now in America, concentrated on the "single-rotor" configuration, with the torque countered by a small rotor at the tail. The first of the new Sikorsky machines, the VS-300, was completed in 1939 when it was tested in tethered form but was not immediately successful. In 1940 the cyclic-pitch control, which allowed the pitch of each blade to be varied as it

rotated for control purposes, was removed, and two small horizontal rotors were added on outriggers for fore-and-aft control and lateral control. Thus modified, the machine flew for the first time on May 13, 1940, and the following year set up a world endurance record for helicopters of 1 hour 32 minutes.

Although he now had a flyable machine, Sikorsky realised that the complication inherent in four rotors was not really practical, and in June he reverted to cyclic-pitch control with a single horizontal rotor at the rear for fore-and-aft control. Finally, in December 1941, the aft rotor was removed and full cyclic pitch control was achieved successfully with the main rotor, the torque of which was countered by a small tail rotor mounted vertically on the side of the fuselage. Thus, in 1942, the VS-300 proved itself to be the first entirely practical and successful "single-rotor" VTOL machine. From it all modern "single-rotor" helicopters have descended.

Although the major credit goes to Sikorsky, his machine embodied important features developed by the Spanish engineer Juan de la Cierva. In 1923 Cierva had produced the first successful autogyro—the important feature of this, so far as subsequent helicopter development was concerned, being the flapping hinges embodied in the rotor blades. (Autogyros are explained on pp. 174-175.)

The need for such hinges arises from the fact that without them the blades of a simple rotor would develop unequal lift as they revolved. This is due to the varying speed of the blades through the air as they revolve into the direction of flight on one side of the rotor and away from it on the opposite side.

When each blade moves forward (advances) it has an airflow equal to its own speed of rotation plus the speed of the helicopter's movement. When moving backward (retreating) it has the speed of rotation less the speed of movement. Since the amount of lift generated by a blade (as with a fixed wing) depends upon the speed of the airflow over it, blades would normally develop more lift on the advancing side than on the retreating side of the rotor, tending to roll the machine over on its back.

Cierva's flapping hinge neatly solved the problems by allowing the advancing blade to pivot (flap) upward as its lift increased. The retreating blade's own weight and centrifugal force brought it down again. Lift was thus balanced out on each side.

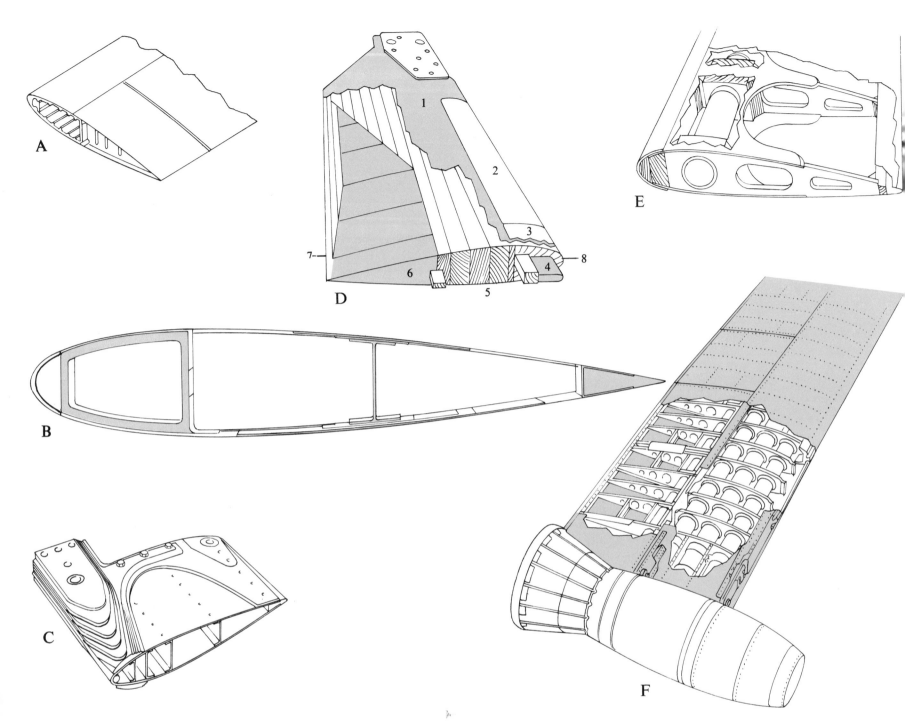

A *Helicopter blade construction. Widely used on Sikorsky helicopters, this particular structure consists of an extruded aluminium alloy D-section nose, to which are attached individual trailing-edge sections.*

B *All-bonded blade, built up around an extruded aluminium alloy spar. This is faced with a brass and aluminium nose block and a stainless steel leading-edge. The trailing-edge section, skins and I-section rear spar are all of aluminium alloy.*

C *This blade is stiffened by a pack of bonded plates where the root is attached to the rotor head hinges.*

D *Composite blade widely used on the early Bell Model 47 series of helicopters. The root-end fitting is bolted through the mild steel core that is used solely for mass balancing.*

1 Glass cloth covering
2 Plastic tape
3 Bonded steel sheath
4 Mild steel core
5 Spruce body
6 Balsa
7 Spruce trailing edge
8 Birch spar

E *Lightweight fabricated blade, built up around a steel-tube spar.*

F *Fabricated blade used on the Fairey Rotodyne VTOL airliner. The machined front spar was slotted to carry an adjustable chordwise balance bar. The complete nose section, housing the tip-drive jet supply ducts, was covered with stainless steel skinning. Aft of the rear spar, construction was of light alloy. At the time of its cancellation the Rotodyne was one of the most promising of all the world's rotary-wing aircraft.*

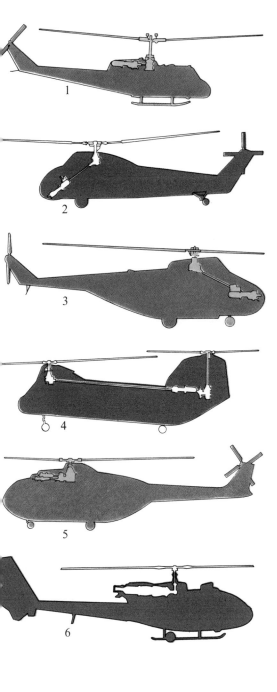

Helicopter engine installations. Turbine engines offer wide flexibility and can be installed virtually anywhere that may be desired or convenient, connection to the rotor head being through a shaft drive.
Bell 204
Westland Wessex
Westland Whirlwind
Boeing Vertol 107
Agusta 101G
  Gazelle

From the VS-300 was developed the R-4B, which went into service with the US Army, this in turn being followed by the further improved R-5, and its civil counterpart the S-51. It is this machine which really ushered in the long delayed era of the helicopter. Powered by a 450-hp engine, it had a cruising speed of 85 mph and an endurance of three hours carrying a pilot and three passengers. It led to the larger S-55, which distinguished itself particularly in the Korean War, when it proved its worth on rescue, ambulance and supply duties. The "Whirlybirds", or "choppers", as they were affectionately known, often landed behind the enemy lines to pick up wounded men. In three years they carried more than 23,000 casualties, half of whom, it has been estimated, would have died without the service offered by the helicopters. Since then helicopters have proliferated, although not quite to the extent anticipated—the day when every householder has a helicopter parked on his back garden has not yet materialised.

On early helicopters, the rotor turned on a shaft driven through gearing by a piston-engine. This type of engine has been largely superseded by shaft-turbine engines, which can be thought of as turboprop engines designed to drive a rotor instead of a propeller.

The development of shaft-turbine engines gave a big impetus to helicopter design. Installation of these new lightweight and powerful engines offered not only greatly improved performance, but resulted in easier and smoother handling qualities. Furthermore, they are small and light enough to mount above the cabin of a transport helicopter. This not only keeps down the size and weight of the transmission shafting, but leaves the fuselage space clear for additional payload. Designers have found that they can more than double the payload of a helicopter by exchanging its bulky piston-engine, mounted in the fuselage, for one or two shaft-turbines mounted above the cabin.

For an aeroplane to take off, it is necessary to produce an upward thrust exceeding its own loaded weight. On conventional aircraft this lift force is generated by the wing deflecting air downward. In fact, all lift, thrust and control forces are obtained by producing a stream of air moving in the required direction. A propeller produces its thrust by drawing in air from ahead and discharging it to the rear at increased speed. A turbojet engine does the same, although the column of air moved is of relatively small diameter and the speed much higher.

The efficiency with which thrust or lift is produced depends upon the relationship between the quantity of air deflected and its speed. In simple terms, the higher the speed of the deflected air, the lower the efficiency—doubling the speed, halves the efficiency.

Thus, the most efficient device for producing lift is the fixed wing of a conventional aircraft. The ability to take off vertically requires the generation of lift without forward movement and this eliminates the most efficient lifting system, the aeroplane wing. As we have seen, on a helicopter the fixed wing is replaced by a rotating wing, known as the rotor.

Although less efficient than a fixed wing, a rotor, in imparting a relatively low velocity to a relatively large quantity of air is more efficient than a turbojet engine, which imparts a high velocity to a relatively small column of air.

However, although the rotating wing is currently the most efficient method of providing vertical and hovering flight, it is not very efficient for normal level flight, nor is it suitable for very high speeds. This has led to a search for other methods of vertical take-off, which in its turn has spurred conventional helicopter development.

In the helicopter field, Lockheed produced the experimental CL-475 with a "rigid" rotor—that is a rotor with the blades fixed rigidly to the rotor head without the usual flapping and drag hinges. This was made possible by the use of a gyroscopic stability technique which prevented the unfortunate effects of early hingeless rotor experiments. The rewards were impressive. Advantages claimed include good stability, high manoeuvrability, low vibration, ease of control, relative mechanical simplicity and high speed.

The rigid rotor concept was carried a step further on the high-performance Lockheed XH-51, which at public displays has been put through its paces as if it were a fixed-wing aircraft, with high-speed fly-pasts followed by steep, nose-up climbs, loops and inverted flying. To obtain information on the high-speed capabilities of the new rotor, one XH-51 was fitted with a 2,600 lb thrust Pratt and Whitney turbojet engine and small wings. In this compound form, the helicopter reached 272 mph, the fastest speed achieved by any rotorcraft at that time. With further refinement, speeds in excess of 300 mph are expected.

Even higher speeds may be achieved by compound or convertiplane helicopters in which the rotor is substantially off-loaded in horizontal flight, the forward thrust being provided independently and a considerable amount of lift being generated by fixed wings. This form of helicopter was pioneered with the British Fairey Gyrodyne and the larger Rotodyne.

On the Rotodyne, the first VTOL airliner in the world, the two powerplants were designed to feed air to the main rotor for vertical flight, the tip of each hollow rotor-blade being fitted with a kerosene-burning pressure jet. Transition to forward flight was achieved by progressively transferring power from the rotor to the propellers. The Rotodyne cruised solely on the thrust of the propellers, lift being provided by the auto-rotating (ie windmilling or free-turning) rotor and the conventional wing, the latter providing over half of the total lift. Britain abandoned this promising inter-city airliner owing to lack of funds, but development of this type of convertiplane has been continued in America by Lockheed and in Russia by Kamov.

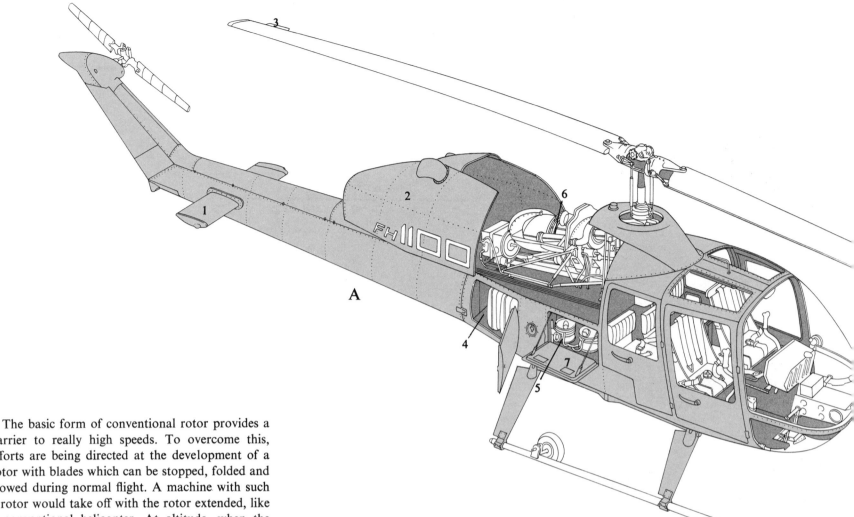

The basic form of conventional rotor provides a barrier to really high speeds. To overcome this, efforts are being directed at the development of a rotor with blades which can be stopped, folded and stowed during normal flight. A machine with such a rotor would take off with the rotor extended, like a conventional helicopter. At altitude, when the forward speed became sufficient for fixed wings to support the machine, the rotor would be stopped, folded and stowed, after which the machine would fly as a conventional aircraft. At the destination, the rotor would be unfolded and a normal helicopter descent made.

Such a solution, even if technically possible, is bound to be complicated mechanically, and several other techniques for VTOL flight are being explored.

As the aim is to produce a machine combining VTOL capability with near-conventional aircraft level flight performance, the idea of using propeller thrust for both vertical and horizontal flight is, perhaps, a natural one for investigation.

One of the first concepts to be tried out was the "tail-sitter" type of aircraft, represented by the Convair XFY-1 and Lockheed XFV-1. These short-wing, turboprop-powered aircraft sat on their tails for take-off, with their noses pointing vertically upward. The propellers acted like a helicopter rotor for take-off, raising the aircraft straight off the ground. Once airborne, the aircraft levelled off and flew conventionally. For landing the procedure was reversed. Speed was reduced until the tail dropped and the nose was pointing straight up, the aircraft then settling down on small wheels at the tips of the wings or tail-fins.

Such a configuration has serious snags and has not been pursued seriously. Even with a special seat,

A *Fairchild Hiller FH-1100 executive helicopter. The basic body is an aluminium alloy stressed-skin structure. A compartment behind the cabin and beneath the engine deck houses oil tanks, radio, electrical and other equipment. The semi-monocoque tail boom, which is cantilevered aft of the fuselage by a four-bolt attachment, has hinged cowlings for access to the tail rotor drive. The powerplant installation is enclosed, the complete cowling sliding aft to expose the engine, accessories, controls and transmission. Built-in maintenance platforms facilitate access to the rotor head. Of all-metal construction, the main rotor blades embody a stainless steel leading-edge spar* and shear web. The trailing-edge section is of aluminium and honeycomb sandwich construction. The blades contain reinforcement doublers at the root section to transfer loads to the point of attachment. Permanent mass balance is provided, and each blade has an aerodynamic tracking tab. The tail rotor blades are of aluminium alloy monocoque construction, bonded to tip ribs and root attachment fittings.

1 Stabilizer
2 Sliding cowl
3 Tracking tab
4 Baggage
5 Oil tanks and equipment
6 Turbine engine
7 Maintenance platform

B *Autogyro flight is fundamentally different from that of a helicopter. On a helicopter the rotor is driven mechanically; on an autogyro it is an independent "free-wheeling" device. Early inventors had assumed that, to develop "lift", a rotor had to be power-driven with the blades set at a high positive angle of pitch. Cierva's basic discovery was that there is a small positive angle of pitch at which a rotor will continue to rotate automatically in an airstream without an engine to drive it, and will develop enough lift for sustained flight. Cierva termed this phenomenon "autorotation". At this angle, the blades will not begin revolving in the* airstream of their own accord, like the blades of a windmill. They have to be set in motion by some mechanical means; but once started, rotation continues automatically provided a continuous airstream is kept flowing through the rotor.
On an autogyro propulsion is obtained by means of an engine driving a conventional propeller. Since the rotor is not power-driven, autogyros are not complicated by problems of torque-reaction. A basic advantage of an autogyro over a conventional fixed-wing aircraft is that it does not stall when flying speed is reduced. It loses height slowly, as the rotor continues

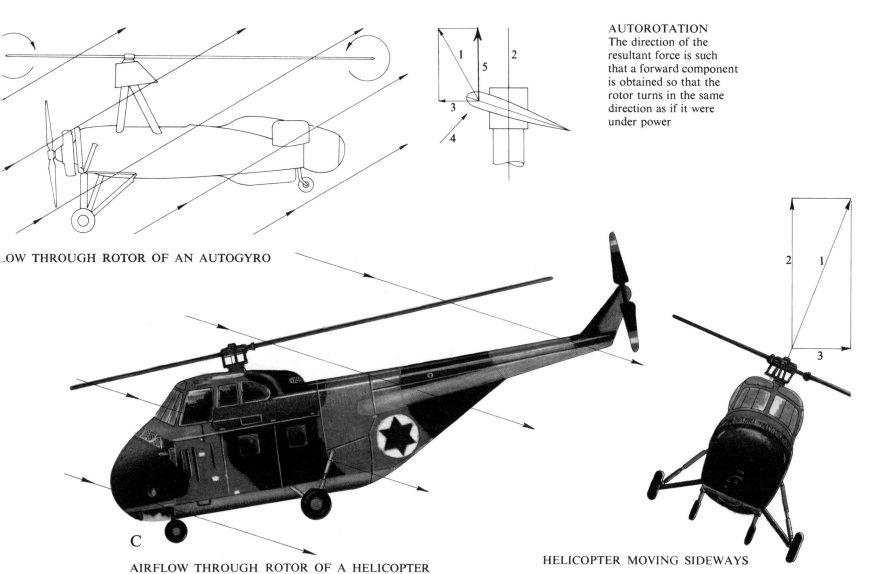

AUTOROTATION
The direction of the resultant force is such that a forward component is obtained so that the rotor turns in the same direction as if it were under power

..OW THROUGH ROTOR OF AN AUTOGYRO

C

AIRFLOW THROUGH ROTOR OF A HELICOPTER

HELICOPTER MOVING SIDEWAYS

..rotate and develop ..t irrespective of ..rward speed. When ..e airspeed is zero, .. autogyro descends ..eeply like a ..arachute, with the ..tor still autorotating. .. this case the flow ..f air up through the ..tor keeps the blades ..volving in exactly ..e same way as does ..e airstream when the ..achine is being pulled ..rough the air in ..vel flight by the ..ropeller.
..ecause its rotor is ..ot power-driven, an ..utogyro has one big ..isadvantage; it ..annot hover in the air .. a constant height, .. can a helicopter. ..arly autogyros ..mbodied conventional ..ircraft-type surfaces— ..ilerons, elevators and ..rudder—for control ..urposes, but roll.

control in particular was poor. A major development in 1932 was the appearance of the "direct control" rotor. This was mounted on a universal joint so that it could be tilted, resulting in a force which tended to pull the autogyro in that direction. The introduction of direct control eliminated the need for ailerons which, with their stub wings, were discarded. Elevators and a rudder are still necessary and these are relatively effective as they are within the slipstream of the propeller.
On some autogyros the rotor can be "spun up" on the ground, permitting vertical take-offs to be achieved. After the desired rotor speed is reached, the drive

system is declutched and the pitch of the blades increased— and up jumps the autogyro.
1 Resultant force on rotor blade
2 Rotor axis
3 Component causing rotation
4 Airflow
5 Lift

C Helicopter flight is quite different from that of fixed-wing aircraft and that of an autogyro. The fundamental advantage of a helicopter is that, having a powered rotor, it can rise vertically and hover—and do so more efficiently than other VTO aircraft. As explained elsewhere in this chapter the swashplate of the rotor head mechanism allows the pitch of individual blades to

be varied as they revolve.
When the pilot moves the control column forward, the swash-plate is tilted forward. The blade control arms are pulled down on the advancing side of the rotor and pushed up on the retreating side. Thus, as each blade approaches the front position of its cycle its pitch is decreased, causing it to fly low. As it continues on its way to the rear, the swashplate control arm increases the pitch so that the blade flies high. The effect is to tilt the whole rotor forward to the same angle as the swashplate, whereupon the helicopter moves forward. This is known as cyclic-pitch control, and the control

column of a helicopter is known normally as the cyclic-pitch change lever.
The rotor tilts in whichever direction the pilot moves the control column, causing the helicopter to move in that direction.
Backwards and sideways flight is possible by moving the control column in the appropriate direction. The fuselage, too, tilts with the rotor, so that it assumes a nose-down attitude in forward flight and a tail-down attitude when travelling backwards. The rotor thus not only provides the lift necessary to sustain a helicopter in the air, but also propels it.
To provide additional lift for take-off and climb, the pitch of all blades is increased

simultaneously by operation of the collective-pitch lever, which works independently of the cyclic-pitch change lever. Another linkage, connected to the collective-pitch lever, automatically opens the engine throttle when the lever is raised. This provides the additional power required by the increase in blade pitch.
Directional control is provided by rudder pedals which are connected to the tail rotor. Pedal movement alters the pitch of the tail rotor blades, and hence their thrust, to turn the helicopter in the desired direction.
1 Total rotor thrust
2 Lift
3 Component force inducing sideways movement

capable of being tilted forward during take-off and landing, the pilot was still partly on his back, which was uncomfortable. The landing manoeuvre was at best tricky and virtually impossible in bad weather. The 90-degree rotation of the fuselage also made the carriage of passengers impracticable.

Many of the disadvantages have been overcome by the tilt-rotor concept, and several aircraft of this type have been or are being tried out. The Bell XV-3 convertiplane, embodying tilting rotors on a fixed wing, made its first "conversion" from vertical to horizontal flight in December 1958. For take-off the two propellers, with their axes vertical, acted as helicopter rotors. At a predetermined forward speed, the "rotors" were tilted slowly forward through 90 degrees, until they acted solely as propellers. During conversion, which took between 10 and 15 seconds, the lift was transferred progressively to the wing.

If the duties envisaged for a VTOL aeroplane are such that it will spend most of its time in level flight, it seems logical that it should be as nearly as possible a conventional aeroplane. This conclusion was reached by Ling-Temco-Vought after an exhaustive survey of VTOL schemes. Such an ideal can be approached by tilting not only the propeller, but the engines and complete wing for vertical flight. In normal flight, with the wings and engines horizontal, such machines are virtually conventional aircraft in all respects.

Biggest example of this type of VTOL aircraft yet flown is the LTV-Hiller-Ryan XC-142A. Designed to carry a payload of 8,000 lb over a combat radius of 200 miles, it has the typical "box" fuselage of a transport aircraft, with a rear loading ramp. Power is provided by four 2,850-shp General Electric T64 turboprops, driving 15 ft 7 in propellers, the slipstream of which effectively blankets the whole wing. The wing, complete with engines and propellers, tilts up to 100 degrees, to give backward movement, or hovering capability in a tailwind.

During vertical and low-speed flight, roll control is provided by differential alteration of propeller pitch, yaw by the ailerons, which are in the slipstream, and pitch control by a horizontal rotor mounted at the rear of the fuselage. During transition, a mechanical mixing linkage integrates the VTOL control system with the conventional ailerons and tail control surfaces in the correct proportions, as a function of wing tilt angle. In normal cruising flight, control is by the conventional control surfaces, with the tail rotor locked. Distinctive leading-edge slats outboard of each engine are used to correct the asymmetry of the slipstream, to preven stalling at the critical transition point.

The XC-142A is of particular interest in that it is part of an operational evaluation programme. During tests a forward speed of 400 mph has been achieved, and a backward speed of 35 mph. A height of 25,000 ft has been reached and both short take-offs and landings and vertical take-offs and landings have been made successfully on the US aircraft carrier *Bennington* while it was under way.

A *A good example of the advantages conferred by turbine powerplants is given by the Kaman Huskie. It was powered originally by a piston-engine which filled the rear of the cabin.*

B *Later versions are powered by a turbine, installed on top of the fuselage and so freeing the whole of the cabin for passengers or freight.*

985   987   988

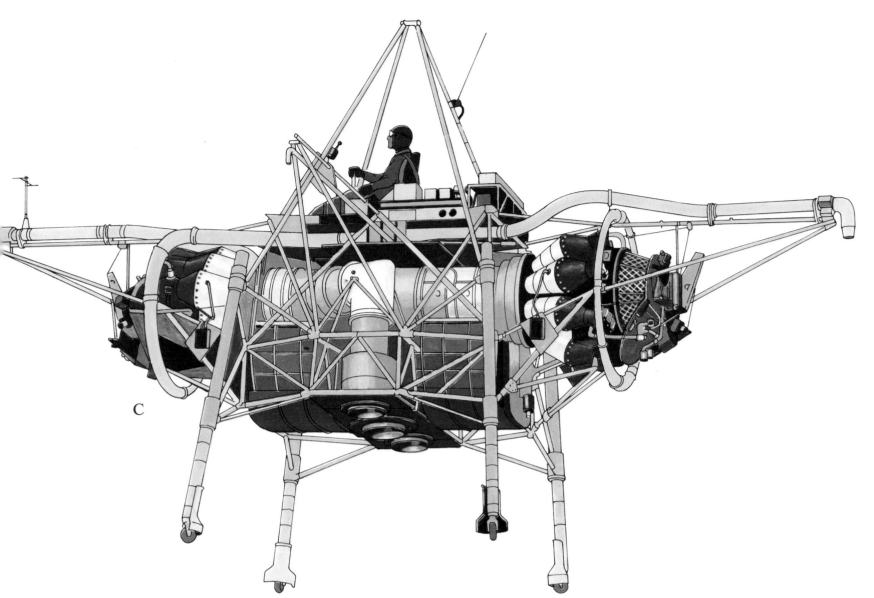

C

During air-drop trials, dummy cargo loads weighing up to 4,000 lb were dropped at heights ranging from 5 ft to 5,000 ft while hovering and at speeds of up to 144 mph. One technique, known as "dump truck" off-loading, involved discharging cargo mounted on rollers from the rear end of the aircraft, with the fuselage tilted upward several degrees, during low-speed flight and while hovering.

Promising as the tilt-wing idea seems, there is a strong school of thought which considers that it is just a step towards the ultimate· in VTOL flight, using pure jet thrust. As such, the critics argue, it should be abandoned and the money so saved devoted to direct jet-lift research. With the example of the losing battle fought by turboprop airliners against turbojet airliners before them, there is a strong case for this point of view.

The first free-flight demonstration of direct jet-lift was made by the Rolls-Royce Thrust Measuring Rig in 1954. Nicknamed the "Flying Bedstead", this successful research vehicle was powered by two Nene jet engines mounted so that their combined thrust of 10,000 lb was exerted straight upward. Lateral control was by compressed air exhausting downward from four nozzles mounted on outriggers

on either side and to the front and rear of the machine.

From the Rolls-Royce Bedstead there evolved the much more advanced Short SC.1 VTOL research vehicle. Delta-winged, this aircraft was powered by five Rolls-Royce RB.108 turbojets each of 2,130 lb thrust. Four of the engines acted as lift engines and the fifth was used for propulsion. The specially-developed lift-jet engines were remarkable for the fact that they developed no less than 8 lb of thrust for every pound they weighed. Later engines develop over 20 lb of thrust for every 1 lb of weight, and this may be compared with less than 5 lb of thrust per lb developed by a conventional modern turbojet engine.

These two machines, the Rolls-Royce Bedstead and the Short SC.1, provide good examples of two fundamental approaches towards obtaining VTOL flight with jet lift. One can use a single engine, or batch of engines, which provide both the lift for vertical flight and the thrust for forward flight, as on the Rolls-Royce machine. Or one can provide separate engines for lift and thrust, as on the Short SC.1. Some disadvantages are inherent in both systems. The separate power plant concept allows

C *Thrust Measuring Rig (TMR). Nicknamed the "Flying Bedstead", this ungainly Rolls-Royce research vehicle ushered in the era of non-rotating-wing VTOL flight. It weighed 3½ tons and was powered by two Nene turbojets, each developing 5,000 lb of thrust, exhausting vertically downward. The rig was stabilized by four downward-pointing compressed-air nozzles, located one each at front and rear, and one on each side. The TMR was flown in the manner of a*

helicopter, that is, by inclining the vehicle in the direction required. Thus, forward flight was achieved by blowing air out of the rear nozzles. This pushed the rear of the rig up slightly, and thus inclined the downward thrust of the engines so that it propelled the vehicle forward as well as sustaining it in the air. Similarly, sideways flight was achieved by blowing air out of one of the side nozzles to raise that side of the machine, so that it moved in the opposite direction.

the individual engines to be designed so that they are ideal for their particular duties. This has to be offset against the fact that, having achieved wing-borne flight, the lift engines are shut down and are then carried as dead weight. The safety conferred by multiple engines is obvious, and will be essential on VTOL airliners.

The dual-purpose engine concept results in a relatively simple aeroplane. Reliability is always desirable and simplicity is one of the best ways of achieving this. With a single engine, the in-flight performance, particularly rate of climb and manoeuvrability, is superior, because of the lighter overall weight. In the case of a transonic strike aircraft, the propulsive thrust requirement is greater than for a subsonic aircraft. With separate lift and thrust engines, it may be necessary to fit afterburners to the propulsion engines to obtain the thrust required—adding weight and complication—whereas with a single engine adequate normal thrust might be available.

Largely because of its simplicity, the single-engine concept was adopted for the Hawker Siddeley Harrier. This tactical V/STOL strike aircraft is powered by a single Pegasus turbofan engine which discharges cold air from the compressor fan through two rotatable (vertical thrust) forward nozzles and hot air from the turbine through two rotatable rear nozzles. The entire thrust can be directed downward for V/STOL operation or rearward for conventional flight. The simplicity of this arrangement, which was pioneered in Britain, is indicated by the fact that the only cockpit control additional to the conventional ones is a lever, alongside the throttle, which is moved backward to rotate the engine nozzles downward.

Other advantages of using rotating nozzles include the ability to make conventional take-offs when the aircraft weight exceeds the thrust available; the ability to do engine power-checks with the nozzles horizontal; and the ability to make short ground rolls when taking off and landing on terrain where debris might be blown up and ingested by the engine.

The simplicity of flying the Harrier is evident from the fact that one pilot completed a programme of conversion in 43 minutes total flying time on the prototype, of which only 16 minutes was hovering and transition time. In due course, history will accord the P. 1127, the prototype of the Harrier, a place ranking with those held by the Wright Flyer and the Sikorsky VS-300 helicopter.

As previously mentioned, a single-engine installation has disadvantages and some of these become more severe with increase in size and when the aircraft is to be used for duties such as passenger-carrying.

Biggest multi-engine VTOL aircraft to reach the hardware state so far is the German Dornier Do 31 tactical VTOL transport. This is powered by two Pegasus lift/cruise engines, with the exhaust ejecting through four rotatable nozzles as on the Harrier, and two wing-tip pods each containing four lift-jet

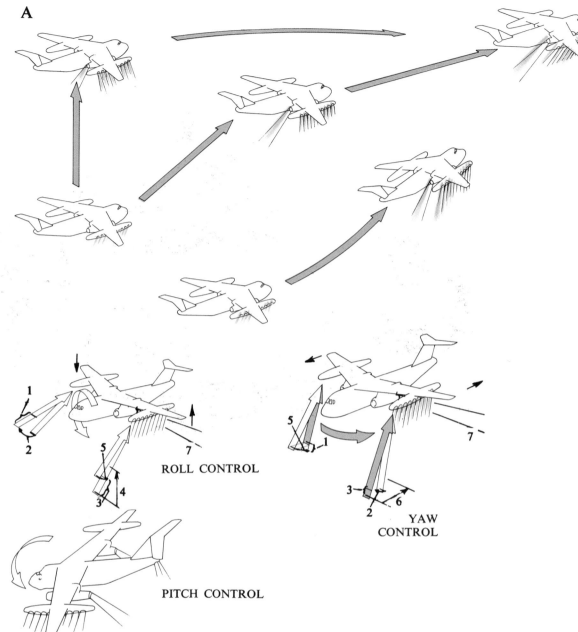

A

ROLL CONTROL

YAW CONTROL

PITCH CONTROL

A *Jet-lift VTOL flight is fundamentally different from that of rotating-wing craft. During hovering and vertical flight, no stability or control can be provided by the aerodynamic surfaces, so that jet-thrust— or reaction type— controls are required. On a mixed power plant VTOL aircraft of the type illustrated above, yaw and roll control is achieved by differential throttling and changing the direction (vectoring) of the lift-engine thrust. During normal flight, control is achieved by movement of the conventional elevators, ailerons and rudder.*

These are designed to operate continuously in response to control movements, so ensuring that smooth transitions occur as a result of the gradual transference of authority from the reaction nozzles to the aerodynamic controls as speed increases. Jet-lift VTOL aircraft can use several techniques when taking off and landing. They can, as is implied, take off vertically. For this, the jet-lift thrust is directed vertically downwards and then, when sufficient height has been gained, is gradually deflected aft to achieve transition to normal

horizontal flight. However, vertical take-offs are not necessary for all missions. When taking off from an airbase with a runway, a "short take-off" technique, involving a relatively short ground run, may be more economical and allow a heavier payload to be carried. For the short take-off technique, the engine thrust is deflected rearwards initially and then partially downwards to augment the aerodynamic lift obtained from the wings. The aircraft depicted above represents a V/STOL transport

projected by Hawker Siddeley Aviation. On this the thrust from the propulsion engines can be partially deflected, through relatively light nozzles. At maximum deflection, there is still some residual horizontal thrust, which has to be counterbalanced by deflecting the lift-engine thrust slightly forward, to achieve a truly vertical take-off. For this aircraft a special technique known as "zero-length" take-off has been evolved. As the term implies, there is no ground run; but for maximum efficiency the lift-engine thrust

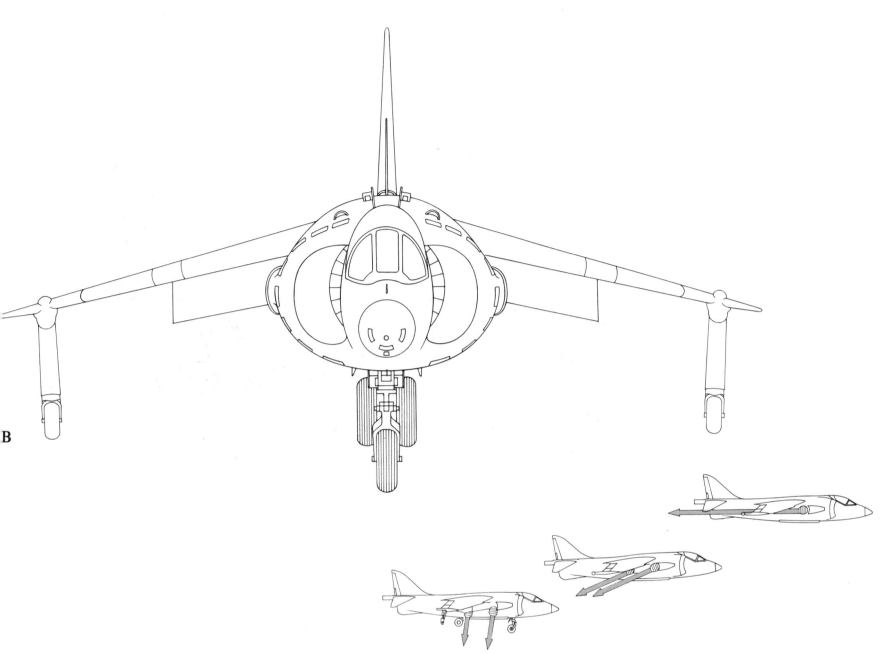

B

is deflected vertically downward, so that after leaving the ground the aircraft moves forward under the residual horizontal thrust from the propulsion engines. The aircraft thus performs its transition in an acceleration climb, at an almost constant angle. This simplifies the pilot's task and the technique is suggested as being the most practical for operation from difficult sites. Various techniques can also be adopted for landing. Near-vertical descents are generally preferred, with some slight forward speed to enable the pilot to

keep his landing point in view until the last possible moment. The lift-engine pods on the Hawker Siddeley project are designed for rapid removal for conventional take-offs and landings where runways are available.

1 Throttle decrease
2 Nozzles forward
3 Throttle increase
4 Differential vertical thrust
5 Nozzles back
6 Differential horizontal thrust
7 Propulsion engine thrust

B *Hawker Siddeley Harrier. Epitomizing the simplicity of the "single-engine" concept, this V/STOL tactical fighter is powered by a Pegasus 6 vectored-thrust turbofan engine of 19,000 lb thrust, with four rotating exhaust nozzles. The air intakes are on each side of the fuselage, aft of the cockpit, and the rotatable exhaust nozzles are under the wing roots. Control at low airspeeds and during hovering is achieved by the use of compressed-air reaction jets (like those of the TMR) at the wingtips, nose and tail. These jets are controlled by*

the conventional stick and rudder pedals. When the nozzles are away from the horizontal position, engine bleed air is ducted automatically to the reaction jets to provide attitude control. The only cockpit control additional to those in a conventional fighter is a lever, alongside the throttle, which is moved rearward to rotate the engine nozzles downward. The technique of variable-thrust vectoring permits both vertical and conventional take-off and landing, or any intermediate degree of STOL operation. The ability to perform

engine runs with the nozzles horizontal also offers advantages.

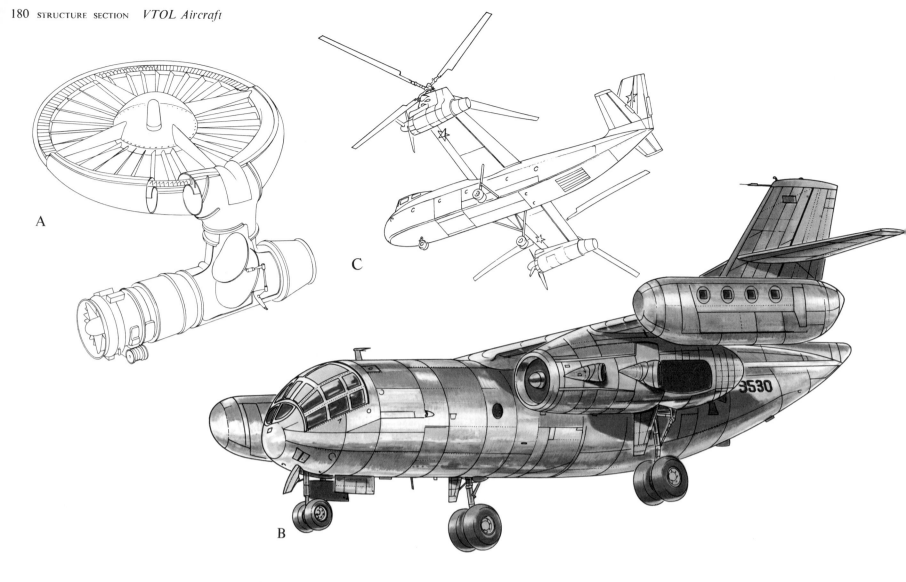

A *Lift-fan installation. For normal flight the turbojet exhaust is directed rearward. When vertical lift is required, the exhaust is diverted to drive turbine blading around the periphery of the fan. The installation, weighing 1,145 lb, produces 7,430 lb of of lift—nearly 5,000 lb more than the turbojet's basic thrust.*

B *Dornier Do 31. This experimental VTOL transport utilizes both deflected thrust and direct jet-lift techniques. The aircraft is powered primarily by two Pegasus vectored-thrust turbofan lift/cruise engines, but has also two wingtip pods each containing four RB. 162 lift-jet engines. In hovering flight, pitch control is obtained by ducting high-pressure air from* the lift engines to the tail of the aircraft, where it exhausts through two sets of nozzles, one set pointing upward, the other down. Roll control is obtained by thrust modulation of the lift/cruise engines. Yaw control is by differential tilting of the lift-engine nozzles. The projected production version would have two lower-powered propulsion engines and ten lift-jet engines in the 5,500 lb-thrust class, the high power of the lift-jets enabling the main engines to be designed for maximum efficiency in cruising flight, instead of having to do two different jobs as well as possible. Clam-shell doors at the rear will permit the ramp loading of vehicles and freight.*

C *Kamov Vintokryl. This twin-rotor converti-plane is powered by a pair of 5,500 shp Soloviev D-25V shaft-turbine engines. As in the earlier Fairey Rotodyne, the engines drive the rotors for vertical and hovering flight, and tractor propellers for cruising flight. Control surfaces are conventional, with hinged flaps extending the length of the wing trailing-edges. The upswept rear fuselage incorporates rear loading doors.*

D *Canadair CL-84. This experimental tilt-wing aircraft is of dual interest in the helicopter development field. It is of importance as representing one line of development in the quest for higher cruising speeds from VTOL aircraft, and the* special tail rotor is of interest as an example of a rotating wing designed to be stopped and restarted in flight. During development flights involving transition manoeuvres, the stresses in the propellers, tail rotor and other major components have been no higher than anticipated. The full range of hovering manoeuvres, both in and out of ground effect, have been explored; and sustained "hands off" flight has proved possible. Flight tests were made with the stability augmentation system de-activated.*

E *Ryan XV-5A. This experimental V/STOL jet aircraft is powered by two 2,660 lb-thrust turbojet engines. Normally, these exhaust under the tail, but for* VTOL flight the efflux is diverted to drive two fans (as in A) housed in the wings. A small fan in the nose provides pitch control. The fans augment the jet thrust by 300 per cent and, therefore, offer relatively economical lift.*

F *The LTV-Hiller-Ryan XC-142 is the biggest VTOL aircraft utilising the tilt-wing principle. Weighing 44,500 lb, the XC-142 is designed to carry a payload of 8,000 lb over a combat radius of 200 miles. Power is provided by four General Electric T64 turboprop engines driving conventional propellers and a horizontally mounted tail rotor through a system of cross-shafting which enables flight to be maintained* on any two engines in an emergency. The wing is able to rotate through an angle of 100°, giving the XC-142 ability to hover in a tail wind. During VTOL flight, roll control is achieved by means of differential collective propeller pitch, yaw control by means of the ailerons working in the propeller slipstream, and pitch control by means of the variable-pitch tail rotor. During transition, a mechanical mixing linkage integrates the VTOL control system with the conventional ailerons and tail control surfaces in the correct proportions required at various wing tilt angles. In normal cruising flight, control is by the conventional control surfaces, with the tail rotor locked.*

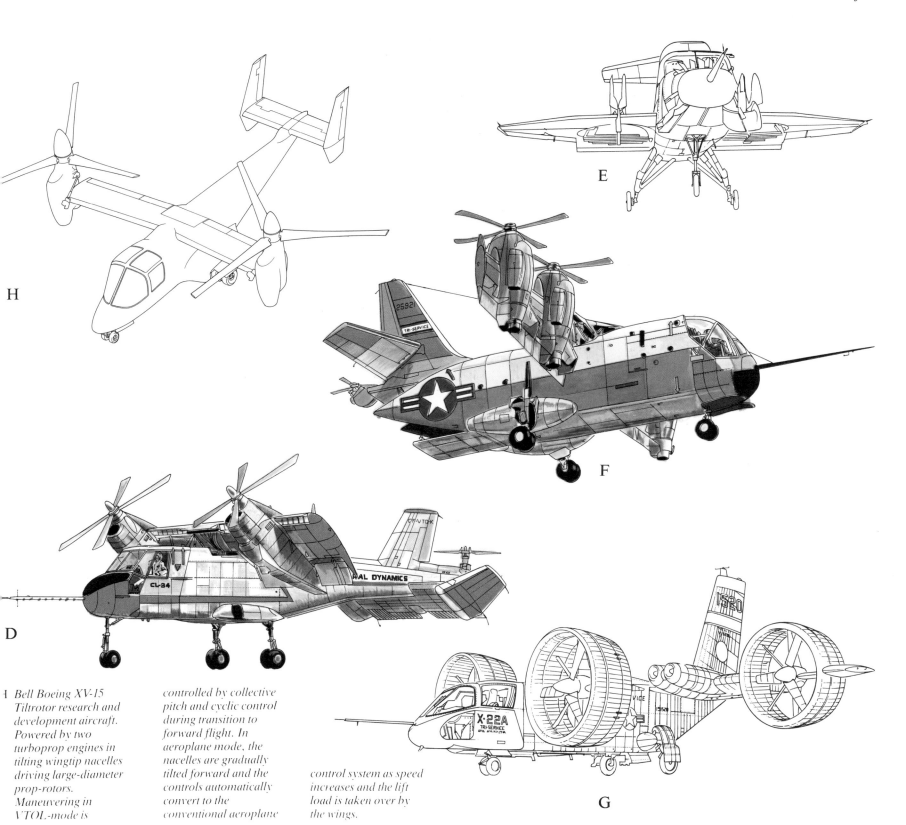

H

E

F

D

G

H *Bell Boeing XV-15 Tiltrotor research and development aircraft. Powered by two turboprop engines in tilting wingtip nacelles driving large-diameter prop-rotors. Maneuvering in VTOL-mode is* controlled by collective pitch and cyclic control during transition to forward flight. In aeroplane mode, the nacelles are gradually tilted forward and the controls automatically convert to the conventional aeroplane control system as speed increases and the lift load is taken over by the wings.

G *Bell Aerosystems' X-22A research vehicle was produced to explore the mechanical and aerodynamic characteristics of the dual-tandem, ducted-propeller V/STOL concept and to assess its military potential. The engine nacelles each house two General Electric T58 turbines which exhaust* at 65° to the horizontal, thus providing additional take-off thrust. Shafts extend from the rear of the engines to the transmission system which embodies inter-connecting shafts so that all four propellers are driven even after the failure of one or more engines. Being intended for research and not for economical day-to-day operation, the X-22A can have a high thrust-to-weight ratio, enabling it to take-off vertically on a hot day with one engine out. This performance margin frees the crew of worry about an engine failure, and permits control and stability investigations to be conducted safely at altitudes which otherwise might be in the "dead man's" zone in the event of engine failure.

Attitude control is obtained by combinations of propeller blade pitch change and elevon deflection. While hovering, pitch is controlled by varying the pitch, and hence the "lift", of the fore and aft pairs of propellers. During transition to conventional flight, when the ducts rotate toward the horizontal, the blade pitch change system is phased out and control by differential fore and aft elevons takes its place. Similarly, roll control is by differential left and right blade pitch change in hover, giving way to differential left and right elevon deflection in conventional flight.

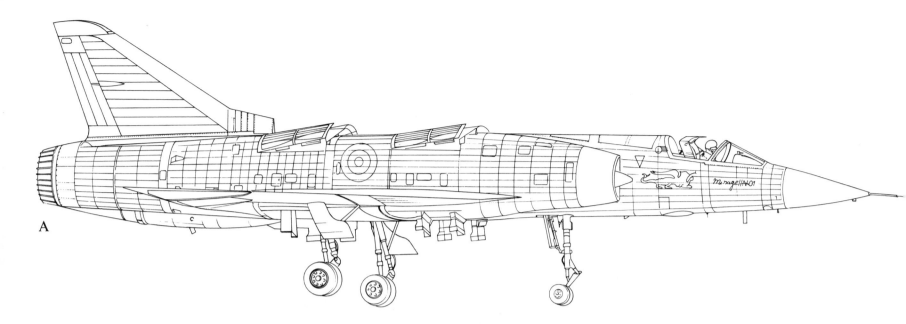

A

engines. For take-off, the Pegasus engine nozzles are rotated to deflect the efflux downward and the lift engines are run. At height, the main engine nozzles are rotated aft and, when sufficient forward speed has been gained, the lift engines are shut down. Thereafter the aircraft flies conventionally.

On the Do 31 prototype, the lift pod engines are mounted vertically, but on developed versions the lift engines would be mounted horizontally and fitted with rotating nozzles. Advantages of this are similar to those evident with the rotating nozzles on the Harrier, including the ability to effect ground checks without ingestion problems, and the fact that the lift engine thrust can be used to assist conventional or short take-offs and landings. In service it is unlikely that vertical take-offs and landings would always be essential, and, where runway conditions permit, it would be more economical to make conventional take-offs and landings, as well as permitting a heavier payload to be carried.

Vertical lift can also be provided by means of devices known as lift-fans. A lift-fan comprises a horizontal rotor, driven by the exhaust from the propulsion engines impinging upon turbine blading around the periphery of the rotor. The huge air flow through the fan makes it a far more efficient lifting device than the turbojet, and shutters can be used to seal it off to preserve the wing profile during normal flight.

An aircraft utilising this form of lift device was the Ryan XV-5A. A two-seater, this had a conventional wing and tail with control surfaces, with the ducted lift-fans incorporated in the wings and a smaller fan in the nose. It was powered by two 2,660 lb thrust turbojets mounted in the top of the fuselage. Normally, the engines exhausted under the tail, but for vertical flight the exhaust was deflected into two ducts leading to the tip-blade drives of the wing and nose fans. The fans augmented the jet thrust by nearly 300 per cent and, therefore, offered relatively economical lift.

In addition to doors over the fan openings there

were deflector vanes to assist control and transition. The nose fan provided pitch control, while the wing fans and their louvres were used to give control in roll and yaw.

Another interesting jet-lift research aircraft was the Lockheed XV-4A Hummingbird. This looked like a conventional jet aeroplane with two close-coupled nacelles and very small wings. Vertical thrust was obtained by diverting the engine thrust into a "mixing-chamber" in the fuselage, where it acted as a jet-pump, thereby inducing some 40 per cent more airflow. For take-off, the fuselage doors were opened, and the two 3,300 lb thrust engines opened up to provide some 8,300 lb lift thrust. When airborne, the nose was tilted down to increase forward speed to 80 knots. The thrust of one engine was then deflected aft and normal horizontal flight was achieved when a speed of 120 knots was reached.

Both of these lift techniques imposed considerable structural problems, making their application to practical aircraft difficult. Ducted lift-fans involve huge cut-outs in the wings, are not easy to house within the wing profile, and encroach upon space needed for fuel; the "mixing chamber" on the Hummingbird occupied a considerable amount of fuselage space.

Because of their relative inefficiency and mechanical complexity, current VTOL aircraft, represented almost entirely by helicopters, are also expensive. Thus, they are generally used only where their ability to rise vertically and to hover enables them to undertake duties not possible with conventional aircraft. The wider use of VTOL aircraft in the future seems to depend upon the development of improved vectored-thrust (rotating-nozzle) turbofans, jet-lift engines and lift-fan devices. Development is likely to be spearheaded by military requirements, with civil applications following later. The benefits of quicker travel promised by VTOL airliners will, however, not be exploited fully until the noise problem has been solved, permitting their operation into the centres of cities.

A *Dassault Mirage III-V. This VTOL research aircraft is powered by a SNECMA turbofan engine of 20,000 lb thrust with afterburning, for propulsion, and eight RB.162 lift engines, each of 4,400 lb thrust, for vertical lift. Special thrust deflector doors are mounted on the underside of the fuselage to reduce ground erosion and possible ingestion problems by turning the lift engine exhaust to the rear, thereby blowing debris and hot gases away from the aircraft. Upon selection of full throttle the deflector doors move automatically so that maximum lift thrust is obtained.*

ENGINES AND EQUIPMENT SECTION

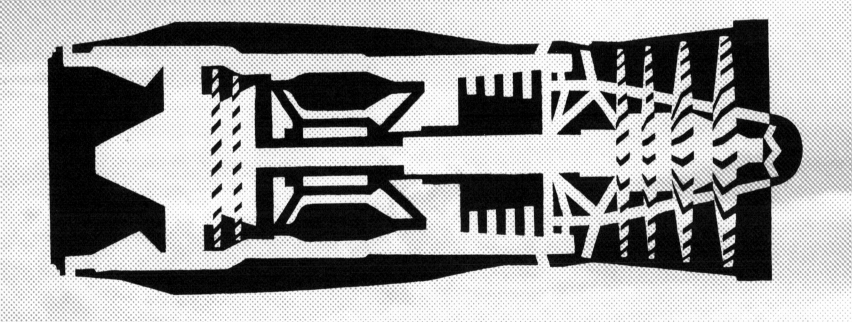

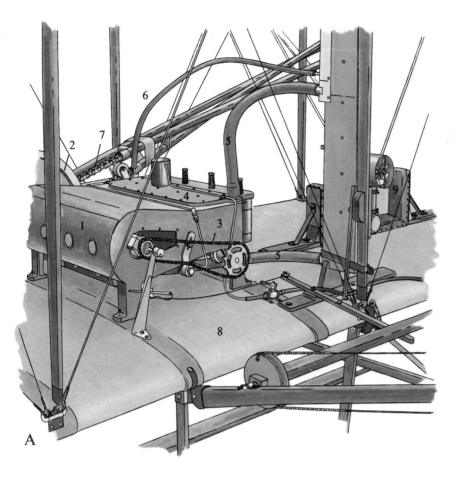

A

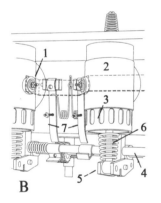

B

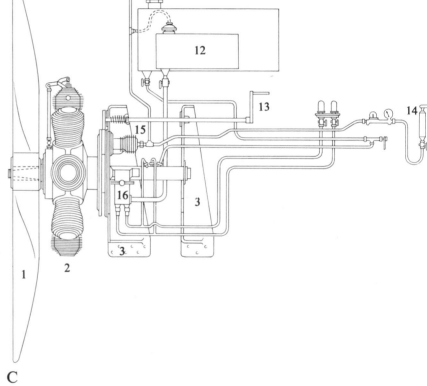

C

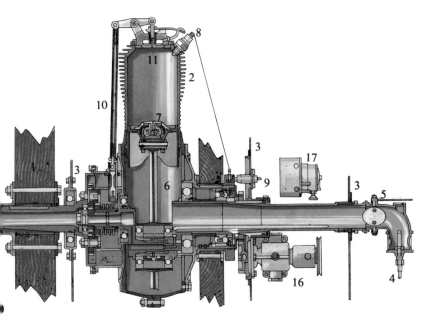

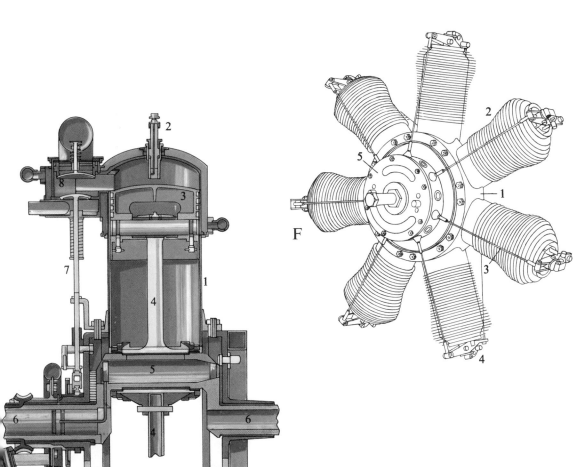

C *The Gnome rotary, in*
D *which the cylinders*
& *rotated with the*
F *propeller around a*
fixed crankshaft, was
designed by the Seguin
brothers in France in
1907-8 and marked the
first complete break
from traditional
automobile engine
design. This cross-
section (D) shows
one cylinder in the
vertically upright
position and
emphasizes the
character—which was
revolutionary in both
senses of the word—of
the seven-cylinder
Gnome of 1909. It will
be seen that the
petrol/air mixture was
admitted to the
crankcase and thence
through a valve in the
head of each piston to
the combustion space,
finally escaping
through the exhaust
valve at the end of the
cylinder. This engine
could be run at up to
1,000 revolutions per
minute and then gave
about 45 to 47 horse-
power. It weighed less
than 170 pounds,
although made almost
entirely of nickel steel.
Gnomes with varying
numbers of cylinders
and giving powers up to
180 horsepower were
made in large quantities
until late in the
1914-18 War. The
complete engine is
pictured on this
page (F).

C & D
1 Wooden propeller
(similar to that
illustrated on page
233)
2 Rotary engine
cylinders
3 Fixed mountings
(forgings or stiff sheet)
4 Air and fuel inlet
5 Throttle valve
6 Combustible mixture
in crankcase
7 Valve in piston to
admit mixture to
cylinder
8 Sparking plug
9 Magneto distributor
contact

10 Valve push rod
11 Exhaust valve
12 Fuel and castor oil
tanks
13 Hand crank for
starting
14 Hand air pump and
gauge
15 Air pump to
pressurize tanks
16 Oil pump
17 Magneto

F
1 Crankcase
2 Seven cylinders
3 Valve push-rod
4 Exhaust valve rocker
5 Propeller hub
mounting

E *By far the most*
advanced internal
combustion engine of
its time,
Charles Manly's five-
cylinder radial
designed for Professor
Langley's 1903
"Aerodrome" did all
that it was designed to
do and, had the
Aerodrome flown
properly, would have
become the first
successful aeroplane
engine. In this
sectional illustration,
only the top, vertical
cylinder is shown; the
four others were
arranged at equal
intervals around the
central crankshaft.
The connecting rod of
the vertical cylinder was
solid but the other four
were hollow, as were
the crankshaft and
several other
components. In
conjunction with the
thin soldered and
brazed construction of
the static parts, this led
to a power/weight ratio
which was not equalled
by any other engine for
20 years.

1 Typical cylinder
2 Sparking plug
3 Piston
4 Connecting rods
5 Hollow crankpin
6 Tubular drive shafts
to port and starboard
propellers
7 Push rod operating
exhaust valve
8 Inlet and exhaust
valves

PISTON ENGINES

The fact that man failed to create a practical powered flying machine until as recently as 1903, despite centuries of thought applied to the problem, can be attributed principally to the lack of a suitable engine. The earliest visionaries who left records of their attempts to build an aeroplane either ignored the problem, and sketched plans for a glider, or else suggested the use of human muscle power in machines which generally were to fly by flapping their wings and in some cases were even pictured with feathers and a beaked head.

In 1809, Sir George Cayley drew attention to the fact that the pectoral muscles of a man, which would be used for flapping artificial wings, are less highly developed than those of a bird and that flight by such means did not appear a possibility.

Cayley, however, realist though he was, felt far from pessimistic about the prospects of ultimately achieving powered flight. "I feel perfectly confident", he wrote, "that we shall soon be able to transport ourselves and families, and their goods and chattels, more securely by air than by water, and with a velocity of from 20 to 100 miles per hour. To produce this effect it is only necessary to have a prime mover, which will generate more power to a given time, in proportion to its weight, than the animal system of muscles".

Cayley was not the first to point out the inadequacy of human muscle power applied directly to flapping wings, but he was, with little doubt, the first to describe in some detail the most practicable alternative: the internal combustion engine. Like many other 19th century inventors he first suggested the use of a steam engine, and proposed to pass the water through a coiled pipe in such a way as to convert it in a single pass from water to steam in the manner of the modern "flash boiler". But even this he regarded as undesirably cumbersome. He emphasized repeatedly the need for lightness, which has characterized aircraft engines ever since, in contrast to the massive machinery used to propel ships and land vehicles. "Lightness is of so much value", he stated, "that it is proper to notice the probability that exists of using the expansion of air, by the sudden expansion of inflammable powders or liquids . . . an engine of this sort might be produced by a gas-tight apparatus and by firing the inflammable air generated with a due portion of common air under a piston".

Cayley actually experimented with engines driven by hot air, and by gunpowder, and left a wealth of material which has only recently been brought to light and published. Most 19th century inventors of flying machines were unaware of his work and proposed the use of steam plant, muscles, twisted rubber, coiled springs, cylinders of compressed air and other sources of power.

In 1889, the Australian Lawrence Hargrave built and tested a small aircraft engine in which three cylinders were arranged radially like the spokes of a wheel. In operation, the engine itself rotated around a fixed crankshaft in the centre, driven by compressed air from a long storage bottle. This mechanical arrangement, known as a rotary engine, was much used for aircraft propulsion a quarter of a century later.

By far the most impressive flying machine actually built before 1900 was the enormous biplane constructed by Sir Hiram Maxim between 1891 and 1894, which appears to have been amazingly sound in design despite its gargantuan proportions (the wings had a span of over 80 feet). Maxim chose to use steam power to drive two large propellers and, with assistance, conceived a pair of compound engines of extraordinarily advanced design. Drawing steam at the unprecedented pressure of 320 pounds per square inch, each of these units achieved an indicated output of no less than 180 hp; yet the total weight of the aircraft laden with coal, 600 pounds of water and three passengers was not in excess of 8,000 pounds. The engines did not blow up or break but worked so well that, on its third trial, the vast machine, with its crew, rose clear away from its guiding rail, at which the steam was immediately shut off. It seems a pity that this aircraft was not tested again.

An even more advanced steam engine was used by Samuel Pierpont Langley, who came within a narrow margin of being the designer of the first successful man-carrying aircraft. By 1896 Maxim had suspended his work, despite its great promise, but in that year Langley achieved successful flights with two model aeroplanes driven by steam engines. Following painstaking research, he perfected the design of the flying machine itself but could not find a suitable engine.

Early in the present century, Langley decided to produce his own power plant, and delegated the task

A *The Liberty engine, a*
B *Vee-12, was designed*
C *for the United States*
& *Army in the three*
D *months following America's entry to the War in 1917. The objective of the design commission, headed by leading engineers of the Packard and the Hall-Scott motor car companies, was to produce a standard engine that could most readily be mass-produced by American industry yet would have the highest possible performance. This aim was met and the Liberty's power of 400-420 horsepower was put to good use in thousands of Allied aircraft.*

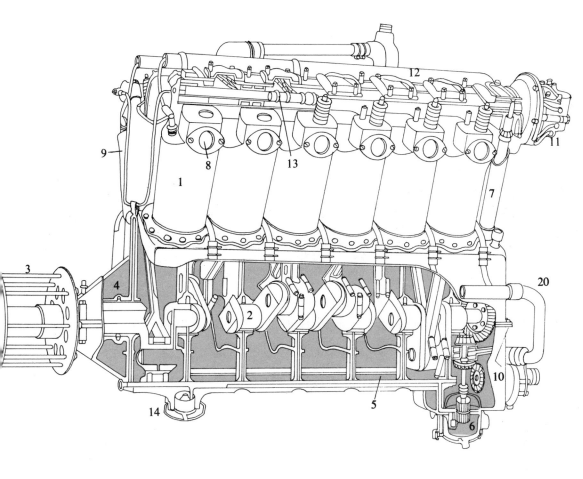

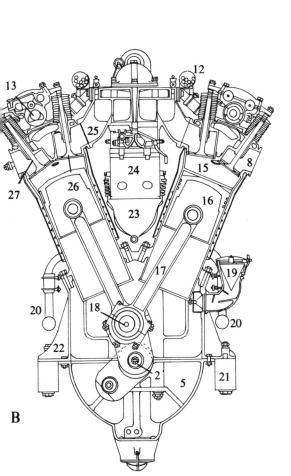

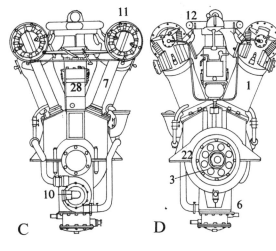

to his assistant Charles M. Manly. Manly's name is known today to only a few, but he created an aircraft engine far ahead of its time and with a performance and ratio of power to weight that was not equalled until well into the 1914-18 War. Basically, Manly's engine was a five-cylinder radial, the cylinders being arranged like a five-pointed star. In contrast to the rotary arrangement, a radial engine has fixed cylinders containing pistons which turn the central crankshaft.

Manly's engine was a true internal combustion engine, with spark plugs to ignite a petrol/air mixture in the cylinders, the latter being cooled by water flowing round them, inside the outer jackets. It thus broadly resembled early car engines, but differed markedly from them in its performance. During a 10-hour test it achieved a measured output of 52.4 hp, despite the fact that it weighed appreciably less than 200 pounds, complete with all accessories, a full petrol tank and the water cooling system and radiator. This amazing engine was installed in Langley's "Aerodrome" so that its crankshaft ran athwartships to drive a propeller on each side of the aircraft. The fact that the machine itself failed to fly was in no way attributable to its outstanding power plant.

Had Langley's machine performed properly on either of its two attempted flights, in October and December 1903, it would with little doubt have been the first successful man-carrying aeroplane. But on both occasions it dived into the Potomac, and the fame of building the first successful aeroplane was won by the Wright brothers.

As related earlier, Wilbur and Orville Wright began by experimenting with gliders, painstakingly improving Lilienthal's and Chanute's concepts by adding control surfaces and various aerodynamic refinements. By 1902 they were gliding so successfully that they felt confident that by adding an engine they would at last realise the ambitions that had proved elusive to mankind for so long.

After carefully looking at all likely means of propulsion, the brothers concluded that the engine would have to be a petrol-burning internal combustion unit. They would have liked to have used an existing car engine as a basis but all existing automobile engines were much too heavy. Ultimately the brothers set to work to design their own engine and during 1903 they completed the design and, helped by their mechanic Mr. Charles Taylor, they constructed it themselves. They did not, as has been widely reported, use the engine from a Pope-Toledo car. When completed their engine weighed about 180 pounds and developed roughly 16 hp.

Although nowhere near so advanced a power plant as Manly's 52-hp masterpiece, the original Wright engine was a robust and reasonably trustworthy unit which achieved all that was demanded of it. The illustration on page 184 shows the way in which its cylinders were arranged horizontally. They were cooled by water flowing through jackets, the water in turn being cooled by

A B C & D

1 First of six port side cylinders
2 Crankshaft
3 Propeller hub
4 Front bearing
5 Lower crankcase
6 Oil scavenge and feed pumps
7 Camshaft drive and oil return pipe
8 Exhaust ports
9 Camshaft oil pipes
10 Cooling water pump
11 Ignition distributor (two)
12 Conduit housing ignition wires
13 Tubular camshaft
14 Oil drain plug
15 Combustion space
16 Piston (first port cylinder)
17 Connecting rod
18 Crankpin
19 Oil filler
20 Cooling water pipes
21 Engine bearers
22 Mounting feet
23 Air intake
24 Carburettor
25 Induction manifold
26 Inlet valves
27 Exhaust valves
28 Vertically mounted electric generator

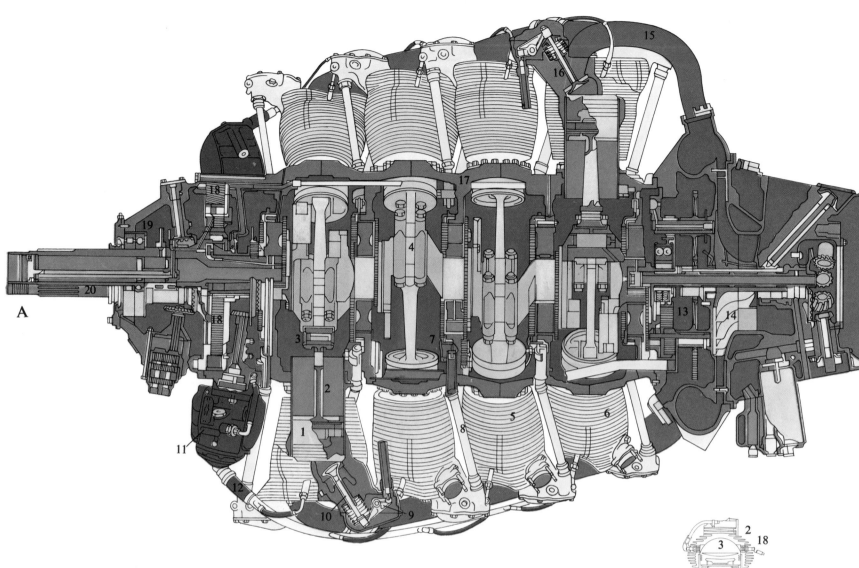

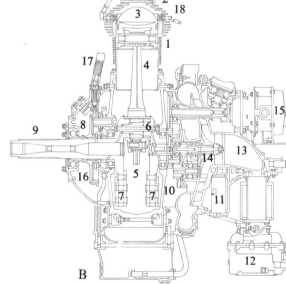

A *On Christmas Eve*
& *1925, a small team*
B *of mechanics in the newly-formed Pratt & Whitney Aircraft company finished the task of assembling their first engine, a nine-cylinder air-cooled radial called the Wasp. It weighed half as much as the Liberty engine but was even more powerful, with a rating of 425 horse-power, and established what is today one of the world's largest aero engine companies. The original engine, seen in section on the right, was developed later to yield 600 horsepower. The company's piston-engines culminated 20 years later in the R-4360 Wasp Major (above), the*

designation indicating "radial engine, 4,360 cubic inches swept volume". In the Wasp Major, Pratt & Whitney boldly joined four rows each of seven cylinders, all fed by a common induction system at the rear and geared to a single propeller shaft at the front. Many such engines remain in use as this book is being written in 1968, having been for 20 years among the most powerful air-craft piston engines. The standard rating is 3,500 horsepower. Compound versions would have given over 4,000 horsepower, but these never saw service as they could not compete with the new generation of gas-turbine engines of the 1950s.

A
1 Piston (in row No. 1)
2 Connecting rod
3 Crankpin
4 Four-throw crankshaft
5 Cylinder barrel in row 3
6 Cylinder in row 4, all rows offset
7 Cam ring and valve push rods
8 Push rod casing
9 Valve rocker
10 Sodium cooled exhaust valve
11 Magneto
12 Ignition harness
13 Step-up supercharger drive gears
14 Supercharger impeller
15 Induction manifold
16 Inlet valve (in row 4)
17 Five section crankcase
18 Propeller reduction gear
19 Roller and ball thrust bearings
20 Propeller shaft

B
1 Forged steel cylinder barrel
2 Aluminium alloy head
3 Forged, hollow crown piston
4 Master connecting rod
5 Connecting rod from a lower cylinder
6 Crankpin
7 Counterweights
8 Ball front bearing
9 Propeller shaft
10 Crankcase
11 Oil filter
12 Carburettor
13 Induction manifold
14 Supercharger impeller
15 Electric starter
16 Valve drive cam ring
17 Valve push rod
18 Sparking plug

assage through a slim vertical assembly of radiator ubes fitting between the wings. The engine was olted to the upper surface of the lower wing, on he starboard side, its weight being counterbalanced y the aviator, who lay on the same wing about the ame distance to port. At the rear end of the ankshaft was mounted a flywheel to smooth ut the pulsations of the four big cylinders and ive a more even drive through a pair of gear-riven chains to the port and starboard pusher ropellers.

After their four flights on December 17, 1903, the rothers set to work to produce a much improved ying machine with a more powerful and more eliable engine. In 1908, their "Flyer No. 3" merged with a power plant which, although asically similar to the 1903 unit, had its cylinders anding upright and incorporated improvements hich raised the output to some 30 hp. This allowed he Wrights to fly with a passenger, and they ere soon making continuous flights of an hour r more.

Up to this time aircraft engines had been the ork of aircraft builders, whose aim was simply to roduce a power plant which gave sufficient horse-ower yet was as light as possible. If the engine ould lift the aircraft it was considered to be 100 er cent successful, and it mattered but little if it ad excessive fuel or oil consumption or was unable o run for more than a few minutes at a time. It was ommon practice to strip an engine down after very flight, or after any prolonged run on the round, and it was considered quite acceptable to ave to replace broken or worn parts. But gradually, s the pioneer aircraft engineers mastered the basic rt of making an engine that would lift an aero-lane, attention turned to improving reliability, uel consumption, control of power, ease of starting nd the life of the components. This process has one on ever since, until today each engine of a arge jet airliner can produce over 30,000 hp t 600 mph, yet remain undisturbed between verhauls for 10,000 hours or more at a time—in hich period the aircraft may complete 1,500 rossings of the Atlantic.

The first engine designed from the start for air-raft that represented a complete departure from ccepted car practice was the famous Gnome rotary. y 1907 the French motor industry, spurred on by ong-distance car racing and the beginning of keen ompetition in Grand Prix racing, had begun to evelop engines superior to those of other nations. ne of their brightest designers, Laurent Seguin, ssisted by his brother, embarked on the task of roducing a completely new internal-combustion otary engine, similar in mechanical layout to that esigned 20 years earlier by the Australian, awrence Hargrave. The Seguins chose the rotary rrangement chiefly because it seemed to offer the est prospect of achieving minimum weight; it also ade the engine act as its own flywheel and, in so oing, simplified the problem of cooling the

cylinders without the complication and weight of a water cooling system.

The Seguins' first rotary, which ran in 1908, had five cylinders and was rated at 34 hp for the out-standingly low weight of about 112 lb, despite the fact that it was constructed almost entirely of steel. It was a very compact and robust unit, and this enabled it to overcome most of the purely mechanical problems which were then afflicting its rival power units. The short, single-throw crankshaft was hollow and served as an inlet tube for a combustible mixture of petrol and air, produced by the simplest possible carburettor controlled by an air throttle. A uniform mixture was thus admitted to the interior of the crankcase and thence, through automatic valves in the piston crowns, to each combustion space in turn. Thus a very even combustion was obtained, in sharp contrast to the erratic burning of contemporary engines, in which the mixture fed to one cylinder did not necessarily bear close relationship to that fed to the others.

The smoothness and sweet running of the Gnome, and its unrivalled low weight, made it the favourite of many aircraft constructors. Its main dis-advantages were its relatively high fuel and oil consumption and the fact that, as it spun around, it acted like a heavy gyro wheel on the airframe and reduced manoeuvrability, especially the ability to turn against the direction of rotation of the engine. In addition, the centrifugal force flung the lubricat-ing oil out through the exhaust (and even through the inlet) valves, and a great deal of power was wasted by spinning the whole weight of the engine and churning up the air. Many thousands of Gnome rotaries were built, however, especially during the early years of the 1914-18 War; but despite its appearance in larger and improved versions, giving up to 200 hp, it began to fade from the scene by 1917 and was soon obsolete.

Although seemingly heavier at first sight, the non-rotating rivals of the Gnome were appreciably more efficient and, when allowance was made for their reduced fuel and oil consumption, they were actually lighter for all except the very shortest flights. Thus, even in the 1914-18 War, many of the engines used in really large numbers were con-ventional spark-ignition, petrol-burning units with a fixed array of cylinders and a rotating crankshaft on which the propeller was mounted. Common configurations included air-cooled radials, air-cooled and water-cooled Vee units with two rows of cylinders in the form of a V seen from the end, and water-cooled in-line engines with a single row of cylinders, usually mounted upright above the crankcase. Numbers of cylinders ranged up to 12 or 14, and maximum output to as high as 300 to 400 hp.

While Britain and France continued to produce engines with all sorts of layouts, the Germans con-centrated on six-cylinder in-line units of basically conservative design. It would have been logical for them to have taken a pair of these cylinder banks and produced a 12-cylinder Vee, but this was first

done by Rolls-Royce in Britain and, slightly later, by the international consortium (mainly American) which is reputed to have designed, built, tested and achieved large-scale production of the Liberty engine in less than three months during 1917. The Liberty had two banks of cylinders set at 45° and was designed for the very high output of 400 hp. Each cylinder was a separate component, assembled from machined steel parts and having a water cool-ing jacket welded on the outside. Bore and stroke were respectively 5 and 7 inches, so that the total volume swept by the pistons in each complete stroke was about 1,650 cubic inches or 27 litres. (At this time car engines were generally much larger in capacity than they are today, and the Liberty was not greatly different in size from the engines of the largest automobiles).

From the start the Liberty was designed for true mass production, on assembly lines of the type then being pioneered in the United States, and everything possible was done to simplify its design and ease of manufacture. Despite these measures, the Liberty engine outperformed every other power plant in the 1914-18 War and achieved not only the highly useful rating of 420 hp but also the unrivalled ratio of weight to power of only 2.1 lb per hp. Thousands were produced during the last year of the war and many were still in service a decade after the Armistice.

Basically, the Liberty and the contemporary Rolls-Royce Falcon and Eagle strongly resembled the water-cooled Vee engines of 1910. Their funda-mental principles were identical, although immense strides had been made in power and weight/power ratio and, to an even greater extent, in reliability. Iron, copper and brass had given way to aluminium alloys and high-strength steels, the rotational speed and cylinder pressures had been approximately doubled, the mixture of petrol and air supplied to the cylinders had been made far more uniform, and a solid basis of knowledge and experience had been laid which made the design of an engine much more exact and predictable.

By 1918 the radial was firmly established as the successor to the rotary. Arranging the cylinders like the spokes of a wheel had several major advantages in an engine designed for driving an aeroplane. The most important was that it enabled each cylinder to be placed in the full airstream behind the propeller, so that, with a suitable arrangement of fins and baffles, the high-speed airflow could serve to carry away all the excess heat and maintain the cylinders at a suitable operating temperature.

In engines with cylinders arranged in axial rows it is much more difficult to achieve satisfactory cooling from direct airflow, and at the time of the 1914-18 War an in-line or Vee engine nearly always had to be water-cooled. This was inevitably more complicated, usually heavier, was more vulnerable to battle damage and often incurred a greater drag penalty because of the need to have a large cooling radiator in the full airstream. In later

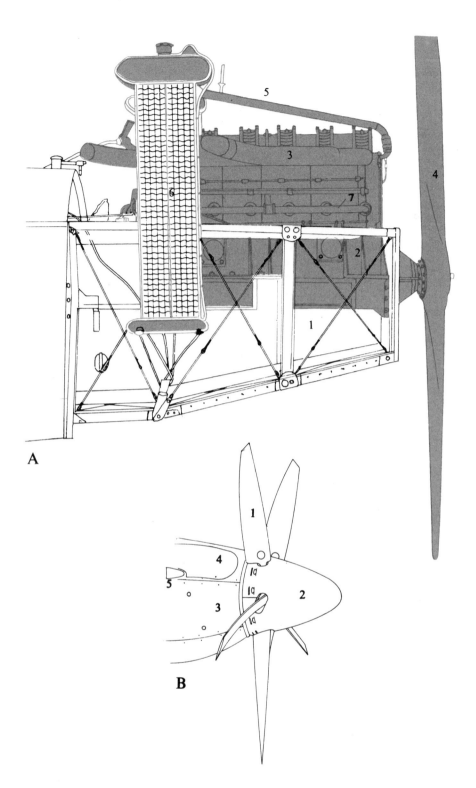

A

B

years the installed drag of both air-cooled a
liquid-cooled engines was steadily reduced, b
both forms had passionate proponents and opp
nents who argued fiercely over which was superio
Such arguments were not resolved until the end
World War II. In that conflict there was no domina
type of engine cooling, but after 1945 the pr
eminent position of the United States as a build
of civil airliners and light aeroplanes for the wor
market meant that the air-cooled engine finally wo
—in radial form for high powers and in the ho
zontally-opposed in-line configuration for ligh
planes, as will be described presently.

A second major advantage of the radial layout
that it enables all the cylinders to be arranged rour
a single crankpin on the output shaft. As a sketc
on page 192 indicates, the connecting rods from
many as nine cylinders can be brought in to th
single crank, with eight of the rods being pivoted
the ninth (termed the master rod). This allows t
engine to be made very short from front to rea
in theory it need be hardly any longer than a singl
cylinder unit. Thus excellent rigidity can be attaine
and the short but stiff crankshaft is unlikely ever
experience the torsional "whip" and vibratic
difficulties which often plagued the long in-line ar
Vee units in the early days of aviation.

Early in the 1920s the British Bristol Jupiter ar
Armstrong Siddeley Jaguar set a high standai
among radial air-cooled engines in the 500-hp clas
The Jaguar was one of the first mass-produce
radials with two rows of cylinders—the second ro
not being directly in line with the first but bein
slightly rotated to expose the rear cylinders to th
airstream passing between the cylinders of the fir
row. The single-row Jupiter was especially succes
ful and formed a staple product of the Brist
company for almost 20 years, besides being man
factured under licence in 14 other countries.

In the United States the 220-hp Wright Whirlwin
set a number of records and was instrumental in th
accomplishment of many notable flights, the mo
famous of which was Charles Lindbergh's non-sto
crossing of the Atlantic from New York to Paris i
1927. Shortly before this dramatic exploit occurre
a small group of engineers in New Jersey, many c
them former Wright employees, designed and bui
a 400-hp radial called the Wasp, and thus starte
the history of what is today America's largest aerc
engine company, Pratt & Whitney. Yet anothe
series of air-cooled engines which founded a lor
chain of successors were the British Cirrus and d
Havilland Gipsy units, both four-cylinder in-lin
engines which were the first in the world to b
designed for the new breed of light private-owne
and club machines.

As engine and aeroplane performance improve
a new problem became apparent and called for a
urgent solution—how to obtain sufficient air for th
engine at high altitudes.

As an aeroplane climbs, the atmospheric pressur
gradually falls away until, at an altitude of abou

A Streamlining was not of
& great importance when
B few aircraft could
    sustain a speed greater
    than about 90 mph,
    but during the 1920s the
    general shape and
    cleanliness of faster
    aircraft underwent a
    great change. The
    engine above, a Hall-
    Scott six-cylinder
    in-line unit dating from
    the period immediately
    following the 1914-18
    War, was installed in
    a way intended to

minimise air drag; the
water cooling radiator
was mounted on the
side, instead of being in
front as in most cars.
Just 20 years later, the
Spitfire engine
installation (bottom)
served to show that the
drag of a liquid-cooled
piston-engine could be
brought very much
lower still.

A
1 Drip tray beneath
  crankcase
2 Front engine mounting
  foot
3 Exhaust pipe
4 Propeller
5 Hot water pipe from
  engine
6 Radiator mounted
  sideways to reduce
  drag
7 Cold water return to
  engine

B
1 Five-blade propeller
2 Spinner
3 Engine cowling
4 Bulge to accommodate
  front of cylinder block
5 First of six
  starboard exhaust
  stubs

,000 feet, the pressure and density of the air around e aircraft have fallen to less than half that at sea el. This reduction in air density causes an exactly oportional loss in engine power. Motorists who oss mountain passes notice the same phenomenon d much the same sort of thing afflicts athletes rforming at high altitudes, as in the Mexico City lympics. The engine rapidly gets out of breath, d something has to be done to pump more air to it.

Even before the end of the 1914-18 War, several pes of supercharger had been tested for this rpose. Among the first were a mechanically-iven centrifugal blower developed by the Royal ircraft Establishment and test flown in an S.E.5, broadly similar Rateau compressor tested on a ispano engine, and a turbosupercharger built by e US General Electric Company which attempted get something for nothing by making use of the aste energy in the engine exhaust to drive a turbine upled to the pumping blower. Turbosuper-argers had been experimented with in Britain and rance before this, but had been abandoned in vour of gear-driven blowers. The United States, owever, persevered with the difficult task of per-cting a turbine driven by the white-hot exhaust gas d finally made scores of thousands of turbo-percharged engines during World War II, as will described later.

The main types of supercharging are outlined in iagrams on pages 192 and 195. In all cases the r is drawn into the eye of a centrifugal blower and en compressed to perhaps two or three times the ressure of the thin upper atmosphere and often to pressure even greater than that of the atmosphere sea level. This enables the downgoing pistons to raw in a much greater weight of air than would therwise be the case. The important factor is that is also means a correspondingly greater weight of xygen, which makes up about 21 per cent of the tmosphere, because it is the quantity of oxygen rawn into the cylinder that determines the amount f fuel that can be burned in it and, hence, the ower produced.

Sometimes a supercharger is used purely to boost e power of an engine, even at low altitudes, as has equently been done with high-performance cars. lowever, this requires that the whole engine be pecially strengthened to bear the increased perating stresses, and it is much more common in ero-engines to use a supercharger to restore ower lost at high altitudes.

Despite the apparently obvious merit of super-harging, engines so equipped were slow to come to use. The first use of a supercharger was in the aguar IVS engine of the Siskin IIIA fighter which ent into service with the Royal Air Force in 1926-7. Small numbers of supercharged engines were rought into military service in other countries efore 1930 in a constant endeavour to improve erformance at altitude. But the basic engineering now-how of that period was still insufficient to pro-

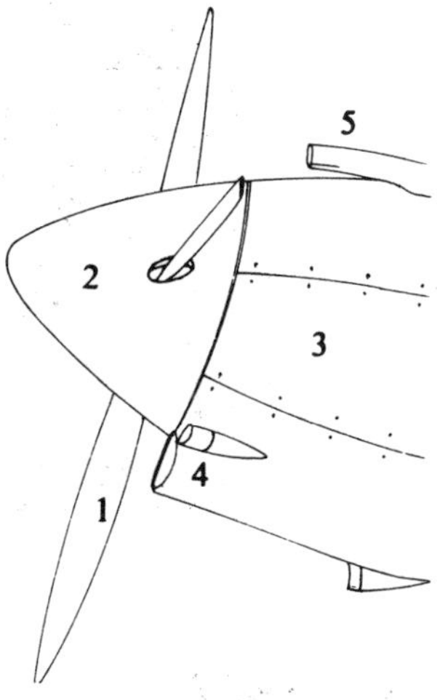

C

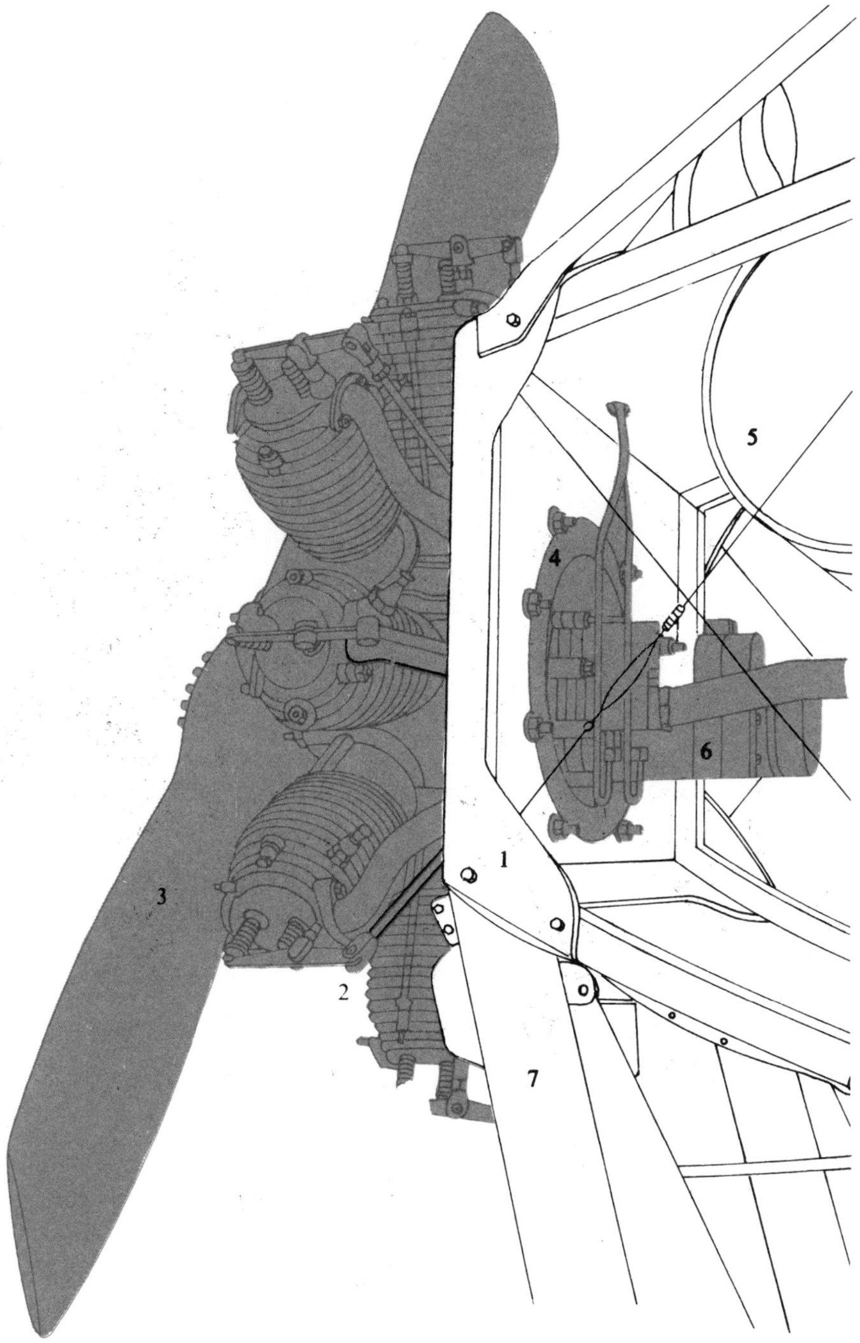

D

C Forming a parallel with & the illustrations
D opposite, these drawings show the advance that was made in the installation of air-cooled radial engines. In most early flying machines the engine was simply fixed securely to the appropriate place and coupled up to a supply of fuel. The example above shows a ten-cylinder Anzani radial (not rotary),

which was bolted to a steel plate mounted square-on to the slipstream on the front of the four strong longerons forming the backbone of the fuselage. Less than 30 years later the Curtiss XP-42 (top) appeared with a Pratt & Whitney Double Wasp engine, with an extension shaft to the propeller and housed in a finely streamlined cowling.

C
1 Three-blade propeller
2 Spinner
3 Engine cowling
4 Cooling air intake
5 Carburettor air intake
D
1 Fuselage nose bolted to upper and lower longerons
2 Fixed radial engine
3 Propeller
4 Support plate and engine mounting ring
5 Oil tank
6 Magneto
7 Undercarriage strut

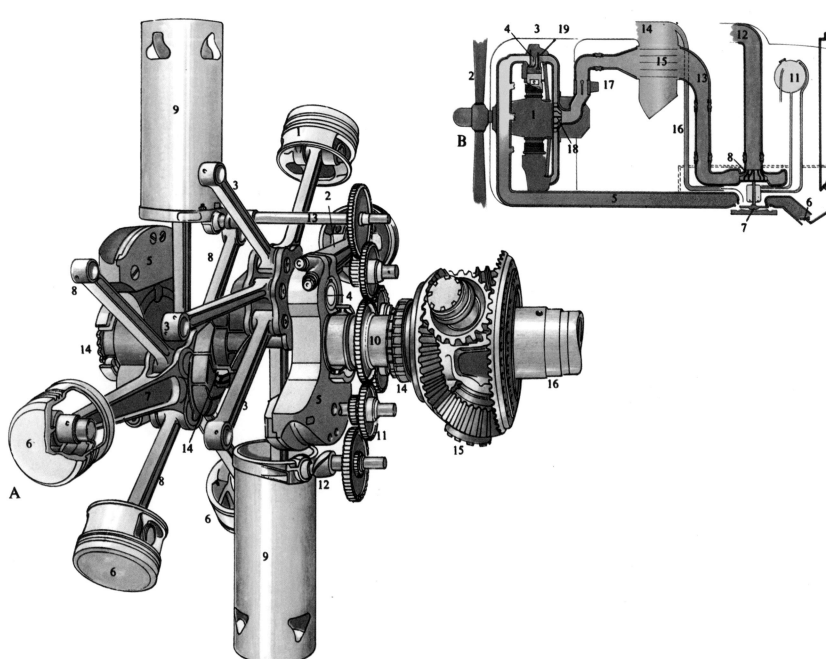

A *From 1925 until 1950, the Bristol Aeroplane Company developed aircraft piston-engines with sleeve valves in place of the more common poppet type. This drawing of the main moving machinery inside the company's Hercules engine shows two of the 14 sleeves used in the two rows of seven cylinders each. The sleeves are driven by cranks so that they rotate and oscillate inside the cylinder, with the piston running inside the sleeve. Charge admission and exhaust expulsion take place as the appropriate ports in the sleeve come opposite*

corresponding ports in the cylinder wall. Also visible are the pistons and connecting rods and the reduction gear (on the right) to the slower-running propeller shaft. Over 57,000 Hercules were delivered during World War 2, and civil and military versions continued in production for several years afterwards, combining powers of around 2,000 horsepower with long life and high reliability.

1 Piston in front row of cylinders
2 Front master rod
3 Slave connecting rod (front row)
4 Front crankpin

5 Counterweight
6 Pistons in rear row (seven)
7 Rear master rod
8 Articulated slave connecting rods (rear)
9 Sleeves (two of 14)
10 Crankshaft and main sleeve drive pinion
11 Front sleeve drive gears
12 Drive crank to sleeve shown in front row
13 Drive shaft to sleeve shown in rear row
14 Rear, centre and front bearings
15 Propeller reduction gear
16 Propeller shaft

B *This drawing is based on one prepared by the General Electric Company (USA) who for 50 years have been world leaders in the art of turbosupercharging aircraft piston-engines. The white-hot exhaust gases are piped to a waste gate through which they can escape direct to the atmosphere in the usual manner. As the aircraft climbs, an automatic boost regulator unit steadily closes the waste gate shutter, so that the gases can escape only by passing through a nozzle box, to drive a turbine. The latter spins at 30,000 to 50,000 revolutions per*

minute, and on the same shaft is a centrifugal blower that super-charges the engine. In this diagram the engine also has a second, mechanically driven supercharger—an arrangement adopted in such aircraft as the Liberator, Superfortress and Stratocruiser. It was General Electric's experience with turbosuperchargers that led to their selection in 1941 as the company to bring the British invention of jet propulsion to the United States.

1 Engine crankcase
2 Propeller
3 Engine cowling
4 Exhaust valve
5 Exhaust pipe
6 Waste gate control valve
7 Gas turbine
8 Supercharger
9 Pilot's boost lever
10 Automatic control system
11 Cooling and lubricating oil
12 Ram air intake
13 Air delivery manifold
14 Airflow through intercooler
15 Cross-flow intercooler
16 Cooling air to turbosupercharger
17 Carburettor
18 Mechanically driven supercharger
19 Inlet valve

uce a reliable engine giving high power at altitudes uch above 10,000 feet. If a centrifugal blower as used, of the type commonly found in a vacuum eaner, it required not only the most careful aero-ynamic design but also the use of step-up gears turn it at perhaps ten times the crankshaft speed. loreover, if the blower was kept in use all the time, pumped too much air into the engine at low ltitudes, or on take-off, and caused it to be over-ressed. This meant that a clutch had to be pro-ided, and neither clutch nor gears could be made ouble-free until well into the 1930s.

As related earlier, much the sharpest spur to ircraft development before 1930 stemmed from the chneider Trophy contests. The objective in these aces was simply to fly round a closed circuit at low ltitude as fast as possible, and this called for use f an engine supercharger not to improve per-ormance at altitude but to boost power low down. hanks to the financial generosity of a private in-ividual, Lady Houston, Rolls-Royce engineers ere able to produce in the final Schneider Trophy ower plant an engine whose performance was not qualled until the end of World War II. This re-narkable unit was the "R" engine, developed rom the Rolls-Royce "H" (later named the luzzard) by a substantial redesign.

Externally, the most obvious alteration was the ddition of a supercharger of exceptional capacity, vhich raised the pressure in the induction manifold o some 12 pounds per square inch above atmos-heric (in other words, it almost doubled it and hus practically doubled the power, although the upercharger itself absorbed several hundred horse-ower). Such a "boost pressure" was previously nheard of. At the same time, the Rolls-Royce esigners stressed the "R" engine not only to bear he enormously increased loads caused by the reater power but also to run at a speed raised from ,200 to over 3,000 rpm. A third major change vas to increase the compression of the combustible nixture inside the cylinders, by leaving a smaller pace above the piston at the top of its stroke. This hange, in particular, meant that special fuel had to e employed. In the 1929 Schneider Trophy race, he "R" delivered about 1,900 hp to the propeller, nore than double the power of the original "H" nit. For the 1931 race the power was raised to ,300 hp, and in the same year a special "sprint" ersion was cleared to deliver 2,600 hp for short eriods. With this engine, the S.6B seaplane achieved world speed record of 407·5 mph. Just as the esign of the Schneider seaplanes exerted a marked eneficial effect on the next generation of military ircraft, so did their engines—and the Rolls-Royce R" in particular led the way to the highly-rated iston-engines of World War II.

There is another problem connected with super-hargers that has to do with the propeller rather than he engine. As we have seen, in an unsupercharged ngine the fall in power as the aircraft climbs s almost exactly related to the fall in air density.

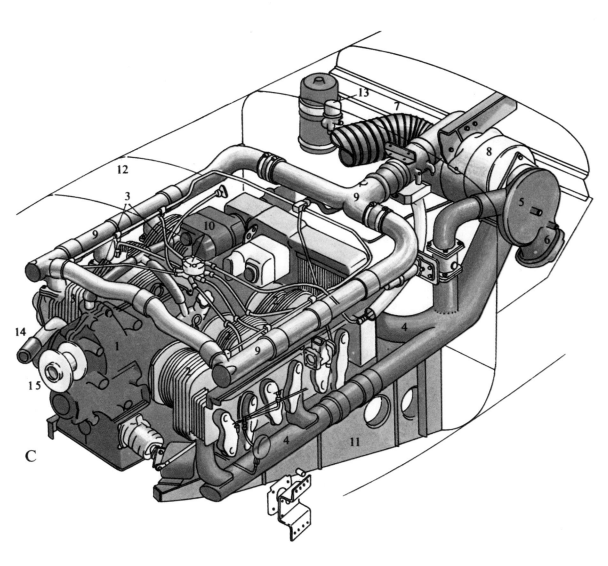

C *The only fields of aviation in which piston-engines continue to be pre-eminent are those of light aeroplanes and helicopters. Undoubtedly, the gas-turbine, described in the next section of this book, will continue to encroach on the piston-engine's last remaining market and is being staved off only by the steady improvement of light piston-engines themselves. Typifying modern lightplane power is the installation of a pair of Continental flat-six (horizontally-opposed, six-cylinder) engines in the Beagle 206 Srs. II. The engine of this high-performance twin is the GTSIO-520C, rated at 340 horsepower; the designation means "geared, turbo-supercharged, direct fuel injection, opposed cylinders and 520 cubic inches swept volume". The induction pipes to the supercharger lie along each side, above the cylinders; the exhaust pipes to the turbine lie below them and the fuel injection distributor is seen in the centre of the top of the engine.*

1 Engine crankcase
2 Port row of three cylinders
3 Starboard cylinders offset further aft
4 Exhaust pipes
5 Gas turbine plenum chamber
6 Exhaust outlet stack
7 Air intake hose
8 Turbosupercharger casing (airframe mounted)
9 Air delivery manifold
10 Fuel injection distributor
11 Engine bearer beam
12 Outline of engine cowling
13 Supercharger control unit
14 Electric generator and ram air cooling intake
15 Propeller hub

The power that the blades of the propeller can impart to the thinning air falls away at the same rate, so the two remain balanced. But with a supercharged engine the power can be held at the full sea-level value up to a height of many thousands of feet, according to the engine design, whereas the power that can be absorbed by the propeller inevitably falls as before.

To keep the propeller properly matched to the engine requires either a variable-ratio gearbox in the drive shaft, or blades of variable camber or area, or else a means for changing the pitch of the blades—their setting to the airflow. The last scheme has been found the simplest to put into practice, and it is also the best answer for matching the propeller to the aircraft both at the minimum flying speed and at the maximum speed of which the machine is capable. Propellers are much more complicated than they appear to be, and are discussed on pages 232-5.

An earlier development, in the years following the 1914-18 War, was the introduction of reduction gearing in the drive to the propeller. This was found increasingly necessary in order to match the characteristics of the engine with those of the propeller. The first aero-engines ran at moderate speed and were well matched to their propellers, but the quest for higher power/weight ratios caused the rpm of crankshafts to rise, while increased power outputs demanded propellers of larger diameter. Both factors tended to raise the tip speed of the propeller and the point was reached in the early 1920s when tip speed neared the speed of sound, causing excessive noise and poor efficiency. The solution was to use a still-larger propeller to handle the power, but to turn it much more slowly by gearing it down from the engine, the latter being allowed to run as fast as practicable.

By 1939, geared drive was almost universal on high-powered engines, but low-powered units continued to use direct drive without overspeeding their small-diameter propeller—an extreme case being the North American T-6 Texan (Harvard) trainer of 550-600 hp, the propeller of which could easily exceed sonic speed and cause loud tip noise. Today even light aircraft frequently have geared propellers, to allow their engines to run at over 3,000 rpm with minimum noise.

One of the more obvious examples of "spin-off" from the Schneider Trophy, and from other forms of aerial racing, was the progressive improvement in streamlining that took place between 1918 and 1939. Both air-cooled and liquid-cooled engines altered their appearance considerably over this period, as drawings on pp. 190-191 emphasize. In the early days, when 100 mph was a good cruising speed for any aeroplane, aerodynamic drag was seldom a prime consideration and it usually took second place to achieving low cost or easy maintenance. But once speeds climbed to 200 mph and beyond it became imperative to reduce drag to the lowest possible level, and the contribution which a properly

faired engine installation can make to this aim is very great.

In the case of the liquid-cooled Vee power plant, the main paths towards minimum drag were to enclose the whole engine in a smooth cowling—if possible making this to follow on naturally from the lines of the "spinner" covering the propeller hub—and to place the cooling radiator in a duct in such a way that, instead of sticking out in the airstream and causing high drag, it actually helped to propel the aircraft by heating the air passing through it, rather like a miniature jet engine. In some World War II aircraft this was actually achieved, two of the best examples being the de Havilland Mosquito and North American Mustang.

At first, an air-cooled radial appeared more difficult to fair smoothly into the airframe because the cylinders have, by definition, to be left in a high-velocity airstream in order to be adequately cooled. But the dual problem of improving cooling and reducing installed drag was largely overcome in the second half of the 1920s by the use of special ring cowlings, resembled a wing, which were wrapped round to form a tube surrounding the engine. The use of an aerofoil-type section enabled the cowling almost to "pull itself along" when placed in the slipstream from the propeller. This installation was further improved by making the trailing-edge of the cowl in the form of a series of hinged flaps which could be opened to maintain cooling at low speeds. These "gills" have been used in many forms to control the cooling airflow, invariably in conjunction with fixed baffles around and between the cylinders.

Reverting to liquid cooling, comparable advances took place in the cooling circuits during the two decades between the wars, to keep these power plants fully competitive. In the earliest aero-engines, the usual liquid coolant was simply water, as it generally is today in a car. But before the end of the 1914-18 War experiments were in hand at the Royal Aircraft Establishment to find better liquids, and this work spread later to the United States and other countries.

The ideal cooling fluid would be one which remained a liquid over the widest possible range of temperatures. Motorists put anti-freeze in their cars, consisting principally of glycol, and this simple hydrocarbon also has the advantage of a high boiling point. In 1935 ethylene glycol was adopted, in an aqueous solution, as the coolant for the famous Rolls-Royce Merlin (page 195), and this decision not only eliminated the problems of freezing and boiling which would otherwise have been encountered by the Hurricane and Spitfire, but also enabled the radiators of these fighters to be made smaller and thus have lower drag. By 1939, glycol or related fluids were standard coolants on every liquid-cooled aero-engine.

The basic mechanical design of piston aero-engines did not change fundamentally between 1918 and the end of the piston era, but it improved

steadily as a result of the introduction of new materials and detail refinements. One of the more significant changes, which did not sweep the board but merely increased variety, was the sleeve valve.

Most four-stroke petrol engines use poppet valves of the form sketched on page 188. One valve is the inlet, which opens when the piston is near the top of its stroke and remains open until the piston has gone well past "bottom dead centre" in order to draw in the greatest possible charge of combustible mixture. The other is the exhaust valve, which opens just before the piston's next passage through bottom dead centre and closes at the end of the exhaust stroke, after the inlet valve has once again begun to open. In many aero-engines there are two of each type of valve in each cylinder.

Valves of this type are driven through fairly complicated arrangements of camshafts or cam rings, push rods, rocker arms and multiple springs. In the mid-1920s the Burt-McCollum sleeve valve appeared as one of the best of the many alternative ideas which were then being put forward. Sleeve valves were at that time being adopted for the Daimler car engine and tests appeared to show that this simpler system could be made to work.

As the left-hand sketch on page 192 shows, the sleeve is a steel cylinder without ends but with carefully profiled ports cut in its sides. It fits inside the cylinder and, in turn, acts as a lining in which the piston slides smoothly. At the bottom of the sleeve is a drive crank which oscillates it in a near-circular motion when the engine is running, so that the apertures in its sides successively uncover the inlet port and the exhaust port. One of the few drawbacks of the sleeve valve is that it is hard to start at low temperatures, since the film of oil on which it slides inside the cylinder becomes viscous in the cold and makes it difficult to move.

By 1926 the Bristol Aeroplane Company was working on sleeve valve engines which led to the famed family of Perseus, Taurus, Hercules and Centaurus. The Napier Sabre, a remarkable 2,200-3,000-hp engine of World War II, used 24 sleeve valves in its two banks of horizontally-opposed cylinders, and the 3,500-hp Rolls-Royce Eagle did likewise.

Sleeve valves for aero-engines were pioneered by Britain, but another development of comparable magnitude was introduced in Germany—direct injection of fuel into the cylinders. In early aircraft engines, as in most car engines today, the fuel was vaporized in a conventional float-chamber carburettor which served to maintain the fuel at the correct level so that a stream of vapour could be drawn off by the incoming air passing through the constriction in the choke tube. The reduction in pressure at this constriction draws off the necessary plume of vapour to turn the airflow into a flow of combustible mixture, but it also has the effect of abruptly lowering the temperature in the choke tube. As a result, choke tubes are prone to acquire ice when the air coming in is already cold and

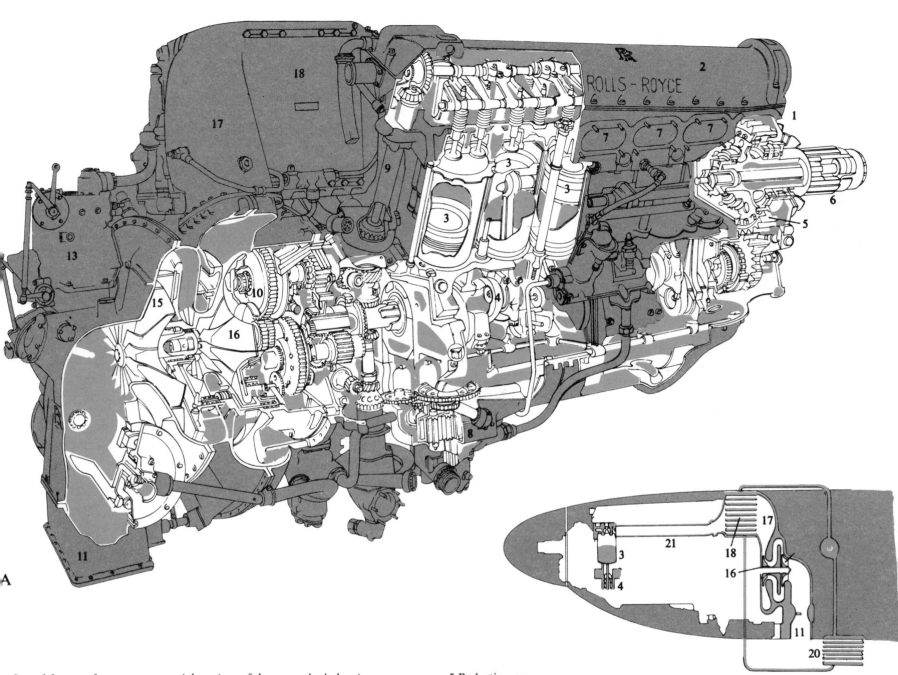

A

One of the most famous engines of all time, the Rolls-Royce Merlin was named not for the legendary wizard but as one of a long series of engines by this company all bearing the names of birds of prey. Starting life at under 900 horsepower in 1935, it was delivering some 1,100 horsepower by the time of the Battle of Britain, in all the RAF's Hurricanes and Spitfires. Later versions gave 1,500 to 1,750 horsepower and were specially designed to offer an enormous increase in power at high altitude. By the winter of 1944-45,

special versions of the Merlin were being run at powers of up to 2,640 horsepower; their brake mean effective pressure was a near-incredible 404 pounds per square inch, and the ratio of weight to power was barely 0.6 pounds per horsepower. All Merlins had 12 liquid-cooled cylinders of 5.4 inches bore and 6 inches stroke, giving a swept volume of about 1,650 cubic inches or 27 litres. This drawing shows one of the later versions, with a two-stage supercharger, the two centrifugal impellers for which are clearly visible in

both drawings.
Over 150,000 Merlins were built, almost all of them during World War 2, for such famous aircraft as the Hurricane, Spitfire, Battle, Defiant, Lancaster, Mosquito, Halifax, Whitley, York and Hornet, and for the American P-51 Mustang and some P-40 Warhawks. A civil version was also much used in Yorks, Lancastrians and Canadair Four "Argonaut" airliners.

1 Starboard bank of six cylinders
2 Valve gear cover
3 Pistons
4 Crankshaft

5 Reduction gear
6 Propeller shaft
7 Exhaust ports
8 Oil pressure and dual scavenge pumps
9 Starboard camshaft drive
10 Supercharger step-up gears
11 Air intake manifold
12 Heated barrel type throttle
13 Automatic injection-type carburettor
14 Fuel injector
15 First stage supercharger
16 Second stage supercharger
17 Air delivery manifold
18 Water cooled intercooler
19 Water pump
20 Intercooler radiator
21 Induction manifold

moisture-laden. In the first 40 years of aviation, thousands of pilots suffered engine failure because of carburettor icing. The obvious palliative was to find a way to keep the choke tube warm, and one way to do this was to keep it fed with a flow of hot oil channelled through pipes in the wall.

Carburettor heating could not, however, overcome an equally basic defect of the conventional carburettor which has already been mentioned—bad mixture distribution. If the mixture is led away through long and tortuous manifolds its consistency tends to vary, so that by the time it reaches the cylinders some parts of the engine receive a stronger mixture than others. In cars this difficulty is sometimes overcome by using two, three or even four carburettors connected to shorter manifolds. In aero-engines this policy has not been followed, partly because the carburettors would all have to discharge their mixture into the single supercharger.

One way of overcoming completely the problem of carburation and mixture distribution—and the drawback that a normal carburettor cannot operate upside down for more than a few seconds—is to adopt the diesel cycle. In this, the fuel is not added to the incoming air, which is taken straight to the cylinder and then compressed to about one-fifteenth of its original volume. This great compression makes the air very hot, and as soon as the oil fuel is injected into it the fine spray of droplets ignites and burns. Thus the diesel is often called a compression-ignition engine. It needs no separate ignition system, but the great compression means that the structure of the engine must be exceptionally robust and heavy and this, in turn, tends to limit the running speed.

Diesels were used widely for airship propulsion, but were little favoured for aeroplanes, except for a unique series of engines produced by the German Junkers company from 1928 onward. In these Jumo diesels, two pistons were mounted in each cylinder, head to head, with the combustion space between them. The engines were of the two-stroke type—in which every downstroke of the piston is a firing stroke and every return an exhaust—as is the case with most small motor-cycles. The only real advantages of the diesel are that, compared with the four-stroke petrol engine, it has higher thermal efficiency and burns what may be a safer fuel from the viewpoint of fire risk. But, in aircraft use, the much greater weight for a given power made it unattractive even in applications calling for considerable range, in which the diesel's fuel economy would have been an advantage.

Germany persevered for a long time with the Jumo diesels, but had virtually abandoned them by 1945. Their greatest effect was to give German aero-engine designers an appreciation of the extremely precise fuel injection systems needed by such power plants. After 1933, when German designers began full-scale work on a broad range of high-power aircraft engines, the attractions of fuel injection for non-diesel units led to a progressive abandonment of the carburettor in favour of various types of direct injection. By 1939, all the main aero-engines for Luftwaffe combat aircraft were fitted with direct injection, the fuel being metered precisely according to the needs of the engine operating conditions and delivered in measured doses by small plunger pumps, each serving one cylinder. The arrangement is inevitably more complex than that with most carburettors, but it is not prone to icing, operates perfectly when inverted and, in many cases, can lead to increased performance.

British trials with direct injection, fitted to a Bristol Pegasus radial, showed no apparent advantage; but in the years following 1945 the few remaining high-power piston aircraft engines were nearly all converted to direct fuel injection. The remaining carburettors were themselves changed in principle so that, instead of having a float chamber and choke tube, they metered the required fuel and injected it at some point in the induction manifold. Today fuel injection is common even on lightplanes and is coming into use on high-performance cars.

During World War II, the great pressure of a fight for survival brought about extraordinary technical progress in aviation. To fight a war, an air force needs proven hardware and not a succession of untried new projects; yet despite this fully-appreciated fact the warring nations virtually doubled the average power of their aero-engines between 1939 and 1945.

In most cases this was achieved largely by a vast increase in the power which could be obtained from the same basic design. The classic instance of this policy will always be the British Rolls-Royce Merlin (page 195), which powered more than 70,000 of the fighters and bombers used by the Allies. The original rating of the experimental Merlins of 1934 was 720 hp. By 1936 the Mark I engine was entering RAF service at 990 hp, and during the Battle of Britain the most common engine of the Hurricanes and Spitfires was the Merlin II of 1,030 hp. By 1945, large numbers of Merlins were in service with maximum powers approaching 2,000 hp and an experimental version surpassed 2,500 hp on the test-bed, using special fuel.

This exceptional rise in power can be attributed largely to a tremendous advance in the technology of supercharging. Following on from their experience with the racing Rolls-Royce "R" engine, the Merlin design staff unceasingly improved the performance of their centrifugal impellers to force ever-greater charges of combustible mixture into the modest-sized (27-litre) engine, and in 1940 they decided to adopt a two-stage arrangement in which the air is compressed first by one impeller and then by a second. The resulting Mark 60 and successive Merlins achieved very much greater powers at high altitude, the more powerful superchargers being used primarily for this purpose rather than for boosting output low down. As a result, the Merlin 61, which went into service with the Spitfire IX in 1941-2, developed double the power at 40,000 feet that could be reached by earlier Merlins at a much lower altitude.

In the United States, two-stage supercharging was common in a different form, in which the first stage of compression of the air took place in a turbo-supercharger. As already stated, General Electric built a turbosupercharged aero-engine during the 1914-18 War, and this was tested at altitude in 1918. After a long period of technical development—much of which was devoted to a search for improved heat-resistant materials similar to that demanded later by gas-turbines—G.E. turbosuperchargers were cleared for service use in 1939, and in that year the first Boeing B-17 Fortress fitted with such engines was delivered to the Army Air Corps (which already had many B-17 bombers of earlier types). The new engines gave a great improvement in altitude performance and all subsequent Fortresses had such power plants.

The engine of the B-17 was the Wright Cyclone nine-cylinder radial. This became famous in 1934 when early 710-hp versions powered the Douglas DC-2 airliner which came second in the England-Australia air race. Wright improved the Cyclone throughout the 1930s, primarily for airline use, in keen competition with rival products from Pratt & Whitney. Although the latter company continued to make enormous quantities of single-row radial engines known as the Wasp and Hornet, most of their development effort was by 1935 concentrated upon Twin Wasps of various kinds. These units, also known as the R-1830 in the American system of designating engines according to their cubic inches of swept volume, were not produced by combining two nine-cylinder Wasps into one engine but by using two rows each having seven cylinders slightly smaller than those of the original Wasp. The first Twin Wasps were rated at about 800 hp when they appeared in 1930, and subsequent versions generally remained rather more powerful than the contemporary Wright Cyclones.

Most Twin Wasps were rated at 1,200 hp, and engines of this output were fitted to the DC-3 and its military variants, as well as to the B-24 Liberator bomber and many other Allied aircraft of World War II. There is little doubt, in fact, that the Twin Wasp was made in greater numbers than any other single basic design of aircraft engine.

The Twin Wasps fitted to the 19,000 Liberators were equipped with a single-stage gear-driven supercharger and an exhaust turbosupercharger similar to that of the engines of the B-17. But by the mid-1930s it was clear that a much larger engine would be needed for both civil transports and combat machines. Wright engineers were forging ahead with the R-2600 (2,600 cubic inch) and R-3350 double-row Cyclones, and to meet this great challenge Pratt & Whitney scrapped many unpromising projects and concentrated upon the R-2800 Double Wasp. Planned initially to give 1,800 hp, it was giving combat powers of up to 3,400 hp by the end of World War 2 and all in all it contributed more

1 *1903 Manly 52 hp*
2 *1908 Anzani 24 hp*
3 *1908 Wright 30 hp*
4 *1909 Gnome 34 hp*
5 *1918 Rolls-Royce Eagle 360 hp*

6 *1929 Hispano Suiza 12 Nbr with reduction gear*
7 *1918 Liberty 400 hp*
8 *1922 Cirrus Hermes 80 hp*

9 *1931 Rolls-Royce R 2,600 hp*
10 *1925 Pratt & Whitney Wasp 400 hp*
11 *1922 Bristol Jupiter 400 hp*

12 *1936-50 Rolls-Royce Merlin 990-2,300 hp*
13 *1942 B.M.W. 801 1,580*
14 *1953 Wright Turbo-Compound 3,500 hp*

horsepower to the Allied nations than any other engine.

As early as 1940, a Double Wasp, then giving about 2,000 hp, had driven a Corsair navy fighter at 405 mph, and the combination of unparalleled power and speed did much to confirm Pratt & Whitney and the U.S. Government in their belief in air-cooled engines.

After the conflict, Double Wasps continued in production for later variants of the same fighter, as well as for the civilian DC-6, Martin 2-0-2 and 4-0-4 and Convair 240, 340 and 440. Meanwhile, Pratt & Whitney reached the pinnacle of piston-engine technique with the R-4360 Wasp Major, in which four rows of seven cylinders, looking rather like a corn-cob, produced 3,500 hp, and more, for the B-36 and B-50 bombers, C-97 and Stratocruiser transports and other large aircraft. But the projected 4,500-hp compound Wasp Major never appeared because its timing brought it into the gas-turbine era.

In contrast, Wright's plan for a compound version of the R-3350 Cyclone did materialize, in the early 1950s. The idea of making a compound engine had appealed to designers from 1910 onwards, because the bulk of the heat released by the combustion of the fuel is wasted in the exhaust. Surely, it was reasoned, it should be possible to harness some of this energy and put it to use in propelling the aircraft? The turbosupercharger offered one method of doing this, as its gas-turbine extracted energy from the exhaust and used it to turn a shaft spinning the supercharger blower. The basic arrangement for a compound engine is to use the same type of gas-turbine not to drive a supercharger but to put extra power straight into the main propeller shaft.

In the Wright Turbo-Compound family of engines, the exhaust gases from the 18 cylinders escape through three "blow-down" turbines arranged at 120° intervals around the rear of the crankcase. Each turbine is mounted on a shaft geared directly to the main engine crankshaft. At sea level, this arrangement increased total power output from about 2,700 hp to an initial rating of 3,250 hp, later raised to 3,500 and 3,700 hp. These increases in power were accomplished with virtually no rise in fuel consumption, the net effect being to extend aircraft range by 20 per cent. It was the Turbo-Compound engine which, in 1952, enabled Douglas and Lockheed to build the DC-7 and later Super Constellation airliners with range sufficient to fly across the United States non-stop. By 1956 the later Turbo-Compounds had been matched with still further-refined airframes in the DC-7C and L-1649 Starliner, both capable of non-stop transatlantic operation without restrictions.

By 1945 Britain also had completed basic studies suggesting that the future lay in compound power-plants, but that country's remarkable advance into the field of gas-turbines caused her engineers to consider a different type of compounding. In the Napier Nomad, developed in 1950-57, the exhaust

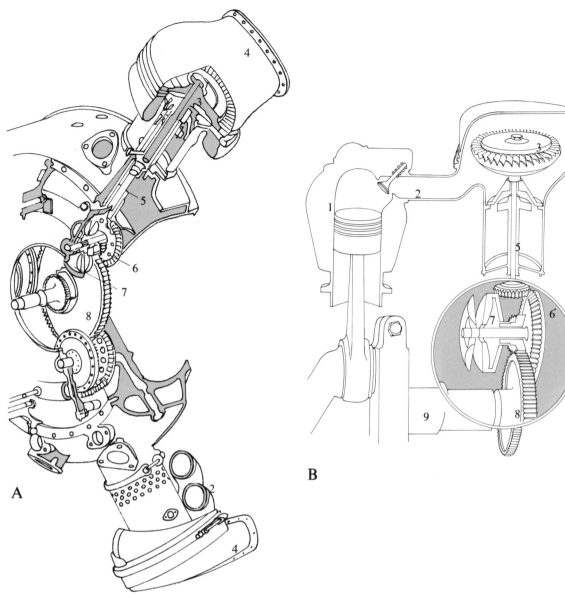

A

B

A Probably the most
& advanced aircraft
B piston-engine ever to go into full-scale service, the Wright Turbo-Compound is the only compound unit ever to have become widely accepted in aviation. Based on the established R-3350 (3,350 cubic inch displacement) Cyclone 18-cylinder two-row radial, used in the B-29 Superfortress and the original types of Lockheed Constellation airliner, the Turbo-Compound recaptured much of the energy that would otherwise be lost in the exhaust gases in much the same manner as is done with turbo-

supercharging. The big difference is that the exhaust gas turbine does not drive a supercharger blower but is geared directly to the main engine crankshaft, so that its power is transmitted straight to the propeller. The simplified schematic drawing at right shows how the "blow-down" turbine spins a fluid coupling to smooth out speed differences in the drive and, thereby, is geared to the rear end of the crankshaft. The more detailed illustration shows two of the three blow-down turbines fitted to the Turbo-Compound engine,

spaced at 120° intervals around the rear. Each puts into the engine some 300 horsepower that would otherwise be lost, raising the maximum power from the 2,700 horsepower of the normal R-3350 to a peak of 3,700 horsepower in some Turbo-Compounds. It was the Turbo-Compound that enabled the DC-7 to fly non-stop schedules coast-to-coast over the USA in 1953 and the DC-7C and L-1649 to operate non-stop Atlantic services in 1956.

1 Engine cylinder
2 Exhaust manifold
3 Exhaust gas turbine
4 Exhaust discharge du
5 Quill shaft driving to crankshaft
6 Reduction gears
7 Hydraulic clutch
8 Main turbine drive pinion on crankshaf
9 Crankshaft

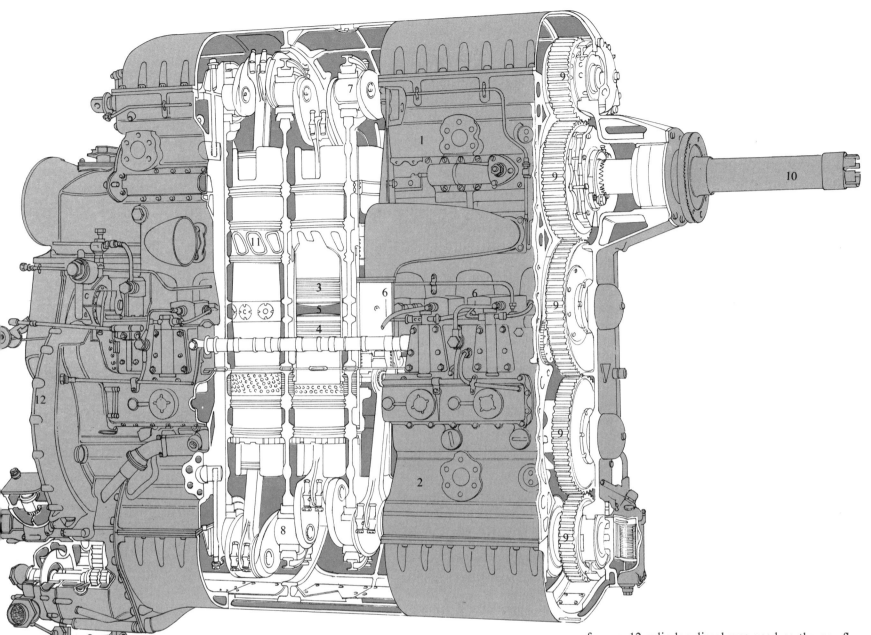

The only compression-ignition (diesel) engines ever to have made a real impact on aviation were the Jumo diesels developed by the German Junkers company from 1928. Operating on the two-stroke principle, they were very high and narrow because the single cylinder block housed six steel liners each having a combustion space in the centre and opposed pistons driving crankshafts at the top and bottom, the propeller shaft being driven through trains of gears. This arrange-ment enabled the pistons to uncover the inlet and exhaust ports, thus dispensing with valve gear, while a supercharger blower blew fresh air through the combustion space between each pair of pistons on each exhaust cycle. This illustration shows a Jumo 205 of 700 horsepower. With a turbosupercharger, the otherwise identical Jumo 207 gave 1,000 horsepower at 3,000 r.p.m., yet weighed only 1,430 pounds. All Jumo diesels had specific fuel consumptions of about 0.35 pounds of heavy fuel oil per horsepower per hour, appreciably below the best attained by conventional aircraft piston engines.

1 Upper bank of cylinders
2 Lower opposed bank of cylinders
3 Upper piston No. 4
4 Lower piston No. 4
5 No. 4 combustion space
6 Oil fuel injection pipes
7 Upper crankshaft
8 Lower crankshaft
9 Propeller drive gears
10 Propeller shaft
11 Upper inlet/exhaust ports
12 Supercharger casing

from a 12-cylinder diesel was used as the gas flow to drive an axial turboshaft unit geared to the diesel by way of a special variable-ratio coupling. The objective was to achieve the lowest possible fuel consumption, but although the Nomad promised to do this it was swept away by the rising tide of gas-turbines.

Even in the field of light aircraft the gas-turbine is beginning to make inroads into what was long regarded as the final safe haven of the piston aero-engine. Most lightplane engines today have horizontally-opposed cylinders arranged on left and right sides of a central crankcase. This gives good balance, easy accessibility, a clean cowling and effective air cooling, and is often used as the basis for an engine with a supercharger (gear-driven or turbo), direct injection and a geared drive to a constant-speed propeller of the type described on pages 234-5. But steady improvements in small turbine engines, combined with a strong wish to "keep up with the Joneses", are making the indefinite continuance of piston-engines in aviation less and less certain.

GAS TURBINES

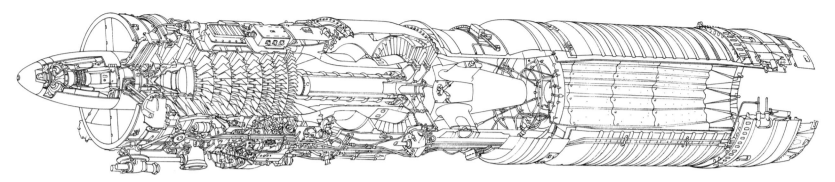

GAS TURBINES exist in many forms. Perhaps the simplest are the small windmills which, turned by the rising column of hot air from an electric lamp, either spin a lampshade or else produce a flickering pattern of light and shade to simulate a coal fire in what is really an electric heater. Gas-turbines for power use were suggested in the 19th century and industrial sets were in production in Switzerland in the mid-1930s. In fact, in the patent literature gas-turbines appear at intervals back to a British proposal of 1791.

In the field of aircraft propulsion, various proposals appeared from about 1910 for the use of some form of turbine power plant instead of a piston-engine. Serious engineering calculations for gas-turbine aircraft propulsion began at the Royal Aircraft Establishment at Farnborough in the mid-1920s, and H. H. Constant and A. A. Griffith—both renowned names in the gas-turbine field for 40 years afterwards—published papers on the characteristics of turboprops, in which the turbine would be used to drive a conventional propeller. But it was the unknown cadet at Cranwell, Frank Whittle, who proposed to combine the use of a gas-turbine with elimination of the propeller to produce the turbojet.

Whittle was called upon to write a thesis in his final term at the Royal Air Force College in 1928, and chose as his subject "Probable future developments in aircraft design". In it he outlined the possibility of creating a wholly new power plant to accomplish jet propulsion. Of course, every powered aircraft is jet propelled, because even a propeller serves only to create a large-diameter jet of air flowing rearward in relation to the still air around the aircraft. But Whittle's jet was to come out of a pipe leading from the heart of the engine; it was to consist of all the hot gases produced in the engine and to move at a speed much higher than the cool air behind a propeller.

As Newton showed in 1680, if a flow of gas is at some point accelerated by the addition of energy, there will be a reactive force acting on the walls of the containing vessel. Whittle's idea was that an engine should be built containing a large-capacity compressor, identical in principle with a piston-engine supercharger, which would draw in fresh air, compress it and feed it to a combustion chamber. There fuel would be injected and burned, convert-ing the airflow into a supply of hot gas of greatly increased energy. This flow of gas would then be allowed to escape by expanding through the blades of a turbine. The turbine would be mounted on a shaft driving the compressor, and in consequence would extract much of the energy from the gas stream. But, reasoned Whittle, there would still remain enough excess energy in the gas stream for useful thrust to be generated when it was allowed to escape to the atmosphere through a suitable form of propelling nozzle. The net result of all the forward and rearward forces acting on the various parts of the engine air/gas duct, on the compressor, combustion chamber and turbine, would be a substantial forward propulsive force.

The beauty of the arrangement was that it did not appear to be prone to sharp limitations in forward speed. The piston-engine and propeller were, in 1928, just capable of reaching 350 mph, and aircraft designers were striving to produce a practical military aeroplane capable of 200 mph in level flight. Even looking far into the future, fundamental difficulties seemed to stand in the way of producing propeller-driven aircraft capable of speeds much above 450 mph, and these difficulties subsequently showed themselves very real in practice. In contrast, the gas-turbine, when used to produce a propulsive jet of hot gas, did not appear to be subject to any limitation. The difficulties, it seemed, would be associated with designing the components for satisfactory efficiency and with making them from materials able to bear high mechanical stress at cherry-red or even white-hot temperatures. In 1928 there were no such materials.

Whittle failed to arouse enthusiasm either within the British Air Ministry or with such renowned companies as Armstrong Siddeley Motors and British Thomson-Houston; but the young Pilot Officer was so convinced of the rightness of his idea that he patented it at his own expense in January 1930. Nobody showed interest for five years. Then, through the agency of private friends and investors, moves were made to put the idea into practice and in March 1936 a company called Power Jets was set up for this purpose, with a capital of £2,000. Whittle, by then a Squadron Leader, began the design of an actual engine, and. the task of making most of it was subcontracted to the B.T.H. company. Eventually, on April 12, 1937, he first Whittle turbojet ran successfully on the est bed. After painstaking development, a modified nit was cleared for flight in the specially-built Gloster E.28/39 and this machine flew for the first ime on May 15, 1941. By this time, the Gloster ompany was already building twelve examples of he twin-engined F.9/40, which became the Meteor ghter and saw active service from the summer of 944 onward.

Whittle's engine was put under full development t Government expense from March 1940, and ndustrial support was provided first by the Rover ar company and then by Rolls-Royce, who quickly uilt up the same outstanding reputation for air-raft gas-turbines that they had earlier established vith piston-engines. In January 1941 the de Havil-and company also had been asked to build a jet ighter with a different design of engine, and this merged as the Vampire in August 1943.

In the spring of 1941, General H. H. Arnold, US Army Air Force Chief of Staff, asked the Bell nd General Electric companies respectively to roduce a jet fighter and its engine. A British Whittle turbojet was shipped across the Atlantic n October 1941, and on October 1, 1942, the first Bell XP-59 Airacomet flew at Lake Muroc (now nown as Edwards Air Force Base). The Allied vartime effort was completed by the amazing per-ormance of Lockheed Aircraft who built the XP-80 ighter prototype around the de Havilland engine in 43 days and flew it on January 9, 1944. As the Shooting Star, the P-80 was soon in production vith an American General Electric engine, based n Whittle's designs.

From 1944 onward, Allied pilots encountered German jet-propelled fighters, and it was soon vident that German engine designers had turned o gas-turbines no later than the British. In fact, lthough Whittle was years earlier in his thinking, nd actually ran an engine first, the German pro-gramme was more purposeful and more than made up for its later start. Thus, although the first German urbojet, Pabst von Ohain's He S1, did not run until September 1937, it was speedily developed into he more practical He S3B and flown in the world's irst jet aircraft, the Heinkel He 178, on August 27, 1939. Heinkel flew his larger twin-jet fighter, the He 280, on April 5, 1941, (it had flown earlier as a glider), but his jet family then came to a halt until he hurried advent of the He 162 *Volksjäger* People's Fighter) of 1944, which was intended to be produced at the rate of 1,000 per month.

Instead, the companies selected by the German Air Ministry to produce operational jets for the *Luftwaffe* were Messerschmitt, BMW and Junkers. The latter firm began gas-turbine work in 1937, started the design of the Jumo 004 late in 1939, ran it on the bench in December 1940 and flew it—slung under a piston-engined Me 110 test-bed—late in 1941. Finally, the Me 262 twin-jet fighter, with much improved Jumo 004B engines, flew in the summer of 1943 and went into production in the

spring of 1944. It entered operational service a few weeks after the Meteor, and from the summer of 1944 onward was a thorn in the side of allied bomber crews, both by day and, in 1945, by night.

By April 1945 there were numerous German programmes involving jet engines and aircraft—more, in fact, than could have been continued even if the war had not ended. In this historical outline, it is also worth mentioning that Germany pioneered in the field of rocket propulsion. Fritz von Opel, the motor tycoon, fitted rockets to an established design of glider in 1929, and a real rocket aeroplane emerged from the Heinkel works ten years later in the form of the He 176, flown in June 1939. Again, however, it was Messerschmitt who produced an operational machine, the tailless Me 163 Komet interceptor, which achieved 623 mph in 1941 on the thrust of a Walter rocket burning a mixture of hydrazine hydrate and alcohol in the oxygen re-leased by decomposing hydrogen peroxide.

The Me 163 was able to climb swiftly towards attacking bombers and then engage them with 30-mm guns; it had very limited range and could serve no other useful function. Since then, rockets have been used mainly for spacecraft, and for re-search aircraft like the very successful North American X-15, which has reached over 4,000 mph and over 300,000 feet altitude. A few other air-craft, including versions of the British Buccaneer and French Mirage III, have auxiliary rocket-engines to speed take-off, climb and combat per-formance.

Today practically the entire field of aviation, apart from light private and sporting machines, relies upon the gas-turbine, used either to drive a propeller (in which form it is called a turboprop) or to provide a propulsive jet (as a turbojet). Increasingly, also, the turbofan is now coming into use, this term covering a range of power plants between the pure turbojet and the turboprop. How these engines work is explained in detail later on; meanwhile, the principle can be understood by considering what happens to the air as it passes through a gas-turbine from front to rear.

In all gas-turbines this flow of air, which becomes a flow of hot gas, is quite steady and smooth, and is thus unlike the intermittent flow of the piston-engine, in which power is produced by a succession of "bangs" occurring in a number of cylinders.

As the air flows through a turbojet—the simplest form of gas-turbine—its pressure, temperature and velocity vary greatly, as illustrated on page 206. It is, in fact, the changing pressure of the flow at different places within the engine that results in there being a large net resultant thrust to propel the aircraft. One of the largest forward thrusts is that exerted on the inside of the forward part of the combustion chamber, but many other surfaces inside the engine make a considerable contribution. When all the forward thrusts are added together they amount to a much greater total than the sum of the thrusts acting in a rearward direction, as so

much of the rearward thrust simply escapes through the jet-nozzle.

Fresh air coming through the intake duct of the aircraft passes straight into the engine compressor. This may be designed in either of two forms, which differ considerably in appearance. All the earliest successful turbojets were designed with centrifugal compressors, identical in concept with those used to supercharge piston-engines. Such a compressor has an impeller of the type sketched on the left of page 211. Often machined from a solid light-alloy forging, this impeller, when rotated at high speed, draws in air at the centre and flings it out radially from its outer edge. Some of the impellers of the early British and American jet engines had blades on both sides of a central disc and thus handled a greater airflow than the single-sided compressor illustrated.

Such a compressor is easy to manufacture, robust, relatively cheap, quite light and is easy to operate without encountering problems of surging or instability. But it has to be turned at very high speed—from 10,000 to 100,000 rpm—and cannot normally achieve a pressure ratio greater than about 5:1. This means that the pressure of the air it delivers to the combustion section is not more than five times the normal outside air pressure at the intake. Such a pressure ratio was adequate to supercharge piston-engines, and most superchargers do not exceed a pressure ratio of 3; but the gas-turbine is vitally dependent upon the compression of the incoming air, because the greater the com-pression the lower the fuel consumption. In addition, the centrifugal compressor tends to have a large frontal area and thus never appeared very suitable for the time when supersonic speeds would become attainable.

The alternative in an aircraft gas-turbine is the axial compressor, in which the air passes more or less straight through a compressor consisting basically of many blades approximately the shape and size of bus tickets arranged radially on the surface of a drum. The blades fit closely inside a surrounding drum, also fitted with blades, in this case projecting inward to lie in rows between those on the central drum. The latter, termed the rotor, is turned at high speed so that the blades it carries behave as miniature wings and compress the air stage by stage. The stator blades fixed in the surrounding case take out the twist imparted to the airflow by each set of rotor blades and feed it at the correct angle into the next rotor stage.

Early axial compressors achieved a pressure ratio of about 3.5:1 by using 14 or 15 stages of blading, and were much heavier and more expensive than the contemporary centrifugals. But the modern compressor with the same number of stages can compress the air by a ratio of as much as 28:1, a figure undreamed of 20 years ago, and has thus transformed the gas-turbine from an interesting concept of limited applications, suffering from a high fuel consumption, to the modern engine which

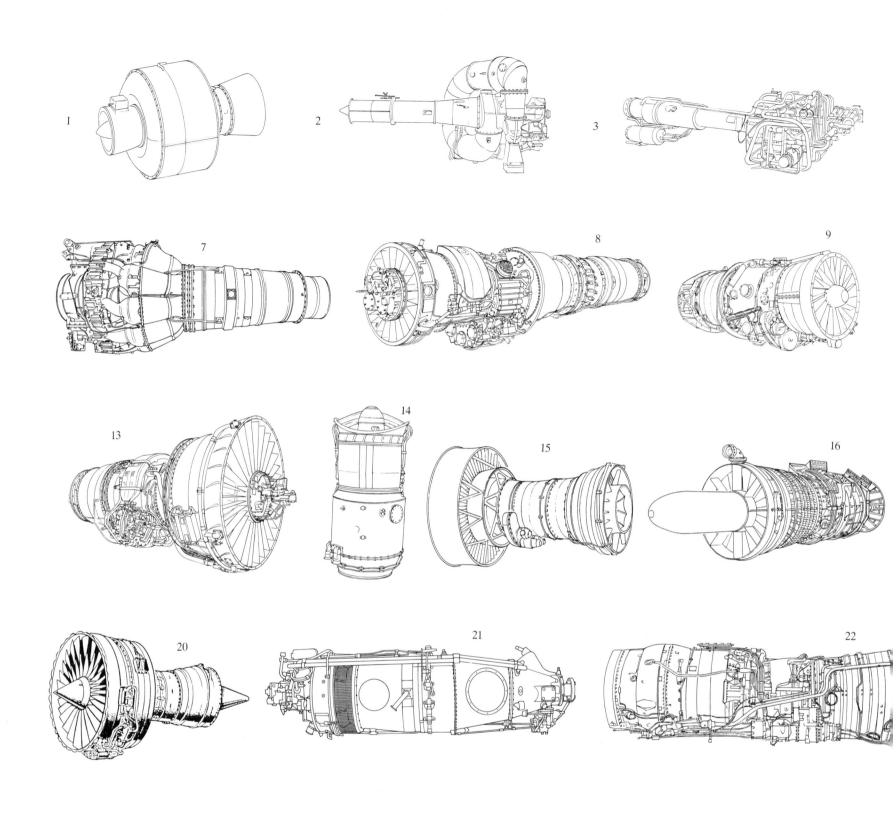

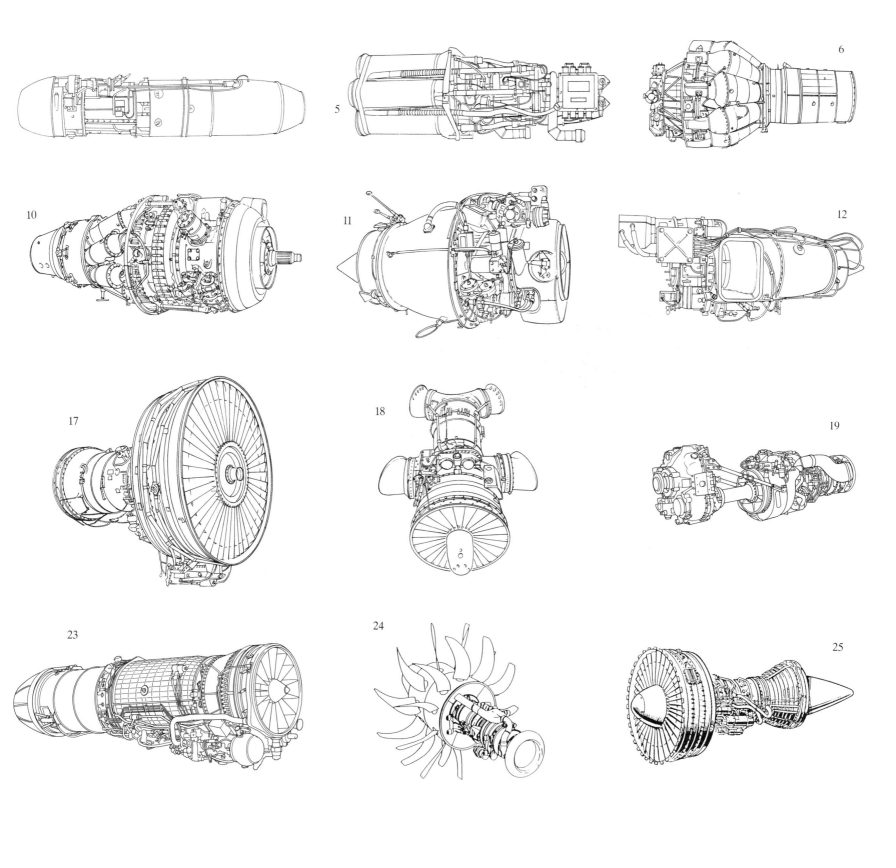

RB.207 53,000 lb
thrust
16 1968 General Electric
GE.4 67,000 lb thrust
17 1967 Pratt & Whitney
JT9D-1 43,500 lb
thrust
18 1968 Rolls-Royce

Bristol Pegasus 20,000
lb thrust
19 1979 General Electric
CT7 1,725 shp
20 1990 Rolls Royce RB.
211-524L 72,000 lb
thrust
21 1979 Pratt & Whitney

Canada PT6A-65A
1,409 shp
22 1967 Rolls-Royce
Bristol/SNECMA
Olympus 593 35,000 lb
thrust
23 1980 General Electric
F404 18,000 lb thrust

24 1986 General Electric
GE36 UDF 22,700 lb
thrust
25 1990 Pratt & Whitney
PW4000 68,000 lb
thrust

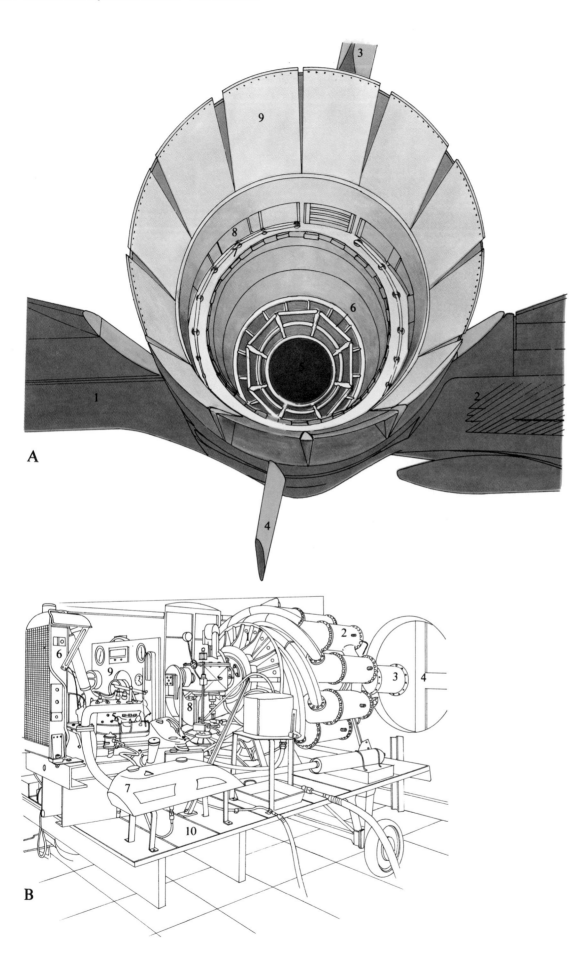

A

B

A *Outside the Soviet Union the most powerful aircraft engine in service is the Pratt & Whitney J58 (JT-11) turbojet, two of which power the Lockheed YF-12A fighter which currently holds the world record for absolute speed by a conventional aircraft, at 2,070 mph. Looking into the propelling nozzle, it is possible to see the fuel injection rings and flameholders which sustain reheat combustion in the afterburner and confer a 300 per cent thrust boost at Mach 3 (2,000 mph). The basic sea level thrust of the J58 at rest is about 34,000 pounds; it is a straight turbojet, rather than a turbofan, and is one of the two American types of air-breathing engine cleared for use with manned aircraft at Mach numbers of 3 or above.*

1 Outer wing panel
2 Inner wing and elevon
3 Port vertical tail
4 Fixed ventral fin
5 Tailcone behind J58 turbine
6 Afterburner fuel manifolds and flame-holder rings
7 Variable area primary nozzle
8 Secondary air intakes inside convergent/ divergent nozzle
9 Variable area divergent secondary nozzle

B *The world first heard the sound of a jet in 1937 at Rugby, Warwickshire. The original Whittle engine was successively modified and rebuilt and by the following year had the appearance shown here. It was by this time known as the W.1X engine (having originally been called the W.1U) and testing had been relocated a few miles from Rugby, at Lutterworth. The engine was mounted on a trolley which was fastened to the rear wall by a device which measured the thrust developed—which, depending on the particular conditions, was of the order of several hundred pounds. The 10 horsepower car engine on the left was used for starting the experimental unit by means of a belt drive. Before each run the doors in the end wall were, of course, opened !*

1 Centrifugal compressor casing
2 Combustion chambers
3 Jet pipe
4 Doors over opening in wall of hut
5 Morris car engine used as starter
6 Car engine radiator
7 Motor cycle fuel tank to serve (5)
8 Belt drive to experimental turbojet
9 Instrument and control panel
10 Chassis held back by thrust measuring device

an rival for economy any other type of prime mover except for the special case of nuclear power.

Even today engineers concerned with marine, rail and industrial power are prone to thinking in terms of high fuel consumptions for gas-turbines which bear no reality to those actually being achieved by the latest aero-engines. This economy stems largely from a very high pressure ratio, but a second source of economy is efficient combustion and, if the engine is a turbofan or turboprop, it is also improved by raising the temperature of the flow to the highest possible value.

From the compressor the air, at perhaps 300 pounds per square inch pressure and already heated by compression to over 200°C, passes straight into the combustion chamber. In Whittle's original engine there was one huge combustion chamber wrapped right round the rest of the engine, because in 1937 nobody knew how to burn fuel at the rate needed without making the burner and chamber very large. Gradually the space needed has been reduced, until today the whole process of burning fuel completely can be accomplished in a volume barely nine inches long and with a cross-section simply made to fit the rest of the engine.

The usual fuels for a gas-turbine aero-engine are various types of kerosene, much like household paraffin; but several grades of petrols (gasolines) and even fuel oils are also used. From the aircraft tanks, the fuel is pumped to the engines through a series of pipes and valves and then passes through a pump on the engine itself, where its pressure is raised to 1,000 pounds per square inch or more. At this high pressure it is forced through a compact unit which determines exactly how much fuel the engine needs, and then enters a manifold pipe encircling the middle of the engine from which numerous short branches lead to the individual burners.

Each burner is designed to atomize the fuel to a precisely controlled spray of vapour and fine liquid droplets. Most of these droplets are only about one millionth of an inch across and the aim is to make all of them as small as possible. The smaller a droplet of fuel is, the quicker it can burn and the smaller the combustion chamber can be made, whilst still achieving complete burning of the fuel. The airflow through the combustion chamber is made as slow as possible to assist burning to take place, but it is still of the order of hundreds of feet per second. Every additional fraction of a second needed for the larger droplets of fuel to be consumed means that the chamber must be made an inch or two longer— and every extra inch on engine length means much greater weight and, possibly, mechanical difficulties caused by "whipping" of a long shaft or instability in a thin-skinned casing. Moreover, if unburned droplets of fuel were to strike the turbine blades they would cause erosion and might even pierce through the hot blade.

In early aircraft gas-turbines it was common to burn the fuel in a number of separate tubular chambers, arranged around the engine as shown in a sketch on page 223. Use of separate small chambers greatly eased the problem of developing the chamber, because enormous quantities of compressed air are needed when running a complete chamber on a test rig and, if 16 separate chambers were chosen, the design could be perfected with only one-sixteenth of the full engine airflow. Inside each chamber was a perforated flame tube, the purpose of which was to allow fresh air to pour in at intervals along the flame from the burning fuel and quickly convert the local regions of intense heat to a uniform flow of gas that would not be too hot for the turbine blades at any point.

Modern chambers are generally of the annular type, in which the burners are arranged around a continuous ring-shape chamber, encircling the engine and filling the gap between compressor and turbine, with the drive shaft passing down the centre. The annular chamber is generally more efficient than the array of separate tubular chambers, and it is certainly much lighter and simpler. Most important of all is that it enables the fuel to be burned in the smallest possible space. In operation, the annular chamber is virtually filled by a single ring-like flame which is fed constantly by the fine fuel jets sprayed from the burners and cooled constantly by fresh air coming in through the many apertures in the flame tube.

At the downstream end of the combustion chamber, the hot gases escape by accelerating rapidly into a converging exit section which terminates in a row of turbine stator blades. These blades, which have to bear the highest temperatures of any metal in the engine, are fixed between inner and outer rings which join the inner and outer parts of the combustion chamber together at the rear. As the gas passes between the white-hot blades it is straightened and directed on to the blades of the turbine rotor.

The turbine is probably the most critical part of the engine, and has always imposed the main limitation upon engine power and development. This is because, while it has to operate in gas temperatures only slightly less severe than those experienced by the stator blades, the whole turbine rotates so fast that centrifugal force exerts a force of many tons on each blade, tending to pull it out of the disc in which it is mounted. The acceleration on each blade may amount to several thousand "g's", so that a blade with a mass of a few ounces may experience an end load of two or three tons and a consequent stress of 30,000 pounds per square inch or more. On top of this, the blade must withstand the very large sideways, or tangential, gas load which makes the assembly rotate, as well as high-frequency vibrations which tend to induce fatigue in the metal.

Special materials have had to be produced to make blades that will stand up to such demands, and gas-turbines for aircraft have been rendered practicable only through the ceaseless efforts of specialist firms which, for 30 years, have continually developed better and better blade alloys. Most of the earliest blades were made of Nimonic 80 and similar alloys containing a high proportion of nickel. Their great advantage was an ability to resist creep, the slow lengthening of a piece of metal when subjected to severe tension under conditions of high temperature. Many metals would run successfully in the form of a turbine blade for a few minutes or hours, but white or even cherry-red heat would weaken them so that they would soon start to rub inside the turbine casing and eventually lengthen so much that they would collapse.

This slow stretching cannot be simulated faithfully in a laboratory and thousands of hours of testing must be completed before improved blading, made of experimental new materials, can be run in an engine. At any given time, thousands of blades are under test in Britain, the United States, the Soviet Union, Sweden, France, Germany and other countries, most of them being held for several years at a precise temperature, as high as 1,165°C, or 1,000°C, or whatever is necessary, while being stretched either continuously or intermittently to a stress of perhaps 35,000 pounds per square inch. There are hundreds of thousands of turbine blades in the air at any given time, all glowing at over 1,000°C, and most of them are fitted to civil airline engines which may not be opened up for inspection for as long as 10,000 hours at a time.

Obviously, anything that can be done to reduce the metal temperature encountered by the blade is worthwhile. In the early 1950s, Rolls-Royce followed on where the German engineers left off in 1945, in attempting to cool the blade by use of some fluid to carry away the heat. The Germans also experimented with ceramics, but ceramics cannot usually carry severe and fluctuating loads. They did pioneer work in making blades which were metal at the root, where stresses are highest, and ceramic towards the tip where temperatures are highest. Today whole new families of "cermets", combining the best qualities of ceramics and metals, are industrially available.

None of the German schemes for cooling the blades was very advanced in concept or development, although most of the blades actually used in German operational jets were made from sheet metal, and had air passages down the centre. Rolls-Royce sought to make far stronger blades, forged or cast in solid heat- and creep-resistant metal but containing integral cooling passages for air. At the immense centrifugal loads experienced by a turbine blade, air allowed to pass down the interior of the blade will do so at very high velocity without the need for pumping. But it can only escape into gas at quite high pressure, so the usual source of cooling air is a "bleed" pipe from some point on the engine compressor. Even if this air is quite hot as a result of compression, it will still have a much lower temperature than the blade. Details of how cooled blades are manufactured may not yet be disclosed but most of the processes involve forging or casting the rough blade with the cooling holes already running the whole length from the root to the tip.

Manufacturing a turbine blade involves a succession of as many as 50 distinct stages, each carried out with the utmost precision. Many conventional parts made from metal are simply pressed to shape when the metal is hot, but the material of the turbine blade is specially chosen so that, even when almost white hot, it retains its strength and will not deform under stress. As a result manufacturing a blade is almost beyond the capability of an ordinary production shop. When de Havilland Engines were building their first turbojet in 1943, all the blades were laboriously finished by hand filing. Today, clever new techniques involving abrasion, electric sparks, chemical erosion and, above all, electrochemical machining are used to produce the exact form required as speedily as possible. Nevertheless, a set of blades that a man could carry in one hand may still cost as much as a Rolls-Royce motor car.

A single row of blades may be inserted into a disc of heat-resistant steel by means of "fir-tree" root fixings or any of several other methods, so that each blade will be completely secure. Each then forms a small vane, often called a "bucket" in the United States, which behaves like a miniature wing. When acted upon by the gas flow from the combustion chamber each blade exerts a tangential force on the disc, causing it to rotate. In a modern engine, a blade no larger than a thumb may generate one hundred horsepower and the total power generated by the turbine may be 50,000 horsepower or more. Usually a complete turbine consists of two or more stages of blades, each inserted into a separate disc, the whole assembly being joined together by through-bolts and splined to the end of the hollow driving shaft. In some very small turbine stages, the whole rotor assembly of disc and blades is machined from a single "cheese" of heat-resistant steel or other alloy, and it is quite common for such a unit to run at 60,000 rpm or more.

To drive the turbine, the gas is redirected by a ring of stator blades interposed ahead of each stage of rotor blading, in the same manner that stator blading is used in the compressor. The stator blades, which are sometimes called nozzle guide vanes, are either inserted singly into the surrounding stator rings or else cast in groups of three, five or even ten to form self-contained segments. Casting the stator blades in groups involves forming them with an integral end shroud, and this is also common with rotor blading, particularly in the case of high-pressure turbines. The latter, which are the smallest in any engine and are operated upon by the hottest and highest-pressure gas, are often fabricated with an end plate which mates with those on the neighbouring blades so that when the rotor is complete the blades are surrounded by a continuous ring shroud which prevents loss of gas round the tips of the blades. Such blades offer a significant improvement in efficiency, but the extra weight on their tips substantially increases the stress they must bear at the root where they join the disc.

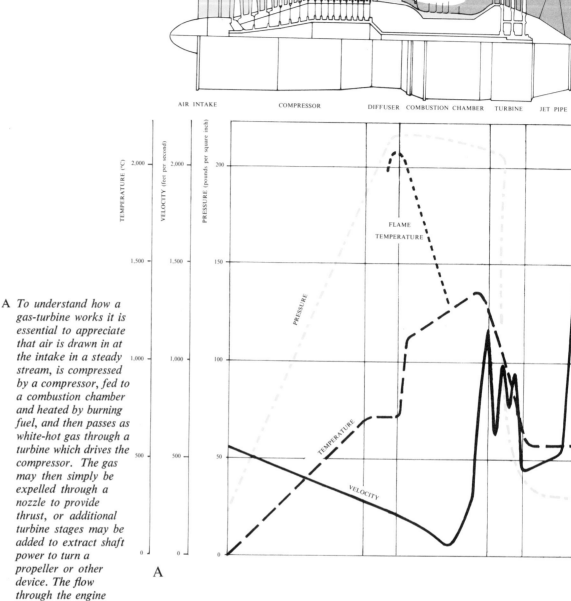

AIR INTAKE   COMPRESSOR   DIFFUSER COMBUSTION CHAMBER   TURBINE   JET PIPE

FLAME TEMPERATURE

PRESSURE

TEMPERATURE

VELOCITY

A

A To understand how a gas-turbine works it is essential to appreciate that air is drawn in at the intake in a steady stream, is compressed by a compressor, fed to a combustion chamber and heated by burning fuel, and then passes as white-hot gas through a turbine which drives the compressor. The gas may then simply be expelled through a nozzle to provide thrust, or additional turbine stages may be added to extract shaft power to turn a propeller or other device. The flow through the engine is always continuous and smooth—unless something is seriously wrong—and the way its pressure, velocity and temperature vary in a typical aircraft turbojet is shown here. The graphical curves are drawn directly below the corresponding parts of the engine to which they refer. The engine depicted is not any particular type, but is simply a straightforward single-shaft turbojet with a 17-stage axial compressor driven by a three-stage turbine. If such an engine had a mass flow of 200 pounds of air per second, it would generate a thrust of about 18,000 pounds.

B All aircraft engines propel aircraft by accelerating to the rear a mass of air, the resulting reaction being in the form of a forward thrust. Propeller-driving engines handle the largest airflow (grey) but impart to it only a moderate acceleration. The turboprop therefore generates a very high thrust at low speeds and is the most efficient type of gas-turbine for aircraft propulsion up to a speed of about 400 mph. But by this time the efficiency of the propeller is falling rapidly and once 500 mph is reached the turbofan is markedly superior. The fan engine handles a smaller airflow than does a propeller of similar static thrust, but it imparts a higher acceleration and does not fall in efficiency so rapidly at higher speeds. A high by-pass ratio turbofan reaches maximum efficiency at about 600 mph and then becomes inferior to the more slender low by-pass ratio unit. Finally, at supersonic speeds, the still more slender turbojet becomes superior, handling a lower airflow but accelerating it to very high velocity as a jet of hot gas (colour). Efficiency of the pure turbojet does not begin to fall until aircraft speed has reached at least 2,500 mph, and then is limited chiefly by the inability of the turbine to stand up to the gas temperatures necessary for such an engine to generate adequate thrust at such speed. For even higher speeds the ramjet is more effective, in which the air compresses itself as it enters the intake and thus needs no compressor or turbine.

Compressor and turbine are joined by a shaft which is always hollow and may be of large diameter, or even of conical form opening out to the full diameter of the rearmost compressor disc. Sometimes the compressor is carried between two bearings —often a roller bearing at the front and a double ball bearing or inclined tapered roller bearing at the rear to withstand any end load. The compressor can then spin by itself and the turbine drive shaft need not be attached rigidly but merely slipped into driving splines. In this case, the turbine may be supported either by a single bearing in front or behind, or by a bearing on both sides. Alternatively, the whole rotating assembly may be coupled rigidly together and supported by only two bearings, one in front of the compressor and one close to the turbine, either in front or behind. Bleed air is taken from various places along the compressor and led through external pipes or through the turbine drive shaft and directed on to the sides of the turbine discs, partly to cool them but primarily so that the different pressures shall cancel out any residual end load imparted by the compressor and turbine blade forces.

Downstream of the turbine the gases are cooled considerably by their expansion through the various turbine stator and rotor blades, but their mean temperature and pressure are still far above those of the atmosphere. In the jet pipe and final nozzle, the flow is accelerated to the greatest possible velocity so that as much energy as possible is converted from heat or pressure to kinetic energy. It is the average change in momentum of each molecule of gas as it passes through the engine that creates the useful thrust, and the ideal jet pipe and nozzle would leave the gas no hotter than its surroundings and travelling at the greatest speed that could be attained. Jet velocity is determined partly by the pressure ratio across the nozzle, so it is higher at higher altitudes provided the internal engine pressures can be maintained; and it depends upon flow temperature.

A simple propelling nozzle is no more than a plain hole at the end of the jet pipe, of exactly the right area for the gas flow. Sometimes small variations to nozzle area are made by inserting fixed fairings, colloquially called "mice", to trim the nozzle area to a particular engine. Clearly, continued reduction in nozzle area will raise jet velocity but it will do so only up to a limiting value at which the nozzle "chokes"; beyond this point no further improvement is possible and, in any case, any reduction in nozzle area raises gas temperature inside the engine.

If nozzle area can be varied in flight, small improvements can be made to overall engine performance; but although this was a feature of early German engines such as the Jumo 004 and B.M.W. 003, no modern engine has an adjustable propelling nozzle unless it has afterburning.

Afterburning, or reheat, consists of injecting fuel into the gas flow downstream of the turbine so that it is fully burned before the flow passes out of the nozzle. At the cost of a considerable increase in fuel consumption, this greatly increases the thrust which

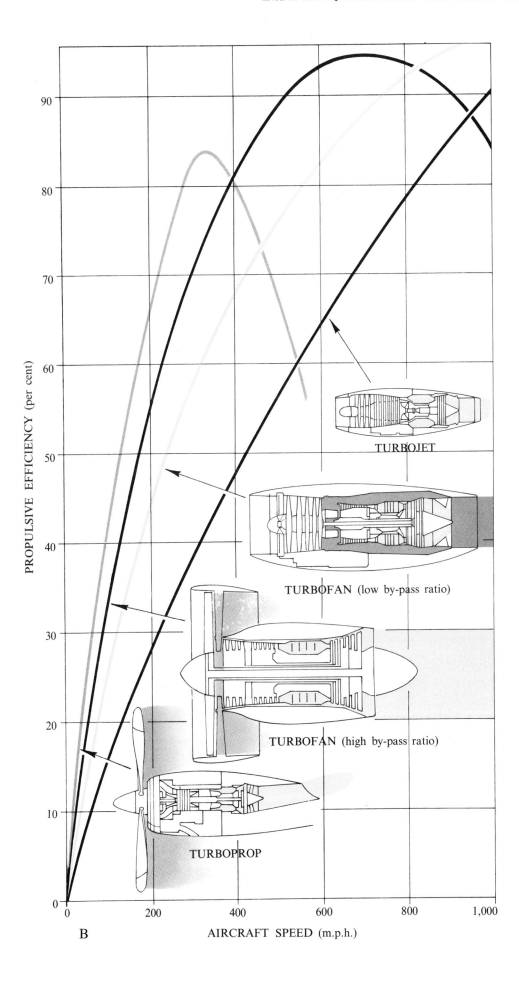

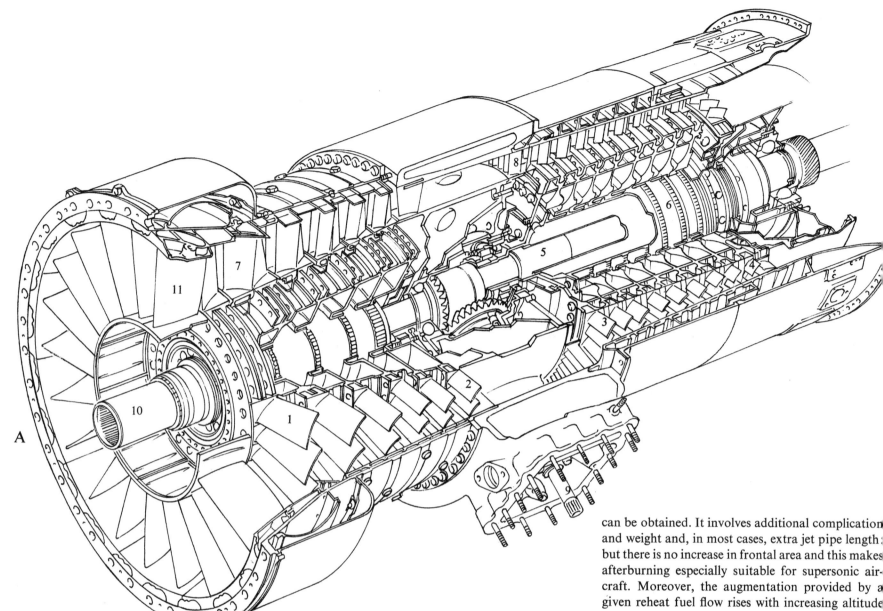

A

11

7

10

1

2

3

4

5

6

8

A *This compressor, for the Rolls-Royce Tyne turboprop, shows the two-spool configuration, in which the work of compressing the air is performed by two mechanically-separate rotors which each adopt their own best rotating speed. The six-stage low-pressure compressor in the Tyne is driven by a three-stage low-pressure turbine which* also drives the front reduction gear for the propeller (*not shown in this drawing but visible on page 226*). The high-pressure compressor has nine stages and runs in its own bearings around the shaft of the low-pressure system; it is driven by a single-stage high-pressure turbine with air-cooled blades.

1 First stage of rotor blading on six-stage low-pressure compressor
2 Last stage of low-pressure rotor blading
3 First stage of rotor blading on nine-stage high-pressure compressor
4 Ninth and last stage of high-pressure rotor blading
5 Low-pressure drive shaft

6 High-pressure drive shaft surrounding (5)
7 First stage of low-pressure stator blading
8 Inlet guide vane ahead of high-pressure compressor
9 Accessory gearbox drive shaft
10 Propeller gearbox drive shaft
11 Inlet guide vanes in air intake

can be obtained. It involves additional complication and weight and, in most cases, extra jet pipe length; but there is no increase in frontal area and this makes afterburning especially suitable for supersonic aircraft. Moreover, the augmentation provided by a given reheat fuel flow rises with increasing altitude and also with increasing Mach number. Thus, an engine having 30 per cent reheat boost under sea level static conditions may offer a boost of 70 per cent under static conditions at 50,000 feet (this can be simulated in a test cell) and 300 per cent at Mach 2.5 at the same height.

Typical of early installations was that of the Rolls-Royce Avon RA.7R, which raised the sea level thrust of the engine from 7,500 to 9,500 lb. The propelling nozzle was a simple two-position unit, with a pair of "eyelid" shutters driven by pneumatic rams. Today's Spey 201 for the F-4K and F-4M Phantom fighters offers an increase in thrust from 12,500 to 20,000 lb at sea level, a boost of almost 70 per cent. At Mach 2.5 a gain of several hundred per cent should be possible, although details are restricted.

Even greater boost can be obtained with turbofan engines, a family of aircraft power plants described presently, in which additional fuel can be burned not only in the hot gas behind the turbine but also in the fresh air delivered by the fan. One of the first duct-burning engines was the TF-306, developed jointly by Pratt & Whitney and the French SNECMA

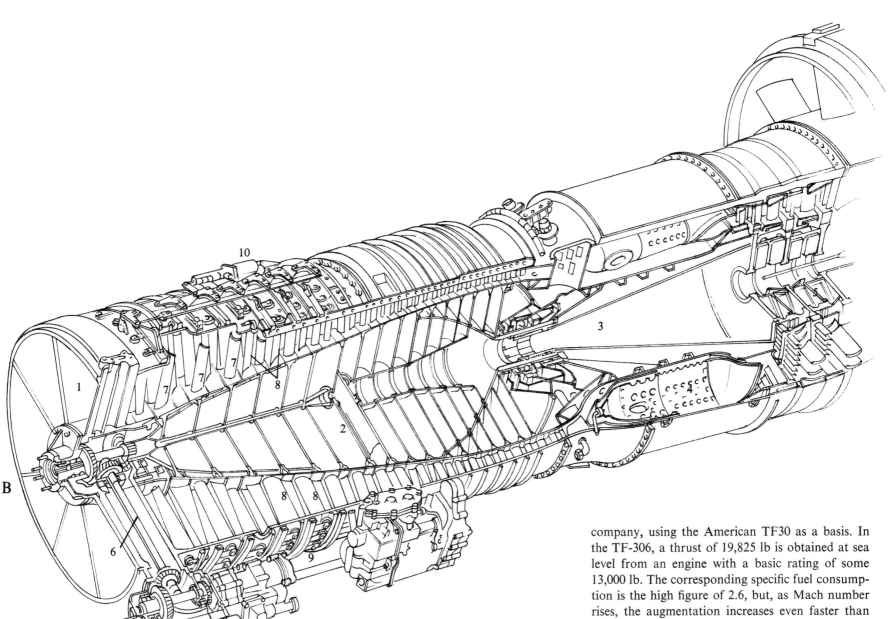

B

company, using the American TF30 as a basis. In the TF-306, a thrust of 19,825 lb is obtained at sea level from an engine with a basic rating of some 13,000 lb. The corresponding specific fuel consumption is the high figure of 2.6, but, as Mach number rises, the augmentation increases even faster than in an afterburning turbojet, offering outstanding performance at great altitudes. A much larger duct-burning engine is the Pratt & Whitney JTF17, mentioned later.

In duct-burning and afterburning engines, provision must be made for varying the area of the nozzle or nozzles to match them to the greatly varying flow passing through them. In all cases the nozzle must be opened up when afterburning is in operation. Such boost can easily accelerate the jet to beyond the speed of sound and, if fuel is still burning in the nozzle, bright shock diamonds can then be seen in the jet for perhaps 20 or 30 feet behind the engine: the shock waves from the nozzle lip cross and re-cross after being reflected at the boundary layer between the jet and the undisturbed air.

Early reheat installations were seldom fully-variable, but today it is usual for continued forward movement of the pilot's throttle lever to result in smooth, steady increase in thrust from idling up to maximum "dry" or minimum "cold" thrust, followed by steady progression through an infinitely-variable afterburning regime right up to maximum reheat or maximum "hot" thrust, without any sudden changes in response. Early jet fighters equipped with afterburners, such as the North American F-100 Super Sabre and LTV F-8 Crusader, brought in their reheat system with a sudden muffed explosion, in an abrupt on/off type system. Another step in the development of the afterburning systems was a possibility to light the afterburner in steps with distinctive stages between the different zones.

Such power plants are extremely noisy and would never be accepted for civil use. In fact, even a plain nozzle is generally too noisy for airline use, without any question of reheat. Since 1950, many laboratories have worked on the problem of reducing jet noise, with minimum harm to engine performance, and almost every jet airliner now in use has a nozzle embodying some feature to suppress noise.

It was discovered that noise varies greatly in character, according to the volume of airflow through the engine and thrust and, especially, jet velocity. Most of the noise stems from the violent shearing action that takes place between the hot gas jet and the cold, still air. If the boundary between the two can be broken up and increased in length this shear can be reduced considerably. The first two nozzles to do this were the Rolls-Royce multi-lobe nozzle of the Comet and the Pratt & Whitney multi-lobe nozzle of the Boeing 707. In both cases the hot gas issues through multiple apertures surrounded by cool air which mixes rapidly with the jets and reduces the shearing action and, thus, the noise. In subsonic aircraft it is doubtful that noise will continue to be a problem because, even in engines of over 40,000 lb thrust, the use of turbofans of high by-pass ratio greatly reduces the noise emitted and, equally important, changes its character from an ear-splitting blast to a deeper sound that is less offensive to human ears.

In many aircraft, especially civil airliners, it is worthwhile making provision to reverse the thrust of the engines in order to shorten the landing run—and sometimes to increase the rate of descent possible in flight without forward speed becoming excessive. Thrust reversal is particularly valuable when a jet lands on an icy or wet runway, when wheel brakes are less effective. The first thrust reversal units for jet engines were complex and did not achieve a braking effect greater than 30 per cent of the full forward thrust. Today, reversers achieve a drag force equal to from 60 to 80 per cent of the maximum forward thrust and are much neater in design; but they remain weighty units because they are large assemblies made of heat-resistant alloys and bear very high and fluctuating loads.

By far the most common reverser design uses a pair of clamshell shutters. These can be swung round by a pneumatic ram system to seal off or at least impede the normal jet, while simultaneously opening large apertures in the side of the jet pipe provided with "cascades" of curved deflectors which turn the gas forward to achieve the greatest possible braking effect. In developing a reverser installation, great

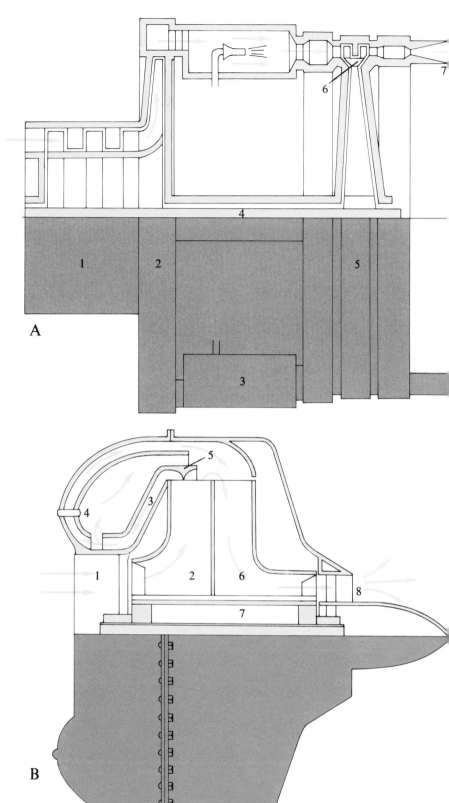

A *Although it bears little relationship in mechanical design to his subsequent engines, the turbojet depicted in the young Frank Whittle's original Patent Specification of 1930 is of great historic interest. He showed the air coming in at the left, being compressed by axial and centrifugal compressors, being delivered through a diffuser with a 90° bend to a combustion system surrounding the middle of the engine, and finally escaping through a two-stage turbine (driving the compressor) followed by an annular divergent propelling nozzle. When his plans reached the hardware stage, he used a single large combustor looped as a "U" round the engine and discharging into a small turbine leading to a single jet pipe.*
1 Casing over axial compressor
2 Casing over centrifugal compressor
3 One of several tubular combustion chambers
4 Drive shaft from turbine to compressor
5 Casing over turbine
6 Turbine with two stages of blades on single disc
7 Annular divergent propelling nozzle

B *Contemporary with Whittle, but unaware of his work, Pabst von Ohain in Germany was throughout the 1930s engaged in the development of a gas-turbine jet engine working in an exactly similar manner. His*

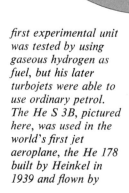

first experimental unit was tested by using gaseous hydrogen as fuel, but his later turbojets were able to use ordinary petrol. The He S 3B, pictured here, was used in the world's first jet aeroplane, the He 178 built by Heinkel in 1939 and flown by

Captain Erich Warsitz on August 27 of that year. But the German government failed to follow up this pioneer line of development and, instead, awarded major turbojet contracts to BMW and Junkers.
1 Air intake
2 Centrifugal compressor

3 Primary airflow to combustion chamber
4 Fuel injector nozzle
5 Airflow splitter with bulk going straight to cool turbine
6 Inwards radial flow turbine backing on compressor
7 Tubular shaft carrying compressor and turbine
8 Propelling nozzle

are is necessary to match it not only to the engine but also to the particular aircraft. The deflected gas streams must not damage any part of the airframe or landing gear, nor pick up stones and other loose objects in such a way that they could be sucked into the engines. Thrust reversers are nowadays a standard feature on all civil airliners, with a few exceptions such as the Fokker F-28. Military transport planes also use the advantage of reverse thrust to minimize landing roll, thus making it possible to operate from short airfields. The only operative combat aircraft to be equipped with thrust reverse is the Swedish SAAB 37 Viggen which has extremely good short-field performance. In some rare cases, the reversing system can be used to back the plane off from the gate of the air terminal; but due to noise problems, the maneuver is generally executed by special push-back tractors.

When a single axial compressor is designed to operate at a high pressure ratio of well over 8:1, the airflow through it cannot be properly matched under all operating conditions. At times the front of the compressor will deliver too much air for the rear stages to handle and at other times too little. There are various ways of improving the performance and flexibility of a high-pressure engine, to achieve not only the highest possible thrust but also rapid and smooth response to the throttle movements. The simplest answer, which is used on many engines, is to fit one or more blow-off valves on to the compressor casing, so that excess air can be discharged automatically. A second method is to make the inlet guide vanes and some of the succeeding stator rows in such a way that the blades can twist in pivoted mountings at each end. All the rows made variable in this way are linked together mechanically and driven by a hydraulic ram so that the incidence of each row of blades is at all times exactly right for the engine speed, atmospheric pressure, aircraft speed and fuel flow. A third method is to split the work of compression into two phases. This involves using a large low-pressure turbine, and surrounding the drive shaft of this assembly by a smaller high-pressure compressor driven through a large-diameter tube by a small high-pressure turbine.

This "two-spool" or two-shaft arrangement was pioneered by Bristol in Britain in 1949 and by Pratt & Whitney in the USA in the following year. Although it makes a gas-turbine more complicated, and requires several additional shaft bearings, it enables a higher pressure ratio to be obtained from a given number of compressor stages and can even result in a lighter engine for a given thrust. It has become immensely important today because, by increasing the size and air-handling capacity of the low-pressure compressor, or by adding enlarged blades to the front of it, it is possible to effect a fundamental improvement in the overall engine efficiency.

Rolls-Royce were the first company to put into production an engine of this more efficient type.

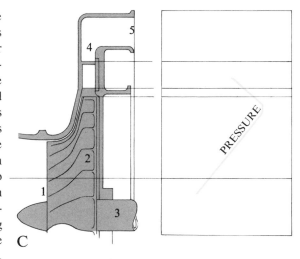

C

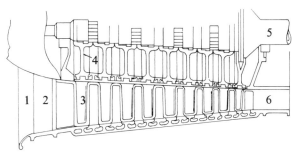

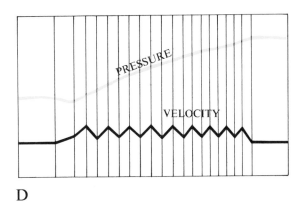

D

C Nearly all the earliest gas turbines used compressors of the centrifugal type, because this form had a long background of experience, and was robust and simple to develop. It comprises a disc carrying radial vanes on either one side or both; for maximum efficiency the inner part of each vane is curved to induce the air. As the air is flung outward between the vanes its velocity and pressure both rise as shown by the graph. In the diffuser the velocity is reduced to increase pressure further.

1 Air intake
2 Centrifugal compressor rotor (impeller)
3 Drive shaft from turbine
4 Diffuser reduces velocity but raises pressure
5 Discharge to combustion chamber

D Capable of generating much higher pressure ratios (ratio of inlet pressure to delivery pressure), at greater efficiencies, the axial type of compressor is more complex than the centrifugal type but is now used universally on all the larger and more advanced aircraft gas-turbines. An axial compressor comprises a rotor running in a casing fitted with stator blades which lie between each stage on the rotor. As the air passes along the compressor its velocity remains substantially constant, as shown, but its pressure rises steadily. This compressor has nine stages, but some have 13 or more and reach pressure ratios in excess of 24:1.

1 Air intake
2 Fixed intake strut
3 First stage of rotor blading of axial compressor
4 First stage rotor disc
5 Drive shaft from turbine
6 Delivery and diffuser section

They enlarged the low-pressure compressor, bypassed the excess air from it through a duct surrounding the smaller high-pressure compressor, combustion chamber and turbine, and finally reinjected the cool air into the jet pipe. They called this type of engine a by-pass turbojet, and the Rolls-Royce Conway showed that such a power plant could be lighter, have much lower fuel consumption and faster acceleration than a comparable "straight" turbojet.

To meet this competition, Pratt & Whitney took their JT3C turbojet and replaced the first three stages of the low-pressure compressor by two rows of huge fan blades, much larger in diameter than the remainder of the engine. They modified the turbine so that it would give enough shaft power to drive the huge fan and the shortened low-pressure compressor, and then cowled the engine in such a way that, while the original jet was unaltered at the rear, the fan delivered fresh air through a short duct discharging along the sides of the cowling over the rest of the engine. They called the result the JT3D turbofan, and it proved to be a vast improvement, for subsonic use, over the original turbojet. Although only slightly heavier, its thrust is increased from 12,000 lb to 18,000 lb, yet it has almost exactly the same fuel consumption as the JT3C.

Since 1960 the turbofan type of engine has emerged as the best possible power plant for most aeroplanes designed to cruise at between about 350 and 640 mph. At below 350 mph the turboprop (described presently) is superior; above 640 mph, the pure jet (the turbojet) gradually becomes superior. The real criterion is the mass airflow handled by the propulsion system. At low relative speeds it pays to handle the largest possible airflow, and for this reason the helicopter rotor is the most efficient way of achieving stationary hovering flight and VTOL (vertical take-off and landing). The conventional propeller handles a smaller airflow but one that is still much larger even than most turbofans. The turbojet handles the smallest airflow for a given thrust and is therefore inefficient for hovering flight but excellent at supersonic speed.

Turbofans can clearly offer a wide range of differing characteristics, filling the gap between the turbojet at one end of the scale and the turboprop at the other. The variable which determines a turbofan's position in this scale is its by-pass ratio—the ratio of the airflow expelled by the fan to that which passes through the smaller high-pressure compressor, combustion chamber and turbine. By-pass ratio in the original Conway was only 0.3:1, which was too low to show much improvement over turbojets. The JT3D boldly raised the ratio to 1:1 which means that as much air was expelled by the fan as passed through the hot part of the engine. In the still later JT9D and TF39, the by-pass ratios are as high as 3:1 to 8:1, so that only a small portion of the total air taken in by the engine issues as the central hot jet.

At about the same time that Pratt & Whitney were developing the first JT3D, in 1960, another US firm,

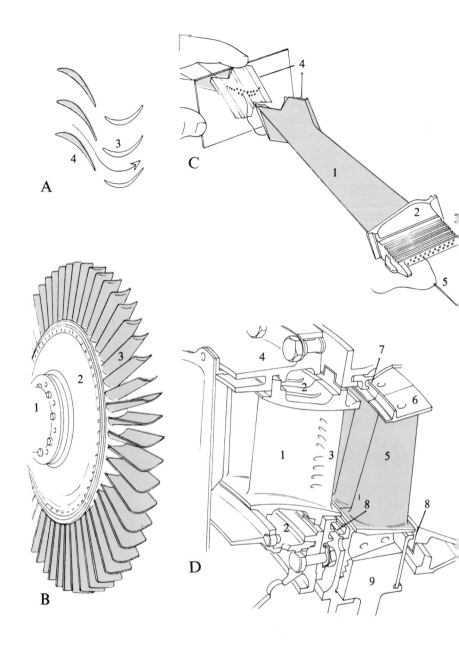

A A gas-turbine consists
& of stationary stator
B blades upstream of moving rotor blades and may be of the impulse type, in which the rotor blades are driven round by the kinetic energy of the impacting gas, or of the reaction type in which the gas expands between the rotor blades and changes its direction to impart a force on the blades as it passes between them. Aircraft gas-turbines are usually half impulse and half reaction, and typical blade profiles and gas paths are as shown here.
1 Turbine drive shaft

2 Turbine rotor disc
3 Turbine rotor blades
4 Turbine stator blades (nozzle guide vanes)

C The hottest and most advanced engines of today are fitted with air-cooled turbine rotor blades. These represent the finest achievements by investment casting technique, which casts the complete finished blade with the necessary cooling air passages. This blade, by Howmet Corporation for an American turbofan engine, has a fir-tree root (foreground) through which the high-pressure cooling air passes via a row of fine holes to

emerge through the ends of the same holes, arranged around the aerofoil section at the tip. The material of such a blade could be neither forged nor machined by any conventional process.
1 Turbine blade
2 Blade root platform
3 Fir-tree root to locate and retain blade
4 Shrouded tip seen in mirror
5 Thread passed through one of the cooling air holes (for demonstration only)

D A portion of a typical turbine rotor, fitted with relatively simple blades inserted into a forged disc
1 Nozzle guide vane
2 High pressure cooling air galleries
3 Cooling air outlet holes
4 Outer retaining and mounting ring
5 Turbine rotor blading
6 Tip shroud on blades
7 Tip seal
8 Root seals
9 Rotor disc

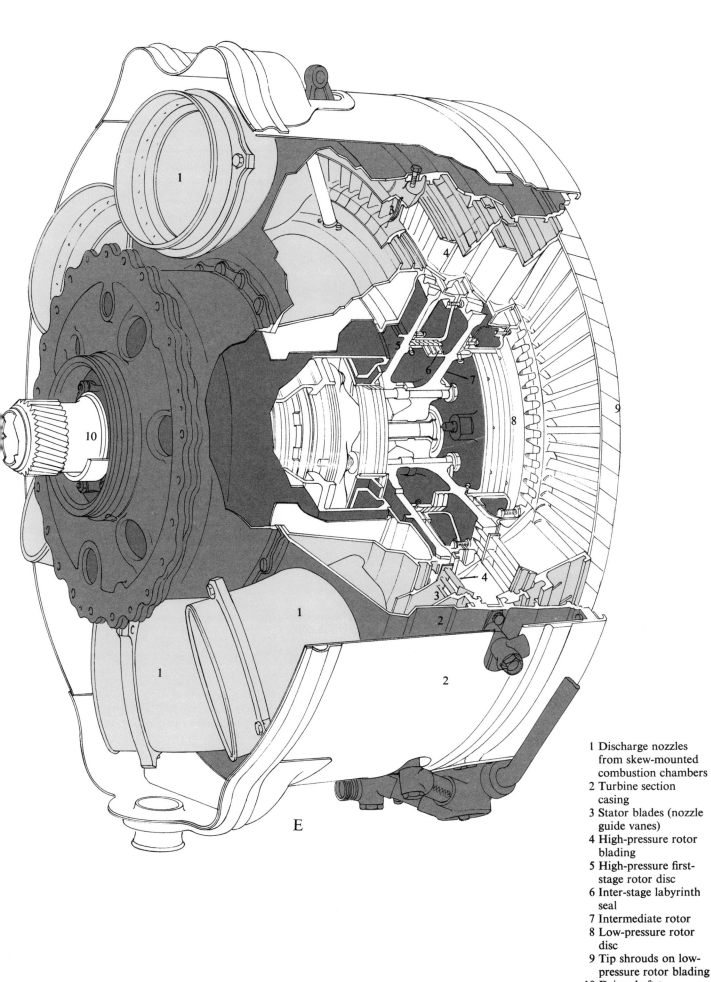

E

1211

E *Typical air cooling flows in a modern turbine are provided by compressor bleed air, supplied along the centre of the engine. Some of the air passes up through a nozzle guide vane, or in from the outer end, and is discharged through rearward-facing holes in its concave undersurface; other air passes outwards through the rotor blading from root to tip. As an example of a complete assembled turbine, the three-stage unit of the Rolls-Royce Dart turboprop is easy to understand and forms an interesting comparison with the same company's later and more advanced turbine for the Tyne turboprop (see page 226). Early Dart engines had a two-stage turbine, but the more powerful versions introduced a third stage to increase efficiency and reduce the loading of each stage despite the increased power. Gas to drive the Dart is generated in seven combustion chambers which, to shorten the engine, are canted helically. The discharge pipes from these can be seen leading to the high-pressure nozzle guide vanes. It will be noted that all three stages of rotor blading have tip shrouds on the blades, and that the three discs are tied together by through-bolts. On the extreme left can be seen the drive shaft to the compressors and the propeller gearbox.*

1 Discharge nozzles from skew-mounted combustion chambers
2 Turbine section casing
3 Stator blades (nozzle guide vanes)
4 High-pressure rotor blading
5 High-pressure first-stage rotor disc
6 Inter-stage labyrinth seal
7 Intermediate rotor
8 Low-pressure rotor disc
9 Tip shrouds on low-pressure rotor blading
10 Drive shaft to compressor and propeller

1212

General Electric, adopted an even more radical solution, but one which had actually been investigated almost 20 years earlier in one of the first families of turbojets—the axial engines by Metropolitan Vickers. This arrangement involves adding a fan at the rear of the basic engine, so that the major machinery of the turbojet need not be altered.

The rear-mounted fan of the Metrovick engine was a complicated multi-stage assembly, but the General Electric CJ-805 aft-fan conversion merely involved adding a single wheel immediately behind the three-stage turbine of the basic turbojet. The disc of this new wheel carries unusual blades of "double deck" type; each has an inner portion which is a turbine blade working in the hot gas stream emerging from the original turbine and an outer part which is a large fan blade working on fresh air in a duct around the rear of the engine (page 221).

The rear-mounted fan concept did not see further development, in contrast to the JT9D-type with front-end fans. Today, this totally dominates the market for heavy jet transports, both civil and military. The by-pass ratio of turbofans has increased steadily and for the current Very High By-Pass engines, the ratio is as high as 12 to 1.

In some ways, the rear fan design can be said to have been brought to life again with the development of propfans, or Ultra High By-Pass engines. The two counter-rotating unducted fans with stubby propeller-like blades are mounted in the rear part of the engine nacelle, and both Pratt & Whitney/Allison and General Electric have made full-scale air tests which are very promising for the future.

The world's most widely used jet engine, the Pratt & Whitney JT8D in all its variants, started as a military project, J52, for the US Navy in 1954. It was a straight turbojet design with five low-pressure and seven high-pressure compressor stages in the 7,500—11,200 thrust range. The J52 was only used in the Grumman A-6 Intruder, and plans to market the commercial version, JT8, were never successful. But the story of JT3, of which the JT3/J52 was a scaled-down distant relative, repeated itself when a redesigned turbofan version became a standard installation on the Boeing 727 and McDonnell Douglas DC-9. JT8D had the low-pressure compressor section changed to six stages driven by a three-stage turbine. Two fans in the front added to the thrust and pressure ratio, which were roughly doubled in comparison to the unfanned original JT8 turbojet. Since its first production in 1960, the JT8D series 200, now in production for the MD-80, has a redesigned fan section with higher by-pass ratio, which brings the thrust up to the 20,000–21,700 bracket. To date, more than 12,000 JT8D's have been built, the highest number of any jet-engine family. The Swedish company Volvo Flygmotor, a subcontractor for JT8D components, has taken the engine as a basis for a military version for the SAAB 37 Viggen. This version, named RM8, is highly modified in the compressor, burner and turbine sections, and is equip-

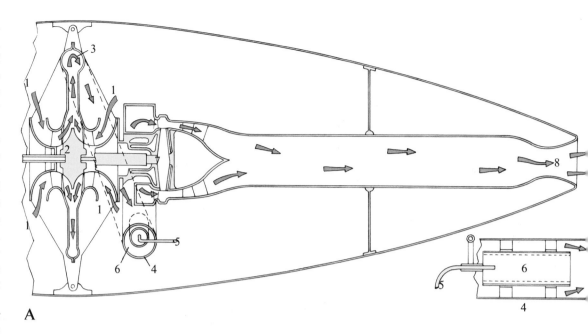

A

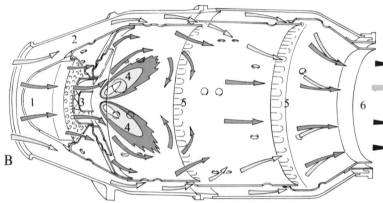

B

A *In the early years of gas-turbine development for aircraft, by far the greatest single problem was the attainment of efficient combustion of petrol or kerosene at intensities far surpassing anything previously attempted. Earlier industrial gas-turbines, of which several were built before 1939, could use enormous combustors, offering fewer problems. The aircraft unit had to achieve very great heat release in chambers only a few inches across; and the combustion had to be regular and complete if the chamber or turbine was not to be damaged. This is the chamber used in the original Whittle engine of 1937: the single fuel pipe discharges through a perforated end tube, into a primary combustion zone in an open-ended tube*

provided with a wire mesh liner to ensure steady burning. The secondary dilution air passes around this tube and mixes with the flame downstream.

1 Air entering double sided compressor casing
2 Double sided centrifugal compressor impeller
3 Diffuser and compressor delivery duct
4 Combustion chamber (see inset)
5 Fuel pipe and injector nozzle
6 Flame tube
7 Single-stage turbine rotor
8 Propelling nozzle

B *A more modern chamber has a primary zone, in which air from the compressor enters through swirl vanes, and a perforated flare to support primary combustion at an air/ fuel ratio of about 5:1. The bulk of the air (over 72 per cent of the total) then enters through apertures farther along the flame tube, to dilute the flame to an air/fuel ratio of from 15:1 to 140:1 and to yield a steady stream of hot gas at a uniform temperature of some 900° to 1,400°C which can be used as the medium both to drive the turbine and, in a jet engine, to provide the propulsive thrust.*

1 Primary air snout
2 Secondary airflow (about 80 per cent of total)
3 Fuel injector here
4 Flame cone from finely divided fuel spray
5 Dilution and cooling air
6 Even temperature hot gas to drive turbine

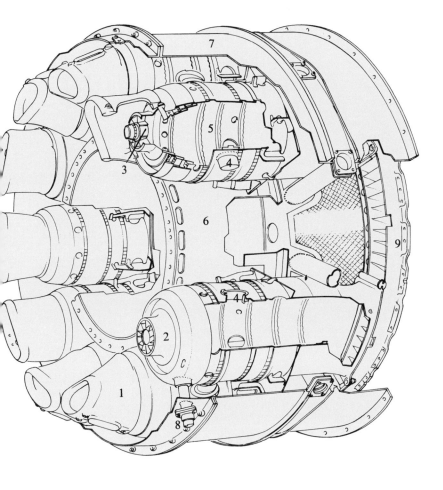

The complete combustion section of a medium-sized turbojet. In this case, a tubo-annular arrangement is used, in which ten individual flame tubes are arranged around an annular chamber between inner and outer casings. Air from the compressor either passes around the flame tubes, to enter them further down through the dilution holes, or else enters a primary air scoop at the front of a tube to sustain the flame burning inside. Interconnecting tubes join all chambers, to ensure that all light up correctly when the engine is started, even though only one (or possibly two) is fitted with an igniter. Downstream, all chambers discharge through the nozzle guide vanes ahead of the turbine.

1 Primary air snout
2 Upstream end of chamber and fuel nozzle carrier
3 Fuel nozzle
4 Interconnectors to transmit flame to all chambers during start
5 Flame tube
6 Combustion section inner liner wall
7 Combustion section outer casing
8 Igniter plug (one or sometimes two)
9 Turbine stators (nozzle guide vanes)

ped with a locally designed afterburner.

All turbofans are rather less noisy than a pure turbojet of similar thrust, and have a specific fuel consumption (fuel burned for a given thrust) up to 40 per cent lower. In their design, there is every incentive to adopt the highest possible pressure ratio. Thus, one of the smaller turbofans, the Rolls-Royce Spey, has a peak temperature of over 1,400°C and a pressure ratio which has been raised from 16:1 in the first version to as much as 21:1 in the most powerful variants now in production. It is difficult to believe that in 1950 designers were striving to get compression better than 5:1.

In fact, it is instructive briefly to compare the best gas-turbine practice of 1950 with what is being achieved today. There were no turbofans in production in 1950, although mention should be made of a tiny French unit called the Turboméca Aspin, produced in the following year, which achieved a specific fuel consumption of 0.63 (i.e. for every pound of thrust the engine developed it burned 0.63 lb fuel per hour). A quite outstanding turbojet of 1950 was the original Bristol Olympus, the first two-spool engine, which was rated at 9,140 lb thrust, weighed 3,520 lb and had a specific fuel consumption of 0.766. These were unparalleled figures in 1950; yet the 1968 Rolls-Royce Trent turbofan was rated at 9,730 lb thrust, weighed only 1,751 lb and had a specific consumption of about 0.45. This is the measure of what was accomplished in 18 years.

The Trent was one of the first of a new generation of three-spool engines with the single-stage fan separate from the compressor and driven by its own two-stage turbine. The first compressor proper had four stages and was driven by a single-stage turbine; and, as the fan had already acted as a first stage of compression, this four-stage spool was called an intermediate compressor. Finally, the high-pressure compressor had five stages and was driven by its own single-stage turbine. The three shafts ran neatly one inside the other and, although there had to be more bearings than in a single-shaft engine, the superb performance and flexibility of the three-shaft layout is actually gained with a shorter compressor. The Trent's ten stages of compression achieve a pressure ratio of 16:1, whereas in a 1952 engine, the civil Avon turbojet, sixteen stages of compression were needed to achieve 9:1.

The "big fan" engines were introduced on the civilian market with the wide-bodied jet airliners. February 9, 1969, marked the beginning of the "Jumbo" era as Boeing 747 took the air for the first time. Designed to carry 500 passengers, it was powered by four Pratt & Whitney JT9D turbofan engines. The design and development of the engine had been going on ever since 1961, at first with a military contract to supply the powerplant for the gigantic transport C-5A. But P&W lost to General Electric, who won the biggest single contract for the company up to that date with its TF39, the first high-bypass-ratio turbofan in the world to go into production. With a

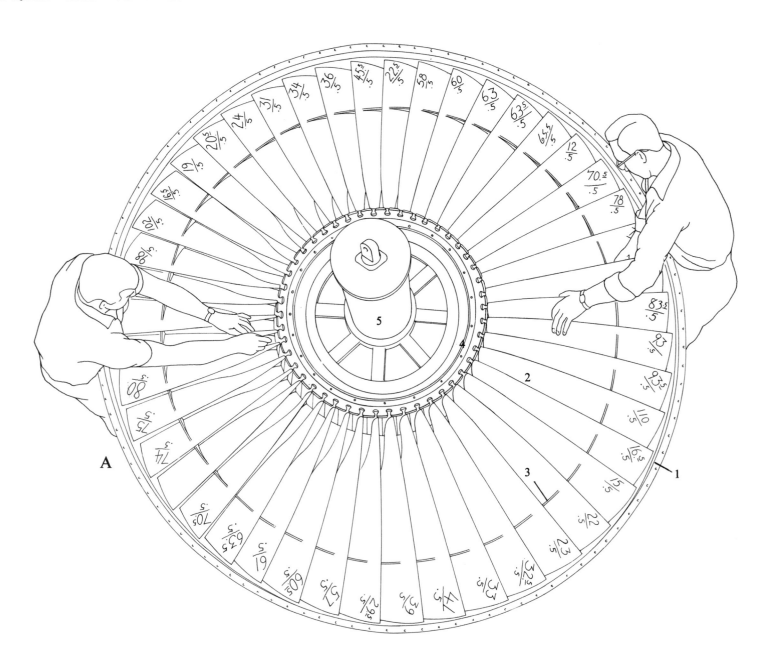

A *Just how far the technology of the gas-turbine has come since the days of Whittle's experiments is emphasized dramatically by watching engineers at the Pratt & Whitney plant in East Hartford, Connecticut, assembling their largest engine, the JT9D. Designed to power the huge*

Boeing 747 jet-liner, the JT9D is an "advanced technology engine" incorporating the latest design thinking on cycle pressure ratios, temperatures and by-pass ratios. As a result, it promises to be the largest single factor in establishing a completely new low level of cost in air transport—principally

by virtue of its specific fuel consumption of 0.34 pounds of fuel per pound thrust per hour. The mighty single-stage fan at the front of the JT9D is almost eight feet across and handles over 1,000 pounds of air per second at full power at sea level.

1 Fan casing
2 Fan blades
3 Blade platform shrouds
4 Fan ring (bolted to conical disc in engine)
5 Mounting and lifting assembly (not part of engine)

ratio 8 to 1, which means that eight times more air is passing the fan than enters the engine, the TF39 pointed the way for future high-performance civil jet engines. Rated at 40,000 lb thrust, a civilian version derived from the TF39 was later turned down as a proposed engine for the 747. It was considered to be not powerful enough and this time the competitor, Pratt & Whitney, won the race with their JT9D giving 43,500 lb.

The most visual significance of the high-bypass-ratio turbofan engines is the enormous front fan, on the JT9D measuring eight feet in diameter. The compressor has 15 stages and is driven by six turbine stages, and the specific fuel consumption is 0.34; in other words, for every pound of thrust it generates, it burns 0.34 lb of fuel per hour. The JT9D is made of titanium high-nickel alloys, stainless steel and other advanced metals; in the latest models, the high-pressure turbine blades are single-crystal and the thrust has risen to 56,000 lb. JT9D has now, in its latest model PW4000, undergone so many modifications and redesigns that it is practically a whole new engine. The number of parts has been reduced by more than 50% and the first of this series, the PW4052 (52,000 lb thrust) entered service on the Airbus A310—300 in mid-1987. There is a built-in growth potential up to 80,000 lb thrust in the most powerful version offered to power the Airbus A330.

General Electric and Rolls-Royce each have developed and manufactured their families of big turbofans for the civil airliner market. GE's first in the series of CF6 engines quietly powered the first DC-10 on its ceremonial rollout in July 1970, and since then the CF6 has gone into almost every type of big jet. The latest, advanced-technology CF6—80 is used in the Airbus A-310 and Boeing 767.

Rolls-Royce's equivalent is the RB211 series. It began life as a launch engine for the Lockheed L-1011 TriStar, but has also been offered as an optional powerplant for other planes. The latest version RB211—524L will have a power output in the 70,000-lb range.

At the other end of the scale, gas-turbines are in use which can be picked up with one hand. Even main propulsion units of aircraft have been built weighing under 100 lb. In such engines, axial blades are smaller than any adult fingernail, and great care has to be exercised in the design of the engine to minimise gas leakage, by tightening up all the fine clearances around the tips of blades. Typical of these tiny units is the Williams Research turbojet which powers the CL-89 pilotless reconnaissance aircraft for the Canadian, British and West German armies. Some of these little gas-turbines spin at speeds of the order of 100,000 rpm.

To conclude this review of current turbojet and turbofan engines, there are two specialist propulsion requirements which tend to call for high-thrust engines: VTOL (vertical take-off and landing) and supersonic flight. The VTOL jet aircraft may use either of two quite distinct methods of getting into the air, in addition to taking off conventionally with a forward run whenever an airfield is available. It may have a "lift/cruise" main engine fitted with deflector valves or swivelling nozzles, enabling the pilot to direct the exhaust vertically downward or at some angle between downward and aft, or it may have an entirely separate battery of lift-jet units pointing down permanently and used only during vertical, hovering and low speed flight.

To produce a lift/cruise engine, there are again two possible solutions. In the first aircraft ever built with jet deflection—a Meteor fighter modified for tests by the Royal Aircraft Establishment—each existing turbojet was arranged to discharge into a valve box. This contained large rotary deflectors capable of directing the gas flow out through an auxiliary nozzle pointing almost straight downward. More recently, Rolls-Royce developed lift/cruise engines with neater and lighter deflectors stemming from their civil airline thrust reversers. Clearly if the jet can be directed forward for braking, the same mechanism may direct it downward for lift. In the Hawker Siddeley 681 V/STOL transport, designed for the Royal Air Force but cancelled in January 1965, four Rolls-Royce Medway turbofans would each have been fitted with a deflector capable of directing the gas flow rearward for cruise, downward for lift or forward for braking. Another Rolls-Royce turbofan, the R.B.153/61, would have been used in a German V/STOL supersonic fighter with long jet pipes coupled to afterburners. Between each engine and its afterburner was inserted a deflector valve which, for the VTOL mode, could divert the gas flow out through downward-pointing pipes exhausting under the rear fuselage.

Another German supersonic jet aeroplane, the VJ-101C, carried four R.B.145 engines disposed in pairs on the tips of its wings, in pods which could be rotated bodily to point either downward or to the rear. It proved the practicability of this seemingly simple scheme in extensive flight trials.

Alternatively, the engine may be a turbofan so arranged that the cool discharge from the fan is expelled immediately behind the fan and the hot discharge from the jet pipe is expelled at the rear, as in the manner of the original American JT3D, already described. This enables thrust to be provided at four points, one on each side of the engine at the front and one on each side at the rear from a bifurcated jet pipe. If the engine is then installed in the centre of the body of a small aircraft, the four nozzles can be made to project through the sides of the fuselage so that, when equipped with deflector cascades, the gas can be directed straight down to lift the aircraft.

The four deflector cascades can then be arranged to rotate in unison to direct the gas jets in any desired direction for lift, thrust or reverse thrust. The pilot needs only a single throttle lever and a single jet deflector lever to effect complete control of the resultant thrust forces imparted to the aircraft.

This is perhaps the simplest jet-lift scheme imaginable and it has been well proven in the transonic Hawker Siddeley P.1127 Kestrel/Harrier. In a more advanced form, as the BS.100 engine with "plenum-chamber burning" to boost the thrust of the forward nozzles, it would have powered the P.1154 strike fighter, cancelled at the same time as the HS.681. The engine of the Kestrel/Harrier is the Rolls-Royce (Bristol) Pegasus, rated at thrusts ranging from 11,000 lb for the prototypes to almost 20,000 lb in the production version for the RAF Harrier strike fighter. The Pegasus is also used in the German Dornier Do 31 twin-engine military transport which first flew in 1966 and, with additional boxes of pure lift-jets, had VTOL performance.

Pure lift-jets are the simplest and lightest gas-turbines that can be made for a given thrust. It is important to make them small, since they are grouped in multiple installations where space is at a premium. Used only in the VTOL phase, they are installed either in streamlined containers under the wings or beside the body, or else inside the airframe where they are normally sealed off by doors over intakes and nozzles. In a de Havilland design of 1962, batteries of lift-jets were installed lying on their sides for minimum drag in underwing nacelles, with deflectors to direct their thrust in the desired direction.

Rolls-Royce pioneered this form of jet-lift, starting with the "Flying Bedstead" of 1954, which was a simple test rig containing two Nene turbojets, and could be flown freely by a pilot seated on top. Also in 1954 the first specialized lift-jet, the R.B.108, appeared; it gave a thrust of 2,300 lb for a weight of 269 lb. In 1960, the much more advanced R.B.162 was in use, giving 4,400 lb thrust for a weight of 275 lb and notable for its extensive use of glass-fibre reinforced plastics throughout the compressor casing and blades and for having a titanium turbine disc and drastically simplified mechanical design. The 162 was later developed to a rating of 5,500 lb thrust and was used in French and German V/STOL military prototypes, being installed in groups of from two to eight, with automatic starting and bleed air systems serving "puffer jet" nozzles to control the aircraft at zero airspeed.

The other specialized propulsion requirement—that for the aeroplane capable of cruising at supersonic speed—demands an engine that is anything but simple. When any flight covers a wide range of Mach numbers, it is difficult to match the engine installation to all the different operating conditions that are encountered, and to attempt to do so demands a great deal of complicated "variable geometry". As described earlier, subsonic aerodynamics are relatively easygoing and forgiving, whereas in supersonic flight seemingly trivial changes of shape can result in catastrophic loads, unacceptable drag or hopelessly impaired economics. In the case of a propulsion system of an air-breathing nature, as distinct from a rocket, the basic engine must be preceded by a long

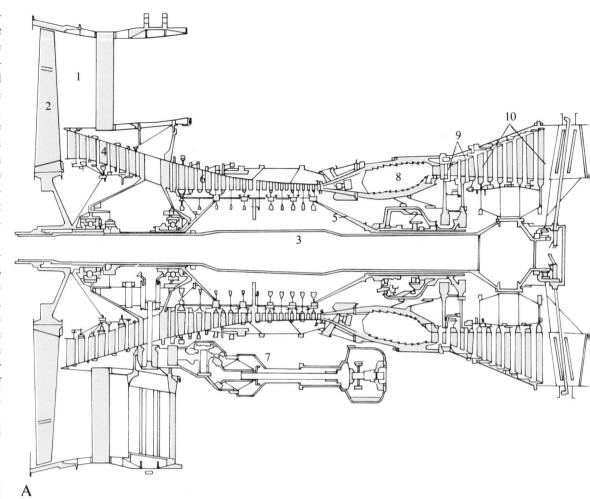

A

A *Although it has an envelope diameter of 96 inches (compared with 53 inches for the earlier JT3D used in most 707 and DC-8 aircraft today), the JT9D is actually shorter than its predecessor—a remarkable fact in view of its 15 stages of compression, achieving a pressure ratio of 24:1. The huge fan and the three-stage low-pressure compressor are driven through a long, large-diameter shaft from the four-stage low-pressure turbine; a separate two-stage high-pressure turbine with cooled blades drives the 11-stage high-pressure compressor. The JT9D-3, which went into airline service in 1970 in the first 747 airliners, has a thrust of 43,500 pounds and weighs 8,608 pounds.*

1 Fan duct
2 Fan blades
3 Fan and low-pressure compressor drive shaft
4 Low-pressure compressor
5 High-pressure compressor drive shaft
6 High-pressure compressor
7 Accessories driven off high-pressure compressor
8 Combustion chamber of can-annular (also called "tubo-annular") type
9 Two-stage high-pressure turbine
10 Four-stage low-pressure turbine

B *In recent years, increasing attention has been paid to minimising the noise of aircraft and particularly the noise of turbojets. Much of the noise stems from the tearing, shearing action that takes place at the boundary of the jet of hot gas, where it passes through the surrounding cold atmosphere. One way to quieten this part of the noise spectrum is to increase the length of the jet periphery and speed up the mixing of the gas with the cold air. In 1958 the first aircraft with quietened nozzles went into airline service. The original Boeing 707 had propelling nozzles consisting of 21 separate pipes; the Comet had the arrangement shown here, in which fresh air is admitted through inward-facing ducts surrounding the single nozzle.*

1 Jet pipe connecting ring
2 Nozzle outer casing
3 Secondary air inlets to break up jet periphery

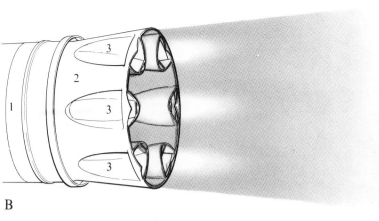

B

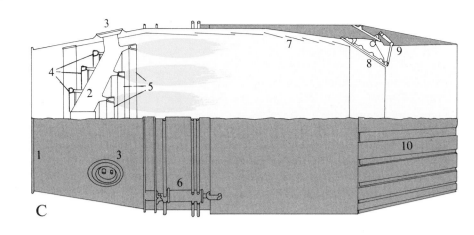

C

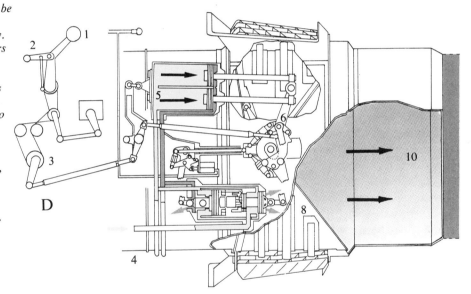

D

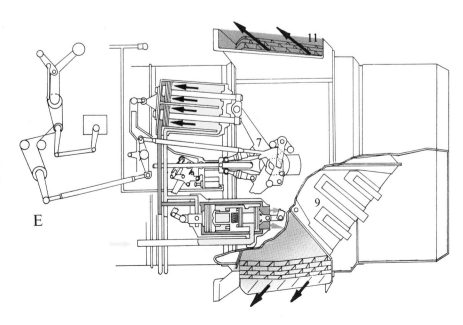

E

Afterburners, or reheat jet pipes, boost the thrust of jet engines by burning extra fuel downstream of the turbine. Early designs were long and clumsy, but today combustion technology has improved so greatly that a reheat temperature of 2,000°C can be achieved within a linear distance of only two or three feet, despite steadily increasing gas velocities. This simplified illustration shows a typical modern afterburner, of the type developed by Rolls-Royce for the Adour reheat turbofan being produced jointly with Société Turboméca for the Anglo-French Jaguar fighter-trainer. It is attached immediately behind the turbines and houses three flameholder rings which locally slow the jet to assist combustion; it terminates on the right with an infinitely-variable propelling nozzle. Cooling air enters through the jet-pipe liner.

1 Flange connecting reheat jet pipe to engine
2 Three radial mounting struts of streamline section
3 Strut anchorages and fuel inlets

4 Three circular fuel injection manifolds
5 Three vee gutter flameholder rings behind (4)
6 Pneumatic actuator driving variable nozzle flaps
7 Cooling air intake rings
8 Variable area primary nozzle
9 Actuating linkage
10 Variable area secondary nozzle

D Although a primitive
& jet thrust reverser was
E flight-tested on a Whittle engine mounted in the tail of a Wellington, there was little incentive to develop such devices until it was realized that they could reduce the runway lengths required by civil jet-liners. They not only enable jets to be brought quickly to a halt even on wet or icy runways, but are used regularly in short-haul service where stops are frequent, greatly reducing tyre and brake wear and spares bills. Several patterns are in use. The standard Rolls-Royce design, shown here, uses a pair of clamshell shutters to seal off the normal nozzle and open reverser exits, containing cascade vanes, to deflect the gas jets forward. For an engine mounted

beneath a wing, it is more usual for the reverser apertures to be on each side rather than above and below. Operation of reversers is by means of pneumatic rams, controlled by systems provided with multiple safeguards to prevent inadvertent operation—although some aircraft, including the Trident, are cleared for reverser operation in flight to enable steep let-downs to be made without reaching excessive airspeeds.

1 Pilot's power lever (throttle)
2 Reverse thrust selector lever
3 Actuation and mixing linkage
4 Jet pipe
5 Twin cylinder pneumatic ram actuator
6 Actuating arm in forward thrust position
7 Actuating arm in reverse position
8 Clamshell reverser doors in forward thrust position
9 Clamshells closed into reverse thrust position
10 Propelling nozzle
11 Forward deflector vanes for reversed efflux

and carefully profiled nozzle.

Taking the intake duct first, it must operate at zero forward speed at the start of the take-off run without reducing the airflow or harming engine performance in any way. This requires the maximum possible intake area and it is common to arrange not only for the area of the intake itself to be opened out fully, by various arrangements of movable centrebodies or sidewalls controlled by automatically positioned actuators, but also to open auxiliary doors in the wall of the intake duct to admit additional air.

As the aircraft accelerates, the intake area is reduced and reaches a minimum at the full supersonic crusing speed. At the maximum Mach number, which is 2.2 in the case of the Concorde and was about 2.6 for the Boeing SST, the geometry of the lips or centrebody at the very start of the intake must also be arranged so that they produce an inclined shock-wave, positioned in such a way that it gives minimum pressure drop to the engine airflow, as well as minimum interference with flow over the rest of the aircraft. Inside the intake duct, the normal shock-wave which reduces the flow velocity to below the speed of sound must also be situated at exactly the right place for minimum pressure drop, and it must be kept there.

The positions of these shock-waves can vary considerably and suddenly, unless the flow through the intake duct can be controlled powerfully and rapidly. An apparently complicated arrangement of interlinked auxiliary intake doors, spill vents and dump doors is used for this purpose in the Concorde engine installation. In addition to these means for disposing of excess air, or taking in additional air, the Concorde intake has a profile which can be altered greatly by means of a hydraulic actuation system that can change the inclinations of the hinged panels forming the entire upper wall of the duct. At low speeds the panels are pulled upwards, giving a maximum duct cross-section; but at supersonic Mach numbers the panels are pushed down into the airflow to create a marked "throat" restriction. The normal shock-wave is positioned at the throat and the divergent duct behind this point acts as a diffuser to continue the process of compression in subsonic flow that was begun by the convergent section upstream in supersonic flow.

To ease the mechanical problems caused by the need to vary the duct geometry, the Concorde intake duct is approximately square in section, with flat walls. The Boeing SST, however, had circular intake ducts and in this case the variable geometry was confined chiefly to the use of a centrebody with flexible walls in a kind of barrel-stave arrangement capable of being altered in profile by internal linkages somewhat like an umbrella frame.

Generally, then, the variable-geometry intake is necessary to ensure that under all flight conditions the intake airflow is exactly the same as the flow required by the engine and, further, to keep the flow velocity at the entry to the engine constant for each flight condition and below the speed of sound.

At speeds well below that of sound an aircraft intake normally has to suck in the air needed by the engine; the air does not come in of its own accord. But at supersonic speed the intake is no longer sucking because the air is "rammed" into the intake by the latter's own movement. The resultant ram pressure built up in the intake rises very rapidly as flight speed rises beyond about Mach 2; at that speed the ram pressure is already as great as that produced by many engine compressors, and at Mach 3 it is markedly greater. At Mach 4 it is pointless to have a mechanical compressor at all in cruising flight, because its effect is insignificant compared with the ram compression. The best air-breathing engine is, therefore, the ramjet, described later, which is basically just a suitable profiled duct.

Even at the cruising speeds attained by the first supersonic airliners, the optimum pressure ratio for the engine's own compressor is considerably less than it is for subsonic transports. The Olympus 593—power plant of the Concorde—is a two-spool engine (actually a very much developed descendant of the first two-spool engine ever built) with seven stages on both its low-pressure compressors. The resultant pressure ratio of 14.8:1 is surprisingly high for a Mach 2.2 engine. The air enters the engine at about 30 pounds per square inch and 150°C, compared with the original pressure and temperature ahead of the intake of about one pound per square inch and –57°C at 60,000 ft altitude. It is then raised to about 85 lb per square inch and 550°C by mechanical compression, before entering the combustion chamber, and finally reaches the turbines at about 1,180°C. The two single-stage turbines both have air-cooled blades.

Downstream of the turbine, the cooling airflows from the turbine, bearings and other parts of the engine mix with the main gas stream and the total flow is accelerated through the propelling nozzle to supersonic speed, falling in pressure simultaneously to that the surrounding atmosphere. Pressure ratio across the propulsive nozzle of a subsonic jet is usually about 2:1 or 3:1, but in the supersonic Concorde a more usual figure would be 15:1. For maximum propulsive efficiency a "con-di" nozzle is necessary, first convergent in order to speed up the initially subsonic gas stream and reduce its pressure and then, at the point at which the jet reaches sonic velocity, changing to a divergent profile to accelerate further the supersonic flow and continue the process of reducing its pressure.

The fact that a supersonic flow can be accelerated and reduced in pressure only in an expanding duct (the converse of subsonic flow) means not only that the propulsive nozzle of an SST engine must be fully variable but also that it must be very large in cruising flight. At Mach 3 the nozzle will be considerably larger in diameter than any other part of the engine, and this effect is accentuated further at Mach 4. In

A *While the JT9D is one of the largest turbofans in the world, the Turboméca Aubisque is the smallest. Built in a thriving factory just north of the Pyrenees, the Aubisque is one of the latest in a prolific family of small aircraft gas-turbines which began at the end of World War II and are used in almost every country. An unusual feature of the Aubisque is that its single-stage fan is driven through a reduction gearbox. If it were driven directly by the shaft on which the small-diameter turbines and centrifugal compressors are mounted, it would reach supersonic speed at the tips of its blades and achieve poor efficiency. As it is, it enables a specific fuel consumption of the order of 0.6 to be achieved for the first time in the 1,000-pound thrust class. First order for the Aubisque was from Sweden for engines to power Saab-105 trainers for the Swedish Air Force.*

1 Intake section
2 Single-stage low-pressure compressor (fan)
3 Step-down gearing
4 Flow divider
5 Axial compressor
6 Centrifugal compressor
7 Vaporizing combustion chamber of annular type
8 Two-stage turbine
9 Jet nozzle
10 Discharge nozzle from fan duct (by-pass duct)

B *An "advanced technology engine" which promises to brin new levels of efficiency and economy to the short-haul airline business is the Rolls-Royce Trent, shown here installed in a cowling pod of the typ in which it would be fitted in a short or medium range transport. One of the Trent's outstanding features is that, like other Rolls-Royce engines for the 1970s, it is a three-shaft engine: the fan is on a shaft of its own and can be throttled on take-off to reduce engine noise without losing thrust. This 9,730-pound thrust engine weighs only 1,751 pounds and is designed for the most economical and easies operation in airline service.*

1 Air intake
2 Single-stage fan
3 Intermediate compressor
4 High-pressure compressor
5 Combustion chamber
6 High-pressure turbine
7 Intermediate turbine
8 Two-stage low-pressure turbine
9 Fan duct (by-pass duct)
10 Jet pipe

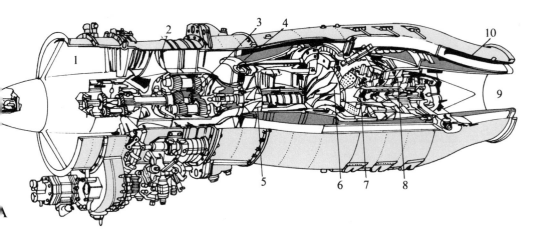

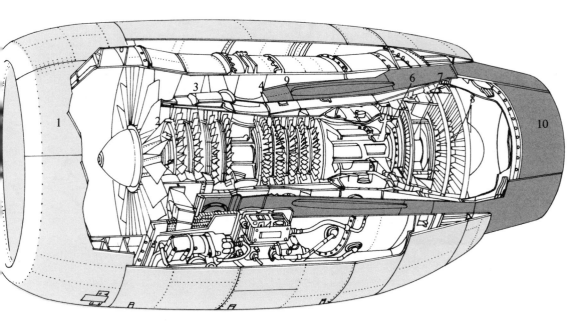

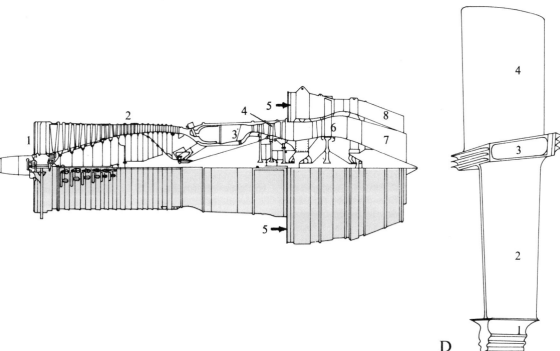

C
& Aft-fan engines have
D not been made in large
numbers, but the
General Electric
Company (USA) has
shown that they can be
made to work well in
small, medium and
large sizes. Their first
turbofan was the
CJ805-23, shown in
the sectioned drawing
at left. Evolved directly
from the J79 turbojet
by adding a new wheel
at the rear with
double-deck turbine/
fan blades, the
CJ805-23 has, since
1961, been serving in
many countries in the
Convair 990 Coronado
airliner. One of its
unusual blades is
pictured below (D).
Known colloquially in
America as a "blucket"
(blade/bucket), it is
made of high-strength
nickel alloy. The highly
stressed fir-tree root by
which the blade is
attached in its disc has
a running temperature
of only 565°C, for the
gas driving the blade
has cooled in
expanding through the
three-stage turbine of
the basic engine. This
565°C temperature is
characteristic of the
whole inner (turbine
blade) portion out to
the shroud, which
forms a continuous
ring when the stage is
assembled; the fan
blade outboard of this
platform has a metal
temperature of 365°C
at the root and 140° at
the tip. Despite its
large moment of
inertia, the fan follows
power changes of the
basic engine with a lag
of only 0.1 second and
can accelerate from 88
to 100 per cent rpm
in 0.7 second. At full
power the thrust of the
CJ805-23 is 16,100
pounds; it weighs
3,765 pounds.

C
1 Air intake
2 Single spool axial
   compressor with
   17 stages
3 Combustion chamber
   of can-annular type
4 Three-stage turbine
5 Aft fan intake
6 Freely rotating fan
7 Jet pipe from engine
8 Cool air duct from fan

D
1 Fir-tree root
2 Turbine blade driven
   by engine hot gas flow
3 Intermediate platform,
   shroud and inter-blade
   seal
4 Fan blade acting as
   propeller in duct
   handling fresh air

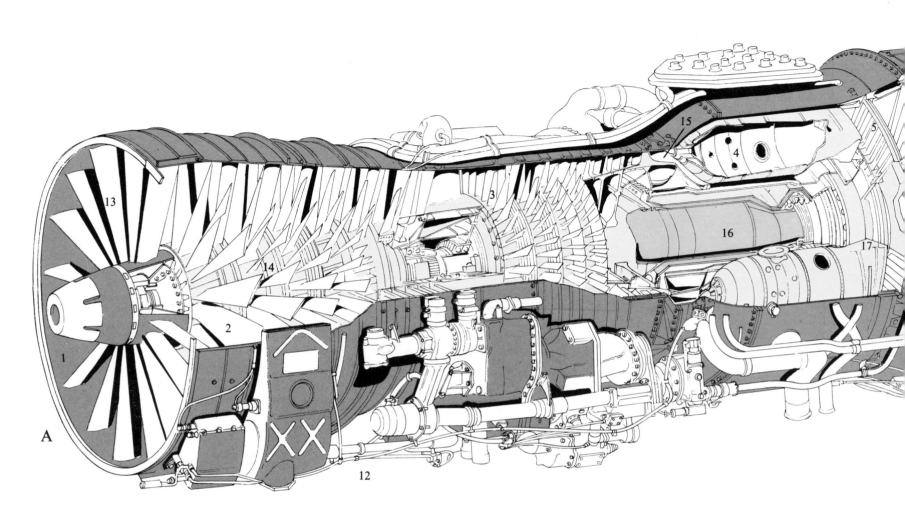

A Twenty years' progress
& in civil airline jet
B propulsion are
exemplified by the
contrast between the
engines of the Comet 1
(right) and those of the
Concorde (above). The
de Havilland Ghost
turbojet of the Comet 1
developed a thrust of
5,000 pounds and
propelled the 105,000-
pound pioneer jetliner
at 490 mph; the
Rolls-Royce/
SNECMA Olympus
593 turbojet of the
Concorde develops
35,080 pounds thrust
and will propel the
376,000 pound super-
sonic jet-liner at
1,450 mph. Although
the range of flight
conditions
encountered by the

Comet 1 was far wider
than that met by any
previous civil aero-
plane, it was not too
great for a fixed-
geometry engine
installation: a simple
ram intake leading
straight to the single-
sided centrifugal
compressor, ten
combustion chambers,
a single-stage turbine
and a simple jet pipe.
But the Concorde flies
over a wider range of
speeds and altitudes so
that there have to be
many ways in which the
engine installation can
be adjusted to meet
changing demands.
The auxiliary doors,
moving ramps and
hinged flaps are
upstream (not shown),
and the

reverser immediately
ahead of the special
SNECMA fully-
variable ejector nozzle
is a particularly neat
piece of engineering.
The simple reheat
system is installed with
practically no weight
penalty and could
boost thrust by 20 per
cent if needed; but the
engine has exceeded
its originally specified
thrust with no reheat
at all. The contrasts
between the volume of
air flow and its
pressures and
temperatures as it
passes through the
engine of 1947 and
that of 1967 are
indicative of millions
of man-hours of
painstaking
development.

A
1 Air intake
2 First stage of low-
pressure compressor
3 First stage of high-
pressure compressor
4 Combustion chamber
of can-annular type
5 High-pressure turbine
driving h-p compressor
6 Low-pressure turbine
driving l-p compressor
7 Simple thrust boosting
afterburner (reheat)
8 Variable area primary
nozzle
9 Variable area
convergent/divergent
secondary nozzle
10 Thrust reverser
clamshells in forward
thrust position
11 Upper and lower
reverse thrust
deflector vanes
12 Atmospheric pressure
about 1 lb/sq in

13 Pressure at 60,000 feet
at Mach 2.2 about
14.5 lb/sq in,
temperature 153°C
14 Airflow approximately
500 lb/second
15 Pressure at compressor
delivery 205 lb/sq in,
temperature 550°C
16 Power required to
drive both compressors
over 100,000 shaft
horsepower
17 Mean gas temperature
over 1000°C
18 With reheat operating,
pressure 100 lb/sq in
at over 1500°C

B
1 Air intake
2 Casing over centrifu
compressor
3 Combustion chambe
(one of ten)
4 Shaft from turbine
driving compressor
5 Single stage turbine
6 Jet pipe extended to
discharge at wing
trailing edge

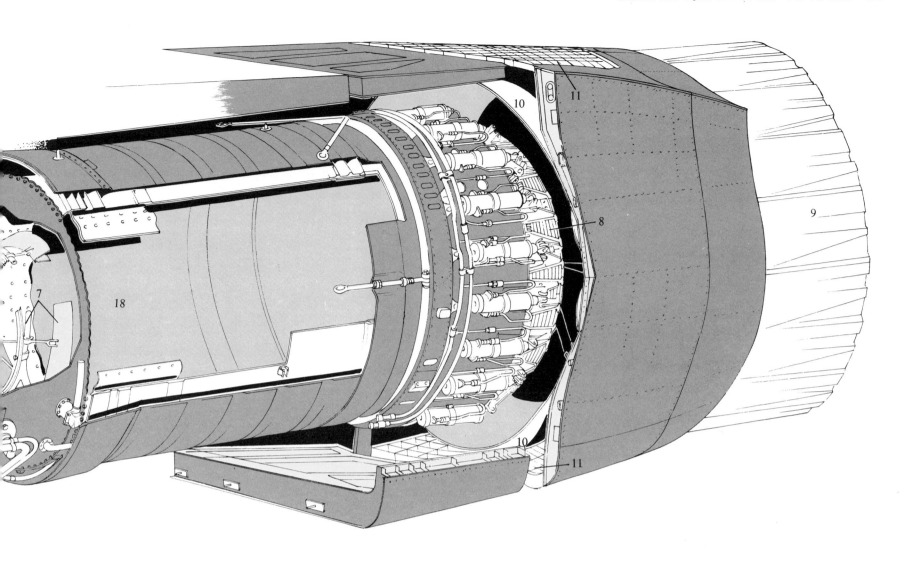

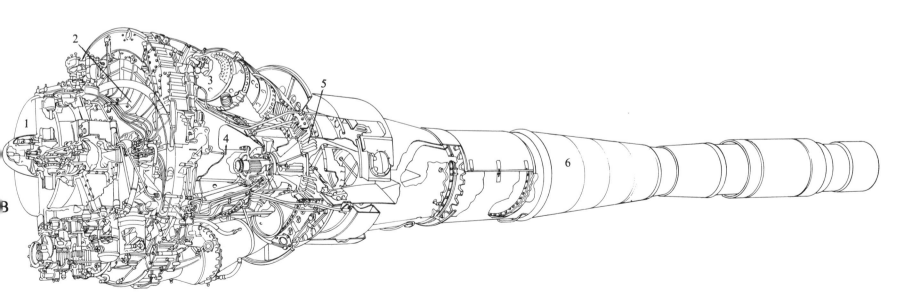

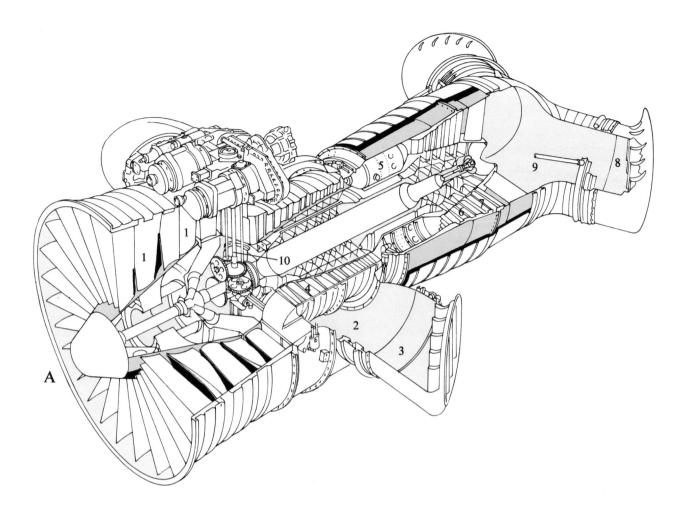

A *In the late 1950s, the great British aircraft designer Sir Sydney Camm discussed with the engine designer Dr Stanley Hooker the possibility of producing a jet engine with rotating nozzles, so that it could not only propel an aeroplane forward but also—if the thrust/ weight ratio of the fully-laden aircraft could be made greater than 1—could lift it off the ground without the need for any forward speed. The unique engine which resulted is the Rolls-Royce Pegasus turbofan. The fan discharges its air through two forward nozzles while the gas jet at the rear is also split into port and starboard flows discharged through two nozzles. The four nozzles project through the sides of the nacelle or aircraft fuselage, the cold nozzles ahead of the centre of gravity and the hot nozzles behind (both equidistant from the c.g. if the thrust is split 50-50 between the front and rear). When the nozzles point downward, the engine can lift the aircraft; once airborne the nozzles can be rotated, under the control of a single cockpit lever, by a bleed-air motor driving cross-shafts and chains which keep the nozzles exactly aligned. Thrust can be directed rearward or even forward, for braking purposes. In the Hawker Siddeley Harrier tactical strike fighter of the RAF, a single Pegasus provides all the power for STOL (short take-off and landing) and VTOL (vertical take-off and landing) as well as for transonic flight.*

1 Two stage fan
2 Fan duct discharging on each side
3 Rotating nozzle on each side for lift, thrust or braking
4 High-pressure compressor rotating in opposition to (1)
5 Combustion chamber of can-annular type
6 High-pressure turbine driving compressor
7 Low-pressure turbine driving fan
8 Rotating nozzle on each side of split jet pipe
9 Nozzle drive ties all four nozzles to turn together
10 Drive to accessories

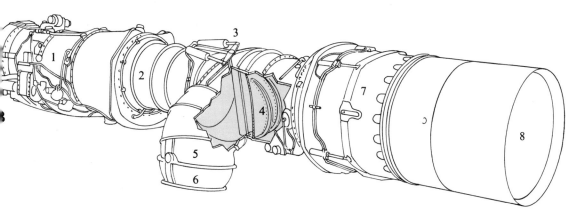

the Concorde the nozzle is of conventional circular section, althow at one time the possibility of using a square nozzle similar to the intake shape was investigated in the hope that it might make variable geometry easier to accomplish.

The shape of the nozzle must be changed to suit varying flight conditions, the pneumatic actuation system being capable of reliable operation at temperatures of the order of 400°C. At low flight Mach numbers the nozzle is closed, and additional air is entrained through auxiliary intakes above and below; but at cruising speed the nozzle is fully open. For reverse thrust after landing, the gas stream can be expelled through conventional upper and lower cascades immediately behind the auxiliary nozzle doors.

In all the gas-turbines described so far, the sole output propelling the aircraft is a jet of hot gas or of gas and atmospheric air. It is possible to build a gas-turbine in which as much power as possible is extracted from the gas stream and made available as torque to turn a shaft. This is the case in virtually all gas-turbines for land or marine traction and for general industrial purposes, and it is also by far the best way of using a gas-turbine to propel a helicopter, a slow aeroplane (up to, say, 350 mph) or an aircraft in which extreme fuel economy or the ability to use very short fields is more important than achieving the maximum level speed.

Clearly the more extreme turbofan engines, in which the by-pass ratio is well over 5:1, are almost gas-turbines driving propellers—albeit propellers with large numbers of fixed-pitch blades encased within a duct. But the true turboprop evolved quite naturally long before the turbofan was thought of, because in the 1920s gas-turbines were considered by several authorities as possible replacements for the piston-engine in driving an aircraft propeller.

Design studies for what are now known as turboprops were pursued at Farnborough: engineers in France and Germany offered proposed designs and a Hungarian named Jendrassik actually built a small engine in the early 1930s. But the first turboprop to fly was the original Rolls-Royce Trent, fitted to a modified Meteor in 1945. Bearing no direct similarity to the 1960s engine of the same name, the Trent was a Derwent turbojet modified to drive a five-bladed Rotol propeller of very small diameter, necessitated by clearance problems on the Meteor; the propeller absorbed about 750 hp and some 1,250 lb of thrust could still be obtained from the jet.

In the three years following World War II, most expert opinion held that the immediate application of the gas-turbine for civil use would be in the form of the turboprop. Not only was this considered to have appreciably better propulsive efficiency than the turbojet, and to promise lower fuel consumption for a given duty, but it also appeared better matched to the range of speeds likely to be achieved by airliners in the then-immediate future. The bold step by de Havilland Aircraft in designing the jetpropelled

A slightly different way of providing thrust vectoring, to enable a jet engine first to lift and then to propel an aeroplane, is to use a switch-in deflector. This device is based upon the Rolls-Royce type of thrust reverser and can be applied to any type of turbojet or turbofan discharging through a single jet pipe. For VTOL flight, the pilot shuts off the normal jet pipe with a single shutter or with twin clamshells, as in an ordinary reverser. The jet then escapes through a side orifice, which can be fitted with a rotating cascade to *deflect the gas downward or in any other desired direction in its plane of rotation. Alternatively—as shown here—the engine can be fitted with an auxiliary jet pipe which is turned downward and terminates in a nozzle capable of deflecting the jet through a small angle to assist aircraft control at low speeds. The powerplant shown is the RB.153 turbofan which was developed jointly by Rolls-Royce and the German MAN company and had an afterburner downstream of the deflector. Normally such an* *engine provides thrust well behind the aircraft centre of gravity and so must be used in conjunction with additional lift-jets further forward.*

1 RB.153/61 turbofan engine
2 Jet pipe
3 Deflector valve actuation package
4 Deflector valve switched in to "lift" position
5 Lift pipe
6 Lift nozzle has limited ability to swivel fore and aft
7 Drive ring for variable afterburning nozzle
8 Propelling nozzle for forward thrust

A *A highly advanced turboprop is the Rolls-Royce Tyne which entered service in Vanguard airliners in 1960 and is now used in addition in military transport and ocean patrol aircraft. In this design, the accent was not only on trouble-free service but also on the highest possible efficiency. The Tyne, therefore, has a two-spool axial compressor with a pressure ratio of 13.5:1, and a high gas temperature necessitating the use of air-cooled turbine blades. The three-stage low-pressure turbine drives both the six-stage low-pressure compressor and the propeller, the latter via a reduction gearbox which incorporates a torquemeter to indicate the precise load on the propeller shaft and thus provide a measure of the power developed. If the engine should fail the propeller can either be feathered or else left to windmill to drive the accessories, which are geared to the low-pressure shaft. The intake is electrically heated for anti-icing protection, and there is a large compressor bleed at the rear of the low-pressure spool, to provide air for cabin pressurization and other purposes. Tynes weigh just over 2,000 pounds each and give from 5,000 to 6,000 horsepower.*

1 Circular air intake with electric de-icing elements
2 Air intake duct
3 First stage of low-pressure compressor
4 First stage of high-pressure compressor
5 Combustion chamber of can-annular type
6 Single-stage high-pressure turbine driving h-p compressor
7 Three-stage l-p turbine driving l-p compressor and propeller
8 Jet pipe
9 Propeller reduction gearbox incorporating torquemeter
10 Propeller shaft

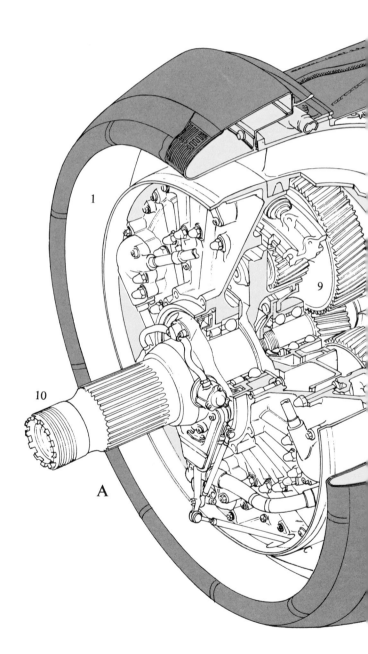

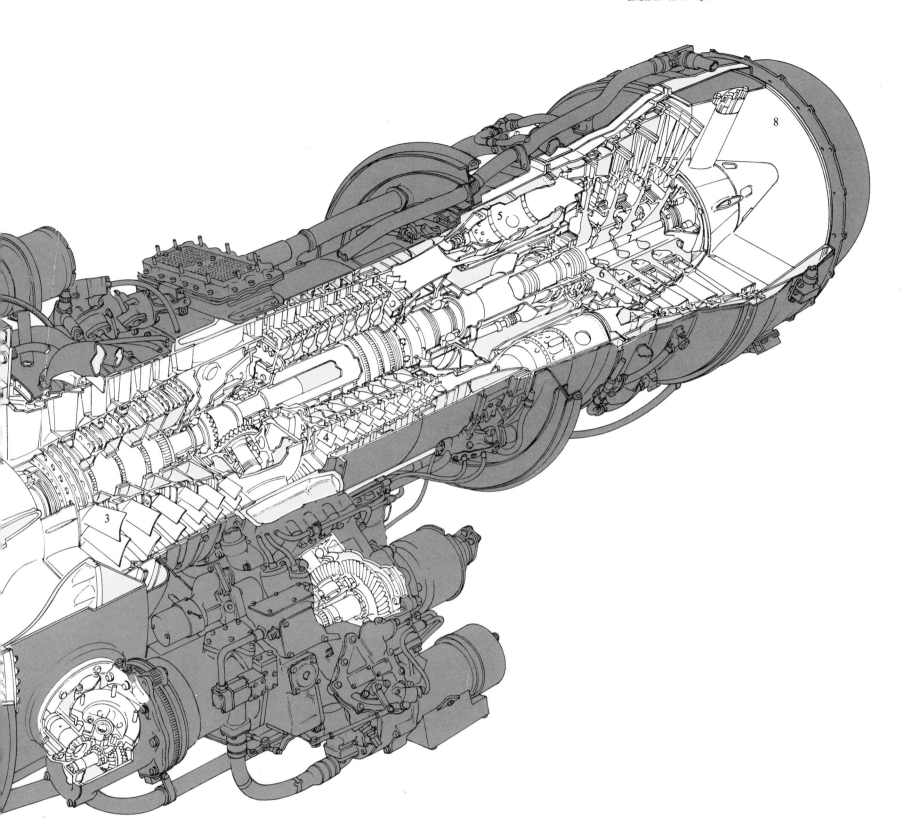

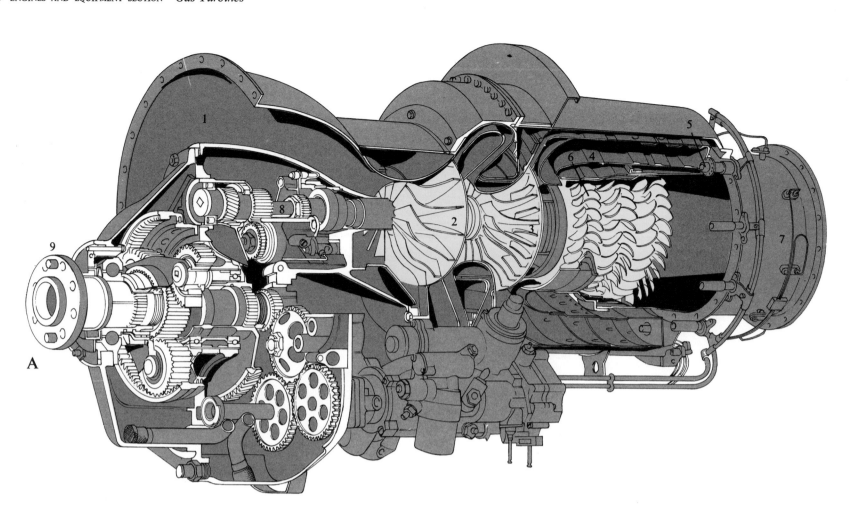

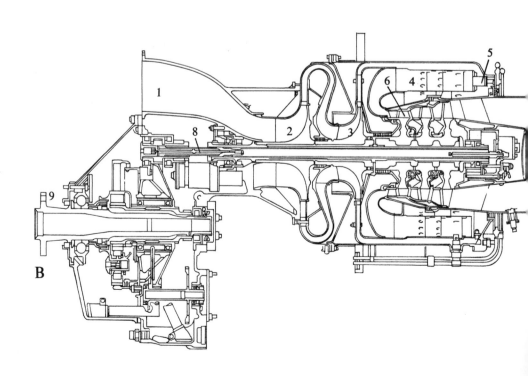

A One of the many small & gas-turbines which are B steadily breaking into the established markets of piston-engines is the AiResearch TPE 331. Known to the US military services as the T76, it is a single-shaft turboprop, built by a company with extensive experience in the field of gas-turbines for auxiliary and industrial applications. It has three turbine stages, each cast in a single piece of high-nickel alloy, complete with blades, connected through a "curvic coupling" directly to a pair of titanium centrifugal compressors in series and thence, via a two-stage reduction gear, to the propeller shaft. The air intake duct passes either above the reduction gear, as shown here, or below it. Throughout the TPE 331, the emphasis is on robust construction for trouble-free operation in trying conditions, yet it weighs only a little over 300 pounds and gives 715 shaft horsepower.

1 Air intake from above propeller spinner
2 First-stage centrifugal compressor impeller
3 Second-stage compressor impeller
4 Reverse-flow annular combustion chamber
5 Fuel injection nozzle
6 First rotor stage of three-stage turbine
7 Jet pipe
8 Propeller gearbox drive shaft
9 Propeller mounting

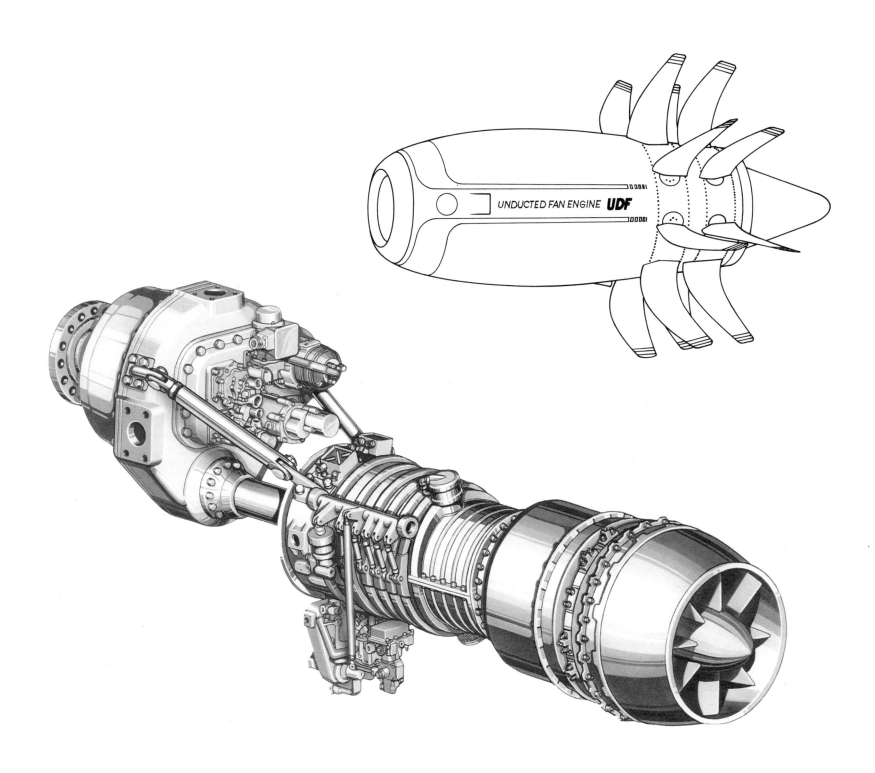

UNDUCTED FAN ENGINE **UDF**

A *The unducted propfan, or ultra-high-bypass engine configuration, has further increased the amount of bypassed air compared to the turbofan, which gives a significant improvement in fuel consumption. Several aeroengine manufacturers have developed different solutions to the propfan concept. General Electric GE36 UDF, pictured, has a turbine-driven, counter-rotating propfan without any gearbox, while the* similar Pratt & Whitney/ Allison design has a mechanical transmission between the gas generator and the propfan. The engines were developed during the 1980s when fuel prices were expected to rise dramatically, and prototyes were successfully flown on Boeing 727 and MD-80. The future market potential for propfans is wholly dependent on the fuel price level.

B *The Allison GMA 2100 is a a civil derivative of the US government-funded T406 for the Bell Boeing V-22 Tiltrotor. Rated 6,000 shp in the basic version, it will power the new high-speed regional aircraft. GMA 2100 turboprop has a growth potential up to the 12,000–13,000 shp range, by increasing the engine operating temperature and raising the pressure ratio by redesigning the* compressor and the turbine. The propulsion package including the engine, gearbox and propeller is about 10 % lighter and considerably more fuel efficient than earlier generations of turboprop engines.

Comet was considered premature; the fact that this jetliner would be in full BOAC service by May 1952, cruising at 490 mph, was not foreseen in 1945, and the British Government viewed the main choice for the propulsion of future airliners as lying between the turboprop and the compound diesel of Nomad type.

But the turboprop was never accepted widely for long-range travel, despite the fine example of the Bristol Britannia and Lockheed Electra with their long-range capacity and low fuel consumption. Their debut coincided with the introduction of the jet airliners Boeing 707 and Douglas DC-8, and the travelling public's attraction to the somewhat faster jets made the turboprop's appearance on the long-haul market a short episode. In the east, however, long-range air traffic for a longer period was flown by turboprops like Tupolev Tu-114 and Ilyushin IL-18, most likely due to the fact that Soviet jet-engine designers concentrated on military engines and it took longer to develop a suitable powerplant for civil airliner use. Even today in 1990, a large portion of the Soviet strategic bomber and reconnaissance fleet still consists of the Tu-95, equipped with four Kuznetsov NK-12MV turboprops rated 14,700 hp, each driving a big-diameter counter-rotating propeller unit.

For short-haul medium-sized airliners, commuters and light business aircraft, the turboprop concept has proved to be the ideal powerplant. Low fuel consumption and maintenance costs, reliability, favourable power/weight ratio, and long intervals between overhauls, offset the relatively higher cost of a turboprop compared to a piston engine.

The outstanding turboprop has been the Rolls-Royce Dart, first run in 1946, and test—flown in the nose of a Lancaster in 1947.

Basically, the Dart stemmed from the two-stage compressor of the 60-series Merlin and Griffon piston-engines. A two-stage turbine was designed to drive it, and the first Dart was rated at 990 shaft hp, driving a 10-ft Rotol propeller. By April 1953 the output had risen to 1,480 equivalent shp—this expression meaning that allowance is made for the value of the residual jet thrust by dividing the thrust in pounds by 2.6 and adding the result to the shaft power.

April 1953 was the month in which the Viscount entered service with British European Airways. This plane, powered by four Darts, was almost devoid of competition for seven years and is still a competitive short-haul transport. Today, the latest Darts are rated at no less than 3,250 equivalent shp. They have a three-stage turbine, but are otherwise almost indistinguishable from the original 990-hp engine. No other type of power plant has ever more than trebled its output purely as a result of detail refinement.

Bristol Aero Engines, later absorbed into Bristol Siddeley and now a division of Rolls-Royce, pioneered long-range turboprop operation with the Proteus, first flown at 3,200 hp in 1949 and used as the engine of the Britannia at 4,450 hp since 1957.

The Dart is a single-shaft power plant, the turbine being connected directly to the compressor and, via a two-stage reduction gear, to the propeller. In contrast, the Proteus is a "free-turbine" engine, the turbine which drives the compressor being mechanically unconnected to the second, low-pressure turbine geared to the propeller. This arrangement makes for a versatile engine and also means that the starting system does not have to turn the reduction gear and propeller; but special precautions have to be taken to prevent turbine runaway in the event of a gear in the propeller drive stripping its teeth. The first American families of turboprops—the T38, T40 and T56 by Allison and the T34 by Pratt & Whitney—are single-shaft engines. Those by Allison are notable in that the propeller gearbox is not attached directly to the compressor casing but is mounted ahead and either above or below on bracing struts which transmit the loads to and from the propeller.

In 1956 several very powerful turboprops were under development, outstanding examples being the 12,000–14,000 hp NK-12M in the Soviet Union and the 15,000 hp T57 by Pratt & Whitney. The Soviet engine entered service as the power plant of the remarkable Tu-95 long-range bomber and maritime-reconnaissance aircraft in 1957, and subsequently entered civil airline use in the basically similar Tu-114. But the T57 was cancelled, as were the British BE.25 Orion and all other high-power propeller engines, partly because of the difficulty and cost of developing corresponding propellers but chiefly owing to the emergence of the high-ratio turbofan as a superions power plant for long-range tranports.

It is revealing to contrast the projected performance of the T57-powered Douglas C-132, planned in 1956, as the future "workhorse" logistic cargo carrier of the US Military Air Transport Service, with the Lockheed C-5A of today's Military Airlift Command. Powered by four General Electric TF39 turbofan engines, the C-5A will carry twice the payload of the C-132 half as far again and at 150 knots higher speed for only marginally greater fuel consumption.

By far the brightest future for the turboprop lies in the range of 3,000 hp and below, with the bulk of sales between 250 and 1,000 hp. Typical of the way in which the market is growing is the fact that between 1959 and 1961 the number of North American aircraft gas-turbines in the 250-hp range rose from nil to three, and in the 500-hp bracket from nil to four. Moreover, in both cases a foreign engine by the French Turboméca company was also made available.

Turboméca is an outstanding company which, by specializing in small aircraft turbines for 20 years, has carved a major niche for itself and gained footholds in 34 countries outside France. Its first engines were jets—including the Aspin turbofan already mentioned—but its products today include turboprops, shaft-turbine engines for helicopters and air compressors for helicopter tip drive, in which air is expelled from jets at the tips of the rotor blades without causing any torque reaction.

Broadly speaking, the turboprop handles a large propulsive mass of air than does any other aircraft gas-turbine and thus has the highest propulsive efficiency. But the fact that propeller tip speeds must be held within certain Mach-number limits imposes an upper limit on forward speed of the order of 450 mph (except in the case of the Soviet Tu-95 and Tu-114 and the weight, cost and maintenance cost of the gearbox and propeller are considerable items which are absent from the balance sheet of a turbofan.

Even the question of noise, formerly a point in the turboprop's favour, is now no longer a major issue between the turboprop and high-ratio fan engine and the latter type of power plant is greatly to be preferred by the passenger or crew inside the aircraft. In general, the turboprop is going to have little success in competing with fan engines at the upper end of the scale of speed or size, but it will continue to find success when competing with piston-engine at the lower end.

In the small-aircraft market, turboprops have now reached the stage where propellers are available for them, their fuel consumption is not excessively costly in comparison with piston-engines, the performance gains are substantial and the added smoothness of flight is a bonus. The only handicap is that a typical turboprop in the 250–500 hp class costs several times as much to buy as does an equivalent piston-engine. But aircraft such as the Pilatus Porter and Short Skyvan, both described earlier, were planned around piston-engines and put into production with turbine power, while American mass-producers of light business aircraft—Beech, Cessna and Aero Commander—have all put into production turboprop aeroplanes having pressurized cabins and cruising speeds in excess of 260 mph.

Handley Page in England has achieved success with a slightly larger executive machine which would have been impossible to produce without the availability of highly-developed turboprops in the 800-hp class. The engine concerned is the Turboméca Astazou XIV, and the aircraft the H.P. 137 Jetstream. The Astazou began life in 1959 as a 460-hp single-shaft engine, only 18 inches in diameter and weighing 270 lb. It had reached 560 hp by 1962, 670 by 1964, 740 by 1965 and 855 by 1967. The Astazou 20 of 1,000 hp come in 1969. Such fantastic increases in power were not possible with piston-engines; in fact a 1,000-hp piston-engine would have weighed more than four times as much as the Astazou 20 and cost more to run in fuel and maintenance.

Fifty Jetstreams equipped with Astazou were produced until the manufacturer, Handley Page, went into liquidation in 1970 and the production lines were closed. At the beginning of the 1980s, however, British Aerospace (BAe) put a slightly modified 19-seater version, Jetstream 31, into production powered by a Garret TPE 331 turboprop engine.

Turboméca was the first company in the world to produce a shaft-turbine engine for helicopters. The as-turbine concerned, the original Artouste of 1950 ated at 220 shp, was installed in the Sud-Aviation louette helicopter which, by 1955, had been developed into a 550-hp machine so far in advance of s piston-engined contemporaries that Sud have ince delivered over 1,000 to some 25 countries. By 960 the piston-engine had been abandoned for all elicopters and VTOL machines except for certain ery small craft and the evergreen Bell 47, Hiller and Hughes families which continued in production simly because they were highly-developed machines, heap to buy and easy for a skilled garage mechanic o maintain.

Apart from these, nearly all helicopters in producion have shaft-turbine engines, ranging from the 00-hp MAN-Turbo 6012 to the 5,500-hp Soloviev ngines of the Russian Mi-6 and Mi-10. In every case he engine is only a fraction of the weight of the best orresponding piston-engine. I aircraft such as the Mi-10 and Sikorsky CH-54A flying cranes the whole oncept would be rendered hardly worthwhile without turbines of great power and light weight. Both ingle-shaft and free-turbine engines are used, the atter having an independent power turbine driving he rotor system of the helicopter and nothing else. o start the engine, the high-pressure turbine and ompressor are run up to speed and the build-up of he gas-flow gradually sets the rotors turning. In a umber of helicopters and VTOL machines, two, hree or four gas-turbines are coupled into a comnon gearbox and mechanical drive system which ontinues to function even if one engine should have o be shut down.

Typical of the engines used in modern helicopters s the American General Electric T64 which has a 14-tage compressor working at a pressure ratio of 13:1. n the Lockheed AH-56A Cheyenne "advanced aeial fire support system" of the US Army, which has performance and firepower to do credit to a fixedwing fighter, the T64-16 delivers 3,400 shp, with pecific fuel consumption of 0.49; yet it is only 30 nches across, 85 inches long and 690 lb in weight.

The final type of gas-turbine to be considered for ircraft propulsion is a special class, used only in elicopters, in which the power is delivered in the orm of a flow of compressed air. Starting with the French Ariel helicopter of 1949, a number of rotorcraft have been built in which the problem of torque reaction is circumvented by making the rotor drive itself from the tips. Hollow blades containing air ducts are used in this rotor, and the drive is obtained either by expelling compressed air from jet nozzles at the tips or, if high rotor power is needed and air pipes to provide such power would not fit within the blade contour, fuel can be burned in "pressure-jet" units on the blade-tips.

Yet another arrangement is to feed the blade ducts with hot gases from the engine itself. This "hot cycle" arrangement can give greater power for a given blade duct section and may find production applications in the near future.

There is another way to drive a helicopter rotor from the tip and this is to use a ramjet. This type of power plant was referred to briefly in the discussion of supersonic propulsion, when it was described as little more than a suitably-profiled duct. At high flight speeds, air is rammed into a forward-facing intake at a substantial pressure ratio, and at flight speeds of the order of Mach 4 the ramjet is far more effective than any gas-turbine. The compression in a suitably designed ramjet intake can exceed the best that can be obtained with a mechanical compressor, and the fact that such a power plant needs no turbine removes the main limitation upon the achievement of higher gas temperatures. But the ramjet cannot start from rest and this has severely restricted its use as the main power plant of aeroplanes.

In the helicopter, however, the difficulty of starting from rest can be partly overcome, either by spinning the rotor by hand or by an auxiliary engine, or by blowing compressed air into tip-mounted ramjets using an external source. About 40 types of helicopter have flown on the power of small tip-mounted ramjets or pulsejets, the latter being a strictly subsonic variation using an arrangement of non-return flap valves over the intake to confer a self-starting capability. Unfortunately the ramjet and pulsejet are extremely noisy and, especially at the relatively low subsonic flight speeds reached by helicopter blades, inefficient and heavy on fuel consumption.

For the highly supersonic aeroplane, however, the ramjet is very attractive. Early experiments using such power units were made with a Lockheed Shooting Star fighter, flown with Marquardt ramjets on its wing-tips in 1950, and a whole family of remarkable aircraft by the French engineer René Leduc, starting in 1949, in which the entire barrel-like fuselage formed a ramjet. The first Leduc aeroplane, the O.10, was carried aloft on the back of another aircraft, but his ultimate intention was to produce a highly supersonic aeroplane capable of being operated in the conventional manner.

Most ramjets built so far have had to slow the air to subsonic speed before it enters the combustion chamber. Extensive research has now shown that supersonic airflows can still be used to support proper combustion of conventional kerosene-type fuels and this is likely to be utilized in future engines which are colloquially called "scramjets"—supersonic-combustion ramjets. Most of the projects at present being studied are based upon enclosed ducts of fairly conventional appearance, but one scheme which seems promising is highly unusual and at first sight appears frightening. The surface-burning ramjet, or external-burning ramjet (ERJ), is hardly an engine at all in the normal sense of the word. On the underside of the aircraft the skin profile is extended downward in the form of a large, flat-sided wedge. Supersonic compression takes place on the windward side of this wedge, fuel is injected along the windward face immediately ahead of the bottom edge and supersonic combustion takes place on the lee side, the increased pressure on the lee face providing both lift and propulsion.

Rolls-Royce have published a study for an aircraft to cruise at an altitude of 200,000 feet at Mach 15 (9,900 mph) in which turborockets (turbojets burning rocket propellents and thus needing much smaller turbines) would be used for take-off, climb and acceleration and external-burning ramjet propulsion would take over at about Mach 4. This would involve the whole aircraft changing shape in full flight at a speed of some 2,500 mph, shutting off the duct to the turborocket engines and lighting up the ERJ system at the same time. Clearly, the aircraft propulsion designers of the future are not going to find life entirely devoid of challenges.

PROPELLERS

Until the era of jet propulsion, the propeller was as universal a feature of power-driven aeroplanes as wheels are of vehicles on land. A propeller was the only truly practicable way in which the shaft power of an engine could be converted into propulsive thrust in a vehicle operating wholly within air. Automobiles could use the grip of a wheel on land to move; ships were able at first to use a wheel equipped with paddles to obtain propulsive force; but for aircraft an adaptation of the screw propeller was the only solution, and hence the one adopted by every successful pioneer of powered flight.

Propellers have a background of service in airships going back to 1852. Most of the earliest designs were based upon tubes or beams which spun around a central axle and carried curved plates at their extremities to deflect the air rearward and so generate thrust. But by 1912 aircraft propeller construction had become standardized, using many laminations of hardwood, usually mahogany, all running from tip to tip except where they were tapered off by the form of the blade. Each lamination was rotated slightly in relation to the lamination below and when the complete assembly had been glued and pinned together it was hand carved to the correct profile.

Until the 1920s there was no means of "designing" a propeller in the sense of an exact process; empirical rules were followed, backed up by rudimentary testing. When two blades could not handle the power, it was necessary to adopt a more difficult configuration having four blades.

Essentially a propeller blade is a miniature wing, and its section profile at any given station is a form of aerofoil. When the propeller is rotated, the air flows past the blades exactly as it would past a wing, although the actual relative speed will clearly increase along the blade from root to tip.

As the propeller turns, it draws in the air ahead and accelerates it to the rear in the form of a large circular jet. If the propeller is free to move forward under the influence of the thrust so created, it will eventually reach an equilibrium condition in which the thrust of the propeller exactly balances the air drag of the aircraft which the propeller has to pull along. When this condition is reached the aircraft might be travelling at 200 mph relative to the undisturbed air, but the "slipstream" behind the propeller might initially be moving past the aircraft

at 300 mph—in other words, the propeller will give each particle of air a rearward velocity of 100 mph which will soon die away in eddies and turbulence. In the case of some very slow aircraft of the 1914-18 War the speed of the slipstream relative to the undisturbed air was much higher than the speed (in the opposite direction, of course) of the aircraft. This was very inefficient; by comparison, in modern turboprop airliners the speed of the aircraft, relative to the still atmosphere, is far higher than that of the slipstream.

Not only is a propeller blade aerodynamically equivalent to a wing but it is also structurally equivalent to a cantilevered beam fixed at the centre and free to bend under the influence of the loads acting on it. By far the greatest loads are two which tend to oppose each other: the thrust created by the rotating blade which tends to bend it forward, and the centrifugal load resulting from the blade's own weight which tends to keep it pulled out straight. The net result can still be a substantial bending load in the forward direction acting across the blade. Like a school ruler, a propeller blade has little ability to resist bending in this direction because a cantilever beam needs depth. The problem was felt only mildly in the early days of thick wooden blades but in the 1920s the quest for speed called inexorably for much thinner blades able to slice through the air faster—just as the same pressure on designers to build faster aeroplanes forced them to adopt thinner wing sections.

Thinner wing sections demanded new forms of construction, and so did thinner propeller blades. The obvious course open to propeller designers was to change to metal construction. The first successful metal propeller was the Reed type, introduced soon after the 1914-18 War and used extensively from about 1926 onward. Reed patented a method of forging a two-blade propeller from a single solid blank of duralumin. The blade profile and twist were introduced by presses and all that remained to be done was to drill the centre to fit on the drive shaft, which in those days usually transmitted the torque through a series of bolts passed through the propeller and a disc fixed firmly to the propeller shaft. This arrangement was necessary with wooden propellers but when metal hubs were introduced it became possible to adopt a simpler and lighter arrangement in which the propeller is threaded on

A *Until well into the 1930-40 period most propellers were made of wood and differed little from this example from immediately before World War I. At the hub there can be seen to be eleven laminations of wood, each chosen to have the grain running longitudinally to bear the centrifugal pull towards the tips. The laminations are glued carefully together and the blades are then carved to have the profile shown in the eight small cross sections.*

1 Front elevation of hub
2 Leading edge (in elevation and in section at same station)
3 Trailing edge (in elevation and in section at same station)
4 Blade root
5 Front of hub in side view
6 Tip
7 Fade-out of a particular lamination (front and side views)

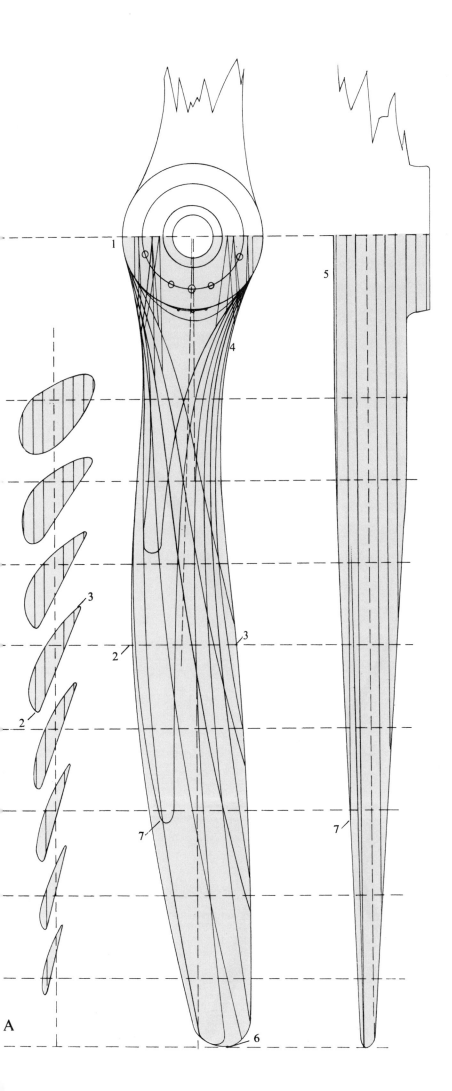

A

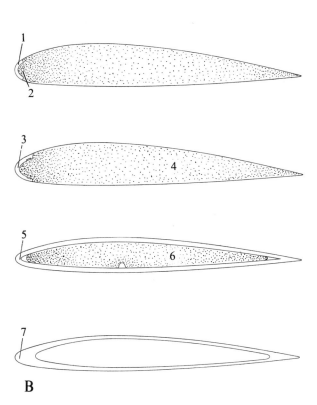

B

B *Four modern methods of making propeller blades (top to bottom): solid 'densified' wood; solid light alloy; hollow steel, welded at leading and trailing edge; and hollow plastics lightweight blade for STOL aircraft.*

1 Protective metal sheath over leading edge
2 Heavy wire to achieve correct blade balance
3 Electric de-icing mat over rebated leading edge
4 Solid light alloy blade
5 Hollow steel shell
6 Lightweight rigid filling to damp vibration
7 Moulded glass fibre reinforced plastics blade

to straight splines grooved in the shaft and secured by a retaining nut.

By the late 1920s the Reed propeller had appeared in a smoother running form with three duralumin blades attached individually to a steel hub. Three was an unusual number of blades in the all-wood era, because it was difficult to achieve adequate strength in the roots of individually attached blades and the linearity of wood grain precluded the attainment of an integral three-bladed propeller. Since 1930, however, three has been the most common number of blades for all aircraft propellers, except on lightplanes.

The angle at which the blades of a propeller meet the air is known as the pitch. In operation, a propeller behaves rather like a nut running on the thread of a long bolt. According to the pitch of the thread, the nut advances a short distance or a larger distance on each rotation. To alter the speed at which the nut moves along a bolt, its rate of rotation must be altered. But this is impracticable with an aircraft propeller, because when it is moving forward most slowly, at the start of the take-off run, is the time when full thrust is needed most vitally and the engine must be run at full speed. To make the propeller give peak thrust at take-off, its blades must be set at a very fine pitch angle (i.e., almost edge-on to the airflow) so that they meet the air at the correct angle even though the engine is running at full speed while the aircraft is moving very slowly. But a fine-pitch propeller of this type would accelerate the aircraft only up to a very modest forward speed of perhaps 100 mph; beyond this point it would overspeed the engine. For peak flight performance the propeller needs to have a very coarse pitch to enable it to take a large "bite" at the air on each revolution and travel forward the maximum possible distance. Clearly a fixed-pitch propeller can only be a compromise, which either restricts forward speed or else imposes poor take-off performance.

The obvious answer is a variable-pitch propeller, which can be set at a fine pitch for take-off and landing and at a coarse pitch in cruising flight. As with so many ideas which are now commonplace, the variable-pitch propeller was first produced at the Royal Aircraft Factory at Farnborough. In 1916 an R.E.8 took off with a four-blade propeller capable of being set to either a fine pitch or a coarse one according to the command of the pilot. It had wooden blades, held in rotary bearings in a steel hub, and weighed about 50 lb more than the usual all-wood propeller. Two-position propellers of this kind were used on many aircraft from about 1928 to 1940, but by as early as 1920 it was clear that a propeller with infinitely variable blades would be only a little more complex and hardly any heavier.

In 1924 the Gloster/Hele-Shaw/Beacham propeller was patented, and this was the first constant-speed design actually to be flight tested, on a Gloster Grebe fighter in 1927. In a constant-speed propeller, a governor unit, usually containing flyweights spun at a rate proportional to engine speed, is connected

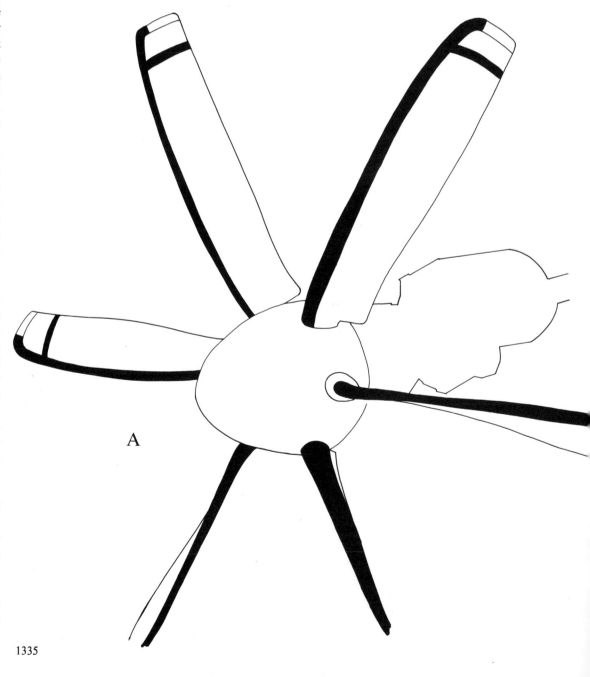

A *Typical of the advanced propellers intended for the new breed of high-speed regional turboprop aircraft, this six-blade unit produced by Dowty Rotol for the Saab 2000 is powered by an Allison GMA 2100 6,000-shp turboprop.*

The lightweight propeller blades are built around a polyurethane foam core sandwiched between two carbon-fibre spars that run from tip to root. The sandwich is covered by a carbon/glassfibre shell and a top coating of protecting polyurethane.

A

to both the propeller and the pilot's propeller control lever. Once the latter has been set to a given operating condition (fine, coarse or some intermediate pitch), it will maintain engine speed approximately constant despite wide variations of airspeed. Thus, the propeller can be set to fine pitch for take-off and the engine opened up to full power. As the aircraft accelerates forward, the constant-speed unit steadily coarsens the pitch of the propeller to prevent the engine from overspeeding, while allowing the aircraft to reach a high forward speed. For full forward speed the pilot must select a coarse pitch; the engine speed will then fall away to a gentle cruising power, while the airspeed continues to increase. If the aircraft is put into a dive, the blades merely assume an extra-coarse setting and the engine speed remains within the desired limits.

Britain was slow to exploit the constant-speed propeller and it was the American Hamilton design that swept the world of aviation in the early 1930s, with a neat hydraulically-operated hub mechanism driving two or three light alloy blades held in roller and ball bearings. Rival systems were the British Rotol, with a hydraulic hub driving blades made of resin-bonded laminations of compressed wood, and the American Curtiss, in which the blade pitch was altered electrically. It appears disgraceful in retrospect that Hurricanes and Spitfires of the Royal Air Force should have had to go to war in 1939 with two-bladed fixed-pitch wooden propellers. However, these were soon replaced by three-bladed constant-speed propellers and both types eventually had to add a fourth blade in order to absorb the increasing power of the Rolls-Royce Merlin engine. The German Fw 190 and Me 109 continued to use three-bladed propellers throughout the war but the Spitfire ultimately appeared with a five-bladed propeller, to transmit the 2,300 hp of the Rolls-Royce Griffon engine, and in 1945 with a six-bladed "contra-prop" which became standard on the Seafire.

The contra-rotating propeller is an assembly of two propellers, usually identical except that one is externally the mirror image of the other, mounted on a common shaft and faired by a single streamlined spinner. As the two units turn in opposite directions, very little residual torque is imparted to the aircraft; nor is there any great corkscrew spiral of slipstream behind it as there is behind a conventional single propeller. Moreover, a contra-prop can absorb high power within the minimum overall diameter. A related arrangement is the co-axial propeller, consisting of independent halves driven by separate engines.

Once it had become possible to alter the pitch angle of propellers in flight, it was only a matter of time before they could be made to rotate their blades through a still larger angle in order to offer two further benefits. First came "feathering" of the propeller, to stop it rotating in flight, by turning the blades until the pitch angle is 90°. The problem here is to twist the blades so as to match the fast-moving airflow at the tips with the slower airflow near the root against the same forward speed at both places; the feathered angle actually achieves a state of balance in which the inboard portions of the blades tend to turn the propeller in one direction, while the middle of each blade is exactly edge-on to the airstream, and the outboard portions of the blades tend to turn the propeller in the opposite direction. Feathering is adopted, either automatically or by pilot command, whenever an engine has to be shut down. Its main purpose is the vital one of reducing the drag of a "windmilling" engine; it is also able to stop an engine from rotating after it has suffered mechanical damage.

The second additional propeller function is to reverse the thrust and so assist in braking the aircraft after landing. The direction of rotation of the propeller is not altered, but the direction of its thrust is reversed by turning the blades right round past the fine pitch position so that their angle of attack passes from a positive angle through zero to a negative angle. The air then, effectively, acts downward on the aerofoil section of each blade so that the resultant force is reversed in direction. The maximum reverse thrust that a propeller can generate is slightly less than the maximum forward thrust, because the aerofoil is less efficient when operating at a negative angle of attack.

Use of reverse pitch inadvertently is normally prevented by special mechanical locks which are disengaged automatically when the weight of the aircraft compresses the landing gear struts after touchdown. The pilot will previously have selected reverse pitch on the propeller control; he then uses his throttles to slow the aircraft, as required. As in the case of thrust reversal of jet engines, reverse propeller pitch is particularly valuable on wet or icy runways. It involves the use of special bearings in the engine propeller shaft, to accept end load in the opposite direction, as well as provision for the engine itself to function with the direction of output torque reversed and with the normal airflow past it or into it disrupted. The latter consideration is especially important with turboprops which, even in the reverse-thrust mode, must not be starved of their proper intake airflow.

Since 1950, turboprops have spurred the development of improved propellers tailored to this new form of powerplant. Most are characterized by small hubs enclosed within a smoothly streamlined spinner, which often carries blade root sections fixed at a positive pitch angle so that, even when the blades themselves are reversed, the airflow past the spinner will continue to feed the engine. Blades are either of solid forged light alloy, hollow built-up steel, hollow light alloy or, in the case of certain special propellers with large and wide blades for V/STOL aircraft, glass-fibre. Spinner and blades may be de-iced by electric heater mats or, in cheaper or older propellers, by a fluid (usually glycol) flung along the blades by centrifugal force.

The operating conditions encountered by a modern propeller are extremely severe. The stresses imposed by thrust, engine torque, centrifugal force and flight manoeuvres are very high and are accentuated by rapid fluctuation and even reversal in some disturbed conditions such as are common in the reverse thrust mode. The air temperature may change from +50°C to -50°C in a single flight. The surface of the whole propeller may be blasted by a sandstorm, rain or hailstones, any of which can erode or even destroy the majority of normal engineering finishes at a 600 mph impact speed.

It is partly because of these severe conditions that propeller blades have so far remained relatively rudimentary, in sharp contrast to wings which are today adorned with slats, leading-edge flaps, triple-slotted trailing-edge flaps, blowing nozzles and other devices. The only fundamentally new concept that may improve propeller performance is the variable camber blade, under development since 1956 by Hamilton Standard. In this pairs of blades are used to make up each complete blade. For take-off and other situations calling for maximum thrust at low forward speed, the two units of each blade are so arranged that they form one blade of large area and a cambered profile, giving high lift. At high forward speeds the rear section of each blade is turned approximately parallel to the first to behave as two separate blades. But it is significant that no announcement of when such a propeller might be put on the market has yet been made.

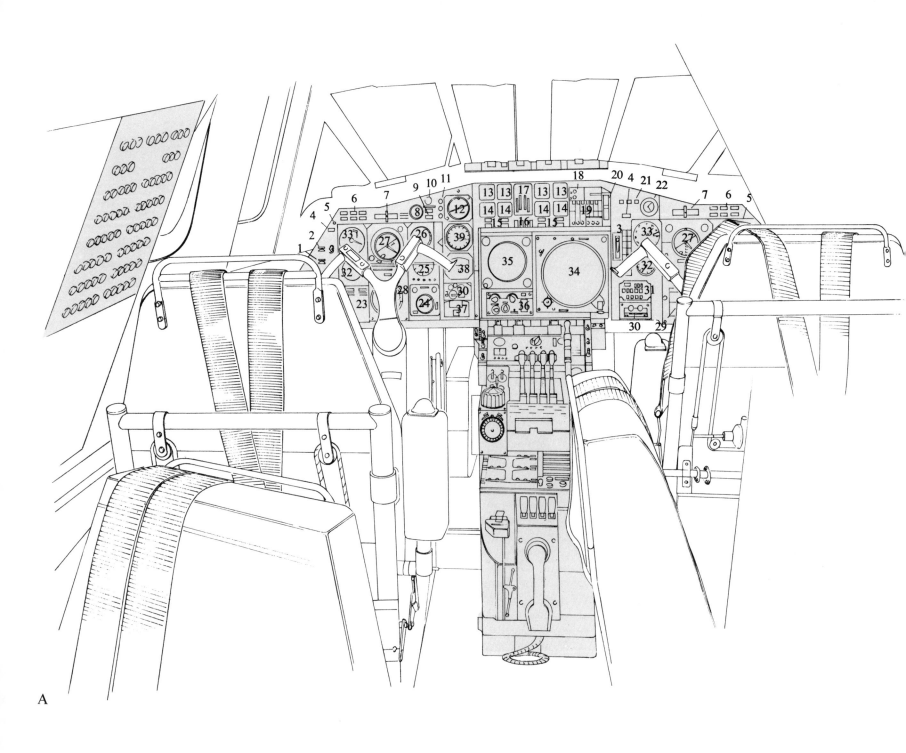

A

A *The flight deck of the Concorde is not large because the fuselage itself is slender and tapers to a pointed nose, but it provides comfortable seats for captain (left), co-pilot (right), third crew member (at a system management panel to the right rear) and a supernumerary (behind the captain). The view on this page shows the captain's outlook in supersonic cruising flight with the nose up and the vizor in the 'up' position. Switches are grouped on roof panels; side consoles contain weather radar display, teleprinter, oxygen and intercom boxes and the central pedestal houses powerplant, radio, navigation computer and other controls. The pilots' panels are drawn on the facing page.*

AIRBORNE SYSTEMS

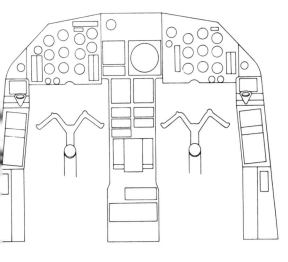

1 Third crew-member
 oxygen
2 Pilot oxygen
3 Angle of attack/
 acceleration
4 Stall warning
5 Ice warning
6 Warnings
 (temperatures,
 pressures, trims)
7 Autopilot state
8 Brake pressure
9 Emergency brakes
10 Marker switch
11 Marker
12 Stand-by horizon
13 Engine h-p speed
14 Turbine temperatures
15 Thrust reverse signals
16 Reheat signals
17 Engine thrusts
18 Warning flasher
19 Control surface angles
20 Centre of gravity
 position

21 Digital clock
22 Cabin altitude
23 Outside air temp.
 (total, static)
24 VOR/DME/ADF
 controls
25 Radio altitude
26 Altimeter
27 Horizon
28 Course control/
 display
29 Yaw
30 3-axis trim
31 Landing gear
32 Machmeter
33 Air-speed
34 Moving map
35 Radar
36 Radar control
37 Brake test
38 Stand-by altimeter
39 Stand-by Mach/
 air-speed

The simplest modern ultralight airplanes cost some £ 5000 to buy. Practically all of this is accounted for by the airframe and propulsion (engine and propeller), and only about £ 500 is needed for accessories and instruments. A medium-size modern jet airliner, on the other hand, may cost some £ 20 million. The airframe accounts for £ 16 million of that, with the avionics costing about £. The interior of the cabin, with furnishings, galleys and passenger seats, will add another £ 300,000.

In the case of a multi-purpose military strike aircraft, the total bill of perhaps £ 20 million may be made up in the following manner: airframe, £ 11 million; engines, £ 3 million; airborne systems and support equipment, £ 6 million. To the cost of the military plane itself, a considerable sum must be added for the armament and weaponry systems.

It is abundantly clear that, in the most advanced and complicated aircraft, the mass of equipment that can be lumped together under the general heading of "Airborne Systems" constitutes the main part of the cost and is therefore responsible for most of the development effort.

It was said of one aeroplane of the 1914 era that if you put a sparrow into the space between the upper and lower wings and released it, the ability of the hapless bird to escape would be a sure test of whether or not there was a loose bracing wire somewhere in the maze of wires and struts holding the aircraft together.

In a present-day supersonic fighter, it is a safe bet that not even a sparrow's egg could be inserted anywhere beneath the airframe's thick skin, except perhaps inside a pipe or duct. Certainly no rifle bullet could pass far into such a machine without damaging a whole series of complex and expensive devices. This trend, which shows little sign of changing, is at once a source of strength and of weakness. It is a source of weakness because there is such a vast array of equipment which can go wrong or be damaged by combat, heat or vibration. Every one of literally thousands of delicate devices is an extra item to buy and maintain, an extra burden on skilled manpower, an extra contributor of weight and occupier of precious space, an extra thing to malfunction and delay a mission, cause an in-flight

emergency and possibly even abandonment of the aircraft. It is a source of strength because every item fulfils a function necessary to the completion of the aircraft's combat mission in any weather and under the most adverse environmental conditions. Every device is tested and checked, and is backed up by other devices which take over if the first goes wrong and by further devices which tell the pilot—and possibly ground staff—just what has gone wrong.

There is a chain-reaction tendency which seems to run the risk of going mad and making an aircraft so complex it would never be serviceable, so packed with gear it could carry nothing else, and so costly it could never be bought. By standards of 1918, or 1939 or even 1950, this stage has long since been passed; but we may be sure that the process will continue and that the aircraft of 1999 will make 1970 airborne systems appear crude.

To fly a light club machine on a fine day, a pilot can manage with an airspeed indicator, altimeter and engine speed indicator—supplemented preferably by a turn/slip indicator to assist in achieving precisely coordinated banked turns and a compass to assist in flying from A to B. All would be immediately understood by a pilot of half a century ago.

To fly a jet airliner, airborne systems of much greater complexity are needed. On page 259 the story begins of what happens on the flight deck of such a machine on a transatlantic flight.

A modern airliner flight deck can vary considerably in size, some having just enough room for three people—captain to port (left), first officer to starboard, and a seat for an observer/instructor in between and slightly behind—to squeeze in. Others are not crowded even with a flight crew of five seated and five more looking over their shoulders under instruction. The example chosen for the drawing on page 236 is the Anglo-French Concorde supersonic airliner. Because of the essential slimness of the fuselage, this is not a large flight deck, but it is comfortable, giving two pilots and an engineer all they need to fly the aircraft and follow accurate predetermined tracks between any two places under all weather conditions—if need be, without help from the ground—and finally land under conditions of very poor visibility.

Some of the basic instruments of the Concorde look very different from those of the light club machine, yet function in exactly the same way. For example, the airspeed indicator has two pointers, one showing the indicated speed in knots and the other a safe limiting speed. A third index mark sets itself automatically around the edge of the dial to show a safe minimum speed, while a digital readout, similar to that of a car mileage indicator, shows the present speed correct to three places of figures. Despite this complexity the instrument is driven by differential air pressures, just like the simple ASI.

Again, the altimeter works in the same way as that of the simple aircraft but has an entirely different presentation. The single needle is used only for fine adjustments and goes right round the dial for every 100-ft change in altitude. The actual indication of aircraft height is again a digital readout similar to the mileage indicator in a car, and the extreme right-hand window indicates 10..20..30 ft in step with the progress of the needle. Also visible is a second digital display indicating the atmospheric pressure in millibars to which the instrument is set, because if atmospheric pressure varies the height indications will also vary. Thus, when the pilot resets the instrument at the start of a flight the reading given should be the height of the airfield above sea level.

An aeroplane like Concorde also has a second type of height indicator, known as a radar altimeter, which bounces radio waves down to the earth and back and measures the time they take to do the double journey. A radar altimeter measures the clear space between the aircraft and the land or sea below, and its reading changes violently during flight over mountains.

The two largest instruments on each pilot's panel are a director horizon and a horizontal situation display, mounted one above the other with a sensitive sideslip indicator between them. The horizon is a very refined and complex descendant of the artificial horizons used in airliners since about 1930. Among many other uses, it gives the pilot an immediate and precise indication of the attitude of the aircraft relative to the Earth. In an aeroplane as fast as the Concorde this must be known accurately all the time, because even a 1° or 2° change in horizontal or vertical planes would soon bouild up a divergence from the planned trajectory measured in miles. The horizontal display is, likewise, a refined and complicated descendant of the plain magnetic compass; it provides a multitude of indications concerning aircraft heading, desired reading, distance to go to a particular place and time at which that place will be reached.

Other instruments have both circular dials and near scales and, following standard practice, those for the engines are grouped in the centre where they ca in be seen by both pilots. The four engine speed indicators give readings in percentage rpm, so at take-off all four would show 100 per cent while at cruising flight a figure of 90 per cent would be more likely. Below them are four dials showing the temperature of the gases in the engine jet-pipes, 800° C being a typical reading. Below are indicator lights for the engine thrust reversers; in the centre, separating each set of four instruments into port and starboard pairs, is a combined vertical scale display showing gas pressure in the engine jet-pipes.

On the right, just inboard of the first officer's flight instruments, are indications of nose and vizor position and of aircraft centre of gravity (c.g.). As described earlier, the Concorde has a hinged nose incorporating a retractable vizor fairing and its c.g. can be adjusted in flight by pumping fuel from one place to another. (The fuel system of the Concorde, as an example of modern practice, has been described earlier.)

The c.g. indicator is quite complicated and, once the crew have told it the "zero fuel c.g."—the position of the centre of gravity without fuel but allowing for all the payload—it automatically computes the c.g. position by integrating the inputs from the various fuel—measuring systems. The crew have a second check on c.g. position by referring to aerodynamic position indicators which confirm that allowance has been made automatically for the major change in trim which takes place during acceleration or deceleration through the transonic regime.

A transport vehicle as "productive" as the Concorde cannot be allowed to waste a single minute on the ground when it could be earning revenue. This pressure to keep aircraft in the air was scarcely felt at all before 1939, but since the start of the jet era in 1952 it has increased continuously. It is reflected partly in the enormous improvement which has been achieved in engineering design and detail and, despite fantastic increases in aircraft complexity, has allowed periods between overhauls and even between routine checks to be extended from a matter of hours to months or years. It is also reflected in the great effort which has been made to equip aircraft with systems of such known reliability that operations can be continued even in the most adverse possible weather.

Protection against icing was a fairly straightforward part of the story of progress towards the all-weather aeroplane; achieving fully blind landings was much more difficult. In the Concorde provision for blind landing and full all-weather operation—which means just what it says, no less—has been incorporated from the outset.

Heart of the Concorde navigation and flight control system is an inertial platform and digital computer. An inertial platform is a small assembly, free to rotate about any axis but carried along with the aircraft, on which are mounted three gyroscopes. Each gyro is of special design and made to an order of accuracy far surpassing anything normally achieved in precision engineering.

The natural oscillation period of an inertial gyro is 84 minutes, the same as that of a pendulum having length equal to the radius of the Earth. By arrangin three such gyros, mutually perpendicular to eac other—one parallel to the aircraft's lateral axis, on to its longitudinal axis and the third vertical—th platform can be maintained in an attitude whic stays exactly constant in relation to the fixed stars. thus acts as a perfect reference, upon which can b mounted three equally precise accelerometers whic sense and record the acceleration of the platform and hence of the aircraft, in the same three plane These accelerations are automatically integrated b the digital computer and the result gives the exac linear displacement from the start of the flight. Sinc the starting point is known, the inertial navigatio system continuously records the exact position of th aircraft without external assistance.

In the Concorde, the inertial platform also serve as an accurate attitude reference for the flight instru ments, the autopilot and the weather radar. In fac there are two inertial platforms and two computers each of which is self-monitoring and also checks th performance of the other.

When they come on board, the flight crew brin with them a tiny cassette. The video tape inside it i projected onto an 8-inch screen in the centre of th instrument panel to show in detail the required rout for the flight and any possible diversions to "alter nate" airports. The inertial navigation system con tinuously monitors the projection of the tape so tha the aircraft position is always in the centre of th screen. The display also indicates aircraft position a indicated by other navigating methods involving ex ternal or self-contained radio and radar systems.

Although the entire route may be fed into the tw computer "memories" before take-off, it is als possible to drive the map display manually until desired future position—a "waypoint"—lies in th centre of the screen. By pressing a button the geo graphic co-ordinates of this point are entered int one of the computer memories. Return to the "navi gate" mode then switches the display back to on centred on present position but with the compas heading, distance and time to the desired waypoin immediately displayed, together with other com puted information such as the fuel that will remai on board at that time. The computer can, in fact, b used to exercise complete control over fuel manage ment as well as to determine the best flight paths i both the horizontal and vertical planes to achiev minimum flight time and minimum fuel consump tion.

Although the twin computers can provide th steering information, the actual flight path of th Concorde is controlled by duplicated Elliott-Bendi autopilots, each of which monitors the other an automatically takes sole charge in the event of th other system malfunctioning or exceeding a permis sible limit of error.

The functions of the autopilot are to provide th following services: attitude hold (guidance to main

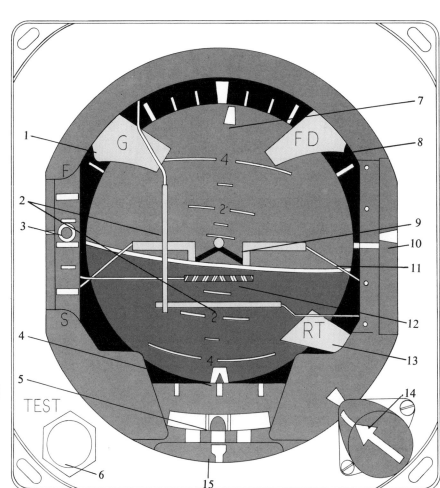

A

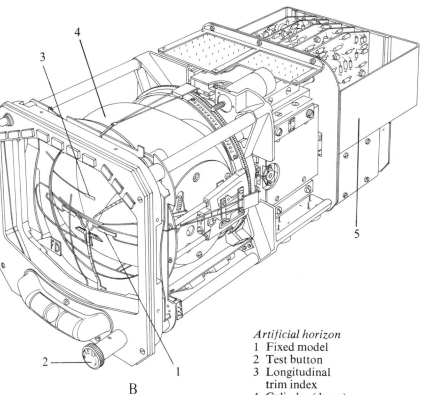

Artificial horizon
1 Fixed model
2 Test button
3 Longitudinal
  trim index
4 Cylinder (drum)
5 Electronic box

B

A *The flight director is
the central, basic
instrument in modern
aircraft and—especially
when the aircraft is
being flown manually
instead of by the
autopilot—it is looked
at more often than all
the others. This
example shows the
attitude of the aircraft
in relation to the
ground (dark blue) and
many other vital
factors as listed on
the right. Another
flight director is
pictured on page 276.*

1 Gyro warning flag
2 Flight director bars
3 Speed command
  display
4 Expanded localizer
  display
5 Inclinometer
6 Self test switch
7 Roll pointer
8 Flight director
  warning flag
9 Miniature aircraft
  symbol: wings;
  fuselage
10 Glide slope display
11 Attitude sphere: above
  horizon; horizon;
  below horizon
12 Radio altitude display

13 Rate of turn warning
  flag
14 Trim knob
15 Rate of turn display

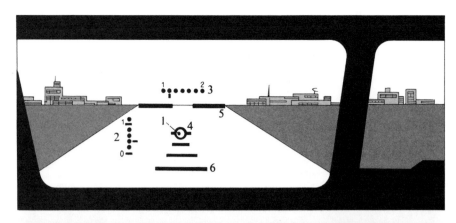

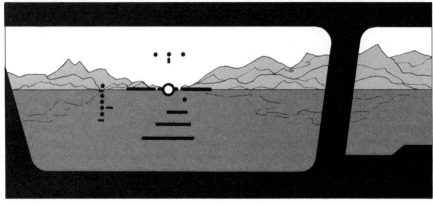

A

A *Now coming into use with civil, as well as with military, aircraft, the head-up display projects essential basic flight information on to the windscreen ahead of the pilot. It is focused at infinity so that the pilot can study the display at the same time as he looks out ahead, without having to readjust his eyes. In the upper view an Elliott head-up display is seen in the approach mode in which it gives basic information on the final stages of an*

approach and landing. The lower drawing illustrates an attack mode against a surface target. The pilot does not need to attend to the display other than to choose the correct operating mode and then adjust its brightness to suit the background against which it is seen.

1 Director dot
2 Altitude scale
3 Angle of attack or IAS
4 Aircraft datum
5 Pitch and roll attitude
6 Perspective track flight director

tain a desired aircraft attitude): altitude selection and hold, or a precomputed vertical profile involving various changes of altitude; Mach number hold, or precomputed variation in Mach number; airspeed hold; vertical speed (rate of climb or descent) selection and hold; ILS radio beam coupling and landing flare control (auto coupling for the optimum approach and landing profile); profile guidance for "missed approach", enabling the aircraft to climb away in the best way without pilot intervention; heading selection, capture and hold (capture means that the aircraft turns exactly onto a desired course and does not overshoot it and then "hunt" each side of it); track selection, capture and hold (flight path over the ground, allowing for wind); VOR (a radio aid) and inertial navigation coupling; localizer capture and coupling, to steer the aircraft in the horizontal plane during a blind landing; and control during the landing run and turn-off, which is essential in truly blind conditions.

Of all the manoeuvres performed by aircraft, landing is the most difficult to achieve in "blind" conditions when the airfield is covered by dense fog right down to the ground. By 1950 such schemes as BABS (blind approach beacon system) and SBA (standard beam approach) had been developed into ILS (instrument landing system) which is now fitted to all runways at major airfields throughout the world.

An ILS installation includes special radio transmitters which send out two narrow beams of signal from near the desired touchdown point on the runway, back along a suitable "glidepath" at an angle of 2,7° to 6°. Control horizonbally is provided by left/right steering information; control in elevation is similar: the correct track of the glidepath is usually shown by cross-pointer indicators on a cockpit instrument. Distance from the runway is indicated by one or two beacons, called fan markers. Modern ILS equipment is provided with OME, continuously showing the distance to go to the runway end.

Blind landing is achieved by coupling ILS signal into the aircraft autopilot. Not only must the auto pilot fly the aircraft but it can also automatically govern the engine power and all other variables. And it must be incredibly reliable. The autopilot is often duplicated or triplicated in such a way that any failure is automatically counteracted by the remaining channel(s), so that the likelihood of a complete failure is less than one in ten million.

Electronics on the Concorde include various radio communications systems, two VOR navigation and ILS instrument landing systems, two distance measuring systems, two radio altimeters, a marker receiver to pick up particular beacons on the ground, and a weather radar in the nose which detects storms, dense cloud and turbulence ahead (as well as mountains, although the radar should not be needed for such a purpose). The flight director computer systems also can control engine power automatically during a blind landing, or at any other time should this be necessary. Finally, the map display described previously can be used to display check lists and emergency drills, thus further reducing (in fact, eliminating) paperwork.

The precision of the ILS, in its most advanced form, CAT III, permits the system to be used for fully automatic landings without assistance from the pilots. The Autoland procedure was introduced in the early part of the 1980s and has become a standard feature on all modern airliners. Most major airports have a CAT III ILS installed on at least one runway and landings can be made with only 75 meters runway visual range. The vertical visibility limit is zero compared to CAT I ILS, where 60 m is a vertical minimum and 600 m runway visual range.

At some airports, heavy terrain problems prevent installation of ILS; but the implementation of a new approach and landing aid, MLS, will make it possible to upgrade any airport to instrument landing standard regardless of the terrain surroundings. MLS is planned to replace the ILS by the beginning of the next century. Instead of two fixed beams, the MLS transmits two beams that sweep back and forth with extremely precise timing. By clocking the exact moment at which each beam strikes the airplane's receiving antenna, a computer on the airplane can determine just where it is with respect to the runway centerline and glide path. Also, the MLS provides a measure of the distance between the airplane and the runway, thereby fixing the plane in space in three dimensions. Unlike the ILS, the MLS allows for the approach paths to be made from any direction within a wide sector instead of lining up along the beams' centres in a straight line.

In military airplanes the degree of complexity and the demand for reliability of airborne systems are comparable with those in a modern airliner, but the role of a combat aircraft calls for somewhat different approach in design. The combat plane must be totally self-contained concerning navigation systems on all levels of flight from tree-top to the highest possible altitude. Also today there is an absolute demand to be able to operate in all weather conditions. The high load manoeuvring during a combat mission is also a factor that must be taken into consideration in the design of the onboard systems equipment in a military airplane.

Clearly, no external navigation assistance can be expected by a combat aeroplane over enemy territory and for this reason inertial systems are relied upon heavily. Nearly all the hard groundwork of experience upon which today's inertial navigation systems have been built was achieved with military aircraft and intercontinental missiles. In aircraft designed for long-distance, low-level penetration of enemy airspace, INS information is continually compared with a preprogrammed terrain profile to determine position and ground clearance with extreme accuracy.

Apart from the problems of navigation and bad-weather landing, the equipment carried depends greatly on the missions the aircraft must fulfil. A fighter, for example, must carry interception radar in its nose capable of detecting other aircraft at a range of 50 to 100 miles and of "lockingon" to the target. A fire-control system must then be brought into action to steer the aircraft automatically to intercept the supposed enemy, as well as IFF (interrogation friend or foe) devices to determine beyond doubt that the target is hostile.

The interception trajectory may be a curve of pursuit, especially when guns are to be used; but with some modern air-to-air guided weapon systems a more efficient tactic is the collision course in which the fighter steers towards a point in space which the enemy will reach in a few seconds' time. This eases the interception task which must be accomplished by the missiles. After the interception has been completed, the automatic procedure may be discontinued and control returned to the pilot—who can, at any time, break into the automatic "closed loop" system and take over control manually should he so wish. The most advanced form of air-to-air weapon is the "fire and forget" active, self-homing missile. It can be launched from any angle in a wide sector at a target from long distances. This smart weapon can be guided by either infra-red sensors or radar and the onboard computer steers the missile without any external information from the attacking plane. All steering data is generated and calculated within the missile itself and the system is very accurate and relatively non-sensitive to different kinds of counter-measure efforts.

Aircraft intended to attack surface targets on land or sea, or within the sea, can use any of the navigation systems already described to reach them—assisted in certain cases by radar systems capable of detecting mental hulls against water or able to picture the terrain below, or by systems detecting the disturbance to the Earth's magnetic field caused by a submerged submarine, or by sonobuoys dropped into the water to pinpoint a submarine by reflected sound waves, or by various other delicate and cunning devices. The actual delivery of the weapon or weapons then also calls for at least an autopilot capable of operating under specified weapon delivery conditions and may necessitate a special system used just for this purpose.

Strategic bombers designed to make deep penetrations of enemy territory, which were thought of as outmoded not too long ago, have been made practicable again by the "stealth" technique, using special shapes and materials to make the aircraft almost invisible to radar and other detection systems. Air-launched "stand-off" missiles can deliver a nuclear or thermonuclear warhead to a target 100 or more miles from where they are launched from the parent aircraft; if they have an inertial navigation system, they need no outside assistance and do not reveal their presence by emitting electromagnetic radiation. But the parent aircraft must know its position precisely, as well as its exact height, speed, heading and acceleration at the moment of release, and feed all these data into the computer of the missile.

For strategic reconnaissance, the conventional manned aircraft are too vulnerable to make deep penetrations of hostile territory and more certain results can be obtained from satellites. But against nations who lack advanced defences or are not an enemy in a "shooting war", it is possible to learn a great deal by using modern reconnaissance aeroplanes that cannot be fully duplicated by a satellite.

The long-established sensor system is the camera taking photographs of more or less conventional type, using either normal film or film sensitive to infra-red (heat) radiation. Some aerial cameras are very large, in order to operate with a focal length measured in feet, while others use slit shutters which sweep the image of the ground across the film at a speed exactly synchronized with that of the aircraft, so that even in a supersonic run at low level the pictures are not blurred.

Modern aircraft photography can yield pictures taken from an altitude of 80,000 ft in which newspaper headlines can be read, salinity of water or its depth can be accurately measured, a car with its engine running can be detected among 1,000 with engines off, the place where a car was recently parked can be seen at once and a real man can be distinguished from a dummy.

Reconnaissance aircraft also make wide use of all kinds of electromagnetic radiation. Radars of various kinds can probe for buildings or metal objects, and certain wavelengths can detect a rifle hidden in a forest. Moving objects can be distinguished from static by the use of the doppler effect in modern radar technology, and by advanced computer-processing of radar signals the specific signature of different objects such as cars, armoured vechicles and aircraft types can be determined. Speed and heading of the target will also be displayed in real-time by the pulse doppler radar. Side-looking aircraft radar (SLAR) can produce a detailed picture of the ground 100 miles inside a country without the carrier aircraft crossing the frontier, while laser devices can measure range to a particular object to an accuracy of within one inch in 100 miles. Special ferreting electronics play a ceaseless game of cat-and-mouse in the secret war of electronic countermeasures (ECM), one of the basic moves in which is to measure every type of radiation emitted by radio and radar equipment in the other country. Other airborne systems can sample the atmosphere and detect a few parts of particular material (a gas or nuclear fission particles) in a billion parts of air.

Tactical strike aircraft spend most of their life at relatively low levels and need a tough airframe to stand up to immense flight loads at high speed in dense air, often after suffering battle damage. Their basic flight profile is "high, low, high", or variations upon this. Their actual approach to the target from perhaps 50 miles out will be right "on the deck", to try to elude defending radars and guided missiles. Too high and they are vulnerable: too low and they

cannot avoid sudden obstacles or hills. Special terrain-following radars are often installed, coupled to the autopilot and a computer to steer the aircraft in both the horizontal and vertical planes in the most efficient manner to avoid striking the ground.

Near the target the flight path of the aircraft may be according to predetermined automatic weapon delivery operation of the autopilot or the flying may be done manually. Various inertial, radar, laser and other systems are used to pinpoint the target and the weapon(s) may then be released either in a straight run, in a zoom climb or in an "over the shoulder" toss. In all cases, some type of digital computer must be used to co-ordinate the various flight activities.

A non-combatant specialist type of aircraft is that used for airborne early warning (AEW). Radars can "see" only along lines of sight and thus, like a human eye, can see a greater distance to the horizon if they are lifted high above the Earth. Powerful radars can detect aeroplanes at distanc of 200 miles or more—in fact, ballistic-missile early warning systems could do so at well over 2,000 miles—but a set installed on Earth cannot make full use of this performance. The AEW aircraft has thus evolved as a machine large enough to house a crew of up to ten or more, plus an enormously powerful search radar with a highlyelliptical dish aerial (antenna) which sweeps the horizon or a large sector of it—unlike other airborne radars which either scan a small sector to-and-fro or else have a conical scan of only 2° or 3° included angle. The AEW scanner is enclosed within a giant "saucer" carried above the centre of gravity of the aircraft. The objects detected by the radar are automatically analyzed and plotted both in the aircraft and on the ground, or in a surface ship or "airborne command post".

Yet another military method which tries to avoid combat was introduced in Vietnam—a forward-area controller (FAC) and psychological warfare aircraft. The duty of the FAC is to examine the ground below in great detail from close range and to find targets, direct attack aircraft or land-force weapons on to them, and observe the results. Light gun and rocket pods are carried, but advanced systems are not used. Likewise the "psycho" aeroplane carries little more than powerful broadcasting apparatus, usually sending out taped messages, and leaflet dispensers. Most of this type of flying is done visually, with navigation assisted by radio beacons or the Decca system.

The first example of a modern multi-mission combat aeroplane was the General Dynamics F-111A. The F-111 was planned under the designation TFX (tactical fighter experimental) in 1960–62 as a replacement not only for the entire "Century series" of fighters, interceptors and strike bombers of the US Air Force but also for the interceptors and some of the strike bombers of the US Navy. In addition to the basic Air Force version, other types have been developed, including the RF-111A for long-range reconnaissance, the FB-111A for strategic bombing and the F-111C for the Royal Australian Air Force.

Among the many major innovations which wer combined in this new-generation aircraft are var able-sweep wings to suit the shape of the airframe t both low-speed and highly supersonic flight, afte burning turbofan engines to achieve good fue economy and very high supersonic performance, a enormous internal fuel capacity, and quite excep tional provisions for rapid maintenance and fault l cation, using built-in self-test systems and quickl replaceable components, accessible through mor than 300 access panels.

The F-111 carries a number of systems designed t detect enemy radars, to determine their frequenc and apply suitable electronic countermeasure (ECM), to create spurious reflections on enem radar screens, to minimize the radar response of th F-111 itself and to take automatic avoiding action t minimize the likelihood of interception by fighters o guided missiles.

The primary flight instrumentation is so extensive and so closely integrated with the other airborne sys tems, that it cannot really be taken separately.
The "interface" between the avionics throughout th airframe and the crew commander's steering in struments is a system coupler which accepts an conditions the incoming electrical signals. Vertica and horizontal steering signals are fed to th commander's flight-director computer, while furthe steering information is provided by pitch, roll an yaw (heading) signals fed to the attitude directo indicator.

A horizontal situation indicator (a new instrumen to most pilots) displays simultaneously headin course, distance and bearing information, as well a selected fixes obtained from airborne and extern navigation systems. The bomb/navigation distance time computer displays the selected mode of opera tion of the avionics systems as well as the time t weapon release or distance to the selected targe The central air-data computer converts data abou the atmosphere around the aircraft—precise pres sure, temperature and other factors—into signal suitable for the other airborne systems.

The airspeed/Mach indicator not only gives spee and Mach number but also the "g" acceleration an the angle of attack of the wings. Associated with is a special device which warns the commander if h tends to exceed the maximum safe Mach number fo any flight condition. The altimeter also indicate vertical velocity (rate of climb or dive) and a wholly new indicator provides exact information on critica temperatures in various parts of the airframe.

The mission and traffic control systems (MTC are basically those which in earlier aircraft wer called communication / navigation / identificatio (CNI). The simplest subsystem is probably th "intercom" and even this can be used betwee flight crew, between them and the ground crew (when the aircraft is on the ground) and betwee various ground crew members. Standard amplitude modulated voice communications are provided by

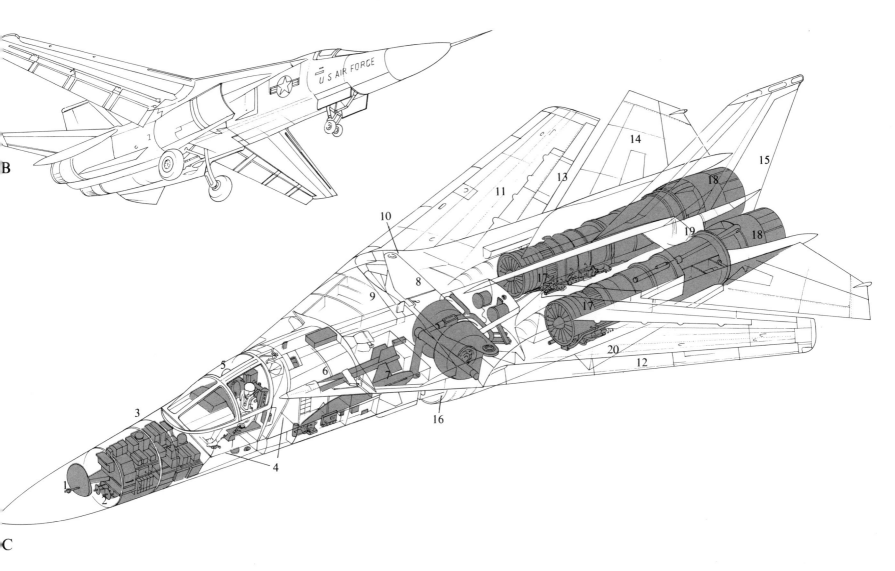

B

C

Typifying modern multi-mission combat aircraft, the General Dynamics F-111A is the first of a family of generally similar designs for the US Air Force and Royal Australian Air Force and, at one time, the US Navy and RAF. All are characterized by having wings whose sweep angle can be varied from 16° for take-off, slow "loitering" and landing (as shown in the inset) up to the 72.5° position illustrated which is adopted for supersonic flight at Mach 2.5. Each of the Pratt & Whitney TF30 after-burning turbofan engines develops up to 20,000 pounds thrust, the crew of two sit in a pressurized "module"

which in emergency can be fired right out of the aircraft and landed by parachute, full-span leading-edge slats and trailing-edge flaps allow the aircraft to land as slowly as 110 knots and come to rest in 2,000 feet, and as many as fifty 750 pound bombs can be carried at once, in an internal weapons bay plus four inboard pylons that swivel during wing sweep plus four-jettisonable non-swivelling outboard pylons. This sketch reveals the strong "bridge piece" to which the swinging wings are hinged, emphasizes how much of the aircraft is occupied by fuel (the basic F-111A can fly 3,300 nautical miles

on internal fuel alone, and can carry six 600 U.S. gallon supplementary tanks on its weapons pylons), and also indicates some of the amazingly complex airborne systems.

1 Main attack radar scanner
2 Terrain following radars
3 Packaged electronics with rapid self-test and replacement
4 Boundary of jettisonable crew module escape system
5 Left and right hinged half canopies
6 Main fuselage fuel tank
7 Internal missile bay
8 Carry through bridge joining wings
9 Fixed part of wing (integral fuel tanks)
10 Wing pivot

11 Variable sweep portion of wing
12 Leading edge drooping slats
13 Trailing edge flaps and spoilers
14 All moving tailplane for pitch and roll
15 Rudder
16 Port engine intake with variable geometry
17 Twin Pratt & Whitney TF30 engines
18 Variable ejector secondary nozzles
19 Rear end of fuel tank between engines
20 Fixed ventral fin

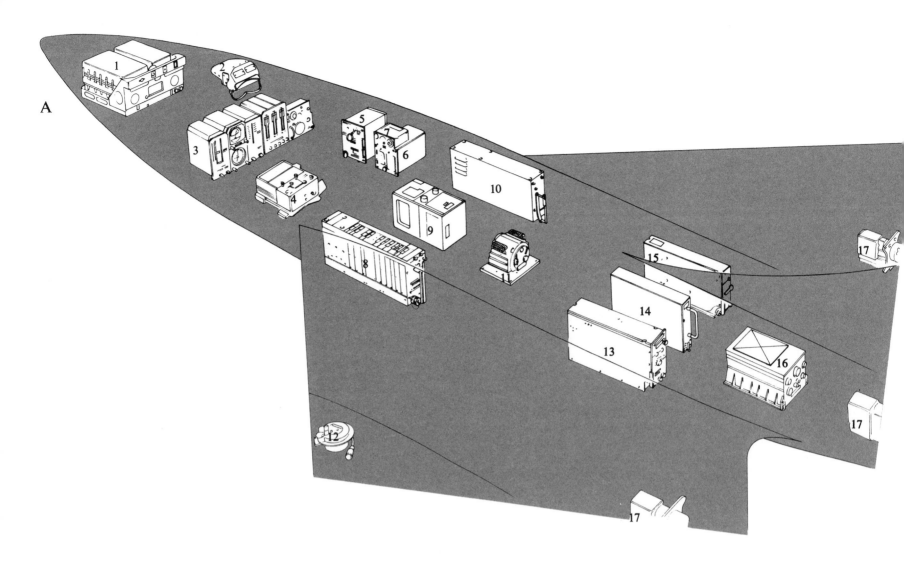

A

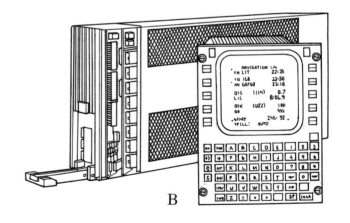

B

A *Integrated flight control and navigation systems for modern military aircraft have become more sophisticated and complex, yet have simultaneously increased in reliability and "MTBF"—mean time between failures. These packages are the products of a single manufacturer, The Bendix Corporation, and are shown as they might be disposed through a supersonic tactical aeroplane. The central air data computer (1) measures and co-ordinates atmospheric parameters and computes output signals for all desired flight profiles to keep the aircraft at the desired altitude or speed; the head-up display (2) is illustrated on page 240; the horizon, director indicator and other flight instruments are on the left of the cockpit panel (3), together with vertical linear scale speed, height and engine presentations, with autopilot and attitude/heading reference system (AHRS) controllers on the right; the navigation map computer (4), navigation computer control (5), and VERNAV vertical navigation controller (6) control the flight path in all three planes, while above the VERNAV is the weapons release controller (7); largest package is the all-weather yaw damper computer (8); the AHRS two-gyro platform (9) and AHRS coupler (10) are just ahead of the autopilot vertical gyro (11) which provides an indication of true vertical; the compass system is fed by a magnetic flux gate (12) in the wing, while three computers serve the VERNAV system (13), navigation (14) and all-weather flare for landing (15); the weapons release programmer (16) times weapons precisely; and the autopilot surface servos (17) drive the flight controls to steer the aircraft.*

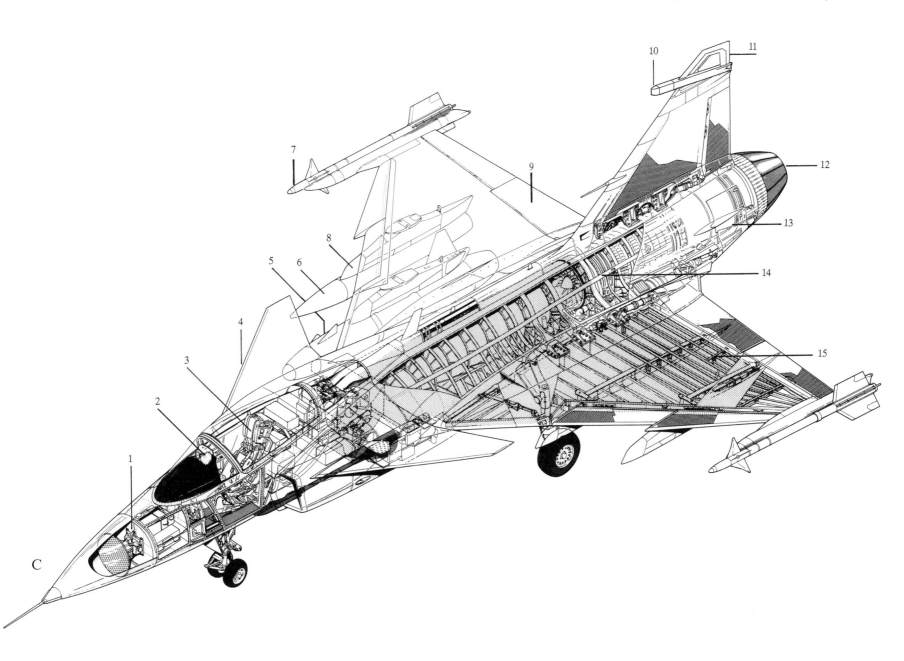

The flight management system is the prime interface between the crew and the aircraft. The system is nowadys a standard installation on most airliners and manages the flight in a most economical way from take-off to landing.

Before the beginnning of the flight, the pilot feeds in the flight plan details such as fuel, distance, load and estimated time of arrival—and the system interfaces with the navigation sensors to guide the aircraft against these parameters. The flight will be carried out in the most efficient way in respect to fuel economy, flight time and aircrew workload.

In the system's database, 52,000 waypoints and airfields are stored, and for each flight plan the pilot can define 256 of them. The multi-sensor navigation system can use a blend of information from VOR/DME, VLF/omega, Loran C, INS and GPS satellites, and decides which system, for the moment, gives the best computed position.

C The Swedish multi-role combat aircraft JAS 39 Gripen (Griffin), a highly maneuverable aircraft with close-coupled delta canard configuration and all-digital fly-by-wire system designed for fighter, attack and reconnaissance missions. In order to attain a high degree of maneuverability and to carry a big variety of external load, the stability of the JAS 39 must be varied in a more flexible way than in earlier types of combat aircraft. This calls for an electrical steering system where the impulses from the pilot and on-board sensors are processed by computers, and the information to the plane's hydraulic servo-assisted steering system is by electric wires with no mechanical connections to the controls. The plane is controlled by two subsystems, one operated by the pilot and the other by an automatic stabilizing system. The pilot's controls, stick pedals and throttle do not affect the plane's rudder positions directly; the computer also gets constant information from the automatic stabilizing system about the plane's attitude. Based on the pilot's inputs on the controls and the automatic stabilizing system, the computer gives the proper signals to the movable steering surfaces to keep the plane in aerodynamic balance during the maneuvers.

1 Multi-mode pulse doppler radar.
2 Wide diffraction optics head-up display.
3 Martin-Baker zero-zero ejection seat.
4 All-moving canard foreplane in composite construction.
5 Saab Bofaros RB 15F air-to-surface anti-shipping missile.
6 Chaff-flare countermeasures pod.
7 Sidewinder AIM-9L air-to-air IR missile.
8 Leading edge flap.
9 Inboard and outboard elevon.
10 Electronics-countermeasure equipment fairing.
11 VHF aerial.
12 Variable-area afterburner nozzle.
13 Airbrake.
14 RM 12/F-404J turbofan engine, 18,000 lb thrust.
15 Carbon-epoxy wing with integral fuel tanks.

uhf (ultra-high frequency) radio systems. An hf (high frequency) set provides for communications over ranges up to 3,000 miles, a TACAN (tactical air navigation) receiver provides position information with respect to a selected ground station and an ILS (instrument landing system) receiver accepts a signal from the destination airfield and automatically steers the aircraft down the glide-path.

Even the Mk 1 avionics of the F-111A provides firepower control which includes subsystems enabling targets to be located and attacked in all weathers. The basic navigation and attack set is used in conjunction with radars in the nose and elsewhere to provide accurate inertial navigation, course computation and automatic bombing ability. It can guide the aircraft from take-off to landing, with continuous presentation of position, attitude, track, velocity and steering commands to either a target or a home airfield. The set also provides everything needed for automatic (radar) blind attacks on surface targets and—a unique feature—an inbuilt ILS, quite separate from the normal one, which enables the F-111 to make instrument landings on runways not equipped with radio or radar landing aids.

The main radar in the nose is the attack system which is designed to "deter" jamming by an enemy and to provide a clear visual presentation of a target. It also performs the functions of ground mapping, air-to-air search and tracking for intercepting other aircraft, navigation fix updating and

radar reconnaissance and photography. Below the nose are two smaller scanners serving the terrain-following radar which provides day/night low-altitude penetration capability by steering the aircraft along the profile of the ground at all speeds within the aircraft's capabilities. It can be set to either manual or automatic operation, checks itself continuously and, should it sense any marginal operation or failure, automatically pulls the aircraft into a climb well away from the Earth.

The terrain-following system incorporates an analogue computer that generates climb/dive signals which are fed to the autopilot to steer the aircraft in both vertical and horizontal planes and are also displayed on a cockpit screen. The commander can thus follow the ground profile together with the radar mapping of the countryside. Ground clearance for such missions is ensured by feeding the terrain-following system with signals from a low-altitude radar altimeter which gives not only terrain clearance but the rate at which this varies.

Finally, the various forward-firing types of armament—which can include 20-mm guns, guided missiles and rockets—can be fired by the F-111's crew-members by using a lead-computing optical gunsight and a missile launch computer. Both systems provide data which are displayed on the windscreen, focused at infinity, in a "head-up display" which enables the pilot or navigator (the USAF use two pilots and the RAAF a pilot and navigator) to read the steering or range data while

at the same time watching the target. The computing display also shows airspeed and dive angle for programmed dive bombing, and additional data can be fed into it from other systems for terrain following, air-to-air attack, radar bombing, blind let-down or instrument landing.

Despite the presence of all these systems, the original requirements for the F-111, which the production aircraft must meet, call for not more than 35 man-hours to be spent on maintenance per hour of flight. Three-quarters of the aircraft with a squadron must always be operationally ready, must be capable of remaining at "alert" status for five days consecutively, must have a reaction time of five minutes or less, must never need longer than minutes to check-out all systems or rectify a fault, must not take more than 30 minutes to turn around between flights and must always be able to fly at least 30 hours a month under combat conditions.

The next generation of military aircraft are certain to be even more complex yet will have to meet even more stringent demands.

Altogether, the widely differing requirements of the private owner, the scheduled airline operator, the major military force prepared to meet a sophisticated enemy and the military force engaged in operations against an elementary and generally hidden foe have brought forth a range of airborne systems and equipment that would have seemed fantastic even ten years ago.

BETWEEN THE WARS

Before 1914 flying was the prerogative of a handful of men generally regarded as lunatics, while in World War 1 it became the temporary, very serious occupation of some quarter million. While infant airlines then struggled into existence in the early 1920s a new sort of aviation arose: the flying club and private owner using aircraft as a means of personal transport or purely for recreation. At first "light aircraft" were often converted military machines, but in 1925 Capt (later Sir) Geoffrey de Havilland produced his famous Moth, designed for this new market. Powered by a new 60 horsepower Cirrus engine, it seated two in tandem, was simple to maintain and could even be towed along a road with its wings folded. Gradually aerodromes arose throughout the world. The needs were small: a few hundred yards of fairly level, flat ground firm enough to take a low pressure pneumatic tyre; a windsock to help pilots to take off and land into the wind; some sort of enclosed building for use as a hangar and office; and, perhaps, a fence to keep out stra animals. Such fields were used both by local aircra based there and also by visiting "circuses", of whic Sir Alan Cobham's was the best known in Europ as well as by lone freelance pilots always prepare to give anybody 'a flip' for a few shillings or cent It was in such surroundings that millions fir experienced the noisy wallowing through air pocke that was light aviation in the 1920s and 30s. Toda many of the grass fields still exist.

1417   1419   1421

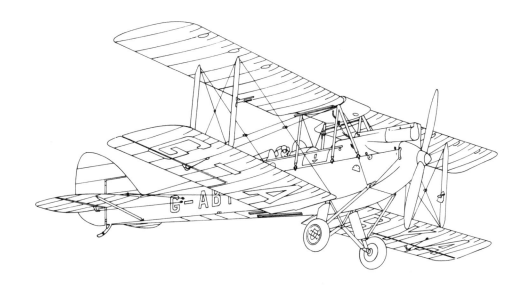

Here four D.H. 60 Moths are operating from an airfield typical of the 1920s. In most parts of the world, flight in open-cockpit aircraft meant aviators had to wear leather jackets or coats and helmets and goggles all the year round.

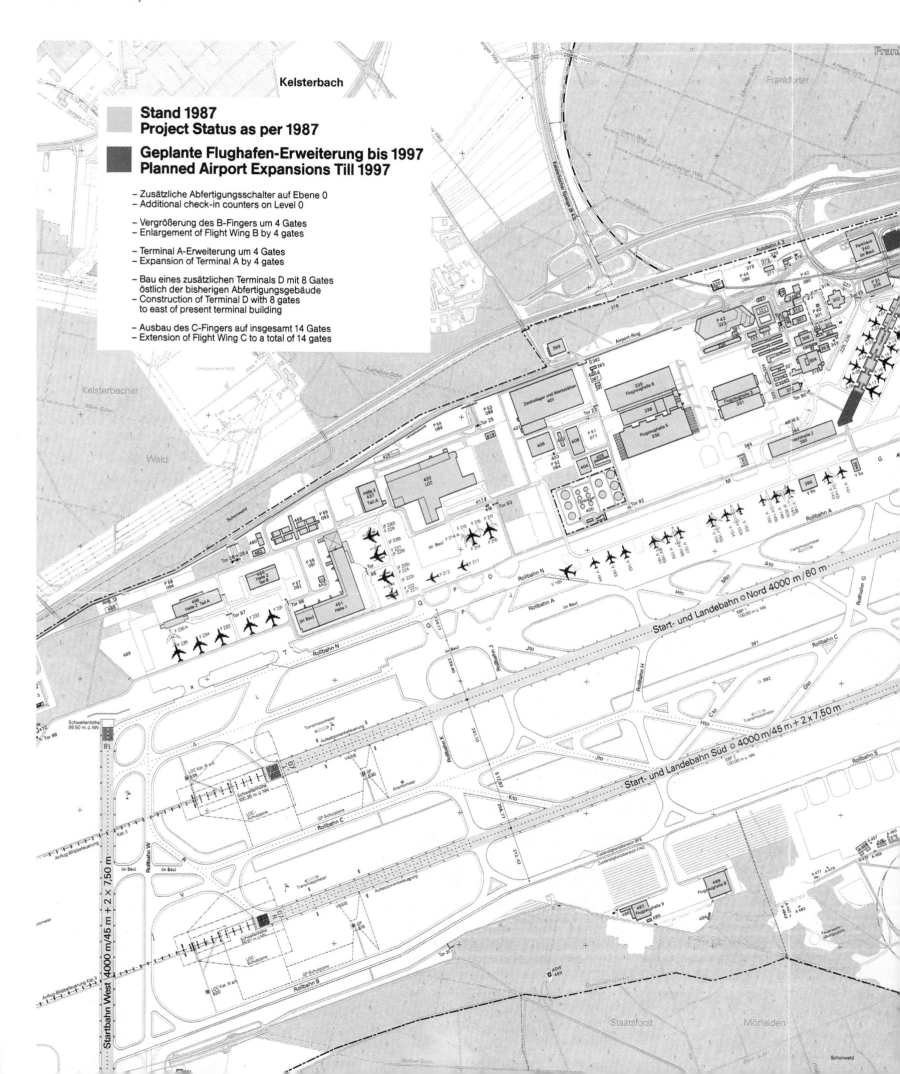

Kelsterbach

Stand 1987
Project Status as per 1987

Geplante Flughafen-Erweiterung bis 1997
Planned Airport Expansions Till 1997

– Zusätzliche Abfertigungsschalter auf Ebene 0
– Additional check-in counters on Level 0

– Vergrößerung des B-Fingers um 4 Gates
– Enlargement of Flight Wing B by 4 gates

– Terminal A-Erweiterung um 4 Gates
– Expansion of Terminal A by 4 gates

– Bau eines zusätzlichen Terminals D mit 8 Gates
  östlich der bisherigen Abfertigungsgebäude
– Construction of Terminal D with 8 gates
  to east of present terminal building

– Ausbau des C-Fingers auf insgesamt 14 Gates
– Extension of Flight Wing C to a total of 14 gates

FRANKFURT AND SCHIPHOL

Two king-size cigar-shaped airships were the main attraction at the official opening of Rheine-Main airport in Frankfurt on July 8 1936. The airport was to serve both as a startingpoint for the pioneering Lufthansa air mail service to South America and the main airship port for the Zeppelins flying between Germany and America.

Initial statistics were modest. In 1936 Frankfurt handled 58,000 passengers and 80 tonnes of freight. In comparison with todays figures, 26,724,430 passengers and 1.1 million tonnes of freight, 1989 Frankfurt exceeds the numbers from 1936 in half a day.

Frankfurt is one of the most important international airports in continental Europe and the main gateway to the Federal Republic of Germany. It is also the main hub of the German flag carrier Lufthansa's national and international traffic network.

The airfield consists of three runways each 4,000 m long. The two parallel runways 25R/07L ("Runway North") and 25L/07R ("Runway South") are both for takeoff and landing operations, whereas Runway 18 ("Runway West") is exclusively designed for takeoff. The parallel runway system permits simultaneous operations and the airport has the capacity of up to 70 starting and landing aricraft an hour under instrument landing conditions. The designated numbers of the runways correspond to the compass course of the approach direction, for example 25R stands for the right runway orientated in 250 degrees. For the last decades, Frankfurt has constantly expanded and the dramatic increase in air traffic during the 80's urgently calls for further expansion and improvement. A new terminal will be ready in 1992 and the situation is not unique to Frankfurt; all major airports around the world are suffering from congestions and travellers can see construction work going on on most of them. It seems leke an airport's growth never stops.

The airport is like a small town and a workingplace for about 45,000 people. Restaurants, supermarket, tax-free shops, hotels, conference center, a chapel and a number of other facilities are situated in the airport area, and the terminal building can be reached by train which stops conveniently at an underground station under the terminal. Frankfurt airport is the third biggest in Europe counting the number of passengers and aircraft movements and ranks first in the volume of cargo handeld. Somewhat smaller in passenger and aircraft movements but with a larger runway structure, Schiphol airport outside the Dutch capital Amsterdam has scheduled air traffic routes to about 200 destinations in over 90 countries.

The four main runways are all equipped with the instrument Landing System (ILS). Two of them permit category 3 operations which means that it allows aircraft to land at a minimum horizontal visibility of

150 meters, this limit will be reduced to 100–75 meters in the future. The category 3 ILS insures a safe operation of the airport regardless of weather conditions. On the ground, aircraft can be directed by the ATC to the parking space in bad visibility with the ASMI, ground radar system. All main airports nowadays have similar types of technical equipment to keep the traffic in operation in almost any kind of weather. The central traffic area at Schiphol is bounded by the four runway systems. Within them is a double ring of taxiing lanes, one for arriving aircraft and one for departing. The "heart" of the traffic area is the terminal building with its piers. The Arrival Hall is on the ground floor, and the Departure Hall one floor above, so that the flows of arriving and departing passengers will not have to cross each other and the processing of passengers can be carried out more swiftly.

Moving walkways makes it passible to reach the exits to the aircraft quicker, which can be essential as transfer times between flight connections sometimes can be short.

To the discomfort of the passengers, unfortunately, security checks have become a part of travelling by air. To minimize the risk of somebody bringing firearms or explosives on board the plane, every passenger must pass the security check where the hand luggage is x-rayed and he/she must pass through a metal detector. At Schiphol, a security check is situated at the entrance to each pier where all departing passengers must pass before reaching the gate. Boarding the planes is made via covered bridges which are directly connected to the plane's door from the gate of the pier.

Situated at Schiphol is also the Fokker aircraft industry and the Aviodome, an aircraft museum where visitors can study the development of aviation on the spot where the first scheduled international air route was opened in 1919.

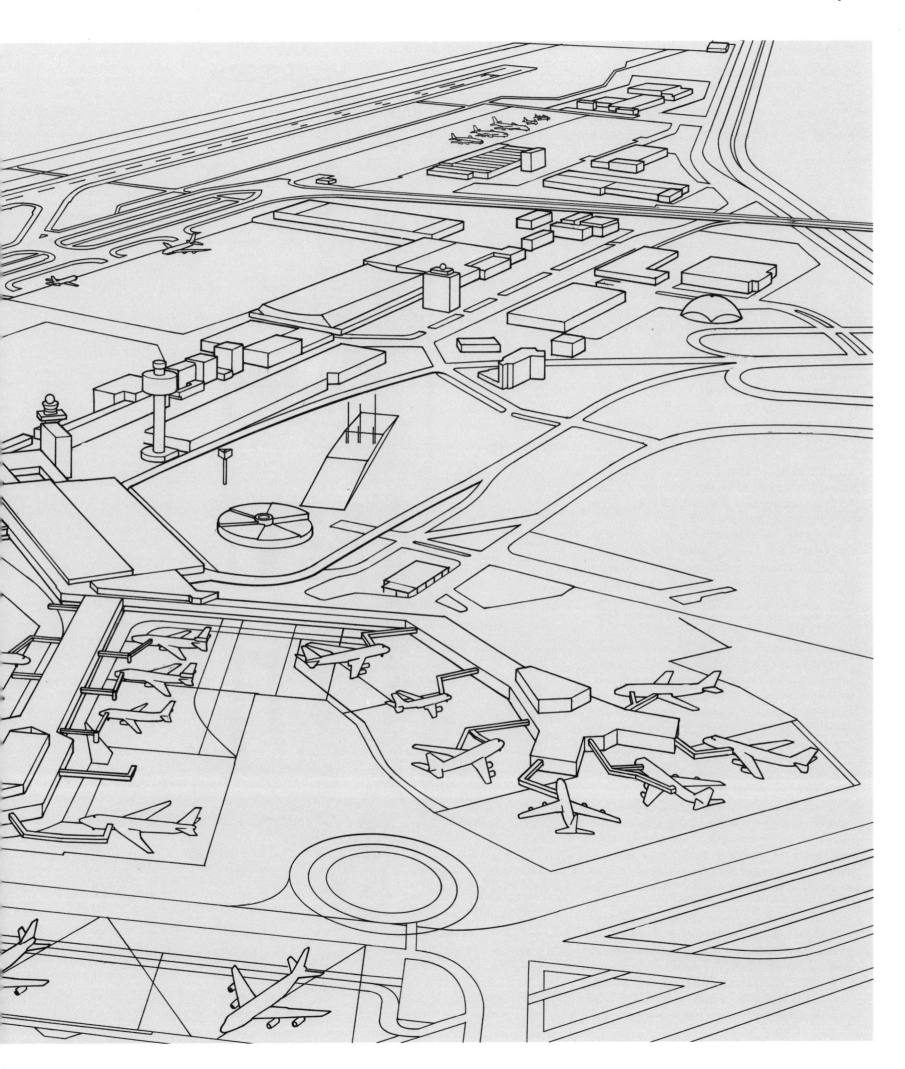

MAN IN SPACE

Although most of this book is concerned with traditional flight within the atmosphere of the Earth, it would be incomplete without a brief look at a much more challenging and expensive form of flight: spaceflight. This new realm of science is fast becoming an established mode of transport which increasingly includes human beings.

On April 12, 1961, the first manned spaceship, Vostok I with Yuri Gagarin, was in orbit around the Earth. However, it was clear that a single astronaut would hardly be able to cope with the diverse and complex programme of a prolonged space flight. Multi-seat, multi-purpose space vehicles were needed. The development of the Russian Vostok series of spacecraft and the American Gemini programme commenced and were completed, to be followed by the Russian Luna and the American Apollo Moon programmes, culminating in the American landing of the first man on the Moon on July 20, 1969.

Astronaut Neil Armstrong, the first man to set foot on a celestial body, coined the immortal sentence: "A small step for a man, one giant leap for mankind." The Apollo programme launched nine expeditions to the Moon. On six of them, 12 astronauts were landed on the Moon surface, before the programme was terminated by the Apollo 17 mission in December 1972.

Meanwhile, the Russians were proceeding with the Soyuz series of space vehicles. The Soyuz consists of a command module on top and a service module at the bottom, with two habitable structures for the crew. The Soviet Union gave up the space race to put a man on the Moon, when Apollo 11 landed with Buzz Aldrin and Neil Armstrong in the summer of 1969. Instead they would concentrate on scientific experiments in manned space laboratories in orbits around the earth. The first in a series of Salyut space stations was launched in April 1971 and it was docked a few days later by two cosmonauts in Soyuz 10 spacecraft. Initial difficulties with the Salyut series led to much redesign, and to ambitious space-station activities when Salyut 6 was launched in September 1977. Salyut 6 was equipped with two docking ports, one in each end, which made it possible to have two spacedrafts docked to the station at the same time for crew exchange, resupplying or rescue work. On December 10, 1978, the two cosmonauts Gretjko and Romanenko set off in Soyuz 26 to encounter Salyut 6 in what was going to be a record stay in space for 96 days and 10 hours. They were visited for short periods by other Soyuz vehicles and a remotely controlled, unmanned supply ship, Progress, which made an automatic docking to deliver food, drink, fuel, fresh air and many other items that made it possible for the cosmonauts to stay in space for so long. The Salyut-Soyuz space-flight procedures became routine,

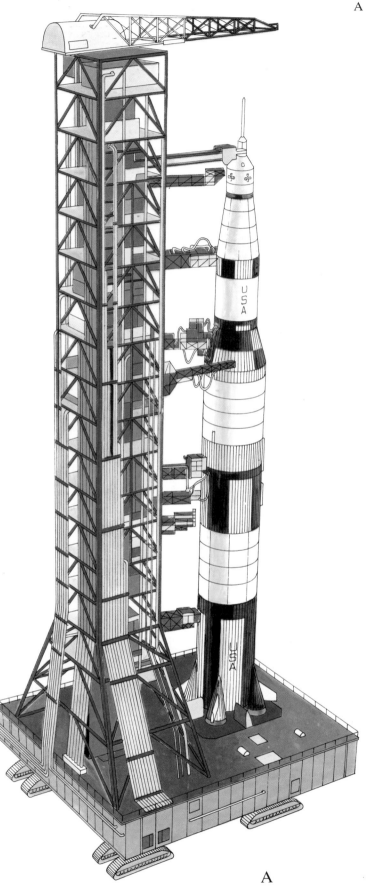

A *The Apollo Moon missions were launched by the three-stage Saturn V booster. The Apollo spacecraft and Lunar Module sat atop the 363-feet-tall rocket, which developed 7.5 million pounds of thrust on take-off. The Apollo/Saturn is standing on a gigantic crawler that took it two miles from the assembly hall to the launch pad.*

A

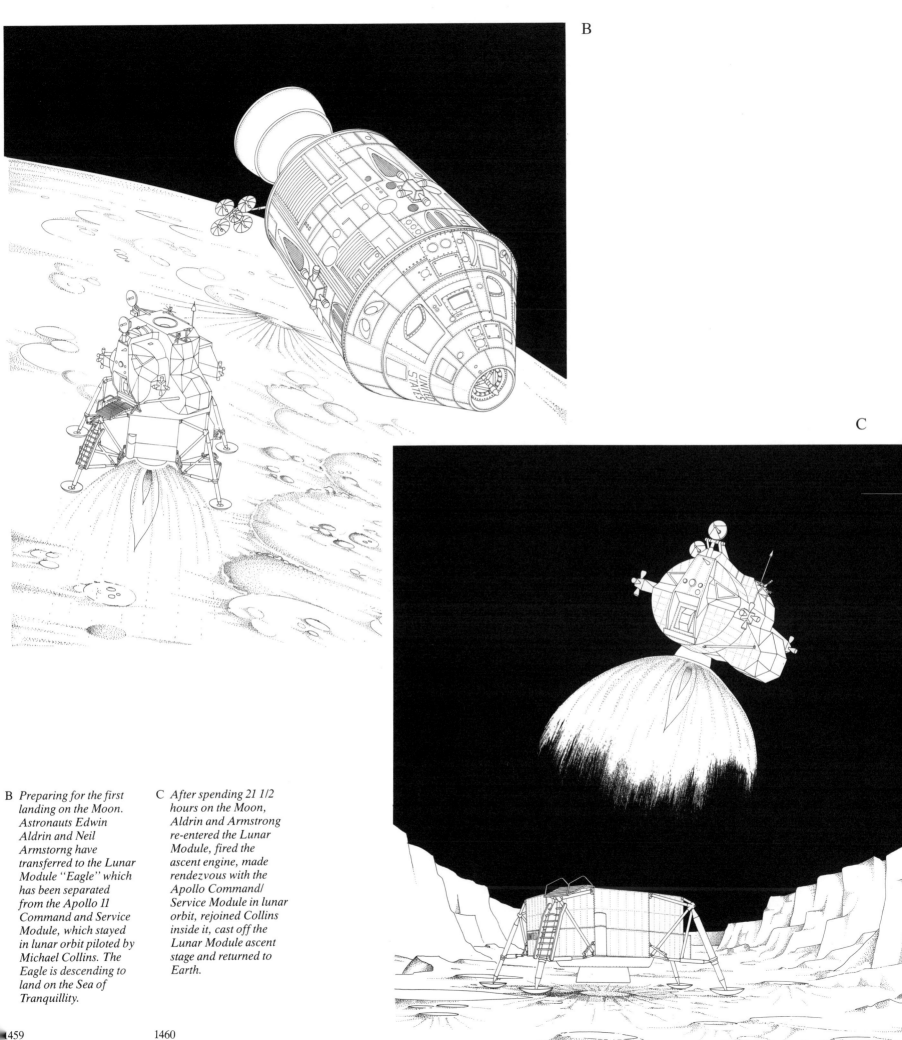

B *Preparing for the first landing on the Moon. Astronauts Edwin Aldrin and Neil Armstorng have transferred to the Lunar Module "Eagle" which has been separated from the Apollo 11 Command and Service Module, which stayed in lunar orbit piloted by Michael Collins. The Eagle is descending to land on the Sea of Tranquillity.*

C *After spending 21 1/2 hours on the Moon, Aldrin and Armstrong re-entered the Lunar Module, fired the ascent engine, made rendezvous with the Apollo Command/ Service Module in lunar orbit, rejoined Collins inside it, cast off the Lunar Module ascent stage and returned to Earth.*

and in 1984 three cosmonauts orbited Earth for 237 days in Salyut 7.

The American Skylab scientific space station was larger than the Soviet Salyut, and during six years in space Skylab circled the earth 34,981 times. Three times Skylab was manned by astronauts and scientists who were ferried to the space station by the Apollo spacecraft. The last mission in Skylab started in November 1973, and lasted for the record time of 84 days; the station was sealed and abandoned after a useful life of less than a year. Five years later the 75-ton Skylab entered Earth's atmosphere and disintegrated, its parts falling in to the Indian Ocean.

The Space Shuttle programme ran fairly problem-free. In the payload bay, which could carry up to 12 tonnes if highly specialized, a carefully chosen load was lifted up to be deployed in orbit or to make observations and experiments on board the Shuttle: scientific instruments, astronomical telescopes, communications relay stations, reconnaissance sensing and reporting systems, nuclear detection devices, weather observation systems, geophysical mapping and measuring systems, navigational systems. Many scientists were given the opportunity to carry out experiments unaffected by the Earth's gravity.

In a series of Shuttle flights, the European Space Agency's Spacelab was carried on board Shuttle Columbia for the first time in November 1983. Columbia was placed in a 150-mile-high orbit; the payload bay doors were opened and for ten days, scientists worked with experiments in five major areas—life sciences, earth studies, space physics, astronomy and material science. Malfunctioning satellites could also be rescued, repaired in space by astronauts wearing pressure suits or retrieved by the Shuttle's remotely controlled manipulator arm.

Tragically, disaster struck the Shuttle programme in January 1986: it blew up shortly after take-off. This was the 25th mission and the programme was not resumed until 1989, when the Soviet Union made public that their first Space Shuttle, Buran, had made a successful flight.

The most spectacular event in the history of manned spaceflight was the historic link-up between the American Apollo and the Soviet Soyuz in July 1975. In a coordinated lift-off sequence, the two spacecraft were launched and guided from the Russin space centre Bajkonur and the American equivalent, Cape Kennedy. At 22,300 miles above Earth, the astronauts from USA rendezvoued with their cosmonaut colleagues from the USSR. The final docking and handshaking could be watched by millions of TVspectators all around the world.

On April 12, 1981, the twentieth anniversary of Gagarin's epic first space voyage in Vostok 1, Pad 32A at Cape Kennedy was the scene of the beginning of a new era in manned space flight. It

A

marked the dawn of the commercial and military space age as the Space Shuttle Columbia lifted off, riding on two booster rockets and an enormous, jettisonable fuel tank. The Space Shuttle would, after its mission, return into the Earth's atmosphere to make a smooth landing like an aircraft. The heat generated when the Space Shuttle re-entered the atmosphere was one of the biggest technical problems overcome during the development phase. It was solved by covering much of the Shuttle's exterior with special ceramic heat-protective tiles.

A *The 85-ton Skylab was the largest assembly ever placed in Earth orbit. Inside, it provided 10,000 cubic feet of space divided into floors by a metal grid. On top is the X-shaped telescope solar array. One of the Skylab's main solar wings was torn off during lauch, but the laboratory was successfully occupied by three crews before it crashed back to Earth on July 11, 1979*

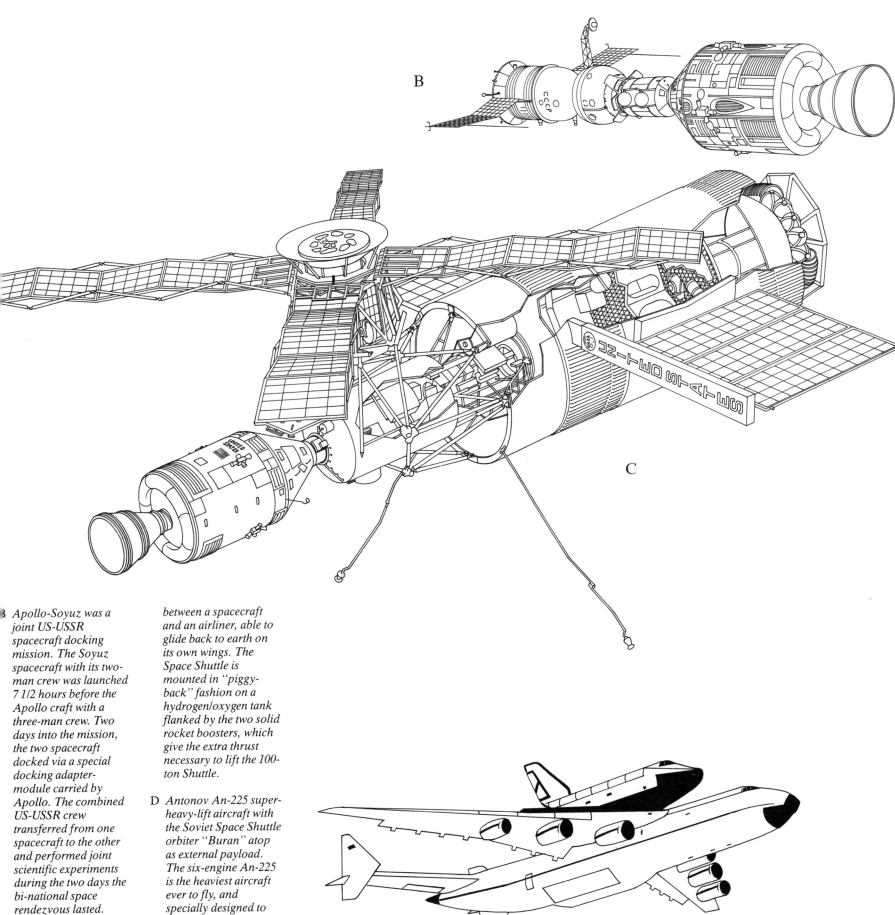

B Apollo-Soyuz was a joint US-USSR spacecraft docking mission. The Soyuz spacecraft with its two-man crew was launched 7 1/2 hours before the Apollo craft with a three-man crew. Two days into the mission, the two spacecraft docked via a special docking adapter-module carried by Apollo. The combined US-USSR crew transferred from one spacecraft to the other and performed joint scientific experiments during the two days the bi-national space rendezvous lasted.

C Space Shuttle, the first reusable spacecraft, being prepared for a mission on launch pad 39A at Kennedy Space Center. The Orbiter vehicle is a hybrid between a spacecraft and an airliner, able to glide back to earth on its own wings. The Space Shuttle is mounted in "piggy-back" fashion on a hydrogen/oxygen tank flanked by the two solid rocket boosters, which give the extra thrust necessary to lift the 100-ton Shuttle.

D Antonov An-225 super-heavy-lift aircraft with the Soviet Space Shuttle orbiter "Buran" atop as external payload. The six-engine An-225 is the heaviest aircraft ever to fly, and specially designed to carry the Buran shuttle orbiter. A 130-ft-long cargo hold can be loaded directly from an aft ramp or the upward-swinging nose. Each engine delivers 54,000 lb thrust.

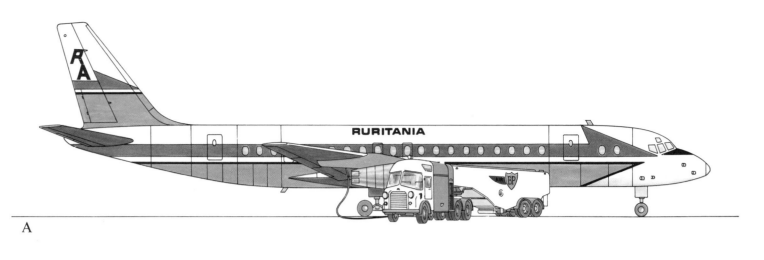

A

B

A *Refuelling: An aircraft that has to fly across the Atlantic may carry about 20,000 gallons of fuel at take-off. From tankers, as in the illustration, or from hydrant pipelines buried underground, fuel is pumped into the tanks through refuelling points beneath the aircraft's wings.*

B *Gauges which indicate fuel contents are located at these refuelling points as well as on the flight deck.*

C *A typical weather chart.*

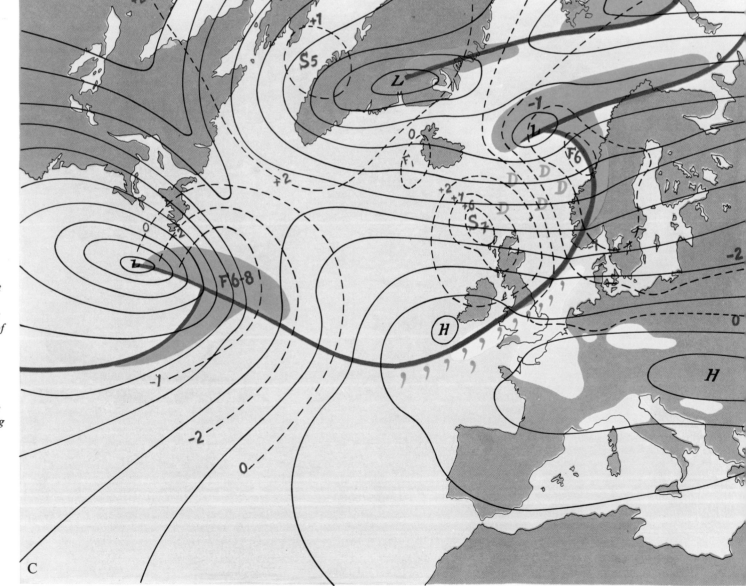

C

A TYPICAL FLIGHT

There is really no such thing as a typical flight. It was once remarked by a pilot that you could fly from Stockholm to Copenhagen once a day for a year and still find something new on each flight. Many scheduled flights are much shorter than that, but many others are very long and non-stop, such as those from Paris to Rio de Janeiro, New York to Buenos Aires, and Los Angeles to London. Nowadays a flight across the Atlantic is considered fairly typical, and in 1989 nearly 30 million passengers crossed the North Atlantic by air.

A normal example would be a flight from Gatwick, south of London, to New York. Suppose we join "Ruritania" flight 887 as it is being prepared for take-off from Gatwick. The passengers are still settling into their seats when a dull murmur indicates that the first of the aircraft's engines has been started. By the time the cabin staff have checked that all seat belts are fastened, a demonstration of the location of the emergency exits and of how to use life vests and the oxygen system is being shown on video sets throughout the cabin. The aircraft is already taxiing toward the runway.

It takes a great deal of preparation by many people, including the crew, to achieve this quick getaway. The Captain and his copilot have had a briefing with the airline flight operations department at least an hour before departure. They will have checked the meteorological conditions for the departure airport, and the weather forecasts for the route and destination. The Captain decides how much fuel to have onboard the aircraft, after consulting with the flight despatcher and the computer-produced flight plan. He will have reviewed the state of the radio facilities along the route, and signed the flight plan.

On reaching the aircraft, the Captain has briefed his crew on how he wants to see the flight carried out. He has made sure that the right amount of fuel is onboard, checked with the technical ground staff about the aircraft's serviceability and components, and signed the load sheet which states that the aircraft is properly balanced. As the passengers boarded, he and his copilot—today often a young lady—were going through a long checklist to verify that all necessary equipment and systems are working and set for take-off.

Now as the aircraft taxis toward the take-off, the cabin staff find their own seats and the passengers settle down to read magazines or newspapers. On the flight deck, the copilot is still reading out the checklist, while he monitors the operation of the hydraulic and electrical systems, the indications of instruments and so forth. At last he calls on the radio: "Gatwick tower, this is Tania 887 standing by for airways clearance."

"Roger, Tania 887 is cleared into position runway 26, cleared for take-off. Wind from 250 degrees at 20 knots, gusting 28. Cleared to New York Newark via standard departure route Daventry 5V, cross Lambourne at 6000." Thus the Captain receives his instructions for the first phase of the flight, the take-off and the route by which he will leave this airport to get on his course.

With one hand on the nosewheel's steering wheel, and the other on the throttles, the Captain manoeuvres the aircraft round the final turn onto the runway, and brings it to a gentle stop by using the footbrakes on the pedals. Quickly the crew runs through the concluding checks before take-off, noting the actual weight and runway conditions to make sure that the available runway length is sufficient.

At low speeds there is no difficulty in keeping the aircraft on a straight track during take-off by using the nosewheel steering. Once the aircraft is fully airborne, it can be tracked along the runway centre-line simply by pointing the nose a few degrees into the wind. But when the aircraft is supported partly by its wings and partly by its wheels, some care must be taken to ensure that it does not "weathercock" into the wind.

TAKING OFF

The Captain advances the throttles—or thrust levers, as they are properly called on a jet aircraft—and releases the brakes. As the aircraft accelerates rapidly along the runway, the copilot confirms that full power has been developed, and calls out as ever higher speeds are reached. The first speed, termed V1, has been calculated before the aircraft started to taxi. If an engine failure occurs before the aircraft has reached this speed, the take-off will normally be abandoned and the aircraft is stopped on the runway. Beyond this speed, the pilot must continue the take-off.

The last speed to be called out may be V2+10. By

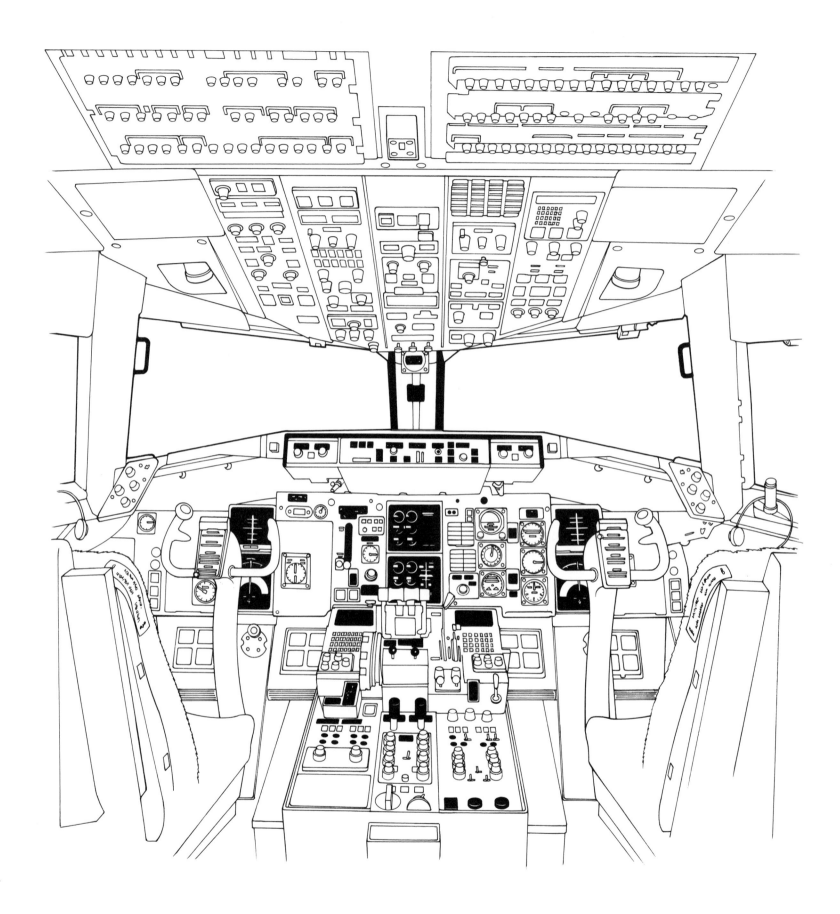

A

A *Cockpit layout: There are so many instruments which must be displayed on the modern flight deck that positioning them* satisfactorily has become of major importance. Each pilot requires separate basic flying instruments and a control column. Most of the flight systems must be duplicated, and warning or emergency lights placed where they will command attention. Apart from these items, everything which may be needed must be within reach of one or other of the flight crew.

1489   1490   1491   1492

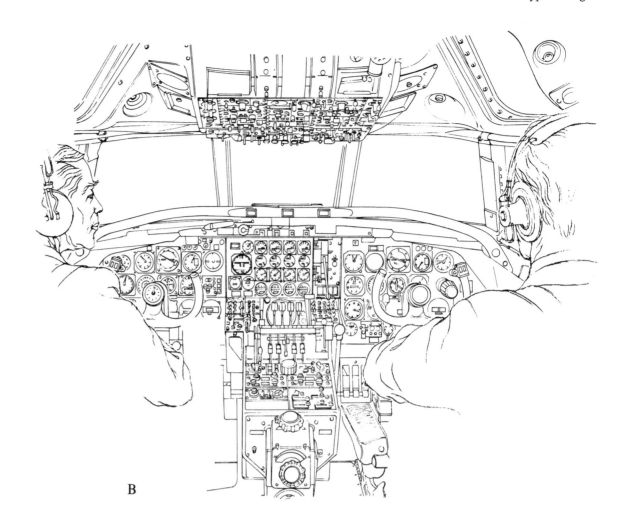

B

AIRCRAFT OPERATIONAL MANUAL 767
NORMAL CHECK LIST

RESTART

1. Circuit breakers	CHECK	P.	
2. Emergency equipment	CHECK		
3. Pins and pitot covers	CHECK		
4. Flight recorder	NORM		
5. SERV INTPH switch	OFF		
6. RESERVE BRAKES & STRG	NORM		
7. BAT switch	ON		
8. STBY POWER selector	AUTO		
9. Hydraulic system	SET		
10. Landing gear lever	DOWN		
11. Flap lever	CHECK		
12. SPEEDBRAKE lever	DOWN		
13. ALTN FLAPS selector	NORM		
14. LE and TE alternate flaps switches	OFF		
15. ALTN GEAR EXTEND switch	OFF		
16. GND PROX FLAP OVRD switch	ON		
17. GND PROX/CONFIG GEAR OVRD switch	OFF		
18. External power	ESTABLISH		
19. IRS selectors	NAV/ALIGN		
20. YAW DAMPER switches	ON		
21. ELEC ENG CONT switches	NORM		
22. HF radios	SET		
23. EVAC COMMAND switch	CHECK		
24. Electrical system	SET		
25. APU selector	START		
26. Cockpit voice recorder	TEST		
27. EMER LIGHTS switch	ARMED		
28. PASS OXY switch	CHECK		
29. RAM AIR TURB switch	CHECK		
30. Engine ignition selector	1 or 2		
31. Ice protection	OFF		
32. External lights	SET		
33. CARGO HEAT switches	ON		
34. WINDOW HEAT switches	ON		
35. PASS SIGNS selectors	AUTO/ON		
36. Pressurization system	SET		
37. EQUIP COOLING selector	AUTO		
38. Pneumatic & air conditioning systems	SET		
39. Fire extinguishing system	CHECK		

AIRCRAFT OPERATIONAL MANUAL 767
NORMAL CHECK LIST

Crew at their stations. (R/P)

40. CDU	SET	1/P.	
41. Nav aids	SET	1/P.	
42. Loose objects and equip	CHECK	P.	
43. Crew papers & Aircraft Log	CHECK	L/P.	
44. VOR/DME	AUTO	P.	
45. F/D switches	ON	P.	
46. EICAS	CHECK	L/P.	
47. EFIS control panels	SET	P.	
48. Crew oxygen	CHECK	P.	
49. INSTR SOURCE SEL	CHECK	P.	
50. Flight instruments	SET	P.	
51. AUTOLAND STATUS annunciators	CHECK	P.	
52. HDG REF switch	NORM/TRUE	R/P.	
53. Parking brake	SET	L/P.	
54. STAB TRIM CUTOUT switches	NORM	L/P.	
55. RESERVE BKS & STRG	OFF	L/P.	
56. Standby instruments	SET/AUTO	L/P.	
57. AUTO BRAKES selector	OFF	L/P.	

When ready to start engines.

58. Load sheet and fuel loading	CHECK	L/P.	
59. BULK CARGO HEAT selector	NORM/VENT	R/P.	
60. CDU/speeds	SET	P.	
61. Cabin report	RECEIVED	L/P.	
62. Doors	CHECKED	L/P.	
63. Start clearance	OBTAIN	P.	
64. Hydraulic system	SET	L/P.	
65. Fuel pump switches	ON	R/P.	
66. Trim	SET	L/P.	
67. Flight controls	CHECK	1/P.	
68. RED anti-collision lights	ON	R/P.	
69. Pack selectors	OFF	R/P.	
70. Check List completed.			

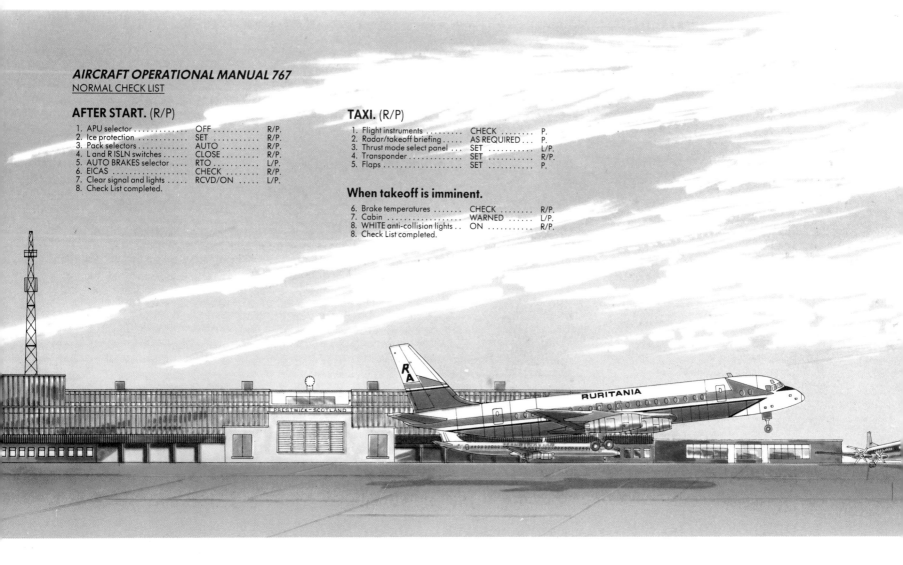

AIRCRAFT OPERATIONAL MANUAL 767
NORMAL CHECK LIST

AFTER START. (R/P)

1. APU selector . . . . OFF . . . . R/P.
2. Ice protection . . . . SET . . . . R/P.
3. Pack selectors . . . . AUTO . . . . R/P.
4. L and R ISLN switches . . . . . . CLOSE . . . . R/P.
5. AUTO BRAKES selector . . . . RTO . . . . L/P.
6. EICAS . . . . CHECK . . . . R/P.
7. Clear signal and lights . . . . . RCVD/ON . . . . . L/P.
8. Check List completed.

TAXI. (R/P)

1. Flight instruments . . . . CHECK . . . . P.
2. Radar/takeoff briefing . . . . . AS REQUIRED . . . P.
3. Thrust mode select panel . . . SET . . . . L/P.
4. Transponder . . . . SET . . . . R/P.
5. Flaps . . . . SET . . . . P.

When takeoff is imminent.

6. Brake temperatures . . . . CHECK . . . . R/P.
7. Cabin . . . . WARNED . . . . . . L/P.
8. WHITE anti-collision lights . . ON . . . . R/P.
8. Check List completed.

AIRCRAFT OPERATIONAL MANUAL 767
NORMAL CHECK LIST

CLIMB.

1. Landing gear lever . . . . OFF . . . . 2/P.
2. Altimeters . . . . 1013, ft . . . . P.
3. External lights . . . . SET . . . . 2/P.
4. Pressurization and
   air conditioning . . . . CHECK . . . . 2/P.
5. PASS SIGNS selectors . . . . . AUTO/ON . . . . . L/P.
6. Check List completed.

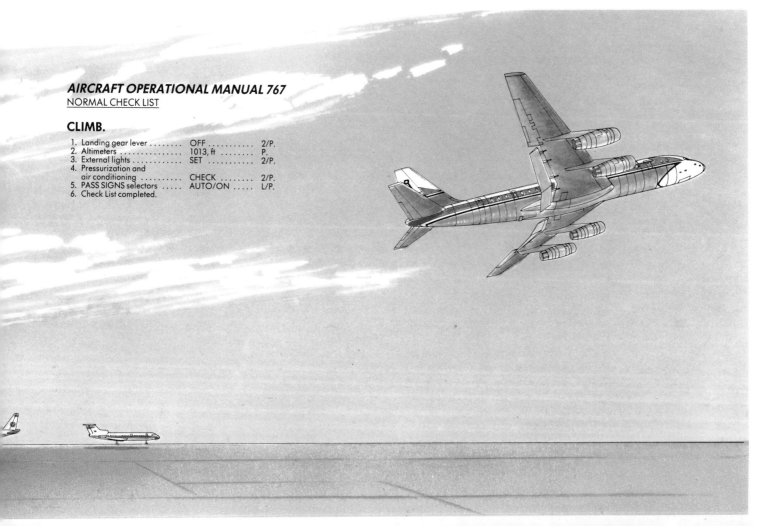

The take-off: Even if there is no cross-wind, and even if runway surface conditions are good, the take-off is one of the most critical stages of the flight. During the rapid acceleration, the crew must finally assess the aircraft's overall fitness to fly, and must be ready for instant action if any emergency arises.

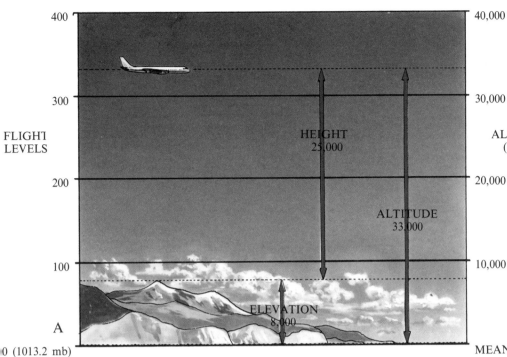

A

FLIGHT LEVELS

ALTITUDE (FEET)

HEIGHT 25,000

ALTITUDE 33,000

ELEVATION 8,000

MEAN SEA LEVEL

0 (1013.2 mb)

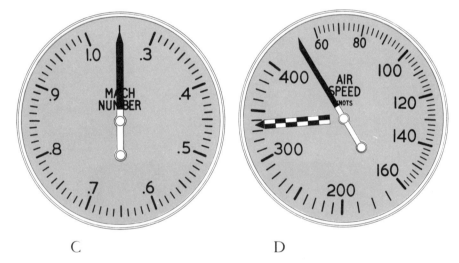

B

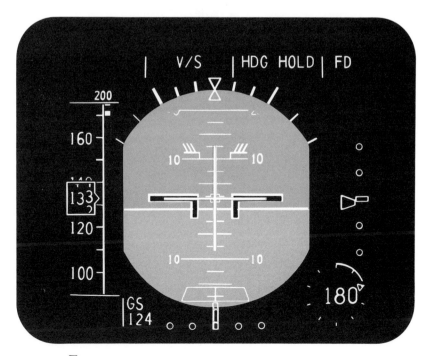

C

D

E

A Definition of altitudes in aviation is, by ICAO standards, made in different ways depending on the circumstances:
ALTITUDE: The vertical distance from a point measured from mean sea level (QNH setting on the altimeter).
HEIGHT: The vertical distance from a point measured from a specified datum (QFE setting on the altimeter).
FLIGHT LEVELS: Surfaces of constant atmospheric pressure which are related to a specific pressure datum, 1013.2 millibars. (ISA or standard setting on the altimeter). The QFE setting is mostly used during approach and landing, as it gives the height over the airfield and indicates ground as zero. QNH is used at altitudes below 3–4,000 ft and for the standard setting at higher altitudes, which gives a common indication for all aircraft; this is essential so that the ATC can separate the air traffic in a safe way.

B The altimeter is essentially a barometer that measures the difference in atmospheric pressures at different altitudes. This example has a "window" at the right that indicates thousands of feet, which corresponds to one full turn of the needle. The small "windows" show barometric settings in millibars and inches of mercury., The altimeter setting is made by turning the knob down to the left.

C The airspeed indicator is used up to 25,000 ft and measures the flying speed of the aircraft. But the indicated airspeed (IAS) has to be corrected depending on the difference in temperature at different altitudes which gives true airspeed (TAS); the difference between IAS and TAS becomes greater, the higher the aircraft goes. For navigation purposes, ground speed (GS) must be calculated by taking tail- or headwind into the speed figure. Speed can be indicated in knots, statute miles or kilometers.

D At higher flight levels, the Machmeter is normallly used. It relates the aircraft speed to that of sound. Most jet airliners cruise at about Mach 0.82 (82 % of the speed of sound).

E The electronic flight instrument system (EFIS) gives the pilot integrated information that was previously presented by many separate mechanical instruments and indicators. This example shows the electronic presentation of the artificial horizon. The angled bars in the middle symbolizes the aircraft's horizontal attitude, in this case level. If the bar were at "10" on the vertical scale, the attitude would been 10 degrees nose-up. The letter symbols on the upper part of the indicator show the setting of autopilot and flight director. The left vertical scale indicates airspeed and the actual IAS is presented in expanded digits. Ground speed, GS, is shown at the bottom right and the small circular symbols to the right indicate ILS localizer position. The square above is the ground indication, which converges with the aircraft symbol on touch-down. The vertical scale to the far right shows the ILS glide slope.

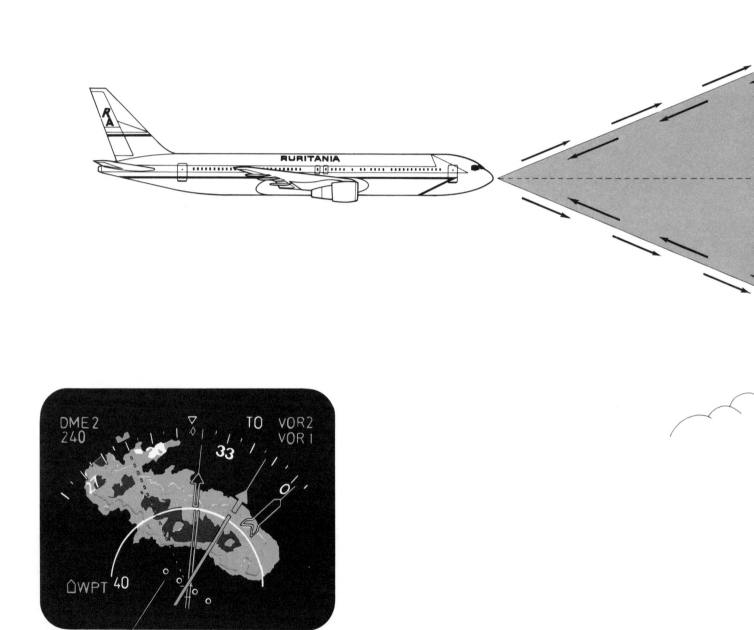

Weather and radar:
A cathode-ray tube is
utilised to display in
the cockpit a picture
of the relative positions
of clouds ahead of the
aircraft, to give
warning in particular
of the severe precipita-
tion associated with
turbulence. The power
of the transmitter may
be judged from the fact
that the set depends
on echoes received
from rain-drops which
may be well over
100 miles away. The
radar scanner in the
nose of the aircraft
can also be tilted
downward, to provide
a radar "map" of the
terrain below the
aircraft, prominent
features such as
coast-lines being
displayed clearly.

then, the aircraft has been rotated, meaning that its nose has lifted and it is climbing away from the runway. The Captain calls for the undercarriage to be retracted. As the speed increases and height is gained, the wing flaps are brought in gradually and the after-take-off check is begun. "Leave the seat-belt signs switched on," says the Captain, "complete the rest of the checklist and have a look at the weather radar."

This is one of the busiest moments of the flight, and everything seems to be happening at once. The Captain has banked the aircraft a little to port, minimizing noise over a nearby residential area. He and the copilot are monitoring the operation of the many aircraft systems which are brought into use during the early stages of the climb. It is bumpy, partly due to convection currents caused by the afternoon sun, and partly since the wind is deflected by rough areas on the ground.

The Air Traffic Controller on the ground is waiting to hear the aircraft's departure message. The aircraft must be navigated to Daventry, which is only 70 miles from Gatwick. But 15 miles this side of Daventry, there are dark clouds building up, with a threat of possible turbulence and icing.

To the crew, the situation is routine, for they have often made similar take-offs. The Captain switches on the flight director, and the information on his TV displays then guides him through the turns enroute to Daventry. The copilot studies the radar and satisfies himself that, on this occasion, the cloud ahead is more innocent than it looks to the naked eye, so the aircraft can fly safely through it.

Next, the crew activates the systems which will protect the aircraft from icing. Meanwhile a call is made to Air Traffic Control (ATC). As the aircraft breaks through the cloud into the golden sky beyond, the Controller's voice comes back: "Tania 887, you are cleared to continue climb to flight level 250. Here is your Oceanic clearance, are you ready to copy?"

"London from Tania 887, leaving flight level 80. Go ahead with the Oceanic clearance." Flight level 80 is equivalent to an altitude of 8,000 feet. Now that the clouds are beneath them and the climb can be continued, the pilots relax slightly. They are offered and accept a particular route and height, which will eventually bring them over St. Anthony at the northern extremity of Newfoundland.

NAVIGATION OVER THE OCEAN

A line on the Captain's TV displays is pointing dead ahead, indicating that the aircraft is homing toward the Daventry VOR (Very high frequency Omnidirectional Range). The Captain sets the figure 288 on a dial, so that he will be able to follow a particular track out of Daventry. As it happens, he has barely finished doing so, when the VOR indicator swings round, showing that he has passed over the ground-based navigation aid.

During the next half hour, there will be some guidance from the short-range navigational aids in Great Britain and Ireland, while ground radar may offer further assistance. But from now on, navigation over the ocean will be performed by the FMS, or flight management system. This is a computerized, self-contained system for horizontal and vertical navigation. It automatically selects the sources from which it derives navigational information, and combines them to get the most accurate possible information about position. In the future, satellite-derived positioning will surely be part of the FMS for greater accuracy.

For over-water navigation today, one source might be INS (Inertial Navigation System). This is taken from space exploration and missile technology. It is completely independent of external sources like ground-based radio beacons, and derives information from built-in gyros that detect changes in speed, direction and attitude. Also commonly used are Omega and Loran, two long-range navigation systems based on radio signals. Combining with computerized interpretation of the received information, they too offer the high precision needed by a modern navigation system suitable for the North Atlantic. Overland VOR information is automatically selected to update the somewhat less exact information from other sources.

During the climb-out, as cocktails are served in the cabin, the passengers notice a change in engine sound and in the aircraft's attitude. After further clearance by ATC, the aircraft has continued up to an assigned cruising height of 31,000 feet, known as flight level 310. Upon reaching this height, the pilots check that the autopilot makes the aircraft level out.

As radio altimeters are used only near the ground, a flight level is normally maintained by referring to conventional altimeters. These measure ambient air pressure, operating like barometers, and have an adjustable scale to compensate for differences in local air pressure, so that they can correctly indicate the height above an airfield. But when an aircraft climbs toward the cruising level, its altimeters are reset to a uniform "standard" setting. This enables different aircraft to be allocated specific "flight levels" in order to minimize the risk of collision.

During take-off and the early stages of the climb, the aircraft's speed was shown on the airspeed indicator, the units of measurement being knots (nautical miles per hour). For speed indication at high levels, the pilots rely mainly on the Mach meter, which is now showing Mach 0.81 (slightly more than 8/10 the speed of sound, or about 8 nautical miles per minute). The figure is maintained exactly, because the Oceanic clearance has stipulated the aircraft's Mach number. This is essential, as another aircraft is on the same route some fifteen minutes ahead, and another may be flying a similar distance behind.

This procedure illustrates the fact that navigating the ocean offers little difficulty to the crews of modern airliners. If there were no other aircraft in the

A *Doppler navigation:
Nearly all civil air-
liners flying the
Atlantic are equipped
with Doppler, which
came into service
about ten years ago.
At present Doppler
offers the most reliable
general method of
ensuring that an aircraft
follows its assigned
route over the ocean.
In the somewhat
unlikely situation
shown on the dial
illustrated, the aircraft
would have a ground
speed of 205 knots and
be drifting 28°
to starboard.*

1 Aircraft's track over
  the ground
2 Aircraft heading
3 Port forward beam
4 Starboard forward
  beam
5 Starboard rear beam
6 Port rear beam
7 Drift angle

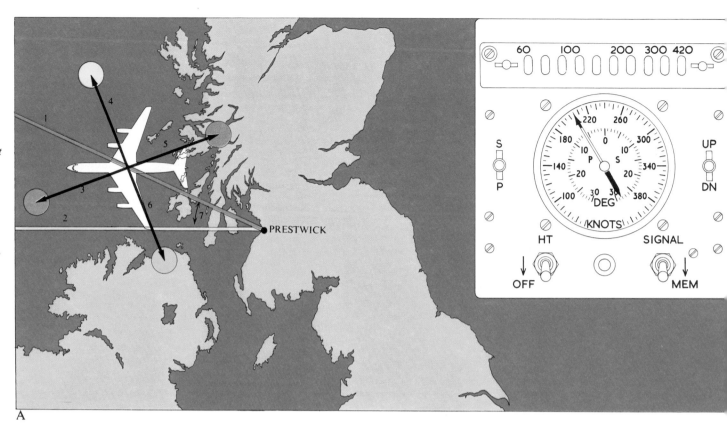

A

B

sky, the pilots could just steer a compass course and be pretty sure of arriving over Canada eventually. But there are many other aircraft about, and up to 250 may cross in one direction between Europe and North America in a day, most of them within a period of about six hours. The airspace is thus relatively congested, and most of the effort which is put into the navigation has the purpose of keeping the aircraft on its cleared track.

Very high winds often blow at these levels over the Atlantic. The strongest ones, called "jet streams", are associated with the low-pressure regions that are constantly moving across the North Atlantic from west to east. The winds move counterclockwise around a centre of low pressure. Naturally all Captains flying the Atlantic want to take advantage of these winds by routing their aircraft through the areas of strong tailwinds. But the traffic makes this impossible, since most aircraft would prefer the same areas of the Atlantic airspace, and the risk of collision would become unacceptably great.

To solve the problem, a set of routes is designed every day for westbound traffic, and another for eastbound. These routes take maximum advantage of the day's wind conditions, while also ensuring safe distances between the aircraft. Such "tracks" are transmitted to the airlines flying the North Atlantic daily, and are presented to the Captain at the pre-flight briefing. They cover the areas with heaviest traffic, which can be described roughly as lying south of the 60-degree latitude. When flying north of the tracks, an aircraft is free to choose its own route—as is always done by flights in the polar region, where winds are very weak.

Each aircraft reports several positions as it passes over them on its way across the ocean, depending on which route it is following. This enables Air Traffic Control to ensure that it does not conflict with other aircraft. When it gets more than about 200 miles from the coast, it is out of range of Very High Frequency radio, and different equipment using High Frequency is used instead for communications. With HF, the pilots can reach any point on earth, although the transmission is usually of much lower quality than with VHF.

A WARNING

At this moment a bell rings on the flight deck and a light is flashing. The aircraft is being called by a ground station near Shannon Airport. Both pilots don their headsets in order to take down the message. "Tania 887, here is a Sigmet issued at 1800 UTM time. A jet stream extends..." In his Irish brogue, the communicator explains the present characteristics of the jet stream and the areas where turbulence is expected.

A jet stream is very different from the winds encountered at medium altitudes. It moves almost like water sprayed from a hose, through the levels at which jet aircraft operate. In its centre, the wind may be extremely fast, occasionally exceeding 200 knots. A few thousand feet below or a few miles to one side, the wind velocity is far less. This variation causes eddies similar to the whirlpools in a fast-flowing river, and as a result there may be turbulence without any associated clouds. Such clear-air turbulence cannot be detected by the aircraft's radar. If encountered unexpectedly, it might have worse effects than spilling some coffee in the aircraft cabin.

In this instance, the Captain has ample time to warn the cabin staff. The seat-belt sign goes on, the cabin is generally tidied, and the Captain's reassuring voice over the loudspeaker system tells the passengers why these measures are being taken. But it was a false alarm: the aircraft drones smoothly onward.

On the flight deck, the Captain is preparing the next position report. He takes increasing interest in the wind information, which has been showing that the wind is strengthening and swinging to come from the southwest. Suddenly a tremor runs along the fuselage. It is barely noticeable in the cabin, but both pilots check that their seat-harnesses are locked tightly. Another tremor, and then the aircraft hits sharp "cobblestone" turbulence.

Some of the passengers look uneasily out of the windows, noticing that the wing-tips are jumping up and down in relation to the horizon. Yet they need not worry, as all aircraft are somewhat flexible, and in certain circumstances it might be normal for the wing-tips to move up and down by eight feet or more. The hostesses are also rather concerned, since they know that some passengers may become airsick if the turbulence continues for too long. After five minutes, the hostesses exchange smiles—already the turbulence is fading, and it now consists mainly of a series of tremors.

A little later the Captain switches off the seat-belt sign, unfastens his own harness, and climbs out of his seat to stretch his legs. The aircraft has passed 30 degrees West, and contact has been established with Gander in Newfoundland, whose ATC Centre is responsible for this part of the Atlantic.

It is interesting to recall how this kind of responsibility was allocated. Each country has certain rights in regard to the airspace above it, and can lay down regulations which must be obeyed. But over the oceans, the circumstances are quite different. In the early years of oceanic flying, there was no ATC, and so few aircraft existed that each could set a course without bothering about the risk of collision. As the traffic increased, however, some degree of regulation became essential. An agency of the United Nations, ICAO (International Civil Aviation Organization), arranges such agreements between the countries involved, and it devised the necessary rules in this area as for other parts of the world.

The great majority of aircraft flying across the ocean are airliners. Most of them belong to companies which are members of IATA (International Air Transport Association), and whose technical

B *VOR: The Very-high frequency Omnidirectional Range, or VOR, is a standard short-range navigational aid. Limited to line-of-sight, in the case of jet aircraft the signals can usually be received over distances of up to about 200 miles. Because of the wave band on which it operates, VOR is vastly more reliable than the non-directional beacon and radio range which preceded it. When following an air route defined by VORs, the crew will often have one set tuned to the VOR which has just been passed, and one to the VOR ahead; the dial which is illustrated might suggest that the aircraft is heading north and is flying just to the east of Prestwick Airport. No 2 VOR is tuned on Prestwick VOR (magnetic bearing 280°) and No 1 VOR is tuned on Skipness (bearing 300°).*
1 Full VOR coverage
2 No coverage

cooperation is far better than that reached on the sea in the early days of maritime development. The countries which border on the ocean accept responsibility for providing ATC and related services near them, and an elaborate network of communications has been set up between the various centres.

Even before Ruritania 887 taxied out at Gatwick, the flight-plan details were received at Gander. Just now the controller who handles the continental airspace is preparing the ATC clearance for this flight from Newfoundland across Canada's Maritime Provinces to New York.

ACROSS THE MARITIME PROVINCES

So the flight proceeds, seemingly always into the setting sun. The change of time between London and New York is five hours, and today the flight is taking 6.5 hours. Thus the jet airliner is almost keeping up with the sun, and indeed may "overtake" it in certain circumstances. But as the aircraft comes within VHF range of Gander, the blue haze below is deepening to purple, and a few faint lights on the ground can be seen. The sky above is still light, and the long silvery contrail of an aircraft at flight level 370 gleams ahead, pointing the way to Montreal.

The crossing of the Atlantic was not made entirely at such a high flight level. When approaching the English coastline, the aircraft was still very heavy, and could only reach a cruising flight level of 290. As fuel is burned, the aircraft becomes ever lighter, and it performs a "step climb"—at first able to reach level 330 and, finally near the Canadian coast, level 370.

The white contrail of an aircraft, actually brown or dark grey as seen from above, consists of fine ice crystals. When the fuel burns, one of the by-products is water, which freezes almost as soon as it leaves the jet exhausts, since the temperature at these levels is likely to be 50–60 degrees below zero Centigrade. If the surrounding air is saturated and unstable, the contrail gradually expands and spreads. Contrails can be quite useful in indicating the presence of an aircraft—for while it is difficult to receive signals from an aircraft more than about six miles away, a contrail can sometimes be seen at twelve times that distance.

There is a medium-wave radio beacon at St. Anthony. This equipment is less sophisticated than the VOR/DME stations located there and along the remaining route to New York—but it serves a similar purpose. The aircraft crosses the Gulf of St. Lawrence, and takes one of the well-marked routes over the border between Canada and the United States.

By now, the crew is planning for the arrival at John F. Kennedy Airport, receiving reports of the weather at that airport, and at alternative ones to which a diversion might be made if the weather at JFK gets worse. Near New York, the weather is not too good: low clouds are drifting in patches over the airport. But as long as it does not turn into dense fog, the landing conditions are acceptable.

Periodically throughout the flight, the pilots have checked the amount of fuel left. Now they compare the fuel needed for possible diversions with the amount that is expected to be left when arriving at JFK. This enables the Captain to estimate how long he can "hold" there (wait in a racetrack-shaped pattern over a designated place near the airport) before beginning a diversion. Moreover, the aircraft is now about thirty tons lighter than it was when leaving Gatwick. Its weight at landing is estimated thirty minutes in advance, so that the approach speeds can be calculated.

The pilots take the JFK landing charts from their route manuals and study these carefully. The pilot who will perform the landing conducts a briefing: he describes the manner of approach, the routes to be flown, the naval aids, minumum altitudes and special regulations which may apply, so that both pilots agree fully about all aspects of the approach. A disagreement, or even a simple question when near the ground, could lead to a missed approach or worse. The necessary charts are fixed to a clip on the steering-wheel column, or beside each pilot, for quick reference—and they are changed as the flight progresses.

Soon after the aircraft passes Boston, ATC calls with clearance to descend along a prescribed route, so as to enter the holding pattern at Deer Park VOR/DME at 11,000 feet. Within a few minutes the aircraft is descending, initially at 2,500 feet per minute. By now the sky is totally dark and, although the lights on the ground are obscured by cloud, the anti collision lights of other aircraft can be seen above and below. The first part of the approach check is completed, and the altimeter settings have been changed to suit the local barometric pressure.

As Deer Park is approached, the pilots listen to the ATIS (Automatic Terminal Information Service), which is a continuous recorded report of the airport weather conditions, runway in use, and status of the relevant navigation aids. This information is updated every half hour. The cloud base is 300 feet, but the visibility below is still more than a mile.

THE FINAL APPROACH

After holding at 11,000 feet for several minutes, the aircraft is successively cleared to lower levels. At last the Approach Controller takes over, giving instructions that the pilots have been waiting for. "Tania 887, leave Deer Park holding in three minutes on radial . . ." Thus the final stages of the flight begin.

JFK is, of course, an extremely busy airport. To prevent midair collisions and "near misses", and to direct aircraft outside the ground-based navigation aids, ground radar is used to check the path of every plane. Each is given a certain speed at which to fly, and the radar controller uses a computer to calculate the spacing between aircraft. Much concentration is needed on the flight deck. The aircraft's flight direc-

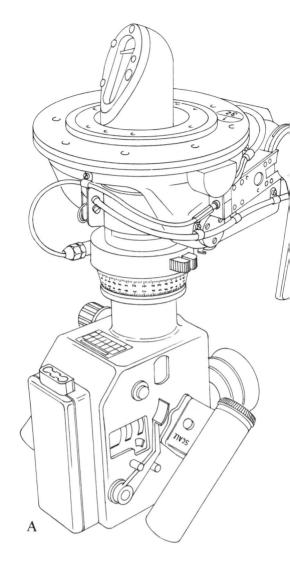

A

A *Periscopic sextant: Before aircraft were pressurised, the navigator stood under an astro-dome in the top of the aircraft, which enabled him to view the entire sky. Nowadays he uses a sextant with a periscopic extension which passes through the pressure hull. The illustration shows the azimuth ring which he will use when checking the aircraft's heading, and the counters which will indicate the altitude of the star or other body which he observes. Normally he will align a bubble on to this star for a period of two minutes, keeping them carefully together; an averaging device incorporated in the sextant will then give him the average for the two-minute observation.*

B *An ancient navigational aid—the map of Vinland. Superimposed on the map is a reconstruction of a ship's compass, of the type probably used by the Vikings. The position of North was given by the shadow cast by the vertical pin at sunrise or sunset; the horizontal pin was adjusted to indicate the course to be steered. Covering the same area as the Loran system chart on the following page, the Vinland map has caused much controversy over who first discovered America. It is claimed to be an authentic map, made half a century before the voyage of Columbus. Vinland(Newfoundland) is shown with the legend: "Island of Vinland, discovered jointly by Bjarni and Leif".*

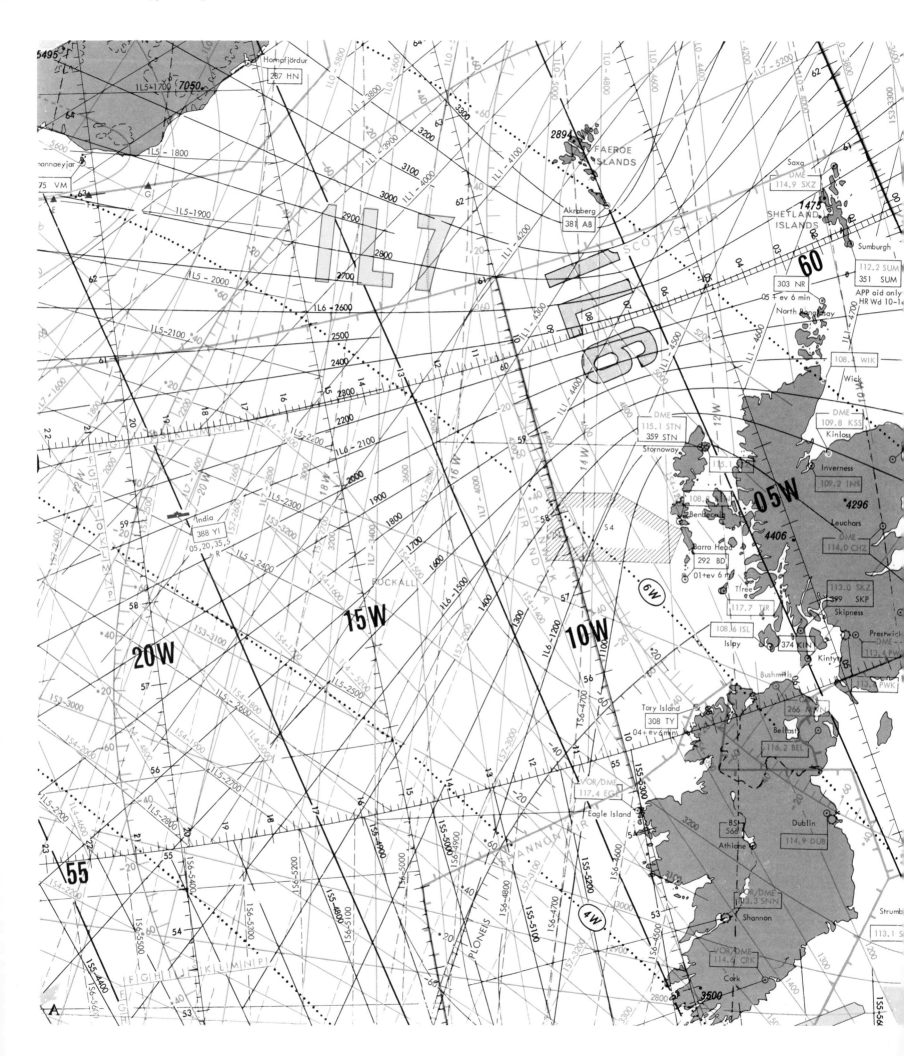

A *Loran: Loran A is the most commonly used ground referenced navigation aid over the ocean. A cathode-ray tube presentation is normal, and enables the pilot or navigator to establish a single position line, or, by reference to several lines, to obtain a "fix"*

B
C *To provide meteorological services for the Atlantic crossing, nine weather-ships were deployed in the North Atlantic. Here a radio-sonde balloon is being launched; it will be tracked by the ship's radar and will provide information on winds, temperature and other parameters, which were reported back to meteorological stations on land.*

The weather-ships have been supplanted by a system of meteorological satellites that gives real-time information directly to the airports' meteorological offices, displayed on a screen or as printouts that can be obtained by the aircraft crew.

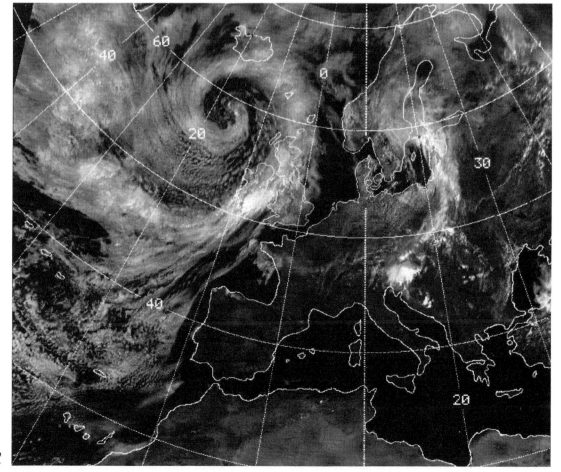

C

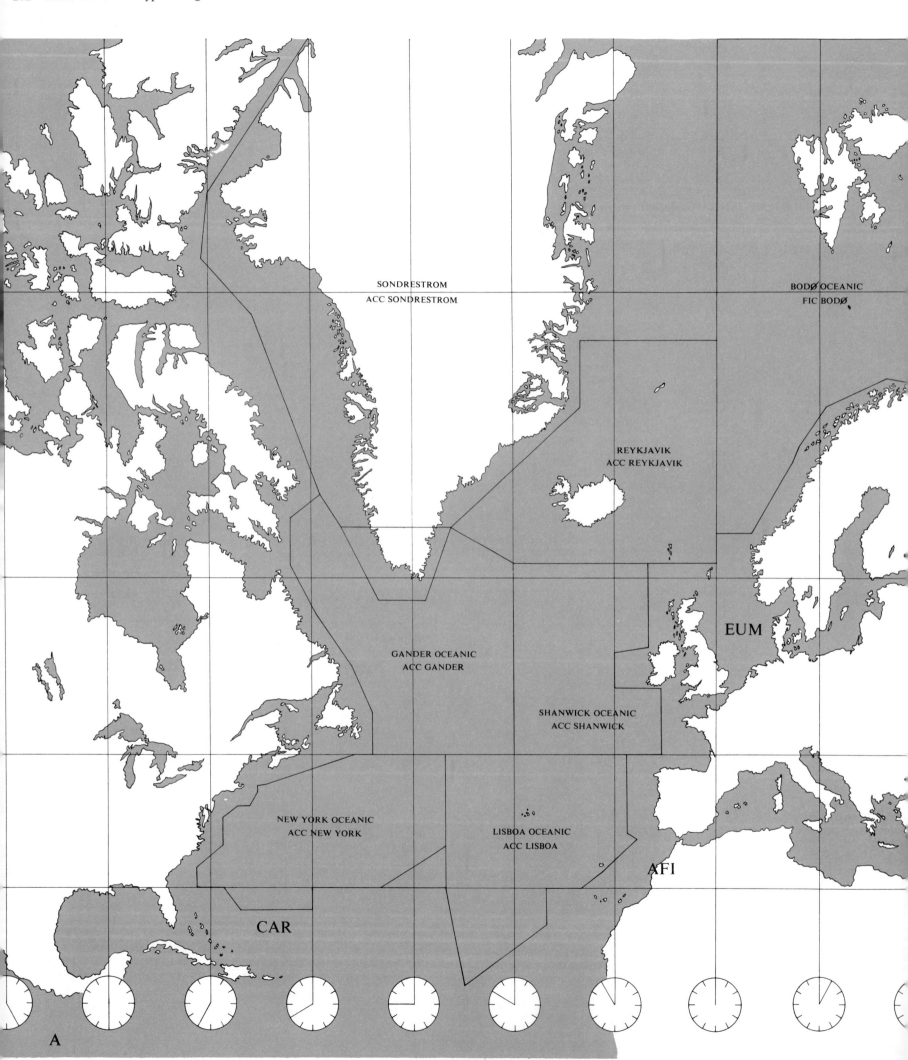

SONDRESTROM
ACC SONDRESTROM

BODØ OCEANIC
FIC BODØ

REYKJAVIK
ACC REYKJAVIK

EUM

GANDER OCEANIC
ACC GANDER

SHANWICK OCEANIC
ACC SHANWICK

NEW YORK OCEANIC
ACC NEW YORK

LISBOA OCEANIC
ACC LISBOA

AFI

CAR

A

tor is reset to the approach mode, when the aircraft turns from the VOR radial to intercept the ILS (Instrument Landing System).

The ILS provides two radio beams. One, the localizer, is rather like a VOR radial, and is usually aligned exactly with the runway centre-line. The other is a glide-slope, which will bring the aircraft over the runway threshold at 50 feet. By following the two beams simultaneously, at the right speed and with correct flap settings, the pilot can reach the runway ready for landing.

Now the aircraft is aligned on the localizer, flying level. As soon as the autopilot is intercepting the glide-slope, the Captain calls briskly, "Landing check!" Undercarriage and flap indications are checked, and as the power is reduced, the aircraft begins to fly down the glide-slope beam. A soft rain is falling, and the windshield wipers swing to and fro. The cloud that enshrouds the flight deck is bright, lit by the lamps below, but no individual lights can be seen yet.

The altimeters read 500 feet, next 400, and then a voice is radioed: "Tania 887, you are cleared to land. After landing, vacate the runway to the right and contact Ground Control on . . ."

This is a tense moment. The aircraft three miles ahead has landed without difficulty, but the sudden rain may have caused the cloud base to sink. If the Captain cannot see the approach lights when he reaches a prescribed height, he must initiate a go-around: open the throttles, reduce flap, bring up the undercarriage, and climb away following a pre-arranged procedure. If, as anticipated, the lights come into view when the aircraft is a little below 300 feet, the Captain must be ready to change from instrumental to visual flying. But he must not do so too abruptly, as there are sometimes further banks of cloud beneath the main base.

Just after the altimeters read 300 feet, the approach lights appear. Their array might bewilder an amateur, but each light has its purpose. Still some distance ahead are the runway lights, marking the edges and centre-line in red, to indicate the threshold. On each side of the threshold are VASIs (Visual Approach Slope Indicators), a set of lights which confirm that the aircraft is on the correct approach slope. Bright lights are flashing in a repeated sequence toward the threshold, while a pattern of approach lights is helping the Captain with guidance as to roll and azimuth.

There is some crosswind, so the aircraft is "crabbing" slightly. The Captain disengages the automatic pilot, which has been doing the flying until now. In his mind's eye, he surveys the power setting, the aircraft speed, and the rate of descent. As the touchdown point rushes toward him at nearly 200 feet per second, he satisfies himself that he is really flying on the centre-line, the runway ahead is clear, the landing lights are illuminated and the landing flaps are fully extended. Then a gentle backward pressure on the elevator controls, a straightening action with the

rudder—and the main wheels start rolling on the runway.

Spoilers extend automatically on the wings, to help decrease aerodynamic lift and to make the wheel brakes more effective. The engines are put into reverse thrust. As the speed drops, the Captain prepares to use the nosewheel's steering wheel, ready for the long taxi to the disembarkation gate. Another routine, uneventful flight has been completed.

"After-landing check," calls the Captain. Easing himself slightly in his seat, he smiles at his copilot. "Where shall we eat tonight?"

AIR TRAFFIC CONTROL OVER THE NORTH ATLANTIC

Before World War II, transatlantic flights were few, but the routes were rapidly developing. The first Oceanic Control Centres were set up at Gander (Newfoundland) and Foynes (Ireland). The Royal Air Force formed an overseas Movement Control at Gloucester, and its Atlantic section moved to Prestwick (Scotland).

In May 1946 the RAF and the British Ministry of Civil Aviation decided to coordinate at an Oceanic Area Control Centre. Similar centres arose at Reykjavik (Iceland), New York, Goose Bay (Canada), Santa Maria (Azores), Shannon (Ireland), Søndre Strømfjord (Greenland), Stavanger and Bodø (Norway), and Casablanca (Morocco). There emerged the finest example of international cooperation yet seen in air traffic control, under the aegis of the North Atlantic Region of ICAO (International Civil Aviation Organization). Prestwick and Shannon collaborated with a service covering the area known as Shanwick.

Lack of navigation aids, and often a lack of communication due to sunspot activity, made it essential to apply separation standards. In the early postwar era, most flights were carried out at night in order to reach destinations in North America during the early morning, as well as to utilize astronavigation and the better high-frequency radio transmission at night. This was consistent with the long time taken by flights—but when jet airliners arrived, companies sought better utilization of the aircraft and tried to complete the round trip between Europe and North America in 24 hours.

Civil aircraft were also rapidly catching up with military planes in performance, and the airspace had to be shared. As navigation aids improved, the need for so many centres decreased, and Stavanger and Goose Bay closed, as did Søndre Strømfjord for a short period. Prestwick retained control of the Shanwick area, where communications continue to be provided by the Ballygirreen radio station of the Department of Transport and Power of the Republic of Ireland. The Azores Centre was renamed Santa Maria (Azores).

The separation standards agreed upon were neces-

Flight information regions: The world's airspace has been divided up by international agreement into Flight Information Regions, which may or may not take account of national sovereignty. Each FIR normally has one Control Centre which is responsible for providing a service to aircraft in flight, and for passing information to adjacent Centres. In some parts of the world this service is very rudimentary, but over the North Atlantic all commercial aircraft are given a positive control service by the responsible Oceanic Area Control Centre.

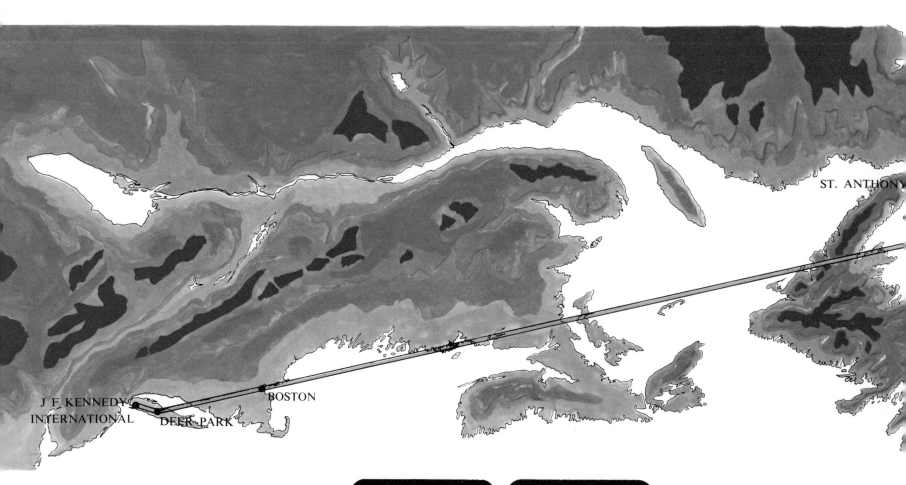

J F KENNEDY
INTERNATIONAL
DEER PARK
BOSTON
ST. ANTHONY

B

C

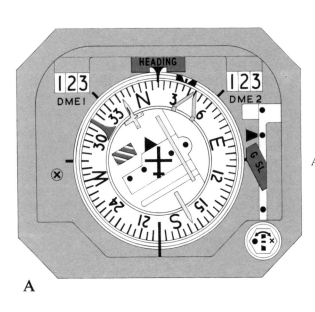

A

A *The flight director: Although the pilot uses his gyro-magnetic compass as a heading reference, this is usually developed into a flight director into which additional information is fed, so that the pilot is presented with "command" information, telling him simply to turn right or left (or climb or descend). The illustration is of a flight director which also incorporates a display of information from DME (distance measuring equipment) which shows that he is 123 miles from the DME station or stations selected.*

B *The electronic Horizontal Situation Display helps the pilot with navigation and to get in the right position for a landing approach. In principle, it gives an overview of the ground from above, with the aircraft marked by a triangle symbol. The aircraft's heading is indicated in digits above the expanded compass scale and the track (heading ± wind) is marked by the line pointing upward from the aircraft symbol. The bar to the upper left of the aircraft is the approach line, and the circles indicate offline to the approach cernterline, which must be superimposed under the aircraft symbol until the plane con turn into the correct ILS approach direction. When the plane is properly lined up, all the symbols have formed a straight line as shown on figure C. The scale to the right indicates the glide path, and below the frequency of the selected radio navigation beacon is indicated.*

D *"Stacked"-aircraft destined for busy airports such as J.F. Kennedy at New York are sometimes instructed by Air Traffic Control to circl over nearby radio facilities. Whilst thus orbiting, they can be cleared in sequence to descend in the holding stack, and eventually directed towards the final approach path.*

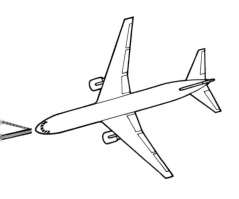

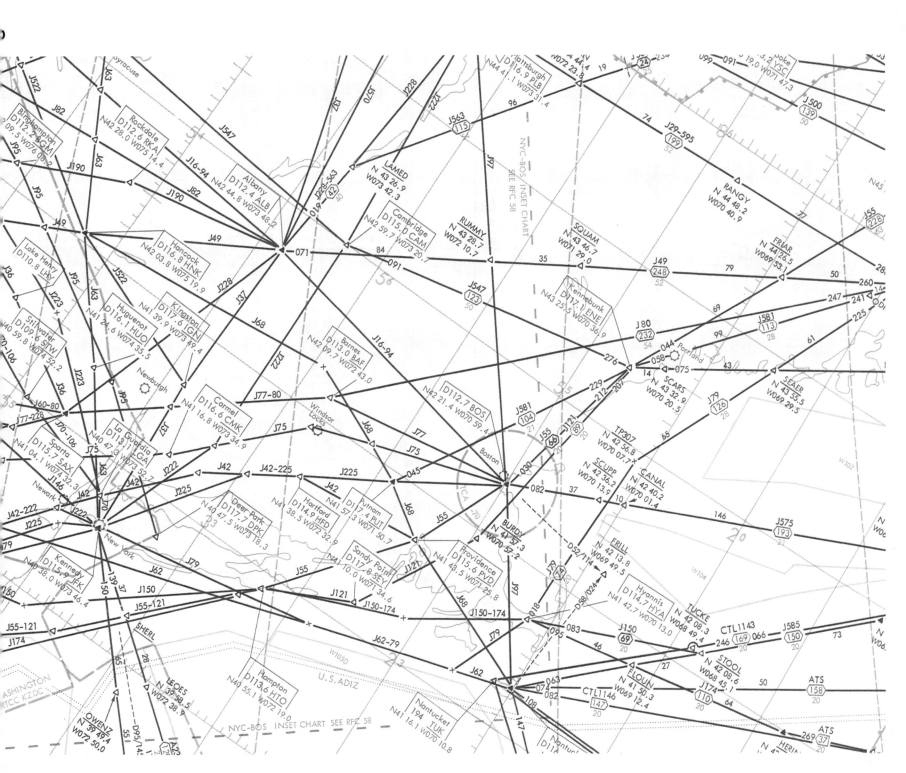

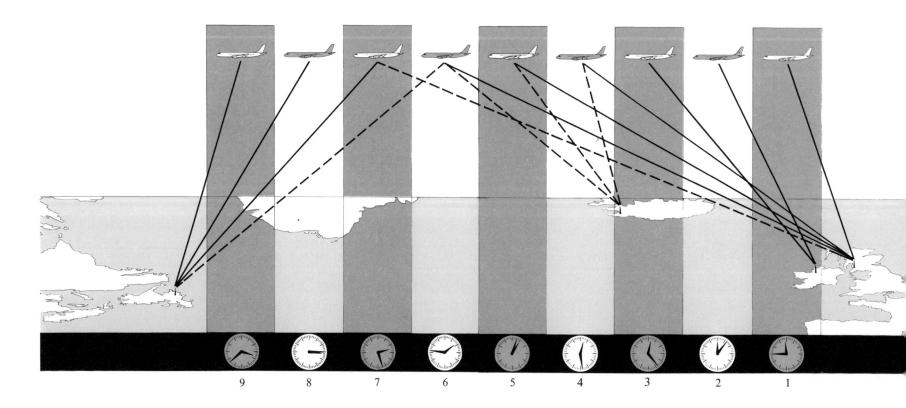

SOME TYPICAL R/T CALLS BY A FLIGHT FROM PRESTWICK TO NEW YORK.
(Not the same flight as that described on pages 258-276).

(start at the extreme right on the facing page, at 1145z—11.45 a.m. Greenwich time.)

9. *1538z*
Tania 887 this is Gander Radio Contact Gander Radar on frequency . . .
1538z
Gander Radar this is Tania 887 over Scad at 1538z Flight Level 310 estimating Cartwright at 1552z. Squawking Alpha 51.
Tania 887 Gander Radar Roger. You are identified position . . . Over.

8. *1515z*
Gander Radio Tania 887 over 54 North 50 West at 1515z Flight Level 310 estimating Scad at 1538z Cartwright at 1552z Over.
Gander reads back.

7. *1427z*
Gander Radio Tania 887 over 55 North 40 West at 1427z Flight Level 310 estimating 54 North 50 West at 1515z. Temperature minus 36. Spot wind at 40 West 210 degrees 15 knots.
Gander reads back. Shanwick Roger.

6. *1346z*
Shanwick Tania 887 over 56 North 30 West at 1346z Flight Level 310 estimating 55 North 40 West at 1427z Over.
Shanwick reads back. Gander Roger. Reykjavik Roger.

5. *1303z*
Shanwick Tania 887 over 5658 North 20 West at 1303z Flight Level 310 estimating 56 North 30 West at 1346z. Temperature minus 38. Spot wind at 20 West 270 degrees 40 knots Over.
Shanwick reads back. Reykjavik Roger.

4. *1229z*
Shanwick Radio this is Tania 887 over 56 North 10 West at 1229z Flight Level 310 Estimating 57 North 20 West at 1302z Over.
Shanwick reads back. Reykjavik Roger.

3. *1224z*
Ulster Radar Tania 887 Level at Flight Level 310. Over.
Tania 887. Ulster Radar Roger. Take the time now. Your position is 5556 North 09 West. Over.
Ulster Radar Tania 887 Roger.
Tania 887 this is Ulster

Radar Squawk Alpha 51 and continue with Shanwick Radio on H/F Route frequencies.
Roger Ulster what are the frequencies?
887 the Primary 5626.5 kcs the Secondary 8913.5 kcs Over.
887 Roger Out.

2. *1206z*
Ulster Radar this is Tania 887 over Skipness at 1206z passing Flight Level 120 for Flight Level 240 estimating 56 North 10 West at 1229z Coding Alpha 24.
Tania 887 Ulster Radar Roger. You are identified. Continue your climb to Flight Level 310 Over.

1. *1145z*
Shanwick this is Tania 887 estimating off Prestwick at 1200z. Request Oceanic Clearance. Over.
Tania 887 Shanwick Standby.
1150z
Tania 887 Shanwick with Oceanic Clearance Over.
Shanwick Tania 887 Go ahead.
Tania 887 Shanwick clears you from 56 North 10 West to New

York via Track Bravo Flight level 310. Mach .81. Code Alpha 51 Read back.
Roger. Tania 887 is cleared via 56 North 10 West, 57 North 20 West, 56 North 30 West, 55 North 40 West, 54 North 50 West Scad Cartwright Airways to New York to maintain Flight Level 310. Mach .81 Code Alpha 51 Over.
Tania 887 Shanwick that is correct. Out.

arily restrictive. It has been said that one aircraft required the airspace equivalent to one pigeon flying alone in London's Trafalgar Square. Originally the separations were 1,000 feet vertically from opposite-direction traffic, 30 minutes longitudinally and 120 nautical miles laterally. Today these rules still apply to a degree, but the separations are only about half as great in the areas of densest traffic.

Jet aircraft have posed problems. They now carry all long-distance traffic, and most of those on the North Atlantic usually fly across that ocean twice daily. But their take-off times are restricted by noise limitations and congestion at most international airports. The requirements of the travelling public further limit the timing of flights. Aircraft must await the arrival of internal "feeder" services, but must not depart too late for the business executive on a tight schedule, or arrive too late to connect with domestic services.

Aircraft weight and temperature limit the ideal flight levels for the crossing. The absence or presence of high winds aloft makes certain routes more attractive than others, restricting the choice of routes. Company flight planning results in large concentrations of aircraft that want similar routes and flight levels during a very short time. Fortunately, the westbound and eastbound aircraft plan routes so far apart that the centres can agree on a "datum line" between the two flows of traffic, and permit the use of all flight levels north or south of the datum line, thus in effect doubling the number of usable flight levels. The main flows of traffic occur at different times of the 24-hour day.

The rules for separation of airliners on the North Atlantic have changed considerably during the 1970s and 1980s, with the advent of new long-range radar stations covering much of the region—and with the introduction of new navigation aids onboard the aircraft, providing greater precision and reliability than the means previously used to determine the position and the progress of the flight.

Foremost among these long-range navigation aids is the INS (Inertial Navigation System), followed by Omega and Loran, as we have already seen. Being managed by the pilots, they render the navigator redundant, and their high accuracy has permitted a reduction in the separation criteria over the North Atlantic, thus increasing the capacity of the available airspace.

To achieve this, a MNPS (Minimum Navigation Performance Specification) area has been established over the North Atlantic—from 27 degrees North to the North Pole, between flight levels 275 and 410. Only aircraft equipped with approved high-precision long-range navigation systems are allowed to enter this area, so that separations can be reduced. Here the aircraft are routed to within 60 nautical miles laterally of each other, while the vertical separation is 2,000 feet (or 1,000 feet below flight level 290). The minimum time difference between two consecutive aircraft is normally 15 minutes.

Exceptionally, the time difference may be reduced to as little as ten, five, or even three minutes, provided that certain other criteria are fulfilled. The first aircraft must have a considerably higher cruising speed than the following aircraft, or the two must have divergent tracks after a common reporting point.

Aircraft that do not meet the standards of MNPS are kept below flight level 275, above level 410, or south of 27 degrees North. Their separation is according to older rules, meaning 120 nautical miles of lateral separation and 20 minutes longitudinally. The rules for vertical separation are identical.

How does the controller ensure this separation? Each morning for westbound flights, or each evening for eastbound, the companies file their flight plans with the centres that handle clearances. All their routes are plotted on a chart. The planning officer computes his own minimum flight path for the day, using the 300-millibar chart. A compromise is then reached between the company tracks, his results, and the possible flow of opposite-direction traffic. The centres confer and agree about tomorrow's track system.

Usually, the most popular track is the central one of three, which lie two degrees of latitude (120 nautical miles) apart at each 10 degrees of longitude—but one degree of latitude within MNPS airspace. These are published for all the centres, airports, radar units and operators concerned. The airlines then refile on one of the published tracks, stating a preferred flight level and/or track.

The time between designated reporting points is calculated for each flight level with the Mach number likely to be used. Separation is based on these times, and is updated enroute by actual position reports received from the aircraft.

Some fifteen minutes before the Oceanic boundary, aircraft make their calls and receive their Oceanic clearance, without which no plane is permitted to enter the Oceanic area. These clearances are relayed to adjacent Air Traffic Control units, so that the aircraft can enter the Oceanic area at the required point and level. This means close liaison between ATC Centres at Gander, Reykjavik, Bodø, New York, Shannon, Paris and Madrid, according to the flight's origin or destination. The enroute controller is responsible for ensuring that separation is maintained, and for liaison with adjacent centres to coordinate the levels, times and positions.

Forecasts of expected standards of High Frequency communications are issued, and the frequency bands used are varied occasionally according to the time of day or season. To avoid the long periods of listening on HF, which was common in the infancy of long-range flight, the SELCAL system is used—enabling ground operators to call up a selected aircraft. Moreover, in areas with radar coverage, secondary radar is employed: every aircraft is given a four-figure individual code, which is displayed on the controller's radar screen along with the aircraft's actual flight level.

Further improvements will be concerned with communications and navigation systems. The static-filled HF pilot-to-controller channel, which is the only long-distance means of exchanging spoken language today, is expected to be replaced by the kind of VHF static-free frequency band that is used everywhere over land when suitable. In addition, communications satellites have been placed in appropriate geosynchronous orbits over the equator. Navigation accuracy will be refined with GPS (Global Positioning System), whose information comes from satellites. A computer system developed in Sweden, enhancing the accuracy so much that satellite-based information can even be used for precision instrument approaches, shows great promise for the future.

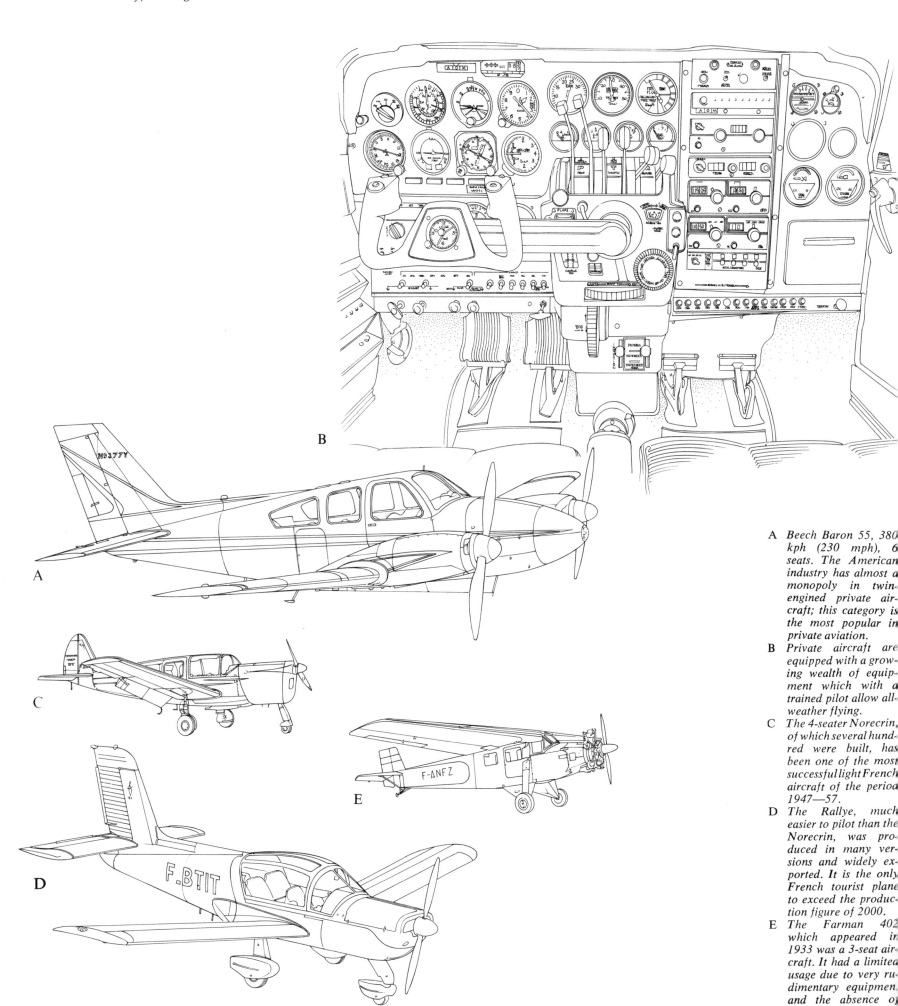

A   *Beech Baron 55, 380 kph (230 mph), 6 seats. The American industry has almost a monopoly in twin-engined private aircraft; this category is the most popular in private aviation.*

B   *Private aircraft are equipped with a growing wealth of equipment which with a trained pilot allow all-weather flying.*

C   *The 4-seater Norecrin, of which several hundred were built, has been one of the most successful light French aircraft of the period 1947—57.*

D   *The Rallye, much easier to pilot than the Norecrin, was produced in many versions and widely exported. It is the only French tourist plane to exceed the production figure of 2000.*

E   *The Farman 402 which appeared in 1933 was a 3-seat aircraft. It had a limited usage due to very rudimentary equipment and the absence of radio.*

FLYING A SMALL AIRCRAFT

In these days of swing-wing supersonics, jumbo jets and airline passengers by the million, it is not generally realised that the great majority of aircraft are small and simple machines. For example, there are over 100,000 privately-owned small aeroplanes in the United States, where they outnumber airliners about a hundred to one.

These "general aviation" aircraft, as they are known, with their piston-engines, fixed landing gears and unpressurized cabins, are a far cry from some of the more sophisticated aircraft featured elsewhere in this work. Their relative simplicity is nowhere more apparent than in the pilot's compartment. Not only does this contain a fraction of the levers and dials found on larger aircraft, but they are smartly styled in the manner of a well-designed, enthusiast's motor car.

Light aircraft, however, share one important feature with their larger counterparts—their flying control systems are fundamentally similar. All types of aeroplane, except for a few unorthodox and research aircraft, are controlled in the air by movable surfaces on the wings and tail. These surfaces are operated by a central control column (or handwheel) and rudder bar, and govern the attitude and actions of the aircraft when airborne.

On big fast aircraft the control surfaces are operated by a complex system of cables, rods, compensators, gearing units, autopilot servo-motors and hydraulic boosters, often duplicated or even triplicated. By contrast, on light aeroplanes the surfaces are operated simply by cables running round pulleys.

Ailerons, on the wings and moving in opposite directions, push one wing up and the other down, causing the aircraft to roll. The elevator, hinged to the rear of the tailplane, pushed the tail up or down causing the aircraft to dive or climb. The rudder, mounted vertically and hinged to the fin, controls the aircraft in yaw—i.e. in rotation about its vertical axis.

These flying control surfaces all work in the same manner—by deflecting the air passing over them to produce a force in the required direction. Varying combinations of the three basic forces on the ailerons, elevator and rudder, together with different settings of the throttle, cause the aircraft to execute its full range of flying manoeuvres.

In order to function, the control surfaces require air to flow over them, and this requirement is met as an aircraft begins to accelerate down the runway for take-off.

Before this can happen, of course, the engine has to be started. On modern light aircraft this operation is little more complicated than starting the engine of a motor car. It is preceded by a simple—but important—series of checks. These are somewhat similar to the checks one makes (or should make) before starting a motor car, such as ensuring that the gear lever is in the neutral position and the brakes on. Before attempting to take off, the pilot should also have filed a flight plan—that is, told someone in authority of his destination and proposed route—and he will also have obtained a weather report. Let us now trace the whole sequence of a flight in a modern light plane with a nose-wheel undercarriage (take-off and landing procedures are slightly different when a tail-wheel landing gear is fitted).

At the take-off point, the engine is run-up to ensure that it is functioning satisfactorily, and a drill of "vital actions" carried out. This drill varies from aircraft to aircraft, but includes actions such as adjusting the elevator and rudder trim, tightening the throttle-friction nut, setting the fuel mixture to rich, checking that the fuel is on and sufficient for the flight, checking the oil pressure, and ensuring that the doors are securely closed and the safety harness done up.

If all is well, and if the pilot is operating from an airport, he can proceed to take-off after obtaining clearance from the control tower.

Lining the aircraft into the wind, the pilot opens the throttle fully and releases the wheel brakes. As the aircraft gains speed, the airflow over the wings starts to generate lift. At the point where the lift almost equals the weight, the pilot eases back the control column. This moves the elevators up, which deflects the air, causing a down load on the tail. This load rotates the nose of the aircraft up thus increasing the angle at which the wing meets the air so that the lift exceeds the weight. The aircraft takes off and starts to climb.

The steeper angle of the wing, in addition to developing higher lift, also generates more drag, making it harder for the propeller to pull the aircraft through the air. So, as soon as the aircraft is airborne, it is levelled off momentarily, to reduce drag

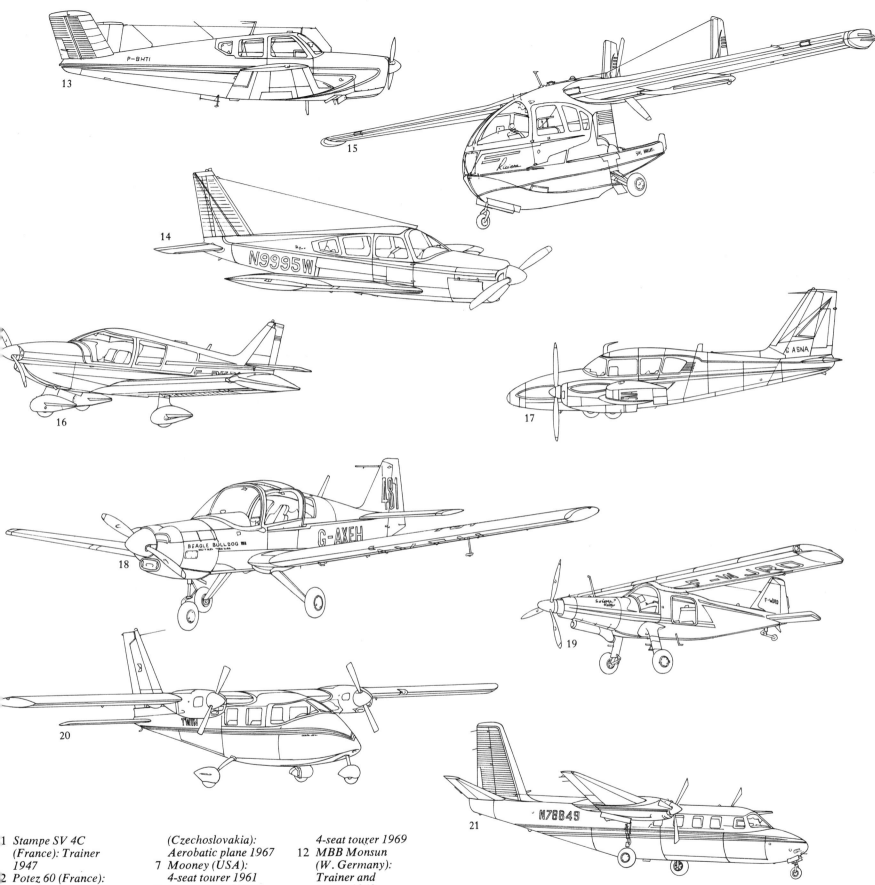

1 *Stampe SV 4C (France): Trainer 1947*

2 *Potez 60 (France): 2-seat tourer 1934*

3 *Piper PA 18 Super Cub (USA): Tourer and observer 1955*

4 *Jodel D.117 (France): 2-seat tourer 1955*

5 *SAAB 91 C Safir (Sweden): Trainer 1960*

5 *Zlin Trener 526 F*

(Czechoslovakia): Aerobatic plane 1967

7 *Mooney (USA): 4-seat tourer 1961*

8 *Cessna 150 (USA): Trainer and tourer 1958*

9 *Fuji Fa.200 Aero Subaru (Japan): 4-seat tourer 1969*

10 *Fournier RF3 (France): Glider plane*

11 *Aerospatiale ST 10 Diplomate (France):*

4-seat tourer 1969

12 *MBB Monsun (W. Germany): Trainer and tourer 1969*

13 *Beech P.35 Bonanza (USA): 4-seat tourer 1963*

14 *Cherokee Piper PA 28 Arrow (USA): 4-seat tourer 1965*

15 *SIAI-Marchetti FN 333 Riviera (Italy): Amphibious 4-seat*

tourer 1963

16 *Robin HR 100-200 B (France): 4-seat tourer 1971*

17 *Piper Aztec PA 23 (USA): Twin-engined tourer 1962*

18 *Scottish Aviation*

Bulldog (Beagle) (Britain): Trainer 1969

19 *Dornier Do 27 Astazon II (W. Germany): Ferry plane 1961*

20 *Partenavia P.68*

Victor (Italy): Twin-engined tourer 1970

21 *Aero Commander Grand Commander 680 F/L (USA): Business and transport plane 1962*

and enable the speed to build up for the climb. On aircraft so fitted, the undercarriage and flaps will be retracted at this stage.

A slightly different take-off technique enjoys wide popularity. In this, the aircraft is held on the ground, if necessary by slight forward movement of the control column, until its speed is such that the lift exceeds the weight. The stick is then eased back and the aircraft leaves the ground and starts to climb. There is no momentary levelling off, and the speed is allowed to increase to the recommended climb speed as quickly as possible.

When the desired climbing speed is attained, the engine power is reduced and the trim adjusted for the climb. The trimmers are small hinged flaps on the trailing-edges of the elevator and rudder, which can be moved to provide a load on the surface, to deflect it. This relieves the pilot of the tiring job of continuously pulling on the control column or pushing the rudder bar, the result being that the aircraft virtually flies itself.

Once the aircraft has climbed to its cruising altitude, the pilot levels off and retards the throttle to obtain the desired cruising speed. While the aircraft is in steady level flight all the forces working on it are in balance. That is, the thrust equals the drag, and the lift equals the weight. This surprises some people, who think that the thrust must be greater than the drag for the aircraft to move along, and the lift greater than the weight for the wings to sustain it in the air. In fact, if the thrust is greater than the drag the aircraft will gain speed, and if the lift is greater than the weight it will start to climb.

The constant matching of thrust and drag, and lift and weight, is not easy. Slight variations in engine power, or in the attitude of the aircraft caused by turbulence, or slight movements of the controls, will upset the balance. To keep in level flight, the pilot has to co-ordinate carefully both the angle of attack and thrust.

For example, suppose the thrust increases slightly —i.e., the engine speeds up momentarily. The speed will increase, and so will the lift. With the lift now greater than the weight, the aircraft will start to gain height.

To return to level flight, the pilot can either close his throttle slightly to regain the original degree of thrust, or he can push the nose down, to reduce the angle of attack and hence the lift. In the latter case, the conditions of flight will have changed. Though now in level flight, the aircraft will be flying faster. Level flight can thus be sustained at varying speeds. At low speed, the aircraft nose will be high and the angle of attack close to the maximum. At maximum speed, the nose will be down and the angle of attack small to keep down the lift generated. Throughout this entire speed range, the lift remains equal to the weight, in spite of the changing attitude of the wing. The attitude itself is dependent upon the engine power.

While in level flight the pilot also has to keep the wings level. Turbulence, or slight differences in lift,

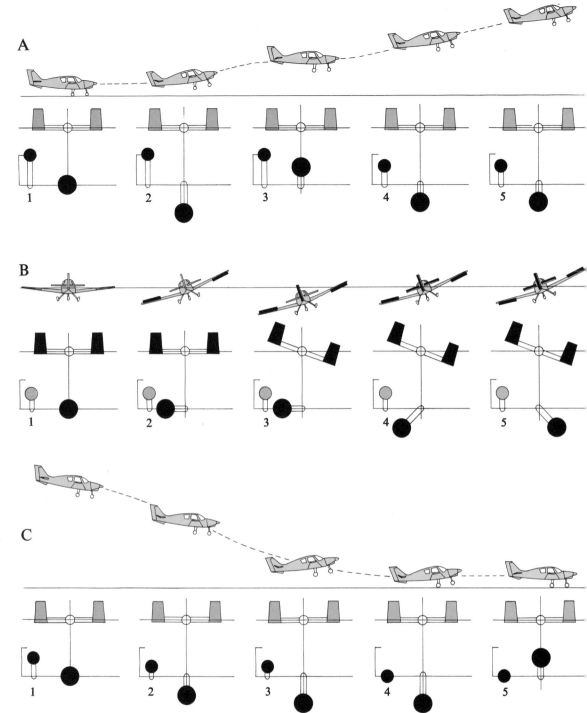

● THROTTLE   ● STICK   ⊏⊐ RUDDER BAR

A *Taking off.*
1 *Stick and rudder bar central. Throttle opened fully. Nose kept aligned on distant object by rudder.*
2 *Stick pulled gently back. Aircraft leaves ground.*
3 *Stick eased forward momentarily to enable speed to pick up.*
4 *Stick eased backward for the climb. Engine throttled back to the climb setting.*
5 *Elevator trim set for*

the climb.
WARNING. *In the unlikely event of engine failure, the pilot must ease the stick forward and glide straight ahead into the wind, turning off the ignition and the fuel supply. He must* never *try and turn back into the airfield. Violation of this rule is disastrous. With insufficient height to make a complete turn, one instinctively*

pulls the stick back to maintain height. This results in a stalled turn and a dive, or spin, into the ground.

B *Turning.*
1 *Before turning, the pilot looks round to ensure that there are no other aircraft in the vicinity.*
2 *The stick is pushed to the left, to put the aircraft into a bank.*
3 *Almost simultaneously left rudder is applied.*

The aircraft turns to the left, but the nose begins to drop.
4 *To prevent the nose dropping, the stick is eased backwards. In a turn the bank tends to increase on its own, as the outer wing is travelling faster than the inner and thus generates more lift.*
5 *To prevent the bank increasing, it is kept constant by pushing the stick to the right.*

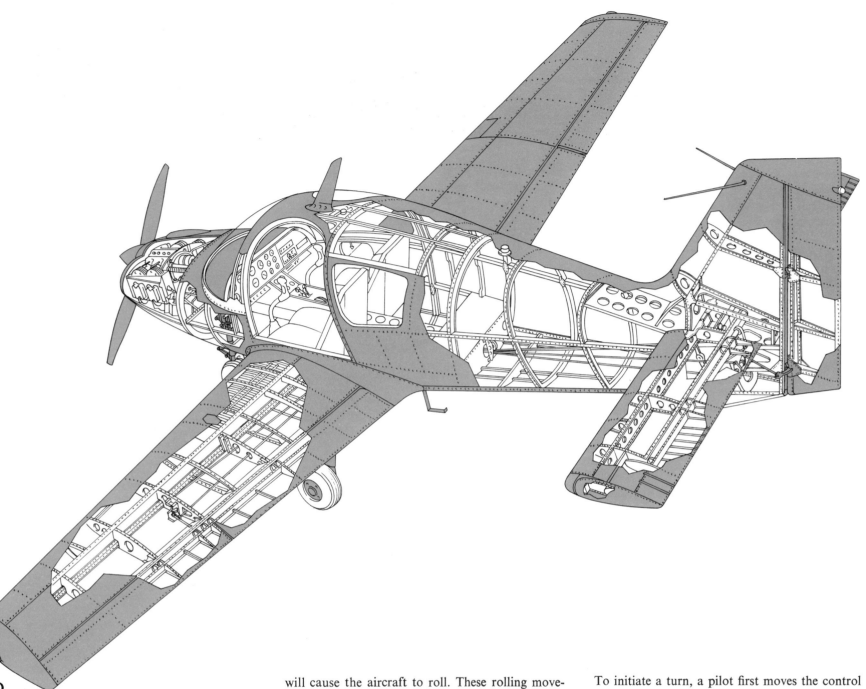

D

Landing

1 *The throttle is closed slightly and the aircraft put into a shallow dive.*

2 *Before turning on final approach power is reduced further and the stick pulled back to obtain the proper approach speed.*

3 *Just above the runway the stick is pulled back; the aircraft "flares out" and floats momentarily just above the ground.*

4 *Touch-down. The throttle is closed for the landing run.*

5 *The nose drops and the aircraft rolls along on its tricycle gear. Slight forward movement of the stick will hasten contact of the nosewheel. While taxying, the flaps are retracted.*

D *Cutaway drawing of Beagle Pup.*

will cause the aircraft to roll. These rolling movements are counteracted by operation of the ailerons. If the left wing drops, the control column is moved to the right. This deflects the left aileron down, generating an upward force, and the right aileron up, generating a push downward, to produce a righting moment. In practice, aeroplanes usually have a degree of inherent stability built into the basic design so that they tend to correct automatically any slight deviations from a straight flight path. The fin at the rear gives the aircraft an arrow-like stability and helps to keep it straight. Lateral stability can be obtained by giving the wings dihedral, or mounting the wing on top of the fuselage to give pendulum stability.

In flight the pilot will also want to turn. This manoeuvre is not executed by simply turning the rudder, as is the case with a ship, but by a combination of aileron, rudder and elevator movement, and adjustment of the engine power.

To initiate a turn, a pilot first moves the control column sideways to deflect the ailerons and cause the aircraft to roll into a banked attitude. With the wings tilted away from the horizontal, the wing lift now provides both a vertical force and an inward force. The vertical force must still equal the weight, and to ensure this the control column is pulled back. This moves the elevator up and places the wings at a higher angle of attack so that more lift is developed. The inward force starts to pull the aircraft into a turn. However, as soon as the aircraft starts to roll, it also starts to yaw away from the turn, because the drag of the down-going aileron on the rising wing is greater than that of the upgoing aileron and tends to retard it.

Some aircraft are compensated for this effect, but on most light aircraft the tendency to yaw must be counteracted by deflecting the rudder towards the turn, to provide the necessary force. Thus, in making a turn, the aircraft is rolled and turned in the same direction with the help of the rudder, so that a balanced change of direction is made without either inward slipping or outward skidding.

One other thing is necessary in a turn. The engine power must be increased to overcome the higher drag resulting from the increased angle of attack, and thus maintain flying speed. As the angle of bank is increased, the turn becomes tighter and the elevator progressively becomes the dominating turning control (the rudder, now nearly horizontal, assumes the role of elevator). To resume level flight, opposite rudder and bank are applied and, when the aircraft is level, the control column and rudder bar are centralised.

While flying the pilot will need to ensure that he is on his correct course; this means that he will have to navigate. To help him, a variety of navigational procedures are available, the four principal methods being map-reading, dead-reckoning, astro-navigation and the use of radio aids. These are used separately or together. For short journeys in daylight, map reading is the answer, and for this purpose a variety of special maps exist which show those features which are most readily identified from the air.

Dead reckoning involves marking the track to the destination on a map, and then calculating the course to be steered in the air to achieve this track, making due allowance for magnetic variations, wind speed and direction. Various major landmarks along the track are selected and the length of time between these is calculated and marked on the map. Many light aircraft can now be equipped with a simple autopilot which holds the aircraft on to a given heading. This relieves the pilot of the task of actually controlling the aircraft in level flight, and leaves him free to concentrate more fully on navigation.

Light aircraft today make considerable use of the complex radio and radar facilities installed for larger aircraft. The equipment which makes this possible has been miniaturized and made suitable for all but the smallest and cheapest of aircraft.

By means of radio the pilot can obtain navigational bearings, listen to weather information, and obtain landing and take-off instructions, which improves flight safety. If necessary, with the help of the radio he can make a controlled descent through cloud and in bad weather may even be "talked" right down to the runway, if the necessary radar is installed at the airfield.

At this point it should be emphasised that these comments apply to light aircraft flying outside the airways—those invisible "motorways" of the air used by airliners. In these busy airways the navigation of airliners is carried out under the strict instructions of air traffic controllers on the ground. To help the pilots of light aircraft, a certain number of "free lanes" are reserved down to ground level through controlled areas, allowing them free access to and from small airfields without interfering with aircraft under direct control.

A

B

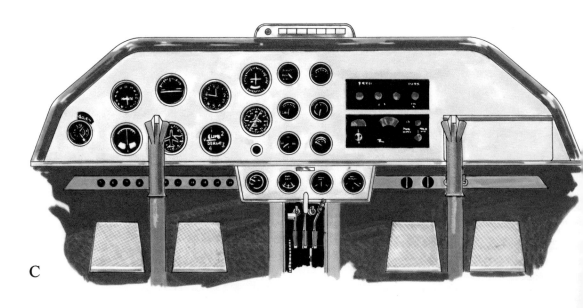

C

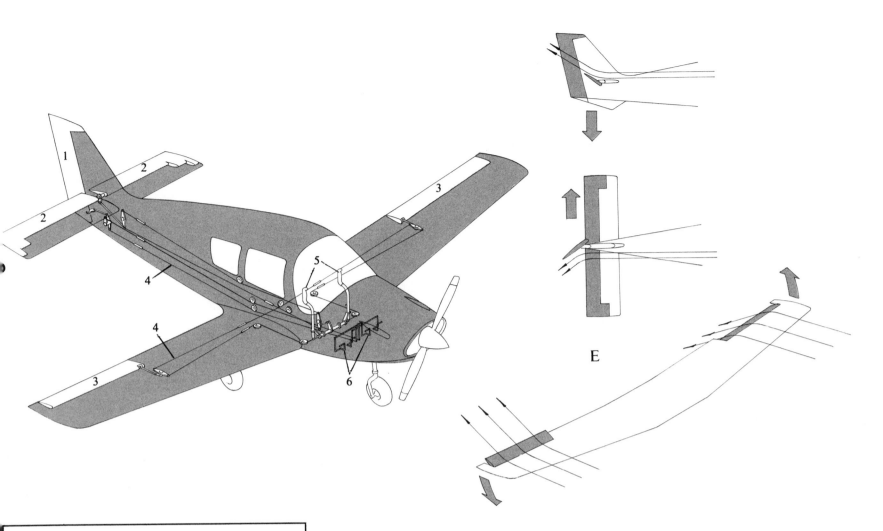

E

F

A *Beagle Pup.*

B *Pilot's Certificate to Fly.*

C *The cockpit of the Beagle Pup is typical of the neat layouts found in modern light aircraft.*

D *Flying controls. The rudder, elevator and ailerons are operated from dual control columns through conventional cable and pulley circuits. The elevator trim-tab is moved by a cable-operated screwjack. An electric motor, mounted under the rear seat, operates the flaps through a series of push-rods.*

1 Rudder
2 Elevators
3 Ailerons
4 Cables
5 Dual control columns (joysticks)
6 Dual sets of rudder pedals

E *Like their larger counterparts, light aircraft are manoeuvred in the air by co-ordinated movements of the ailerons, rudder and elevators. Each surface works in the same manner—air is deflected to generate a force in the opposite direction. The ailerons move differentially (i.e., in opposite directions), causing one wing to rise and the other to drop.*

F *Check list. These are the things the pilot must check and do before taking off.*

However, if a pilot has the necessary training and his aircraft carries the required equipment, he can fly on the airways system and make full use of all the control facilities just like an airliner. He can then fly at night or in weather too poor for normal map-reading.

Near the destination, the pilot can put his maps away, and call the control tower by radio for permission to land. When this has been given, the pilot turns down wind, parallel to the runway, and carries out a cockpit check to ensure all is correct, if necessary lowering the undercarriage and extending the flaps. He then turns across wind and later turns again so that he is in line with the runway, on the final approach. Power is reduced, and the control column pulled back to increase the angle of attack and maintain the lift. Near the ground the control column is eased back further and the aircraft levels out. As speed drops off, the control column is pulled back still further and, with the wings about to stall, the aircraft touches smoothly down on its main wheels. As speed is reduced, the nose drops and the pilot raises the flaps and taxies to the parking area. Putting the brakes on, he switches off the engine.

Most of the time, the task of flying a light aeroplane is easier than driving a car, less strenuous than riding a horse and requires less skill than fishing for trout. It does, however, require constant alertness and any lapses of concentration can be serious.

1 *Boeing 747–400 (USA) 412-seat four-jet airliner.*
2 *Ilyushin Il-62 (USSR): 186-seat four-jet airliner*
3 *Boeing 767 (USA) 250-seat twin-jet airliner.*
4 *BAC/Sud-Aviation Concorde (UK/France): 136-seat four-jet supersonic airliner*
5 *McDonnell Douglas MD-11 (USA) 320-seat four-jet airliner.*
6 *British Aeropace BAe 146 (UK) 85-seat four-jet airliner/transport.*
7 *Fokker 100 (Netherlands) 100-seat twin-jet airliner.*
8 *Boeing 737–300 (USA) 140-seat twin-jet airliner.*
9 *Fokker F.27 Srs 200 Friendship (Netherlands): 52-seat twin-turboprop airliner*
10 *Aerospatiale-Aeritalia ATR 72 (France & Italy) 70-seat twin-turboprop regional airliner.*
11 *Shorts 360 (UK) 36-seat twin-turboprop regional airliner*
12 *Britten-Norman BN-2 Islander (UK): 10-seat twin-engined light transport*
13 *De Havilland DHC-6 Twin Otter (Canada): 20-seat twin-turboprop light transport*
14 *British Aerospace ATP (UK) 64-seat twin-turboprop airliner.*
15 *LET 410 UVP-E (Czechoslovakia) 19-seat twin-turboprop regional transport*
16 *Dassault Fan Jet Falcon (France): 14-seat twin-jet business transport*
17 *SAAB 340 (Sweden) 35-seat twin-turboprop regional airliner.*

5

7

6

8

13

14

17

15

16

25

23

24

26

31

34

35

32

33

Some of these liveries were changed recently. See pp. 356—359 for current company symbols.

18 Learjet 31 (USA) 10-seat twin-jet business transport.
19 Aero Commander (USA): 11-seat twin-engined business transport
20 Beechcraft Queen Air B80 (USA): 8-seat twin-engined business

transport
21 Piper PA-46 Malibou (USA) 6-seat business aircraft.
22 Cessna 337 Super Skymaster (USA): 6-seat twin-engined (tandem) business aircraft
23 Piper Comanche

(USA): 4/6-seat light business aircraft
24 Pilatus Turbo-Porter (Switzerland): 11-seat single-turboprop utility transport
25 Morane-Saulnier Rallye Commodore (France): 4-seat light aircraft

26 Piper PA-18 Super Cub (USA) 2-seat light aircraft
27 PZL-104 Wilga 3P (Poland): 4-seat utility aircraft
28 Piper Cherokee 180 (USA): 4-seat light aircraft
29 Cessna 172 (USA):

4-seat light aircraft
30 SAAB/MFI 15/17 (Sweden) 2/3-seat utility aircraft (Lincence manufacture in Pakistan)
31 Canadair Cl-215 (Canada): Twin-engined water bomber for fire-fighting duties

32 Piper Pawnee 235 (USA): Single-seat agricultural sprayplane
33 Sikorsky S-61N (USA): 29-seat amphibious helicopter airliner
34 Mil Mi-8 (USSR): 28-seat helicopter airliner
35 Bell Jet Ranger (USA): 5-seat light helicopter

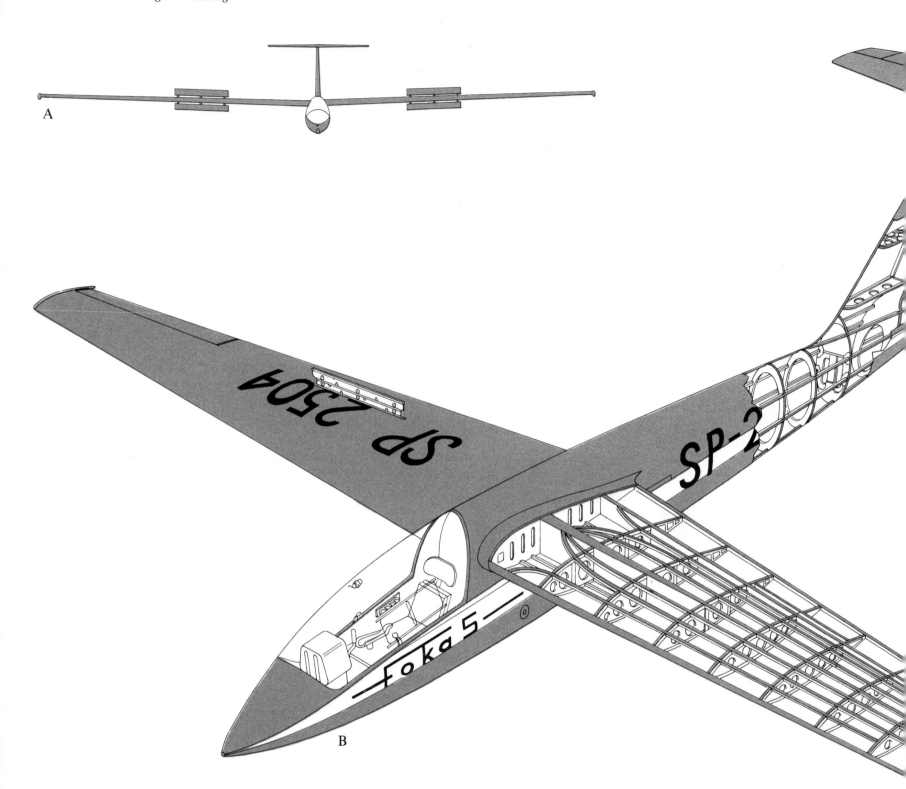

A

B

Foka 5

SP 2504

SP-2

A *Airbrakes. Consisting of flat plates extending above and below the wing, these destroy lift and create drag, steepening the angle of glide. They are used primarily during landing.*

B *Typical of today's high performance sailplanes is the Foka 5 (Seal) produced by the Experimental Glider Establishment in Poland. The single-seat Standard Class craft is a development of earlier Fokas which have achieved notable successes in several World Championships. Major improvements of the Foka 5 include a roomier cockpit, and*

the tailplane is mounted on top of the fin to prevent damage when landing in long grass. The best glide ratio is 34 to 1. The craft is of all wood construction. The wing is a multi-longeron spar-less structure and the fuselage is a semi-monocoque structure of oval section. The landing gear comprises a fixed mono-wheel, to

comply with Standard Class requirements. The cockpit contains a fully-reclining seat, with adjustable back rest and rudder pedals. Blind flying instrumentation is fitted, and provision is made for radio and oxygen equipment.

1657   1658   1659   1660

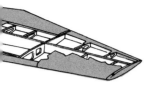

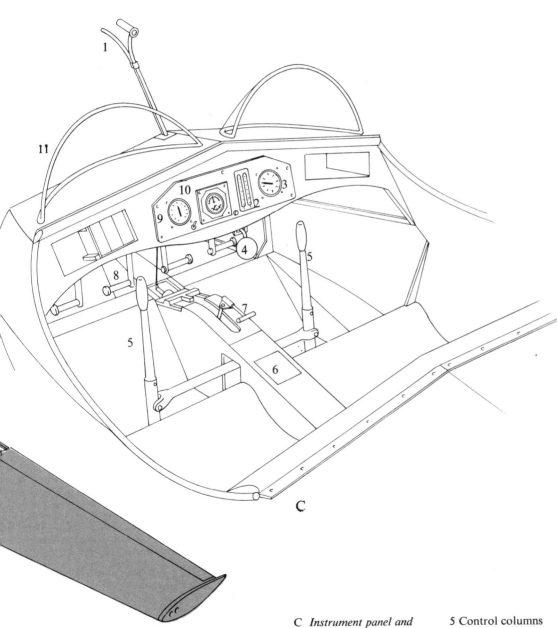

C *Instrument panel and controls of a dual control training glider (the Slingsby T-21).*
1 Pitot head and venturi assembly
2 Cobb-Slater variometer
3 Altimeter
4 Cable release control
5 Control columns
6 Flying limitations data card
7 Spoiler (airbrakes) control
8 Rudder pedals
9 Air speed indicator
10 Turn and slip indicator
11 Wind shield

Gliding and soaring have never been more popular than they are today. This is understandable because, in an age when powered flying is a costly pastime, they provide an exciting way of gaining air experience and a knowledge of aeronautics at comparatively low cost. Even youth groups can afford the simpler forms of gliding, while the more advanced sport of soaring provides both a challenge for the dedicated enthusiast and relaxation for professional pilots who earn their living by flying huge civil airliners or complex military aircraft.

Basically, gliding is a sport, with few restrictions, licences or inspections. It is somewhat akin to the sport of sailing in the world of seaborne vessels; in fact, the two pastimes have much in common.

Being unpowered, a glider needs some assistance to launch it into the air. The dominating methods today are powered-aircraft tow and winch launch. Car towing and bungee-catapult launch are methods that, for practical reasons, have gone out of fashion. So too with winch towing, it seemed some years ago; but especially in central Europe, the winch has again become a popular way to get gliders into the sky. Early winches consisted of old cars with their rear wheels jacked up, one wheel being replaced by a drum. When the engine was put in gear and its clutch released, the drum spun round, winding up the cable in the process. Today, special winches equipped with as many as eight drums are available, permitting many planes to be launched in quick succession. Modern winches embody a guillotine-like device for cutting the cable in an emergency, and a cage in which the operator sits. The cage protects him in case the cable breaks or must be cut, and whips back over the winch.

The other widely used method of launching is aero-towing. Virtually any light aircraft of sufficient power can be used, and this method has the big advantage of being able to take the glider up to far greater heights than is possible with ground launches. In addition, the glider can be towed to height-gaining thermals before release. The drawback is that it is much more expensive than other methods, but it is possible to take two or even three gliders on tow, mainly for cross-country tranportation. The minimum engine power for a towplane must be around 150 hp, and Piper Super Cubs have been the real workhorse of this trade. Nowadays, converted agricultural planes like the 235-hp Piper Pawnee have proved to be very suitable for glide towing.

To ensure the safety of the gliders, launching cables are fitted with a short length of cable, known as

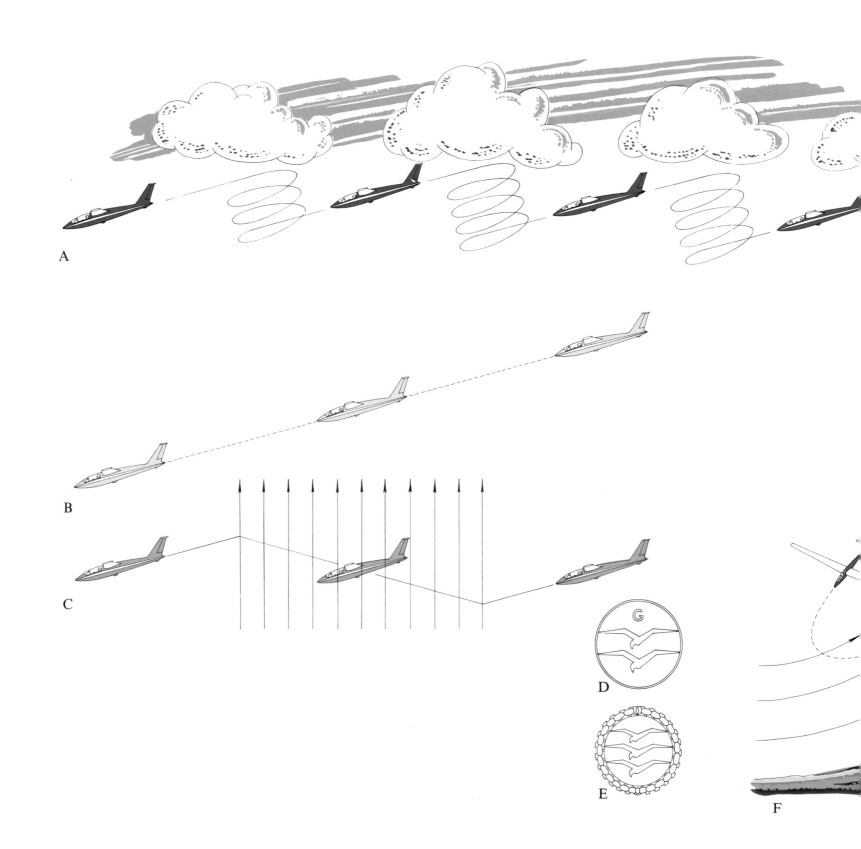

A

B

C

D

E

F

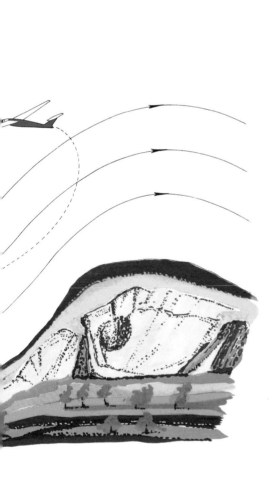

the weak link, which breaks before the glider can be damaged if something goes wrong. The cables also have a small parachute at the end, to lower them gently to the ground after release, and also facilitate their location on the ground.

A flight in a glider is both thrilling and relaxing, while the sensation of soaring has to be experienced to be appreciated fully. For people used to powered flight the initial difference is pleasant—and terrifying.

The difference becomes apparent as soon as you approach the take-off area. Almost certainly you will hear voices from overhead as the pilots in twoseaters talk to each other. This would, of course, be impossible if they were flying powered aircraft.

Before take-off a few simple, but vital, rules are observed. The harness is checked and the airbrakes extended and retracted to ensure that they function correctly and retract flush. The quick-release under the nose of the glider is examined for wear and the ring on the end of the towing cable is then hooked into position. A runner holds one wingtip to keep the wings level. After glancing to ensure that it is "clear above and behind", a signal to "take up slack" is flashed to the winch operator. Slowly the cable pulls through the grass as it tightens up. Soon it is taut and upon the signal "all out" the winch is put into top gear.

The glider gathers speed quickly and in a few seconds is airborne and climbing rapidly. The steepness of the climb is, perhaps, the characteristic which most surprises the pilots of powered aircraft. Their first climb terrifies them, as at that angle a powered craft would stall and crash. Not so the glider which, pulled on by the cable, climbs steadily. However, glider pilots must resist the temptation to climb too quickly, in case the cable breaks and the nose has to be put down quickly to prevent stalling. At about 1,000 ft the craft is at the top of the launch and the cable is released. If by any chance the pilot should forget to release the cable, or should his release jam, the release is so designed that it will automatically throw off the cable as soon as the glider starts to overfly the winch.

Free of the hook the glider flies on like a giant bird. While flying, of course, it loses height; in fact a glider is always diving slightly relative to the air. If a glider gains height, this means that it is in a volume of air, known as a thermal, moving upward at a greater rate than the glider's own rate of sink.

Because of the gliders' excellent aerodynamic form and fine finish, they can cover surprisingly long distances, even in still air. For example, a high-efficiency sailplane, in calm weather, could glide about 15 miles from a height of only 2,500 ft. Gliding distances are normally extended by the pilots' finding thermals in which height can be gained. Long distances can be covered by gliding from one thermal to another. Cross-country trips of hundreds of miles have been made in this manner, the record being over 1,000 miles.

To help the pilot locate thermals and upward-moving air, gliders are fitted with a variometer, a sensitive instrument which indicates the vertical speed of the air around the plane. The most modern types, called total-energy or compensated variometers, measure only the true vertical outside airspeed, which compensates for the plane's own vertical manoeuvres.

When circling, the variometer may indicate gains in height during only part of each successive circuit. This means that the glider is half in and half out of a thermal. By manoeuvring the glider, the centre of the thermal can be located and the maximum increase in height gained.

In normal gliding flight the controls act conventionally, as on powered aircraft. Gliding is utterly relaxing, the smooth whistle of the air flowing past providing a musical backdrop. This air noise is a distinctive feature of gliding. It is very susceptible to change, and experienced pilots can tell precisely what their glider is doing from the note of the airflow.

When coming in to land, another difference between gliding and powered flight is apparent. When on the final approach the airfield boundary can be crossed at a height of a hundred feet or more. In a powered aircraft, this would almost certainly result in an overshoot, and the pilot would have to go round again. In a glider, by extending the air-brakes—flat plates which rise above and below a section of the wing, destroying lift and creating drag—the glide angle can be steepened dramatically, resulting in a rapid loss of height. A glider, of course, cannot go round again; but in the event of an undershoot the glide-path can be extended by retracting the airbrakes.

Nearing the ground, the pilot checks the descent and continues to ease the control column back to keep the glider just off the ground while the speed dies away. Owing to the drastic effect of the airbrakes, this happens quickly and soon the glider is on the ground. Most craft have only a single wheel, and the ailerons are used to keep the wings level as long as possible during the short landing run, so that when one wingtip finally touches the ground, the speed is as low as possible.

Glider pilots can earn a number of badges, indicating their degree of proficiency. The first certificate is the "C", requiring a soaring flight of over five minutes above the point of cable release. After a great deal more experience, the pilot can then attempt to obtain the International Silver "C" Certificate. This necessitates a flight of 32 miles, a climb of 3,200 ft and a flight of five hours. A pilot does not have to do all three on one flight; they may be achieved at intervals, on separate occasions.

The next certificate, the Gold "C", is far more difficult to get. For this a pilot has to cover a distance of nearly 200 miles and a climb of over 10,000 ft.

A really top pilot can add Diamonds to his Gold "C"—by gliding 300 miles to a designated destination and by climbing at least 16,000 ft.

A popular misconception regarding gliders is that they are slow. While most of them do fly at, say, 40 to 50 mph—their low speed is half their attraction—many gliders can dive safely at speeds of up to 150 mph. Nor are gliders limited when it comes to aerobatics. Properly-designed sailplanes can execute most of the aerobatic manoeuvres of their powered counterparts, including the bunt, or outside loop. An outside loop begins with the aircraft in level flight; then it dives, flies upside down, and climbs back to level flight. It is one of the most strenuous manoeuvres to which an aircraft can be subjected.

Sailplanes can also reach surprisingly high altitudes—far above the ceiling of most propeller-driven craft. The absolute official world record is held by Robert Harris, USA. On February 17, 1986, he reached 45,531 ft. But already in February 1961 a height of 46,267 ft was reached by Paul F. Bikle, Director of the NASA Flight Research Center. Taking off from Lancaster, in the rim of the Mojave Desert in California, Bikle was towed to a height of 10,000 ft and then released. For more than an hour he manoeuvred his high-performance sailplane, specially equipped with an oxygen system, into rising thermals, climbing higher and higher, until he had reached almost 9 miles above the Earth. He could possibly have gone even higher but, as he was not wearing a pressure suit, he could remain above 40,000 ft for only five minutes.

The gliders of today stand for the most efficient aerodynamic shape of all designs in the aviation world. An aircraft in the open class, whose only limitation is max. 750 kg, can soar a distance of more than one mile while losing only 100 ft of height in calm air. Theoretically the most extreme gliders would be aerodynamically able to fly at supersonic speed, but structural design does not permit speeds exceeding 200 mph. Plastic composite material was first used in the construction of gliders, which gave the designers much more freedom to create aerodynamically very efficient shapes. Also, it meant lower structural weight and higher strength. The sleek wings of a modern glider have a span of 75 ft, but it doesn't weigh more than can be lifted by one person.

A new trend in gliding is the motorized glider, which can come off the ground without any outside help, and can reach the home airfield even if the height loss has been too much to allow gliding back. The engine is mounted on a folding pylon aft of the pilot's seat, and when not in use it is covered by shaped panel-doors that do not affect the aerodynamics of the plane. The performance of the motorized glider is equal, or just slightly inferior, to a "clean" gliding plane with a glide-to-sink ratio of 42:1 for a 50-ft-span plane. A 43-hp engine adds about 50 kg to the weight, but enables the plane to take off after 500 ft of ground roll.

In short, gliding provides a relaxation, a knowledge of the weather, adventure and a camaraderie seldom associated with other forms of flying.

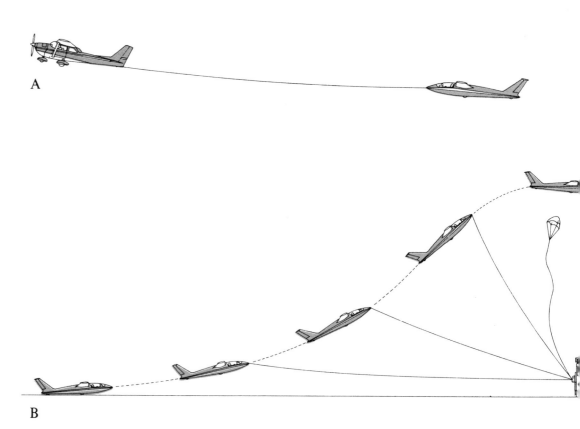

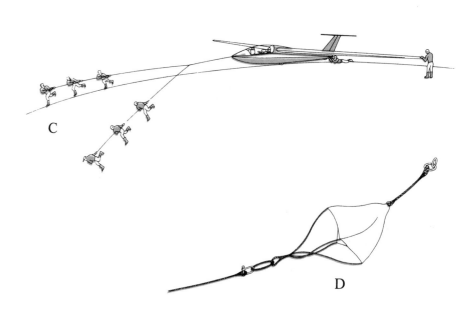

A *Launching by air-towing. This has the advantage that the glider can be released at great heights and can be towed into a thermal.*

B *Launching a glider by winch. The cable is released just before the craft overflies the truck. Car towing is similar, the car replacing the winch.*

C *Diagram of a bungee cord launch. When this method is used, the launch is usually from a hill-top.*

D *Launching cable wea link. This is designe to break before any excessive loads can imposed on the glide*

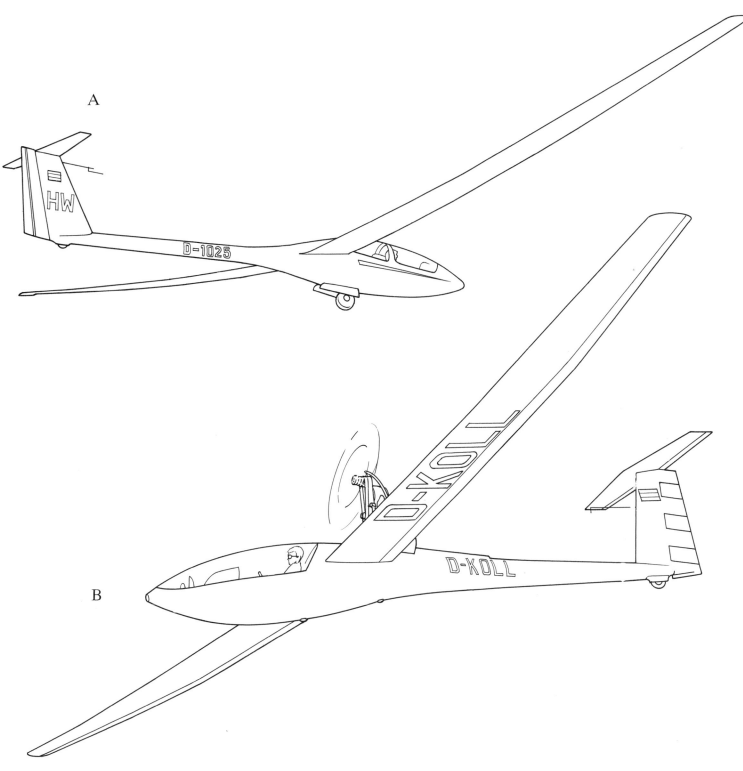

A *The queen among gliders, the ASH-25 is the most advanced high-performance glider with a glide to sink ratio of 58:1. The twoseater has a span of 75 ft and is built in composite material, mostly glassfibre epoxy and carbonfibre/epoxy. To increase speed, watertanks in the wings can be filled with 180 litres. Flaps along the*

trailing edge of the wing can be used in different settings fron 6 degrees negative to 38 degrees positive, to achieve the best performance depending on the flying conditions. ASW-25 is equipped with a computer that gives the pilot continuous information about the most effective speed related to thermal winds.

B *The DG-400 is a high-performance glider equipped with a two-stroke 43-hp engine for take-off and climb, which makes the plane independent of outside assistance to get in the air. The engine can be tilted back into the fuselage and covered by clamshell doors. With the engine stowed, the DG-400 behaves exactly like a glider.*

ENCYCLOPÆDIC INDEX

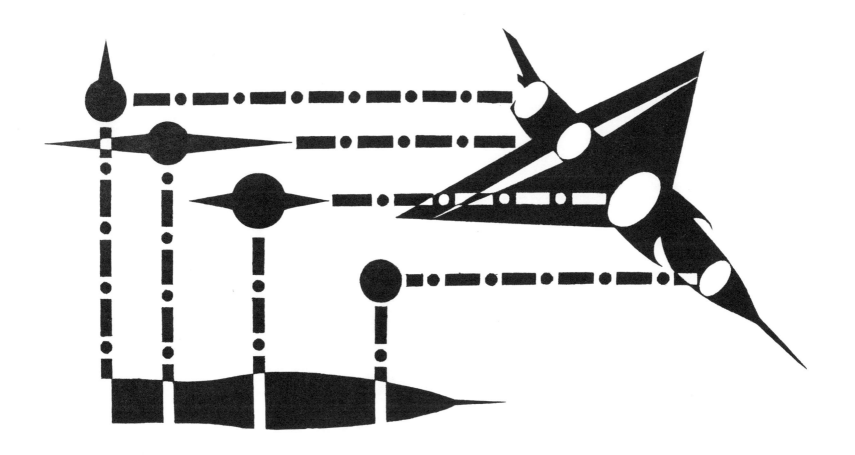

ENCYCLOPAEDIC INDEX

This section of the book is more than just an ordinary Index. Its contents have been written and arranged so that it forms an important and valuable complement to the main part of the book, comprising more than 1,550 entries, totalling over 120,000 words, and many unique illustrations.

Thus, as well as forming a comprehensive index to all important references in the main text, it also provides (for example) much additional information about the most important pioneering figures in the development of aviation, and a glossary of the principal aerodynamic and structural terms and their applications.

Hundreds of additional entries have been included to provide salient data about the world's leading airframe and aero-engine manufacturers and their products; the principal air forces and their equip-ment; more than 100 of the world's major airlines, their equipment and principal routes; and over 40 of the world's most important airports. A unique feature is the inclusion of chronologies of the development of aviation in the leading countries of the world, in many cases illustrated with photographs and drawings never before published.

USING THE INDEX:

The figures appearing immediately after a heading in the Encyclopaedic Index refer to the *column* numbers in the main sections of the book where reference to that subject can be found. Additional cross-references to the main book are indicated by the words "see text".

Where the abbreviation "*q.v.*" is used, this indicates a cross-reference between one heading and another within the Index itself.

A-5 Vigilante, see North American

A 101G, see Agusta

Ader, Clément, 181, 184

The Eole of Clément Ader

After being involved in developing the telephone, this French engineer turned his attention in the eighteen-seventies to heavier-than-air flight. His *Eole* —which, it is claimed, hopped off the ground in 1890—was a bat-winged steam-powered aeroplane. He later developed the *Avion III*, but this did not accomplish so much as the *Eole*.

Aerial Steam Carriage, see Henson

Aerial survey

The production of maps from aerial photographs (photogrammetry) has become an indispensable part of modern surveying. Its applications are wide, the uses of such data being of inestimable value in such fields as town planning, civil engineering, communications, agriculture and geology. Frequently, aerial survey can provide such data for regions where making a surface survey would be arduous or even impossible.

The practice of aerial photography is as old as flying itself; the first aerial pictures were taken from balloons, and the aeroplane was used for photographic as well as visual reconnaissance before the 1914-18 War. Serious commercial application of the art really got under way in the nineteen-twenties, as a result of pioneers like Britain's Sydney Cotton, but the real impetus was provided by World War Two with its tremendous strides forward, not only in photographic equipment but in the highly-specialized art of interpreting aerial photographs. Probably the best-known illustration of the value of such

activity was the discovery of the V1 flying-bomb centre at Peenemünde by this means.

A typical aerial survey today involves the making of a "mosaic", of perhaps hundreds of vertical photographs, which can be virtually as accurate as a fully-detailed map of a large area of terrain. This may serve as a quick and comparatively cheap alternative to plotting and charting an unknown area, or, with the aid of oblique photographs of the same area, may be used to interpret (say) areas of forest or moving herds of wild animals. Terrain relief may point the way to the best route for a new road or railway, or the disposition of geological formations leading to new archaeological discoveries.

Nor is the camera the only weapon in the aerial surveyor's armoury, especially when mineral prospecting from the air is undertaken. Instruments such as magnetometers and scintillation counters have been specially developed to measure rock structures or to detect radio-active mineral deposits.

Aer Lingus—Irish International Airlines

Aer Lingus Teoranta operates services from the Irish Republic to points in the UK and Europe. Its first services were flown in May 1936, between Dublin and Bristol, with a de Havilland Dragon biplane. Transatlantic routes are

Aerial survey: a vertical by Fairey Air Surveys of Kariba Dam, Rhodesia

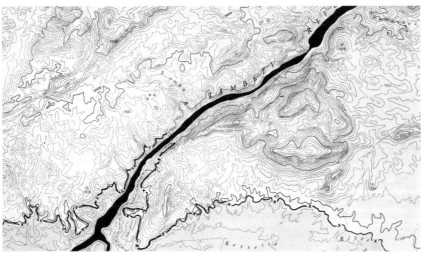

The corresponding map produced by automatic photogrammetry and stereo plotting

flown by its sister airline Aerlinte Eireann (formed after World War Two). The two airlines have separate membership of IATA, but operate together under the joint title of Aer Lingus-Irish International.

Headquarters: Dublin.

Aermacchi M.B.326, 430-433

This aircraft was developed by Aeronautica Macchi SpA in 1957 as a tandem two-seat jet trainer for all-through jet training. It became a standard trainer for the Italian Air Force, entering service in 1962, and has also been sold to the air forces of Tunisia, Australia, Ghana and South Africa (where it is known as the Impala), and the Argentinian Navy. It is licence-built in South Africa and Australia. The Italian airline, Alitalia, has used

it for jet training. Ground-attack versions have also been built, and the latest such version, the M.B. 326GB, uses an uprated Viper 20 engine of 3,410 lb st.

Aero

One of the three major aircraft manufacturing companies in Czechoslovakia prior to the outbreak of World War Two was Aero Tovarna Letadel Dr Kabes.

Founded in 1919, its first product was a copy of the wartime Austrian Phönix (Brandenburg) two-seater. The company's own designs included the A 11 reconnaissance and light bomber aircraft of 1923, which set up several international records; the A 18 fighter; and the A 30, A 100, A 101 and A 304 reconnaissance-bombers. In the late nineteen-thirties Aero also built under licence the French MB 200 twin-engined bomber.

Aero O/Y, see Finnair

Aeroelastic deflection, 445

Aeroflot, 312

Aeroflot, the State-controlled domestic and international airline of the USSR, was formed in 1923 under the name of Dobrolet. In 1932, after absorbing other

ambulance, fishery protection, survey and rescue work and crop spraying in addition to its primary function as a passenger and freight carrier. Although not a member of IATA, Aeroflot is by far the world's largest airline, currently carrying well over 6 million passengers a year. It is responsible for all airport services, training and navigation facilities in the USSR, and a special division known as Avia-Artika exists to carry out operations inside the Arctic Circle. In 1956 Aeroflot became the second airline in the world to operate scheduled services by jet, when it introduced the Tu-104 into service. Recent equipment includes the four-turboprop Tu-114, the four-jet Il-62 and the twin-jet Tu-134. A vast internal network of routes is flown and international services operate to Africa, Cuba, Japan, UK and USA. Aeroflot will have in service, within the next few years, the Tu-144 supersonic airliner.

Headquarters: Moscow.

Aerofoil, see Flight, principles of

Aerolineas Argentinas

Formed by the Argentine Government in 1949 to combine the activities of several smaller operators, Aero-

Mi-4P helicopter of Aeroflot leaving Moscow central post office

pioneer Soviet operators, it assumed its present name, and today is responsible for

lineas Argentinas has an extensive network of routes within Argentina, and to other South

American countries, the US and Europe. The 6,400-mile route between Buenos Aires and Madrid is among the longest in the world.
Headquarters: Buenos Aires.
Member of IATA.

Aerolineas Peruanas

Began operations in Summer of 1957, offering low-price services from Peru to the US. Currently provides internal and regional passenger and freight services in South America and jet services to the US.
Headquarters: Lima.
Member of IATA.

Aero-medical services

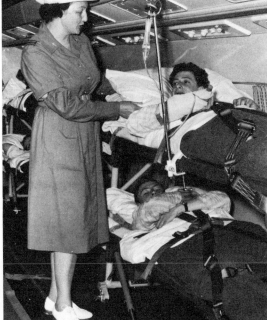

Aero-medical services: a flight sister of Princess Mary's RAF Nursing Service

Not to be confused with aviation medicine (*q.v.*), which is essentially a preventive activity, aero-medical services are those that provide assistance, with the aid of aeroplanes, for the sick or injured.
The best-known such service is the Royal Flying Doctor Service, which originated in Australia in 1928 using one small, single-engined D.H.50 which had no radio; today, with aircraft developed specially to operate in remote areas, it has regional divisions serving every state in Australia. Patients in isolated homesteads can call up the doctor by radio link, secure in the knowledge that his flying ambulance can bring him practically to their doorstep within a few hours if necessary. Similar organizations have grown up in recent years in other parts of the world where the population is scattered widely over vast areas—notably in several African states and in the large island groups of the south-west Pacific.
On the military side, aero-medical services are provided as a matter of course for treating casualties of war and, where necessary, evacuating them from forward areas to a convenient base hospital.
In both the civil and military fields, helicopters and STOL lightplanes are the major types of aircraft used for such work.

Aeronaves de Mexico

In addition to an extensive domestic route network, Aeronaves flies international services to the US, Canada and Europe. Operations began in September 1934 with a service linking Mexico City and Acapulco, and after World War Two Aeronaves expanded steadily by absorbing other smaller Mexican operators. In recent years Douglas aircraft have predominated in the fleet.
Headquarters: Mexico City.
Member of IATA.

Aero Spacelines

Aero Spacelines is a cargo-carrying company specializing in the transportation of outsize missile and spacecraft components on behalf of NASA. For this work it has a small fleet of three specially-modified Boeing Stratocruisers known as Mini-Guppy, Pregnant Guppy and Super Guppy.
Headquarters: Santa Barbara, California.

Aerostat

Term used to denote a lighter-than-air craft, e.g. a balloon or airship.

AEW (Airborne Early Warning), 1381

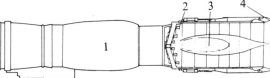

1 Hawkeye aircraft
2 Airborne radar horizon
3 Surface radar horizon
4 Line of sight of airborne radar
5 Line of sight of surface radar

A radar system designed for installation aboard an operational aircraft, which detects the presence of aircraft or ships approaching the patrol area. Signals received by the AEW systems are relayed to surface stations or to other patrolling aircraft equipped with the means of intercepting the intruder. It is generally housed either in a large ventral radome, as on the Fairey Gannet AEW Mk 3 and Douglas Skyraider, or in a large disc-shaped radome mounted above the fuselage on struts, as on the Grumman Tracer and Hawkeye.

Afterburner, 481, 1153-1154, 1171-81, 1243-1244, 1303

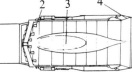

1 Engine
2 Afterburner fuel
3 Reheat flame
4 Variable-area nozzle

A gas-turbine engine jet-pipe specially designed so that additional fuel can be burned in it, to give extra thrust at the expense of a much higher fuel consumption. Afterburning is seldom resorted to except with a pure jet engine (turbojet) or turbofan intended for use at supersonic speeds. In such power plants it is especially desirable to achieve maximum jet velocity and maximum thrust per unit frontal area, and afterburning achieves both. There is ample free oxygen remaining in the hot gases issuing from the engine turbine to support combustion of additional fuel; and such combustion can be effected if the "reheat fuel" is injected as a finely atomized spray upstream of flameholders or baffles designed to cause local turbulence and reduction of flow velocity, so that each droplet of fuel is burned before it passes out of the propelling nozzle. The gas temperature may be allowed to rise to at least 2,000°K (1,727°C) as there is no turbine imposing a limitation on temperature, and the jet is hot enough to be capable of accelerating itself to well beyond the speed of sound in a properly designed nozzle. The nozzle area must be capable of being varied, from a minimum in the "cold" (no reheat) condition up to a maximum with full reheat. In a power plant designed for Mach 2 or above, the nozzle becomes the largest-diameter part of the engine, and at Mach 4 the full nozzle area is more than twice that at the engine intake. An afterburner is much more than a mere pipe; in a large engine it may weigh over 2,000 lb.

Aft-fan engine, 1201, 1213, 1259-60

The aft-fan engine is a turbojet engine with, at the rear, a large-diameter fan which not only is driven by the hot gases passing through the turbines but also moves a mass of cold air ducted around the outside of the engine. It has been found that such a layout has the advantages of reducing noise level and giving a better fuel consumption, although, since it was introduced, the general tendency has been to put the fan in front of the engine.

Agricultural aviation

Large areas of growing crops can be fostered and protected by dusting or spraying them with pesticides and/or fertilizers carried in containers aboard a small aeroplane flying above them and distributing the chemical evenly by means of spray-bars or other forms of dispenser attached to the aircraft. The primary requisite of such an aircraft is, therefore, the ability to fly at a very low height and speed under complete control while dispensing the aerial dressing. Special equipment for this job has been designed for fitting to existing types of aircraft, but specialized types such as the Cessna Agwagon, Grumman Ag-Cat and Piper Pawnee have also been designed specifically for the agricultural rôle.
As might be expected, because of the vast collective farming areas under cultivation, the USSR leads the world in the use of the aeroplane for this purpose, and reportedly has nearly a hundred and fifty million acres under aerial treatment. Second largest acreage under treatment from the air is in the United States (about sixty million acres), and

Agricultural flying: spraying a wide swath

The use of aircraft in the control and regulation of crops has expanded considerably during the past two decades, and, with the continuing deterioration in the ratio between world population and world food production, this application of the aeroplane is likely to assume an even greater importance in years to come.

among countries following the practice on a lesser scale are Canada, Czechoslovakia, Australia and New Zealand.

Agusta A 101G, 967

The A 101G is a three-engined helicopter for heavy duty rôles, able to carry 35 passengers or up to 11,025 lb of freight. It is powered by three Rolls-Royce (Bristol) Gnome shaft-turbines of 1,250 shp each, driving a single, five-blade rotor of 66 ft 11 in diameter. The first A 101G flew in October 1964.

AH-56A Cheyenne, see Lockheed

Airacomet, see Bell

Airacuda, see Bell

Air Afrique

Air Afrique has been operating since 1961. It represents the interests of 12 former French African colonies that are now independent, and serves most countries on the African continent and several points in Western Europe.
Headquarters: Abidjan, Ivory Coast.
Member of IATA.

Air Algérie

The present Air Algérie was formed in 1953, the Algerian Government being the major shareholder. Mixed jet- and piston-engined fleet, operating services in North Africa and to Europe, the Middle East and USSR.
Headquarters: Algiers.
Member of IATA.

Air America

Formed in 1950, Air America provides transportation in support of civil constructional and military operations in south-east Asia, including supply-dropping and search and rescue facilities for the US forces in Vietnam.
Headquarters: Washington, DC.

Airborne Early Warning, see AEW

Air brake 615 ff, 1657

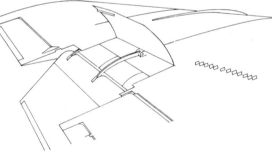

Simple flat-plate type of air brake on the North American Super Sabre.
A device which can be raised into the airflow to create drag to reduce speed. Sometimes known as "speed brakes", air brakes are necessary on

high-speed aircraft to enable them to reduce speed to that required by air traffic control and to increase descent rates. On military aircraft, air brakes assist formation flying and combat manoeuvres.

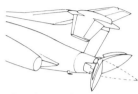

Petal-type air brake on the Hawker Siddeley Buccaneer.
Air brakes usually take the form of hinged flat plates or perforated fingers, arranged as far as possible so that the drag from the brakes when open passes through the aircraft's centre of gravity, to minimize the effect on trim. Various types of parachute are also used for air-braking, and on some aircraft the landing gear can be extended for this purpose.

Air brake-cum-spoiler, used to assist aileron control on the Boeing 707.
On some aircraft the air brakes, mounted in the wings, are used differentially to assist the ailerons for lateral control. Ailerons alone are sufficient for controlling roll at high speeds, but some assistance may be desirable at low speeds, particularly during turbulent conditions when landing

and taking off. It is in crosswind landings and take-offs that the differential action of the air brakes becomes most important, as the ailerons alone may not provide sufficient rolling moment. In such cases, the aileron system is usually arranged so that when the up-going aileron has moved beyond a certain setting the air brake on that wing extends proportionately to supplement aileron movement. When so used, air brakes are sometimes known as "spoilers" as, in addition to creating drag, they are "spoiling" the lift of the wing.

Air Canada

Originating in 1937 as Trans-Canada Air Lines, this company was established as a national passenger and air mail carrier before World War Two. Transatlantic services to the UK were established as part of Canada's war effort, and these were reopened on a commercial basis in 1948. In addition to its domestic routes, Air Canada flies services to the US, Caribbean and Europe.
Headquarters: Montreal.
Member of IATA.

Air Ceylon

State-owned national airline of Ceylon, formed in 1947. Operates services internally and, in association with BOAC, to London and Singapore.
Headquarters: Colombo.
Member of IATA.

Air charter, see Charter flying

Air Congo

National airline of the Congolese Democratic Republic, the Government of which has a majority interest. In-

ternal and international services to other African states and to Europe.
Headquarters: Kinshasa.
Member of IATA.

Air-cooled engine, 1043, 1077

A piston-engine in which excess heat from the cylinders is dissipated directly by air flowing past them. As in the case of most motor-cycles and some cars (notably the Volkswagen), the majority of piston-engines used in current aircraft are air-cooled and have special cylinders whose exterior consists of closely-spaced, deep fins providing a large radiative area. The shape and position of these fins is designed so that cooling is achieved effectively by an airflow guided past the cylinder(s) by a series of deflectors and baffles. The complete engine installation is sometimes so arranged that the cooling air enters through a forward-facing ram intake, picks up heat energy from the engine and is then expelled through a propulsive slit or nozzle to give useful thrust. The airflow is usually under the pilot's control, according to the position of movable hinged gills or flaps.

Aircraft carrier, 217

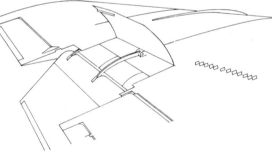

The birth of the aircraft carrier dates from 14 November 1910, when Eugene Ely took off, in a Curtiss biplane developed from the Golden Flyer,

from a platform on the deck of the cruiser USS Birmingham. On 18 January 1911 he made a landing on a similar platform, this time on the USS Pennsylvania, the aircraft being brought to a halt by engaging hooks on the undercarriage in a series of cables held on the deck by sandbags. Similar experiments followed, notably in Britain and Germany, and both of these countries developed ship-borne aircraft for a variety of rôles.
An early variation of aircraft carrier was the seaplane tender, used by both sides in the 1914-18 War. The first ship built from the outset as an aircraft carrier was HMS Furious of the Royal Navy, on which the first British deck-landing trials (with a Sopwith Pup) were carried out in August 1917.
The aircraft carrier came into its own during World War Two, sizeable carrier task forces being operated by Britain, the USA and Japan. Carrier-based aircraft carried out the attack on Pearl Harbor in December 1941, and it was the eventual superiority of the US carrier fleet and its aircraft that turned the tide of the war in the Pacific. Since World War Two the aircraft carrier, with its attributes of mobility and versatility, has fre-

quently been urged as the basis for a more viable defence system than one dependent upon large, fixed bases on land; but, although Australia,

Britain, France, India and Brazil still operate modest numbers of carriers, most of these will have vanished by the early nineteen-seventies. The only sizeable carrier fleet today is that operated by the US Navy.

Aircraft wheels, 527, 529ff, see also Bogie landing gear

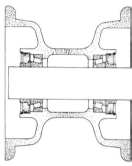

Well-based wheel.
Constructed in one piece, this type of wheel is provided with a deep well for tyre fitting and removal. Although light, it is rarely used now due to the difficulty of tyre removal, particularly on larger sizes with stiff beads. The inside of the rim is usually knurled deeply to assist tyre bead adhesion.

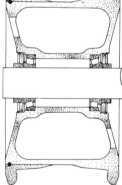

Detachable flange wheel. *On this type of wheel one flange is made readily removable to facilitate tyre fitting and removal. The flange is retained by a circular ring engaging a groove in the wheel. The torque between the loose rim and the hub is taken up by a key on the rim fitting in a slot in the hub. Such wheels suffer from inherent weakness in fatigue of the retaining ring grooves, which*

are subjected to stress concentrations. To overcome this, on some detachable flange wheels the flange is bolted in position.

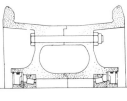

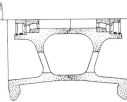

Divided wheel. *Known alternatively as a split wheel, this type is constructed in two halves bolted together to facilitate tyre fitting and removal. Generally used for tail-wheels and nose-wheels.*

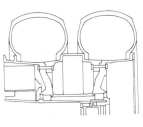

Twin-tyred wheel. *As the name suggests, this is a single wheel fitted with two tyres. Such wheels embody a single pair of bearings and two detachable flanges.*

Air data computer, 705, 1393

AiResearch TPE 331 engine, 1297-1298

Air Forces, see under country of origin

Air France

International French flag carrier, formed in 1933 by the amalgamation of several smaller airlines, some of which had been operating since 1919. Among its early claims to fame was the pioneering of air mail services in and to the South American continent, and mail-carrying is still a prominent part of the airline's extensive activities. By the outbreak of World War Two the fleet numbered nearly 100 aircraft, and following the occupation of France in 1940 Air France continued to operate restricted services from bases in North Africa—a connection which it still retains by a financial interest in several North African airlines.

Air France has the largest fleet of Sud-Aviation Caravelles

At the end of the war Air France was nationalized, and resumed inter-continental operations in 1946. It was among the first customers for the Viscount turboprop airliner, and later became the first, and largest, operator of the rear-engined Caravelle jet transport. A trans-Polar service to Tokyo was opened in 1958, and today Air France's quarter-of-a-million-mile route network extends to all parts of the globe. In the nineteen-seventies it will be one of the first to operate the supersonic Concorde.

Headquarters: Paris. Member of IATA.

Air Guinée

Formed in 1961 to operate internal and regional African services as the national airline of the Republic of Guinea.

Headquarters: Conakry. Member of IATA.

Air-India

The national airline of India was formed in 1948 and has been entirely State-owned since 1953; but its experience dates back to the 1932 air mail services inaugurated by its present Chairman, Mr J. R. D. Tata. Today it flies services to Europe, Africa, the USSR and the Far East, and with BOAC and Qantas provides the "Kangaroo" route from Sydney to New York via Singapore, Bombay and London.

Headquarters: Bombay. Member of IATA.

Airline insignia, see colour plates on pp. 360-363

Airline operators, see under individual operators' titles.

Air mail

The first recorded occasion when an aeroplane was used to deliver letters to their destination occurred in India on 18 February 1911, when a sack of mail was flown some 6 miles from Allahabad to Nairi by the French pilot Pequet. In this, the coronation year of King George V, Britain was quick to follow the example, and the Royal Mail Aerial Postal Service between London and Windsor in September 1911 was a huge success. During the

Pioneer days of air mail: Jupiter-engined de Havilland 50 about to leave Australia (Archerfield) for London, 1931

three days of its operation over 100,000 letters and postcards were carried. In the US, Earle Ovington ran a series of aerial postal flights from Long Island, NY, in September/October 1911, and similar flights in several other countries took place in the years preceding the 1914-18 War. With the establishment of commercial air transport in the 'twenties, the delivery of mail and newspapers by air became an important part of an operator's activities, and several of today's leading airlines owe their position at least in part to the air mail contracts with which they started business. This was particularly true in the United States, where aircraft such as the Boeing Monomail (*q.v.*) were evolved with such services in mind. Almost all international mail can nowadays be sent by air anywhere in the world, reaching its destination in a matter of hours compared with the days or even weeks taken by surface transport.

Air Malawi

Formerly a subsidiary of Central African Airways, the national airline of Malawi became autonomous at the beginning of 1968. Services are flown internally and to neighbouring African states.

Headquarters: Blantyre. Member of IATA.

Air Mali

National airline of the Mali Republic, formed in 1960 with assistance from Aeroflot and BEA. Services internally, to surrounding African states and to Europe with a mixture of Soviet, Czechoslovak and US aircraft.

Headquarters: Bamako. Member of IATA.

Air New Zealand

Known from 1940-65 as TEAL (Tasman Empire Airways Ltd), Air New Zealand was originated, with British assistance, to provide services connecting New Zealand with Australia. It became wholly New Zealand-owned in 1961. Until 1954 it operated Short Solents on its network, and was one of the last major airlines to keep flying boats in service. Nowadays it provides services to Singapore, Hong Kong and across the Pacific to Honolulu and the US.

Headquarters: Auckland. Member of IATA.

Air Rhodesia

A subsidiary of the former Central African Airways Corporation until UDI, the Rhodesian carrier was renamed Air Rhodesia Corporation on 1 September 1967. Its modest-sized fleet at present operates internally and to Malawi, Mozambique and South Africa.

Headquarters: Salisbury.

Member of IATA.

Air routes, see International air routes

Air/sea rescue, see Rescue services

Air shows and exhibitions

The world's first exhibition of aeroplanes was held in December 1908, when statically-displayed aircraft formed a substantial section of the *Salon de l'Automobile et de l'Aéronautique* in Paris. A year later, it became a *Salon de l'Aéronautique* in its own right, being held annually until 1913 and biennially between the wars and since World War Two. Today it is generally accepted as the world's leading international aerospace show, and the 29th *Salon* was staged at Le Bourget in 1969. A similar, though smaller and less truly international exhibition is that staged in the UK by the Society of British Aerospace Companies, which is now also biennial and alternates with the Paris *Salons*. Originally held at Radlett, Hertfordshire, its location was moved to Farnborough, Hampshire, shortly after World War Two.

The Olympia Aero Show (*q.v.*) was its early British counterpart, and was held annually from 1910-14. The first international flying meeting was the *Grande Semaine de l'Aviation* at Rheims in 1909 (see Competitive flying).

In the nineteen-sixties, other international air shows have begun to assume growing importance. Chief of these is that held biennially at Hanover; others are held in Turin, Venice and Tokyo, and a special exhibition for light aircraft is held at Cannes. Also important, although not international in nature and not held at regular intervals, are the Aviation Day displays held at Tushino airfield near Moscow. These are predominantly military in content.

Static park at the 1947 show of the Society of British Aircraft Constructors at Radlett, Hertfordshire (the SBAC show is now held at Farnborough, Hampshire, every two years)

Air shuttle, see Commuter air services

Airspeed indicator, 675, 689, 695, 720, 1355, 1533, 1666

The airspeed indicator is one of the most vital instruments the pilot uses. It is a pressure instrument—that is, it relies on variation in air pressure to obtain a reading. The airspeed indicator relies on the difference between the static and dynamic pressure of air to obtain the required value. Air is drawn from a pressure (pitot) head into the instrument. The pitot head contains a sealed static head of air which feeds one side of a diaphragm, while an open dynamic tube in the pitot head feeds the other side of the diaphragm. The movement of the diaphragm in relation to the varying pressures is linked to a pointer in front of a dial in the cockpit. By careful design, the indicators can be made very accurate, although they are subject to two errors. As in all pressure instruments, there is a lag in the reading between the actual and the indicated, although this is small and not often serious in the ASI.

Secondly, there is position error: wherever the pitot head is installed and however carefully it is positioned in order to obtain the most undisturbed airflow, there will always be conditions of flight in which it will be at an angle to the line of airflow. This is especially so at high angles of attack, as presented by some modern aircraft in the landing configuration, and results in a difference between the True and the Indicated Air Speed.

Air Vietnam

Domestic and regional airline, flying since 1951 services formerly operated by Air France, which still has a minority holding.

Headquarters: Saigon. Member of IATA.

Alaska Airlines

Known as Star Air Lines from 1937-44, Alaska Airlines flies domestic passenger services with Boeing 727s. In addition, it performs cargo-carrying and other services in support of oil exploration in Alaska, Ecuador and Peru, and charter operations for Military Airlift Command of the United States Air Force.

Headquarters: Seattle.

Albatros, 361

The Albatros-Werke GmbH of Johannisthal, Germany, was one of the foremost manufacturers of military aircraft for the German Air Force in the 1914-18 War. Building both single- and two-seat biplanes of advanced design, it maintained a steady flow of operational types of which the D.III and D.V fighters achieved particular fame. With finely streamlined monocoque fuselages, thin wings and unbraced tailplanes, they were used by many of the leading German aces of the 1914-18 War.

Albatross (D.H.91), see de Havilland

Alcock, Captain John W., 253, 255, 813, 815

Alcock and Brown memorial at London (Heathrow)

Alexander the Great, 112

Alia (Royal Jordanian Airlines)

Named after the daughter of HM King Hussein, Alia was established in 1963 and is now entirely State-owned. Services are mainly internal and to other Middle Eastern countries, but routes to Paris, Rome and London are also flown.

Headquarters: Amman. Member of IATA.

Albatros D.Va captured by Anzac troops

Alitalia

Formed jointly in 1946 by British and Italian interests, Alitalia took over the routes of its major competitor, LAI, in 1957 and is now the No. 1 internal and international passenger carrier of Italy. Alitalia still maintains one of the largest Caravelle fleets in operation, as

well as more than a dozen DC-8s. In addition to passenger services to all continents of the world, it provides all-cargo jet services to New York.

Headquarters: Rome. Member of IATA.

Allegheny Airlines

A pioneer US air mail carrier, Allegheny was founded in 1938 as All-American Airways. It changed to its present title in 1953, four years after beginning domestic passenger services, and is now one of the principal US local service carriers, operating between several of the north-eastern states and to Toronto.

Headquarters: Washington, DC.

Allison, 819, 822, 1235, 1309

The Allison Engineering Company entered the aero-engine industry in the nineteen-twenties. It specialized in the design and construction of reduction gears, air-cooled cylinders and superchargers, and converted the 400-hp Liberty engine into an inverted air-cooled engine, marketing it as the Allison VG-1410. It became a subsidiary of the General Motors Corporation, and in 1935 expanded to put into production the 1,000-hp V-1710 twelve-cylinder liquid-cooled Vee engine. A whole family of V-1710s was produced during the nineteen-forties, and powered many US military aircraft. Some versions were supercharged and one, the E11, was built with remote propeller

drive at the end of a shaft, being used in the Bell P-39/P-63 series of aircraft. An experimental engine, the V-3420, was built as a 24-cylinder W-shaped liquid-cooled engine and was virtually two V-1710s linked together. It was not put into production. After World War Two, the company became the Allison Division of General Motors and has since produced General Motors gas-turbines.

Allison T38 engine, 1309

The T38 was an axial-flow turboprop development engine, built for the US Navy and flown in a Convair-liner. It achieved a performance of 2,925 hp.

Allison T40 engine, 1309

The T40 turboprop was a development of the T38, consisting of two T38 axial-flow power sections united to a common reduction gearing to drive two three-blade co-axial contra-rotating propellers, producing 5,520 hp. The T40 was fitted to three US Navy experimental aircraft, the Douglas XA2D-1, the North American XA2J-1, the Convair XP5Y-1, and the Convair R3Y Tradewind flying boat which went into limited production.

Allison T56 engine, 1309

The T56 was the first Allison turboprop to be widely adopted for US aircraft. It first flew in 1954, mounted in the nose of a Boeing B-17, and was then used to power an experimental Convair-liner for the US Navy. Having been successfully proved, the T56 went into production for the Lockheed C-130 Hercules and Grumman E-2A Hawkeye. A commercial development, the Model 501, was also built in quantity for the Lockheed Electra airliner.

The T56 is an axial-flow turboprop developing between 3,750 and 4,910 eshp, according to the model.

Alouette, see Sud-Aviation

This vital instrument is the means by which a pilot calculates the height at which he is flying. The standard altimeter is a pressure instrument, and its indication is given as a result of the difference between the outside air pressure and a standard capsule of air in the instrument. A diaphragm of one form or another moves according to the pressure differential, and this operates a series of levers which turn the needles on the face of the altimeter. It is important to note that the altimeter does not necessarily give the height of the aircraft above the ground, but above a fixed datum pressure, and it is necessary to know the ground level pressure and set the instrument in relation to this before the height above ground level can be known. It is also a feature of pressure instruments that there is a lag between the actual reading and the indicated reading due to the time taken to adjust to a change of pressure. Since World War Two great advances have been made in the development of radio altimeters, which obviate the disadvantages of the pressure altimeter, and most high - performance aircraft are now equipped with both types.

American Airlines

Although it operates within the North American continent, American Airlines has one of the largest aircraft fleets and longest route networks of any airline in the world. It was formed in 1934, although its constituent companies' experience dates back eight years before this. Substantial orders from American Airlines played a large part in the success of such types as the Convair 240 and 990, Douglas DC-7 and Lockheed Electra, and

its total fleet today is in the region of 200 aircraft, almost all of them jet-powered.
Headquarters: New York.
Member of IATA.
Amsterdam Airport (Schiphol), Netherlands
Lies on the site of the former Haarlem Lake, reclaimed in 1852. A military airfield was built at Schiphol in 1917, and it has been a commercial airfield since 1920.
Distance from city centre: 8 miles
Elevation: 13 ft below sea level
Main runways:
09/27 = 11,330 ft
06/24 = 10,663 ft
01/19 = 9,563 ft
05/23 = 6,618 ft
14/32 = 5,906 ft
Passenger movements in 1967: 3,263,300
A term formerly much

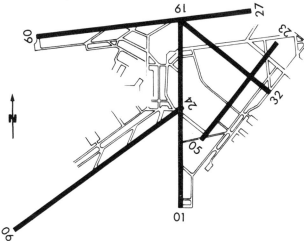

1 Angle of incidence
2 Chord line
3 Aircraft horizontal datum

misunderstood and misused, the angle of attack is the angle formed between the chord line of an aerofoil surface and a line representing the

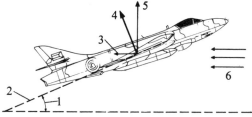

1 Angle of attack   4 Total reaction
2 Chord line   5 Lift
3 Drag   6 Direction of airflow

undisturbed airflow relative to that surface. It is, therefore, strictly an aerodynamic term and should not be confused (though it often is) with "angle of incidence", which is a design or constructional term with a different meaning.

Angle of incidence
Often confused with

"angle of attack" (q.v.), the angle of incidence is that formed between the chord line of an aerofoil surface and the datum line, or longitudinal axis, of the body (e.g. fuselage) supporting that surface.

ANIP (Army/Navy Instrumentation Program), 707
Ansett-ANA
Ansett Airways was founded by Mr R. M. Ansett in 1936, grad-

ually expanding after World War Two until, with the acquisition in 1957 of Australian National Airways, it became the biggest privately-owned airline in Australia. It is now a component of Ansett Transport Industries, which in turn has absorbed several other smaller Australian operators. Ansett-ANA's network covers all the Australian states and includes routes to New Guinea and Papua. A mixed fleet includes jet-, turboprop- and piston-engined airliners and several helicopters.
Headquarters: Melbourne.
Associate member of IATA.
These devices are used to permit full wheel braking power to be utilized without locking the wheels. Wheel locking not only extends the landing run, but also leads to the serious risk of burst tyres and subsequent damage to the wheel and, possibly, the aircraft. Such devices became essential with

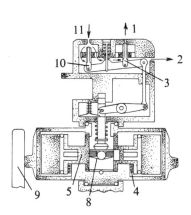

1 To reservoir
2 Delivery to brakes
3 Exhaust valves closed
4 Drive ring and spring
5 Drum driven by drive spring
6 Flywheel driven by drum
7 Apex of cam profile
8 Cam profile
9 Maxaret track on aircraft wheel
10 Inlet valve open
11 Supply from brake control

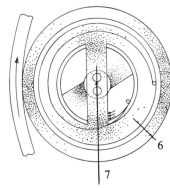

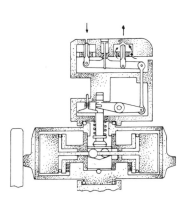

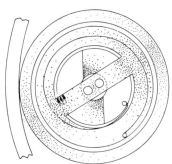

Principle of Maxaret anti-skid device

the development of bogie-type landing gears, where some weight transference on the wheels usually occurs, leading to ready locking.
Anti-skid devices are usually flywheel or centrifugal mechanisms which release the brake hydraulic pressure the moment a wheel starts to lock. The basic principle of current mechanical

devices is to rotate a small fly-wheel with the aircraft wheel, both being spun up at touchdown, and to use the tendency of the fly-wheel to "overrun" the wheel, when the latter is decelerated, to release the brake. The tension between the fly-wheel and the aircraft wheel can be adjusted to regulate the deceleration rate at which control begins. An electric anti-skid device has been developed. On this the wheel unit is a small DC generator, the output of which is proportional to the speed of the wheel. The generator is connected to a condenser and relay housed in a control box. When the wheel accelerates, the generator output increases, and a current flows into the condenser through the relay, causing it to operate. When the wheel stops accelerating (i.e., on reaching the ground speed) the voltage from the generator stabilizes and the current flow in the condenser ceases.

When the wheel begins to decelerate, the generator output falls off and, consequently, the condenser discharges slowly. If the wheel decelerates at a rate sufficient to cause a skid, the generator output will fall off rapidly, resulting in a discharge rate from the condenser sufficient to operate the relay. This energizes a solenoid in the brake supply line to release the brake pressure.

Aircraft fitted with anti-skid devices are often described as having automatic braking.

Antoinette, 209, 229, 523

Antoinette of Hubert Latham down in the English Channel, July 1909

Antoinette was the trade name of the engines and aeroplanes built by Léon Levavasseur, who named them after his partner's daughter. The engines, first used in 1905, eventually produced 50 hp and became much sought after for use in aircraft in the period up to 1910, before the rotaries came into their own. The aircraft, similar in conception to Blériot's monoplane, were more attractive in appearance but more cumbersome

in operation; they were popular on both sides of the Channel in 1908-9. The Antoinette gradually disappeared after 1911. The Antoinette could easily have become as famous as the Blériot, for Hubert Latham used one to make two attempts to cross the Channel, one on 19 July 1909 and another on 27 July 1909, as soon as he heard that Blériot had crossed and the weather was favourable. Both his attempts failed.

Antonov, Oleg Konstantinovich

Oleg Antonov, one of the USSR's senior aircraft designers, was little known outside the Soviet Union prior to 1947, when he had been known primarily as a designer of gliders and sailplanes. In 1947 his first piston-engined aircraft to enter large-scale production appeared, the single-engined An-2 general-purpose biplane which in its latest form is still in production today, both in the USSR and in Poland.

It is as a designer of multi-engined transport aircraft that he is best known today; the range of Antonov types in current production include the twin-engined An-14 and An-24, and the four-engined An-10, An-12 and An-22.

Antonov An-12 ("Cub"), 789

The An-12 was developed in parallel with the commercial An-10 four-turboprop airliner. This involved modification of the rear fuselage to include a rear-loading ramp and a tail gun-turret. The An-12 is a high-wing monoplane, similar in configuration to the Lockheed Hercules, and has

been built in large numbers for the Soviet Air Force. In addition, it has been supplied to the air forces of Algeria, Egypt, India, Indonesia and Iraq.

Antonov An-22 ("Cock")

In 1965 a very large aircraft, of similar layout to the An-12 except for its twin fins and rudders, made its first public appearance at the Paris *Salon*. Designated An-22 and named *Anteus*, it was at that time the world's largest aeroplane, with a length of 189 ft and a wing span of 211 ft. It is powered by four Kuznetsov NK-12MA turboprop engines of 15,000 shp each, and has a huge hold for the carriage of bulky items of freight or military equipment. An unusual feature is the location of crew and passenger doors in the wheel fairings.

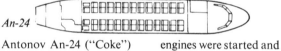

An-24

Antonov An-24 ("Coke")

Soviet counterpart to the Fokker Friendship and Handley Page Herald (*qq.v.*) twin-turboprop airliners, the An-24 was flown for the first time in April 1960 and, in modified form, was introduced on Aeroflot's internal services in September 1963. Since 1964 it has been sold to several European and Middle Eastern airlines, and to operators in China and Cuba. Current version is the An-24V, in which the seating capacity is increased from 48 to 50. This version is powered by two 2,550 ehp Ivchenko AI-24 turboprops (Series I) or two 2,550 ehp AI-24As (Series II) which give it a cruising speed of around 300 mph at 19,700 ft.

Anzani engines, 195, 217, 229, 1077

The Anzani engines, which were used in many early aircraft, were all of the radial type, as opposed to the more commonplace rotary engines of later. It

was a 25-hp Anzani which powered Blériot's monoplane for the Channel crossing in 1909. Anzani continued to build radial engines during the 1914-18 War, the most powerful engine achieving 200 hp. A British Anzani company was formed to build its products in the UK.

Apollo spacecraft, 340-342, 503-4

APU (Auxiliary Power Unit), 943

Many large modern aircraft need a considerable amount of power to keep their various systems operative. Until fairly recently external power aids were needed on the ground only before take-off; so it was the practice to use ground compressors and generators which plugged into the aircraft but which were disengaged when the engines were started and engine-run generators and compressors could take over the power load.

Nowadays, most modern airliners and large military aircraft have their own built-in power unit, usually in the tail of the aircraft. It is customarily a small gas-turbine, similar to the light turbojet engines which power small jet aircraft, and supplies all the power necessary for the aircraft systems on the ground and in the air. This has the added advantage that the aircraft is independent of the need for ground trolleys carrying generators and compressors.

Area rule

An aerodynamic technique used in the design of high-speed aircraft. In general terms, it means that the wings, fuselage, tail and all other appendages have to be considered together when working out the streamlining. This is necessary so that the cross-sectional area of successive "slices" of an aircraft from

nose to tail conform to those of a simple body of ideal streamline shape.

Without area rule, it is obvious that the greatest total cross-sectional area comes where the wings are attached to the fuselage. In such cases the cross-sectional area continues to increase to a point well back from the nose, when it should ideally be reducing to give the familiar streamlined shape.

It can thus be appreciated that, to "area rule" an aircraft, the cross-section of the body must be reduced to compensate for the presence of a wing, nacelle or any other protuberance. The most common way of achieving this is to reduce the diameter of the fuselage where the wings join it or enlarge the fuselage ahead of and behind the wing, giving it a "wasp-waist", or "coke-bottle" shape.

The area-ruled Buccaneer

The area rule is now accepted universally in the design of transonic and most supersonic aircraft. A good example of an aircraft completely designed to transonic area ruling is the Hawker Siddeley Buccaneer naval strike aircraft.

Argentine Republic: Air Forces

Fuerza Aerea Argentina
Established as an autonomous force in January

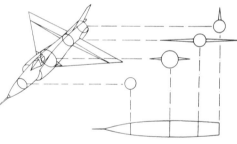

The area rule applied to the Convair F-102 fighter. The fuselage diameter is reduced where the wing frontal area is greatest, so that the total cross-sectional area is not more than that of an ideal streamline shape at the same relative position

At the time when aircraft were beginning to fly at the speed of sound, several were redesigned to conform to the area rule. The classic example is the Convair F-102 Delta Dagger delta-wing fighter. When first built, this could not fly faster than sound, even with its turbojet engine on reheat, giving a total thrust of 15,000 lb. It was extensively redesigned to conform to the area rule, after which it easily reached a speed of 800

mph. Not only was performance substantially increased, but valuable additional storage space was gained in the enlarged sections of the fuselage.

1945, although military aviation in the country dates from 1912. It made little progress until after the 1914-18 War, and during the nineteen twenties was largely Italian- and French-equipped. After the Military Aircraft Factory was established at Cordoba in 192? domestically-designed aircraft began to enter service, but shortly before World War Two licence-built aircraft of foreign design were reintroduced. Foreign aircraft have constituted the major equipment ever since, although the nationally-designed Huanquero and Guarani transports are currently in service. Present combat aircraft include Sabre and Skyhawk jet fighter-bombers.

Aviación Naval. The aircraft carrier *Independencia* has been in service with the Argentine Navy since 1958. The carrier-based Panther jet fighters are still retained, but the wartime Corsair fighter-bombers have recently been replaced by NAA T-28S Fennecs, and Aermacchi M.B. 326GB jet light attack aircraft.

Land-based naval types include Neptune anti-submarine patrol aircraft and Catalina flying boats. Late in 1968 the carrier *Karel Doorman* was purchased from the Netherlands Government.

A small air arm is also operated by the Argentine Army; this consists principally of light transport and observation types.

Argonaut, see Douglas DC-4

Argosy (biplane), see Armstrong Whitworth

Argosy (monoplane), see Hawker Siddeley

Argus engines, 234

The Argus Motoren GmbH was established in 1900 to build components for the nascent automobile industry. It produced the first German water-cooled aero engine, which was used almost exclusively on German aircraft by 1912 and also powered the Russian Sikorsky *Russkii Vitiaz.* It declined subsequently until the 1914-18 War, when its engines were used extensively. In 1918 it ceased production and did not return to aero engine manufacture until 1926. It settled down to producing engines for light aircraft between the wars but by the outbreak of World War Two had become part of the Junkers combine.

Ariel helicopter

A two-seat experimental helicopter built in 1949 by SNCASO in France. It was powered by a piston engine driving a compressor (built by Turboméca) which delivered compressed air to jet units at the rotor tips,

whence the driving power of the rotor was established. This eliminated the problem of torque. A developed version, the Ariel III, used a tail-mounted jet for directional control.

Arlanda Airport, see Stockholm

Armstrong Siddeley Double Mamba engine, 829

The Armstrong Siddeley Mamba was one of the first axial-flow turbojet engines to go into production. The Double Mamba was developed to meet a Royal Navy requirement for twin-engined versatility in a single-motor installation, and is virtually two Mambas side-by-side, geared to drive two contra-rotating propellers. The combined unit produces 2,640 shp and 810 lb static thrust. The Double Mamba was so conceived that one engine could be shut down, enabling the aircraft to cruise on the other, thus saving fuel and giving a longer endurance. It was installed in the Fairey Gannet anti-submarine aircraft, which entered service with the Royal Navy in 1953 and was followed by an airborne early warning version, the Gannet AEW Mk. 3.

Armstrong Siddeley Jaguar engine, 1073, 1075

Armstrong Siddeley were among the foremost manufacturers of radial engines in the UK between the wars, and in the early nineteen-thirties had no fewer than eleven types in production. The Jaguar was one of these; it was a two-row fourteen-cylinder air-cooled radial in ungeared, geared or supercharged versions, all producing between 400 and 460 hp. It was used mainly in the medium-sized twin-engined civil transports and heavy single-engined biplanes of the day.

Armstrong Whitworth, see also Hawker Siddeley

Sir W. G. Armstrong

Whitworth Ltd was formed in 1920, and in 1921 produced its first aircraft design, the single-seat Siskin biplane, which became a standard RAF fighter of the nineteen-twenties. This was followed by the Atlas two-seat Army co-operation aircraft, the three-engined Argosy airliner for Imperial Airways and then, in 1932, by the company's first monoplane design, the four-engined Atalanta airliner.

Other fighter designs appeared during the 'thirties, but the next significant product was the twin-engined Whitley bomber, first flown in 1935. This was one of the standard RAF bombers at the outbreak of World War Two, although it was transferred to mine-laying, reconnaissance, transport and glider towing duties as more modern bombers became available. Other Armstrong Whitworth types to see wartime service included the Albemarle transport and glider tug, and the pre-war Ensign airliners, which gave much useful service with BOAC during this period.

The company's first

Armstrong Whitworth-built Sea Hawk 101 of the West German Kriegsmarine

post-war civil venture, the four-turboprop Apollo airliner, was overshadowed by the Vickers Viscount and did not go into production. Armstrong Whitworth then assumed responsibility for continued development of the Gloster Meteor as a night fighter and of the Hawker Sea Hawk naval fighter. Last product to appear under the company name was the Armstrong Whitworth (now Hawker Siddeley) Argosy four-turboprop freighter

which was built in some numbers for the RAF, BEA and other civil operators.

Armstrong Whitworth Argosy biplane, 724

The Argosy was built as a result of the establishment of Imperial Airways in April 1924. This unified British airline placed orders with three manufacturers for airliners and the Argosy was Armstrong Whitworth's contribution. It was a three-engined biplane, seating 20 passengers. Power was supplied by three Armstrong Siddeley Jaguar engines, the Mk II having slightly more powerful Jaguar IVAs compared with Jaguar IIIs in the Mk I.

The first Argosy appeared in 1926, and six further machines were built, the type being used principally on the European routes, although two were based in Cairo for the South African route at a later date.

Armstrong Whitworth Atalanta, 540

The Atalanta four-engined airliner was, in a sense, ahead of its time. It was produced for the tropical routes of Imperial Airways in 1932, when many leading British authorities were still biplane-minded; despite this, eight were ordered and used on overseas routes. The Atalanta was a high-winged monoplane with four Armstrong Siddeley Serval engines. It was extremely clean aerodynamically for its day, and carried nine passengers in comfort. During its six years of commercial service, three Atalantas were lost in accidents; the remaining five were used for coastal anti-submarine work in

India during World War Two.

Armstrong Whitworth Siskin, 1075

A fighter aircraft developed during the nineteen-twenties which eventually went into production and service with the RAF in 1924. It was the first fighter of all-metal construction to enter squadron service with the RAF. The Siskin III was a single-bay biplane, powered by a Jaguar III engine of 325 hp, giving it a top speed of 134 mph. It was succeeded by the Siskin IIIA, which was a much better aircraft with a supercharged 450-hp Jaguar IVA, a top speed of 156 mph and a service ceiling of 27,000 ft. The IIIA entered service in 1927 and remained a standard RAF fighter until 1932.

Armstrong Whitworth Whitley, 1101

The Whitley was developed from the A.W.23 troop transport and made its appearance in 1936. It was well-established in service when World War Two broke out. Originally, the Whitley was powered by Armstrong Siddeley Tiger radial engines, but the versions in most widespread operational service were the Merlin-

Armstrong Whitworth Whitley IV, 1939

engined Mks IV and V. The Whitley had the longest range of any RAF bomber in service at the outbreak of war, and was thus the first type to bomb Berlin and the first used on raids to Italy. It was a somewhat slow aircraft, and by the end of 1941 was too vulnerable for the bomber offensive. However, it found a further operational career for some time with Coastal Command

on convoy and anti-submarine duties.

The Whitley also served as a glider-tug and as an operational trainer, although it was virtually extinct at the end of the war.

Army Aircraft Factory, see Royal Aircraft Factory

Arrow, see Avro (Canada)

Artificial horizon, 685, 696

A combined flight instrument, using a gyroscope aligned to the level of the Earth and connected to the instrument, which gives a pictorial representation of the aircraft and the horizon. When the aircraft climbs, the model aircraft rises above the horizon and when it dives the model falls below the horizon. It is graduated in the same way as a turn indicator and the gyroscope, aligned in the same way, shows the model aircraft banking as the aircraft banks.

Thus, on one instrument there is given a pictorial record of the aircraft's attitude at all times. This represented a considerable advance on all other flight instruments at the time it was introduced, initially by the Sperry Gyroscope Company.

Artouste engine, see Turboméca

"Ash" missile, 895

Little is known about the air-to-air missile code-named "Ash" by NATO, of which four are carried beneath the wings of the Tupolev "Fiddler" supersonic all-weather fighters of the Soviet Air Force. It is operational in both infra-red and radar homing versions.

Aspin engine, see Turboméca

Astazou engine, see Turboméca

Atalanta, see Armstrong Whitworth

"Atoll" missile, 895

"Atoll" is the code-name given to a standard air-to-air missile in large-scale service with the Soviet *bloc*. It is carried by the MiG-21 fighter, has a solid-propellant rocket motor and infra-red homing head.

Aubisque engine, see Turbomeca

Australia: Air Forces
Royal Australian Air Force. Created on 1 April 1921, having been preceded by the Australian Flying Corps (founded 1913) and Australian Air Corps (from 1919). Fought in the Middle East during the 1914-18 War and re-equipped by "Imperial Gift" with war-surplus D.H.9s and S.E.5s in the early 'twenties. Expansion was very slow until the mid-nineteen-thirties, and even by late 1939 it had less than 250 aircraft and less than 3,500 personnel. After the outbreak of World War Two, there was a considerable build-up of men and equipment, and RAAF units played a valuable part in the Allied war effort, in the Pacific area especially but also in Europe and the Middle East. Most aircraft were of British or American origin, although the Australian-designed Boomerang fighter also served with distinction. The peace-time RAAF was not to enjoy peace for very long, and became committed in the Korean War of 1950-53 and, later, to the US campaign in Vietnam. The present-day RAAF is a substantial force, whose combat aircraft include Australian-built Mirage III and Sabre fighters, F-111 strike aircraft, and Canberra light bombers, with Lockheed Orions and Neptunes for maritime patrol and Hercules for transport.
Royal Australian Navy. Australian naval avia-

Australian-built Canberra B.20 of RAAF No 1 Sqn based at Amberley, Queensland

tion dates from 1948, when it was formed to operate from two new light fleet carriers, HMAS *Melbourne* and *Sydney*. Since then it has been committed operationally in a similar manner to the RAAF, and is at present equipped chiefly with anti-submarine aircraft (twin-engined Trackers and Wessex helicopters) and a few Skyhawk fighter-bombers.
Australian Army Aviation Corps. Comparatively small force, consisting mainly of Turbo-Porter light transports and Sioux helicopters.

Australia: Chronology
1851. Dr William Bland, an English naval surgeon transported to Sydney, sent drawings and a model of his Atmotic Ship to England. These were exhibited at the Crystal Palace in 1852. The "ship" was to be a balloon 200 ft long, driven by a steam engine, lifting five tons in a suspended cabin, and able to cruise at 50 mph. Bland claimed it would be possible to start regular five-day services between Sydney and London.
1894. Lawrence Hargrave, Australia's first great aviation pioneer, lifted himself from the ground to a height of 12 ft, using four box-kites. In 20 years of experiments, Hargrave developed the box-kite to the stage at which, in 1905, Gabriel Voisin could base his first aircraft upon it. It was adopted also by the US Meteorological Bureau. Hargrave also invented, and made models of, a rotary engine — a design "rediscovered" by French engineers years later.
1909-1910. Three flights

are credited as the first heavier-than-air flights in Australia. On 9 December 1909, racing driver Colin Defries made an uncontrolled flight of 5½ seconds in a Wright biplane at Victoria Park, Sydney. On 17 March 1910, Fred Custance stayed airborne for 5½ minutes in a Blériot "too scared to try landing". On 18 March 1910, Harry Houdini, the famous magician, made a properly controlled series of flights in a Voisin at Digger's Rest, Victoria, and he is generally credited with the first successful powered flight in Australia.
16 July 1910. John Duigan flew the first Australian-built machine at Mia Mia, Victoria.
13 June 1911. H.R. Busteed became the first Australian to secure a flying brevet, the RAC certificate No. 94.
5 December 1911. W. E. Hart, a dentist of Wyalong, New South Wales, qualified for the No. 1 Australian flying licence.

Australia: Deperdussin (1914) and Commonwealth Mirage III-O in the RAAF Museum

July 1914. Maurice Guillaux, a French pilot, flew the first air-mail and freight, on the first Melbourne-Sydney flight.
17 August 1914. The first Australian flying course began at Point Cook, near Melbourne, with four students. One

graduate was sent to German New Guinea with an aircraft.
April 1915. Australia became the first Dominion to send a flying unit abroad when, at the request of the Indian Government, a half-flight was sent to Mesopotamia.
1919. Ross and Keith Smith made the first England-Australia flight in a Vickers Vimy in 27 days 20 hours, arriving at Port Darwin on 10 December. This flight won the prize of £10,000 offered by the Australian Government, and was the first flight made across the globe by an aircraft of any nation.
1921. The Royal Australian Air Force was formed on 1 April, with a total of 20 officers and 120 airmen. The first regular domestic airline service was inaugurated by West Australian Airways Ltd over a route of 1,211 miles between Derby and Geraldton, in December. The aircraft used were Bristol Tourer Coupés, and two of the first three aircraft crashed on their first flight.
2 November 1922. Qantas began operations on a regular basis between Longreach and Charleville, Queensland.
6 April to 19 May 1924. Wing Commander S. J. Goble and Flight Lieutenant I. E. McIntyre, flying an RAAF Fairey IIID seaplane, made the first round-Australia flight, a distance of 8,500 miles, in 90 flying hours.
25 September-7 December 1926. Group Captain

R. Williams (now Air Marshal Sir Richard Williams, RAAF CAS), and I. E. McIntyre made the first flight from Australia into the New Guinea area in an RAAF D.H.50A. They flew a route Point Cook—New Guinea—Solomon Islands and return in 126 hours flying time.
10 January 1928. New Zealand pilots Hood and Moncrieff lost their lives in the first attempt to fly the Tasman Sea, from Sydney to New Zealand. The first successful crossing was made by Charles Kingsford-Smith and Charles Ulm, between Sydney and Christchurch, on 10-11 September 1928.

"Southern Cross" memorial at Brisbane

31 May-9 June 1928. First successful crossing of the Pacific Ocean, from America to Australia, was made by Kingsford-Smith, Ulm and the Americans Harry Lyon and James Warner, in the Fokker Trimotor *Southern Cross*.
1 January 1930. Kingsford-Smith's Australian National Airways started a Sydney-Brisbane service with Avro Tens. This was the first major attempt to create a modern air service with up-to-date aircraft on the east coast of Australia.
24-25 June 1930. Charles Kingsford-Smith, Evert Van Dyk, Saul and Stannage, made a successful east-west crossing of the North Atlantic. By flying from New York to San Francisco on 2-4 July, the *Southern Cross* linked up with the starting point of the earlier Pacific flight, and was the first aircraft to make a round-the-world flight which crossed the Equator.
20 October-3 November 1934. Charles Kings-

ford-Smith and Capt P. G. Taylor, in the Lockheed Altair *Lady Southern Cross*, made the first successful Pacific flight from Australia to America.
10 December 1934. Qantas Empire Airways Ltd, formed on 18 January by an agreement between Queensland and Northern Territory Aerial Services Ltd and Imperial Airways, began to operate the Singapore end of the London-Sydney service.
1936. Australia's first major aircraft production venture began with the foundation of the Commonwealth Aircraft Corporation at Fishermen's Bend, near Melbourne. A number of large Australian industrial groups provided the capital for its first product, the RAAF's Wirraway fighter.
3-21 June 1939. P. G. (later Sir Gordon) Taylor and Richard Archbold made the first crossing of the Indian Ocean in Archbold's PBY Catalina (which was chartered to the Australian Government). They left Sydney on 3 June and, flying via Batavia, Cocos, Diego Garcia and the Seychelles, landed at Mombasa on 21 June.
2 July 1950. The RAAF's No. 77 (Fighter) Squadron began operations over Korea, the first non-American assistance of any kind given in the Korean War. Their Mustang aircraft were replaced later by British-built Meteor 8s.
14 January 1958. Qantas began the world's first scheduled round-the-world service with Super Constellation aircraft.

Austrian Airlines
Beginning services in 1958 with chartered Viscounts, Austrian Airlines currently flies jet and turboprop services to points in eastern and western Europe, the USSR and the Middle East.

Two of the factors which most disturb the safe and successful operation of commercial airliners are the high proportion of total aircraft accidents that occur during take-off and landing, and the cost and inconvenience incurred when an airliner is diverted from the airport of its destination owing to impossible weather conditions. To combat these factors, considerable research has been expended in the past two decades to evolve automatic landing equipment which can land an aeroplane safely in thick fog, snowstorms or other conditions where a visual landing would be impossible.

First in the field was the Blind Landing Experimental Unit of the British Royal Aircraft Establishment, which began its experiments in 1948. At a time when the average failure rate of a simple autopilot was approximately once in every 25 thousand landings, the British aim was, and is, the achievement of an automatic landing system with a failure rate of less than once in every 10 million landings. Such a degree of reliability is not achieved overnight, and it is only within the past few years that automatic landings of scheduled passenger flights have been introduced. Even these are still restricted to use in certain weather conditions (q.v.), and can be used only at airports with ILS (q.v.) equipment; but continued improvement will eventually make automatic landing in the future a matter of routine. The major systems in use in 1968 were Bendix-Elliott (as in the VC10), Smiths Autoland (Trident), Sperry (Boeing 727) and Sud-

Lear (Caravelle), but none of these had then been cleared for full operation down to Category 3C (zero visibility) weather conditions.

Automatic landing comprises three basic stages. The first of these is autoflare, in which automatic throttles control the approach and landing speeds, while radar altimeters measure accurately the aircraft's altitude as it descends until at the appropriate moment the engines are flared out and the aircraft's attitude is changed automatically from the approach angle to the correct landing attitude. Additional equipment "kicks off" any drift during the second stage and aligns the aircraft correctly with the runway immediately prior to touchdown. The third stage keeps the aircraft straight on the runway and can guide it, if necessary, along the taxiways right up to the terminal building.

Autopilot, 1361, 1373, 1385

Since about 1910, various mechanical systems capable of controlling an aircraft in flight without human assistance have been developed, primarily to minimise pilot fatigue on long flights. The first practical system was evolved by the American Elmer Sperry and flown on a Curtiss flying boat in 1912. This used a single gyroscope which, as a result of the inherent property of any spinning mass, tended to resist any change in the plane of its axis of rotation. Departures by the aircraft from its original attitude thus caused a small force to be applied to a spring connected to one end of the gyro axis and this, magnified mechanically, was used to effect a restoring movement of the aircraft controls.

All simple autopilots since that time have used similar principles, although for control

Automatic Landing

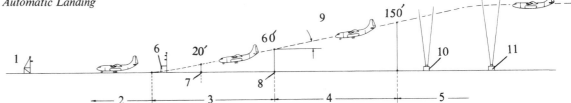

1 ILS localizer
2 Roll-out on localizer
3 Radio altimeter
4 Constant attitude
5 Approach on ILS
6 ILS glide path
7 Drift kick-off
8 Start of flare-out
9 2½° nominal defined
10 Middle marker
11 Outer marker

about all three axes—pitch, roll and yaw—three gyros are needed mounted on mutually perpendicular axes. By 1930 the British Royal Aircraft Establishment and several private companies had developed refined autopilot systems which were gradually introduced to both military and civil aircraft. At first, simple gyro stabilization was accepted, but soon the magnetic compass was coupled in as a heading reference. Later, during

Autopilot instruction at RAF College, Cranwell

World War Two, much more complicated autopilots were produced in which, by simply moving a control knob or similar manual input, the aircraft could be held on a steady heading at a constant altitude, or made to describe a steady rate of turn in either direction, or even change altitude in a precise manner.

From here it was a short step to the auto-approach coupler in which signals from an instrument landing system were made to control an aircraft automatically during approach to land down the glide slope/localizer beams. Further very considerable refinement, increase in accuracy and the adoption of duplicated, triplex, twin self-monitoring pairs and other fail-operative sys-

tems has enabled auto-pilots to be used which can be relied upon to bring an aircraft right on to the runway and keep it on the centre-line and turn off on to the taxiway, all in totally "blind" conditions. Such an autopilot must have a statistical likelihood of failure of less than one in ten million landings. Aircraft today use the autopilot as part of an integrated flight control system linked by a digital computer to an air data system, ground transmission and receiver stations and various kinds of navigational systems. Only by this means has the burden of the high-speed work load on the flight crew been kept at a reasonable level in the latest aircraft.

Autorotation, 977, 983

Auxiliary power unit, see APU

Avia

The Avia works at Prague-Cakovice was responsible for producing most of the fighter aircraft used by the Czechoslovak Air Force during the nineteen-twenties and 'thirties. Its BH-3 of 1921 was the first Czechoslovak fighter to go into series production. The most extensively-used fighter was the B 534, which was standard with most

fighter squadrons at the outbreak of World War Two; this had an Avia-built Hispano-Suiza engine of 860 hp, a top speed of 250 mph, and was armed with four machine-guns. A later fighter, the B 35, appeared too late for use against the Nazi invaders. After the war, Avia built German Messerschmitt Bf 109s from captured components; these served with the post-war Czechoslovak Air Force under the designation S 199 until replaced by Soviet MiG-15 jet fighters in the early 'fifties.

Aviaco

Aviaco is the operating name of Aviacion y Comercio SA, which was formed in 1948 and began scheduled passenger services in 1950. It serves routes within Spain and regionally to the Balearic Islands, Canary Islands and North and West Africa. A car-ferry service is also operated.
Headquarters: Madrid. Member of IATA.

Avianca

Avianca (Aerovias Nacionales de Colombia SA) has a history dating from 1919, when a Junkers floatplane opened a service between Barranquilla and Giradot, and is thus the oldest surviving airline in the Americas. It was predominantly financed

by Pan American from 1931, and Pan Am still retains a substantial (though no longer controlling) interest. The airline took its present title in 1940 and today has a substantial network within Colombia as well as international services to Central and North America and to Europe. It also owns Aerotaxi, a domestic charter company, and Helicol, a local helicopter charter operator.
Headquarters: Bogotá. Member of IATA.

Aviation medicine

Aviation medicine is that branch of medicine devoted, in a preventive rather than a curative sense, to helping human beings to withstand the various stresses which flying can place on the body. It is a branch that has made outstanding advances since World War Two, especially since the advent of supersonic flight and space travel.

Aviation medicine: subject entering RAF centrifuge

The main subjects embraced by aviation medicine include biophysics, biochemistry, psychology, physiology, and the effects of such phenomena as acceleration and deceleration and climate. Research into these fields yields improvements in aircrew clothing and escape and rescue apparatus. Principal aviation medicine research centres are

Avia 534 single-seat fighter

at Farnborough in the UK, at Wright-Patterson Air Force Base in the USA, and in the USSR.

RAAF partial pressure-g-suited Mirage III-O pilot

One of the most common applications of aviation medical research is the "g" suit worn by every pilot of high-speed military aircraft today. The term "g" is used to express the rate of acceleration, due to gravity, of a body in free fall. When, for example, a supersonic fighter is put into a tight turn at high speed, pressures of several "g" are exerted on both the aircraft and the pilot. Beyond certain limits these could cause the pilot to lose consciousness and his aircraft to break up, but even before reaching these limits he would "black out" were it not for his "g" suit, which, by applying pressure to the lower part of the body, prevents the blood draining from the brain.

Aviation Traders Carvair, 791

Aviation Traders Ltd of Southend, England, produced an unusual modification to the standard DC-4 in 1961, when it built a completely new nose on to one of these aircraft. The cockpit was raised above the fuselage, leaving an unobstructed cargo hold, with the whole nose swinging aside as a door. This had obvious applications for car ferry services, and several aircraft were modified for this purpose. A production line for converting DC-4s was set up and Carvairs were sold to

operators in various other parts of the world where their easy-loading capabilities were of use.

Avon engine, see Rolls-Royce

Avro, see also Roe, Alliott Verdon; see also Hawker Siddeley

Avro was the shortened title of A. V. Roe & Co, formed in 1910 as one of the earliest British aircraft manufacturing companies. Perhaps its most famous product of all was the Avro 504 two-seat biplane, more than 10,000 of which were built during and for long after the 1914-18 War The 504 was one of the best-loved training aircraft of any generation, and during the war was used for many other duties including day bombing and night fighting: in the 'twenties it was a mainstay of the numerous flying clubs and air "circuses" that sprang up when the war was over.

Avro F cabin monoplane, 769, 775

Built in 1912, the Avro

Type F cabin monoplane was novel in featuring a completely enclosed cabin. It might have been successful if it had not been underpowered.

Avro G cabin biplane, 769

The Avro Type G cabin biplane appeared in 1912 and was one of the most reliable aircraft taking part in the Military Trials of that year. It was powered by a 60-hp Green engine and seated two men in tandem, enclosed completely in a narrow cabin, with celluloid windows, which filled the whole gap between the wings and so avoided the necessity for centresection strutting.

Avro Lancaster, 885, 1101, 1281

The Lancaster bomber, most famous of all British World War Two bombers, is a good example of a brilliant success resulting from an apparent failure. The failure was the Avro

Avro 652 prototype Anson

Throughout the period up to World War Two, Avro's main activities were concerned with the production of military aircraft for the British services. Notable among these were the Rota (Cierva C.30A) autogiro, the Tutor trainer and the Anson twin-engined patrol aircraft. Wartime products included the celebrated Lancaster bomber, from which were developed the post-war Lincoln and the Shackleton maritime patrol aircraft. Last aircraft to appear under the company's own name were the Avro (now Hawker Siddeley) Vulcan delta-winged bomber and 748 twin-turboprop transport.

Avro F cabin monoplane, 769, 775

Built in 1912, the Avro

Manchester bomber powered by two Rolls-Royce Vulture engines. Rushed into production, the engine was not fully proved and failed the aircraft badly; Avro hit upon the idea of replacing the two Vultures by four Rolls-Royce Merlin engines, and built a prototype. It was soon found that this combination proved more effective than the two officially-ordered four-engined bombers, the Stirling and Halifax. The Lancaster was accordingly ordered into production and entered service by the beginning of 1942. It began operations in March 1942 with minelaying, but the following month took part in a long, low-level daylight raid on Augsburg for which the CO

of the leading squadron, Sqn Ldr Nettleton, gained the VC.

From then on the Lancaster force grew quickly, and soon became the major portion of the nightly offensive against Germany. The type was used on special raids such as the dam-busting attack in May 1943 and it was also modified to take bigger and bigger bombs culminating in the 22,000 lb "Grand Slam" (*q.v.*).

A total of 7,366 Lancasters was built during and just after the war. The type remained in service with Bomber Command until 1953 (the last unit being a reconnaissance one) and with Coastal Command, having been modified for maritime reconnaissance duties after the war, until 1954.

After the war 65 were modified as transports with accommodation for nine passengers. These were known as Lancastrians. As well as serving with the RAF they also served BOAC, British South American Airways and Trans Canada Airlines as interim airliners until more economical aircraft could be acquired.

Avro Shackleton, 817

This four-engined long-range maritime reconnaissance aircraft originated from the use by the RAF of the Lancaster bomber in this rôle, allied to development of the Lincoln bomber. A replacement for the Lancaster was needed, so a new fuselage was designed for the Lincoln, capable of taking the necessary equipment for the rôle. Sufficient room was allowed for a considerable increase in the amount of equipment carried.

The Mk 1 Shackleton entered service with RAF Coastal Command in 1951 and was immediately found to be satisfactory. A long line of development took place to keep the Shackleton abreast of

equipment and operational development, resulting in the Mk 2 which first flew in 1952. This had a redesigned fuselage with a lengthened nose. The Mk 3 followed in 1955, involving a considerable redesign, with a tricycle undercarriage and increased tankage, including tip-tanks. The Mk 3

Avro (Hawker Siddeley) Shackleton MR.3

is capable of staying airborne for 24 hours, and has been progressively up-dated with the latest anti-submarine equipment. This version now has Viper turbojets installed in the outboard engine nacelles, supplementing the Griffon engines for take-off, attack and evasion conditions.

Avro York, 1101

The York began as a private venture during World War Two but did not come into large-scale production until the end of the war. Basically, it used the Lancaster's wings and tail with a different fuselage twice the cubic capacity of the Lancaster's. It was a high-wing monoplane and a central fin was added to the twin fins to counteract the extra keel area. It carried a crew of five and 24 passengers, and after the war Yorks became part of the fleets of BOAC and British South American Airways.

The York was used extensively in the Berlin Airlift and remained in substantial RAF service until superseded by the Hastings in the early nineteen-fifties. It continued to be used by some charter airlines for some years after this.

Avro (Canada) Arrow (CF-105), 841

The Avro CF-105 Arrow was a large, twin-engined delta-wing all-weather interceptor

fighter designed wit the problems of Canada's northern de fence lines in mind. reached the prototyp stage and flying report were encouraging, bu the whole project wa cancelled in favour o missile defence—a de cision hotly conteste at the time.

Axial compressor, 1139 1195, 1199, 1253, 1260 see also Compressor

B-17 Fortress, see Boein

B-24 Liberator, see Con solidated

B-29 Superfortress, se Boeing

B-36, see Convair

B-45 Tornado, see Nort American

B-47 Stratojet, see Boein

B-50 Superfortress, se Boeing

B-52 Stratofortress, se Boeing

B-58 Hustler, see Genera Dynamics

B-70 Valkyrie, see Nort American

B.206, see Beagle

BABS (Blind Approac Beacon System), 1373

A development of th SBA System (*q.v.*) de veloped in the UK t assist the returnin bombers of RA Bomber Command. Th landing aircraft interro gates equipment on th ground, which send back signals giving radar picture in th aircraft showing posi tion relative to th runway.

BAC (British Aircraf Corporation)

BAC is one of the tw major airframe manu facturing groups in th British aircraft industry It came into being i February 1960 and in corporates the aircraf and guided weapon interests of the forme Bristol Aeroplane C Ltd (now BAC Filto Division), The Englis Electric Co Ltd (now BAC Preston Division and Vickers Ltd (now BAC Weybridge Divi sion). Military and civ aircraft work is no conducted by the Pre ston and Weybridg Divisions respectively except for the BA share in the Concord programme, which is the responsibility of the Filton Division.

The Lightning became Britain's first supersonic interceptor in July 1960. It was developed, via the P.1B, from the original English Electric P.1A supersonic development aircraft. Powered by two Rolls-Royce Avon 300 Series turbojets, mounted one above the other in the fuselage, it has a maximum speed of Mach 2 plus and a service ceiling of 60,000 ft. In its latest versions (F Mk 3 and F Mk 6) it provides fighter defence of the UK, Cyprus and Singapore; the F 2A version is used in Germany both as interceptor and for ground attack. The ground attack Lightning has also been exported to Saudi Arabia and Kuwait.
Interceptor armament comprises two Firestreak or Red Top air-to-air missiles; various combinations of armament can be added for ground attack duties.

One-Eleven Series 500

This aircraft started life as a project by Hunting Aircraft Corporation, which became part of British Aircraft Corporation. It underwent a long period of development in project form and emerged as the BAC One-Eleven short/medium-range jet transport, powered by two Rolls-Royce Spey turbofan engines. It was sponsored by an independent airline, British United Airways, which bought the first production aircraft, and it has been sold subsequently, in developed versions, worldwide. First developments were concerned with increasing range (Series 300) and then embodying the requirements of US operators (Series 400); the

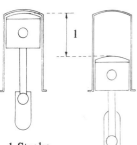
BAC One-Eleven 400 "Astrojet" of American Airlines

latest version in production is the long-fuselage Series 500, first ordered by British European Airways to carry 99 passengers and a crew of two.

Super VC10

A Vickers design, before that company became part of British Aircraft Corporation, the VC10 is a four-engined intercontinental jet airliner built primarily for British Overseas Airways Corporation, although it has been sold to other airlines and serves also with the Royal Air Force.
It is powered by four Rolls-Royce Conway RCo.42 turbofan engines of 21,000 lb st each, grouped in pairs each side of the rear fuselage; the tailplane is carried high on top of the fin. There are two basic versions, the VC10 and Super VC10, the latter carrying up to 174 passengers compared with 151 in the standard VC10.

Concorde

The Concorde is the most ambitious of the first group of international collaboration programmes, as it is pioneering a completely new field for both countries involved, France and Great Britain. The go-ahead to produce this

supersonic airliner in prototype form was given in November 1962 and since then work has progressed to the stage where both prototypes, one at Toulouse in France and one at Filton in England, were taken up for their first flights in the spring of 1969. As the Concorde is likely to be the only such airliner available outside the Soviet Union for several years, there is already a reasonably large order book, for delivery in the early 'seventies. It is intended that the aircraft will fly at Mach 2.05 for 4,250 miles with a max payload of 128 passengers.

Designed and built in 1944, the German Bachem Ba 349A *Natter* (Viper) was a cheap, semi-expendable interceptor fighter. It had basically a rocket motor (Walter HWK 109-509A) with a pilot's compartment in front, small stub wings and a cruciform tail, with a battery of rockets in the nose.
The *Natter* was intended as a crude weapons system, to be flown to the bombers by radio/radar link before the pilot took over. After firing his rockets, he was supposed to eject, the rocket motor returning to the ground by parachute.
Twelve launchings were

made, only the last being piloted. All were unsuccessful.

This officer of the *Regia Aeronautica* became Italy's Air Minister and led two highly successful long-distance trans-atlantic formation flights by *Regia Aeronautica* crews in S-55 flying boats. These flights made a world-wide impression in the middle 'thirties, and Balbo's name entered into RAF slang at that time, any large formation of aircraft being called a "Balbo".

General Italo Balbo

Popular title describing the air fighting in defence of Britain during the Summer of 1940. Its beginnings were in the attacks made by the German *Luftwaffe* against convoys in the English Channel in July 1940. These attacks were stepped up and then, in August, the *Luftwaffe* mounted heavy raids against Fighter Command airfields. The RAF, though heavily outnumbered, fought back more successfully than the German Command anticipated and by September it was apparent that the *Luftwaffe* had not gained the upper hand. It

turned its attention to attacks against the civil population, chiefly in London, but these failed in their objective of reducing morale. After the biggest air battles of the time, in mid-September, the mass raids were withdrawn and the *Luftwaffe* concentrated more and more on night raiding which, while very wearing on the civil population, did not achieve its original object, namely to prepare the way for the invasion of England. Whatever else was accomplished at the time, the Battle of Britain was the pivotal factor in World War Two which stemmed the victorious progress of Nazi Germany throughout Europe and forced her leaders to look for expansion elsewhere, to their ultimate destruction.

1 Stroke

In a piston-engine, the lowest position reached by the piston and crankpin of any particular cylinder. This occurs when the crankpin is at the "6 o'clock" position, directly below the centre of the cylinder, and the connecting rod is vertical *(assuming an engine with the cylinder vertically above the crankshaft). At BDC, piston acceleration is at its maximum but piston velocity is zero.*

The B.E.2 was one of the basic designs produced in 1912 which went into mass-production and subsequent development in the 1914-18 War. It was a two-seat tractor biplane, designed by Geoffrey de Havilland at the Royal Aircraft Factory, Farnborough. It was found to have very practical flying qualities and, as it was Government policy at that time to order aircraft for the British forces from Factory designs, the B.E.2 was put into immediate large-scale production for the Royal Flying Corps. In the meantime, the aircraft was progressively refined and, in 1914, it was the B.E.2a which was the first British aircraft to go to war in France. The variant produced in greatest numbers was the B.E.2c, which became the standard mount of the Corps squadrons in France. It was the task of these squadrons to co-operate with the Army in spotting for the guns and making reconnaissance flights. Because the B.E.2c was inherently stable, it was not easy to manoeuvre quickly; furthermore, it had the disadvantage of having the pilot in the rear cockpit and the observer forward, hampering de-

B.E.2c of the RFC "presented by the Indian Nobles"

fensive fire. Consequently, it was easy to shoot down and heavy casualties were encountered by B.E.2c squadrons. This and later developments of the B.E.2 were used for other duties, including night fighting in the UK.

BEA (British European Airways Corporation), 284, 307, 734-736, 1309, 1439

British European Airways came into being as a result of the British Government's Civil Aviation Act of 1946. It began with continental routes, using Dakotas inherited from BOAC, supplemented by the first of a fleet of Vickers Vikings. In February 1947 it took over the small airlines which had been running domestic services, and soon established a large internal network. In accordance with its policy of buying British, when BEA ordered replacement aircraft it chose first the Ambassador and then the Viscount, being the sponsor of this important turboprop airliner which achieved success throughout the world.

Vickers Viscount 701 airliner in service with BEA, April 1953

The Viscount enabled BEA to expand and capture a large slice of the continental traffic, and from that time BEA has expanded steadily. Subsequent aircraft built to BEA's requirements include the Vanguard and Trident.
Headquarters: London. Member of IATA.

Beagle B.206 Series II, 1090

Beagle Aircraft was formed in 1962 to fill a gap in the British aircraft industry by producing competitive light aircraft. It had the backing of the Pressed Steel Company and absorbed Auster Aircraft Ltd. Four years later it was taken over by the British Government to ensure the continuation of a British light aviation industry.

The first product to make any impact on the market was the Beagle B.206, a twin-engined low-wing monoplane with acommodation for five to eight persons and available in either unsupercharged (Series I) or supercharged (Series II) versions. The Series II has a maximum cruising speed of 227 mph, service ceiling of 25,400 ft, and range of 1,600 miles. This version has been sold to many countries. The version produced as the Basset CC Mk 1 communications aircraft for the RAF can transport an entire V-bomber crew and their equipment,

this being part of the RAF's requirement.

Beagle Pup, 1639-1642

The Pup is a fully-aerobatic two/four-seat light monoplane, powered by a 100-hp Rolls-Royce/Continental O-200 or 150-hp Lycoming O-320 engine. It has a tricycle undercarriage and is in large-scale production

for UK and export customers. It cruises at between 112 and 147 mph, according to the engine fitted, the range varying between 540 and 760 miles.

Beardmore Inflexible, 549

In 1924 the shipbuilding firm of William Beardmore & Co Ltd, of Glasgow, acquired licence rights in the Rohrbach methods of metal construction for aircraft and subsequently designed and built the prototype Inflexible, resembling an enlarged Rohrbach Roland. It was a huge three-engined monoplane of 150 ft span, weighing 15 tons, and was the largest metal monoplane to have flown successfully in its day (1927). While it provided much information on large metal aircraft, no development potential was foreseen and Beardmore built no further aircraft.

Beirut International Airport, Lebanon

Distance from city centre: 10 miles
Height above sea level: 85 ft
Main runways:
 18/36 = 10,663 ft
 03/21 = 10,433 ft
Passenger movements in 1967: 1,018,000

Belfast, see Short
Belgium: Air Force

Force Aérienne Belge. The Belgian Air Force has existed under the above title only since

Republic F-84Fs of Belgian Air Force

October 1946, but it is in fact one of the oldest in the world, having originated as the *Compagnie des Aviateurs* in 1913, which two years later was renamed *Aviation Militaire*. Its early Blériots, Nieuports, Spads and Voisins played an important part in the air campaigns of the 1914-18 War, and after the peacetime run-down the force was equipped chiefly with British or French aircraft, many of them eventually built in Belgium. It became a component of the Army in 1925 with the new title *Aéronautique Militaire* and by the outbreak of

World War Two was equipped with Fox and Gladiator fighters and a handful of Battle day bombers. Escapees from occupied Belgium served with the RAF for the remainder of the war, and later formed the nucleus of the new post-war air force. Belgium now contributes an important part of the NATO air defence forces, and the

FAB's front-line combat types include F-104 Starfighter and F-84F Thunderstreak fighter/strike aircraft, with a substantial number of Mirage 5s on order to replace the latter.

Belgium: Chronology

1901. The Aero-Club of Belgium was founded, with the basic aim of gathering together those people involved or interested in ballooning.

1906. During the first years of the century, Belgian forces had shown interest in lighter-than-air flying, and a specialized corps was founded in 1906. Known as the *Aérostation Militaire*, it began using balloons for reconnaissance purposes.

1909. Baron de Caters qualified for the first Belgian pilot's licence. In the same year, a number of air shows were organized in Belgium, in which several national pilots participated.

7 July 1910. Foundation date of Belgian military aviation, on the authority of the War Minister, General Hellebaut. It

Laminne, a civilian who had been flying Farman for some months. The first squadron was formed later the same year 1912. Even at this early date, defensive guns had been mounted on some aircraft, and Belgium most probably had the first air-gunners in the world.

1914-1918. Belgian military pilots developed a number of aerial bombing techniques.

10 February 1919. The second known international commercial flight in the world was made between Paris and Brussels, a Caudron aircraft carrying five fare-paying passengers.

1919. The *Syndicat National d'Etudes du Transport Aérien* (SNETA) was formed to study the possibilities of commercial aviation. Prime object was to ascertain whether commercial aircraft could operate from Brussels to terminals in foreign countries. Some trials were made in the Belgian Congo. As a result of the studies the *Société Anonyme Belge d'Exploitation de la Navigation Aérienne* (more popularly known, then and now, as Sabena) was founded.

1925. Edmond Thieffry made the first flight from Belgium to the Congo, piloting a three-engined Handley Page airliner, the *Princesse Marie-José*. In the following year a military

"Princess Marie-José" Handley Page W8f of Sabena

was not until after World War Two that a separate Air Force came into being. First military pilots were trained by the Chevalier de

crew made a similar flight in a Breguet XIX 1939-1945. Belgian military aviation had few opportunities to enter into combat with the

German *Luftwaffe* in May 1940, due to lack of suitable fighter air-aircraft. Many pilots managed to join the British Royal Air Force and, by the end of the fighting in Europe, had accumulated a total of over 100 victories scored against the enemy.

1950. Sabena inaugurated the first scheduled helicopter postal network, using Bell 47s. Three years later it started helicopter passenger services with Sikorsky S-55s, subsequently changing to S-58s. At one time, this network included 12 scheduled stops in four countries, with unduplicated routes totalling 652 miles.

Sikorsky S-55 of Sabena at Brussels

Bell Model 47, 963, 1313
The Bell 47 was evolved from the original Model 30 that the Bell Aircraft Corporation built in 1943. The prototype appeared in 1945 and received its commercial licence on 8 March 1946. Production was begun that year and US Army and Navy orders ensued. The Model 47 is still in production more than 23 years later, although it has been modified to suit various rôles. It is produced not only in America but by Agusta in Italy and Westland in Great Britain.

It is basically a three-seat light helicopter, powered by a 305 hp Lycoming VO-540 engine; and most versions have a steel-tube, uncovered structure aft of the engine, and skid undercarriage.

Bell Model 204, 967
The Bell 204 helicopter won a US Army design competition in 1955, and has been in continuous production ever since to meet civil and military orders, well

over 5,000 having been built by 1968. The latest versions have twice the engine power and carrying capacity of the original XH-40 prototypes, the first of which flew on 22 October 1956. Since redesignated UH-1 and named Iroquois, the military version has become one of the major workhorses of the US war in Vietnam and has also been supplied to numerous foreign air forces. The majority of commercial Bell 204s have been built by Agusta in Italy and Fuji in Japan. Agusta

Bell UH-1B (204) of Japan Ground Self Defence Force

have also supplied military 204s to the Italian, Austrian, Netherlands, Saudi Arabian, Spanish, Swedish and Turkish forces. The commercial 204 is a 10-seater and has a larger rotor and slightly longer tail-boom than the military Iroquois. Military roles include troop transport and casualty evacuation; armed patrol versions can be fitted with machine-guns and unguided air-to-ground rockets.

Bell Airacomet (XP-59A), 293, 1135
Fourth nation to fly a jet aircraft was the USA, the aircraft in question being the Bell XP-59A Airacomet. Three prototypes were built, from designs initiated in September 1941, and the first of them flew on 1 October 1942, with two Whittle-type turbojets, designated General Electric I-A. In July 1943, two I-16 engines were fitted and, subsequently, the three aircraft were used for developing other engines. They were fol-

lowed by 13 pre-production YP-59As and 20 production P-59As, the latter powered by the General Electric J31-GE-3 turbojet, conferring a top speed of 413 mph.

Bell Airacuda (XFM-1), 819

Bell XS-1 (X-1), 297, 300
Soon after World War Two, the USA projected a research aircraft to explore the transonic regions of flight. It was built by the Bell Aircraft Corporation as the XS-1, and the method of flight was to air-launch it from a B-29 mother-plane, whence it accelerated and climbed to its operational height of approx 70,000 ft. Here it completed the required test and then glided to land at Edwards Air Force Base, California. The XS-1 was powered by a Reaction Motors XLR11-RM-5 motor, comprising four independent rocket chambers which could be used individually or together. The programme was to fly the aircraft at higher and higher Mach numbers until the transonic region was reached and passed. Severe buffeting was encountered at Mach 0.94, but the eventual transition to Mach 1 was otherwise uneventful. The first supersonic flight was made on 14 October, 1947, the pilot being Captain Charles Yeager, USAF.

The aircraft was later developed into the X-1A, with which trials were continued, and eventually these aircraft reached a speed of Mach 2.5 and a height

of 90,000 ft.

Bell XV-3, 985
Two Bell XV-3 convertiplanes were built as part of a joint US Army-Air Force contract to investigate the possibilities of tilt-rotor aircraft for VTOL operation. The first flew in August 1955 and made several conversions from horizontal to vertical flight before being badly damaged in a crash. The programme was continued with the second aircraft. The XV-3 was 30 ft long and had a wing span of 31 ft. At the end of each wing was a pod containing the rotor/propeller, which was tilted from horizontal to vertical by an electric motor. It was powered by a 450-hp Pratt & Whitney R-986 engine located in the fuselage behind the four-seat cockpit. Speed range was from 15 to 180 mph and a ceiling of 12,000 ft was achieved.

Bell Aerosystems X-22A, 1015-1020

Benoist flying boats, 258

Benz engines, see Daimler-Benz

Bert, Paul, 925

Bf 109 and Bf 110, see Messerschmitt

BHP (Brake horsepower)
The power developed by a piston-engine calculated from knowledge of the actual work done at its output shaft. The output shaft is attached to a dynamometer, a testing device which imparts a known torque tending to prevent the engine shaft from turning. The known brake load multiplied by the crankshaft speed gives the bhp. For a four-stroke engine, the bhp is equal to bmep (*q.v.*) multiplied by swept volume multiplied by crankshaft speed divided by 792,000. Brake horsepower multiplied by 100, divided by the indicated horsepower (obtained from an indicator diagram), gives the mechanical efficiency, which allows for work done on the in-

duction, compression and exhaust strokes and also for energy wasted in friction and in driving accessories.

Bienvenu, 145, 161

Blackburn, see also Hawker Siddeley
The name of Blackburn is associated indissolubly with the manufacture of a long line of aircraft types for the British Royal Navy, although Robert Blackburn was a pioneer aviator in his own right for some years before his manufacturing company was formed.

The company's first official order came from the Admiralty in May 1914, when it was asked to build a dozen B.E.2c biplanes for the RNAS. Later in the 1914-18 War it built the GP seaplane, from which was developed the Kangaroo land-based bomber. In the inter-war years the company specialized in torpedo-bombers and fleet spotter aircraft, including such well-known types as the Dart, Blackburn, Ripon, Baffin and Shark biplanes. Blackburn

then built the Fleet Air Arm's first operational monoplanes—the Skua dive-bomber and Roc turret-armed fighter—and, as World War Two ended, the Firebrand torpedo strike fighter was about to enter service. The line of Fleet Air Arm combat aircraft ended with the Blackburn (now Hawker Siddeley) Buccaneer transonic strike aircraft, powered by two Rolls-Royce Spey turbofans and capable of carrying a 16,000-lb weapon load internally and under the

wings.

Non-naval Blackburn products included the Lincock fighter, Bluebird and B.2 lightplanes between the wars, and the four-engined Beverley troop and freight transport which served with the RAF from 1956.

Blackburn Bluebird, 538
The Blackburn Bluebird was designed and built in the nineteen-twenties to the classical biplane formula for light aircraft which had been pioneered by de Havilland with the Moth. It featured side-by-side seating and, in its later forms, metal construction. It was produced in quantity by Saunders-Roe Ltd. Several notable long-distance flights were made by Bluebirds.

Blackburn monoplane, 630
Robert Blackburn of Leeds, Yorkshire, was inspired to build aircraft in 1908 as a result of a visit to the Paris *Salon* of that year. In 1909 he built his first monoplane, which owed much to the Blériot, but

Blackburn Ripon torpedo bomber (500-hp Napier Lion)

achieved little success with it. His second aircraft, built in 1910, bore greater resemblance to the Antoinette. It was more successful and several "production" models were built, as two-seaters. One of them made a tour of the west of England in 1911 and a double crossing of the English Channel. From these beginnings stemmed the long line of Blackburn aircraft, culminating in another two-seat monoplane, the Buccaneer strike aircraft of today.

Blanchard, Jean-Pierre, 115

Born in 1753, Blanchard was not only a designer of "aerial carriages" but a balloonist of note. His original idea was the "flying boat" designed in 1781. In 1784 he first took to the air in a balloon and subsequently made the first ascents in Germany, Holland, Belgium, Switzerland, Poland and the United States. At the same time, he was intensely interested in developing the parachute and dropped many animals from his balloons. He never made a parachute descent himself.

Blériot, Louis, 193, 195, 209, 217, 221, 222, 229

Originally a pilot of aircraft made by Voisin, Louis Blériot determined to make aircraft to his own design and settled on the idea of the monoplane when most others were constructing biplanes and triplanes. His first aircraft, the "Canard", was built in 1907, and by the end of that year he had built three different designs. He also developed flying controls essentially as we have them today, with a control column actuating the elevators and wing-warping (ailerons) and a rudder bar for the rudder. He gained world wide recognition and fame on 25 July 1909 when he flew his monoplane No. XI from Baraques on the French Coast to Dover — a flight which brought with it the realization that political and social changes were inevitable as a result of perfection of a practical aeroplane. He continued to produce further aircraft and his monoplanes went to war in 1914 with both the French and British forces. He formed a company for aircraft construction and this became one of the main French aircraft manufacturing concerns between the wars, until it was nationalized in 1937.

Blériot 1909 monoplane, 193, 195, 209, 217, 524

See also Blériot, Louis

Blind Approach Beacon System, see BABS

Blind flying panel

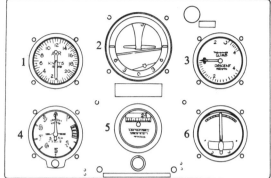

1 Air speed indicator
2 Artificial horizon
3 Vertical speed indicator
4 Altimeter
5 Gyro compass
6 Turn and slip indicator

The blind flying panel, introduced into the Royal Air Force in the nineteen-thirties, was a standard panel carrying six basic flight instruments, comprising altimeter, air speed indicator, vertical speed indicator, turn and slip indicator, gyro compass and artificial horizon. This standard panel was made common to all aircraft in RAF service, and all the instruments were grouped in the same place on each panel. This provided familiarity for pilots, in that they knew where to find the basic flight instruments in relation to one another, whatever type of aircraft they were flying; also, all the basic instruments were within easy eye-span of one another.

This innovation improved the standard of instrument flying in the RAF and paved the way for night and bad weather operations in World War Two.

Bluebird, see Blackburn

Blue Steel missile, see Hawker Siddeley

BMEP or bmep (Brake Mean Effective Pressure), 1100

In a piston-engine, the mean effective pressure produced in the cylinders that is put to useful work. Numerically it is equal to Indicated Mean Effective Pressure (IMEP) multiplied by the engine's mechanical efficiency; it is also equal to the brake horsepower divided by the swept volume multiplied by engine speed. BMEP is effectively the specific brake horsepower—in other words, the power produced per unit of swept volume per unit of crankshaft speed; it is thus a yardstick by which all piston-engines may be judged. In normally aspirated engines, BMEP varies from 120 to 160 lb/sq in; in supercharged engines it varies from 150 to 330 lb/sq in.

BMW (Bayerische Motorenwerke AG)

German engine manufacturing company, formed in 1916. Its first aero engine was produced toward the end of the 1914-18 War and was a six-cylinder in-line unit, intended for high altitude use, with oversized cylinders and a high compression ratio. The company was purchased shortly after the Armistice by BFW (Bayerische Flugzeugwerke), but retained its original title until 1939, when it became known as the Bayerische Flugmotorenbau GmbH. It took over in 1939 the Brandenburgische Motorenwerke GmbH, manufacturer of the Siemens-Halske Sh-14 radial engine, and renamed that company BMW Flugmotorenwerke Brandenburg GmbH. Engines manufactured by the latter company were marketed under the name Bramo (or Bramo Fafnir), while those of the original company were given BMW prefixes. Principal BMW engine of World War Two was the BMW 801 14-cylinder two-row radial of 1,580 hp, fitted *inter alia* to many versions of the Fw 190 fighter and the Ju 88, Ju 188 and Do 217 bombers.

BOAC (British Overseas Airways Corporation), 307, 1283, 1439

BOAC was formed on the eve of World War Two to accomplish the nationalization of Britain's external airlines. It took executive control of Imperial Airways and British Airways on 1 April 1940, its first task being to provide essential air transport links under the difficulties of a World War. It gained much experience during the next five years, and under the Civil Aviation Act of 1946 went ahead to develop Britain's long-distance routes, having fostered the reopened European air routes which it then handed

Boeing 707-436 of BOAC on flight line at Renton

over to British European Airways.

Since then BOAC has expanded in step with the growth of world air transport and has become one of the world's leading inter-continental carriers.

Headquarters: London. Member of IATA.

Boeing

Originated by William E. Boeing as the Pacific Aero Products Corporation, this leading US manufacturer became The Boeing Airplane Company on 26 April 1917; the word "Airplane" was dropped from the title in May 1961.

The first aeroplane designed and built by Boeing, in collabora-

William E. Boeing

tion with Conrad Westervelt of the US Navy, was the two-seat B & W twin-float seaplane of 1916. Two were built and sold to the New Zealand Government, and from them was developed the Model C, built in quantity for the US Navy and in smaller numbers for the Army. First commercial aircraft design was the B-1, a single-pusher-engined flying boat which later developed into the more successful Model 204. Between the wars, Boeing will be remembered chiefly for the Monomail and other single-engined mailplanes; for the Model 247 twin-engined all-metal airliner with retractable undercarriage; and for the P-12 and P-26 series of Army and Navy fighters.

In 1935 came the Boeing Model 299, forerunner of the B-17 Fortress and B-29 Superfortress bombers of wartime fame; and in 1938 the first of the giant Model 314 "Clipper" flying boats, built for Pan American's services across the Atlantic and Pacific. Three of these were later sold to

BOAC, who operate them across the Atlant during the early part World War Two. An other pre-war airline to serve usefully as war transport was th Model 307 Stratoline built for PAA an TWA.

From the B-29 Supe fortress stemmed th military Stratofreight and commercial Strat cruiser after the wa followed by Boeing first jet bombers, th B-47 Stratojet and th eight-engined B-5 Stratofortress. The Boe ing 707, ordered orig nally in flying tanke form to refuel the B-47 and B-52s of Strateg Air Command, was co currently developed in a passenger-carryir aircraft—a step whicl when Pan American pt it into service over th North Atlantic in 195 launched the biggest ai line spending spree i the history of aviatio In addition to the 70 "family" of jet trans ports, Boeing has sinc launched the tri-jet 72 the twin-turbofan 73 and the huge 74 "jumbo jet", and is cu rently engaged upon th design of the first U supersonic transport.

Vertol Aircraft Co poration was taken ove in March 1960 and now the Vertol Divisio of Boeing. Current hel copter products includ the Model 107 and th CH-47 Chinook tran ports.

Boeing 247, 279, 287

It was the Boeing 24 which initiated the grea step forward in air tran port in the mid-'thirtie A sleek, twin-engine low-wing monoplan with retractable unde carriage and contro able-pitch propellers, offered a cruising spee of 184 mph. United A Lines was first to put into service, with whor it achieved great su cess. Many were sold but its true significanc was overshadowed b the Douglas DC-2 whic followed hard on i heels.

Boeing 707, 309, 399, 413, 469-471, 576, 579, 613, 637, 791, 837, 852, 1189, 1201, 1237, 1243

Boeing 707–420

Boeing, first off the mark with a jet transport in the United States, has since reaped a huge reward. The prototype of the Boeing 707 appeared in 1954, making its first flight in July. The USAF almost immediately ordered a developed version for service as a tanker aircraft, designated KC-135. In the following year, this became the basis for a commercial version and, under the impetus of an initial order by Pan American World Airways, the 707 and similar 720 virtually scooped the board as regards worldwide jet airliner orders.

The first production commercial 707 flew on 20 December 1957 and the type went into transatlantic service in October 1958. From then on, variations in powerplant and fuselage length have enabled 707s and 720s to be made to suit any transcontinental or transocean airline

and orders have poured in. By August 1969, 836 had been sold, outnumbering all the other long-range jet airliners built outside Russia. There is still no sign of production terminating.
Boeing 727, 579, 791, 837

In 1959 The Boeing Company decided to enter the short/medium-range jet airliner field with a three-engined aircraft, following the layout established for the Hawker Siddeley Trident two years earlier. This aircraft was the Boeing 727, and featured a rear-engined layout with a T-tail, one engine being placed in the rear of the fuselage and the

Boeing 727–200

other two on pylons beside it, leaving a completely free, uncluttered wing. The first 727 flew in February 1963 and it has subsequently proved most successful, being developed into cargo and cargo/passenger versions, some models with a longer fuselage. Cruising at 596 mph, the basic 727-100 has a range of 3,430 miles with a 25,000 lb payload and a service ceiling of 36,000 ft.
Boeing 737, 616, 754, 791, 839

The Boeing 737 represents The Boeing Company's entry into the short-haul jet airliner market. Decision to go ahead was given in February 1965, and the first aircraft flew in April 1967. The design went against the current trend of rear engines and T-tails, by suspending the two Pratt & Whitney JT8D turbofans under the low, swept wing—thus, in effect, producing a miniature, twin-engined Boeing 707. The 737 has proved very popular and a Series 200 version with extended fuselage flew in August 1967, earning

Boeing 737–200

more orders than the original model. It can carry between 88 and 113 passengers over 2,000 miles at a cruising speed of 573 mph at 20,600 ft.
Boeing 747, 315, 317, 617-18, 763-5, 1223, 1225-8, 1237-8, 1439, 1441

Boeing 727 of Pacific Southwest

of layouts enables anything from 363 to 490 passengers to be carried at a cruising speed of Mach 0.89 for a range of 4,600 miles (with maximum payload).
Boeing Air Transport, see United Air Lines
Boeing Clipper (Model 314), 745, 761
Only a few Boeing 314s were built—six for Pan American and three for BOAC, but they represented a big step forward in air transport because of their long-range capabilities (3,100 miles with 40 passengers). The 314 was one of the first really practical trans-ocean aircraft and in 1939 PAA introduced three on their Atlantic service and

three on their Pacific route.
The 314s provided a high standard of comfort for their passengers. The three sold to BOAC gave fine wartime service, and transported Prime Minister Winston Churchill on several of his historic journeys. They were withdrawn from service in 1948.

Boeing Fortress (B-17), 299, 397, 569, 885, 1109

B-17F Fortress; this model introduced the clear moulded plastic nose and "chin turret"

A major heavy bomber used by US Forces in World War Two, the B-17 played a large part in the air war over Europe from 1943 onward, being used chiefly in massed daylight raids. Casualties were high, but by virtue of numbers the raids were successful.
The B-17 was developed considerably, having been initiated in 1935 with the Boeing Model 299. By 1941 the B-17C version was current, and twenty of these were supplied to the RAF for high-altitude bombing raids. They were found to be ineffective and such raids were abandoned, after which the Fortress, in this and its later versions, served the RAF mainly with Coastal Command. The

USAAF, learning from the RAF's operations, produced the B-17E, a more heavily-armed version. With this, the Americans themselves began daylight bombing operations in Europe and gradually built up the 8th Air Force with this and the later B-17F and B-17G variants. In the meantime, the B-17 had not been idle in the Pacific area, where it had begun operations soon after the Pearl Harbor attack; soon it was in service wherever the US Army Air Force was operating. Production was prodigious, at its height over 500 being built each month; and in August 1944 the US Army Air Force had a total of 4,574 on

Boeing 747

Following Boeing's great successes with the 707, 727 and 737 jet airliners, the 747 heralds a new class of aircraft, presently described as "jumbo-jets". The first 747 flew early in 1969 and orders for 183 had been placed by August 1969. With a wing span of 195 ft 8 in and length of 231 ft 4 in, a variation

Some of the first 2,000 commercial jetliners built at Renton: Boeing 707s in final assembly

Boeing 747, supported on air cushions while having compass systems swung

strength. The B-17 continued in service, throughout the war and for some time afterward, in a multiplicity of variants.

Boeing Monomail, 731

Boeing Monomail (Model 200), May 1930

Designed and built in 1930, this aircraft was well ahead of its time. It was a sleek monoplane of 59 ft span, with a fuselage containing three mail compartments and a centrally-positioned cockpit for one pilot. The engine was a Pratt & Whitney Hornet of 575 hp, giving the aircraft a maximum speed of 158 mph—a figure contributed to by the then-revolutionary retractable undercarriage. The Monomail did not find favour with those airlines with US Mail contracts, and its further development was discontinued.

Boeing Stratocruiser, (Model 377), 733, 739-740, 747, 757, 1085, 1109

The Stratocruiser grew out of the military transport aircraft built for the USAF as the C-97. It became the first luxury-class airliner on the transatlantic routes after World War Two and was used by Pan American, BOAC and American Overseas as well as by United and Northwest Airlines within the USA.

It was a large aircraft for its day, with a wing span of 141 ft, and was powered by four Pratt & Whitney R-4360 Double Wasp four-row radial engines providing, with water injection, 3,500 hp each for take-off. A pressurized cabin with a downstairs cocktail bar provided comfortable accommodation

for 55-100 passengers.

Boeing Stratofortress (B-52), 577, 581, 585, 637, 837, 851

Although similar in layout to the B-47, the B-52 is a completely different aircraft. It is an eight-jet long-range heavy bomber which first flew in prototype form in 1952 and entered service in 1955. Its eight Pratt & Whitney J57 turbojets give it a maximum speed of over 650 mph and it has a range of over 9,000 miles. It has been developed extensively and can carry Hound Dog air-to-surface missiles. Production ceased in 1962 with the 744th aircraft but the B-52 still forms a large proportion of the USAF Strategic Air Command's strength.

Boeing Stratofreighter (C-97), 757, 1117

The C-97, or Boeing Model 367, grew out of the B-29 Superfortress. A new fuselage, of "double-bubble" cross-section and pressurized, brought new capabilities to the USAF in terms of cargo and personnel lift over long distances, and the C-97 was ordered into mass-production, the first aircraft being delivered in October 1949. As well as serving in the transport rôle, it was also built in quantity for the tanker rôle, being the standard refuelling aircraft for Strategic Air Command Wings until replaced by the KC-135 tanker version of the Boeing 707 jet airliner.

Boeing Stratojet (B-47), 577, 835

The B-47 was built by The Boeing Airplane Company just after

World War Two as a very advanced, high-performance medium bomber. It was the first service bomber with thin, narrow, swept-back wings and jet engines in under-wing pods. It also incorporated a bicycle undercarriage and all these features established a reputation for the aircraft of being difficult to land. The first prototype flew on 17 December 1947 and the type entered service in the early 1950s, equipping a large part of the USAF's Strategic Air Command. The B-47 was developed extensively and saw service in the tanker rôle and as a long-range high-altitude reconnaissance aircraft.

Boeing Stratoliner (Model 307), 755, 925

This aircraft has its place of fame in the history of civil flying as the first airliner to enter scheduled service with a pressurized cabin for the passengers. Pan American and TWA were the two airlines that introduced the type in 1940. With a service ceiling of 23,300 ft, the Stratoliners enabled flights to be made "over the weather", providing a smoother, more comfortable journey for passengers. They were built to carry 33 passengers, and had a maximum range of 2,340 miles.

One or two Stratoliners are still in service in various parts of the world, which bears testimony to the soundness of their design.

Boeing Superfortress (B-29), 300, 885, 887, 893-4, 1085, 1119

The Boeing B-29 Superfortress will always have its niche in history as the aircraft which dropped the atomic bombs on Japan in 1945. It was conceived as a replacement for the B-17 and B-24 bombers, with which the United States had built up its world wide strategic bombing force in 1943-45. Apart from being stretched in almost every parameter,

the B-29 was also pressurized, enabling much higher altitudes to be flown. The prototype first flew on 21 September 1942, but development was held up when this aircraft caught fire and crashed in February 1943. Further aircraft were flying by July, and development continued unhindered. By the time the B-29 was in service, the obvious operational field for it was the Far East and the first Wing, the 58th (B) Wing, began operations against Japan in June 1944. In January 1945, it was used at night instead of by day, with increased effectiveness. Even then, preparations were being made for the use of atomic weapons, and 6 August 1945 brought the climax of this effort when B-29 44-86292 *Enola Gay* dropped the bomb on Hiroshima; three days later the second bomb was dropped on Nagasaki. After the war, the B-29

Boeing SB-29 Superfortress of US Air Force Air Rescue Service

remained in service with the USAF for many years and was also used by the RAF as the Washington B.1. Three B-29s which landed short of fuel in Russia were confiscated and, in due course, Russia produced its own copy, the Tu-4. B-29s were also used as engine testbeds, for flight refuelling, and as motherplanes for rocket development aircraft, as well as serving on air-sea rescue duties and for weather reconnaissance.

Boeing Superfortress (B-50), 1117

A 75 per cent rebuild of the Boeing B-29 Superfortress, the B-50 followed it into service with the USAF's Strategic Air Command in

1948. With a range of over 6,000 miles and an operating height of over 40,000 ft, it was a formidable long-distance bomber for the first decade after the war. It was relegated to tanker duties in the late 'fifties, providing a three-point tanker system which enabled three fighter aircraft to be refuelled at once; this KB-50J version was used in support of Tactical Wings of the USAF for their ferrying journeys about the world.

Boeing Trimotor (Model 80-A), 737

The Model 80-A was built by Boeing and introduced by its own airline, Boeing Air Transport, in 1929. It was a large biplane, powered by three Pratt & Whitney Hornet engines, each producing 525 hp, and could carry 12 passengers in luxury or 18 in standard accommodation. The 80-A served in this capacity for some time but was soon made obsolete by the advent of such twin-engined monoplanes as the Boeing 247 and the Douglas DC-2.

Boeing-Vertol 107, 967

This helicopter was developed by Vertol in 1956 as a twin-turbine transport for civil and military applications. It

employs tandem rotor and is powered by tw General Electric T5 shaft-turbines. It is 8 ft long overall, and ha accommodation for 2 passengers. The first 10 flew in April 1958 an the design was deve loped into productio versions for the U Navy and Marin Corps; it also serve with civil helicopte operators and with th Canadian, Japanese an Swedish forces. Com mercial and non-US military 107s are buil in Japan by Kawasaki

Bogie landing gear, 571 573, 579, 581, 582, 58

The bogie landing gea came into widesprea use following Worl War Two, with th need to absorb th greatly - increasing all up weight of aircraft i a unit capable of fittin into an aircraft and a the same time copin with the high landin and take-off speeds tha became the norm. Usu ally, the undercarriag leg terminates in a hori zontal beam supportin any number of axle with the landing whee attached. Thus, the loa of the aircraft is sprea over four, eight or mor wheels, reducing th weight taken by any on wheel and also reducin the load to be carried b the concrete apron o runway. It also mili tates against damag caused by tyre failure enabling the load to b carried by other whee if one tyre should burst Bogie gears have be come standard for al large modern aircraft.

Civil Boeing Vertol Model 107 helicopter (New Yor Airways)

Bombay Airport (Santa Cruz), India

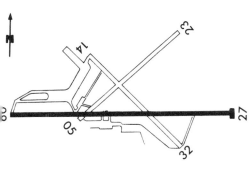

Distance from city centre: 18 miles
Height above sea level: 27 ft
Main runway:
  09/27 = 10,500 ft
Passenger movements in 1967: 1,076,000

Bonding, 441-3, 463, 466, 976

Bonmartini, 553

Boost pressure, 1075, 1087
In a supercharged piston-engine, it is possible to use the supercharger not merely to restore power lost as a result of increasing altitude, but also to boost power above the maximum which could be achieved by a normally-aspirated engine. This is done by using the supercharger to raise the pressure of the combustible mixture in the induction manifold to a "boost pressure" greater than the pressure in the surrounding atmosphere. In the atmosphere at sea level the pressure is about 14.7 lb/sq in; the highest boost pressures reached by highly-developed piston aero-engines are of the order of 25 lb/sq in and, assuming there is no great fall in engine efficiency, this represents almost twice the power of a normally-aspirated engine. Boost pressure can usually be regulated by an automatic control system to ensure that the engine cannot become overstressed, and to keep fuel feed continuously matched with bulk airflow. In many engines boost pressure is restricted at sea level by a capsule sensitive to atmospheric pressure, and is allowed to reach its full value only at the

"rated altitude," beyond which height the supercharger would be unable to prevent the pressure in the induction manifold from falling.

Bore
In a cylinder of a piston-engine, the internal diameter. The piston is a loose fit but is equipped with rings which fit the bore tightly and minimize loss of gas or lubricating oil.

Bore sighting
The process by which fixed guns carried by a military aeroplane are aligned accurately. They may be harmonized to converge at a point a given distance in front of the aircraft or may fire in parallel. Most bore sighting techniques involve the use of mirror-equipped bore-scopes or fibre optics in order to observe a target through the barrels of the guns.

Boston, see Douglas

Bottom Dead Centre, see BDC

Boulton, M. P. W., 625
This far-seeing Englishman predicted, in 1868, the need of and use for the aileron system of control for banking the wings of an aircraft. He took out British Patent No 392 for his invention—a patent which was overlooked even when practical aeroplanes had become commonplace.

Boundary layer control
When an aircraft is in flight, the layer of air in contact with the skin is actually at rest—that is, it moves with the surface. At some distance from the skin the air flows freely at the speed of

the aircraft. The intermediate layers, composed of turbulent and sluggish regions sliding over adjacent layers of air, constitute what is known as the boundary layer.

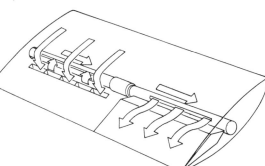

Boundary layer control. The thin layer of air next to the wing skin is sucked inside and carried through a duct to the flaps or ailerons to increase their efficiency.

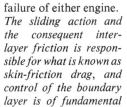

Boundary layer control on the Buccaneer. This aircraft employs extensive boundary layer control to increase lift and improve control. The tailplane has a boundary layer control slot underneath the leading-edge; the flaps and ailerons have powerful BLC slots blowing air across the top of each surface to increase lift. In addition, the leading-edge of the wing has a boundary layer control slot to prevent flow separation, which

could result from the extreme lift pressure created by the flap boundary layer control slots. Duplicated supplies and shut-off valves keep all slots fully blown after failure of either engine.
The sliding action and the consequent inter-layer friction is responsible for what is known as skin-friction drag, and control of the boundary layer is of fundamental importance to the achievement of a high lift coefficient. This control can be obtained by sucking away the turbulent layer through narrow slits in the wing surface or by injecting a high-velocity airflow tangentially across the wing from aft-facing slits close to the leading- or trailing-edges or, in a rather primitive manner, by employing vortex generators.
Boundary layer control can be used to improve the efficiency of flaps by ejecting a jet of air over the top surface. Such a

system may also be used to increase the effectiveness of control surfaces, by preventing breakdown of the airflow at large surface deflections.

Brake, disc, see Disc brake

Brake, drum, see Drum brake

Brake horsepower, see BHP

Brake Mean Effective Pressure, see BMEP

Bramo Fafnir engine, 911, see also BMW
This nine-cylinder air-cooled radial engine was used in World War Two in early versions of the Dornier Do 17 bomber, in the Fw 200 Condor maritime reconnaissance aircraft and in the Henschel Hs 126 army co-operation aircraft. Its maximum output was 985 hp.

Braniff International Airways
Formed in 1928 by the brothers Thomas and Paul Braniff as the Tulsa—Oklahoma City Airline, this company was renamed Braniff Airways in 1930 and Braniff International

Airways in 1948. It serves a large area of the central US and since 1948 has flown international services to Cuba and South America; the former South American airline Panagra (Pan American-Grace Airways) was absorbed in 1967, and Braniff's total route network is now well over 20,000 miles.
Headquarters: Dallas, Texas.
Member of IATA.

Brazil: Air Forces
Força Aérea Brasileira.
Came into being in January 1940 to combine the former air arms of the Army and Navy, which had taken an interest in aviation since just before the 1914-18 War. Prior to 1940, equipment was predominantly of US or British origin, a trend which continued throughout World War Two. During the war (which Brazil entered in August 1942), the FAB contributed transport services across the south Atlantic as well as maritime patrol services within Brazilian waters; and Brazilian pilots served with the US Twelfth Air Force in Europe in 1944-45. After a post-war rundown the FAB was re-equipped from 1947 with Thunderbolt fighters, Mitchell and Fortress bombers and Texan trainers from the US, to which was added in 1954 a substantial force of British Meteor jet fighters. The Texans, with later T-28Cs, now serve in the light ground attack rôle, while Douglas Invaders are now the standard light

Westland Whirlwind 10 (R-R Gnome) of the Brazilian Navy

bombers; the Meteors remain in service.

Marinha do Guerra. As indicated above, early Brazilian naval aviation was integrated with that of the Army in 1940; but following the acquisition of the aircraft carrier *Minas Gerais* a new, small naval air arm came into being, whose aircraft and crews are supplied by the FAB. Aircraft at present in service include Tracker and Neptune ASW types and Gnome-Whirlwind helicopters.

Breguet *Gyroplane Laboratoire* (1935), 957
The Breguet *Gyroplane Laboratoire* of 1935 was one of the first really controllable rotary-wing craft. It had an open-girder fuselage, with the pilot in an exposed cockpit at the front. The engine was mounted amidships, and conventional tail control surfaces were fitted. To overcome the torque problem, two counter-rotating rotors were used, mounted coaxially one above the other. The mechanical engineering was complex. As a result the machine was overweight and its lifting performance was poor, although it still represented a considerable advance upon other helicopters produced up to that time.

Breguet, Louis, 529, 957
The name Breguet is synonymous with the development of French aviation through the years. Louis and his brother Jacques first turned their attention to helicopters, and it is to them that the credit goes for building the first full-sized helicopter actually to leave the ground. This was in 1907 with a device employing twin biplane rotors. At that time it appeared that the helicopter was a "dead-end" development, for materials and powerplants were not light enough nor powerful enough. So Louis turned to conven-

tional aircraft and soon adopted metal construction for his biplanes, which went into quantity production before and during the 1914-18 War. He also developed the oleo strut (*q.v.*) at a time when most constructors were satisfied with rubber cord springing. The Breguet 14 bomber was the mainstay of French two-seater squadrons in 1917-18 and many were issued to the Americans in France. The type continued in service until 1930 and its post-war descendant, the Breguet XIX, was used in the Spanish Civil War. The XIX was also used for several long-distance flights in the mid-'twenties. During this period, Louis Breguet produced mainly single-engined biplanes, but in the early nineteen-thirties he concentrated on single and twin-engined sesquiplanes, mostly with a military application. He entered the flying boat field with his Saigon and Bizerte three-engined biplanes, and produced the *Gyroplane Laboratoire* in the mid-'thirties in association with René Dorand. With the nationalization of the French aircraft industry, Breguet stayed outside as an independent constructor and, just before World War Two, received a quantity production order for the Br. 690 twin-engined monoplane. Many were built before France fell. During the war, his company was forced to build for the *Luftwaffe* and the Vichy Government, and also developed the Br. 500, a twin-engined commercial aircraft. With the return to peace it was again active in design and development. First post-war product was the Type 76 series of large four-engined transports for Air France and the French Air Force. Louis Breguet himself died on 4 May 1955, with

his company prosperous and engaged in the development of the Atlantic anti-submarine aircraft, production of the Alizé carrier aircraft and development of the 940 and 941 STOL transports.

Breguet Br 941, 577
The Br 941 is a practical STOL transport, re-

Breguet Br 941S STOL transport of the Armée de l'Air *at Toulouse-Blagnac*

sulting from Breguet's many years of developing the deflected-slipstream technique. This involves blowing the slipstream from four propellers over the entire wing span and utilizing an extensive slotted-flap system on the wing trailing-edge. An engine synchronization system ensures a constant airflow over the entire wing, with the result that low landing speeds and short take-offs and landing distances are achieved. The Breguet 941 is a four-engined high-wing monoplane, with accommodation for 57 passengers, 40 troops or 24 stretchers. The prototype flew in 1964 and has been followed by four pre-production Model 941S aircraft. The design is sponsored in the USA by McDonnell Douglas as the McDonnell Douglas 188.

Bristol Aeroplane Company, 85, 1081-1083, 1097, 1283
This company originated as the British and Colonial Aeroplane Company before the 1914-18 War, building its own aircraft and maintaining one of the early flying schools. During the war it produced a series of light

scouts and a fighter monoplane, all of which saw service in small numbers. It also developed the Bristol Fighter which set the pattern for two-seat fighters and was one of the most useful aircraft in the RFC and RAF in that war. The "Brisfit" remained in service in large numbers for more than a decade after the Armistice, as a general-purpose aircraft, both at home and abroad. Between the wars the Bristol Company built a large number of aircraft, its most successful being the Bulldog fighter and the Blenheim bomber. The latter served in World War Two, being the RAF's most effective light bomber at the outbreak of hostilities. The same general confi-

guration led to the Beaufort torpedo-bomber, and the Beaufighter night fighter which played a significant part in the night defence of the UK from 1940 onwards. It was also developed as a strike fighter-bomber for Coastal Command, with great success.

After World War Two, Bristol produced the Freighter, with which Silver City Airways launched its highly-successful cross-Channel car ferries, and the vast Brabazon. This was followed by the Britannia, still in service today, and a number of helicopters. The helicopter division was taken over by Westland at the end of the nineteen-fifties. Since then Bristol has become part of British Aircraft Corporation and has tackled the major British share of the Concorde programme.

In addition, the Bristol Company has long been a producer of aero engines. In 1920 it began producing radial engines and scored an outstanding success with the Jupiter series, not only selling it well in the UK but also overseas. It was followed

by further engin "families", the Pegasu and Mercury in parti cular powering many o the aircraft with whic Great Britain entere World War Two. Fo some time the Compan had been applying th sleeve-valve principle t aircraft engines and th first complete engine o this type, the Perseu appeared in 1932. It wa followed by the Hercule and Taurus, both o which went into pro duction. Developmen of the Hercules con tinued during and afte the War and it powere a number of post-wa military aircraft and air liners. The final sleeve valve engine from th Bristol stable was th Centaurus (*q.v.*). Bristo were not slow to ente the turbine field, turnin their attention at firs to turboprop engine for large aircraft. Th Theseus was the first t appear, in early 1947 followed by the Proteu producing up to 4,44 ehp. This was develope into the Couple Proteus for the Princes flying-boat. In th turbojet field, Bristo produced the Orpheu and began the Olympu family which power the Vulcan bomber and

Bristol Boxkites in the Filton erection shop, 1911

in a highly-developed form, the Concorde supersonic airliner.

Britannia Series 312

The Britannia was developed by the Bristol Aeroplane Co as a long-range turboprop airliner. It was built initially for BOAC, who used it in a short-fuselage form (Series 102) on their African and Eastern routes and in long-fuselage form (Series 312), with accommodation for 133 passengers and range of 4,268 miles, on trans-Atlantic routes. The later version had an export appeal, seven other airlines buying it. The Britannia was also built for RAF Transport Command with whom it has served in the long-distance trooping rôle. By the time it was in service, the world's airlines were turning more and more to turbojet airliners, and so its development and popularity were not as great as its capabilities warranted. It proved to have one of the best safety records of all airliners in scheduled service.

Bristol Centaurus engine, 1097

From experience gained with the Bristol Hercules engine came the Centaurus. Biggest of Bristol's sleeve-valve engines, it was an eighteen-cylinder two-row engine with a stroke of 7 in. In its most developed version, it produced a take-off power of 2,810 hp at 2,900 rpm. The Centaurus was in production at the end of World War Two, and powered the Vickers Warwick for Coastal Command and the F Mk II version of the Hawker Tempest. Post-war, it was built in quantity to power the Firebrand torpedo-fighter and Sea Fury

fighter for the Fleet Air Arm, the Beverley heavy-lift transport for the RAF and the Ambassador airliner for British European Airways.

Bristol Fighter (F.2B), 877

The Bristol Fighter began life as a replacement for the reconnaissance B.Es. which were being shot down in great numbers on the Western Front in 1915-16. Additionally, it was thought that it might be useful on scouting patrols. The first prototype flew in September 1916, followed quickly by a second, and they proved so promising that a production batch was ordered without delay. The first squadron, No 48, went to France in March 1917 and flew into action in April; at first it seemed that the Bristol Fighter would be unsuccessful, for within ten days 12 had been lost. However, as soon as the aircraft was used as though it were a single-seat fighter, the position was reversed and it quickly became a formidable combat aircraft. Lt McKeever, one of the Bristol's finest exponents, achieved most of his 30 victories in this type. It operated throughout the last years of the war with great success and on into the 'twenties, when it became a general-purpose aircraft serving particularly well in the Middle East and India for more than a decade.

Bristol Freighter (Type 170), 775

The Bristol Freighter was conceived at the end of World War Two as a freight or passenger transport aircraft, of simple basic design, for carrying large payloads over comparatively short stages. It was a high-wing monoplane with fixed under-

carriage, powered by two 2,000 hp Bristol Hercules engines. A large number were built, with production extending into the late nineteen-fifties. Their nose loading doors, which opened like two large jaws, enabled bulky loads, the size of the cross-section of the freight hold, to be put aboard quickly. Availability of the

Bristol 170 Freighter used as cross-Channel car ferry

Freighter led to the development by Silver City Airways of the car-ferry service over the Channel. By employing the aircraft in conjunction with specially designed loading ramps, cars could be driven to an airfield near the coast (Silver City eventually built its own airfield at Lydd), loaded, flown across the Channel in 20 minutes and unloaded in France. The operation proved extremely successful and tens of thousands of cars were transported in this way, the business only beginning to fall away when the shipping companies built really competitive and cheaper ship ferries in the late nineteen-sixties. Besides this facet of the Freighter's operations, many were sold throughout the World for tasks requiring a large-capacity hold for a short-distance flight and for general "mixed traffic" work. Four air forces also used the Freighter.

Bristol Hercules engine, 1081, 1097

The Bristol Hercules two-row fourteen-cylin-

der radial engine, operating on the then-new sleeve-valve principle, first appeared in 1936. With a bore of $5\frac{3}{4}$ in and stroke of $6\frac{1}{2}$ in, it had a capacity of 2,360 cu in and provided a take-off power of 1,290 hp at 2,800 rpm. It found its first applications in flying boats, including the Short "Golden Hind" class, developed from the "Empire" 'boats and intended for the Atlantic run, and the Saunders-Roe Lerwick coastal patrol flying boat. By the outbreak of World War Two later versions of the Hercules were developing 1,375 hp at 4,000 ft, and had been fitted to the Mk III version of the Vickers Wellington bomber. The engine was ideally suited to this airframe and the Mk III became the principal version of the Wellington to serve with Bomber Command. It also operated with this Command as the power plant of the first of the four-engined heavy bombers, the Short Stirling; by then, the Hercules XI had appeared, developing 1,600 hp. It was also used on Halifaxes, Lancasters and Beaufighters and powered many of the immediate post-war civil aircraft such as the Viking, Bristol Freighter and Short Solent flying boat, giving 2,000 hp in some versions.

Bristol Jupiter engine, 1073

The Jupiter engine was the progenitor of the wide range of Bristol air-cooled radial engines. It was really a family of nine-cylinder

single-row 400-500 hp radials, tailored to meet a variety of operating conditions. In large-scale production in the nineteen-twenties and early 'thirties, it was used by practically every important air force in the world and by a large number of commercial airlines.

Bristol Orion engine (BE 25), 1309

The Orion benefited from Bristol's experience with the Proteus and Olympus, being a supercharged turboprop. It was intended to produce 4,000 shp and to have an extremely low fuel consumption. However, it was cancelled at an early stage.

Bristol Pegasus piston-engine, 1107

The Pegasus range of radial engines was introduced by the Bristol Aeroplane Company in 1931, when the Jupiter range was reaching its limit of development. In effect, it took over from the Jupiter and was used extensively during the 'thirties and in World War Two. The original Pegasus (a nine-cylinder radial) produced 550 hp, but by the outbreak of war the later versions were producing 830 hp. Swordfish, Harrow, Hampden, "Empire" 'boat, London, Sunderland, Stranraer, Walrus, Wellesley and Wellington are among the well-known aircraft which were powered by the Bristol Pegasus. The name Pegasus was revived for the Bristol Siddeley BS.53 vectored-thrust turbofan engine that powers the Hawker Siddeley Harrier strike aircraft; this engine is described under the Rolls-Royce (Bristol) heading.

Bristol Perseus engine, 1097

The Perseus was the first Bristol sleeve-valve engine to go into production, in 1932. It was used in the Mk IV version of the Vickers Vildebeeste biplane torpedo-bomber in the RAF, and gave much

valuable operational service. It was a nine-cylinder radial engine producing, in its most developed Mk XIVc form, a take-off power of 890 hp. The Perseus was used in the Botha trainer, Flamingo airliner, Skua dive-bomber and Roc fleet fighter, modified "Empire" flying boat and Mk II Lysander.

Bristol (Prier) monoplane (1911), 664

Bristol Pullman, 729

In 1920, when the Bristol Aeroplane Co found that its four-engined triplane bomber, the Braemar, was not required by the Royal Air Force, it took the second prototype and converted it into an airliner. With luxurious seating for six passengers and an enclosed cabin for two pilots, it was a great advance on contemporary types but went no further. Its performance was good, with a maximum speed of 135 mph.

Bristol Scout, 873

The Bristol Scout first appeared at the 1914 Olympia Aero Show. It was a single-bay biplane, powered by an 80 hp Gnome mounted at the front of a slim single-seat fuselage. It became the first Bristol scout, or fighter, aircraft to be put into production and later versions were equipped with a single machine-gun firing through the propeller. The Bristol Scout served on the Western Front for the first two years of the 1914-18 War, two or three examples being issued to each reconnaissance squadron. Early versions had a top speed of 95 mph, increased to 110 mph in later aircraft.

Bristol Taurus engine, 1097

The Taurus was an enlarged development of the Perseus in the successful range of Bristol sleeve-valve engines of the nineteen-thirties. It was a two-row radial with fourteen cylinders, giving a power output of

1,090 hp for take-off.
It was used principally in the Bristol Beaufort torpedo-bomber and the Fairey Albacore fleet torpedo - spotter - reconnaissance aircraft during World War Two.

Bristol Siddeley BS.100 engine, 1233
The BS.100 was projected as a very large turbofan lift/thrust engine for supersonic use, employing swivelling nozzles as in the vectored-thrust Pegasus engine. It was intended to have a thrust in the 35,000 to 40,000 lb range, and the first engine was built and run in November 1964. With the cancellation of the P.1154 supersonic VTOL fighter, for which the BS.100 was intended, development of this engine was abandoned.

Bristol Siddeley Coupled Proteus engine, 831
The Coupled Proteus installation was developed specially for the Saunders-Roe Princess flying boat and the Bristol Brabazon, although it never flew in the latter aircraft. The Proteus engines were mounted side-by-side and coupled together by a train of double helical pinions so that the combined output of both engines, totalling 6,640 ehp, was delivered through a single shaft to a further gearbox where the power was divided to two contra-rotating coaxial propellers.

Bristol Siddeley Proteus engine, 1309
The Bristol Proteus was a large turboprop engine designed and built by the Bristol Engine Company in 1949 and intended for the Britannia airliner. It was developed as the 700 Series which went into production for this aircraft, with a maximum power output of 3,320 shp plus 1,200 lb static thrust. Two Proteus were coupled together to form the Coupled Proteus which was used in the Princess flying boat. Further development of the Proteus engine in

the Britannia brought a final take-off rating of 4,445 ehp.

Britannia, see Bristol

British Aircraft Corporation, see BAC

British Deperdussin Seagull seaplane, 361

British Deperdussin Thunderbug landplane, 361

British Eagle International Airlines
Known as Eagle Airways from 1953-60 and as Cunard Eagle Airways from 1960-63. Business included scheduled passenger services and charter and trooping flights within the UK and to Europe and North Africa, but the Company went into voluntary liquidation in November 1968.
Headquarters: London.
Formerly a member of IATA.

British European Airways Corporation, see BEA

British Overseas Airways Corporation, see BOAC

British United Airways, see BUA

British West Indian Airways, see BWIA

Bromma Airport, see Stockholm

Brown, A. W., see Whitten-Brown

Browning machine-gun, 879
The Browning machine-gun had its inception early in the 1930s when the British Air Ministry approached the Colt Automatic Weapon Corporation to find out whether the US company could adapt its 0.30-in calibre automatic gun to British 0.303-in rimmed cartridges. This was done, and licence arrangements were agreed between Colt and BSA. This weapon had a rate of fire of 1,200 rounds per minute, which was superior to anything else in RAF use at the time. The Browning became the gun that won the Battle of Britain, for it was installed in both the Hurricane and the Spitfire, as well as in all other British fighter aircraft using a fixed-gun

installation, until the general adoption of much heavier 20-mm Hispano guns in 1942. It was also adopted for use in many of the gun-turrets which formed the defensive armament of Britain's bomber and coastal reconnaissance aircraft, and served throughout the war in this way. The US forces used a 0.5-in calibre version in fixed, hand-held and power-driven turrets, and this much heavier pattern was also used by the RAF.

BS.100 engine, see Bristol Siddeley

BTH (British Thomson-Houston Company), 1131
An industrial company in the UK, which began experimenting with gas-turbines in 1933 in order to harness them to electrical generators. It was sub-contractor to Whittle's Power Jets company for the drawings and parts required for the first Whittle jet engine. This was first run, on 12 April 1937, in the BTH factory, the engine running away and not responding to throttle control. The cause of this was speedily traced. BTH later developed its own version of the Whittle W.1A engine.

BUA (British United Airways)
BUA, the largest British independent airline, was formed in 1960 by merging Airwork and Hunting-Clan, and now operates over routes within the UK and to Europe, Africa and South America. Operations include the all-freight Africargo service and charter and trooping flights.
Headquarters: London.
Member of IATA.

Buccaneer, see Hawker Siddeley

Buenos Aires Airport (Ezeiza), Argentine Republic
Distance from city centre: 22 miles
Height above sea level: 66 ft
Main runways:

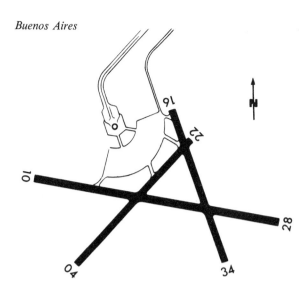

Buenos Aires

10/28 = 10,825 ft
04/22 = 7,200 ft
16/34 = 6,890 ft
Passenger movements in 1967 : 1,731,000

Bulgarian Air Transport
This Bulgarian airline was formed in 1947 as TABSO (Bulgarian Civil Air Transport) and from 1949-54 was operated with Soviet assistance. The fleet still consists entirely of Soviet-designed aircraft, but the airline has been fully Bulgarian-owned since 1954. It operates services internally and to major cities in eastern and western Europe and the Mediterranean area.
Headquarters: Sofia.

Bullpup missile, see Martin Marietta

Business aviation
This area of aviation activity has risen sharply in importance and scope during the past two decades, and today represents a booming industry in its own right. Any sizeable company whose management executives need to travel—and this means most of them—can find the operation of its own aircraft useful and profitable. There are aircraft of the right size and price for almost any company's budget, although the most widely used are the light twin-engined types seating from four to six people. Over distances of up to a thousand miles these can offer travel on terms, and at

speeds, that compare favourably with those of the regional airlines, with the added advantage that they can use either established airports or smaller airfields where the larger airlines cannot go. In this category the American Beech, Cessna and Piper range of light twins reign supreme, despite the existence of other useful and attractive types such as the

Business aviation: Cessna "Turbo System" Skyknight Executive

Beagle B. 206 (q.v.).
Higher up the scale are the bigger medium twins, which are really semi-airliners seating anywhere up to 20 people. They include the Handley Page Jetstream, Mitsubishi MU-2, de Havilland Canada Twin Otter and the Aero Commander range, as well as comparable types from Beech, Cessna and Piper. Among the most promising recent models is the 16-seat Piper Pocono, with two 500-hp turbo-supercharged engines. Jet types in this category include the Dassault Fan Jet

Falcon, Hawker Siddeley 125 and Lear Jet. At the most expensive level the largest propeller-driven business "twin" is the 24-seat Grumman Gulfstream, while even airliners of the size of the BAC One-Eleven, Boeing 737 and Douglas DC-9 are available with luxury executive interiors for the ultra-wealthy customer.
At the other end of the scale are a wide choice of small helicopters such as the FH-1100 and Jet Ranger, ideal for metropolitan commuting or inter-city business travel.

BWIA (British West Indian Airways)
Controlled since 196— by the Trinidad and Tobago Government, BWIA was founded in 1940 and was from 194— a subsidiary of British South American Airways and later of BOAC. Its network covers the Caribbean area and also extends to the eastern seaboard of the US and southward to Guyana

and Surinam.
Headquarters: Port of Spain.
Member of IATA.

By-pass ratio, 120—

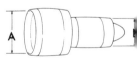

In a turbofan engine the ratio of the airflow (A) handled by the fan alone, that does not pass through the combustion chamber and turbine, to the airflow (B) that passes through the high-pressure compressor, combustion chamber and turbine. The first turbofans had by-pass ratios in the range 0.3-1.3, but more recent units intended for aircraft with speeds ranging from 380 to 640 mph have by-pass ratios between 3 and 8. In the latter, the total airflow through the engine is divided in the ratio of eight parts through the fan duct to one part through the turbine.

by-pass turbojet, 1195 A turbojet having a compressor which delivers more air than is needed by the combustion chamber and turbine. The excess air is discharged either directly to atmosphere through propulsive nozzles or, in the usual arrangement giving the engine its name, through a duct by-passing the rest of the engine but discharging the compressed air into the hot gas stream downstream of the turbine. In most engines of this type, the compressor is made in separate spools and the front spool is larger than the high-pressure assembly. Essentially, a by-pass turbojet is the same as a turbofan of the front-fan variety, and it is an anachronism to cast a distinction between them.

-5A Galaxy, see Lockheed
-82 Packet, see Fairchild
-97 Stratofreighter, see Boeing
-119 Packet, see Fairchild
-124 Globemaster, see Douglas
-130 Hercules, see Lockheed
-132, see Douglas
-133 Cargomaster, see Douglas
-141 StarLifter, see Lockheed
-160, see Transall

AAC (Civil Aviation Administration of China).

CAAC, the airline of Communist China, was formed in 1952 with Soviet assistance, but has been wholly Chinese controlled since 1954. Like its Soviet counterpart, Aeroflot, CAAC is responsible for internal air taxi, survey and agricultural activities as well as for domestic air services and international routes to other countries in south-east Asia.

Headquarters: Peking.
Cairo International Airport, Egypt

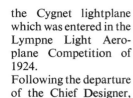
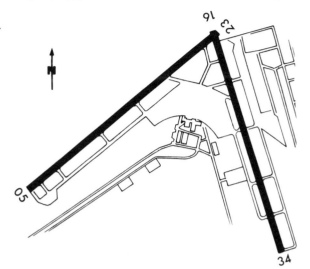

Distance from city centre: 16 miles
Height above sea level: 366 ft
Main runways:
  05/23 = 10,827 ft
  16/34 = 10,278 ft
Passenger movements in 1966 : 1,262,000
Callao Airport, see Lima
Camm, Sir Sydney, 1273

Sir Sydney Camm

One of the most famous of British designers, Sir Sydney Camm started with model aircraft before the 1914-18 War, and then joined the Martinsyde Company, with which he gained aircraft engineering experience before entering the design department. He joined the H. G. Hawker Company in about 1923 and it is his work with this company and its successors for which he will always be remembered. The first aircraft he designed was the Cygnet lightplane which was entered in the Lympne Light Aeroplane Competition of 1924.

Following the departure of the Chief Designer, W. G. Carter, in 1925, Camm was given this post. His first products were mostly adaptations of the Woodcock fighter; his first production success was the Hawker Horsley bomber. A prolific series of aircraft flowed from his design office, but his first real winners came in 1929 when he produced the Hornet single-seat fighter and Hart two-seat bomber, both powered by the Kestrel engine. The Hart entered service first and was faster than any RAF fighter sent up to intercept it in exercises. It remained in production for nearly a decade in its many forms, begetting a whole family of two-seat biplanes. The Hornet, renamed Fury, achieved equal fame as an RAF fighter in the 'thirties. Although preoccupied with a welter of variants of the Hart, Sydney Camm turned his attention to one of the most significant aircraft of its time. This was a monoplane fighter with retractable undercarriage and the new Merlin engine. Fitted with eight machine-guns, it entered service with the RAF as the Hurricane in 1937 and bore the major part of the German onslaught in the Battle of Britain. The Hurricane was a war-winner, but Camm did not rest on his laurels. Taking the new and more powerful Sabre and Vulture engines, he drew up fighter designs around them. The Sabre-engined Typhoon was eventually to be the scourge of German armour during the 1944 invasion of France. Camm, meanwhile, developed and refined the basic Typhoon theme through the Tempest and Fury monoplane fighters until the turbojet rendered further development pointless.

It was with his jet aircraft that Sydney Camm's eye for beauty of line came to the fore, with the Sea Hawk naval fighter and then the Hunter, one of the most successful jet fighters ever produced. But his most imaginative design was the P.1127 VTOL fighter which, as the Harrier, pioneers a new fighter concept and is now in squadron service with the RAF as the first operational VTOL fighter in the world, and a worthy epitaph to the late Sir Sydney Camm.

Camshaft, 1036, 1038, 1054, 1097, 1102

Canada: Air Forces
Canadian Armed Forces. All branches of the Canadian armed services now operate under this collective title, the use of separate titles having been discontinued in the mid-nineteen-sixties. An embryo Canadian Air Force was reorganized in 1920 and received the prefix "Royal" in April 1924. It was equipped largely with aircraft of US and British design, many of which were built under licence in Canada during the inter-war period. The RCAF became independent of Army control late in 1938, and expanded rapidly (as did the domestic aircraft industry) after the outbreak of World War Two. Units of the RCAF served in all theatres of the war, and Canada also made a tremendous contribution to the Commonwealth Air Training Plan, training nearly 132,000 aircrew for the Allied war effort during 1939-45. After the war the RCAF contributed a component of the initial Occupation Force in Germany and still maintains an important contribution to the NATO air defences of Western Europe. Principal types of combat aircraft in service include CF-5s, Voodoos and Starfighters for fighter/strike duties, and Argus maritime patrol aircraft.

The Canadian Navy has provided an air arm of Canada's defence forces since shortly after World War Two (although a small Naval Air Service existed briefly from 1918-19). The modern naval air arm has been centred mainly around the carriers HMCS *Magnificent* (now retired) and *Bonaventure*. Aircraft aboard the latter include Trackers and Sea Kings for anti-submarine operations. The Canadian Army also maintains a comparatively small air arm, equipped mainly with observation and transport aircraft, including Boeing-Vertol 107 helicopters.

Canada: Chronology
31 July 1879. First balloon flight in Canada.
August 1893. First aerial photographs of Canada taken over Halifax citadel by Capt H. Elsdale of the Royal Engineers, using an automatic-release camera suspended from a captive balloon.
12 July 1906. First airship flight in Canada, by Lincoln Beachey at Montreal.
1907. Frederick W.

Canadair CL-28 Argus 2 of Maritime Command, Canadian Armed Forces

("Casey") Baldwin and John A. D. McCurdy, Toronto University graduates in Engineering, helped Alexander Graham Bell to form the Aerial Experiment Association (*q.v.*).
12 March 1908. Frederick W. Baldwin flew in *Red Wing* at Hammondsport, NY,

J. A. D. McCurdy airborne over the Canadian prairie in 1909—first flight by a Commonwealth citizen in the Commonwealth

F. W. Baldwin shortly before flying the "Red Wing" in the USA in 1908

becoming the first Canadian to fly a heavier-than-air machine. The *Red Wing*, powered by a 40-hp air-cooled Curtiss engine, was designed and built by Glenn H. Curtiss.

23 February 1909. John A. D. McCurdy flew the *Silver Dart* from the frozen surface of Bras d'Or Lake, Nova Scotia. He was the first person to fly anywhere in Canada. This was the first successful controlled flight in a powered heavier-than-air machine at any point in the British Empire.

2-13 August 1909. First military demonstrations of aircraft in Canada, carried out at Petawawa by J. A. D. McCurdy and F. W. Baldwin with the *Silver Dart* and Baddeck No. 1 biplanes. Both aircraft wrecked in accidents.

2 June 1917. Capt W. A. Bishop of the RFC won Canada's first air Victoria Cross on an intruder mission to an enemy aerodrome. He subsequently became the top fighter pilot of the RAF, with 72 victories.

24 June 1918. Capt Brian A. Peck of the 89th Training Squadron, Leaside, Ontario, accompanied by Cpl Mathers as engineer, flew from Montreal to Toronto, with stops at Kingston and Camp Desoronto. This was the first mail-carrying flight in Canada.

7 August 1919. Capt Ernest C. Hoy made the first flight over the Canadian Rockies, from Vancouver to Calgary, with stops at Vernon, Grand Forks, Cranbrook and Lethbridge. Elapsed time 16 hr 40 min, of which 12 hr 15 min were spent in the air.

17 October 1920. Completion of first trans-Canada air mail flight, from Halifax to Vancouver. Relays of pilots flew a Fairey floatplane, a Curtiss HS-2L flying boat and two D.H.9As. July 1923. Vickers

Viking went into service. Six of the eight aircraft acquired were built in Canada—the beginning of the post-World War I aircraft industry in Canada.

1924. Ontario Provincial Air Service organized. This grew to be the largest organization of its type in the world.

11 September 1924. First regular air mail service in Canada, to Rouyn goldfields area.

October 1924. Prototype Vickers Vedette launched in Montreal—the first Canadian-designed and Canadian-built aircraft to enter service with the RCAF.

9-10 October 1930. First successful Canadian crossing of the Atlantic, by J. Erroll Boyd and Harry P. Connor in a Bellanca monoplane.

1937. Aviation Section of the Royal Canadian Mounted Police formed.

10 February 1949. Orenda jet engine tested. It was a development of the Chinook (which ran successfully but was never installed in an aircraft) and became the first Canadian-designed jet engine to go into production. It was used extensively in Avro Canada CF-100 and Canadair F-86 Sabre fighter aircraft.

10 August 1949. First flight of Avro Canada Jetliner—the first jet transport to fly on the North American continent.

19 January 1950. First flight of Avro Canada CF-100—the first jet-powered military aircraft designed and built in Canada. Subsequently considerable numbers served with NATO forces in Europe.

23 February 1959. On the 50th anniversary of the first flight in Canada, Wing Commander Paul Hartman flew a replica of the *Silver Dart* from the surface of Bras d'Or Lake, watched by J. A. D. McCurdy, whose feat half a century earlier he was emulating. The replica was constructed by an RCAF

de Havilland Canada Beaver floatplane

crew under the direction of LAC Lionel McCaffrey.

Canadair

Canadair Ltd of Montreal is the Canadian subsidiary of General Dynamics Corporation of New York, and has been developing and building military and civil aircraft since 1944. Current or recent products include the CL-28 Argus anti-submarine aircraft, the CL-41 jet trainer/ground attack aircraft, the CL-44 swing-tailed transport, the CL-84 tilt-wing V/STOL aircraft and the CL-215 water-bomber. In the nineteen-fifties, Canadair built the North American F-86 Sabre jet fighter under licence for the Canadian forces.

Canadair CL-44

The Canadair CL-44 was the first large swing-tail aircraft to enter production. The design was based largely on that of the Britannia airliner, with Rolls-Royce Tyne turbo-props. The fuselage length was extended to 136 ft and the whole tail end of the aircraft was hinged forward of the fin at the point where the fuselage became parallel-sided.

Deliveries to two US freight airlines began in June 1961 and a third also bought some later. The aircraft has served on intercontinental freight duties since then, its range of 3,260 miles with a 61,664-lb payload enabling it to cover most journeys required of a freight aircraft. A version built for the RCAF as the Yukon, has more conventional side-loading fuselage arrangements.

Canadair CL-84, 1011-1012

Like the XC-142A, the Canadair CL-84 is an experimental VTOL aircraft in which the whole wing assembly, together with its power plants, is made to rotate between the vertical and horizontal. It was first flown in May 1965, in the hovering mode, with the wings in the vertical position. By the end of that year, the CL-84 began flight tests as a conventional aircraft, with the wings horizontal, and the first transition took place in January 1966. Since then the type has been fully proved, and search and rescue operations already have been simulated, as one of the applications for this type of aircraft. The CL-84 is a high-wing monoplane, powered by two Lycoming T53 shaft-turbine engines mounted midway along the 33 ft span stub-wings and driving propellers which provide lifting airflow across the whole wing. The propellers are interconnected, so that both are powered by either engine in the event of an engine failure. In addition, the engines drive a horizontal tail rotor behind the fin and rudder for control purposes.

Canadair Four *Argonaut*, see Douglas DC-4

Canadair Tutor (CL-41), 709

Canadian Pacific Airlines Set up in 1942 by the Canadian Pacific Railway, the original network of CPAir was confined to the western half of Canada, but international services across the Pacific were opened in 1949 and CPAir now flies regularly to Honolulu, Australia, New Zealand, Japan and Hong Kong. Other routes extend south-

ward to Mexico and South America, and eastward (either via the North Pole or directly across the Atlantic) to points in Europe.
Headquarters: Vancouver.
Member of IATA.

Caproni

One of the two largest aircraft and aero-engine manufacturing groups in Italy until the end of World War Two, the Società Italiana Caproni was founded in 1908 by Count Gianni Caproni and became

Count Caproni

widely known at an early date for the range of large, multi-engined biplane and triplane bombers built during the 1914-18 War.

Between the wars the group expanded to include several other Italian companies, chief among which were the Bergamaschi, Reggiane and Vizzola airframe constructors and the Isotta-Fraschini aero

engine company. The Caproni range of aircraft was improved considerably after 1934 when the veteran designer Rodolfo Verduzi joined the company. His first noteworthy products included the Ca 133 transport based on the same configuration as the Fokker trimotors of the period, and the single-engined Ca 161 biplane which regained the World Altitude Record for Italy in May 1937 by flying to a height of 51,361 ft. A further modified Ca 161 raised this to 56,046 ft on 2 October 1938—a record which still stands 30 years later for height achieved by a piston-engined aircraft.

The Caproni-Campini N.1 (*q.v.*), Italy's first jet aircraft, was flown on 28 August 1940, but the only other noteworthy types to serve during World War Two were the Reggiane R 2000-2005 series of fighters and a range of Caproni twin-engined light reconnaissance and transport aircraft, the Ca 310-314.

Caproni-Campini N.1 (CC2), 29

Caproni Ca 42 bomber triplane (central nacelle, tail carried on booms)

Caproni Ca 161bis, with pressure cabin, holder of world altitude record

This aircraft, often referred to incorrectly as the CC2, was the second jet aircraft to take the air and the first to make a cross-country flight. The engine it employed was something of a

Caproni-Campini N.1 ground running with tail unit removed; there was no turbine, the compressor being driven by a piston-engine

hybrid, being an Isotta-Fraschini piston-engine driving a three-stage fan and afterburner. Only one N.1 was built and it flew for just eight months (first flight on 27 August 1940). Its performance was generally unsatisfactory and Italy's next jet aircraft, the Fiat G80, did not appear until after World War Two.

apacitance gauge, see Fuel Contents Indicating System

aravelle, see Sud-Aviation

arburettor, 1054, 1060, 1097, 1105

ar ferries, see Vehicle ferries

argomaster (C-133), see Douglas

aribair

This is the trading name of Caribbean Atlantic Airlines, also known as Lineas Aereas de Puerto Rico. As its name suggests, Caribair operates extensively throughout the Caribbean area. It was formed in 1939. Headquarters: Isla Verde, Puerto Rico. Member of IATA.

arvair, see Aviation Traders

asualty evacuation, see Aero-medical services

AT (Civil Air Transport) CAT was brought into being in 1946 by Major General Chennault of the American Volunteer Group in China to provide famine relief to the Chinese population after World War Two. It was evacuated to Formosa from mainland China in 1950 and currently operates passenger, cargo and charter services in the south-west Pacific area. Headquarters: Taipei, Formosa. Member of IATA.

Cathay Pacific Airways
Founded in 1946, Cathay Pacific incorporates the former Hong Kong Airways and currently flies all-jet services to all the major countries in south-east Asia.
Headquarters: Hong Kong.

Cayley, Sir George, 138, 145, 161, 162, 163, 164, 165, 521, 522, 527, 608, 609, 622, 1045, 1047.

Centaurus engine, see Bristol

Centrifugal compressor, see Compressor

Ceskoslovenské Aerolinie (Czechoslovak Airlines), see CSA

Cessna Super Skymaster, 815, 817
A rarity among modern twin-engined lightplanes is the Cessna Super Skymaster, as the tail, with twin fins and rudders, is carried on tail booms projecting from the wings. The engines are in line, one forward and one aft of the cabin, so that twin-engined safety can be provided and, even if one engine fails in flight, the aircraft remains as easy to fly as a normal single-engined aircraft. The Super Skymaster is an all-metal four/six-seater, aimed at the business market in the USA, and has two 210 hp Continental engines which give it a cruising speed of around 200 mph.

CH-54A, see Sikorsky

Chance Vought Corsair (F4U), 1117
The Chance Vought F4U Corsair single-seat fighter was delivered to the US Navy in prototype form in 1940. It had a protracted development programme, the first pro-

Chance Vought F4U-4 Corsair (1944)

duction aircraft flying in June 1942.
An interesting feature of the aircraft was the inverted gull-wing, which lifted the engine well off the ground for adequate propeller clearance and, at the same time, enabled a short retractable undercarriage to be fitted, simplifying stowage problems.
The Corsair proved to be a very fine fighter, being used extensively by the US Navy from 1942 onward and also by the British Royal Navy. It remained in production until the mid-fifties, served in the Korean War and was supplied also to the French and Argentine Navies. In its developed form it carried four 20-mm cannon in the wings and optional racks under the wings for ten 5-in rockets and two 1,000-lb bombs, or a total of 1,600-lb of bombs. Its maximum speed was 450 mph.

Chance Vought F-8 Crusader, see LTV

Chandler, Captain, 863

Chanute, Octave, 169, 176
Octave Chanute had already become a highly respected engineer in his native America before turning his attention to aviation in his sixties. Following Lilienthal's lead, he began building gliders, but not before he had published *Progress in Flying Machines*,

a book containing as much technical information from all over the world as he could garner. He produced the most successful glider to appear before those of the Wright brothers, and gave considerable advice and assistance to the Wrights in the design and construction of their early machines.

Charles, Professor J.A.C., 128, 129, 209
A French physicist of the 18th century who was commissioned by the Paris *Académie des Sciences* to carry out investigations into lighter-than-air flight following the successes of the Montgolfier brothers. He quickly decided that the newly-discovered hydrogen gas would provide a lifting agent superior to "Montgolfier gas" and designed and built hydrogen balloons with marked success, his first flying in 1783.

Charter flying
An air charter operation is one in which an individual or corporate body charters one or more aeroplanes to make a specific flight or series of flights with an agreed type of load, as distinct from an airline service which runs to a pre-announced timetable.
It is a profitable branch of aviation business, and typical applications of air charter are the operation of inclusive holiday tours, using aircraft chartered by the sponsoring travel agency; the large-scale transportation of troops by air, still carried out for many military services by charter operators; or, on a smaller scale, the operation of "aerial taxi" services.

Chemical etching, 441, 443

Cherokee Six, see Piper

Cheyenne (AH-56A), see Lockheed

Chicago Airport (O'Hare), Illinois, USA

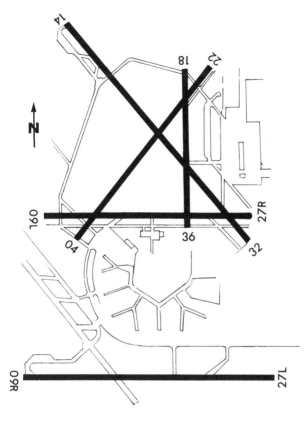

World's busiest airport, both in terms of passengers handled and of total aircraft movements per year, although by no means all of the latter are airline movements.
Distance from city centre: 22½ miles
Height above sea level: 667 ft
Main runways:
  14L/32R = 10,000 ft
  09R/27L = 10,000 ft
  04/22 = 7,500 ft
  09L/27R = 7,416 ft
  18/36 = 5,334 ft
Passenger movements in 1967: 27,552,800

Chicago Helicopter Airways
Originally founded in 1949, as Helicopter Air Service, this operator began business with a local air mail service. It assumed its present title with the introduction of regular passenger services in November 1956, linking Chicago's Midway and O'Hare airports, but these were discontinued in 1965.
Headquarters: Chicago, Illinois.

Associate member of IATA.

China: Air Force
The association of China with practical aeroplane flying—indeed, with flight in any form—is one of the oldest in the world, and the first flying school was set up in the country in 1913.
Attempts were made after the 1914-18 War to put military aviation on an organized footing, and practically the whole of the 1919-39 period was occupied by a succession of Air Missions to the country from Britain, France, Italy, the US and the USSR. Many of the aircraft supplied as a result of these missions found themselves in combat with one another during the frequently-recurring conflicts between local war-lords in feudal China, and also against the air forces of Japan during the nineteen-thirties.
Very shortly after Pearl Harbor the establishment of the American Volunteer Group in China brought large numbers of Lend-Lease

and "transferred" US aircraft into the country, and these or their replacements—also of US origin—formed the basis of the immediate post-war Chinese Air Force under the Chiang Kai-Shek administration. Most of them had to be left behind when this government was ousted by the Communists.

China (Formosa): Air Force
Nationalist Air Force: When the CNAF was re-formed in Formosa, it was re-equipped with US aircraft, and today still maintains a

such as the MiG-15 and Il-28, began to arrive in the early 'fifties, and a few years later the Communist State Aircraft Factories in China began to build Soviet aircraft with or without

"flap" up and down as they advanced and retreated, so balancing the lift forces and keeping the machine stable in flight.

Cierva also discovered the basic principle of

In 1926, Cierva established his own manufacturing company in the UK, and Autogiros of the basic Cierva type were subsequently built in considerable numbers in France, Ger-

As this field was then expanding, the new engine found a ready market and further variants of this basic layout were

CL-41, CL-44, CL-84, see Canadair
CL-475, see Lockheed
Clamshell, 1010, 1189, 1245, 1264

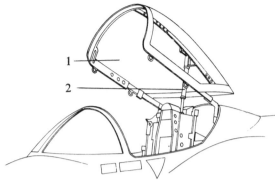

1 Canopy 2 Ram

Il-28 jet bomber, probably manufactured in the Chinese People's Republic, in service with the CPAFAF, 1963

Curtiss P-40D Tomahawk of the American Volunteer Group/Chinese Air Force in Burma, about 1942

considerable strength. Present equipment of the CNAF includes Sabres, Super Sabres, Voodoos and Starfighters for fighter/strike/reconnaissance duties.

China (People's Republic): Air Force
Chinese People's Armed Forces Air Force: After their involvement in World War Two the air forces of mainland China were extensively reorganized and re-equipped with American Mustang and Thunderbolt fighters, Liberator and Mitchell bombers, Lightning reconnaissance aircraft and Commando transports. Most of these were left behind when the Chiang Kai-Shek administration withdrew to Formosa, after which they began to be supplemented by war-surplus piston-engined fighters and bombers from the USSR. Jet replacements,

a licence. Among the types known to be in service are MiG-17, MiG-19 and MiG-21 fighters, An-2 general-purpose aircraft and Mi-4 helicopters.

Although, except for the MiG-21, the CPAFAF is not equipped with especially modern aircraft types, it must be regarded, in terms of numerical strength alone, as one of the largest in the world at the present time.

Choke tube, 1097
Cierva, Juan de la, 322, 959, 977
Juan de la Cierva, a Spanish engineer, produced or influenced the design of almost every successful Autogiro to appear in the nineteen-twenties and 'thirties. The first of these was flown near Madrid on 9 January 1923, and owed its success to the articulated hinges that allowed the rotor blades to

autorotation, showing that there is a small positive angle of attack at which blades can be set to ensure that a rotor will continue to rotate automatically in an airstream, without an engine to drive it, and still develop enough lift to sustain flight.

Cierva C.8L Autogiro

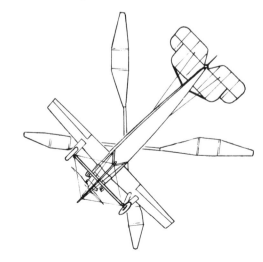

many, Japan and the USA. In 1933, Cierva evolved the first successful "jump-start" Autogiro and designed perhaps his most successful type, the C.30A, in which the engine was geared directly to drive the rotor blades before take-off. The C.30A was the first production Autogiro in the world to employ this arrangement. Cierva was killed in an airline accident at Croydon in December 1936.

Cirrus engines, 1073
Cirrus Aero-Engines Ltd were pioneers of the four-cylinder air-cooled inline aero engine. The Cirrus Mk I was produced in 1925 and immediately found favour with the builders of two-seat light biplanes for training and club work.

developed. In 1934, the company became part of the Blackburn organization, which produced a range of four light aircraft engines, the Cirrus-Hermes Mk II of 110 hp, an upright engine; the Mk IVa, an inverted engine of 120 hp; the Cirrus Minor of 80 hp; and the Cirrus Major of 150 hp. This range continued up to and during World War Two, powering training and liaison types. After the war it was used extensively in the Auster range of light monoplanes. A new engine, the 180-hp Bombardier, was introduced and the other engines in the range were refined.

The Cirrus range went out of production in the nineteen-fifties when Blackburn acquired licences for the Turbomeca range of light turbine engines.

Civil Air Transport, see CAT
Civil Aviation Administration of China, see CAAC
CJ-805 engine, see General Electric

Descriptive adjective applied to any part of an aircraft having a convex profile and capable of being rotated by a hinge axis along one end or edge. Frequently the resemblance to the real shell of a clam is not immediately apparent. Some "clamshell" cockpit canopies are very long and slender and so curved as to be almost circular in cross-section. Similarly, the "clamshell" jet deflectors on some thrust reversers are of almost cylindrical profile; but, as is also the case with the variable nozzles of some early afterburners, they do at least consist of similar halves which open and shut in the manner of a natural clamshell.

Clipper, see Boeing
Collective pitch, 983-984, 1014,
see also Cyclic pitch and Swashplate
The collective pitch control in a helicopter is that used to vary the rate of vertical ascent or descent. It is achieved by a control linkage which enables the angle of pitch of the

Cierva C.8L Mk II (in the Musée de l'Air, Paris)

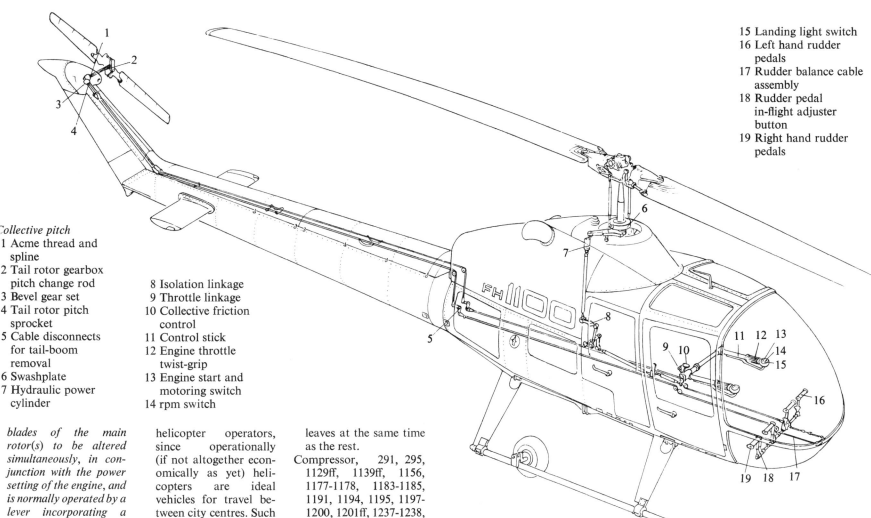

15 Landing light switch
16 Left hand rudder
 pedals
17 Rudder balance cable
 assembly
18 Rudder pedal
 in-flight adjuster
 button
19 Right hand rudder
 pedals

Collective pitch
1 Acme thread and
 spline
2 Tail rotor gearbox
 pitch change rod
3 Bevel gear set
4 Tail rotor pitch
 sprocket
5 Cable disconnects
 for tail-boom
 removal
6 Swashplate
7 Hydraulic power
 cylinder

8 Isolation linkage
9 Throttle linkage
10 Collective friction
 control
11 Control stick
12 Engine throttle
 twist-grip
13 Engine start and
 motoring switch
14 rpm switch

*blades of the main
rotor(s) to be altered
simultaneously, in con-
junction with the power
setting of the engine, and
is normally operated by a
lever incorporating a
twist-grip throttle con-
trol. Together with cyclic
pitch (q.v.) it forms the
main directional control
system for the helicopter.*

Combustion chamber,
 1159, 1161, 1215-18,
 1219-20, 1238

Comet, see de Havilland

Commercial Airways
 Known as Commercial
 Air Services until 1967,
 this carrier has been
 operating since 1949 on
 local service routes in
 South Africa.
 Headquarters:
 Johannesburg.
 Associate member of
 IATA.

Commuter air services
 The realization of an
 "airbus" type of aero-
 plane which will carry
 up to 300 or so passen-
 gers on a close timetable
 akin to that operated by
 leading bus services is
 still some years away. In
 the meantime, however,
 a few operators have
 initiated services which
 are geared to the needs
 of local commuters as
 distinct from long-
 distance travellers.
 Several of these are

helicopter operators,
since operationally
(if not altogether econ-
omically as yet) heli-
copters are ideal
vehicles for travel be-
tween city centres. Such
services by BEA in the
UK and Sabena in
Europe in the late 'fifties
were discontinued on
economic grounds, but
are still operated by
New York Airways,
Los Angeles Airways,
Chicago Helicopter Air-
ways, among others, in
the US; and by Aeroflot
in the USSR. Another
similar service, though it
uses fixed-wing aircraft,
is the Hagerstown Com-
muter service to New
York.
 A major step forward
toward an aerial "bus
service" is the Air
Shuttle, inaugurated in
1961 by Eastern Air
Lines. This connects
New York, Washington
and Boston with a
service that needs no
prior reservation of a
seat yet guarantees a
seat to every passenger.
Thus, if (say) a 50-seat
aircraft is due to leave at
12 o'clock and there are
51 passengers waiting,
an additional aircraft
will be put on to the
service to ensure that
the "excess" passenger

leaves at the same time
as the rest.

Compressor, 291, 295,
 1129ff, 1139ff, 1156,
 1177-1178, 1183-1185,
 1191, 1194, 1195, 1197-
 1200, 1201ff, 1237-1238,
 1253, 1262, 1264-1265,
 1285-1286

A device for raising the
pressure of a fluid, in-
variably of a gas or
mixture of gases such
as air. Where a small
flow is to be raised to
very high pressure a
reciprocating compres-
sor may be used, as is
the case in some high-
pressure pneumatic
systems served by a
multi-stage piston/cyl-
inder compressor look-
ing like a motor-cycle
engine. Where a large
flow is to be raised to a
more moderate pres-
sure, a rotary compres-
sor is the preferred
choice. Some cabin pres-
surization systems use
intermeshing-lobe com-
pressors, such as the
Roots type, or the inter-
meshing-screw Godfrey-
SRM units of the VC10
airliner. Gas-turbine en-
gines relied initially
upon centrifugal com-
pressors of the type
found in most domestic
vacuum cleaners; these
are robust and simple
and are still found on

small engines of low
cost. Their main draw-
backs are their inability
to achieve a pressure
ratio (ratio of output
pressure to input pres-
sure) greater than about
5.5:1 and their rapid
rise in diameter and
weight in large engines.
Today nearly all gas-
turbine aero engines use
axial compressors, in
which the air flows
approximately axially
along an annular space
between an outer casing,
carrying successive cir-
cumferential rows of
fixed stator blades, and
an inner rotor carrying
successive circumferen-
tial rows of moving
rotor blades. Such a
compressor can achieve
a typical pressure ratio
of 13:1 in eight stages
and 24:1 in 15 stages
(these figures are for
two Pratt & Whitney
engines, the JTF17 and
JT9D respectively). For

such high pressure ratios
to be achieved without
losing ease of control
and flexibility of
handling, over a wide
range of operating con-
ditions, it is necessary
either to arrange for the
stator blades, or a
number of stages of
them, to have their
angular setting variable
and accurately control-
able by an automatic
system or else to split
the compressor into two
or even three mechani-
cally independent sec-
tions, each of which can
run at its own best
rotational speed (see
"Two-spool engine").

Compressor blade
 A small rectangular
blade having a highly
cambered aerofoil sec-
tion and behaving es-
sentially as a miniature
wing. In an axial com-
pressor the air is acted
upon by successive rows
of blades, some rotating

on the rotor and the
intermediate rows being
stationary stator blades.
With few exceptions, all
are cantilevered from
one end, those on the
rotor projecting out-
ward and those on the
stator casing projecting
inward so that the alter-
nate moving and fixed
rows lie head-to-
tail with very small
clearances between
them. Blades are usu-
ally of steel, alumin-
ium alloy, titanium,
aluminium-bronze, or
glass-reinforced plastics.

Compressor bleed
 A system of pipes, internal ducts, valves or
other devices for ex-
tracting a flow of air
from a chosen point on
the compressor of a gas-
turbine. In an axial
compressor there may
be several bleeds at
different stages, each
chosen so that the ex-
tracted airflow has the

minimum adverse effect on engine performance, while still having adequate pressure and mass flow to achieve the desired purpose. Bleeds may be provided for balancing pressures inside the engine, cooling the turbine, protecting the intake against icing, pressurizing various parts of the aircraft including the cabin, driving air turbines, heating various components such as fuel filters and many other duties.

Compressor disc
A strong but light disc or ring on which are mounted the rotor blades of an axial compressor. To assemble the complete rotor the discs are either joined by a number of peripheral through-bolts or else stacked on a splined central shaft.

Compressor rotor
The moving part of a rotary compressor. In a centrifugal compressor the rotor is a single component usually having the form of a disc with radial blades on one side; an axial compressor rotor is a built-up assembly of discs with inserted blades and, usually, spacer rings between each stage to leave room for the stator blades.

Compass, 666, 671, 674, 676, 681, 720
A compass has been a necessity for aircraft use ever since aeroplanes first reached the stage of flying across country instead of within the confines of an airfield. The ordinary magnetic compass required some modification to make it suitable for use aboard an aircraft for, mounted in a liquid such as alcohol, it was very sensitive to the movements of an aircraft. Various means of damping the swinging of the magnet and pointers were used, such as fitting two magnets in parallel and building a grid of damping filaments to act with a braking effect on the swirling alcohol in which the whole was

immersed. The magnetic compass remained as the basic navigational instrument until relatively recently, although the gyro-compass was introduced between the wars. The gyro-compass, based on the directional properties of a gyroscope, is unaffected by some of the shortcomings of the magnetic compass (i.e., variation according to position on the Earth, deviation according to the amount of metal in the aircraft, lag due to swinging in response to aircraft movement, etc). However, the ordinary gyroscope is prone to "precession", a gradual moving away from its original directional heading, and so the first gyro-compasses had periodically to be reset against a magnetic compass.

A system was therefore evolved whereby a master magnetic compass was installed in the aircraft, usually at the most favourable position magnetically, and repeater compasses in the various cockpit positions (on a large aircraft) were gyro-compasses reading off from the master compass.

Since World War Two the gyro-compass has been considerably refined, and is now the normal navigational instrument. Another compass type is the radio compass, by which the pilot can read his heading off on a compass tuned to a directional bearing.

Competitive flying
The competitive element quickly became apparent once practical flying was accomplished in the US and Europe, and most of the famous international competitions were held during aviation's first quarter-century. First of these was the *Grande Semaine de l'Aviation* held at Rheims, France, in August 1909, which attracted such early pioneer aviators as Blériot, Curtiss, Farman (*qq.v.*) and

many others. A contemporary report declared: "It taxes the English tongue to the utmost of its reaches adequately to picture the epoch-making events that have marked the progress of the flying races at Rheims".

The first great aviation meeting ever held: Rheims, France, 1909

Rheims was also the venue for the first Gordon Bennett Aviation Trophy race, held five times before the 1914-18 War, each race taking place in the home country of the previous year's winner. In the 1913 event, also at Rheims, the world speed record was shattered by Maurice Prévost in the Deperdussin racer, which flew at an average speed of 124.6 mph (200.5 kmh).

the Sopwith Schneider floatplane at an average of 86.78 mph (139.66 kmh). The Trophy was hotly contested during the 'twenties and early 'thirties until, having been won three times in succession by British competitors, it passed into British keeping permanently in 1931. The final event was won by Flt Lt J.N. Boothman of the RAF in the Supermarine S.6B (2,300-hp Rolls-Royce "R" engine) with a speed of 340.08 mph (547.3 kmh). Other major competitive events, other than individual attempts upon world performance records, have mostly been on a national rather than an international basis. Among

The first international air race to be held at Hendon, England, in 1913, was won by Claude Grahame-White, here seen finishing

The Schneider Trophy races achieved their greatest prominence after the 1914-18 War, but the first two events were held during the Monaco seaplane meetings in April 1913 and April 1914. The second of these was won by

these should be mentioned the National Air Races and those for the Lockheed, Pulitzer and Thompson Trophies in the US, and the King's Cup Air Race in the UK, all of which are still competed for annually.

Concorde, see BAC/Sud
Connecting rod, 1044,

Compound engine, 1047 1117, 1119-22
The first use of a compound engine in aviation was by Sir Hiram Maxim, who developed steam compound engines to power his 1894 biplane. They were, in fact, too powerful and effective for his aircraft. In a sense, a return to the compound idea was made with the advent of the supercharger (*q.v.*), which uses the exhaust gases to power a supercharger blower. However, the Wright Corporation took this idea one step further in the early nineteen-fifties by establishing a true compound engine. In this, the exhaust gases are blown down turbines arranged around a radial engine, the turbines being connected to a shaft geared to the crankshaft; thus the extra power makes a direct contribution to the engine power driving the propeller shaft. In its final fully developed form, this increased the horsepower from a basic 2,700 to 3,700. Engines of this type were used to power the ultimate piston-engined airliners of the Super Constellation and DC-7C series. However, the civil turbojet was by then a reality and these extremely complicated piston-engines were developed no further.

1054, 1059, 1060, 1073
In a piston-engine, the rod linking the piston to the crankshaft. The small end of the connecting rod is pivoted to the piston pin (gudgeon pin) of the piston and has a purely linear oscillatory movement. The big end of the connecting rod incorporates a bearing running on the crankpin of the crankshaft and thus has a purely circular rotary movement. The rod is usually a forging in high-strength steel, highly polished and free from sudden changes in section or discontinuities which could cause fatigue. Over most of its length it is usually a bar of "H" section. In horizontally opposed or Vee engines the connecting rods from two cylinders must both drive the same crank, and the most common solution is for one to have a double, forked big end within which the other big end can fit (see picture of Merlin engine on page 396). In a radial engine the connecting rods from as many as nine cylinders must all drive the same crankpin; in this case it is usual for one rod to be a "master rod" terminating in an enlarged big end incorporating bearings for the smaller articulated connecting rods from the other cylinders.

Consolidated B-24
Liberator, 1085, 1109
The US Army ordered this four-engined, long-range bomber in March 1939, and the prototype flew on 29 December that year. Production began in 1940 for the USAAC and the British and French forces; after the fall of France the latter order was diverted to the UK. Six went into service with BOAC on the Atlantic Ferry, and a further 20 as Liberator Mk Is with RAF Coastal Command, for whose purpose their 2,400-mile range made them ideal for hunting U-boats in the Atlantic. The RAF

One of the 18,000; a B-24C Liberator, fresh off the San Diego line early in 1942

also operated Liberator Mk II bombers in the Middle East and Mk III patrol aircraft with Coastal Command.

The B-24D (equivalent to the British Mk III) was the first version put into bombing service with the USAAF: it had increased armament and turbosupercharged Twin Wasp engines, and served both in the European Theatre of Operations and in the Middle East.

Further improvement in defensive armament led successively to the B-24G, H and J, distinguishable by their nose gun turrets, and these were built in large quantities for the USAAF, US Navy, Canada and other Commonwealth air forces. They continued to be used in attacks over Europe, but it was in the Pacific area in particular that their tremendous range was of especial advantage, in both the bombing and the transport rôles, and the total production of all versions exceeded 18,000—approximately one and a half times as many as were built of their famous contemporary, the Boeing B-17 Fortress (*q.v.*). During World War Two, Liberators were credited with delivering nearly 635,000 tons of bombs on targets in Europe, North Africa and the Pacific war zones.

Consolidated NY-2, 687
The Consolidated NY-2 trainer was a development of the NY-1 Husky manufactured by the Consolidated Aircraft Corporation for the US Navy as a primary

trainer in the mid-'twenties. It was a tandem two-seat biplane with a Wright Whirlwind engine.

Constant, H.H., 1129
Constantinesco, Georges, 877
Constantinesco gear, 877
The Constantinesco gear was a British interrupter gear for firing fixed machine-guns through the propeller arc. It was the work of a Rumanian, Georges Constantinesco, who used hydraulics under pressure to perform the task.

Constant-speed propeller, 1333, 1339

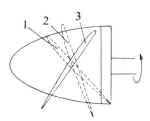

1 Cruise pitch
2 Take-off pitch
3 Reverse pitch

A propeller in which the pitch of the blades is varied automatically by a governor, which may be activated mechanically, hydraulically or electrically so that the engine may maintain a constant rotational speed as set by the pilot.

Constellation, see Lockheed
Continental Air Lines
Inaugurated as a division of Varney Speed Lanes in 1934, Continental adopted its present title in 1937. Since World War Two it has absorbed other, smaller operators and grown into one of the foremost regional carriers in the middle and

western United States. It also undertakes civil charters to Europe and contract flights to the Vietnam theatre for Military Airlift Command of the USAF. Headquarters: Los Angeles, California. Associate Member of IATA.

Continental engines, 1090-2
Continental Motors
This company first entered the aircraft field with a sleeve-valve engine in 1928, but this was not to be the main line of development for the company's aero engines. In 1931, a flat-four engine was put on the market and this became extremely popular for light aircraft, over 8,000 having been produced by 1939. After the war a flat-six range, producing from 115 to 140 hp, was introduced. Each engine comprised three sets of horizontally-opposed cylinders, set one behind the other, the whole engine being air-cooled. This basic range has been developed progressively and improved until today there are thirteen different models in Continental's flat-six range, varying from a simple 210 hp engine to a turbosupercharged 375 hp version.

Continental engines are also the subject of a licence agreement with Rolls-Royce, who market them as their light aircraft engine range.

Continental Aviation and Engineering Corporation is a wholly-owned subsidiary which for several years has been manufacturing small gas-turbines (mainly turbojets) under licence from the French Turboméca company.

Contra-rotating propellers, 822, 829, 831, 1339, 1341
Convair 880 and 990, 579, 789, 1259

Convair's bid for the first-generation jet airliner market took the form of a four-engined, medium-range aircraft with what was to become the classic American design formula of four podded engines slung underneath a swept wing. First announced, as the Model 600 Skylark, in April 1956, it was later redesignated Model 880 and made its first flight on 27 January 1959. Both the domestic Model 880 and the intercontinental 880-M have maximum accommodation for up to 124 passengers, the latter having greater fuel capacity, more powerful engines and an enlarged tail-fin. Sixty-five of both versions were built, most of them for US domestic air lines, but the market had largely been captured by the Boeing 707 and Douglas DC-8 (*qq.v.*).

The Model 990 was developed from the 880 to meet the requirements of American Airlines, and the first flight on 24 January 1961 was made by the first machine of this order. The 990 has a 10-ft longer fuselage, which seats 108 passengers, and is powered by 16,050 lb st General Electric CJ805-23B turbofan engines. It has a range of 3,800 miles and, with a cruising speed of Mach 0.84, was the fastest jetliner then in service. Thirty-six 990s were built, for American Airlines (20) and four other operators. These have since been modified to Model 990A standard, with shorter engine pylons, full-span Krueger leading-edge flaps and four aerodynamic anti-shock fairings projecting from the wing trailing-edges. These fairings also serve

as additional fuel tanks. The 990As in service with Swissair are known by the name Coronado.

Convair B-36, 553, 573, 815, 817, 1117
The B-36 was conceived as a global bomber for the USAF soon after World War Two, the first prototype flying in August 1946. It began life as a six-engined aircraft, with Pratt & Whitney R-4360-41 engines, buried in the wings, driving pusher propellers behind the trailing-edge. Later aircraft also had four General Electric J47 turbojets in two pods, one under each wing, and eventually all B-36s were modified to this ten-engined configuration. The B-36 was a large aircraft, having a wing span of 230 ft and length of 162 ft, and had a designed range of 10,000 miles, with a maximum speed of over 435 mph. It served for some time as a bomber, but more extensively as a long-range strategic reconnaissance aircraft with the USAF. It was also the subject of experiments to carry Republic RF-84F fighter-bombers into the air to provide a stand-off capability. The B-36 was intended to carry the RF-84F a long distance and then launch it. The RF-84F could perform either a reconnaissance or bombing mission and then return to the B-36 which retrieved it and returned to base. Although flight tests were successful the idea was not adopted operationally.

Convair Hustler (B-58), see General Dynamics
Convair-liners (240, 340, 440 Series), 307, 1117
The Convair-liner, developed from the Model 110 of 1946, was produced by Consoli-

dated Vultee in 1947 as a DC-3 replacement. The original Model 240 was a twin-engined low-wing monoplane with tricycle undercarriage and seating for 40 passengers. Its two 2,400-hp R-2800 radial engines gave it a cruising speed of 272 mph and range of 920 miles. It was put into quantity production, with orders from the American domestics and from airlines in Europe, Australia, India, China and South America. It was used also on a small scale by the USAF and US Navy. The basic airframe was considered capable of "stretching" and so was evolved into the Model 340 with greater wing span, more powerful engines and a longer fuselage for 44 passengers. This version entered service in 1952 and many of the airlines which had used 240s ordered the 340. Accommodation was further increased to 52 passengers in the Model 440 Metropolitan, but by the time this version was in service competition from the turboprop Viscount and Friendship was severe. Several Convair-liners have in recent years been re-engined with Allison or Rolls-Royce Dart turboprops, thus considerably extending their useful lives.

Convair "Pogo" (XFY-1), 325, 326, 973
This aircraft evolved from a US Navy design competition in the early nineteen-fifties for a vertical take-off fighter. One prototype was built, making its first vertical take-off on 1 August 1954 and its first horizontal take-off in the following November. The intention was to produce a fighter with a competitive performance which could take off from the decks of quite small vessels, including merchant ships. After considerable development flying the design was abandoned.

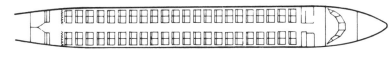

Convair 880

Convair Sea Dart
(XF2Y-1), 583, 603
The Sea Dart was produced in 1953 as an experimental twin-jet fighter seaplane for the US Navy. Two prototypes were built. The aircraft incorporated a delta wing and its undercarriage took the form of one or two retractable hydro-skis, under the fuselage.

Convair Tradewind
(R3Y), 798
Although the end of the transport flying boat was in sight after World War Two, several companies produced designs, and built them for continued use. One of the most feasible was the Consolidated Vultee P5Y-1 Tradewind, which, powered with four Allison T40 turbo-prop engines, was intended for long-range maritime patrol and assault landing duties. For the latter task, it had an upward-hinged nose portion, allowing it to land near the shore and run up the beach to discharge troops and/or equipment. It was found, however, that this was practical only in calm weather conditions. In the event, only a small production order for this aircraft came from the US Navy who later redesignated it the R3Y.

Conway engine, see
Rolls-Royce
Coolant, 1095
Copenhagen Airport
(Kastrup), Denmark

Opened in 1925, when it handled 5,000 passengers. Has expanded rapidly in the past ten years.
Distance from city centre: 6½ miles
Height above sea level: 17 ft
Main runways:
04/22 = 10,827 ft
12/30 = 10,072 ft
09/27 = 5,945 ft
Passenger movements in 1967: 4,511,200

Cornell, see Fairchild
Cornu, Paul, 321, 957
Although the helicopter designed by Louis and Jacques Breguet (q.v.) in 1907 was the first full-sized aircraft of this type to leave the ground vertically, the first to do so without any assistance or guidance from, or connection with, the ground was the machine designed and built by Paul Cornu. At Coquainvilliers, near Lisieux, this aircraft (which resembled nothing so much as a bicycle with rotor blades) lifted itself a foot off the ground for about 20 seconds on 13 November, 1907, under the power of a 24 hp Antoinette engine. Although of historical importance, the machine was not really a practical design and after a few subsequent tests its further development was abandoned.

Corsair (F4U), see Chance Vought
Coupled Proteus engine, see Bristol Siddeley

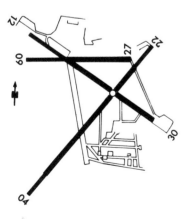

Cowie, 565
Cowling, 109ff

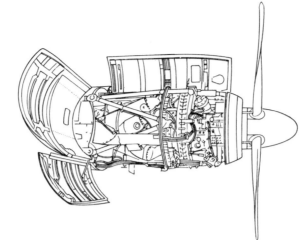

The streamlined fairing over a portion of an aircraft exposed to the passing airflow; especially, the fairing over an externally mounted engine. In piston-engines there was originally no cowling at all. By 1914 a cowling was usually provided to enclose the crankcase of an inline unit, while a rotary was usually housed in a cylindrical cowling with a rounded, inward-turned leading-edge leaving a central aperture for cooling air (and often with the lower part uncowled). By 1925 liquid-cooled engines were entirely cowled in all but relatively slow machines, while radials usually had only the air-cooled cylinders projecting through a streamlined enclosure, which often had as its leading-edge an exhaust collector ring. By 1940 all engines were fully enclosed, radials having a long-chord cowling with cooling controlled by adjustable gills or shutters around the trailing-edge and sometimes assisted by an engine-driven fan in front. Gas-turbines have invariably been fully cowled from the outset except in such machines as the Alouette and Scout/Wasp helicopters in which accessibility is more important than low drag. Modern cowlings are precision assemblies in

glass-fibre, light alloy, steel or titanium sheet, *arranged to hinge open in a few seconds by releasing special fastening catches.*

Coxwell, Henry, 915, 925
Crankcase, 1037, 1043, 1044, 1059, 1060, 1065, 1071, 1086, 1092, 1127
The hollow casing enclosing the crankshaft of a piston-engine and providing a sealed box within which an oil-mist atmosphere may be maintained. All the cylinders are secured directly to the crankcase, and the latter generally serves as the foundation of the engine upon which the complete powerplant is assembled and by which it is mounted in the aircraft. It is usually made of an alloy of aluminium and/or magnesium.

Crankpin, 1044, 1054, 1059, 1060, 1082
Crankshaft, 1043, 1044, 1047, 1053, 1055, 1059, 1083, 1087, 1101, 1117, 1120, 1122, 1123, 1125
The shaft, incorporating one or more crankpins, by which the work done in a piston-engine is converted to a continuous rotary motion. In a single-cylinder engine, or a single-row radial, there is only one crankpin and the shaft may be very short and robust. When the cylinders are arranged in a line, the crankshaft must have the same number of throws (crankpins) as there are cylinders:

straight-eight engines have often been used in cars, but six-throw crankshafts have seldom been exceeded in aero engines. The shaft is usually built up or machined from a single forging in high-strength steel and incorporates balance weights counteracting the rotating masses of the crankpin(s) and the big end of the connecting rod(s). In most modern piston aero engines the rear end of the crankshaft drives trains of gears to accessories, while the front end either carries the propeller directly or terminates in a straight or helically-toothed spur gear through which the drive is stepped down to a propeller shaft running at lower speed.

Coxwell, Henry, 915, 925
Creagh-Osborne, Captain, 661
Creep, 445, 1163
This is the descriptive name given to the gradual lengthening of a metal part that is both stretched and heated. When any solid material is subjected to a tensile load it stretches slightly. Flabby substances such as putty are unable to return to their original dimensions but elastic materials such as most metals, have the ability to stretch slightly under load and then recover their original form exactly once the load is removed, provided the load is not so great that it exceeds the "elastic limit" for the piece of metal concerned. At ordinary temperatures the elastic limit is high enough for metals to be very useful structural materials which can be loaded repeatedly without any permanent "set" or change in dimensions, but if the environmental temperature is raised the elastic limit falls. By the time a dull red heat is reached, light alloys are totally useless, and even steels are weakened and rapidly acquire a permanent deformation under load. The phenomenon of

creep is seen very acutely in gas-turbine rotor blades, which must bear a sustained and intense tensile load while at bright orange - white temperature. Ordinary metals would creep so rapidly as to rub on the outer casing within a second or two, but the latest refractory alloys used for such blades can withstand such conditions for 1,000 hours or more. But creep is always present in any component subjected to high tensile load under high temperature, and this slow stretching always limits the life of a hot turbine blade.

Crop-dusting and spraying, see Agricultural aviation
Crusader, see LTV
Cruzeiro do Sul
Cruzeiro is one of the principal South American airlines, flying services to most countries on the continent, including those of the former Panair do Brasil. These incorporate a shuttle service (see Commuter air services) between Rio de Janeiro and São Paulo in which it participates with Sadia, VARIG and VASP. It was founded in 1927 and adopted its present title in 1942.
Headquarters: Rio de Janeiro.
Member of IATA.

CSA (Ceskoslovenské Aerolinie—Czechoslovak Airlines)
CSA was re-formed in 1946, succeeding its similarly-named predecessor which was formed in 1923 and operated up to the time of the Nazi occupation; it was nationalized in 1948. It now flies to most countries in Europe, to the USSR and the Middle and Near East. Activities include air taxi and agricultural work in addition to airline services.
Headquarters: Prague.
Member of IATA.

Cubana
The present-day Empresa Consolidada Cubana de Aviación

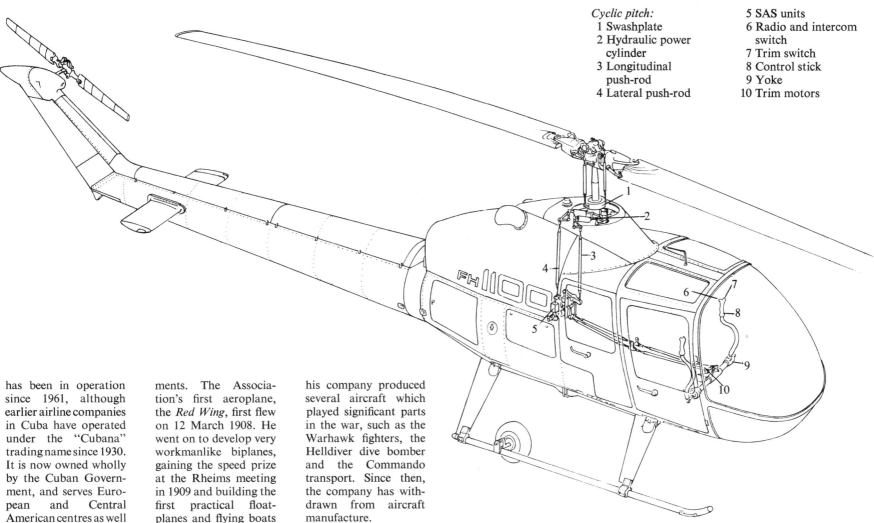

Cyclic pitch:
1 Swashplate
2 Hydraulic power
   cylinder
3 Longitudinal
   push-rod
4 Lateral push-rod
5 SAS units
6 Radio and intercom
   switch
7 Trim switch
8 Control stick
9 Yoke
10 Trim motors

has been in operation since 1961, although earlier airline companies in Cuba have operated under the "Cubana" trading name since 1930. It is now owned wholly by the Cuban Government, and serves European and Central American centres as well as providing an internal network.

Headquarters: Havana. Member of IATA.

Glenn Curtiss

Curtiss, Glenn H., 217, 221, 222, 565

Glenn Curtiss came into aviation as a result of his successes in building light engines and motor-cycles and riding the latter. When the Aerial Experiment Association was formed in America, in 1907, he was appointed Director of Experi-ments. The Associa-tion's first aeroplane, the *Red Wing*, first flew on 12 March 1908. He went on to develop very workmanlike biplanes, gaining the speed prize at the Rheims meeting in 1909 and building the first practical float-planes and flying boats in 1911-12. The float-planes were used for the earliest aircraft carrier (*q.v.*) experiments. Be-fore the 1914-18 War, his company produced a flying-boat for an attempted transatlantic flight by Lieutenant John Porte, RN. During the war, many of his flying boats saw service with the British Royal Naval Air Service, for coastal patrol work in the UK and elsewhere. His greatest contribu-tion to that war was the "Jenny", which was used as an elementary trainer for many years. After the war, a Curtiss flying boat made the first crossing of the Atlantic, in several stages. In the 'twenties and 'thirties, Curtiss built up one of the major aircraft com-panies of the USA, producing aircraft for the US Navy and other military and commer-cial customers. When World War Two came,

his company produced several aircraft which played significant parts in the war, such as the Warhawk fighters, the Helldiver dive bomber and the Commando transport. Since then, the company has with-drawn from aircraft manufacture.

Curtiss biplane, 217, 671
Curtiss flying boats, 217, 253
Curtiss Warhawk (P-40), 1101

When it was found that the Curtiss P-40A and P-40B Tomahawks which the US had sup-plied to the RAF were not up to the latter's required battle perfor-mance for European fighting conditions, Curtiss re-engined this single-seat fighter with a more powerful Allison and another version with a Packard-Merlin. These types were known as the Warhawk to the USAAF and the Kitty-hawk to the RAF. The US and Chinese forces used them successfully in the Pacific area and Mediterranean, and the RAF in the Western Desert and Italy.

They were much used as fighter-bombers, with a 1,500-lb bomb load in addition to up to six 0.50 in guns in the wings. The P-40 was the first

US fighter to be pro-duced on a really large scale: over 14,000 were built of all versions. Al-though not a brilliant design, particularly dur-ing the early part of its career, the P-40 was a strong, dependable aeroplane which per-formed a useful job when little else of com-parable performance was available to the Allies in sufficient quantities.

Curtiss XP-42, 1078-79
Cyclic pitch, 982-984
   see also Collective pitch and Swashplate.

The cyclic pitch control in a helicopter is that which regulates the for-ward or other directional movements of the air-craft other than those of vertical ascent or descent. It is achieved by altering the angle of attack of the main rotor blades, by tilting the rotational plane of the entire rotor in the direction desired, and is effected from the cockpit by a control lever operating in a similar manner to the control column in the fixed-wing aeroplane. Tilting the rotor thus provides both a component of lift from the vertical thrust, and a propulsive component from the horizontal thrust; to provide a good forward speed, the angle of tilt is greatest in that direction.

Cyclone engine, see Wright
Cylinder, 217, 1038, 1043-44, 1047, 1053, 1055, 1057-60, 1075, 1092, 1095, 1101, 1122, 1123-25

In a piston-engine, the cylindrical vessel in which the piston

oscillates. Combustion of the fuel/air mixture takes place in the en-closed volume between the top of the piston and the inside of the cylinder head, the latter con-taining inlet and ex-haust ports which are sealed off by the closed inlet and exhaust valves during the period of combustion. The open-ended base of the cy-linder is attached rigidly to a corresponding aperture in the crank-case, either by being bolted or screwed in, or by a "force fit" produced by making the cylinder oversize and cooling it until it can just be inserted. In most piston aircraft engines there are from 2 to 18 cy-linders. They may be arranged in a straight line, or in parallel banks

1903   1904   1905   1906   1907   1908

with an angle of from 45° to 180° between them (45° would be a Vee or inverted Vee layout, while 180° would be a flat or horizontally-opposed engine), or in a radial form in which they are arranged like the spokes of a wheel. In more powerful radial engines there are two (usually identical) rows of cylinders, arranged one behind the other with the rear cylinders usually staggered to lie between those of the front row, to improve cooling; one very powerful engine still in use has four rows of seven cylinders each. In virtually every modern engine, the centre line axis of every cylinder passes through the axis of the crankshaft driven by its piston. Some engines had two crankshafts, notable examples being Napier "H" and "flat H" engines and the Junkers Jumo two-stroke diesels, in which the cylinders were double-ended and housed two pistons with the combustion space in the centre.

Cyprus Airways

Formed in 1947 by the Cyprus Government (which now has the majority holding), BEA and private investors. Its small fleet of Viscounts operate to several countries in the Mediterranean area; there are no internal air services.
Headquarters: Nicosia.
Member of IATA.

Czechoslovak Airlines, see CSA

Czechoslovakia: Air Force

Ceskoslovenské Letectvo (Czechoslovak Air Force). Came into being as an Army Air Force shortly after the Czechoslovak Republic was created in October 1918, when it was equipped with war-surplus French, German, Italian and Russian aircraft. Due to an expanding domestic industry these were largely replaced by more modern types, of Czechoslovak design,

Aero A 24 bomber of the Czech Air Force

The all-Czech L-29 Delfin, selected in 1964 as the standard jet trainer of Warsaw Pact powers

from the mid-'twenties; and by the time of the Nazi invasion in 1939 Czechoslovakia was one of the leading aviation countries of Europe, both industrially and militarily. After their own country was over-run, Czechoslovak aircrew served throughout the war with the air forces of France, Poland, the UK and the USSR.

A major reorganization and re-equipment programme followed the seizure of power by the Communist regime in 1948, since when the domestic industry has built several types of Soviet warplane under licence for the Air Force, and many of these Czechoslovak-built machines have been supplied to other air forces in the Soviet bloc. Principal front-line types are MiG-19 and MiG-21 interceptors and Su-7 ground attack aircraft, together with large numbers of the Czechoslovak-designed Delfin jet trainer.

Czechoslovakia: Chronology

1770. Legend tells of a countryman named Fucík, of southern Bohemia, reputed to have made several flights with wings of his own construction.
1780. A monk, Brother

The Kaderávek flapping-wing proposal of 1860

Fucík is reputed in Bohemian legend to have flown in about 1770

Tsyprian, was reported to have made a safe landing after a gliding flight from a high rock, using wings of his own construction.
18 February 1784. Tadeas Hanka, a world renowned physician and botanist, launched a small Montgolfier-type unmanned balloon.
31 October 1790. First Czech citizen to become airborne, Count Joachim Von Sternberg, accompanied by the French balloonist J. P. Blanchard in an ascent in a Montgolfier balloon.
1860. First heavier-than-air experiments in Bohemia. Václav Kaderávek of Prague experimented with ornithopter models powered by small electro-magnets.
1892. F. Stepánek, piloting a primitive glider designed and built by himself, made a number of flights near Prague of 200-330 ft (60-100m).

1895. Ing G. V. Finger projected a fixed-wing aircraft which he named *Cyklon*. Of particular interest was his intention of using two or three engines driving propellers in shrouded ducts.
1904. Igo Etrich of Trutnov, in north-west Bohemia, and Franz Wels of Vienna, flew a glider of their own design which featured inherently stable characteristics. They moved subsequently to Vienna, where they later developed such well-known Etrich aircraft as the *Taube*.
1907. At an automobile exhibition in Prague a 15-hp rotary engine, the first engine of its type, was demonstrated.
1910. Ing Otto Hieronymus, chief designer of the motor-cycle and automobile factory of Laurin & Klement (now Skoda Automobiles), flew a monoplane powered by a 50-hp engine, the first of his own design.
13 May 1911. Ing Jan Kaspar, flying a machine designed

This Gnome-powered monoplane named the "Rapid" wa[s] the work of Eugen Cihák about 1913

jointly with his cousin, Eugen Cihák, completed a flight of 75 miles (120 km) from Pardubice to Prague.
13 July 1911. Mr Cermák, a member of the Aeronautical Co-operative Society of Pilsen, in western Bohemia, became the first Czech pilot to receive the official Austrian Pilot's Diploma.
28 October 1918. Czechoslovakia emerged as an independent republic. On the following day the Czechoslovak Air Arm was founded.
27 April 1919. First Czechoslovak-designed and built aircraft flew at Pilsen. This was a small two-seat biplane, the B-5, powered by a pre-war NAG engine of 32/45 hp.
April 1920. First Czechoslovak-designed and built military aircraft, a light bomber/

reconnaissance biplan[e] the Smolík S 1, powere[d] by a Hiero L engine [of] 230 hp, was tested i[n] Prague.
1 March 1923. The Sta[te] airline, Ceskoslovensk[á] Státní Aerolinie (CSA) inaugurated its first ser[-]vice between Prague an[d] Bratislava with Aer[o] A 14 biplanes. In th[e] same year Dr Lhota wo[n] first prize in a Belgia[n] rally, flying an Avia B[H] 5 monoplane, a develop[-]ment of the BH [?] powered by a 60-h[p] Anzani engine.

1924. Third Internat[-] ional Aeronautical Ex[-]hibition in Prague. O[n] show was the Aero A 2[4] night bomber, first twin[-]engined biplane produ[-]ced in Czechoslovakia During the year Aer[o] A 11 and A 12 re[-]connaissance biplane[s] gained internationa[l] records confirmed by th[e] FAI.
1925. First Czechoslo[-]vak aircraft to be selec[-]ted for licence construc[-]tion abroad, the Avi[a] BH 21 biplane fighte[r] was built by SABCA fo[r] the Belgian Air Force.
1926. The *Coppa d'Itali[a]* won by an Avia BH 11 piloted by Bican.
1927. Ceskoslovensk[á] Letecká Spolecnos[t] (CLS), air transpor[t] company founded, a subsidiary of the Skod[a] works. A year later CL[S] started a service from Prague to Dresden, Ber[-]lin and Vienna, with BH 25 biplanes havin[g]

The first Czech-designed and built military aircraft wa[s] the Smolik S.1 of 1920

accommodation for a crew of two and five passengers.

1927. Letov started production of its first all-metal biplane, the S 16, many of which were built for export.

1935. Notable for the production of the Beta and Bibi monoplanes designed by Benes, and the Zlín Z XII light monoplane. In this year CSA commenced its longest international service—Prague to Moscow—with Airspeed Envoy airliners.

1936. Many successes in international competitions, particularly with the Benes Be 501 and Be 502 monoplanes, together with new international records in light-plane categories confirmed by the FAI.

15 March 1939. Czechoslovakia occupied by Nazi forces; her aircraft factories came under German control.

1939-1945. Many hundreds of military air and ground personnel, together with other Czechoslovak civilians, escaped to Poland, Yugoslavia and France in the first months of the occupation. After September 1939 they fought in the Polish Air Army and the French *Armeé de l'Air*, and after the fall of France escaped to Britain. During the war years all national aeronautical activities ceased.

October 1945. The aircraft factories were nationalized and production began again. Preparations made to recommence civil air services.

1 March 1946. The national airline, CSA (now incorporating the former airline CLS), recommenced civil air services.

8 March 1946. First post-war nationally-designed aircraft, the M1A Sokol designed by M Rublic, flew for the first time.

September 1947. The first Zlín Z 26 Trener was test-flown. This aircraft was subsequently produced in large numbers, and the latest versions are still in production today.

1949. CSA purchased its first Soviet airliner, an Ilyushin I1-12.

1951. Czechoslovakian aircraft industry began licence production of Soviet MiG-15 fighters and Ilyushin Il-10 attack aircraft and their engines.

1957. CSA began regular services with Avia 14 twin-engined airliners (Ilyushin Il-14s produced under licence). In November Tupolev Tu-104As joined CSA for use on international routes, making CSA one of the first European airlines to operate turbojet airliners.

1958. In August an Aero Ae 45S crossed the South Atlantic between Natal and Dakar, first transatlantic crossing by a Czechoslovak-built aircraft.

5 April 1959. First flight of the prototype L 29 Delfín two-seat jet-powered basic and advanced trainer. The Delfín was the first jet aircraft of national design and construction, and was powered by a 1765 lb st M 701 engine also of Czechoslovak design.

1960. Series production began of the HC 102 Heli-Baby light training helicopter.

1961. Mr Bezák flying a Zlín Z 226T became World Champion in the first World Aerobatic Championship at Brati-

Czech Z-37 Cmelák agricultural aircraft

slava. A Z 226T took the championship in 1962 and a Z 326 completed the hat-trick in 1963 at Bilbao.

June 1963. The Z 37 Cmelák, first specialized agricultural aircraft of Czechoslovak design, made its first flight.

1967. CSA initiated a transatlantic route to Canada with Il-18D turboprop airliners.

1968. CSA hired an Ilyushin Il-62 for intercontinental services.

Daimler-Benz, see also Mercedes

The Daimler Motoren Gesellschaft of Stuttgart was, prior to 1911, exclusive supplier of petrol engines for the Zeppelin airships, and was a primary manufacturer of aeroplane engines during the 1914-18 War. The Benz company, similarly, provided most of the remaining power-plants required by the German aircraft of the latter period.

Both companies' aero-engine activities were halted by the Treaty of Versailles in 1919, but in 1926 the two amalgamated to form the Daimler-Benz AG, and the new company began to evolve a range of air-cooled and liquid-cooled petrol and diesel aero engines. It achieved especial prominence in 1938 with the appearance of the DB 600 twelve-cylinder inverted-Vee engine, comparable in performance with the Rolls-Royce Merlin (*q.v.*) This was subsequently developed into the DB 601, DB 603 and DB 605 range which powered

many pre-war record-breakers and such leading *Luftwaffe* aircraft as the Messerschmitt Bf 109 and Bf 110, and the Italian Macchi C.202 and C.205 fighters. A specially-boosted DB 601, developing 2,300 hp for a short period, powered the Me 209V1 when it set a World Absolute Speed Record of 469.22 mph in April 1939.

The Company made a brief entry into the gas-turbine field in 1943, and has recently begun again to evolve small turbine engines of its own design.

Daimler-Benz DB 601 engine, 909

Daimler-Benz DB 624 engine, 1114-1116

Dakota, see Douglas DC-3

Dallas Airport (Love Field), Texas, USA

Dart sailplane, see Slingsby

Dassault

Avions Marcel Dassault is one of the two major aircraft constructing organizations in France today. It was formed by Marcel Bloch (chief of the pre-war Bloch company) on his return from imprisonment during World War Two, when he adopted the name Dassault, by which he had been known in the French Resistance movement during the war.

The company's products have been predominantly military in nature, beginning with the first French production jet fighter, the Ouragan, from which were developed the transonic Mystère and the supersonic Super Mystère. Its

given rise to the swept-wing Mirage F1 and the swing-wing Mirage G. The Mirage III and 5 fighter/strike and reconnaissance aircraft have been the most successful European combat aircraft since the Hunter in terms of export orders, and have been ordered by several air forces the world over.

In the commercial field, another Dassault success is the Fan Jet Falcon (Mystère 20) executive jet, and in the Autumn of 1968 the prototype was flown of a new twin-turboprop utility transport and training aircraft, the Hirondelle.

Dassault Fan Jet Falcon

This attractive and successful jet executive aircraft flew for the first

From the rear: Mystère II-C, Mystère IV-A, Super Mystère IV-B2, Etendard, Mirage III-C, Mirage III-R and Mirage IV-A

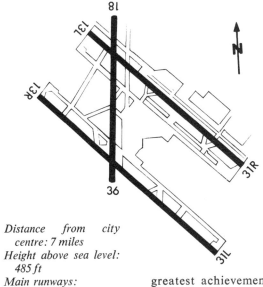

Distance from city centre: 7 miles
Height above sea level: 485 ft
Main runways:
13R/31L = 8,800 ft
13L/31R = 7,751 ft
18/36 = 6,146 ft
Passenger movements in 1967: 3,807,000

Dart engine, see Rolls-Royce

greatest achievement has been the delta-winged Mirage range of aircraft, which includes the multi-mission Mirage III and 5, and the Mirage IV-A nuclear bomber; these have also

time on 4 May 1963. It is known in France as the Mystère 20. Its success has been achieved mostly in the highly competitive US market, where the Business Jets Division of Pan American World Airways is its sponsor, and altogether over 200 examples have been sold in a little over three years. The standard production version of the Fan Jet Falcon is a 10-seater and is powered by two rear-mounted CF700-2D turbofan engines, each of 4,250 lb thrust, which give it a cruising speed of 534 mph. In addition to corporate and private owners, Falcons are also in use as transports by the RCAF and as crew trainers by Air France. The latest version,

Sokol monoplane, designed by M. Rublic

Mystère 20 executive jet, otherwise known as the Fan Jet Falcon

called the Falcon F, entered production in 1969; this has improved take-off and landing performance and a range of nearly 2,000 miles with maximum payload.

Dassault Mirage III, 1137
The Mirage III French delta-winged Mach 2 all-weather fighter and reconnaissance aircraft is in production and service with several of the world's major air forces. It first flew in November 1956, becoming operational with the French Air Force in 1961. Versions are also used by the air forces of Israel, the Lebanon, Pakistan, Peru, South Africa and Switzerland; and it was licence-built in Australia for the RAAF. It saw action most effectively with the Israeli Air Force during the "six days' war" in June 1967.

The standard III-E version is powered by a SNECMA Atar 09C turbojet delivering 13,670 lb thrust with afterburning and may be fitted optionally with a 3,307-lb thrust jettisonable SEPR rocket for assisted take-off. Maximum speed at 36,000 ft is 1,430 mph, it can climb to 50,000 ft in 4 min 40 sec and has a combat radius of 470 miles. It can be armed with a Matra R 530 air-

to-air missile, with provision for two Sidewinder missiles and two 30-mm cannons in addition. The Mirage 5, a simplified multi-mission export version, is in production for the air forces of Belgium and Peru.

Mirage 5, specially developed to appeal to overseas air forces

da Vinci, Leonardo, 113, 117, 145, 159, 607, 609
Leonardo da Vinci's contribution to the science of aviation was great, especially considering the period in which he lived. His scientific intellect enabled him to invent and design the parachute, ornithopter and helicopter in the fifteenth century although he made no attempt to construct them, apart from a small model of his helicopter.

Dayton-Wright, R.B., 567
This racing monoplane was built for the 1920 Gordon Bennett Aviation Trophy race. To reduce drag, the pilot was totally enclosed within the fuselage, to

which a retractable undercarriage was fitted. The pilot's position severely restricted the forward view, and windows were provided in the sides of the fuselage to offset this. A special feature was the variable-camber wing. Powered by a 250-hp Hall-Scott Special engine, the RB was entered for the race, but the variable-camber gear change gave trouble and the aircraft force-landed.

DB 601 and DB 624 engines, see Daimler-Benz

DC-1, DC-2 etc, see Douglas

Decca Navigator, 1383

son with VOR/DME and similar systems which merely give bearing or distance to a ground station. The Decca system is based upon a carefully surveyed group of ground transmitting stations (usually one "master" and two "slaves") which emit a continuous wave signal in unison. Each station sends out its emission like ripples from a stone dropped into a pond, and the "ripples" from the three stations intersect each other along hyperbolic lines. Originally the pilot had to read three instruments, each of which indicated the position of the aircraft on a particular line of a special map; but by 1958 the system had been completely automated and made to give an immediate position on a roller map on which are also shown airfield runways, beacons and other information. Decca can be used anywhere in a geographical region served by the system; the aircraft does not have to follow any particular route. It can be used right down to sea level and is also practicable for all-weather helicopter operation. Extensions of the system are the Dectra long-range system, suitable for trans-oceanic routes, the Omnitrac airline guidance system and the Harco traffic control system for Europe and other regions with dense

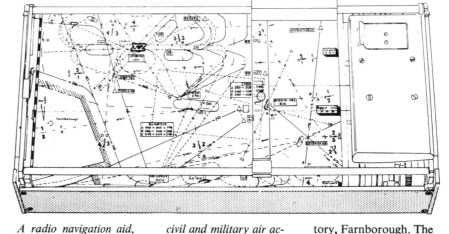

A radio navigation aid, developed in England in the late nineteen-forties, which gives full area coverage and has marked advantages by compari-

a Jesuit priest. He designed and patented the "Passarola", which had wings and a "parachute" sail. It is believed that a model was built and may have flown in Lisbon.

de Havilland, Sir Geoffrey, 565
Geoffrey de Havilland was an early participant in aviation, making his

de Havilland D.H.9a of 39 Squadron RAF

earliest successful flight in the second aircraft which he designed and built, on 10 September 1910. This aircraft was bought by the British Government and he began designing aircraft at the Royal Aircraft Fac-

Sir Geoffrey de Havilland

civil and military air activity.

de Gusmão, Bartholomeu Lourenço, 121, 126
Born in Santos in 1686, de Gusmão later became

tory, Farnborough. The B.E.2 series was conceived under his leadership, but shortly after the outbreak of the 1914-18 War he left Farn-

borough to serve in the RFC. He was soon seconded to the Aircraft Manufacturing Company Ltd (Airco), and began his long series of D.H. designs. The D.H.1, D.H.2 and D.H.5 fighters, D.H.4 and D.H.9 bombers and D.H.6 trainer all entered large-scale service during the War, the D.H.4

and D.H.9 being the mainstays of the British day bombing force. The D.H.4 also equipped a considerable part of the American Expeditionary Force in Europe in 1917-18. After the war, Airco became the de Havilland Aircraft Company and turned its attention to refurbishing and "civilianizing" wartime D.H. aircraft and building de Havilland's new designs. Several of these met limited success in the sparse days of the nineteen-twenties, but it was with the D.H.60 Moth that Geoffrey de Havilland made his mark. It was the first truly practical and economical private aircraft, achieving world-wide fame and sales, and its configuration was copied by many other manufacturers. It was one of the greatest aircraft of all time, and made possible many record-breaking flights. With the Moth the flying clubs in the UK multiplied and the light aircraft movement was born.

Other winners included the Comet Racer monoplane, which won the 1934 England-Australia air race; the Dragon, Dragon Express and Dragon Rapide short-haul biplane airliners, which made possible domestic airlines in all parts of the world (the Rapide still serves in this rôle), and the Mos-

Matra R 530 air-to-air missile carried by Mirage III

The original Comet: the D.H.88 of 1934 (this is the actual machine flown by C. W. A. Scott and T. Campbell-Black to win the Mildenhall-Melbourne race of 1934)

D.H.84 Dragon of Jersey Airways

quito (*q.v.*) of World War Two. By now the de Havilland enterprise had grown, with manufacturing companies in the Dominions, its own engine and propeller companies and subsidiary companies such as Airspeed Ltd. As well as the Mosquito, in production for most of World War Two, the company built thousands of Airspeed Oxford twin-engined trainers. With the war over, Geoffrey de Havilland left the designing of the new generation of aircraft largely in the hands of his staff, and from them came the first jet airliner in service, the Comet, the second jet fighter in RAF service, the Vampire, and a succession of Vampire variants and developments. Meanwhile, the Dove took up where the Dragon Rapide had left off and was joined by the Heron four-engined short-haul airliner. For the Royal Navy, the Sea Vixen all-weather fighter was produced; and the second generation of jet airliners was heralded

by the three-engined Trident, coincidental with the company losing its separate identity when it was embodied in the vast Hawker Siddeley Aviation combine.

four-engined transport in 1936, as the D.H.91 Albatross. This aircraft was produced for Imperial Airways, who used it for fast services to Paris, Brussels and Zurich. Its career was cut short by the outbreak of war in 1939, after it had been in service just under a year. Powered by four D.H. Gipsy Twelve engines, the Albatross had a maximum speed of 222 mph and a range of 1,040 miles. Seven were produced, but all were destroyed before the end of the war.

de Havilland Comet (D.H.106), 307, 311, 579, 581, 787, 835, 851, 1189, 1261, 1283

Comet 4

Conceived by the Brabazon Committee before the end of World War Two, the de Havilland Comet pure-jet airliner flew for the first time in 1949 and began the first scheduled jet passenger services in the world, with BOAC, in May 1952. The aircraft went into series production

in its Mk 1 and 2 forms, only to suffer a severe setback in 1954, when two crashes revealed serious fatigue problems. Intensive research enabled the fatigue problem to be cured and the Mk 2s previously scheduled for BOAC went into highly successful service, with rebuilt fuselages, with RAF Transport Command. Meanwhile, the much-improved Mk 4 was produced for BOAC and operated the first transatlantic jet services in October 1958, slightly ahead of Pan American's Boeing 707s. The Mk 4 and developed Mk 4B and 4C versions are still in service with many airlines and with the RAF and have provided the basis for the world's most advanced maritime patrol bomber, the Hawker Siddeley Nimrod.

de Havilland Albatross (D.H.91), 569

Benefiting from experience gained in building the D.H.88 Comet Racer, the de Havilland Company produced a

D.H.2s, and the type soon proved successful against the Fokker *Eindecker* which was proving a thorn in the flesh of the Allied air forces. For most of 1916 the D.H.2 was used successfully in action, but with the advent of the Albatros fighters its days were numbered, and by June 1917 it had disappeared from the fighting. With a maximum speed at ground level of 93 mph, it had a ceiling of 14,000 ft and an endurance of 2 hr 45 min, enabling long patrols to be made over the fighting lines.

de Havilland Flamingo (D.H.95)

A twin-engined high-wing short-range airliner, produced by the de Havilland Company in 1939.

de Havilland Ghost engine, 835, 851, 1261ff

The Ghost was the sec-

A D.H.95 Flamingo in wartime livery

de Havilland D.H.2, 815

The D.H.2 was the first single-seat fighter designed by Geoffrey de Havilland. When it was conceived, there was no British interrupter or synchronizing mechanism whereby machine-guns could be fired between the propeller blades, so de Havilland designed and built a pusher biplane. The nacelle contained the engine at the rear, driving a pusher propeller, and the pilot in front with a machine-gun.

The aircraft was ordered into quantity production, and deliveries to Corps squadrons began in January 1916. No 24 Squadron of the Royal Flying Corps was the first to go to the Western Front fully equipped as a fighter squadron, with

ond jet engine to be put into production by the de Havilland Engine Company. It was intended as the power plant for the Comet airliner and powered the Mk 1 version which became the first jet airliner in service in the world (with BOAC in 1952). The engine also powered the Venom fighter for the RAF and Royal Navy. It followed the same basic layout as de Havilland's original Goblin, having a single-sided centrifugal compressor with ten large combustion chambers. Its basic output was 5,000 lb static thrust.

de Havilland Gipsy engine, 1073

The success of the Cirrus engine (*q.v.*), and its obvious suitability for light aeroplanes, led the

The second of the tailless D.H.108 Swallow research aircraft

de Havilland Engine Company to set about designing its own engine for the series of lightplanes it was building and planning. In July 1927, the first experimental Gipsy engine was run. It produced 135 hp and was installed in the little Tiger Moth monoplane which established a light aircraft speed record of 186 mph.

Subsequently, the Gipsy I went into production, followed by the II and III; these had considerable success, not only in de Havilland aircraft but in other types, many from foreign constructors.

From them was developed the Gipsy range—including the Gipsy Major four-cylinder inverted in-line air-cooled engine of 120 hp and the six-cylinder Gipsy Six of 185/205 hp. These engines were developed during the 'thirties and were standard light aircraft engines of the period in many countries. The Gipsy Queen was added to the range just before World War Two. It was a 250-hp six-cylinder engine which, after the War, was developed to give 330 hp.

This range of engines is still in widespread use throughout the world.

de Havilland Hornet (D.H.103), 1101

The Hornet was de Havilland's last piston-engined operational aircraft, being designed for the Far East war but not coming into service until World War Two had ended.

It was a twin-Merlin-engined single-seat fighter with a maximum speed of 472 mph and a maximum range of 3,000 miles. It was armed with four 20-mm cannon in the nose and rockets or bombs under the wings.

In its F Mk 3 version, the Hornet saw service in the Far East against Malayan terrorists in 1951. Here it was used

mainly for strike duties, armed with rockets. It was also developed, as the Sea Hornet, for the Royal Navy, in the day fighter, night fighter and fighter-reconnaissance rôles (with two seats).

de Havilland Mosquito (D.H.98), 361, 1101

One of the outstanding aircraft of World War Two, the Mosquito stemmed from the de Havilland Aircraft Company's experience of producing advanced wooden aircraft for high-speed duties, such as the Comet Racer of 1934 and the Albatross airliner which had entered service with Imperial Airways just before the war.

After initial government disinterest, the Mosquito was put into production and entered service with the RAF in three distinct rôles, as a photo-reconnaissance aircraft, bomber and fighter. The first operations were flown in the PR rôle, beginning in September 1941 with a daylight reconnaissance over the French Atlantic ports. Within nine months, as many as ten sorties a day were being flown and for the rest of the war successively-improved versions were employed continuously on photographic sorties, not only over Europe but over the Mediterranean and Far Eastern theatres of operation.

As a bomber, the Mosquito first appeared in its Mk IV version and equipped a Wing at Marham in May 1942. It was employed initially on low-level daylight raids against specific targets, being almost immune from fighter interception due to its high speed. Later, this type of bombing was taken over by the Mosquito Mk VIs of 2nd TAF, and the Bomber Command Mosquito squadrons, by then equipped with Mk IXs and Mk XVIs, were employed increasingly on the specialized task of pathfinding for the

main bomber stream.

In the fighter rôle, May 1942 was the operative date when the first two night-fighter squadrons of Fighter Command re-equipped with Mk II Mosquitos. The type was extremely successful as a night-fighter, especially in its later versions with more advanced radar, and by the end of the war RAF Fighter Command's entire night-fighter force, which also included intruder squadrons, was Mosquito-equipped. The Mosquito also joined Coastal Command, and the Mk VI, equipped with a combination of cannon, bombs and rockets, augmented and began to replace the Beaufighter in the shipping strike rôle. Other fighter Mosquitos, seconded from Fighter Command, joined the bomber streams over Germany at night, attacking the German night-fighters in their home territory. The Mosquito was also built in Canada and Australia, and when production ended the total that had been produced was 7,781.

de Havilland Vampire (D.H. 100), 1135

The Vampire represented de Havilland's first approach to jet fighter design. It followed the Meteor into service with the RAF, the prototype flying in September 1943, with a 2,700-lb st de Havilland Goblin jet engine. The aircraft was ordered into production in 1944 and entered squadron service in 1946.

It was a single-seat mid-wing fighter, with the engine directly behind

the pilot and the tail unit carried on twin booms, to enable a short jet-pipe to be used and so cut down power loss. The first production version had a maximum speed of 540 mph at 20,000 ft, with an initial rate of climb of 4,200 ft/min, but its range was only 730 miles. This was improved to 1,220 miles in the final single-seat version, the FB Mk 9. The Vampire F Mk 1 was followed into service by the F Mk 3, which was used for the first transatlantic flight by a jet fighter squadron. The most produced version was the FB Mk 5, which saw service in the UK in Fighter Command, in Germany, and in the Middle East and Far East. The NF Mk 10 side-by-side two-seat night fighter also equipped several RAF squadrons.

A two-seat trainer version was produced as the standard jet trainer for the RAF and several overseas air forces, and the Royal Navy also used the Vampire for jet familiarization and experiments in deck-landing.

de Havilland Canada Twin Otter (DHC-6)

Twin Otter Series 100

Currently one of the world's most widely used STOL transports, the Twin Otter was designed in 1964 and made its first flight in May 1965. Three years later the total number of orders was around 300, with production scheduled to continue into the nineteen-seventies. Many of these are

for corporate owners or military use, but many more are for small local-service and feeder-line operators, for whom the Twin Otter is especially suitable, due to its short take-off and landing capabilities and economical operation costs. The aircraft seats up to 19 passengers and is powered by two 579-eshp PT6A-20 turbo-prop engines. In 1968 the Series 200 and 300 were introduced; both have increased baggage space in a lengthened nose, and the Series 300 has 652-eshp PT6A-27 engines.

De-icing, see Anti-icing

Delta Airlines

Delta, one of the largest US regional carriers, has an unusual origin, having first been founded in 1925 as Huff Daland Dusters, the first company in the world to be formed to carry out crop-dusting from the air. It began its first passenger services in 1929. Its present route network of some 14,000 miles reaches some 600 points, mostly in the eastern half of the US. Headquarters: Atlanta, Georgia.
Member of IATA.

Denmark: Air Force

Flyvevabnet (Royal Danish Air Force). Has existed as an independent force since October 1950, when it was formed to combine the former separate air services of the Navy and Army, whose histories date from 1911 and 1912 respectively. After the 1914-18 War the Navy's aircraft consisted mainly of ex-German types while the chief Army types were Bristol Bulldog fighters and Dutch-built Fokkers. Denmark's air strength was decimated in the Nazi invasion of 1940, but the domestic industry began to provide its own training aircraft after World War Two, and other types were supplied by

Britain and the US. The RDAF has selected in subsequent aircraft almost exclusively from these two countries ever since, the major exception being an order in 1968 for Drakens from Sweden. Other front line aircraft at present in service include the British Hunter and the American Star fighter, Super Sabre and Thunderflash.

Denmark: Chronology

Although the Danish Navy had started training pilots in 1911, it was not until 25 March 1912 that it received its first aircraft. The Army was a little slower off the mark, with its first flying school starting operations in 1912. During the period 1914-19, the Orlogsvaerft (Navy Ship Yard) built and supplied flying-boats to the Navy, and although production continued after 1919, it was mainly of foreign aircraft built under licence. A flying school was established at Kastrup in 1921.

Denmark established a civil airline as early as 1918, the Det Danske Luftfartselskab (DDL), but scheduled services to Germany and Sweden did not commence until August 1920. This company was integrated into the Scandinavian Airlines System (SAS) in 1946. In 1937, Skandinavisk Aero Industri began to build aircraft at Alborg and following that date a number of different types were produced including K.Z.II, III, IV, VIII and X.

When German forces invaded Denmark in April 1940, obsolete equipment prevented her Air Force from offering any effective resistance. Many Danish pilots escaped to Britain, however, and continued the fight in RAF squadrons. Many other Danes were trained by the RAF with Norwegian units formed in Canada, and some Danish pilots

One of the most numerous Vampires: an FB Mk 5

Danish-produced K.Z.VII Laerke of the RDAF

were trained by the Swedish Air Force. Now equipped with modern aircraft, Denmark is a member of NATO.

Deperdussin, 361, see also British Deperdussin

de Rozier, François Pilâtre, 127, 208, see also Montgolfier

Derwent engine, see Rolls-Royce

DETA (Direcçao de Exploraçao dos Transportes Aéreos) Owned by the Government of Mozambique, by whom it was established in 1936, DETA provides mainly an internal network, although international services are flown to points in Rhodesia, South Africa and Swaziland. Its somewhat outdated fleet has been slightly improved in recent years by the acquisition of Fokker Friendship airliners. Headquarters: Lourenço Marques. Member of IATA.

Deutsche Lufthansa, see Lufthansa

Deutschland II, 253, see also Zeppelin, Ferdinand von The first of five Zeppelins operated by Delag, the German airship company, between 1910 and 1914 from Lake Constance to other cities, including Berlin.

Dewoitine, Emile, 565 Emile Dewoitine founded his aircraft company in 1922 and specialized in building small all-metal military aircraft. He built a long series of parasol monoplanes which saw service with the French and other air forces, later entering the commercial field and also producing a series of low-wing fighters with fixed,

spatted undercarriages. These entered production both for France and for export. In 1936 his company was nationalized as the SNCAM but it continued to build aircraft to his design and on the outbreak of War the Dewoitine D.520 was the French Air Force's most advanced fighter.

D.H.2, see de Havilland

Differential aileron, 1641

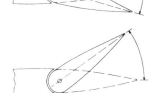

During normal movement of the ailerons, the down-going surface, while increasing lift, tends to increase drag to the point where it can turn a light aircraft in the opposite direction to that required. This effect can be countered by making the up-going aileron move through a larger angle than the aileron which moves downward. This ensures that the up-going aileron causes most drag and decreases lift, while the down-going aileron, owing to its smaller movement, does not cause so much drag.

Diffuser, 1198, 1216 Term usually applied to a portion of a duct in which the cross-section area increases. If such a divergent passage is used to transmit a subsonic flow of air or other compressible fluid, the velocity of flow will decrease and its pressure will increase. This is desirable upstream of any major heat exchange process. For ex-

ample, diffuser passages are provided in the intake duct to a ram-air-cooled heat exchanger (radiator), between the intake and the subsonic combustion chamber of a ramjet and between the compressor and combustion chamber of a turbojet. (*Note*: a divergent passage has the opposite effect on a supersonic flow, which becomes accelerated and rarefied. For this reason rocket engines always have a divergent nozzle, ideally a curved bell shape.)

di Mendoza, Francisco, 119

Dipstick, 913 Graduated rod used to measure, *while it is stationary*, the contents of a tank containing a liquid.

Director horizon, 1357

Disc brake, 567

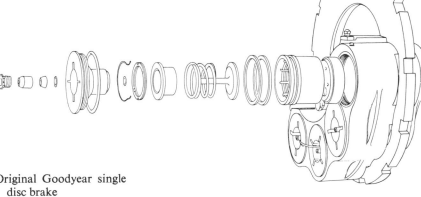

Original Goodyear single disc brake

In this type of brake the drum is replaced by a flat disc or plate. The inherent advantage of a disc is that it expands in a direction normal to the braking forces, and thus does not affect the pressure displacement. Some distortion of a disc inevitably occurs, as it is virtually impossible to heat and cool any piece of metal repeatedly without eventual distortion, but it is much less than with a drum.

In disc brakes the pressure can be applied either by annular pistons or by a series of individual circular pistons housed in the main frame of the brake. The pistons act on

a pressure plate which carries the brake friction lining material and which is keyed to the backplate. A single disc, or a series of discs keyed to the wheel, can be utilized, depending upon the severity of the brake duty. A backplate carries the final set of linings.

Distortion can be further minimized by making the disc of separate segments, articulated together by links and keyed to rotate with the wheel. Articulated discs eliminate distortion troubles and also provide paths within the brake to get rid of the dust generated during each stop. The paths also assist in cooling the brake after a stop. Such brakes can be parked while hot without causing brake seizure or brake lining deterioration.

Dissipation of brake heat is a major problem. The kinetic energy of the aircraft, gained during take-off by the thrust of powerful jet engines operating for thirty seconds or more, has to be absorbed in 10 to 15 seconds by relatively diminutive brakes housed within the confines of the landing wheels. Almost all of the energy is converted into heat, as there is negligible energy loss to the atmosphere during the actual stop. Disc temperatures of 500° C are common, and a fan is often necessary to assist cooling after a stop. Even so, cooling can take half an hour or more before the

brakes can be used again.

Distance Measuring Equipment, see DME

DME (Distance Measuring Equipment), 705-7 A radio device for indicating distance. It resembles radar in that the aircraft transmits a radio pulse and immediately receives back another signal, but the received signal is not a mere reflection. Instead the emission from the aircraft is picked up by an unmanned ground station which is thereby triggered and sends out a different signal which is picked up by the aircraft. The time interval between transmission from the aircraft and receipt of the ground response gives an accurate measure of distance to the ground station. DME is usually combined with VOR

(*q.v.*) so that the ground station gives both distance and bearing.

Doolittle, J. H., 671, 685, 687 J. H. ("Jimmy") Doolittle, US pioneer pilot, was the first man to fly "blind" successfully, on instruments alone, on 24 September 1929. Doolittle also took part in many early air races, including the 1925 Schneider Trophy contest, which he won at a speed of 232 mph, and the 1932 Thompson Trophy race which he won in the over-300-mph Gee Bee Super Sportster. Doolittle achieved further fame for his exploits during

World War Two, including the bombing of Tokyo from the aircraft carrier *Hornet* in 1942.

Doppler 1523, 1537

Doppler effect Named after its discoverer, the 19th-century German mathematician Christian Doppler, the phenomenon known as Doppler effect is the apparent variation in the frequency of vibrations (eg sound or radio waves) when the listener and the source of the vibrations are in relative motion to one another. As the distance between observer and observed is increased or decreased, so do the frequencies appear to decrease or increase in proportion.

Dornier The Dornier Metallbauten was formed in 1922 to succeed the former Zeppelin Werke GmbH at Lindau, which had built aircraft designed by Professor Claude Dornier since the 1914-18 War.

Prof Claude Dornier

These included the Komet and Merkur high-wing light transports and a wide range of seaplanes and flying boats. Among the best-known marine aircraft of the period were the 1922 Dornier Wal flying boat, built in considerable numbers (including licence manufacture in Italy, Japan, the Netherlands, Spain, Switzerland and the USA); and the huge Do X of 1929, which had twelve engines mounted in pairs above the 157½-ft-span wing and a range of 1,740 miles cruising at about 120 mph.

In the nineteen-thirties came the famous "Flying Pencil", the ultraslim Do 17 medium bomber which, with its developed version, the Do 217, served widely as a bomber and as a night fighter during World War Two. Several more modern flying boats also appeared during this period, but the most radical wartime Dornier design was the "push-and-pull" Do 335 fighter, which was only just too late to see operational service.

The reconstituted postwar Dornier GmbH was established initially in Spain, but is now based in Friedrichshafen, where its products include the Do 27, Do 28 and Skyservant generalpurpose/light transport aircraft and an advanced experimental V/STOL type, the Do 31.

Dornier Do X, 811

The Dornier Do X of 1929 was remarkable on several counts. First of all, it was fitted with twelve engines, a greater number than any other type before or since. They were originally Siemens Jupiters, but these were replaced by 600-hp Curtiss Conquerors. Secondly, it was a monoplane of vast proportions in a period when almost all large aircraft still required the strength of a biplane structure. It was also a double-decker aircraft, with a crew deck and a passenger deck, the crew compartment closely resembling the bridge of a ship. The Do X made many long-distance proving flights to various parts of the world but was too far ahead of its time to become a production possibility.

Dornier Do 17, 911

Dornier Do 31, 997, 1009-1010, 1021, 1233

Dornier Do 335, 817, 820-821

Professor Claude Dornier conceived the tandem-engine layout in 1937 and patented it, but was not able to make use of it until 1942. The

result was the Do 335, which began its flight trials too late for any possibility of seeing action in World War Two. It was powered by two Daimler-Benz DB 603E twelve-cylinder engines of 1,800 hp each, giving the aircraft a maximum speed of 430 mph. The aircraft combined the advantage of twin-engined reliability with single-engined manoeuvrability, and was intended for service as a day fighter/bomber or as a two-seat night fighter.

Dornier RSI, 367

Dorval Airport, see Montreal

Double Mamba engine, see Armstrong Siddeley

Douglas, 387, 737

Donald W. Douglas, US aircraft designer, founded in 1920 a company that was destined to have a great influence on world aviation. At first, most of its designs were for the US Army Air Corps, and through the nineteen-twenties and early 'thirties a succession of Douglas aircraft flowed from the factory. It was in the mid-thirties that Douglas produced his DC-2 airliner and then the DC-3 which became the world's leading airliner for the next twenty years or so. It was followed by the DC-4, DC-6 and DC-7 series of four-piston-engined airliners and then by the DC-8 and DC-9 jets. In the military field, Douglas produced the majority of the transport aircraft used by the USA and Britain in World War Two, and also supplied aircraft such as the Boston bomber for the RAF and USAAF and the Dauntless dive-bomber for the US Navy. Since World War Two, the Douglas company has continued to produce military aircraft, principally for the US Navy, including the Skyraider general-purpose bomber and Skyhawk light attack aircraft. It is now part of

the McDonnell Douglas Corporation.

Douglas Boston, 573

The Boston originated as the DB-7 light bomber, which was ordered by the French Air Force before World War Two. Defeat prevented the French from taking delivery, and it was diverted to the Royal Air Force. It first served with the British in the night-fighter and intruder role, as the Havoc; but a later version, the Boston III, was used as a light bomber, both from the UK and in the Middle East theatres, with great effect. The type also went into service with the USAAF as the A-20 in various configurations, for both bombing and attack duties in the European, Middle East and Pacific theatres.

Douglas C-132, 1309

Douglas Cargomaster (C-133), 573, 789

The C-133 Cargomaster was conceived specifically as a transporter for the American forces' IRBM and ICBM surface-to-surface missiles. Thus its fuselage was tailored to fit these weapons and all else was designed accordingly. In the event, the aircraft has had many other uses as a long-distance heavy-lift transport aircraft. It is a high-wing aeroplane, powered by four Pratt & Whitney T34 turboprop engines. Normal cruising speed is 327 mph and range 4,400 miles with typical load. The C-133 first went into service in 1957.

Douglas DC-1, 737, 743, 745

This was the first of the long line of successful Douglas Commercial monoplanes. Only one was built, to the requirements of TWA, and it saw service with that airline until the DC-2s were ready for service, after which it was used by Howard Hughes as an executive aircraft. It went to Lord Forbes in the UK in 1938, but he sold it in France four

months later and it went from there to Spain, where it crashed in December 1940. Virtually indistinguishable from the later DC-2, the DC-1 had accommodation for two crew and ten passengers, and cruised at 190 mph with two 875-hp Wright Cyclone engines.

Douglas DC-2, 287, 387, 745, 761, 1109

Douglas DC-3 (Dakota), 287, 387, 550, 571, 745, 887, 1109

Most famous commercial aeroplane ever built, the DC-3 was evolved from the Douglas Aircraft Company's DC-1 prototype and the DC-2 which had followed the Boeing 247 into airline service and overtook that aircraft in popularity. The DC-3 found immediate acceptance with airlines in America and Europe, American Airlines being the first to use it commercially. With the outbreak of World War Two it went into immediate massproduction and served in thousands with the US services and with the RAF, as the C-47

The Douglas DC-3

Skytrain and Dakota. The end of the war brought a ready market for surplus DC-3s, and few airlines and air forces throughout the world have not used DC-3s at some time or other. They still outoutnumbered any other type of transport aircraft in service in 1968. The C-47 acquired a new lease of life in Vietnam when, equipped with a battery of Miniguns, it became a ground-attack aircraft for the first time in its long existence.

Douglas DC-4, 307, see also Aviation Traders Carvair

Following the success of the DC-3, the Douglas Aircraft Company produced a bigger, longer-range transport with four engines. This was given the USAAF designation C-54, and the name Skymaster, and entered service towards the end of World War Two. The civil version received the Douglas designation DC-4.

When the war ended, the embryo airlines were shopping for suitable aircraft and the DC-4 was soon in demand by airlines the world over. It was widely used on the air routes across the Atlantic from 1946 onward, and has continued to serve in both military and civil guise to this day. In 1947, Canadair acquired a licence to build the type in Canada and reengined it with Rolls-Royce Merlins to improve its performance. As the North Star, this version served with the RCAF, and with Canadian and British airlines. In 1961, Aviation Traders in the UK made substantial modifications to a DC-4, building on an entirely new nose section to make the aircraft suitable for car-ferry and bulky freight transport. Known as the Carvair, this conversion has been supplied to several operators.

Douglas DC-6, 1117

Benefiting from the experience of building the DC-4 as a wartime fourengined transport and converting it back into a civil airliner post-war, Douglas set about designing a larger fourengined airliner specifically for long-range airline use. This aircraft

was the DC-6.

It had a longer fuselage, more powerful engine installation (1,800-hp Double Wasps), pressurization, reversible pitch propellers and thermal de-icing for the wings. As well as being popular with airlines at the end of the nineteen-forties, it also went into production as the C-118 transport for the US services. From it were developed in 1949 the DC-6A and DC-6B with a 5 ft longer fuselage. The DC-6A was a freight aircraft and the DC-6B its passenger derivative; the latter became the most popular of all the DC-6 series. Powered by 2,500-hp Double Wasp engines, it had a cruising speed of 311 mph and a normal range of 3,560 miles.

Douglas DC-7, 1117, 1121

Taking advantage of the greater power of the Wright Turbo-Compound engine, Douglas made a final "stretch" of their basic four-engined airliner formula. The result was the DC-7, which went into service in 1953. It was capable of carrying up to 95 passengers at a maximum cruising speed of 360 mph at 22,000 ft, and had a range of 3,905 miles. The DC-7B was a cargo version and the DC-7C an improved passenger-carrying version with increased wing span of 127 ft and more fuel, increasing the range to 5,000 miles. This version entered service in 1956 and for two years provided the most advanced standards of airline operation. However, in 1958 the advent of the big jets brought the DC-7's supremacy to an end and it did not remain in front-line service for long afterwards.

DC-7C long-range transport

Douglas DC-8, 309, 579, 791, 837, 1201, 1237

DC–8–30

Douglas's contender for the jet airline market in the late 'fifties was the DC-8. It appeared later than the Boeing 707, and although it has not achieved such great popularity, several hundred have been built and sold. A wide variety of versions is available, with the DC-8 Super Sixty series providing long-range and high-density versions that pointed the way to the airbus or "jumbo jet" concept.

Douglas DC-9, 679-84, 837

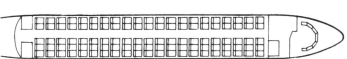

DC–9–30

Design of this short-haul jet airliner began in 1962. In layout it is almost identical to the BAC One-Eleven, with two rear-mounted Pratt & Whitney JT8D turbofans and a T-tail. The first DC-9 flew in February 1965, since when the type has continued in quantity production for many airlines and for the USAF. Several different versions are available, with variations in seating, fuselage length, power and fuel capacity.

Douglas DC-10, 839

Douglas Genie missile, 895
Douglas commenced development of this air-to-air missile in December 1955, and it entered service on F-89J, F-101 and F-106 aircraft after successful trials in July 1957. It is unusual in being an unguided missile and in having a nuclear warhead. It has a speed of Mach 3 and a range of 6 miles.

Douglas Globemaster (C-124), 789, 797

Douglas Mixmaster (XB-42), 822

Drache, see Focke-Achgelis Fa 223

Drag Hinge, 971

1 Drag hinge
2 Drag angle

Hinge used on non-rigid helicopter rotors to allow a blade to swing backward and forward slightly in the plane of rotation. The need for such hinges arises partly out of the blade's flapping motions during rotation. In steady forward flight, although the overall speed of rotation remains constant, the blade on the advancing side of the rotor tends to accelerate while that on the retreating side tends to decelerate. If drag hinges were not fitted, the blades would bend and impose high stresses. Too much freedom of movement about the drag hinge would cause excessive vibration and to prevent this a friction damper is usually fitted to restrict the movement.

"Dragonships", 887
A name given to Doug-las C-47s (DC-3s) modified with Miniguns for the ground-attack rôle in the Vietnam War in the mid-nineteen-sixties.

Dripstick
This is basically an inverted dipstick, mounted in the underside of a fuel tank. When withdrawn, fuel drips through the hollow "stick" when an orifice reaches the level of the fuel, thereby indicating accurately the contents.
Improved under-wing indicators are available which eliminate fuel spillage. These consist of a circular float which slides on a sealed tube containing a calibrated inner tube. Two magnets are embodied in the assembly, one in the float and one in the inner tube. To measure the tank contents, the calibrated tube is released, when it falls to the point where the inner magnet links magnetically with the one in the float, the contents being read directly off the protruding tube. It is used to measure contents of a fuel tank in a stationary aircraft.

Dripstick
1 Magnet
2 Float
3 Magnet
4 Contents calibration marks
5 Fuel

Drum brake, 567

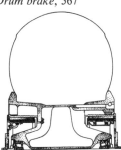

Wheel with two expander tube brakes. On early wheels the brake drum was a tight fit inside the wheel. The wheel illustrated has a detached drum taking the heat away from the tyre beads.
The simplest type of drum brake employs two rigid shoes, as do those on road vehicles, and can be operated mechanically or hydraulically. Use of drum brakes is now confined to very light aircraft, although in the past they were used successfully on quite large aircraft, such as the Junkers Ju 88.
The bag-type or "expander-tube" drum brake was developed to a high state of efficiency. On this, a bag is inflated pneumatically or hydraulically to expand a series of circumferential brake blocks, keyed to the brake frame, against the brake drum secured to the wheel. Although providing powerful braking for a low weight, the tendency for the drum to expand outwards, releasing the brake, and the high operating temperatures which caused severe distortion of the brake drum, proved to be a serious disadvantage. On hydraulically operated brakes, there is often an appreciable lag on release, making taxying difficult. Drum distortion was eased, giving bag-brakes a further lease of life, by the use of bi-metal drums, made by casting iron alloy into a pressed steel casing by a centrifugal process. The result was a mechanically strong drum which did not distort on cooling owing to the advantageous properties of the cast iron. The energy absorption of this type of drum was extraordinarily high, reaching its zenith on such aircraft as the Boeing B-47 Stratojet. This type of brake was used widely until about 1950, when increased landing speeds and weights brought about the introduction of disc brakes (q.v.).

DTA (Direcçao de Exploração dos Transportes Aéreos)
Known alternatively as Angola Airlines, DTA should not be confused with the Mozambique airline DETA (q.v.). DTA, formed in 1938, was originally a division of the Angola Ports, Railways and Transport Administration and began services in 1940. Present services are internal and to nearby African states.
Headquarters: Luanda, Angola.
Member of IATA.

Duct-burning engines, 1181, 1187
A duct-burning engine is basically a turbofan engine in which fuel is also injected into the cold air passing around the combustion area and ignited to provide additional thrust.

Dulles International Airport, see Washington

du Temple de la Croix, Félix, 159

Dyott, G. M., 666
G. M. Dyott designed a small, single-seat monoplane in 1913 as a result of his experience of demonstration flying in the USA, Mexico and Central America. It was built by Hewlett and Blondeau of Clapham, England, and had a 50-hp Gnome rotary, giving it a maximum speed of 75 mph. Dyott took it to the United States, where it performed well. Later he returned and raced it in the UK until, on 6 November 1913, when taking part in a night race from Hendon, he found himself over Eastbourne and landed on Beachy Head, where the monoplane overturned.
In 1914 he designed a large twin-engined bomber for the Admiralty, which flew in 1916. Although heavily armed with five Lewis guns—and, in one form, with a large cannon—it never went into service.

E.28/39, see Gloster

Eagle engine, see Rolls-Royce

East African Airways
Established in 1946, EAA is now responsible for fostering the growth of air transport in the states of Kenya, Uganda and Tanzania. Internal services within and between these three states are flown, as well as longer routes to some

Three Minigun "Gatling" pods delivering a broadside from an AC-47 "Dragonship"

other African countries, to Europe (including the UK), the Middle and Near East. Some of the domestic services are operated by EAA's subsidiary, Seychelles-Kilimanjaro Air Transport. Headquarters: Nairobi. Member of IATA.

Eastern Air Lines

A pioneer US air mail carrier, Eastern was formed in 1927 as Pitcairn Aviation, changing its title to Eastern Air Transport in 1930 and to the present one in 1938. During its existence, Eastern has absorbed several smaller US airlines, both before and since World War Two, and its domestic network, of nearly 20,000 miles, alone serves more than 100 points in the eastern half of the US. Foreign routes include services to Canada, Mexico, the Bahamas and Bermuda. It was Eastern Air Lines which, in 1961, introduced the air "shuttle" (q.v.) commuter service between New York, Boston and Washington. Headquarters: New York.

Member of IATA.

Eastern Provincial Airways

Newfoundland airline, formed in 1949 and incorporating (since 1963) the former Maritime Central Airways. Its mixed fleet consists of piston- and turboprop-engined aircraft of various sizes and includes landplanes, seaplanes and helicopters. Scheduled services are flown to points in Newfoundland, Labrador, Nova Scotia and Quebec. Headquarters: Gander, Newfoundland.

Associate member of IATA.

Edison, Thomas Alva, 317

Egg, Durs, 139

Egypt: Air Force

The first air force in Egypt was formed, with British assistance, in 1930 and became independent of British influence in 1937, although it continued to operate with British aircraft. It became the Royal Egyptian Air Force in 1939, and when re-equipping after World War Two again chose British fighters and bombers for the purpose. The "Royal" prefix was dropped in 1953 after the deposition of the monarchy, and two years later a trade agreement with Czechoslovakia marked the beginning of supplies of Soviet-designed military aircraft to Egypt. A high proportion of the Egyptian Air Force's strength was lost during the Suez conflict of 1956 and again in the "six days' war" with Israel in 1967, but after each of these incidents the Air Force was restocked with aircraft from the USSR. In 1969 the principal combat types included MiG-17 and MiG-21 fighters, Su-7 ground attack aircraft, and Tu-16 jet bombers.

Eindecker, see Fokker

Ejection seats

During World War Two, with the speeds of piston-engined aircraft increasing to over 400 mph and the imminent introduction of even faster jet-driven types, it became essential to develop some form of assistance to enable aircrew members to escape from their aircraft in an emergency. Little was then known of the physical or the mechanical problems involved, but the diligent research of James Martin, of the Martin-Baker Aircraft Co, gave rise to the first ejection seats, the design and

Martin-Baker seat for F-104G Starfighter

development of which has been improved and advanced continuously ever since. The name of Martin-Baker is today synonymous with ejection apparatus, although mention must also be made of valuable early work carried out by Folland, Saab and the SNCA du Sud-Est (now Sud-Aviation).

In essence, an ejection seat is one which is fired from the aircraft by explosive charge in an emergency, complete with its occupant; once clear of the aircraft, parachutes are deployed automatically to bring the seat slowly down to earth after first separating it from its occupant, who descends by parachute in the ordinary manner. All major combat aircraft (and most jet trainers) are fitted with ejection seats as standard equipment. Variations on the basic theme include underwater ejection seats; downward-ejecting seats (through the floor of the aircraft); conventional zero-speed, zero-altitude seats (from which a pilot can eject, for example, if an emergency arises during take-off or landing); and complete ejection "capsules", such as are fitted to the production F-111A, whereby the entire cockpit unit or nose section of the aircraft is separated from the main airframe.

El Al Israel Airlines

El Al came into being in 1948, soon after the formation of the state of Israel itself, and the Israeli Government still retains the majority interest. It began operations with routes into Europe in 1949, followed by services to South Africa and the USA in the following year. The first jet services began in 1961, and the fleet now consists entirely of jet-driven types. El Al operates no domestic services itself, but has a half-share in Arkia, the Israeli internal airline.

Headquarters: Tel Aviv. Member of IATA.

Electra, see Lockheed

Ellehammer, J. C. H., 186 Jacob Christian Hansen Ellehammer was a gifted Danish inventor who, during his lifetime, took out over 400 patents for a wide variety of mechanical inventions. His first successful aeroplane was the "semi-biplane" which he built in 1906; although tethered to a mast, it flew round a circular track on the tiny island of Lindholm for about 140 ft at a speed of some 35 mph. The 1907 triplane developed from it, although fitted with a more powerful engine, was less successful, and Ellehammer then concentrated his efforts on developing his aero engines up to and during the 1914-18 War. (His first, a three-cylinder piston-engine developed in 1903, was probably the world's first radial engine.)

Replica of Ellehammer's 1912 helicopter

In 1912 Ellehammer also tested with some success a helicopter, powered by a 36-hp engine which drove a conventional tractor propeller as well as the rotor system, and having a rudimentary form of cyclic pitch control (q.v.). This was destroyed in 1916, and Ellehammer produced no new helicopter designs until the nineteen-thirties, when he built working models of two types driven by compressed air, one of them having retractable rotor

blades. Neither of these, however, was built as a full-sized aircraft.

Ely, Eugene, 217, see also Aircraft carrier

Embakasi Airport, see Nairobi

"Empire" flying boats, see Short

Engine speed indicator, 661, 664, 666, 678, 683, 690, 696, 720

This was one of the first instruments used in the cockpits of aircraft, being a gauge showing the number of revolutions per minute at which the engine was turning. This gave a very good indication of the power output of the engine and enabled a pilot to see speedily if there was the slightest variation in the engine's power. The instrument has remained an indispensable part of every aircraft, up to the jet aircraft of the present day; it is known by various names, such as tachometer, rev counter, rpm indicator etc.

Originally it was a mechanical instrument, worked by a set of governors on a centrifugal principle and geared to the rotating part of the engine. Nowadays the rpm indicator is usually driven electrically by means of a generator.

Esnault-Pelterie, Robert, 529, 625

One of the foremost Frenchmen involved in the birth of heavier-than-air flying in Europe, Robert Esnault-Pelterie built

and flew a Wright-type glider in 1904, in a misguided attempt to prove that the Wright's theory of lateral control was unsound. Three years later he got off the ground with a monoplane of his own design. Although not a great success, it was structurally sound and was the forerunner of several successful monoplanes, some of unconventional design. One of his later monoplanes was the first aircraft to be built by Vickers in the UK. Early on he used metal construction and incorporated advanced ideas, being the first to employ hydraulic damping in undercarriages, utilizing oil. He was also one of the first to evolve the control column as a primary means of control; was one of the first to introduce ailerons as a means of controlling an aeroplane laterally; and was among the foremost early manufacturers of aero engines in Europe.

Etévé speed indicators, 661, 665

Ethiopian Airlines

Ethiopian Airlines is owned wholly by the Ethiopian Government, although it continues to receive assistance from TWA, as it has done since its foundation at the end of 1945. The mixed fleet ranges from single-engined Cessna and Piper light planes to four-jet Boeing 720s, and these aircraft are used for the internal network and for services to neighbouring African states, to Europe, the Middle East, India and Pakistan.

Headquarters: Addis Ababa.

Member of IATA.

Euler, August, 861

Executive aircraft, see Business aviation

Exhaust valve, 1044, 1054, 1059, 1086

Experimental Fighting Biplane No 1, see Vickers "Gunbus"

External-burning ramjet, 1319

Ezeiza Airport, see Buenos Aires

abre, Henri, 217, 221

Henri Fabre holds his place in aviation history as the pioneer of the seaplane. He began to study the work of contemporary French aviators in 1905, after qualifying as an engineer, and designed and built his first *hydravion*, or seaplane, in 1909. This did not fly, but his second machine, powered by one of the first examples of the new seven-cylinder Gnome rotary engine, made a successful take-off from the harbour at La Mède, near Marseilles, on 28 March 1910. On the following day it made a flight of 3¾ miles and proved to be very stable in the air despite its apparent frailty. In fact it was less frail than it appeared, due chiefly to the then novel lattice-girder construction of the fuselage and main wing spar. The seaplane crashed on 18 May 1910, but it was later rebuilt and improved, making further flights until wrecked in March 1911. Lacking the money to continue building complete aircraft, Fabre then concentrated, very successfully, on the design and manufacture of floats for other aircraft. At the Monaco seaplane meeting in 1913, every winning machine was fitted with floats designed by Henri Fabre. His original seaplane has since been restored and is now exhibited in the *Musée de l'Air* in Paris.

FAI (*Fédération Aéronautique Internationale*)
The FAI, which was founded in 1905, is the only internationally-accepted body qualified to formulate and en-

force rules governing the conduct of sporting air meetings and international air records. Its headquarters is in Paris. The first speed record to be recognized officially by the FAI was the flight at Bagatelle, France, by Alberto Santos-Dumont on 12 November 1906, when he reached a speed of 25.06 mph.

"Fail-safe", 387, 407, 432, 439, 463, 473, 476-477, 935

Fairchild Cornell, 573
The Cornell was built in quantity from 1941 onward as a tandem two-seat primary trainer for the US Army Air Corps (as the PT-19, -23 and -26), the Royal Canadian Air Force and, as the Cornell, for the Royal Air Force; it served with the RAF in Rhodesia under the Empire Air Training Scheme. It was a low-wing monoplane with fixed undercarriage and various alternative power plants.

Fairchild Packet (C-82 and C-119), 573
This military transport was first built towards the end of World War Two by Fairchild for the USAAF. The first C-82 Packet flew in September 1944, but production aircraft did not enter service until after the war had ended. It was a twin-engined, high-wing monoplane of 106 ft span with a fuselage in the form of a large nacelle with rearward-opening doors. The tail was carried by two tail-booms extending rearward from the engine nacelles. It later saw service in the Korean War. A developed version, the C-119, was even more extensively produced and entered service with several other air forces, including those of Canada, Belgium, India and Italy. It still serves in several parts of the world. Those of the Indian Air Force now have an improved performance, resulting from the addition of a

small podded jet engine above the roof of the centre nacelle.

Fairchild "Packplane" (XC-120), 601
This aircraft, one prototype of which was built for the USAF, was a logical development of the C-119 Packet. The XC-120 had a fixed "skeleton" airframe, with a detachable cargo compartment comprising the lower half of the fuselage nacelle; it could fly with or without its cargo pack, thus enabling it to act like a lorry picking up and dropping palletized loads as required. In the event, the XC-120 did not replace the C-119, but the principles on which it was designed can today be seen in large transport helicopters, notably the Sikorsky S-64 Skycrane and the Mil Mi-10K.

Fairchild Hiller
The former Fairchild company has expanded considerably within the last decade, during which period it has absorbed a number of other important concerns. Among the most prominent of them were the former pioneering helicopter company of Hiller (*q.v.*), and Republic Aviation, producers of the celebrated Thunderbolt fighter of World War Two and a range of jet fighter/reconnaissance types culminating in the powerful F-105 Thunderchief in service in Vietnam. Fairchild itself was previously noted chiefly for its training and transport aircraft, such as the Cornell primary trainer, and the Argus light transport and the C-82 and C-119 Packet "flying boxcar" freighters. Fairchild Hiller Corporation today has a wide range of aviation, space and other industrial interests. Aircraft production includes licence manufacture of the Dutch Fokker Friendship airliner and the Swiss Pilatus Turbo-Porter, the latter also in armed configuration for the COIN (counter-in-

surgency) rôle; and the FH-1100 five-seat turbine-powered helicopter.

Fairchild Hiller FH-1100, 975

Fairey
The Fairey Aviation Co Ltd was established in 1916, and until its absorption by Westland Aircraft Ltd in 1960 was known primarily for the wide variety of successful military aircraft which it produced, mainly for the Royal Navy, up to the end of World War Two. Among the many types to achieve renown for their design, performance and combat capabilities were the Royal Navy's Campania, Fairey IIID, Flycatcher, Swordfish, Firefly and Barracuda; and the Fox, Gordon and Battle of the RAF. In 1928, Fairey produced the Long Range Monoplane for attempts on the World Distance Record. Two were built, and the second of these achieved success in February 1933 by flying 5,309 miles from Cranwell, England, to Walvis Bay in South-West Africa in just under 57½ hours flying time.
After World War Two, Fairey entered the field of rotary-winged aircraft, producing first the experimental Gyrodyne and then the highly-promising Rotodyne convertiplane (*q.v.*), which had reached an advanced state of development before being abandoned in 1962.

Fairey Delta 2 (F.D.2), 297, 302
The Fairey F.D.2 was designed with the intention of exploring flight problems at high subsonic speeds. Two aircraft were built, the first flying in October 1954. The design enabled the aircraft to fly with very little disturbance in regions where previously severe buffeting had been experienced. As a result of its high subsonic performance, the aircraft went on to explore

speeds above Mach 1, and on 10 March 1956 it became the first aircraft officially to exceed 1,000 mph under FAI rules for an official World Speed Record.

Fairey Delta 2 shortly after breaking world speed record

Subsequently, the first aircraft was extensively rebuilt, as the BAC-221, with an ogival wing form for high-speed development work in connection with the Concorde.

Fairey Gannet, 829-830
The last in a long line of Fairey aircraft for the Royal Navy, the Gannet was developed, along with its Double Mamba engine, as an anti-submarine hunter/killer aircraft. The Double Mamba, with its ability to shut down one engine for cruising yet having twin-engined power for take-off and operational necessity, gave the aircraft great flexibility and its deep fuselage contained adequate room for a large weapons bay and search radar. It was a three-seat aircraft with pilot, navigator and radar operator.
The Gannet entered service in 1953 and was the main Fleet Air Arm anti-submarine aircraft for many years. The Gannet AEW Mk 3 was developed for AEW (Airborne Early Warning, *q.v.*) with a large "guppy" radome under the forward fuselage and this version still serves with the Royal Navy.

Fairey Gyrodyne, 971
The Gyrodyne was Fairey's first helicopter, flying in December 1947 and embodying several new ideas in helicopter design. With a central main rotor, the anti-torque component was provided by a small forward-facing propeller on the starboard stub-wing which also

provided a forward thrust component; it also provided the ability to convert to autogyro configuration upon loss of power and ensured that the main rotor blades were always capable of autorotation.
In 1948 the Gyrodyne set up an International Helicopter Speed Record of 124.3 mph. The aircraft gathered practical experience for the design of the Rotodyne: in 1954 it was rebuilt with a large, two-bladed rotor and the Alvis Leonides engine powering a compressor which drove compressed air to small jets at the extremities of the rotor blades.

Fairey Rotodyne, 964, 971
The Rotodyne was an advanced design for an inter-city VTOL transport, and made its first vertical take-off in November 1957. Two 3,000-ehp Napier Eland turbo-prop engines, suspended from the stub wings, gave forward propulsive power and also drove the rotor by supplying air under pressure to the tip-jets at the extremities of the rotor blades. The Rotodyne was a true convertiplane in that, once it was airborne, the wings and Eland engines took most of the load in forward flight, the rotor then autorotating. It was a large aircraft, with a rotor diameter of 90 ft and fuselage length of 58 ft 8 in.
In 1959 the Rotodyne established a closed-circuit speed record for rotorcraft of 191 mph, which was 29 mph faster than the existing *absolute* helicopter speed record. Later that year an enlarged version, able to carry up to 75 passengers or 18,000 lb of cargo and powered by Rolls-Royce Tyne en-

Fairey (later Westland) Rotodyne VTOL transport

gines, attracted orders from BEA and New York Airways and interest from the RAF. Firm orders failed to materialize, however, and although Westland (which took over Fairey's activities in 1960) continued its development for a time, this was suspended early in 1962 without the enlarged version ever being built.

Falcon engine, see Rolls-Royce

Falcon missile, see Hughes

Fan Jet Falcon, see Dassault

Farman, Henry, 188, 209, 266, 815

An Englishman living in France, Farman purchased a Voisin and in due course became one of the most accomplished of pilots, making a number of record

horn" and "Shorthorn", which appeared in 1912, became the standard trainers in France and Britain by the beginning of the 1914-18 War. The company prospered and continued after the war, specializing in large biplanes. Farman Goliath airliners played a large part in putting France on the airline map in the nineteen-twenties.

Farman 1913 biplane, 349-352, 355

Letov/Avia Farman Goliath of CSA (Czech Airlines)

Henry Farman flying one of his early biplanes

flights. In 1909 he set up in business on his own and developed what was basically the Voisin biplane into a much more practical aircraft; improvements included the use of ailerons and the new Gnome engine, and the elimination of side curtains. This aircraft, the Farman III, was built for several customers. By 1912 Henry had been joined by his brother Maurice who, although he designed his own aircraft independently, manufactured them jointly with Henry; his "Long-

Farman Goliath, 266

The Farman Goliath was the most important twin-engined aircraft produced in France after the 1914-18 War. It equipped various commercial companies which came together to form Air Union and was used to pioneer many of Air Union's European routes. It was developed as a bomber for the *Armée de l'Air*, and also achieved some success in the export field.

Farman (Maurice) "Longhorn", 815

Fatigue, 409ff, 457ff

It is easy to design a structure to withstand a known steady load; but aircraft are subjected to widely varying loads in flight, and many components even experience load reversals. For example, the wings outboard of the main landing gear sag downwards on the ground, so that the upper skin is in tension and the underside in compression; but in flight the wing is lifted by the aerodynamic forces acting on it and thus bends upwards, causing a complete reversal of the loads. Unfortunately, any metal part subjected to a wide variation or reversal of structural loads is gradually weakened. The most obvious example of this weakening is that

this fatigue process can be used to break off the lid of a tin can or any other piece of metal by bending it to and fro: eventually it breaks without the need to apply any large tensile load.

Early aircraft were designed purely according to the magnitude of the highest static loads likely to be encountered, but since 1945 civil airliners have had to be designed for total flight lives of the order of 40,000 hr or more, and this necessitates a close study of how the airframe will stand up under repeated loads. The subject was vividly highlighted in 1954 when fatigue of the pressure cabin, a hitherto unthought-of menace, caused the break-up in flight of two Comet 1 airliners.

To avoid such a catastrophe the airframe

must not only be made strong enough for the basic static loads, but must also avoid any sudden discontinuities in section or profile, cuts, nicks, notches caused by damage (such as a dent from a slipped screwdriver) or any other seemingly trivial place where loads might be concentrated. Holes must be cleaned free from burrs, machined surfaces must be polished, corners must be rounded to a large radius and, as a basic philosophy, all primary members must, wherever possible, be duplicated so that failure of one will still leave the other to bear the load. The primary structure must also be capable of being inspected, so that if any fatigue crack should develop it will be seen before it can spread far. All new airframe designs are now fatigue-tested by being subjected to thousands of load reversals; but the proof really comes in actual service, when fail-safe design (such as duplication of load paths) allows fatigue to be seen and rectified before structural failure can cause break-up of the aircraft in flight. Some aircraft still break up in flight, but this is almost invariably the result of gross overloading, pilot error or extraordinary storm turbulence.

Faucett (Compañia de Aviacion Faucett)

One of the earliest airlines to be formed within the South American continent, this company was named after its founder, Elmer J. Faucett. From its formation in 1928 until 1946, all Faucett's services within Peru were operated by single-engined Stinson aircraft, which from 1934 onward the airline staff built themselves. Some of these veteran aircraft remained in service until the beginning of the nineteen-sixties, but the current fleet consists almost entirely

of Douglas piston-engined types and one Boeing 727 jet. Braniff (*q.v.*) has a small interest in Faucett.

Headquarters: Lima, Peru.

Feathering, 1341

When an engine stops in flight, the propeller continues to turn, owing to the action of the airflow on the propeller blades. As the propeller is no longer being driven through the air, its action in turning or windmilling produces a drag component on the aircraft known as windmilling drag, which is detrimental to the flight profile. To avoid this, propellers have been designed to feather: that is, to turn the blades at 90° to the airflow by an extension of the variable-pitch mechanism; with the blades end-on, the airflow no longer turns the propeller and the windmilling drag is eliminated. Another feature is that, if the engine is damaged, the propeller no longer keeps the engine turning, so eliminating the risk of further damage to the engine.

Fédération Aéronautique Internationale, see FAI

Fellowship, see Fokker

FH-1100, see Fairchild Hiller

Fiat

The aeronautical activities of the Fiat group of companies can be traced back to early 1916, when production began of the biplane scouts produced by Ing O. Pomilio. After the 1914-18 War, its name was changed

Fiat G91T/1, transonic trainer and tactical strike aircraft

to Aeronautica Ansaldo SA, but its fame really began to be established on a world-wide basis with the appearance in the mid-'twenties of the single-seat biplanes designed by Celestino Rosatelli. Most famous of these were the C.R.30, C.R.32 and C.R.4 fighters—both for their outstanding performances at the air displays of the period and for their combat record during the Abyssinian campaigns, the Spanish Civil War and even, despite their obsolescence, during World War Two. Other, though generally less prominent, aircraft types produced during this period included the B.R.20 bomber, the R.S.14 reconnaissance seaplane and the G.12 transport; the company also produced, in large numbers, aero engines of its own design.

Since World War Two Fiat has evolved several new designs including Italy's first production jet aircraft, the G80 trainer, and the G91 strike fighter for the Italian and German air forces. Current pro-

jects include the G222 medium-range V/STOL transport.

Fiji Airways

Fiji Airways, with its comparatively small fleet, flies services between the Fiji Islands and to other island groups in the Southern Pacific. It was founded in 1951 by Harold Gatty, who was Wiley Post's companion during his round-the-world flight in a single-engined Lockheed Vega monoplane in 1931. The present ownership is divided equally between Air New Zealand, BOAC, Qantas and the Government of Fiji.

Headquarters: Suva, Fiji.

Ansaldo (Fiat) SVA.9, 1918

inland: Air Force *Ilmavoimat* (Finnish Air Force). Formed in 1918, only a few months after the declaration making Finland an independent republic, the Finnish Air Force has remained one of modest size to the present day. Russian- or German-designed marine aircraft constituted a major part of its early equipment, followed from the mid-'twenties onward by aircraft of various nationalities built under licence at the newly-created State Aircraft Factory. By 1939, Dutch Fokker types predominated; during World War Two the Finnish-designed Myrsky fighter and Pyry trainer made their appearance, although neither was built in great numbers. The air fighting in Finland was conducted with a mixture of American, German and Soviet combat types. Of these, the Messerschmitt Bf 109G remained as standard fighter immediately after the war until Vampire jet fighters were acquired in the mid-'fifties. Principal operational aircraft in 1968 were Finnish-built Magister jet trainers, and British Gnat and Soviet MiG-21 jet fighters.

Folland Gnats of the Finnish AF

Finland: Chronology
A Swedish-built Albatros was the first military aircraft operated in Finland, and this was delivered to Kokkola on 25 February 1918. A small number of additional aircraft were obtained after that date, and were used on reconnaissance and bombing operations

during the civil war. After 1919, the Finnish Air Force was equipped almost entirely with British or French aircraft. A Government Aircraft Factory, established in 1920, not only built a number of different foreign aircraft but produced many of original design.

The Finnish civil airline Aero O/Y was formed in 1923, and inaugurated its first scheduled service to Reval in 1924. Towards the end of the 'thirties the company had a number of international routes, and in 1938 several domestic routes were started. At the end of World War Two, the Stockholm route was the first to be reopened, in 1947, and four years later the company adopted the name of Finnair, by which it is known today. When Russian forces attacked Finland in 1939, her Air Force comprised only a small number of bomber and fighter aircraft. These were quickly reinforced by machines supplied from England, Germany, Italy and America. A Swedish volunteer unit also operated in Finland against the Russian forces.

The Armistice of 1945 compelled Finland to limit her Air Force to a strength of 60 first-line aircraft, but today it is equipped with machines which include the Mikoyan MiG-15, MiG-21, Folland Gnat and Agusta-Bell Jet-Ranger.

Finnair
Finnair (Aero O/Y), one of Europe's oldest airlines, began operating in 1924 and for many years had an all-seaplane fleet; its first landplanes were two de Havilland Rapides acquired in 1936. Largely State-owned since 1947, it now operates passenger and freight services to points in Europe, the UK, the USA, the USSR and the Far East.

Headquarters: Helsinki. Member of IATA.
Fir-tree root, 1165, 1204, 1259-60

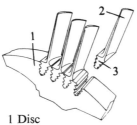

1 Disc
2 Blade
3 Fir-tree root

A type of blade root commonly adopted to secure a turbine rotor blade to its disc. The blade is a cantilever, subjected to great and varying stresses, especially in an outward radial direction. Its root is first forged to have taper and is then accurately machined, either by a precision broach or by electro-chemical machining, to have a series of deep grooves, so that when the blade is held root upward and viewed from the end it has an appearance reminiscent of a fir-tree. The root fits exactly into matching slots cut diagonally into the edge of the turbine disc.

Fiumicino Airport, see Rome
Flame tube, 1216, 1220
Flamingo (D.H.95), see de Havilland
Flap, 613-618, 641, 645
A hinged surface, usually at the trailing-edge of a wing, used to increase the lift of the wing at slow speeds; to steepen the glide path; or to act as an air-brake during approach and landing.
Flaps can take various forms, the simplest being the hingeing of the entire trailing-edge of the wing. A more sophisticated form involves the split flap, where only the lower surface of the trailing-edge is hinged, the upper surface being fixed and maintaining a normal airflow. Slotted flaps, when opened, allow an airflow over the upper surface of the flap, thus increasing lift; the flap

is lowered into the airstream through guides. These are the basic forms, although several variations of them are in use on the complicated airframes of today.
Flapping hinge, 955-6, 959, 971

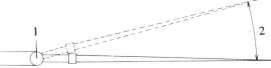

1 Flapping hinge   2 Flapping angle

Hinge used on non-rigid helicopter rotors to allow the blades to move up and down and so relieve the severe bending loads. When initially introduced on the Cierva Autogiro, flapping hinges allowed the blades to rise and fall automatically under natural forces and so balance the unequal lift that otherwise would have been generated by the advancing and retreating blades, owing to their different speeds through the air.
Flapping wings, 113, 117, see also Ornithopter
Flechettes, 873
Flight, principles of

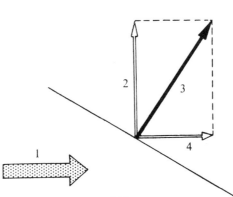

1 Air flow   4 Drag
2 Lift   5 Inclined plane
3 Resultant force

Aerostats are supported against the acceleration due to gravity by the fact that their overall density is initially less than that of the surrounding atmosphere, so that they rise. When they reach a height at which local air density is the same as their own, they stop rising. Further change in altitude is effected by releasing low-density gas (hydrogen or helium) to descend, and releasing

ballast (such as sand) to rise.
Aerodynes are supported in flight by the reaction of air flowing past their wing(s). A wing (q.v.) has a cross-section known as an aerofoil. In the earliest aircraft the section was little more than a shallow, curved plate—convex on top, concave underneath—with rounded leading-edge and sharp trailing-edge. Later, cantilever monoplanes appeared, with thick wings to give adequate depth to support the lift forces, and these initially had nearly flat undersides and high, arched upper surfaces leading back from a bluff, rounded leading-edge. By 1943 laminar-flow wings suited to speeds up to 450 mph were introduced; these had sharper leading-edge curvature and maximum thickness at least half-way back (i.e. at 50 per cent chord or more).
Supersonic aircraft have very thin wings, with a ratio of thickness to chord of 3 per cent or less, with both leading- and trailing-edges very sharp and a roughly equal, gentle curvature on top and bottom surfaces.
Jet VTOL aircraft are supported by the thrust of their engines, which exhaust downward. In-

cluded in such aircraft are ballistic missiles and all spacecraft so far launched. Ultimately, of course, spacecraft may enter orbit round the Earth or some other body and then cease to need engine thrust in order to remain in orbit; their centrifugal and centripetal forces are exactly balanced, as they are for a weight hurled round and round on the end of a string fixed at the other end.
Most wings generate lift by virtue of the fact that the distance the air has to travel in passing across the top of the wing is slightly greater than in passing just beneath it. The air is therefore accelerated to a higher speed, relative to the aircraft, across the top of the wing and thus is correspondingly reduced in pressure. The fact that most of the lift comes from this reduction in pressure across the top of the wing can easily be seen by blowing across the top of a piece of paper shaped in an arched, curving form like the upper surface of a wing; the airflow tends to lift the paper into a straight line behind the leading-edge. A further contribution to lift is made by increased pressure on the under-surface.
Lift depends on the area of the wing, the local air density, the square of the speed of the wing through the air and a factor known as the lift coefficient. The latter depends on the cross-sectional shape and also on other details of the aircraft which may raise or lower it. Modern wings have variable cross-section: flaps can be lowered from below the rear of the wing to increase lift coefficient (perhaps from 1.2 to 2.5, for example) to allow the wing to continue to give lift equal to the weight of the aircraft at much lower speeds (for example, an aircraft might be able to fly at 70 knots with full flap but

could not safely slow below 120 knots in "clean" condition). Lift coefficient also depends on angle of attack (q.v.), the angle at which the air meets the wing. Most wings give zero lift at a slightly negative angle of attack, the air then approaching the wing from slightly above. Lift coefficient then rises steadily up to an angle of attack of 14-17°. Then the airflow suddenly breaks away from the upper surface of the wing; the wing leaves a massively turbulent wake and the aircraft drops almost like a stone until the pilot can restore the angle of attack to a low value, typically by pushing down the aircraft nose and increasing airspeed. This is known as the stall, and it can be postponed by fitting slats to the wing leading-edge to keep the airflow "attached" across the upper surface; and the onset of turbulence should warn the pilot of the impending stall. (Advanced aircraft actually measure angle of attack and provide a positive warning in the cockpit.)

In steady cruising flight the lift of the wing equals the weight of the aircraft and the thrust of the propulsion system equals the total aircraft drag. If these four forces do not all pass through the same point, then one pair must tend to tip the aircraft nose-down by exactly the same amount as the other pair tend to tip it nose-up. Flight of the aircraft is controlled by deliberately spoiling this balance by applying disturbing forces far from the centre of gravity. In nearly all aircraft these forces are imparted by moveable aerodynamic surfaces—elevators to move the nose up and down to cause a climb or dive, rudder to yaw the aircraft left or right or cancel the asymmetric thrust caused by a dead engine on one side, and ailerons or spoilers near the wing-

tips to roll the aircraft. But jet-lift VTOL machines must be able to fly at speeds too low for such surfaces to be effective and their control forces are generally provided by compressed air jets at the extremities of the body and wings.

Other aspects of the principles of flight are dealt with in the body of the book (Flying a small aircraft, page 283).

Flight directors, 1391

Flight recorder, see Flight safety

Flight refuelling

The maximum effort required of an aeroplane's power plant is that needed at the very outset of a mission, to get it off the ground and into the air. Once in the air it is capable of carrying a greater load than the maximum imposed by its take-off requirements.

For this reason, the idea of replenishing and increasing its fuel load once it is in the air is one that has attracted interest since the closing stages of the 1914-18 War. In the late 'twenties and early 'thirties, in-flight refuelling was often employed to boost the capabilities of aircraft attempting to set up new distance or endurance records, and the military value of the practice for extending the range of a well-loaded bomber or a reconnaissance aircraft was obvious. In the US, an endurance flight in 1929 of 150 hours by the Fokker trimotor *Question Mark* was considered an outstanding achievement—until the monoplane *Greater St Louis* was kept in the air for an incredible 647½ hours by being repeatedly refuelled in flight.

For a long time, such achievements were regarded more as stunts than as a serious contribution to aviation progress; but after the formation in 1934 of Flight Refuelling Ltd by Sir Alan Cobham— one of the foremost

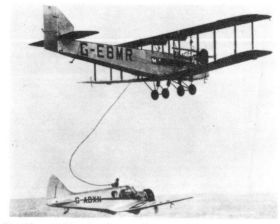

Sir Alan Cobham's Courier being refuelled by Handley Page W.10 in 1934

pioneers and exponents of aerial refuelling—the idea began to attract serious civil and military consideration. In the Autumn of 1939 two "C" class flying boats of Imperial Airways, with a Harrow tanker aircraft, introduced the first air-refuelled service across the Atlantic. In 1944 Cobham's company was asked to convert 1,200 Lancaster and Lincoln bombers for the Tiger Force to bomb Japan from south-east Asia. In the event this proved unnecessary when advanced bases, within the Lincoln's normal range, became available; but military recognition of the value of flight refuelling had been achieved, and in the years since World War Two it has become standard practice in the principal air forces of the world.

It is no longer confined to global bombers or other large aircraft: even single-engined types with modest ranges can nowadays refuel one another, using the "buddy" system of small, jettisonable fuel tanks carried under the wings or fuselage.

The two principal systems of refuelling in current use are the US boom method and the British probe and drogue system.

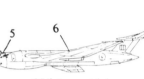

1 Victor tanker
2 Ram-air-driven windmill drives pump and drogue
3 "Buddy" pack
4 Drogue
5 Probe
6 Victor receiving aircraft

Flight safety
The steady rise in the number of passengers travelling by air, the increasing density of air traffic and the increasing size of the airliners themselves must lead to a corresponding risk of an increase in aircraft accidents unless an even greater increase can be achieved in the safety measures designed to prevent them.

The safety of the passenger while in transit has always been a paramount factor in air travel, and the present loss rate among jet airliners, of approximately one per 400,000 flying hours, leaves no room for complacency. A rate

of one per million flying hours is technically and operationally feasible, and advances made in recent years should do much to help realize that goal. All commercial aircraft are already subject to stringent regulations imposing a fatigue life on an airframe or a TBO (Time Between Overhauls) on its engines, and great advances have also been made in the use of failsafe (q.v.) constructional techniques. From the design or structural aspects, it is doubtful if much more could be done to make aircraft intrinsically much safer than they already are; it is in the operation of

them that the significant improvements must come.

Of all aircraft accidents, a high proportion occur during take-off or landing. In recent years certain types in or entering service have been fitted with automatic landing (q.v.) equipment designed to take over complete control and bring them down in complete safety in conditions where the pilot cannot make a visual landing or where it would involve risk for him to do so. The employment of such equipment will increase considerably during the nineteen-seventies, and will eventually become mandatory for all passenger aircraft.

Additionally, a more thorough understanding of the accidents that do occur is possible by installing an instrument called a flight recorder. (Alternative titles are AIDS=Airborne Integrated Data-recording Systems or ADR =

Accident Data Recorders.) As their name implies, these instruments record automatically every action, flight path and other manoeuvre taken by an aircraft whether under manual or automatic control. Under-water ejection and recovery systems are also being developed so that, when an accident occurs, the survival of the recorder is virtually guaranteed and the data contained in its memory can be evaluated during the subsequent investigation. From each investigation something is learned which reduces the risk of another accident from the same cause, and eventually it is likely that the installation of flight recorders will also become mandatory.

Flight simulators
Simulators are a large and essential part of the training schedule of any major airline or air force. Any device that simulates some or all of the conditions of actual flight may be defined as a flight simulator, but the term is most frequently used to denote ground training apparatus representing the flight deck and instrument layout of an aircraft, which is used to train aircrew in the use of flight systems or in operating procedures. Nowadays the tendency is toward the use of a separate procedure trainer for the latter purpose, since these devices can be made more simply and cheaply than fully-instrumented simulators.

Probably the best-known and most widely-used basic flight simula-

Rigid boom flight refuelling technique. J57-powered KC-135A fuelling a TF33 fan-engined EC-135C

tor is the Link trainer. Made by Link Aviation Inc of the US, this is in effect a small "dummy" aeroplane which simulates the control reactions of a real aircraft, and can reproduce banking, diving, spinning and other manoeuvres.

Flugfelag Islands, see Icelandair

Flyer, see Wright Flyer

"Flying Bedstead", see Rolls-Royce

Flying doctors, see Aeromedical services

Flying Tiger Line

The Flying Tiger Line is among the world's principal air cargo carriers, operating a coast-to-coast freight service within the US as well as charter contracts to all parts of the world. The latter includes a regular cargo service to southeast Asia on behalf of the USAF's Military Airlift Command. Flying Tiger took its present name in 1946, the year after its formation, and was the first all-cargo airline to be set up in the USA. It was also the first, and still is the major, operator of the Canadair CL-44 swing-tailed freight transport, which has been in service since 1961. Headquarters: Los Angeles, California. Associate member of IATA.

Focke-Achgelis Fw 61, 954, 957

While Cierva (q.v.) was gaining some measure of success with his Autogiros in the United Kingdom, Professor H. Focke, of the Focke-Wulf Flugzeugbau GmbH in Germany, designed and built the first reasonably practical helicopter. It could be flown and manoeuvred within a small space, and was demonstrated inside the *Deutschland Halle*, Berlin. The aircraft used a normal Siemens radial engine for forward thrust, at the front of a conventional single-seat fuselage. Drives from the engine went to each of two rotors

mounted side-by-side on outriggers from the fuselage. The aircraft first flew in 1936, and in June 1937 gained five world helicopter records. The following October Fraulein Hanna Reitsch flew the aircraft a distance of 68 miles, and it later set a distance record of 143 miles in June 1938 and an altitude record of 11,243 ft in January 1939.

Focke-Achgelis Fa 223 *Drache* (Dragon), 954

A development of the Focke-Achgelis Fw 61, this helicopter first flew in 1940. A 1,000-hp Bramo 323 engine drove two non-overlapping rotors through reduction gearing, enabling the aircraft to lift a useful load of 1,760 lb.

Focke-Wulf

The Focke-Wulf Flugzeugbau GmbH was established in 1924, and in 1931 absorbed the former Albatros Flugzeugwerke (q.v.). In the years before World War Two it was noted chiefly for the production of training aircraft (such as the Fw 44 *Stieglitz* and the Fw 58 *Weihe*) and transports, chief among which was the four-engined Fw 200 *Condor* which became a formidable maritime patrol type during the war. Another outstanding product during the nineteen-thirties was the Fw 61, one of the first really practical helicopters, which set up several new international records in 1937-39.

Focke-Wulf will be best remembered, however, for its production of the excellent Fw 190 single-seat fighter designed by Dr Kurt Tank, who is now with Hindustan Aeronautics Ltd in India, where he has been responsible for the design of India's first jet fighter, the Marut. After a short period post-war under its own name, Focke-Wulf Gmbh joined with the Weser Flugzeugbau GmbH in 1963 to form

the VFW (Vereinigte Flugtechnische Werke) GmbH, which is developing the VAK 191B VTOL strike fighter (in association with Fiat in Italy), the VFW 614 short-haul transport and the VFW-H3 helicopter.

Focke-Wulf Fw 190, 1339

The Fw 190, Kurt Tank's brilliant fighter design to supplement the Messerschmitt Bf 109, first flew in June 1939. It went into production powered by the BMW 801 radial engine rated at 1,700 hp for take-off, and in July 1941 fought in action over the English Channel, establishing immediate superiority over the Spitfire V, then Britain's leading fighter. It was subsequently developed in a variety of versions and carried out fighter-bomber duties with bomb-racks fitted under the fuselage centre-section and, in some cases, under the wings. With a top speed of 408 mph and a rate of climb at sea level of 2,350 ft/min, the Fw 190 had good handling qualities, and gave its pilots immense confidence. A difficult opponent for Allied fighters, its sphere of influence spread to the Russian Front, the Mediterranean and North Africa, and wherever it went it was a success. A change to the Jumo 211 engine came with the Fw 190D, a considerable redesign with a longer fuselage and new tail. This was even faster and followed the earlier series on the production lines. It was in turn developed into the Ta 152, which was in production but had not entered service when the war ended.

Fokker, Anthony H. G., 369, 565, 875

The foremost Dutch designer, Anthony Herman Gerard Fokker was already designing outstanding aircraft before the 1914-18 War. When the war began he offered his designs to both combatants and the Ger-

Fokker F.VIIA of KLM, member of a family used and licence-built all over the world

mans took up his offer. Thenceforward he designed a series of formidable warplanes, specializing in single-seat scouts. His 1915 monoplane, with an interrupter gear fitted to its forward-firing machine-gun, gained early air superiority for Germany and became a formidable aircraft in skilled hands until the Allies produced a new generation of armed scouts. In 1917 Fokker produced his Dr.I triplane, and this was followed by the D.VII biplane and the D.VIII parasol monoplane, all of which were used by the German Air Force to good effect.

After the war, Fokker returned to his native Holland and built up his own manufacturing company, reconditioning many of his wartime-built aircraft for the Dutch Air Force and for export and designing a steady succession of new aircraft. For the airlines he designed a series of high-wing monoplanes which became more and more popular, the best-known example being the three-engined Fokker F.VII/3m. This aircraft, which could be powered by any suitable radial engine, was exported widely, licence-built and copied the world over. It was refined and deve-

loped until overtaken by the low-wing monoplanes with retractable undercarriages.

Fokker continued to build warplanes and the Dutch Air Force was equipped with many Fokker aircraft at the outbreak of World War Two. The Fokker factories were destroyed during the Nazi occupation, but after the war Fokker's successors built the present-day Fokker company which, after several tentative designs, made a breakthrough into the feeder-line market with the twin-turboprop Friendship, recently joined by the twin-jet Fellowship now in production.

Fokker *Eindecker*, 243, 246, 248, 625, 875, 877, see also Fokker M.5K

The Fokker *Eindecker* (monoplane) became the first dominant military aircraft during the 1914-18 War. Developed from earlier and less successful monoplanes, the E.I version introduced in 1915 was the

only scout aircraft over the Western Front fitted with an interrupter gear enabling its gun to fire through the propeller disc. In a short

time, with the improved E.II and E.III versions, it accomplished a temporary mastery in the air, building up a high score of victories over its British and French opponents, and not until a new generation of French and British front-gunned scouts appeared was it challenged. By the end of 1916, the day of the Fokker monoplane was over.

Fokker F.VII/3m, 263 see also Fokker, Anthony H. G.

Towards the end of the nineteen-twenties, this three-engined high-wing monoplane built by Fokker in Holland and licence-built as far afield as Japan and the USA, became the standard airliner of the time and set the pattern for many copies by other companies. It was used in several long-distance flights, notably by the Australian Kingsford Smith in the Pacific, and was developed continuously into the 'thirties until eventually superseded by aircraft with retractable undercarriages and variable-pitch propellers.

Fokker F.X trimotor, 267

Fokker Fellowship (F.28)

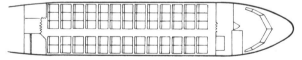

Fellowship

To succeed the highly-successful Friendship twin-turboprop airliner, Fokker evolved in the early nineteen-sixties the F.28 Fellowship, a 40/65-seat short-haul transport powered by two aft-mounted Rolls-Royce Spey Junior turbofans each developing 9,850 lb thrust. The Fellowship prototype flew on 9 May 1967, and the first delivery was made in mid-1968. With 60 passengers the Fellowship has a range of 637 miles at a cruising speed of 521 mph.

Grounded scourge: the 100-hp Fokker E-type monoplane

Fokker Friendship (F.27), 787, 791

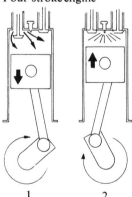

Friendship Series 200

The Fokker Friendship represented the re-entry of the Fokker company into commercial aircraft production after World War Two. The prototype of this high-wing twin-turboprop short/ medium-haul airliner flew for the first time at Schiphol in November 1955. Flight trials were encouraging, and very soon a large order book was building up. Deliveries began in November 1958, and production still continues. Fairchild Hiller in the USA acquired the licence to build the Friendship in that hemisphere, and production there has been on a comparable scale. By Autumn 1969, well over 500 had been sold by both companies and a family of military, combined cargo/passenger and long-fuselage Friendship variants was in being.

Fokker M.5K monoplane, 877

The Fokker M.5K was the short-span version of two military scout monoplanes built by the Dutch pioneer aircraft designer Anthony Fokker. Powered by a 50-hp Gnome engine, it first flew in April 1914. This particular aircraft was used by Fokker for his early interrupter gear experiments, becoming, in effect, the prototype for the famous *Eindecker* (*q.v.*) which, with its machine-gun firing forward through the propeller disc, became the scourge of Allied aircraft over the Western Front during the summer and autumn of 1915.

Ford "Tin Goose", 737

The Ford Trimotor was based on the general layout of the highly-successful Fokker F.VII/ 3m (*q.v.*) three-engined high-wing monoplane, except that the Ford was all-metal—even to its corrugated metal skinning which earned it the nickname "Tin Goose". It was a flexible design which could be fitted with a variety of power plants in the 300/400-hp power bracket. Many were built and used in different parts of the world, especially in the less-developed areas, for which the Ford was ideal. With the depression in the early 'thirties, however, the Ford Company ceased to make aircraft.

It was originally designed by William B. Stout, who, in 1966, reintroduced the type for production as the Bushmaster 2000, having modernized the design to meet modern-day requirements in terms of safety, comfort and operational facilities as a simple and cheap transport aircraft for use from small grass fields.

Fortress (B-17), see Boeing

Four-stroke engine

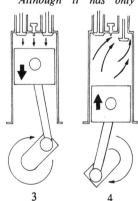

1   2   3   4

A reciprocating (piston) engine operating on the four-stroke cycle first described by Beau de Rocha in 1862 and then modified by Otto and others and today known as the Otto cycle. There are four piston strokes (thus, two complete revolutions of the crankshaft) to each complete cycle. The cycle begins with a downward induction stroke in which the descending piston draws combustible fuel/ air mixture (or, in a diesel engine, air alone) into the cylinder through the open inlet valve. The inlet valve then closes so that on the next upward compression stroke, the piston compresses the mixture in the sealed upper part of the cylinder. Near the top of this stroke the mixture is fired: in a spark-ignition petrol engine this is done by a spark discharge timed at the appropriate point; in a diesel engine the combustion is automatic, as the high compression raises the air temperature to such a level that when a metred dose of fuel oil is injected it burns as it enters. The next downward stroke is the power stroke in which the expanding hot gases perform useful work. Finally, the fourth stroke is the upward exhaust stroke in which the piston expels the spent gas through the opened exhaust valve. Although it has only half as many firing strokes in a given time as a two-stroke engine running at the same speed, the four-stroke unit is appreciably more efficient.

France: Air Forces

Armée de l'Air. At the end of the 1914-18 War, France's *Aviation Militaire* (created 1912) was easily the best-equipped air force in Europe, and most of its aircraft were of national design. Re-duced in size, it was kept busy in French North Africa in the nineteen-twenties. A separate Air Ministry was set up in 1928, and in 1933 the *Armée de l'Air* was created as an independent force. Re-equipment with new fighters began some two years later, but the bomber force was not comparably modernized, and by the outbreak of World War Two France still had virtually no up-to-date medium or heavy bombers in service. When France succumbed to the Nazi onslaught in 1940, those of her aircraft and aircrew that escaped—mostly to Britain or North Africa —continued to fight as the FAFL (Free French Air Force) on the Allied side; while those remaining in France continued as the Vichy Air Force in support of the Axis powers. The post-war pattern repeated that which followed the 1914-18 War, with unrest in the French North African Territories that did not end until the late 'fifties. In addition, similar troubles in French Indo-China (part of which is now Vietnam) engaged the *Armée de l'Air* operationally until 1954. Despite these commitments, France has until comparatively recently made a major contribution to the NATO defence network in Western Europe, and is currently one of the world's leading countries in all aspects of aviation. The *Armée de l'Air* of today is equipped with Mirage III, Super Mystère, F-84F and F-100D fighter-strike aircraft, Mirage IVA nuclear bombers, and Vautour light bombers. In a few years it will have the supersonic Jaguar strike aircraft in service.

Aéronautique Navale (usually abbreviated to *Aéronavale*). French naval aviation also began in 1912 and, al-though secondary to the military air arm, it possessed over 1,200 aircraft by the end of the 1914-18 War. Its first aircraft carrier appeared in 1920, and in 1925 an *Aéronautique Maritime* was created, for operation from both ship and shore bases. Like the *Armée de l'Air*, the naval squadrons also were short of modern equipment at the outbreak of World War Two. Eventually, most *Aéronavale* personnel joined the fighting on the Allied side, playing an important part in hunting down U-boat packs in the North Atlantic. *Aéronavale* carriers and aircraft were well to the forefront during the war in Indo-China; and, as the domestic aircraft industry re-established itself in the nineteen-fifties, new naval aircraft of French design began to replace the British and US types hitherto employed. Today the only major non-French types in French Navy service are the F-8 Crusader and the P-2 Neptune; the Etendard fighter, Alizé and Super Frelon anti-submarine aircraft and the Atlantic maritime patrol aircraft are all of French origin.

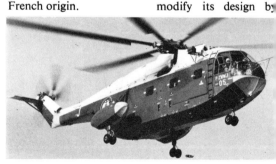

Sud Super Frelon three-turbine helicopter of the French Navy

Aviation Légère de l'Armée de Terre (ALAT). This force, created in 1952, is of substantial size, although the majority of its aircraft are for liaison, observation and communications duties rather than combat types. Newest types include the SA 330 Puma assault helicopter and SA 341 Gazelle observation helicopter.

France: Chronology

21 November 1783. Pilâtre de Rozier and the Marquis d'Arlande rose to a height of 305 ft in a Montgolfier hot air balloon and, carried by the wind, crossed Paris in 25 minutes.

1 December 1783. Charles and Robert made the first ascent in a hydrogen balloon leaving Paris at night and landing near Nesles from where Charles, unaccompanied, made another ascent, rising to 9,150 ft.

28 April 1784. Launoy and Bienvénu demonstrated to the *Académie des Sciences* their toy helicopter, powered by a bow-string mechanism, which flew up into the cupola of the building. A description of the apparatus—the first heavier-than-air device to fly under its own power—and a report on the experiment appeared in the proceedings of the *Académie* for 1 May 1784.

22 October 1797. Garnerin made the first parachute jump from a balloon at the height of 2,300 ft. The oscillation of his parachute during descent led the inventor, advised by the astronomer Lalande, to modify its design by making a hole in the centre of the canopy.

7 October 1870. Léon Gambetta and an American war correspondent escaped from Paris, while it was besieged by the Prussian Army, in the balloon *Armand Barbès* and *Georges Sand*. During the siege 66 balloons carrying 168 people, 400 carrier pigeons and

nearly 11 tons of mail left the city, representing the world's first air mail service.

9 August 1884. The airship *La France*, built by Captains Renard and Krebs and powered by an 8 hp electric motor, was the first airship to be steered in flight back to its point of ascent, the *Parc d'Aérostation Militaire* at Chalais-Meudon.

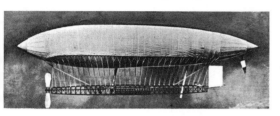

"La France", non-rigid airship of 1884

9 October 1890. At Armainvilliers, Clément Ader claimed to have made a powered (but uncontrolled) hop of 50 yards in his machine *Eole*. In 1897, he made another attempt before a military commission with a new machine called the *Avion III*, but this failed to leave the ground. Both machines were powered by steam engines.

13 November 1907. Paul Cornu, an engineer from Lisieux, succeeded in leaving the ground in a helicopter of his own design powered by a 24-hp Antoinette direct-injection engine.

13 January 1908. Henry Farman, piloting a Voisin biplane, won the Deutsch-Archdeacon prize at Issy-les-Moulineaux for the first officially-observed circular flight of 1 km, under the auspices of the Aéro Club de France. A similar flight had been accomplished on 9 November 1907, but was not observed officially.

30 October 1908. The first cross-country flight between two towns was made by Henry Farman in a Voisin biplane, taking 20 minutes to complete the 16¾ miles between Bouy and Rheims, at a speed of 45 mph and at an average height of 130 ft.

December 1908. The world's first air show, the *Salon de l'Automobile et de l'Aéronautique*, was held in Paris. Its successor, the present-day *Salon International de l'Aéronautique et de l'Astronautique*, is presented at Le Bourget Airport every two years.

19 July 1909. Henry Farman made the first night flight of 1 hr 23 min in an aeroplane built by himself, but which he damaged while making a rough landing in the dark.

25 July 1909. Taking off from the French coast, near Calais, Louis Blériot crossed the English Channel in his Blériot XI monoplane, and landed at Dover after a 37-minute flight, thereby winning the *Daily Mail* prize offered to the first aviator to achieve this feat.

22 August 1909. The world's first flying meeting began at Rheims. All the leading aviators of the day took part, including Blériot, Curtiss and Farman, and in the course of the meeting all existing air records were broken.

8 March 1910. The Baroness de la Roche became the first woman to obtain a pilot's licence (No 38). She learned to fly in a Voisin aircraft, and made her first solo flight on 22 October 1909.

28 March 1910. The engineer Henri Fabre flew the world's first seaplane, near Marseilles, making a total of three flights on that day. The aircraft, powered by a 50-hp Gnome rotary engine, was supported on three floats.

15 October 1910. Henry Coanda exhibited, at the second *Salon de*

l'Aéronautique in Paris, the first full-sized turbine-driven aeroplane to be built. A Clerget in-line piston-engine, driving a compressor unit, was mounted at the front of the aeroplane.

18 February 1911. The pilot Pequet flew the first aeroplane mail, from Allahabad to Nairi in India, flying a Sommer aircraft which was powered by a 50-hp Gnome rotary engine.

21 September 1913. Adolphe Pégoud, the most famous aerobatic pilot of his day, became the first person after the Russian Nesterov to perform the feat known as looping the loop, in a Blériot XI-2 at Buc. Subsequently, other aerobatic manoeuvres were evolved.

23 September 1913. Roland Garros made the first non-stop flight across the Mediterranean, from St Raphaël to Bizerta, in a Morane-Saulnier monoplane powered by a 60-hp rotary engine.

29 September 1913. Prévost won the Gordon Bennett Trophy at Rheims in a Deperdussin monoplane powered by a Gnome rotary engine of 160 hp. He attained a top speed of nearly 127 mph and broke 12 records during the flight.

5 October 1914. Flying a Voisin aircraft, Sergeant Frantz, accompanied by his engineer Quenault, achieved the first aerial victory in history against a German Aviatik, crewed by Sergeant Schlichting and his observer Lieutenant Von Zangen.

24 December 1918. The industrialist Pierre Latécoère, on board a Salmson aeroplane piloted by Cornemont, opened the first section —between Toulouse and Barcelona—of the route which was to link France and Chile a few years later.

8 February 1919. A Farman F.60 Goliath, piloted by Lucien Bossoutrot, and carrying

twelve passengers, made the first flight on the London-Paris route, with a return flight being made the next day. On 10 February, a Caudron R.II carrying five passengers inaugurated the Paris-Brussels route.

1 April 1921. Adrienne Bolland, a French woman pilot, crossed the Andes from Mendoza to Santiago in Chile in 3 hours 15 minutes, flying a Caudron G.III aircraft which dated from the beginning of the war. An altitude of 13,750 ft was attained during the flight.

Caudron G.III, a mass produced type of the 1914-18 War period

11 May 1930. Air mail carried from Toulouse to Saint Louis in Senegal. On 12 May, it was taken to Natal in Brazil in a Latécoère 28 seaplane piloted by Mermoz, arriving on 14 May. This mail was distributed in Rio de Janeiro and Buenos Aires on the same day, and at Santiago in Chile on 15 May.

2 September 1930. Costes and Bellonte, flying the Breguet XIX Super TR *Point d' Interrogation*, took off from Le Bourget and landed at Curtiss Field, Long Island, 37 hr 17 min later, thereby achieving the first east-west non-stop flight between Paris and New York.

25 December 1934. Delmotte set a landplane speed record at a speed of 313 mph, flying a Caudron 460, powered by a 380-hp Renault engine. In 1936, flying the same aircraft, he won the Thompson Trophy and the Greve Trophy at the National Air Races held at Los Angeles.

26 June 1935. The Breguet-Dorand *Gyroplane* made its first flight, piloted by Maurice Claisse. During the next two years this helicopter held all the major rotary-wing aircraft records of its day.

12 December 1936. A few days after the disappearance of Mermoz in the South Atlantic, Maryse Bastié made a similar flight from Dakar in Senegal to Natal in Brazil, flying a Caudron Simoun aircraft powered by a 220-hp Renault engine.

21 April 1949. The Leduc 010, the first French aircraft to be powered by a ramjet engine, made its first powered flight, piloted by Jean Gonord. The conception of such an aircraft dates from 1913, and design of the Leduc 010 from 1936. It was partly-constructed in secret during the Occupation, and made its first unpowered tests on 21 October 1947.

14 July 1949. The Fouga Sylphe, powered by a Turboméca TR 3 turbojet of 200 lb st, was flown for the first time by Léon Bourriau. This gave France a lead in the field of low-thrust gas turbine engines.

25 February 1959. André Turcat, flying the Nord Griffon, powered by a combined turbojet/ramjet powerplant, set a new world speed record over a 100 km circuit of 1,018 mph at Istres. On 31 October 1958, the Griffon had attained a maximum level speed of Mach 2.05 at 55,775 feet.

25 November 1965. France launched her first earth satellite, *A1 Asterix*, by means of a Diamant four-stage rocket. All stages of the rocket were of French manufacture.

Nord 1500 Griffon supersonic ramjet research aircraft

Flight deck of Caravelle VI-N No 203 (VT-DUH of Indian Airlines)

Frankfurt Airport (Rhein/Main), Germany

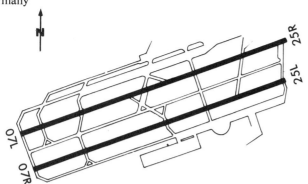

First commercial flying from Frankfurt, by the predecessors of today's Lufthansa, started in 1925; it was the scene of many Zeppelin operations in the 'thirties, and was prominent in the 1949 Berlin Airlift.
Distance from city centre: 6 miles
Height above sea level: 368 ft
Main runways:
07L/25R = 12,795 ft
07R/25L = 12,303 ft
Passenger movements in 1967: 6,129,900

Free-turbine engine, 1309
Freight

Air freighting has come into its own only since World War Two, although the carriage of non-human payloads such as air mail (q.v.) or merchandise has always been a recognized part of the general air transport scene. The scale on which supplies and equipment had to be air-lifted during 1939-45, and the aircraft evolved to carry them, led to expectations of a boom in the air freight business once the war was over.

The expected growth in business did not really reach appreciable proportions until the middle nineteen-fifties, when large piston-engined passenger aircraft that might otherwise have become obsolete were turned into freighters once their future was overshadowed by the new jet airliners. Now there are jet freighters as well, such as the DC-8F and the military StarLifter, and specialized freight-carriers have been developed in

Palletized cargo going aboard a McDonnell Douglas DC-8F freighter

the form of the swing-tailed CL-44 and vehicle ferries (q.v.) like the Carvair/DC-4 conversion. Most bizarre of all are the "Guppy" conversions of the Boeing Stratocruiser, whose special freight consists of components of the giant Saturn rockets used to launch the US Apollo spacecraft.
The carriage of more conventional merchandise by air is increasing steadily, and many of the well-known passenger airlines of today in fact occupy a higher position in the table of freight ton-miles per year than they do in terms of human transportation.

Friendship, see Fokker
Friendship International Airport, see Washington
Frise ailerons

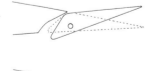

The aileron-balancing system invented by the British engineer, L. G. Frise, involves a set-back hinge and a shroud on the upper surface of the wing, so that the aileron

as it moves downward, presents a smooth contour with the top surface of the wing and thus creates a minimum of drag. The nose of the opposite aileron, as it moves upwards, projects below the bottom surface, upsetting the airflow and assisting the wing to drop, as well as creating drag, so tending to turn the aircraft in the direction of the roll.
The need for such a system is due to the severe yawing effect caused by normal ailerons in the opposite direction to the roll. On some early aircraft, this aileron drag was sufficiently powerful to turn the aircraft in the opposite direction to that intended.
As the effect of Frise ailerons may not be sufficient, they are often combined with differential operation.

Frontier Airlines

A major US inter-state carrier, Frontier Airlines was formed in 1950 by the amalgamation of three smaller companies, the oldest of which had began operating in the Autumn of 1946. Since its merger with Central Airlines in 1967, Frontier's network has been increased

to nearly 14,000 miles, serving well over 100 cities in 15 of the western United States. A fleet of some 50 aircraft includes Boeing 727s and 737s, Convair 580s and Douglas DC-3s.
Headquarters: Denver, Colorado.

Fuel, 913

Apart from the special case of man-powered aircraft, all powered aircraft are propelled by putting to mechanical use the heat energy liberated by burning fuel. In a nuclear-powered aircraft the energy would be obtained by nuclear fission, but such a machine has not yet been built—despite very extensive study in both the United States and Soviet Union since 1953. All other powered aircraft use energy liberated by chemical combustion. Rocket-powered aircraft carry within their tanks all the materials needed for their propulsion, and are thus independent of the atmosphere. Typical fuels used by rocket aircraft include: HTP (high-test peroxide) used either by itself or to provide oxygen for the burning of kerosene; calcium permanganate and hydrazine; liquid oxygen and either ethyl alcohol or ammonia; and fuming nitric acid and furfuryl alcohol. The remaining forms of propulsion, which account for more than 99.99 per cent of all powered aircraft ever built, are all air breathers: they combine their fuel with oxygen from the atmosphere.

The earliest powered aircraft that flew successfully used ordinary petrol (gasoline) as fuel, burned in engines fundamentally similar to those of cars. By 1940 aviation engines were calling for special petrols better able to resist detonation "knocking" and having this resistance expressed in the form of an octane number. Although 87/91 octane is more common in modern lightplanes,

100/130 and even 115/145 octane is used by the few remaining high-power piston-engines. All gas turbine aircraft burn kerosene (paraffin) or a blend of kerosene with petrol. Plain kerosene is often designated Avtur (aviation turbine fuel) or JP-1; blends with petrol are often designated Avtag (aviation turbine gasoline) and JP-4. Carrier-based aircraft of the Royal Navy are cleared to use ships' diesel fuel. Apart from crude trials with rockets in 1929, solid fuel has not been used since 19th-century steam-driven experiments, although a German (Lippisch) project of 1944 envisaged the use of a ramjet duct in which hydrocarbon liquid fuel was sprayed over a fixed mass of white-hot charcoal.

Fuel Contents Indicating System (Capacitance type), 913

This consists of a series of capacitance-type units immersed in the fuel and connected via an amplifier to an indicator. The capacitors consist basically of two concentric tubes extending the depth of the tank. As the fuel level varies, the medium between the plates changes from liquid to mixture of air and fuel vapour and the capacitance of the units varies. This variation is sensed and a signal proportional to the difference is transmitted via the amplifier to the indicator.
Capacitance systems, in addition to indicating volume of fuel, can also measure the mass of fuel. This is a most useful measurement for fuel system management, and capacitance systems are used almost universally on turbojet and turboprop aircraft.
This type of system is practically unaffected by changes of fuel temperature, and by suitable positioning more than one tank unit in a tank and connecting them in parallel, can be made independent of changes in aircraft attitude.

Fuel Contents Indicating System (Potentiometer type)

This comprises a float arm-operated potentio-

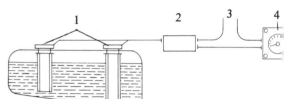
Schematic diagram of capacitance type fuel content indicating system.

1 Tank units	3 Electrical system
2 Amplifier	4 Indicator

meter-type transmitter connected to an indicator. As the fuel level changes, the float arm rotates the transmitter wiper arm. This varies the voltage and current combinations, the changes being transmitted to the indicator the pointer of which.

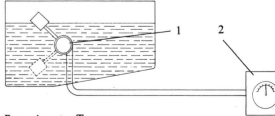

Potentiometer Type

1 Potentiometer transmitter

2 Indicator

follows the resultant magnetic field and so reflects the changing fuel level.

This type of gauging system, which measures the volume of fuel, is not suitable for indicating the combined contents of several tanks and is affected by changes in temperature and aircraft attitude. Its use is generally confined to light aircraft.

Fuel injection, 1097, 1105, 1107

Fury, Hawker, see Hawker

G, see Aviation medicine

Gagarin, Yuri, 329

Yuri Gagarin before lift-off, 12 April 1961

Major Yuri Gagarin of the Soviet Air Force has a secure niche in history as the first human being to orbit in space, a feat which was accomplished on 12 April 1961 by making one circuit of the Earth in the Russian spacecraft *Vostok I*. He was killed in an aircraft accident on 27 March 1968.

Galaxy, see Lockheed C-5A

Galeao Airport, see Rio de Janeiro

Gannet, see Fairey

Garros, Roland, 209, 875

Garuda Indonesian Airways

Garuda, the national airline of Indonesia, was formed jointly by KLM (*q.v.*) and the Indonesian Government in 1950, and has been completely State-owned since 1954. In addition to internal services, it flies internationally to other major cities in India, Pakistan and south-east Asia, as well as to Amsterdam, Cairo and Rome.

Headquarters: Djakarta. Member of IATA.

Gas turbine, 295, 1129ff

The gas turbine has become the basic means of propulsion for all but light aircraft since its introduction at the end of World War Two. It works on the principle whereby air is drawn into a tube in which it is compressed and ignited, giving it increased energy, and exhausted through a tapering nozzle increasing its velocity. To this there is a resultant reaction which propels the tube in the opposite direction. This same basic principle applies also to ramjet and pulse-jet engines, but these have control problems for aircraft use.

A normal gas turbine engine for aircraft use consists of an intake, a compressor (which is mounted on the same shaft as the turbine which drives it), a combustion chamber (or number of chambers), a turbine (through which the ignited gases pass and which drives the compressor) and a jet pipe, through which the gases exhaust to atmosphere. The early turbo-jet engines had centrifugal compressors in which the air drawn in was fed from the central compressor disc to the periphery by centrifugal force, where it was collected and fed into the combustion chambers. Later, axial compressors were introduced, comprising rows of aerofoil-shaped blades on a concentric shaft, by means of which the air is progressively compressed before passing directly into the combustion chamber.

These basic forms have been improved upon during the development of newer and better engines with such refinements as two-spool compressors, reheat, water injection and the ducted fan.

Gatwick Airport, see London

GCA (Ground Controlled Approach), see also Take-off and Landing Techniques.

A system devised to enable a controller on the ground to assist a pilot to approach and land under conditions of poor visibility. Throughout the procedure the aircraft concerned is observed by the controller on radar screens which give a precise indication of its position. At the start of the approach the aircraft captain hands over control, and responsibility, to the ground controller, who then uses voice commands to guide the aircraft left/right and up/down (the latter by calling for changes in rate of descent) to keep it on the centreline of the optimum glide path. At a height of 100 ft the controller tells the pilot to look ahead and land. If the runway or its approach lighting cannot be seen the pilot must overshoot and, probably, divert to an "alternate" airport.

GE.4 engine, see General Electric

Gee Bee Super Sportster, 539

The Gee Bee series of racing aeroplanes was built by Granville Bros Aircraft Inc in the USA in 1930-31. They were the outcome of the National Air Races which comprised mostly aircraft of home design; the Gee Bees were built for the use of racing pilots to enable them to make use of a proved and successful design. All were single-seat, low-wing monoplanes, highly streamlined and polished, and every drag-reducing refinement was introduced in the final design for the 1932 races.

Gemini spacecraft, 337, 503-4

The second of America's human space research programmes was Project Gemini, which utilized modified Titan II rockets as launch vehicles and space capsules once again made by McDonnell. After two unmanned firings, the first manned operation in this programme took place on 23 March 1965, when Gemini 3 took Virgil Grissom and John Young on a three-orbit flight. Three months later Gemini 4 accomplished the longest mission to that date when James McDivitt and Edward White flew a 63-orbit mission lasting four days. On 21 August 1965 this time envelope was doubled, with Gemini 4 and astronauts Gordon Cooper and Charles Conrad. The programme continued until the last flight by Gemini 12 on 11 November 1966, which included outside activity and some excellent space photography.

General Dynamics

The giant General Dynamics Corporation has 14 operating divisions; of these, three (at Fort Worth and Pomona, Texas, and San Diego, California) are concerned with aerospace activities. They were formerly components of Convair (previously Consolidated). Principal current aircraft products are the F-111 variable-geometry multi-purpose combat aircraft, the GD/Martin RB-57F, the GD/Convair F-102 and F-106 fighters and the conversion of Convair-liners (*q.v.*) to turboprop power as the GD/Convair 600/640 series.

General Dynamics F-111, 596, 1383, 1399-1402, 1405, 1407, 1409

The F-111 is the first application of the "swing-wing" principle to an operational aircraft. It won the contract for the US Dept of Defense's TFX requirement for a tactical fighter and was developed for both the USAF and the US Navy, although the latter variant was later cancelled. The F-111 was scheduled to become operational in 1967 but many development problems had to be overcome, although the general principle of the swing-wing formula has been established. The F-111A is intended as the standard USAF strike fighter and will eventually replace the various other types currently in use in this role. The aircraft has also been ordered for the Royal Australian Air Force; the Royal Air Force order for 50 was subsequently cancelled. The F-111A is powered by two Pratt & Whitney TF30 turbofans of approximately 20,000 lb thrust each with after-burning, giving it a maximum speed at height of Mach 2.5 and at low level of Mach 1.2, and a range with maximum internal fuel of 3,800 miles.

General Dynamics Hustler (B-58), 481, 579, 582, 841, 887

Selected as the USAF's first supersonic strategic bomber, the Convair B-58A entered operational service in 1961, 116 aircraft being built. It was conceived as a "weapons system", in which aircraft, weapons and operational task are all developed under one programme. The Hustler was one of the first aircraft to enter service with the ability to cruise for long periods at supersonic speed and its design incorporates many features designed to combat associated problems, especially kinetic heating (see text).

General Electric Company, 817, 887, 1075, 1084-1086, 1109, 1135, 1201, 1213, 1259-1260

The General Electric Company of the USA had, among its other accomplishments, worked in the gas-turbine field as long ago as 1895. With the advent of the aircraft piston-engine the company became involved in producing some of the earliest turbosuperchargers, and its experience in this field led the company to pioneer in the United States the construction of turbojets based on the early Whittle and Rolls-Royce Welland engines. These were installed in the first US jet-propelled aircraft, the Bell XP-59A, in 1942.

Such pioneering development put the company in a good position as the turbojet field opened up after World War Two, and General Electric is now one of the foremost turbojet producers in the world. The Company is also concerned in many aspects of the ancillary side of aviation.

General Electric CJ-805 engine, 1213, 1259-1260

The CJ-805 is the civil version of the J79 turbojet engine (*q.v.*), and is used in the Convair 880 and 990A airliners General Electric developed this engine, as the Model 23, as an aft-fan by-pass engine by fitting a free-floating fan behind the turbine to move a large mass of cool air into the tail-pipe, thus reducing the noise level and improving the specific fuel consumption by 8-12%.

General Electric GE.4 engine, 1251, 1303-1305

General Electric J79 engine, 1184-1186

The J79 was America's first high-compression variable-stator turbojet; development began in 1952. Selected for use in the Convair B-58 Hustler supersonic bomber, it became the first production engine for sustained speeds of Mach 2. It has achieved great popularity and is licence-manufactured in Canada, Germany, Italy and Japan, principally for powering the F-104 Starfighter.

A large number of variants has been produced, including the commercial CJ-805 version, and these have ratings, without afterburning, varying from 9,600 lb to 11,870 lb st.

General Electric J85 engine, 709

The J85 is a lightweight turbojet for use in small training and ground attack military aircraft or executive type civil aircraft. It is an axial-flow turbojet with an optional afterburner (in some versions) and powers the Northrop T-38A, Canadair Tutor and Saab 105XT trainers, the Northrop F-5A, Fiat G91Y and Cessna A-37B fighters and the Bell X-14A and Ryan XV-5A VTOL research aircraft. It is licence-built by Orenda in Canada, and by Alfa Romeo in Italy.

General Electric T64 engine, 1313
The T64 is an adaptable gas-turbine engine, designed and built under the auspices of the US Navy, which can be used either as a turboprop or as a shaft-turbine for helicopters. Its most recent application is in the Lockheed Cheyenne compound helicopter, although it is also used in the DHC Buffalo, LTV/Ryan/ Hiller XC-142A VTOL transport, Sikorsky CH-53A helicopter, Kawasaki P-2J patrol bomber and the Shin Meiwa PS-1 anti-submarine flying boat.

General Electric TF39 engine, 1201, 1213, 1311
The TF39 is General Electric's contribution to the new generation of turbofan engines, and powers the Lockheed C-5A Galaxy heavy transport aircraft.
It has an 8 ft diameter, two-stage front fan providing a by-pass ratio of 8 : 1 and delivers 41,100 lb static thrust in its initial form.

Genie missile, see Douglas

Geodetic construction, 399, 412

The basket-like geodetic form of construction devised by the British engineer Sir Barnes Wallis was used on the Vickers Wellesley, Wellington and Warwick aircraft.
The construction comprises a series of continuous helical members, each of which has an opposite member running contrariwise, the two members being secured to each other at the points of intersection. Under load stresses the two opposing helices tend to counterbalance each other, the compressive load in one member being equal to the tensile load in the member running in the opposite direction. Any geodetic member tending to bow outwards under compression is balanced at its centre by the tendency of its opposed member to bend inwards under tension. The geodetic form of construction provided an interior free from subsidiary structure, and, spreading the loads in numerous members, was able to withstand heavy combat damage without failure.

Germany: Chronology
1888. Otto Lilienthal published his book *Bird Flight as the Foundation of the Art of Flying*, a major contribution to early literature on the technique of flight.
1891. First entirely successful gliding flight made by Lilienthal, over a distance of 15 metres, in Berlin.

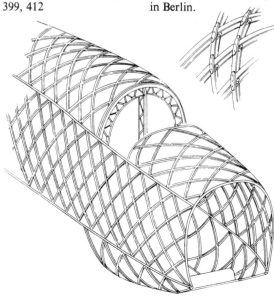

1895/96. Lilienthal made successful gliding flights over distances of up to 350 metres.
9 August 1896. Lilienthal crashed and died. At the time, he was planning to fit an engine to one of his gliders.
3 July 1900. Graf von Zeppelin flew his first big rigid airship over Lake Constance.
18 August 1903. Karl Jatho of Hanover made hops of up to 18 metres in his 9-hp tail-less triplane. After modification, the aircraft is said to have hopped 60 metres in November.
1909. At Johannisthal, near Berlin, Hans Grade won the *Lanzpreis der Lüfte* in a monoplane of his own construction, with a 24-hp air-cooled two-stroke engine.
1909. August Euler awarded the first German pilot's licence.
1912. Hellmuth Hirth set up a world height record of 14,830 ft in an Etrich *Taube* monoplane.
1913. First passenger flight between Berlin and Paris by an Etrich *Taube*, piloted by Alfred Friedrich with Dr Elias as passenger.
1914. Founding of Dornier Flugzeugbau as a branch of the Zeppelin Werke.
1914. Distance record of 777 miles set up between Berlin and Egri Palanka by Max Schüler (pilot) and Wilhelm Seekatz (passenger) on a DFW biplane with 150-hp Benz engine.

Junkers E.I, a pioneer cantilever metal aircraft

1915. Hugo Junkers built the first all-metal aeroplane.
1918. Junkers built the first all-metal passenger aircraft, the F.13, which established a world height record of 22,146 ft in 1919.

Junkers F.13 corrugated-skin transport

1920. First German international air service opened by Deutsche Luftreederei between Copenhagen and London, via Hamburg and Amsterdam.
1922. First sailplane flight of more than one hour's duration (66 minutes) by Arthur Martens of Germany.
1924. Junkers built the G 24, the first three-engined aircraft in commercial service.
1926. Deutsche Lufthansa AG founded.
1927. Total of 33 records set up by Junkers aircraft and aero-engines in Germany and abroad, including a world duration record of 52 hours 27 minutes and a world distance record of 2,895 miles.
12 April 1928. First east-west crossing of the North Atlantic by Köhl, von Hünefeld and Fitzmaurice in a Junkers Ju 33 with Ju L5 engine.
1929. Round-the-world flight by the airship *Graf Zeppelin*, com-

manded by Hugo Eckener.
1929. Flights to North and South America by the twelve-engined Dornier Do X flying-boat.
1930. Wolfgang von Gronau flew from Europe to America in

a Dornier Wal twin-engined flying boat, via the Faroes, Greenland and Labrador.
1931. Elly Beinhorn flew a Klemm 25 light aircraft 7,500 miles to Africa and back.
26 June 1936. First flight of Professor Focke's twin-rotor Fw 61, one of the first practical helicopters.
1938. The Focke-Wulf Fw 200 Condor four-engined airliner made a non-stop flight from Berlin to New York and back.
26 April 1939. World's absolute speed record of 469.22 mph set up by Fritz Wendel in a Messerschmitt Me 209 at Augsberg. This record was not beaten by a piston-engined aircraft for 30 years.
27 August 1939. First jet aircraft to fly was the Heinkel He 178, piloted by Erich Warsitz. In the same year, Warsitz also flew the first rocket-plane, the He 176.

Dornier Do-X, powered by 12 Curtiss Conqueror engines mounted in pairs

1941. In a Messerschmitt Me 163 rocket fighter, Heini Dittmar became the first human being to travel at over 1,000 kmh.
1941. First flight of the Messerschmitt Me 262, first jet fighter to enter series production.

Germany (Democratic Republic): Air Force
Luftstreitkräfte. The present-day air force of East Germany grew out of the small air branch of the *Volkspolizei*, formed in 1950, and has existed under its present title since 1955. Its

principal aircraft are c Soviet origin; many c these were built unde licence until the closin down of the domesti aircraft industry a fe years ago, since whe Soviet-built aircra have been supplied di ect. Main strength lies i the air defence squad rons, equipped wit MiG-17, MiG-19 an MiG-21 fighters. Obso lescent MiG-15 an Il-28 jets constitut the ground attack/ligh bombing force, but substantial transpor force is maintaine which includes moder types such as the four turboprop Il-18 and th twin-jet Tu-124.

Germany (Federal Repub lic): Air Forces
Luftwaffe. The origina German air force gre up around the Zeppeli airships in the earl years of this century, i first heavier-than-a craft being adopted i 1910 and a Militar Aviation Service for med in 1912. Wit some excellent ind vidual types of aircraf this Service acquitte itself well during th 1914-18 War, and cap tured German aircraf formed the initial equip ment of several new an smaller air forces in othe countries after the Arm stice. The 1919 Treat of Versailles forbade th post-war manufactu of combat aircraft, bu this was circumvente by building "sporting monoplanes and fa "mailplanes" whic were but thinly-di guised fighters an bombers such as th Heinkel He 51 an Dornier Do 17. Hitle officially created th new *Luftwaffe* in 193 although it had existe in effect for some tw years before this, an the Spanish Civil Wa of the late 'thirties wa siezed upon as an idea opportunity to "blood its new aircraft unde operational condition In 1939-40, howeve the *Luftwaffe* quickl found that its Civ War tactics were muc

less effective against the better-equipped RAF, and more modern types were quickly developed in the attempt to regain air superiority. These included such excellent designs as the Fw 190 fighter and the Ju 88 and Do 217 bombers, joined toward the end of World War Two by Germany's first jet types, including the Me 262 fighter and the Ar 234 bomber. Reconstitution of a post-war *Luftwaffe*, as a part of the general NATO defence organization, began in 1955 with the introduction of an extensive aircrew training programme followed by the acquisition of transport and tactical combat aircraft. Today's *Luftwaffe* consists chiefly of fighter/tactical strike units of Starfighter and Fiat G91 aircraft, and a substantial transport force equipped mainly with the Noratlas and the Transall C-160. *Marineflieger*. The first German Naval Air Service was established in 1912, and during the 1914-18 War it played a substantial part in the overall German war effort. Subsequent naval air activity was for the most part under the control of the *Luftwaffe*, but with the re-creation of the latter force in 1955 a new *Marineflieger* was also formed, which today has a creditable strength including Starfighter interceptor/strike aircraft, Breguet Atlantic maritime patrollers and Grumman Albatrosses for air/sea rescue.

Heeresflieger. Brought into being in 1955, this is numerically a large force, though its aircraft are primarily transport, liaison and observation types for the Army. They include Bell UH-1D and Sikorsky H-34 helicopters and Do 27 light planes.

Ghana Airways
In 1958, shortly after the Gold Coast became the newly-independent state of Ghana, the services formerly provided in and to that country by West African Airways Corporation were transferred to a newly-established national airline, Ghana Airways. For the first two years, however, WAAC continued to carry out the domestic and regional services under contract, while intercontinental services to the UK and Europe were operated by Ghana Airways in pool with BOAC. Attempts at expansion, begun in 1963 with Soviet assistance, proved too ambitious to be successful economically; but since 1966, when the Soviet aircraft were replaced by Viscounts and VC10s, a firmer operating basis has been achieved.
Headquarters: Accra. Member of IATA.

Ghost engine, see de Havilland

Giffard, Henri, 133, 148

Gipsy engines, see de Havilland

Gladiator, see Gloster

Glaisher, James, 915, 925

Glenn, John, 330,
see also Mercury spacecraft

Globemaster (C-124), see Douglas

Gloster E.28/39, 290, 295, 1135
It was not until the outbreak of World War Two that the go-ahead was given to the Gloster design office for an aircraft to flight-test the Whittle W.1 turbojet. Two were built, the design comprising a pilot's cockpit on top of a long jet-pipe with a small, low wing and an empennage attached above the rear of the jet-pipe. The first aircraft was flown from Cranwell on 15 May 1941 and in due course both aircraft were used to flight-test several turbojets. The second aircraft crashed on 30 July 1943 due to aileron jamming, but the original E.28/39 survived the war and was presented to the Science Museum, London.
With the original engine a maximum level speed of 338 mph had been obtained, but with the Rover W.2B engine, by accidentally exceeding maximum revs, a speed of 466 mph was reached on one occasion.

Gloster Gladiator, 387, 549
The Gladiator was the last of the long line of RAF fighter biplanes, and with a maximum speed of 257 mph at 14,500 ft and four machine-guns it represented the ultimate in biplane fighters with a fixed undercarriage.
A development of the open-cockpit Gauntlet biplane, it entered service with the RAF in February 1937, No 72 Squadron being the first to receive Gladiators. By the outbreak of War there were eight operational Gladiator squadrons, four in the UK and four in the Middle East, despite re-equipment with Hurricanes and Spitfires.
The Gladiator also went to sea, as the Sea Gladiator, and served with six Fleet Air Arm squadrons between 1939 and 1941. It had little opportunity for sustained combat, although three Sea Gladiators became possibly the most famous of all. These were in Malta, crated, at the time of the heavy air raids. They were uncrated and used by the RAF in the fighter defence of Malta. Thus arose the legend of "Faith", "Hope" and "Charity". In addition, the Gladiator was sold to several other air forces; Belgium, Latvia, Portugal, Sweden, Norway and Finland all received them, the last two countries using them operationally against Germany and Russia respectively.

Gloster Grebe, 1333
A single-seat biplane fighter introduced into the RAF in 1923, the Grebe was powered by a 400-hp Armstrong Siddeley Jaguar IV radial engine, giving it a maximum speed of 152 mph. It was armed with two Vickers machine-guns. Grebes served with six RAF fighter squadrons and remained in service until 1928. In 1926 two Grebes were modified to fly from the airship R.33 in flight.

Gloster Meteor, 294, 295, 817, 1135, 1223, 1283
The Gloster Meteor was a logical development from the original British jet aircraft, the E.28/39, via the range of F.9/40 development prototypes. Once a suitable combination of power plant and airframe was found, very little re-design was needed to provide a viable fighter. The combination was the F.9/40 airframe and the Welland engine and the Meteor F.1 was the result. This type went into production and eventually entered operational service with No 616 Squadron, who used it briefly against the V.1's in July/August 1944 with some success. In January 1945 a flight was sent to Belgium, where it flew interdiction sorties against German transport; but the hoped-for encounters with Me 262's never came, as the *Luftwaffe* was by then virtually grounded. However, the Meteor began a long course of post-war development: the F.3 was re-engined with Derwents and became Fighter Command's standard fighter, and from this many single and two-seat versions were built and exported. The most successful export models were the T.7 trainer and the F.8 fighter. Final evolution of the Meteor was carried out by Armstrong Whitworth, which produced the Mks. 11-14 series of night fighters with interception radar in their lengthened noses. The Meteor remained in service as a fighter until the late nineteen-fifties, and still serves with the RAF and other air forces in subsidiary rôles.

Gloster / Hele-Shaw / Beacham propeller, 1333

Gnome rotary engines, 229, 232, 248, 811, 1039, 1047

Godwin, Bishop, 114

Goliath, see Farman

Golubev, I.N., 159, 169

Gordon Bennett, see Competitive flying

Gotha
The Gothaer Waggonfabrik AG of Germany was founded before the 1914-18 War, and achieved prominence chiefly for the multi-engined bomber aircraft which it produced during that period—indeed, to the uninitiated all large multi-engined German aircraft were "Gothas". The company then disappeared from the aviation scene until 1934, when it at first engaged in building light commercial and training aircraft. Subsequently it built substantial numbers of Dornier Do 17 bombers for the wartime *Luftwaffe*, and modest numbers of the Gotha Go 242 transport glider and its powered counterpart, the Go 244.

"Grand Slam" bomb, 885
The "Grand Slam" was the ultimate in size among conventional bombs and was specially tailored to the Avro Lancaster bomber, which was the only type able to carry it in World War Two. Development of this 22,000-lb bomb began after D-Day, and its principle of operation was that of creating a minor earthquake. The method was to drop it from as high as possible (from 40,000 ft it penetrated one hundred feet before exploding). The shock waves from the explosion would then rock to pieces any structure in the vicinity. No 617 Squadron RAF (the "Dam-busters") was the first to use it, dropping 41 "Grand Slam" bombs in all.

Gravity, effects of, see Aviation medicine

Grebe, see Gloster

Greece: Air Force
Royal Hellenic Air Force. Established under this title in November 1935, combining the former Army Air Force and Naval Air Service which had existed, under various titles, since before the 1914-18 War. Aircraft in service in the mid-'thirties were mainly of British or French design, to which had been added Polish PZL P-24 fighters by 1940. During 1940-43 Greece's chief opponent in World War Two was Italy, while in 1943-44 many RHAF personnel fought alongside the RAF in the Western Desert. After the war, rehabilitation was slow, due to various factors, including the need to rebuild nearly every airfield as well as to replenish both manpower and machines; but steady progress was sustained after Greece's entry into NATO in 1952. Today Greece has a modest-sized but modern air force equipped with aircraft mostly US in origin. These include F-104G Starfighters and Northrop F-5As as well as the F-84F and F-86D.

Green, Charles, 205
Born in 1785, Charles Green became a professional aeronaut in Britain in the eighteen-twenties, making more than 500 balloon flights during his lifetime. His experience and skill enabled him to improve the operation of balloons, and it was Green who brought into normal use the trail rope, which acted as automatic ballast. He used coal gas in his balloons from the start and before long this became the standard gas in use. Many of his flights were notable, especially his 480-mile voyage in the "Great Balloon of Nassau".

Griffith, A. A., 1125

Griffon engine, see Rolls-Royce

Griffon fighter, see Nord

Ground Controlled Approach, see GCA

Ground cushion, 488
Term used to describe an aerodynamic condi-

tion effected by certain types of wings.

The undersurface of a wing, when proceeding through the air, tends to deflect the airflow passing under it in a downward direction. This is more noticeable when the wing is set at a high angle of attack, as is the case with many modern fixed-wing aircraft and with helicopters in the landing approach configuration. When such an aircraft nears the ground on landing, a considerable body of air is deflected downwards on to the ground from where it rebounds to form a buoyancy "cushion" under the aircraft's wings or rotors, increasing the pressure and thus the lift under the wings or rotors and automatically lessening the rate of descent in the last stage of landing.

Ground effect

The effect of the "pushing" of air compressed against the ground by a landing aeroplane or helicopter (ground cushion, *q.v.*). This effect can be turned to especial advantage by helicopters when required to land on high plateaux or mountain-sides, their proximity to high terrain then enabling them to operate at heights greater than that of their normal hovering ceiling over low ground.

Grumman

Grumman Aircraft Engineering Corporation was formed at the end of 1929 and for some 40 years has been predominantly a producer of various types of military aircraft for service with the US and foreign navies. Its first biplane

fighter, the FF-1, was succeeded by several other biplanes and then by the F4F Wildcat and F6F Hellcat monoplane fighters of World War Two fame. The line of production piston-engined fighters ended with the F8F Bearcat, after which appeared the straight-winged Panther and swept-wing Cougar and Tiger jet fighters of post-war years. Grumman Tigers were the mounts of the US Navy's "Blue Angels" aerobatic team. The A-6 Intruder low-level strike aircraft is being used successfully in Vietnam, as is the US Army's OV-1 Mohawk observation aircraft. The company was also responsible for development of the US Navy version (now abandoned) of the F-111 "swing-wing" strike fighter.

Grumman has in addition produced in recent years several multi-engined carrier-based aircraft, the most important of which are the S-2 Tracker anti-submarine aircraft, the E-2 Hawkeye AEW patroller and the C-2A Greyhound transport.

Grumman JF-1, 567-9

The JF-1 was a two/three-seat single-engined amphibian biplane built for the US Navy. It originated in 1933 and had a large metal central float underneath the fuselage, into which the undercarriage retracted upward and inward. Twenty-seven were supplied to the US Navy and nine to the US Coast Guard, the latter aircraft being equipped

A classic US Navy fighter: the Grumman FF-1

for rescue work.

Guggenheim, Daniel, 671
Guggenheim, Harry, 671

President of the Daniel Guggenheim Fund for the Promotion of Aeronautics, and Chairman of the Daniel and Florence Guggenheim Aviation Safety Center at Cornell University, USA. This centre was established in 1950 to foster the improvement of aviation safety through research, education, training, and the dissemination of safety studies to the industry and of air safety information to the general public.

Guynemer, Georges, 883
Gyro-compass, see Compass
Gyrodyne, see Fairey
Gyroplane Laboratoire, see Breguet
H-43 Huskie, see Kaman
Haenlein, Paul, 133
Halifax, see Handley Page
Hall-Scott, 1050, 1069

The Hall-Scott Motor Car Co Ltd produced, in 1917, a 125-hp six-cylinder in-line engine for aircraft use. It subsequently developed this engine, but did not remain in the aviation business for long after the 1914-18 War.

Hamilton propellers, 1339

It was the Hamilton hydraulically-operated, variable-pitch propeller which, in the early nineteen-thirties, brought this type of propeller into world-wide operation. Employed on such new monoplane transports as the Boeing 247 and Douglas DC-2, it was seen in many parts of the world and set the pattern for other propeller manufacturers.

Handley Page, 361, 537, 733

One of the early British pioneers, Sir Frederick W. Handley Page formed his own manufacturing company before the Olympia Aero Show of 1911, at which he exhibited a monocoque fuselage. During the 1914-18 War he was asked to build a long-range heavy bomber for the RNAS.

Sir Frederick Handley Page

His resulting O/100 and O/400 twin-engined bombers played a large part in the success of the Independent Air Force, the strategic bombing element of the newly-formed RAF in 1918.

From these bombers, Handley Page evolved a long line of biplane airliners after the war finished, and set up his own airline, Handley Page Air Transport, which operated for some years before being absorbed into Imperial Airways (*q.v.*).

Another of Handley Page's interests was improving aircraft control at the slow end of the speed range, and his developments in this direction led to the introduction of the leading-edge slot, a device which postponed the stalling point of a wing. This he introduced and patented in the late 'twenties.

During the inter-war years Handley Page Ltd concentrated on large aircraft, both airliners and bombers; two separate families of aircraft evolved, culminating respectively in the H.P.42 four-engined airliners (which enabled Imperial Airways to put on the most comfortable and luxurious service of the 'thirties) and the Heyford bomber. Thereafter, Handley Page turned to monoplane designs and provided two famous bombers for the RAF in World War Two: the twin-engined Hampden medium bomber and the Halifax four-engined heavy

bomber. The latter was second only to the Lancaster in numbers used by Bomber Command, and from it evolved the Halton transport of the early post-war years.

Handley Page continued to specialize in large aircraft, among them the Hastings and Hermes transports and the four-jet Victor bomber; but in the last decade its attention has been directed more towards small-sized airliner and executive aircraft, of which the Jetstream is the latest and most promising. When in the late 'fifties the British Government put pressure on aircraft manufacturers to amalgamate, Sir Frederick resisted such pressure, and Handley Page remained an independent company. He died on 21 April 1962.

Handley Page Halifax, 1101

Second of Britain's four-engined monoplane bombers of World War Two, the Halifax originated in 1936 with an Air Ministry requirement for a twin-Vulture-engined bomber. An anticipated shortage of Vulture engines led to an aircraft with four Merlins being developed instead, and the prototype flew in October 1939.

The first production Halifax flew almost a year later, and the type was in RAF service just over a month after that. After several months of working-up, it became operational in March 1941. The Halifax was used originally to replace the Whitley bomber in No. 4 Group, Bomber Command. It was soon undergoing further development with regard to its defensive armament: dorsal turrets at first were added, later removed and finally replaced, and the separate nose turret was deleted and replaced by a hand-operated gun. After two versions had been produced with Merlin engines, new

variants appeared with Bristol Hercules radial engines; and of these the Mks III, VI and VII took over the bombing offensive from 1943 onwards. The type was also extensively used by the Canadian Bomber Group (No. 6). Towards the end of the war, the Halifax entered upon two new rôles those of meteorological survey and general reconnaissance with Coastal Command, and glider-tug and troop transport with Transport Command—rôles in which it continued to serve in the immediate post-war period. It was also developed into the Halton interim airliner for BOAC and some charter airlines.

Handley Page Herald

In its original form the Herald medium-range airliner was powered by four Leonides Major piston-engines, but in 1958 a twin-turboprop power plant was adopted instead. In this form the Herald Series 100 went into service with BEA and the lengthened Series 200 with Jersey Airlines (now BUA (*q.v.*)); the latter version is now operated by several airlines and a trooper version serves with the Royal Malaysian Air Force. The Series 200 passenger version seats up to 56, is powered by two 2,100 ehp Rolls-Royce Dart engines and has a range of 1,725 miles cruising at 275 mph.

Handley Page H.P.42E and H.P.42W, 726, 731

The Handley Page H.P.42 appeared in November 1930. It was designed specifically for Imperial Airways, with the keynote on unsurpassed standards of passenger comfort. Eight aircraft in all were built, four H.P.42Ws for European routes and four H.P.42Es for Eastern routes. The aircraft were each powered by four Bristol Jupiter engines and had a wing span of 130 ft. An unusual feature was the installation

Handley Page 42, typifying the Imperial Airways philosophy of safe, majestic travel

A rare photograph of Lawrence Hargrave flying box-kites

of two engines on the upper wing and two on the lower wing. Maximum speed was 127 mph and cruising speed 95 mph.

Apart from one aircraft burnt out in a hangar fire, all continued in service up to the outbreak of World War Two, earning an enviable reputation for safety and reliability. Put to various war uses for which they were not designed, the aircraft did not survive long under wartime conditions, but they finished their service with the reputation of never having harmed a single passenger in peacetime.

Handley Page Jetstream (H.P.137), 1313
Powered by two Turbomeca Astazou XIV engines, this low-wing turboprop executive aircraft was first flown in August 1967. Many orders were received "off the drawing-board", and deliveries began in 1968. It has a crew of two and the cabin can seat between four and eighteen passengers; there is a freight hold at the rear. Maximum cruising speed is 306 mph; range with a 3,300-lb payload is 250 miles, or 2,000 miles with maximum fuel load.

Handley Page O/100 and O/400, 861
Designed at the request of the British Admiralty for a long-range bomber, the twin-engined O/100 first flew on 18 December 1915. It had a wing span of 100 ft and was powered by two 250-hp Rolls-Royce Mk II engines (later known as the Eagle).

Delivery to the RNAS began in September 1916, and in November the first aircraft went to Dunkirk; unfortunately, the third aircraft landed behind the German lines, and the secret was out. Build-up was slow at first, but as the O/100s went into action they began to make their mark, particularly for their night raiding, despite an almost total lack of night-flying aids. They were used increasingly during the last year of the 1914-18 War, and most squadrons were incorporated into Trenchard's Independent Air Force, the first strategic bombing command in history, after the assimilation of the RNAS into the RAF in April 1918. The O/100 led to the O/400, which served to and beyond the end of the war.

Handley Page Victor, 579, 835
The Victor was the third V-bomber to enter Britain's nuclear deterrent force and was based on Handley Page's research with the crescent wing form. It entered service in November 1957, and equipped four squadrons of Bomber Command in its B Mk 1 version. A new version with 20,600-lb st Conway engines in place of the four 11,000-lb Sap-

phires emerged in 1959, entering service in 1962 with two squadrons as the B Mk 2; it was equipped with the Blue Steel missile. Subsequently, all B Mk 1s were converted into refuelling tanker aircraft and form the RAF's tanker force; and the B Mk 2 has taken on the strategic reconnaissance rôle. The Victor has a 110 ft span and a cruising speed of Mach 0.92.

Haneda Airport, see Tokyo

Hanover Air Show, see Air shows and exhibitions

"Hap"
Allied code-name allocated to the Japanese Mitsubishi A6M3 fighter of World War Two which, when first sighted, was thought to be a new design. Later contacts showed it to be a development of the Zero, with a more powerful engine and the folding wing-tips removed to increase the top speed. Later the code name was altered to "Hamp".

Hargrave, Lawrence, 187, 1047, 1063
Although cut off from the main stream of aeronautical invention in Australia, Lawrence Hargrave began his experiments in 1884 and made industrious records of the results in the form of papers to

Handley Page Victor B.2 carrying Blue Steel stand-off missile

the Royal Society of New South Wales. His main contributions to aeronautics were the invention of the rotary engine, which played a vital part in providing light aero engines for the experimenters of the early nineteen-hundreds, and the box-kite. He first turned his attention to kites in 1893, and his box-kite became the standard design followed in many parts of the world. Hargrave realized that his box-kite formed the basis for a lifting device, although Voisin first exploited the design fully.

Harrier, see Hawker Siddeley

Hawk, 175,
see also Pilcher, Percy

Hawker, Harry G., 579
Harry Hawker qualified for his Aviator's Certificate in September 1912, on a Farman biplane at Brooklands, England. He immediately began aiming for the British Empire Michelin Trophy No 1, an award for the longest flight in an all-British aeroplane without touching the ground. This he won on 31 October 1912 by staying aloft in a Sopwith Wright-type biplane for 8 hr 23 min. Soon after this he was appointed test pilot to the Sopwith company, a post he retained until his death.

Within nine months of learning to fly, Hawker captured four new records in one day (16 June 1913), on a Sopwith biplane with an 80 hp Gnome: three altitude records and an endurance record. The following month he flew the Sopwith Bat Boat amphibian over the Solent, to win the Mortimer Singer seaplane prize. He also attempted the *Daily Mail* seaplane trial, involving a round-Britain flight of 1,540 miles. Flying much farther than his only competitor, he left Southampton and flew round and up the East Coast as far as Cromarty and then down to Oban and across to Dublin, where he crash-landed. He later demonstrated the new Sopwith Tabloid with great verve, taking it to his native Australia in 1913 and returning the following year to resume his test pilot duties.

Harry Hawker

Throughout the 1914-18 War Hawker bent his energies to testing Sopwith products, and he

became probably the finest exponent of the Sopwith Camel. After the war he returned to record attempts, and in May 1919 he and Mackenzie-Grieve set off for England from Newfoundland, only to be forced down and picked up by a Danish ship. The following year the Sopwith company was renamed H. G. Hawker Engineering Co Ltd in his honour; but he was not long to be part of the company, for on 12 July 1921, while testing the Nieuport Goshawk, he suffered a hæmorrhage and died in the ensuing crash.

Hawker, Captain Lanoe G., 873

Hawker, 385, 387, 403, 405.
Hawker Aircraft Ltd was born out of the Sopwith (q.v.) company which built many famous fighting aircraft in the 1914-18 War. The Hawker company concentrated on military aircraft: towards the end of the nineteen-twenties it combined its particular type of metal construction with the new Rolls-Royce Kestrel engine to produce the Hart family of two-seaters and the Fury and Nimrod single-seater fighters. From these developed the Hurricane (q.v.) eight-gun monoplane fighter which played a prominent part in World War Two. All these

bore the stamp of chief designer Sydney Camm (*q.v.*), who also designed the Typhoon and Tempest, which played significant parts later in the war.

In the post-war period Hawker made the transition to jet aircraft, still under Sir Sydney Camm's direction, and produced first the Sea Hawk and then the Hunter, one of Britain's most successful post-war aircraft. Hawker Aircraft became part of the Hawker Siddeley Group in 1935. This has brought to the production stage the world's first VTOL fighter, the Harrier, which was Camm's last design before his death in 1967.

Hawker Hurricane, 285, 287, 879, 883, 885, 909, 911, 1095, 1099, 1107, 1339

The Hurricane began life in 1934 as a private venture by Sydney Camm, Hawker's chief designer. It made its first flight on 6 November 1935, and was ordered for the RAF the following June—the first monoplane fighter since the Bristol M.1C of 1917 to see RAF service. No 111 Squadron at Northolt received the first Hurricanes to enter service in December 1937, and became the first unit to possess eight-gun fighters. The Hurricane was soon deployed on the outbreak of World War Two: four squadrons went to France, but the majority were delivered to Fighter Command. In May 1940 Hurricanes went to Norway, in August to Malta and in October to the Western Desert. By then the aircraft had borne the brunt in the Battle of Britain, being the most numerous (and thus the most heavily engaged) type in the Battle.

During the Winter of 1940 Hurricanes were used in night-fighting, and in 1941-42 found considerable success in intruder sorties over German airfields in France at night. In 1941 they went even further afield, to Singapore and Burma, and there also began the mass supply of Hurricanes to the Soviet Air Force. By now the Mk II was the standard type in use, with either twelve machine-guns or four 20-mm cannon, or as a fighter-bomber. A tank-busting version with 40-mm cannon was very successful in the Western Desert, and in Burma the final version was the Mk IV, which had a "universal" wing that could take long-range tanks, bombs or rocket rails. The Hurricane also went to sea, at first being catapulted from merchant ships and later as the Sea Hurricane, equipping Fleet Air Arm carrier-based squadrons and serving from 1941-43. By the end of the war Hurricanes had almost left operational service, although one or two units in the Middle and Far East still used them and they did not finally cease front-line service with the RAF until October 1946. Altogether 14,230 Hurricanes were produced, of which 1,451 were built in Canada by the Canadian Car and Foundry Co.

Hawker P.1040, 817, 835

The P.1040 was the first essay of Hawker Aircraft Ltd into the jet-fighter field and showed a basic understanding of the new advantages conferred by the jet engine, which was buried in the fuselage. The P.1040 employed bifurcated jet-pipes and intakes for the engine, so that the aircraft could be kept a clean shape aerodynamically, and it was one of the most handsome of the early jet fighters.

It was conceived in 1944, but development to the prototype-ordering stage was slow because of conflict of requirements between the RAF and the Royal Navy, and the proto-

type did not fly until September 1947. It performed well and was eventually developed into the Sea Hawk for the Royal Navy. No order came from the Royal Air Force, but the P.1040 began the long line of development and refinement which led to the Hawker Hunter.

Hawker Sea Fury, 835

The Sea Fury was Hawker's ultimate piston-engined fighter; it stemmed from the Typhoon, through the various versions of Tempest and Fury. The first prototype flew in February 1945, and production began the following year; over 650 were built for the Royal Navy, whose standard fighter it remained until the advent of the Sea Hawk. It entered service in May 1948 and was in action against the MiG-15 over Korea from June 1950 until the fighting ceased.

The Sea Fury could reach a maximum speed of 460 mph at 18,000 ft and, with two 90-gallon drop-tanks, had a range of 1,040 miles. As well as four 20mm cannon it had under-wing points for fuel tanks, bombs and rockets according to need. It was powered by a 2,480-hp Bristol Centaurus 18 radial engine.

Hawker Sea Hawk, 828, 835

The Sea Hawk was basically a "navalized" development of Hawker's first jet aircraft, the P.1040. It was ordered into production in 1950, and the first aircraft entered squadron service in March 1953. The Sea Hawk became the Navy's standard interceptor fighter for nearly a decade, being progressively refined and adapted for the fighter-bomber and ground attack rôles. It went into action during the Suez conflict of 1956, in which five squadrons took part. Sea Hawks were sold to West Germany, the Netherlands and In-

dia, and its excellent handling qualities endeared it to those who flew it. Powered by a Rolls-Royce Nene engine, it had a maximum speed of Mach 0.84 at 36,000 ft and a ceiling of 44,500 ft.

Hawker Tempest, 292, 385

A wartime development of the Hawker Typhoon with semi-elliptical high-speed wings and other modifications, the Tempest entered service in 1944. Its first main rôle was to attack the V1 flying-bombs being sent against south-east England; at the time it was the fastest Allied fighter, with a maximum speed of 427 mph at 18,500 ft. Altogether, Tempests destroyed over 600 V1s. It was used in the final assault in Europe, both

for ground-attack and as an air superiority fighter, with the special task of engaging

The lines of a Camm thoroughbred: Tempest VI at Langley, 1944

the German Messerschmitt Me 262 jet fighters. It had destroyed 20 of these by the end of the war.

After the war a version with a Bristol Centaurus radial engine appeared, serving both in Europe and in the Far East, where it was used operationally, armed with rockets, against the Malayan terrorists.

Hawker Typhoon, 385, 885

Developed by Sydney Camm (*q.v.*), the Hawker Typhoon surmounted initial teething troubles to become

a formidable weapon when used in the interdiction role against German transport and tanks. Many squadrons took part in the assault on Europe in 1944-45. It was successful against the V1 pulse-jet flying-bombs, and in the final offensive against Germany, during which it attacked what was left of the German fighter force.

Hawker Siddeley

Hawker Siddeley Aviation is one of the seven major divisions of the Hawker Siddeley Group (other than those in Australia and Canada). It operated originally in three divisions but has been a single organization since April 1965. Included in the division are the former separate aircraft manufacturing companies, Armstrong

Whitworth, Avro, Blackburn, de Havilland, Folland, Gloster and Hawker (*qq.v.*).

Hawker Siddeley 125, 457-468, 692-696, 787, 793, 837

One of the most popular executive jets, this aircraft came from de Havilland, now a part of the Hawker Siddeley Group. It first appeared in 1962 and has so far been produced for

world-wide service in four versions, the second being as a navigational trainer for the RAF. It is powered by two Viper turbojets which give it a cruising speed of 500 mph at 30,000 ft. With maximum fuel and payload its range is 1,710 miles.

Hawker Siddeley 681, 1233

This was a project for a four-engined high-wing V/STOL monoplane transport, with swept wings and a large fuselage for military transport duties. It was to be powered by four Rolls-Royce Medway turbofans with deflectors for directing the lift downwards for V/STOL operation. The aircraft was under development for the RAF when cancelled in January 1965.

Hawker Siddeley 748

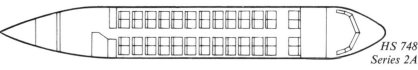

HS 748 Series 2A

This twin-turboprop transport originated as the Avro 748, the prototype flying for the first time on 24 June 1960. It first entered service, with Skyways Coach Air, in the Spring of 1962. The initial Series 1 model has been superseded by the Series 2 and 2A with more powerful Dart engines (2,105- or 2,230-ehp, compared with 1,880-ehp in the Series 1), and these predominate in service. Hawker Siddeley 748s serve with several air lines around the world, with the RAF Royal Flight and the Brazilian Air Force, and are being built under licence by Hindustan Aeronautics Ltd for Indian Airlines and the Indian Air Force. In

Hawker Siddeley (de Havilland) 125 business jet

addition to the standard 748, which is known as the Andover CC Mk 2 in RAF service, a modified freighter version also serves with RAF Air Support Command as the Andover C Mk 1. The HS 748 Series 2 seats up to 56 passengers and has a range of 704 miles with full payload.

Hawker Siddeley Andover
The Andover is a military development of the Hawker Siddeley 748 (*q.v.*) twin-turboprop short/medium range airliner. It is in service with the Royal Air Force in two versions: the C Mk 1, which has a modified fuselage with a rear-loading ramp, facilities for parachuting and a "kneeling" undercarriage (see text) enabling the rear ramp to be levelled to any required height for loading and unloading. Thirty-one of these aircraft entered service with the RAF from 1966. The second version is the CC Mk 2 which is in all outward respects similar to the HS 748. Two are used by the Queen's Flight and others for VIP transport generally.

Hawker Siddeley Argosy, 789
The Argosy was designed as a private venture by Armstrong Whitworth in 1959 to fill the growing need for a short/medium-range freight aircraft. It has four Rolls-Royce Dart turboprops on a high wing from which two tail-booms carry the empennage. The fuselage is basically a large freight compartment (with the flight deck above it) having loading doors fore and aft through which cargo can be unloaded and loaded simultaneously. A modified version was built for the RAF, with which it still serves; the civil Argosy is used by BEA and two American freight lines.

Hawker Siddeley Blue Steel missile, 896
Royal Air Force Strike Command has this large missile in service with its Hawker Siddeley Vulcan Mk 2 squadrons. It is a stand-off bomb which the parent aircraft can launch outside the target's defences. The missile then proceeds to the target under its own inertial guidance system. It is powered by a Bristol Siddeley Stentor twin-chamber liquid-propellant rocket motor and has a thermonuclear warhead.

Hawker Siddeley Buccaneer, 1137
Designed by the former Blackburn company, to fill a naval requirement (NA 39) for a subsonic strike aircraft for operation from Royal Navy aircraft carriers, the Buccaneer first flew on 30 April 1958, powered by two Gyron Junior turbojets.
After intensive flying of twenty development aircraft, the type was ordered into production. The production Buccaneer S1 first flew in January 1962 and became operational in July 1962. Three Royal Navy squadrons were equipped with this version before it was replaced by the S2, powered by Rolls-Royce Spey turbofans. This type has become the standard strike aircraft for the Royal Navy; it also serves with the South African Air Force as the S50. The Royal Air Force is to have Buccaneers as interim strike aircraft: eventually it will take over those at present in Royal Navy service.

Hawker Siddeley Harrier, 326-327, 577, 646, 997, 1006-1008, 1233, 1275
The Hawker Siddeley Harrier represents eight years of development from the original P.1127, which began hovering tests in October 1960. There were six P.1127 prototypes, each incorporating various modifications to prove the concept of using a lift/thrust ducted-fan engine with swivelling nozzles. A tripartite squadron of RAF, USAF and *Luftwaffe* pilots was formed to evaluate a modified version, the Kestrel, and as a result of their experience an even further-developed version was ordered for the RAF as the Harrier ground-attack fighter. The first pre-production Harrier began test-flying in 1966 and the aircraft entered service in 1969. The Harrier is the first fully-operational fixed-wing VTOL aircraft in the world, and has an operational performance superior to that of the Hunter, which it has replaced in RAF service.

Hawker Siddeley P.1127, see also Hawker Siddeley Harrier, 581, 852, 1233

Hawker Siddeley Trident, 591-595, 641-644, 787, 837, 852, 931-932, 1245
The Trident, as the de Havilland D.H.121, was designed around British European Airways' requirement for a short-haul airliner for the nineteen-sixties. As the Hawker Siddeley Trident, two dozen were originally built for BEA, entering service in 1964. A few other airlines also bought the Series 1 Trident; this was followed by the Series 2, also serving with BEA, which has more powerful Spey turbofan engines and longer range. A Series 3 version is now being built for BEA, with a longer fuselage permitting up to 146 passengers. The Trident pioneered the three-rear-engine layout with a "T" tail, subsequently adopted successfully by Boeing in the 727.

Hawker Siddeley Vulcan, 579, 835, 852
The second of Britain's three V-bombers to enter production, the Avro Vulcan (as it then was) first flew on 30 August 1952. It was built on the tail-less delta principle, with a tall fin and rudder and four jet engines buried in the thick roots of the wings. The Vulcan was first produced in B Mk 1 form, 45 of this version entering service with the RAF. It was followed by the B Mk 2, with more powerful Olympus 201 (later 301) engines, the latter rated at 20,000 lb st. Wing span was increased and the B Mk 2 is able to carry the Blue Steel stand-off bomb. This version is the mainstay of the RAF's nuclear deterrent force. Originally built for a high-level rôle, the Vulcan has since been strengthened and adapted for low-level under-the-radar attack as well.

Hawker Siddeley/Matra Martel missile, 896
This missile is a joint product of SA Engins Matra of France and Hawker Siddeley Dynamics of Britain. It comes in two versions, the first in the anti-radar rôle with an ECM head. In this rôle it has a long range, giving the aircraft a "stand-off" capability. The second version is a television-guided weapon. This version flies to the target automatically, but the final attack is guided by a weapon operator aboard the parent aircraft, working from a direct-vision picture.

Head-up display, 1369-1370

Elliott head-up display during evaluation in a McDonnell Douglas DC-9

A display of signals, symbols, numerical data and other information in the form of illuminated characters projected on to the windscreen in front of the pilot or other crew member. The display is focused at infinity, so that the pilot can see it and the sky or ground beyond simultaneously without having to refocus his eyes.

Such displays are used primarily by military aircraft performing various attack, interception or landing functions, especially at low level or in bad weather when to keep switching pilot vision from the cockpit instruments to the outside world would be hazardous. The data contained in the display are generally concerned with aircraft attitude, speed, angle of attack, range to a target or runway, steering information to achieve a desired objective, and altitude.

Heathrow Airport, see London

Hawker Siddeley Argosy of BEA being loaded with palletized freight

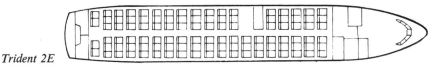

Trident 2E

Hawker Siddeley Vulcan B2 carrying Blue Steel stand-off missile

Heinkel

The name of Dr Ernst Heinkel first came into prominence during the 1914-18 War, when, as chief designer to the Hansa-Brandenburgische Flugzeugwerke, he was noted mainly for the design and development of a wide variety of marine aircraft. Heinkel formed his own company in 1922 and produced a steady output of successful aircraft. These included such revolutionary designs as the rocket-engined He 176; the He 178—the world's first jet-powered aircraft to fly; and the first twin-jet aircraft to fly, the He 280. Despite these pioneering achievements Heinkel's first production jet aircraft, the He 162 *Volks-jäger* or Salamander, was operationally a failure, as was the piston-engined He 177 heavy bomber of World War Two. Probably the best-known of Heinkel's aircraft is the He 111 medium bomber, variants of which were in service throughout the war, serving on torpedo-bombing, mine-sweeping and glider towing duties as well as those for which it was originally designed.

The company continued to operate under its own name for a while after World War Two. Then, in 1959, it combined with Bölkow and Messerschmitt to form the Entwicklungsring-Süd GmbH to undertake development of a Mach 2 VTOL fighter. However, at the end of 1964 Heinkel withdrew from this consortium and became, with Focke-Wulf (*q.v.*) and Weser Flugzeugbau, a constituent company of VFW (Vereinigte Flugtechnische Werke).

Heinkel He 111, 901, 909

Designed ostensibly as a civil transport aeroplane, the He 111 did not maintain this illusion for long. It was in service with the *Luftwaffe* in 1936, and was soon one of the combat types sent by Germany to fight in the Spanish Civil War. Pre-war versions included the He 111B, He 111D and He 111P, and it was the third of these models, with its distinctive offset nose turret, that characterized all later versions of this adaptable bomber. Principal development changes concerned increases in the defensive armament, after the type sustained heavy losses against the RAF during the early part of World War Two. Most common wartime model was the He 111H, with 1,340-hp Junkers Jumo engines, a maximum speed of 258 mph at 16,400 ft and a range of 1,740 miles. Later in the war, the He 111 was used increasingly for alternative duties, including those of torpedo-bomber, paratroop transport and glider tug.

Heinkel He 162 *Volks-jäger*, 1135

The He 162 was built to a crash programme in 1944, as an interceptor fighter for production by semi-skilled personnel with non-strategic materials. It was intended to produce thousands a month, but in the event only 116 were built.

It was a tiny shoulder-wing single-seat monoplane, of 23-ft span and 29-ft length, with a BMW 003A turbojet mounted above the fuselage and exhausting over the dihedral tailplane which had twin fins and rudders.

It attained 522 mph at 19,700 ft and had a range of 410 miles; the war ended before it could become fully operational.

Heinkel He 176

This aircraft was designed to exploit the promising potential of the rocket motor. Work began in 1937 on this advanced aeroplane, which measured only 16 ft long and had a 16-ft wingspan. It had a retractable tailwheel undercarriage, and the pilot sat in an enclosed cockpit in the nose in a semi-reclining position. The He 176 began trials with a Walter HWK R.1 rocket motor, and left the ground for the first time in June 1939. Further flights were made, but development was dropped on the outbreak of war.

Heinkel He 178, 289, 295, 1135, 1192

As in Britain, as soon as the existence of a turbojet engine became a probability, design work was put in hand in Germany for an aerodynamic vehicle to get it airborne. The aircraft in this case was the Heinkel He 178, design work on which began in 1938; the engine was the Heinkel-Hirth He S 3B centrifugal turbojet. The aircraft was, in essence, a cockpit above a jet-pipe, with a shoulder wing and a normal empennage. The fuselage was of duralumin and the wings of wood.

The He 178 first hopped off the ground on 24 August 1939, the first true flight following on 27 August—although it resulted in an emergency landing. Thereafter it carried out a full development programme. It was the first turbojet aircraft in the world to fly.

Heinkel He 280, 1135

This aircraft had the distinction of being the first twin-engined jet aircraft to fly (on 5 April 1941) and also the first jet aircraft designed for operational use as a fighter. Powered by two He S8 engines of 1,320-lb thrust each, its top speed was 577 mph and initial climb rate 4,920 ft/min.

However, the original turbojets were not sufficiently reliable; the He 280 airframes were used as engine test-beds and the aircraft never entered production.

Heinkel He S 1 & S 3B engines, 1135, 1191-1194

Germany's first jet engine, the S 1, was designed by Pabst von Ohain. It was run in Sept. 1937, giving a static thrust of 550 lb, but was purely a bench engine. A further development, for installation in an aircraft, was the S 3 (and later the S 3B) which gave double the thrust of the S 1.

The S 3B was installed in the Heinkel He 178 which, on 27 August 1939, became the world's first turbojet-powered aircraft to fly.

Helicopter tip drive

Some helicopters have been built in which the rotors are driven by means of compressed air being piped along the

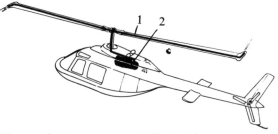

1 Hot gas duct 2 Gas turbine

blades to the tips and either ejected under pressure or ignited in a basic pulse-jet. This method reduces the torque action on the whole airframe and simplifies the mechanism of the rotor head. It has not, however, found general favour.

Henschel Hs 130E, 1112-1113

The Hs 130 was developed from the Hs 128, an earlier experimental twin-engined monoplane which was used for developing the pressure cabin. The Hs 130 series was intended as an operational development of this experimental airframe, intended for high-altitude bomber and reconnaissance duties. Three versions appeared: the Hs 130A, C and E, they did not become operational, but in effect served as development aircraft for German supercharger development, the most advanced being the Hs 130E fitted with Daimler-Benz's HZ *Anlage* (*q.v.*) installation.

Henson, William Samuel, 157, 166, 609

Henson Aerial Steam Carriage, 157, 166, 609

Hercules (C-130), see Lockheed

Hercules engine, see Bristol

Heston Phoenix, 550, 569

First and most successful product of the Heston Aircraft Co Ltd, the Phoenix was an advanced five-seat commercial aircraft. It was almost a sesquiplane (*q.v.*), being a high-wing monoplane with a lower stub-wing faired into the wing struts, the stub wing doing duty as a housing for the retractable undercarriage. This represented an advanced feature for an aircraft of this size at that time (1934). Five aircraft in all were built.

Hiller, 1313; see also Fairchild Hiller

Hiller Industries Aircraft Division was established in 1942 to develop and produce co-axial helicopters in the United States. The first such aircraft was flown in 1944, followed by the UH-4 in 1946. With a new control system the Hiller 360, with a tail-rotor for directional control, was built soon after and entered production as a three-seat utility helicopter. It soon achieved success, and through the years has been modified in many forms, over two thousand being built. It has served with armed forces in many parts of the world and is particularly useful for *ab initio* helicopter training and for crop-spraying.

Hindenburg, 253, 725, see also Zeppelin, Ferdinand von

Hispano

Hispano Aviación is the airframe design and manufacturing division of the Hispano-Suiza (*q.v.*) aero engine company. During World War Two it was engaged chiefly in the manufacture of foreign warplanes under licence, and also produced for the Spanish Air Force the H.S.42 two-seat trainer of its own design. The HA-1109 was a licence-built version of the Messerschmitt Bf 109G, with Rolls-Royce Merlin engine; this was the standard post-war Spanish fighter for many years, and a number are still in service. Other post-war products have included the HA-100 Triana trainer (to replace the H.S.42) and the HA-200 Saeta jet trainer. The now-abandoned HA-300 jet fighter, developed in Egypt, was designed at Hispano Aviación by Prof Willy Messerschmitt, but the project was sold to Egypt in 1960.

Hispano-Suiza, 883, 1075

This well-known company has been in the aero engine industry since the 1914-18 War. It began as a Spanish company in Barcelona, manufacturing cars and engines, and soon found a ready market for its aircraft engines in French and British aircraft.

After the 1914-18 War the aero engine division of the company set up factories in France, and manufactured a long series of successful engines in collaboration with the Wright Corporation of America.

It revived after World War Two and entered the turbine field by licence-building the Rolls-Royce Nene and Tay turbojet engines. It has also played a considerable part in building aircraft accessories and armament, and its cannon were greatly used in Allied aircraft during and after World War Two.

HK S-3, 625

This was a West German single-seat high-performance glider of the type which won the 1958 Gliding Championship held at Leszno, Poland. The wing loading of 5.2 lb/sq ft was high for its day, but the fine wing gave the creditable glide ratio of 37 to 1. An unusual feature was that wing-warping (*q.v.*) was utilized for lateral control and for flaps.

Honeycomb sandwich, 474, 481, 485, 502, 503, 618, 976

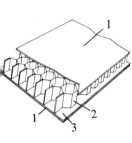

1 Skins
2 Honeycomb core
3 Layer of brazing material, ensuring a joint at all edges of the honeycomb

Form of construction, consisting of two skins between which is interposed a honeycomb core, so named because of its similarity to the combs made by bees. Honeycomb sandwich panels give high strength and great rigidity for low weight, and may be made of wood or metal, or a combination of both. Panels intended for use on high-performance aircraft are usually fabricated from stainless steel, and consist of a honeycomb core built up from strips of thin stainless steel brazed together. The cores are then machined, often while stabilized by a resin-type filler, and placed between pre-formed inner and outer skins. After light tack-welding, the assembly is placed in a jig of the precise contour of the finished part and then heated in an inert atmosphere until the

brazing joints have been made.

Hong Kong International Airport (Kai Tak)
A Hong Kong airport has existed since 1929, but the present one dates from 1954, when work

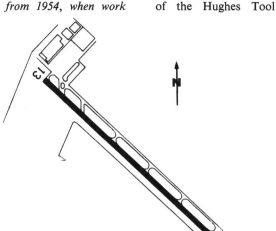

started on building its 8,350-ft runway out into Kowloon Bay—a feature unique among world airports.
Distance from city centre: 4½ miles
Height above sea level: 15 ft
Main runway: 13/31 = 8,350 ft
Passenger movements in 1967: 1,229,000

Honolulu International Airport

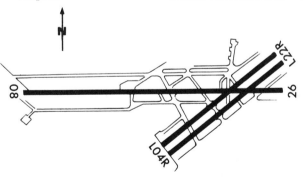

Distance from city centre: 9 miles
Height above sea level: 13 ft
Main runways: 08/26 = 12,380 ft
04R/22L = 7,005 ft
04L/22R = 6,950 ft
Passenger movements in 1966: 3,526,000

Hooker, Dr Stanley, 1273
Horizontal situation display, 1357
Hornet (D.H. 103), see de Havilland
Hotchkiss machine-gun, 875

Hound Dog missile, see North American
Houston, Lady, 1087
H.P. 42E and H.P. 42W, see Handley Page
Hughes, 1313
The Aircraft Division of the Hughes Tool Company built its Model 269A (now Model 300) helicopter as a private venture in 1955. This is a three-seat light utility helicopter which has been built in substantial numbers for service with the US Army and foreign armed forces as a primary helicopter trainer. About 250 have also been produced for civilian customers. Further Lycoming-powered helicopter variants were built before Hughes turned over to turbine-powered helicopters, of which the OH-6A/Model 500 is the current example. Military production alone of the OH-6A Cayuse was well over 1,200 by early 1969.
Hughes Falcon missile, 895
Developed by the Hughes Aircraft Company, the Falcon has

been produced in many versions, one being the first air-to-air guided missile with a nuclear warhead. It is standard equipment for the F-102 and F-106 fighters of North American Air Defense Command and is propelled by a Thiokol M60 solid-propellant rocket giving it a speed of Mach 2 and a range of 5 miles.
Hughes Phoenix missile, 895
An advanced American missile originally developed for the now-cancelled F-111B variable-geometry strike aircraft for the US Navy. It is a radar-homing missile with a Rocketdyne solid-propellant rocket motor.
Hummingbird (XV-4), see Lockheed
Hungary: Air Force
Magyar Legierö (Hungarian Air Force). Formerly a part of the Austro-Hungarian Empire, which was allied to Germany in the 1914-18 War, Hungary became an independent republic shortly after the Armistice. It was subject to the Versailles Treaty of 1919 which forbade the production of combat aircraft, but a small air force was formed in 1936 with aircraft acquired from Germany and Italy. It became a German satellite again early in 1939, and the Hungarian Air Force fought alongside the *Luftwaffe* in World War Two until Hungary signed a separate Armistice with the Allies in January 1945. Again the surrender terms forbade her military aircraft, but following the Soviet occupation, the Air Force was reorganized and re-equipped with Soviet aircraft. At the time of the 1956 uprising it had a large force of MiG-15 fighters and a few MiG-17s, and a tactical force of Il-10, Tu-2 and Il-28 attack/light bomber types. After a period of "good behaviour" following the suppression of the

uprising, the Hungarian Air Force was gradually re-equipped with modern Soviet-designed aircraft. Those at present in service include MiG-17 and MiG-21 fighters and Su-7 ground-attack aircraft.
Hurricane, see Hawker
Huskie (H-43), see Kaman
Hustler (B-58), see General Dynamics
Hydraulic pumps

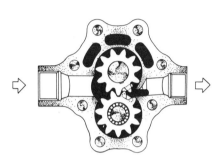

Hydraulic gear pump

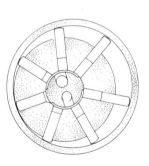

Hydraulic pump. Rotary, radially disposed pistons arrangement

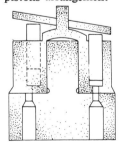

Hydraulic pump. Arrangement with axially disposed pistons and swashplate

These convert mechanical energy into fluid energy by using the

power input to change the pressure level of the fluid passing through a hydraulic system.
The simplest pumps are of the gear type, similar to those used in general engineering. On these, two gears are located in a housing with inlet and delivery ports. One gear is driven, the drive being transmitted to the other by the teeth. As the teeth revolve, fluid is carried round to the other side of the pump, where it is forced out of the delivery port, since it cannot return to the "inlet" side as the teeth intermesh at the centre. Gear pumps are rarely used today, the majority of hydraulic pumps in aircraft utilizing pistons, these either being arranged radially and operated by an eccentric, or parallel and operated by an inclined rotating swashplate. On pumps with radial pistons the cylinders rotate; on pumps with parallel pistons the cylinders are normally stationary.
A feature of aircraft hydraulic pumps is that for most of their operating life they are not required to do any actual work, as for example, during normal cruising flight, when the landing gear is retracted and flaps are in. Because of this, off-loading is necessary to prevent overheating of the oil, which would occur if the pump

continued to pump its full output at maximum pressure through a relief valve. Other reasons for off-loading pumps include the need to reduce pump wear and tear, and relieve the power output from the engine.

Two basic methods of off-loading are used, one involving pressure reduction and the other flow reduction. On early aircraft, off-loading by pressure reduction was effected by a manually-operated valve located between the main pressure and return lines, which was closed when power was required and opened when power was not required. This early arrangement has been superseded by the "open centre" system and the "automatic cut-out" system, in which the pump circulates fluid at a nominal zero pressure unless a service is being operated.

Where off-loading is effected by reducing the flow, the pump displacement is reduced to a figure that just makes good the pump's internal leakage and that of the circuit to which it is connected. Alternatively, the delivery is reduced to a small pre-determined flow which is passed through a relief valve at the full off-loaded pressure. Generally, however, off-loading using the flow reduction technique is effected by means of a pressure-controlled, variable delivery pump, or by a constant delivery pump with an automatic unloading valve.

Hydraulic system, 935

HZ *Anlage,* 1112-1113

IATA (International Air Transport Association), 1537

IATA was founded in 1946 as an international trade association of the principal airlines. Within a year it gained 63 member airlines, and began to expand its activities into all fields in which international co-operation between the airlines would benefit them all. It now has over 100 members, handling about 90 per cent of the world's air commerce. Since 1946, IATA has maintained a world-wide pattern of rates, fares and other matters relating to airline traffic, at the request of various governments. It has divided the world into nine areas, and a Traffic Conference for each area meets to establish fares and rates in that area, together with related subjects such as relationships with travel agencies, cargo agencies, etc, together with more general development of such practical matters as cargo handling.

In the financial field it has established the IATA Clearing House in London, by means of which inter-airline payments can be offset and the actual transactions reduced to a minimum. It also maintains financial sub-committees. IATA maintains a Legal Committee, which keeps an eye on international legal conventions which affect airline operations; and a Medical Committee, which investigates medical problems of airline operatives and passengers. See also pp. 360-363.

Iberia

Operating title of Lineas Aereas de España SA. The original Iberia was formed in 1927, subsequently being absorbed by successive operators which were in turn taken over by the present company on its formation in 1938. The network was then almost entirely internal, but after World War Two an international network was built up steadily until today Iberia flies services to the US, Central and South America and Africa, as well as to Western Europe and the UK. The company is wholly State-owned.
Headquarters: Madrid.
Member of IATA.

ICAO (International Civil Aviation Organization), 1537, 1591

In December 1944 the representatives of 54 nations met in Chicago to discuss matters of mutual interest and collaboration concerning post-war civil aviation. One outcome of this conference was the establishment, with effect from June 1945, of a Provisional International Civil Aviation Organization (PICAO); this became a permanent body as from April 1947, and by the end of that year 46 nations had become members. Today the membership is well over 100 countries and the declared objectives of ICAO are: "To promote the safe and orderly development of commercial and private international civil aviation with equal opportunity for all its member nations; the establishment of world-wide standards for safety, reliability and regularity of air navigation; the economic development of aviation; the reduction of formalities of Customs, immigration, health and currency control and the continuous evolution of international air law."
The ICAO headquarters is in Montreal, Canada. In addition, it has regional offices in Mexico City, Lima, Paris, Cairo, Dakar and Bangkok.

Icarus, 112

Subject of a Greek legend, Icarus was the son of an architect (Daedalus) who made wings of feathers for himself and his son to escape imprisonment by King Minos of Crete. Icarus perished in the attempt. It is possible that this legend was based on an early attempt to fly.

Icelandair

Operating title of Flugfelag Islands HF, the main Icelandic airline. It began operations in 1937 with a domestic seaplane service, adopting its present title in 1940. International services, first to Glasgow and later to Copenhagen, were started in 1945 with Catalina flying boats and, a year later, with Liberator transports. Present day international services are to Greenland, Denmark, Norway and the UK.
Headquarters: Reykjavik.
Member of IATA.

IFF (Identification, Friend or Foe), 1377

ILS (Instrument Landing System), 647, 703, 1373, 1569, see also Take-off and Landing Techniques

One of the most important radio aids developed since World War Two, ILS is now in world-wide use by both military and civil aviation. It can be used both for pure navigational purposes and also as an approach and landing aid at night or in bad weather. It comprises two radio transmitting units at the airfield and a receiver in the aircraft linked to a display instrument in the pilot's cockpit.

In the basic ILS, one airfield transmitter emits a radio beam along the axis of the runway, known as a localizer. The other transmitter alongside the runway emits a beam along the desired glide-slope to land on that runway. The localizer beam is received in the aircraft and interpreted by a vertical needle in the ILS instrument which must be aligned with a vertical line on the dial. Similarly, the glide-slope beam is interpreted by a horizontal pointer which must be aligned with a further line on the dial to ensure a correct approach to the runway. From this basic instrument have come refinements which give the pilot more and more information (see text) to show him what he must do to align his aircraft on the glide-path on the runway. In some cases the ILS display is co-ordinated with a basic flight instrument such as an artificial horizon, to inform the pilot not only how to get on to the runway but what attitude his aircraft is taking while he is doing so.

Ilya Mourometz, see Sikorsky

Ilyushin

The design bureau headed by Sergei Vladimirovich Ilyushin, one of the USSR's most versatile and experienced aircraft designers, came into prominence during the early part of World War Two. It was then responsible for the twin-engined DB-3F (Il-4) medium bomber and for perhaps the most celebrated ground-attack aeroplane ever built, the Il-2 *Stormovik.* In 1950 appeared the twin-jet Il-28 tactical bomber, which was built in very large numbers and remains in service with the air forces of some Soviet *bloc* countries today.

Ilyushin's chief post-war contribution has been to the civil aircraft market, beginning with the Il-12 and Il-14 twin-engined airliners which were Soviet counterparts to such Western types as the Convairliner and the Vickers Viking. In 1957 appeared the four-turboprop Il-18, now used by several airlines outside the USSR, and still in production in 1969. The only other airliner in the world, apart from the BAC VC10, to be powered by four rear-mounted turbine engines is also an Ilyushin design. This is the Il-62, which went into service with Aeroflot in 1967 and in 1968 inaugurated the airline's service between Moscow and New York.

Ilyushin Il-2 *Stormovik,* 885

Designed by Sergei Ilyushin, the Il-2 *Stormovik* was a two-seat (originally single-seat) assault aircraft or dive-bomber. It was widely used by the Soviet Air Force in World War Two, during which over 30,000 were produced. A low-wing monoplane, the Il-2 was powered by a 1,300 hp AM-38 twelve-cylinder liquid-cooled Vee engine, which gave it a maximum speed of about 280 mph. The main landing gear retracted rearward into two distinctive bulges under each wing. The pilot's cockpit provided a good view, and armour-plate was fitted beneath and behind the seat, and on the sides and top of the cockpit cover. Armour-plate was also fitted under the engine. Armament consisted of two 20-mm cannon and two 7.62-mm machine-guns mounted in the leading edge of the wing, firing outside the propeller disc, and one 12.7-mm gun firing aft from the rear cockpit. Eight 56-lb rocket-impelled fragmentation bombs were carried on special guide-rail type racks, four under each wing; alternatively, up to 880 lb of bombs could be carried underwing. Wing span was 48 ft and the length 38 ft 6 in.

Ilyushin Il-18 ("Coot"), 787

Malév (Hungarian Airlines) uses the Ilyushin Il-18 turboprop airliner on international routes

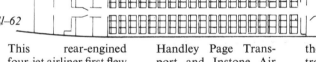

Il–18

The Il-18 has become the standard medium-capacity domestic and continental turboprop airliner of the Soviet *bloc*. First aircraft flew in July 1957 and production aircraft entered airline service with Aeroflot in April 1959. The type has been produced in large numbers (over 500) and has been supplied to twelve airlines outside the USSR; it is also operated by several military organizations, including the Polish Air Force. In standard form it has capacity for 110 passengers, a range of 3,230 miles and a cruising speed of 419 mph.

lyushin Il-62 ("Classic"), 837

Mainstay of the IAF bomber force is the English Electric Canberra

I–62

This rear-engined four-jet airliner first flew in January 1963, and since entering service in 1967 has taken over some of the long-distance routes of Aeroflot. It has accommodation for 186 passengers, and a range of 4,160 miles with maximum payload of 50,700 lb. Cruising speed is about 550 mph.

mmelmann, Max, 658, 877

Max Immelmann was one of two brothers from Dresden who volunteered for the Aviation Corps on the outbreak of war in August 1914. He first went into action in April 1915, flying LVG's on artillery observation. Later that year he understudied Boelcke in the Fokker Monoplane. On 3 August 1915 he had his first combat in a Fokker *Eindecker* and brought down an Allied aircraft. He soon became a formidable exponent of the Fokker, and built up a total score of 17 enemy aircraft shot down.

Most of these were slow reconnaissance biplanes, for the Allied air forces had little to match the Fokker at this time. Immelmann was killed in June 1916 when his aircraft broke up in the air.

His name will be remembered for the aerobatic manoeuvre called the "Immelmann Turn" which he evolved for use in combat. Basically it was a stall turn by which, when he was being followed, he was able to turn quickly and face his opponent.

Imperial Airways, see also BOAC

Four British air transport companies—British Marine Air Navigation, Daimler Hire, Handley Page Transport and Instone Air Line—amalgamated in 1924 to form Imperial Airways. The amalgamation gave the new airline a fleet of 18 very assorted aircraft and a 1,760-mile network of routes which were mostly cross-Channel or shorter domestic ones. From these modest beginnings Imperial Airways set about pioneering the development of routes to the farthest points of the then extensive British Empire. First new equipment for the Empire routes were D.H. 66 Hercules three-

engined biplanes. In 1927 the airline opened a service from Cairo to Basra, followed progressively by others to Karachi (1929); Lake Victoria (1931); Nairobi and Capetown (1932); Calcutta, Rangoon and Singapore (1933); and Brisbane (1935). The Singapore-Brisbane leg of the last-named route was operated by Qantas (*q.v.*).

In 1937 the Short flying boats *Caledonia* and *Cambria* paved the way for the first commercial air services across the North Atlantic. In July 1938 the Short-Mayo composite aircraft *Maia* and *Mercury* achieved the first return trip across the Atlantic (to Montreal) ever made by a payload-carrying aeroplane, and in the following year an experimental transatlantic mail service was inaugurated by the flying boats *Cabot* and *Caribou*, which flew 50,000 miles, without incident and with a high degree of regularity, before the service was discontinued on 30 September 1939. Under the British Overseas Airways Act a new corporation, BOAC (*q.v.*), was created with effect from 24 November 1939, and on 1 April 1940, BOAC formally took

over the undertakings of Imperial Airways and another smaller operator, British Airways.

India: Air Forces

Bharatiya Vayu Sena (Indian Air Force). Created as a small force on 1 April 1933 with assistance from the RAF, which had been established in the country for over a decade. Operational against frontier tribesmen in 1937 and 1939, but had expanded little before 1941, when the Pacific war made India an important strategic base. At the same time the establishment of a domestic aircraft industry both increased the strength of the IAF and made it less dependent upon outside assistance. The IAF achieved a notable contribution to the Allied war effort; in recognition of this it was granted a "Royal" prefix in March 1945, but India became an independent state in mid-1947 and the prefix was dropped early in 1950.

Since then the Indian Air Force has selected most of its combat aircraft from Britain or France, although more recently a few Soviet types have also been introduced. Of the types at present in service, the Gnat and MiG-21 fighters have been licence-built in India while imported types include British Hunter fighter-bombers and

Canberra jet bombers, French Mystère IVA fighters and Soviet An-12 heavy transport aircraft. The HF-24 Marut fighter, designed in India by Dr Kurt Tank (*q.v.*) is also in limited service.

Indian Naval Aviation. The formation of the Indian Navy's air arm dates from January 1950, when India became an independent republic within the British Commonwealth, but its first combat aircraft were not introduced until after the acquisition in 1957 of the ex-British aircraft carrier *Hercules*. This is now in commission as the *Vikrant*, on which are based the Indian Navy's Sea Hawk fighters and Alizé anti-submarine aircraft.

Indian Airlines

Indian Airlines Corporation is the Indian domestic and regional airline, and was formed for this purpose in August 1953. It flies services to points within India, and to Afghanistan, Pakistan, Nepal, Burma and Ceylon. Regular freight services, and a nocturnal air mail link between Bombay, Calcutta, Delhi and Madras.

Headquarters: New Delhi.

Member of IATA.

Indicated horsepower (ihp)
In a piston-engine, the total power developed, calculated from knowledge of the rotational speed and of the gas pressure achieved inside the cylinder during operation. The pressure is obtained by attaching an "indicator" to the cylinder (any cylinder in a multi-cylinder

engine, assuming all function equally) producing an indicator diagram showing the variation in pressure during each firing cycle. The useful power available externally is less than the ihp because the latter does not allow for the power lost in friction and in the induction, compression and exhaust strokes of each four-stroke cycle. Indicated hp is simply the indicated mean effective pressure in the cylinder multiplied by the total area of the piston(s) multiplied by the engine speed.

Indonesia: Air Forces
Angkatan Udara Republik Indonesia (Indonesian Republican Air Force). Established in 1946 and expanded and re-equipped following the transfer of sovereignty from the Netherlands Government in December 1949. The initial equipment of captured or salvaged Japanese aircraft was then replaced with war-surplus US types until the formation of the first jet unit with British Vampire trainers in 1955. Jet combat aircraft did not appear until after Indonesia concluded an agreement with Czechoslovakia in March 1958, since when all front-line units have been re-equipped with aircraft of Soviet origin. In 1968, the AURI was a sizeable defensive and tactical force possessing MiG-17, MiG-19 and MiG-21 fighters, and Il-28 and Tu-16 bombers.

Angkatan Laut Republik Indonesia (Indonesian Naval Air Force) is a shore-based service

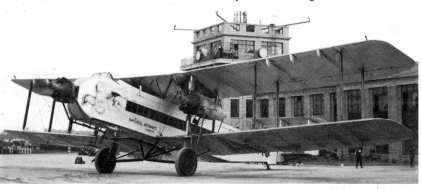

A familiar sight at Croydon in the early 1930s was the Armstrong Whitworth Argosy used on Imperial Airways' European routes

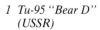

19

20

21

22

23

24

25

26

27

29

28

31

30

32

33

34

35

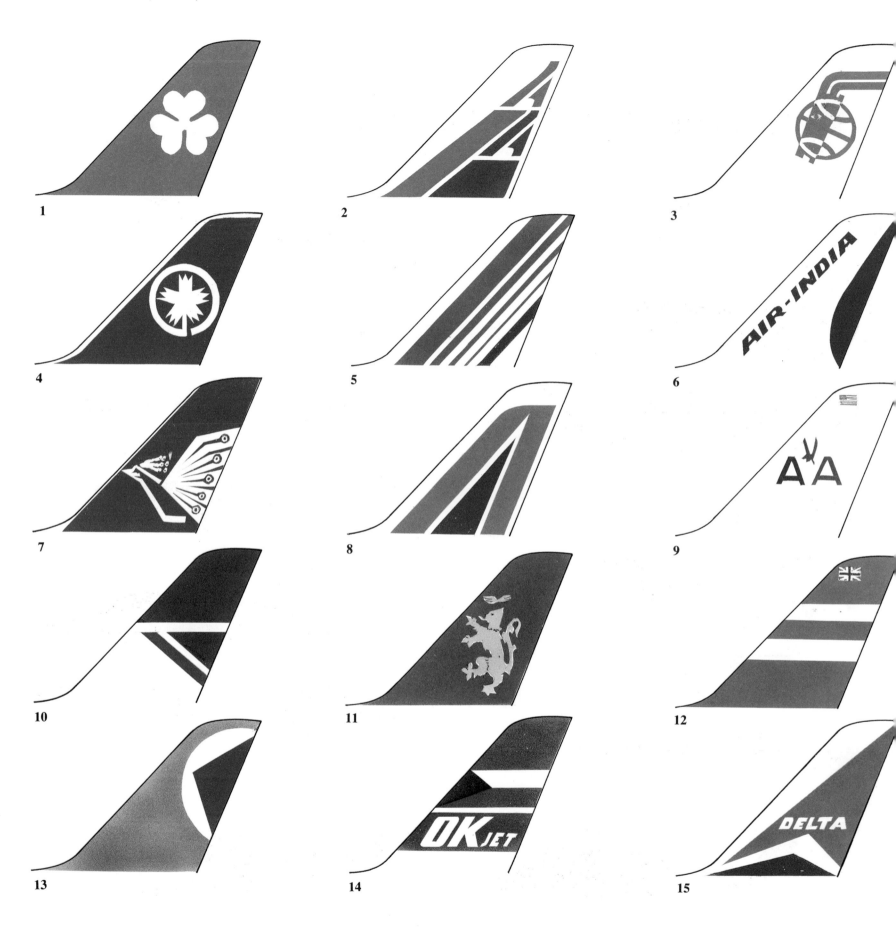

1

2

3

4

5

6

7

8

9

10

11

12

13

14

15

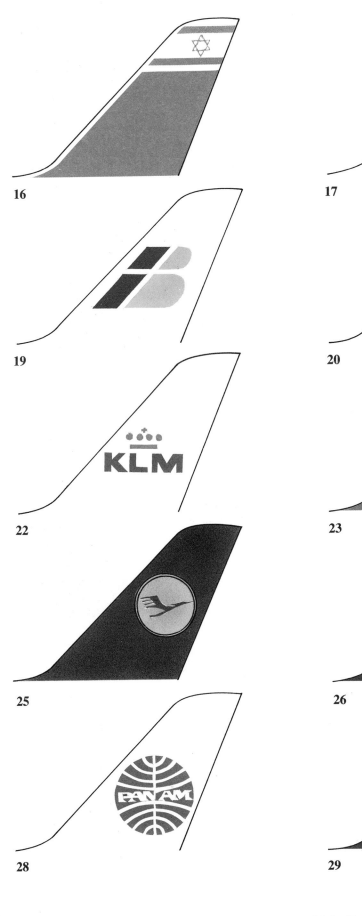

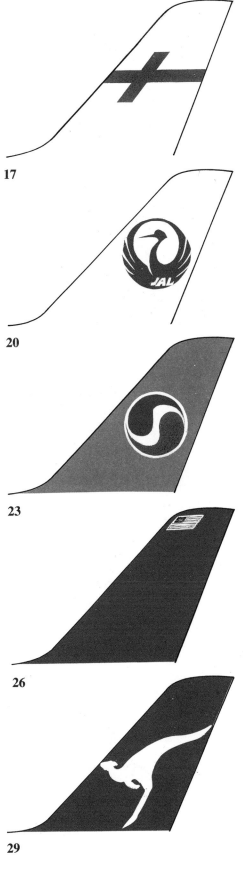

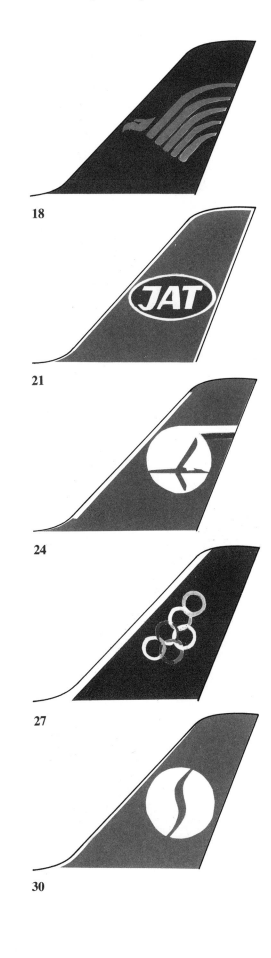

16

17

18

19

20

21

22

23

24

25

26

27

28

29

30

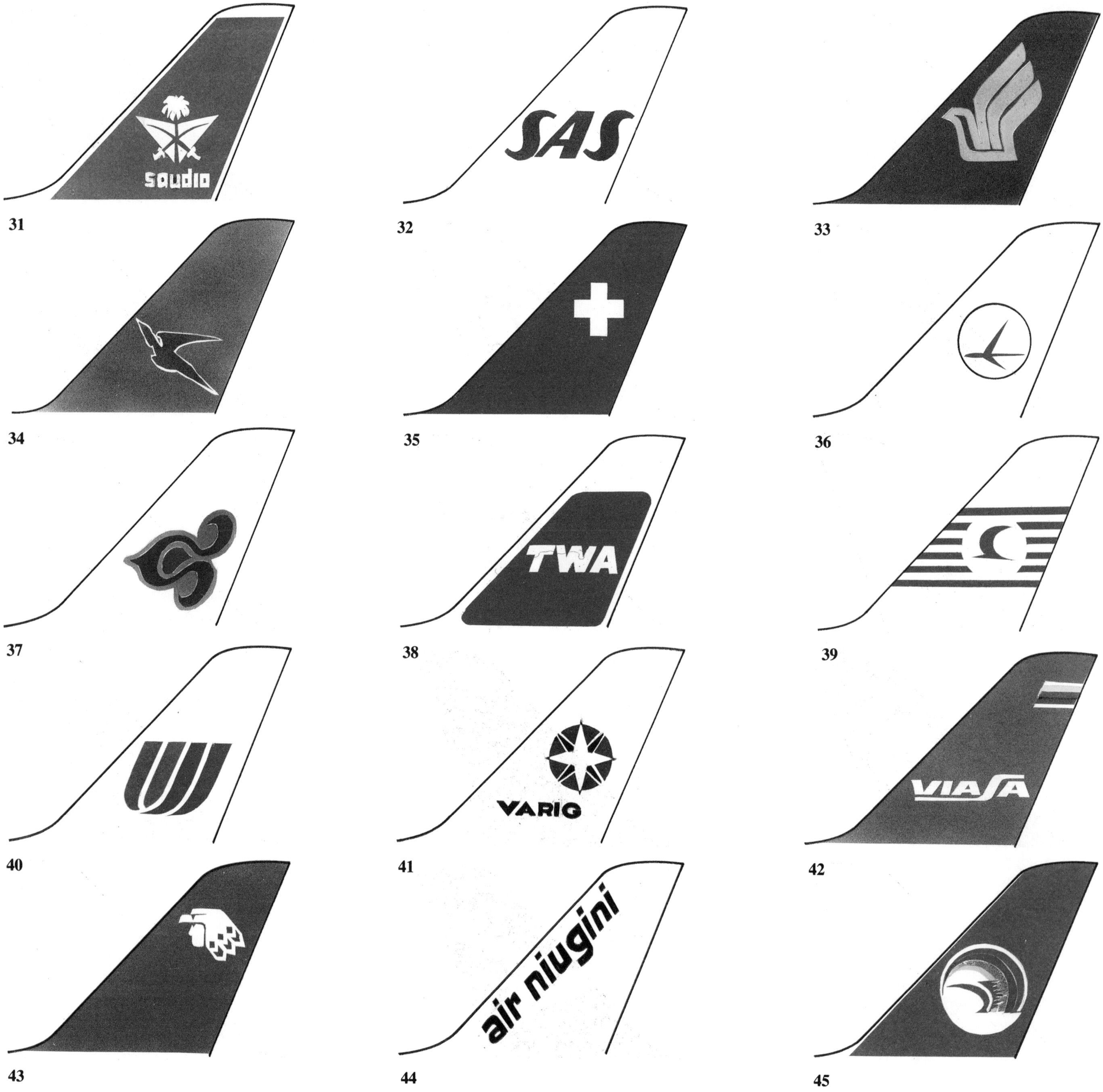

31 Saudia
32 SAS – Scandinavian Airline System
33 Singapore Airlines – SIA
34 South African Airways
35 Swissair
36 Tarom
37 Thai International
38 Trans World Airlines – TWA
39 Turk Hava Yollari·THY
40 United Airlines
41 Varig
42 Viasa
43 Aero Mexico
44 Air Niugini
45 Air Pacific

whose principal combat aircraft are anti-submarine Fairey Gannets. It was formed in 1958 and also operates a small number of helicopters and general purpose aircraft.

In 1959 Fairey supplied Gannet AS.4s to the Indonesian Navy

Induction manifold, 1054, 1059, 1060, 1087, 1102, 1107

Inertial Guidance System, 707

Inertial Platform, 1359, 1361

Inflexible, see Beardmore

Infra-red missiles, 887
Missiles equipped with infra-red sensing equipment linked to their control system. They are usually air-to-air missiles, and work on the basis that the infra-red equipment will "home" on to the hot engine exhaust of the aircraft being attacked. The ad-vantage of such a system is that it is less affected than others by jamming devices.

Inlet valve, 1044, 1054, 1059, 1086, 1097

In-line engines, 909, 911, 1065, 1067, 1069, 1073
Aero engines in which the cylinders are placed one behind the other, linked by a common crankshaft. For greater power, in-line engines have been built with two banks of in-line cylinders either hori-zontally-opposed or in Vee shape and with four banks either in "H", "X" or "W" shape.

Instrument Landing System, see ILS

Integral machining, 443, 445, 485

Integrally-stiffened skin, 441-443

Interflug
Operating title of the East German inter-national airline, which was formed in 1954. In September 1963 the name was changed from the original one of Deutsche Lufthansa, to avoid confusion with the West German airline of the same name, and Interflug's full title is now *Gesellschaft für Internationalen Flugver-kehr mbH*. Scheduled operations began with a Berlin-Warsaw service in February 1956, and an internal network was opened in June 1957. Interflug, in addition to charter and agricultural work in East Germany, flies international services to all Eastern European capitals and to other points in the Mediterranean, the Middle East and North Africa.
Headquarters: Berlin-Schonefeld.

INTERNATIONAL AIRLINES: The members of the International Air Transport Association (IATA)

AIRLINE	COUNTRY	PASSENGERS CARRIED*	CORRECT TONNE-KILOMETRES PERFORMED*	AIRCRAFT FLEET†
Aerial Tours	Papua	—	—	— Various
Aer Lingus	Ireland	1,473,308	223,307	4 One-Eleven,
Aerlinte	Ireland	295,773		6 B737, 2 B747, 6 B707, 2 B720
Aerolineas Argentinas	Argentina	1,159,356	247,108	8 B707, 3 Caravelle, 8 H.S.748, 8 B737
Aeronaves de Mexico S.A.	Mexico	1,481,943	188,799	6 DC-8, 10 DC-9, 5 H.S.748, 1 Twin Otter
Aerovias Nacionales de Colombia S.A.	Colombia	1,947,442	207,807	2 B707, 4 B720, 7 B727, 2 H.S.748, 13 DC-4, 8 DC-3
Air Afrique	Ivory Coast	368,825	168,963	3 DC-8, 2 Caravelle, 2 YS-11A, 3 DC-4, 1 DC-3
Air Algerie	Algeria	683,013	60,225	2 B727, 1 B737, 4 Caravelle, 4 CV-640, 1 DC-4
Air Canada	Canada	7,207,483	1,295,185	29 Viscount, 12 Vanguard, 38 DC-8, 36 DC-9, 3 B747
Air Ceylon	Ceylon	116,047	12,093	1 DC-8, 1 Trident, 1 H.S.748, 2 DC-3
Air France	France	6,314,276	1,361,015	5 B747, 33 B707, 17 B727, 39 Caravelle, 5 DC-4, 15 F-27
Air Guinée	Republic of Guinée	—	—	4 An-24, 3 Il-18
Air India	India	391,707	268,018	9 B707, 2 B747
Air Malawi	Malawi	66,337	4,160	2 Viscount, 2 H.S.748, 2 Islander, 1 One-Eleven
Air Mali	Mali	38,958	8,440	1 B727, 1 Il-18, 1 An-24, 1 An-2, 1 DC-3, 1 Caravelle
Air New Zealand	NZ	379,050	141,401	6 DC-8, 2 Electra, 1 H.S.748, 1 DC-4
Air Siam	Thailand	12,637	35,821	1 DC-8, 3 DC-4
Air Vietnam	Vietnam	1,051,345	40,124	1 B707, 3 B727, 4 DC-6, 10 DC-4, 15 DC-3, 4 Cessna 185
Air Zaire	Congo	325,939	64,406	4 DC-8, 7 DC-4, 8 F-27, 2 Caravelle
Alia-Royal Jordanian Airlines	Jordan	124,943	19,833	2 B707, 3 Caravelle
Alitalia	Italy	5,776,637	1,081,503	22 DC-8, 36 DC-9, 4 B747, 18 Caravelle
Allegheny Airlines	US	—	—	31 DC-9, 31 One-Eleven, 17 F.H.227, 40 CV-580
American Airlines	US	19,244,730	3,369,923	16 B747, 5 DC-10, 98 B707, 100 B727
Ansett-ANA	Australia	2,695,486	261,212	6 B727, 12 DC-9, 3 Electra, 12 F-27, 1 Twin Otter, 1 DC-4, 3 Carvair
Ansett-ANA	Papua	320,000	10,512	5 F-27, 10 DC-3, 2 Twin Otter, 2 Skyvan
Ariana Afghan Airlines	Afghanistan	—	—	2 B727, 1 CV-440, 1 DC-3
Austrian Airlines	Austria	582,071	43,235	7 DC-9, 4 Caravelle
Avna	Republic of South Africa	—	—	—
Braniff International	US	6,014,823	779,218	40 B727, 7 DC-8, 1 B747, 8 One-Eleven, 5 B720, 7 B707
British Caledonian Airways	UK	791,762	120,222	7 B707, 4 VC-10, 18 One-Eleven
British European Airways	UK	8,865,934	533,632	18 One-Eleven, 49 Trident, 10 Vanguard, 2 Heron, 8 Merchantman, 21 Viscount
British Overseas Airways	UK	1,967,651	1,542,108	27 VC-10, 8 B747, 28 B707
British West Indian Airways	Trinidad and Tobago	314,338	73,197	6 B707
C.P. Air	Canada	1,459,848	440,675	11 DC-8, 7 B737, 4 B727

*Traffic statistics are for calender year 1971; tonne-kilometres, a measure of the work done by the airline, are given in thousands.
†Fleet as on December 31, 1971, including aircraft leased.

AIRLINE	COUNTRY	PASSENGERS CARRIED*	CORRECT TONNE-KILOMETRES PERFORMED*	AIRCRAFT FLEET†
Ceskoslovenske Aerolinie	Czechoslovakia	1,087,398	97,463	21 Il-14, 7 Il-18, 4 Tu-104, 2 Tu-124, 1 Tu-134, 4 Il-62
Chicago Helicopter	US	19,188	51	3 S-58, 3 Bell 206A, 3 Bell 47G, 1 Bell 47J
China Airlines Limited	Republic of China	920,816	135,656	3 B707, 3 B727, 1 Caravelle, 1 YS-11, 3 DC-4, 1 C-46
Commercial Airways	Republic of South Africa	21,394	868	4 DC-3
Compania Ecuatoriana de Aviacion S.A.	Ecuador	—	—	3 Electra, 4 DC-6, 1 DC-4, 1 B23
Compania Mexicana de Aviacion S.A.	Mexico	1,460,054	146,477	4 DC-6, 10 B727
Continental	US	5,530,068	875,382	27 B720, 12 B707, 4 B747, 1 B727, 19 DC-9, 1 DC-6
Cyprus Airways	Cyprus	150,162	19,006	2 Trident
Delta Airlines	US	16,917,773	1,757,793	41 DC-8, 77 DC-9, 16 CV-880, 6 B747, 3 Hercules
Deutsche Lufthansa	West Germany	7,029,186	1,365,168	4 B747, 20 B707, 23 B727, 28 B737
DETA	Mozambique	182,411	16,284	3 B737, 3 F-27, 2 DC-3, 1 Viscount, 1 Aztec
DTA	Angola	170,587	9,963	5 F-27, 6 DC-3
East African Airways	Kenya, Tanzania, Uganda	564,229	106,352	5 VC-10, 3 DC-9, 4 F-27, 4 DC-3, 4 Twin Otter
Eastern Airlines	US	22,581,746	2,526,452	28 DC-8, 82 DC-9, 101 B727, 3 B747, 19 Electra
Eastern Provincial	Canada	331,630	25,407	3 B737, 3 Herald, 3 Carvair, 3 DC-3, 1 Twin Otter
East-West Airlines	Australia	311,163	11,491	6 F-27, 2 DC-3, 1 Twin Otter
Egyptair	Arab Republic of Egypt	513,358	107,614	4 Comet, 4 B707, 4 Il-18, 3 An-24, 1 DC-6, 3 Il-62, 1 Cessna 207
El Al	Israel	622,752	383,818	2 B747, 8 B707, 2 B720
Empressa Consolidada Cubana de Aviacion	Cuba	979,188	55,313	4 Britannia, 4 Il-18, 5 An-24, 1 Il-14, 2 An-2
Ethiopian Airlines	Ethiopia	263,051	49,874	1 B707, 2 B720, 2 DC-6, 10 DC-3, 2 Cessna 180, 1 Bell 47
Finnair	Finland	1,335,967	99,214	3 DC-8, 6 DC-9, 2 DC-6, 8 Caravelle, 7 Metropolitan
Flugfelag	Iceland	195,658	16,803	2 B727, 2 F-27, 2 DC-6, 2 DC-3
Flying Tiger	US	—	833,239	17 DC-8
Garuda	Indonesia	837,335	101,166	2 DC-8, 3 DC-9, 2 Electra, 11 F-27, 3 F-28, 3 DC-3
Ghana Airways	Ghana	135,569	16,222	1 VC-10, 2 Viscount, 1 DC-3, 1 H.S.748
Iberia	Spain	6,563,143	721,633	2 B747, 14 DC-8, 25 DC-9, 19 Caravelle, 12 Metropolitan, 7 F-27, 3 F-28
Indian Airlines	India	2,133,868	163,223	7 B737, 7 Caravelle, 7 Viscount, 12 F-27, 13 H.S.748, 8 DC-3
Iran Air	Iran	761,163	83,187	2 B707, 4 B727, 3 B737, 4 DC-6
Iraqi Airways	Iraq	161,875	21,302	3 Trident, 3 Viscount
Japan Air Lines	Japan	6,848,237	1,393,002	47 DC-8, 7 B747, 12 B727, 12 Beechcraft H-18
JAT	Yugoslavia	1,398,990	90,231	2 B707, 8 DC-9, 6 Caravelle, 7 CV-440, 2 DC-3
KLM	Netherlands	2,580,739	1,008,508	7 B747, 27 DC-8, 19 DC-9, 1 F-27
Kuwait Airways	Kuwait	342,177	71,684	3 B707, 2 Trident
Libyan Arab Airlines	Libya	259,386	26,952	2 B727, 2 Caravelle, 2 F-27
Linea Aerea del Cobre	Chile	82,724	8,981	3 DC-6, 2 DC-3, 1 Beechcraft Baron, 1 Beechcraft Queen Air
Linea Aerea Nacional de Chile	Chile	590,296	127,894	2 B707, 5 B727, 3 Caravelle, 9 H.S.748, 8 DC-6, 9 DC-3, 1 Cessna 310
Malta Airlines	Malta	110,816	16,475	Operated by BEA aircraft
MEA—Airliban	Lebanon	648,009	126,591	3 B707, 8 B720, 1 Comet, 1 Caravelle, 2 CV990
Mount Cook Airlines	—	—	—	2 H.S.748, 2 DC-3, 1 Twin Otter, 2 Islander, 4 Widgeon, 1 Fletcher Fu-24, 3 Cessna 180, 15 Cessna 185
National Airlines	US	5,492,182	775,112	18 DC-8, 3 DC-10, 38 B727, 2 B747
New York Airways	US	334,662	857	4 S-61
NZ National Airways Corporation	NZ	1,494,351	68,313	4 B737, 5 Viscount, 13 F-27
Nigerian	Nigeria	227,061	34,189	1 B707, 1 B737, 5 F-27, 1 F-28
Northwest Airlines	US	6,089,273	1,187,568	56 B727, 13 B720, 33 B707, 15 B747
Olympic Airways	Greece	2,021,355	239,266	6 B707, 6 B727, 6 DC-6, 8 YS-11
Pakistan International	Pakistan	1,085,609	253,265	8 B707, 3 B720, 8 F-27, 5 Twin Otter
Pan American World Airways	US	10,449,667	3,980,056	27 B747, 97 B707, 9 B720, 24 B727, 4 Falcon Jet
Polskie Linie Lotnicze	Poland	1,031,196	61,703	8 Il-18, 5 Tu-134, 14 An-24, 5 Il-14
Qantas Airways	Australia	718,286	553,533	21 B707, 4 B747, 2 DC-4
Quebecair	Canada	293,372	16,923	2 One-Eleven, 4 F-27, 3 DC-3, 1 C-46, 3 Otter, 5 Beaver, 2 D-25, 5 Cessna 185, 1 Beechcraft Queen Air
Sabena	Belgium	1,401,988	455,999	2 B747, 12 B707, 5 B727, 7 Caravelle, 1 F-27
Saudi Arabian Airlines	Saudi Arabia	613,609	68,561	2 B707, 2 B720, 3 DC-9, 8 DC-3, 8 CV-340
SAS	Scandinavia	5,679,864	757,096	2 B747, 13 DC-8, 32 DC-9, 13 Caravelle, 9 Metropolitan
Servicos Aereos Cruzeiro do Sul S.A.	Brazil	907,802	80,869	3 B727, 7 Caravelle, 8 YS-11, 7 C-47, 1 C-82, 3 FH-227
South African Airways	Republic of South Africa	1,637,448	345,884	8 B707, 2 B747, 9 B727, 6 B737, 7 Viscount, 3 H.S.748
Sudan Airways	Sudan	139,514	17,550	2 Comet, 4 F-27, 3 Twin Otter
Suidwes Lugdiens	Republic of South Africa	12,820	1,158	2 DC-3, 1 Cessna 402, 2 Cessna 310, 2 Aztec, 4 Twin Commanche, 1 Cessna 205, 1 Commanche

*Traffic statistics are for calender year 1971; tonne-kilometres, a measure of the work done by the airline, are given in thousands.
†Fleet as on December 31, 1971, including aircraft leased.

AIRLINE	COUNTRY	PASSENGERS CARRIED*	CORRECT TONNE-KILOMETRES PERFORMED*	AIRCRAFT FLEET†
Swissair	Switzerland	3,894,984	640,152	2 B747, 8 DC-8, 22 DC-9, 7 CV-990
Syrian Arab Airlines	Syria	166,552	22,682	4 Caravelle, 3 DC-6
Territory Airlines	New Guinea	—	—	1 Cessna 180
Trans-Australia Airlines	Australia	3,074,206	270,146	6 B727, 12 DC-9, 17 F-27, 11 Twin Otter, 2 Electra, 9 DC-3
Transportes Aeros Portugueses	Portugal	1,209,386	296,175	8 B707, 7 B727, 3 Caravelle
Trans-Mediterranean Airways	Lebanon	—	167,002	3 B707, 5 DC-6
Trans World Airlines	US	14,033,568	3,715,094	19 B747, 103 B707, 35 B727, 25 CV-880, 19 DC-9
Tunis Air	Tunisia	308,858	29,933	1 B707, 5 Caravelle, 1 Nord 262
Turk Hava Yollari	Turkey	1,475,995	84,267	3 B707, 9 DC-9, 1 Viscount, 7 F-27
Union de Transports Aeriens	France	387,140	328,991	11 DC-8, 2 Caravelle
United Airlines	US	25,512,953	4,311,247	5 DC-10, 12 B747, 111 DC-8, 29 B720, 150 B727, 72 B737, 1 CV-580
Varig	Brazil	1,540,072	437,812	11 B707, 4 B727, 1 DC-8, 9 H.S.748, 10 Electra, 3 FH-227, 5 DC-3, 1 CV-990
Venezolana Internacional de Aviacion S.A.	Venezuela	—	—	5 DC-8, 1 DC-9
VASP	Brazil	938,021	64,180	5 B737, 4 DC-6, 2 One-Eleven, 6 YS-11, 4 Viscount, 8 DC-3
Zambia Airways	Zambia	214,487	39,519	2 DC-8, 2 One-Eleven, 3 H.S.748

*Traffic statistics are for calender year 1971; tonne-kilometres, a measure of the work done by the airline, are given in thousands.
†Fleet as on December 31, 1971, including aircraft leased.

INTERNATIONAL AIR ROUTES

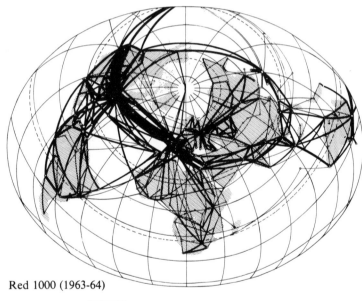
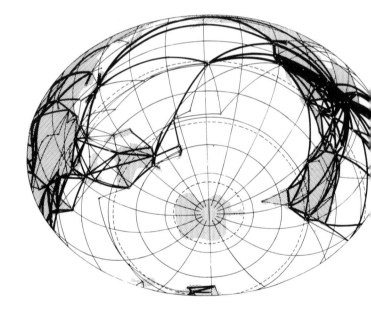

Annual passenger traffic   Red 1000 (1963-64)

Under 30	
30—100	
101—500	
501—1000	
1001—2000	
Over 2000	

International Air Transport Association, see IATA

International Civil Aviation Organization, see ICAO

International phonetic alphabet

The phonetic alphabet introduced in 1956 by ICAO and NATO for international radio call-signs utilizes the undermentioned words to represent the letters of the alphabet:

Alpha   Bravo
Charlie   Delta
Echo   Foxtrot
Golf   Hotel
India   Juliet
Kilo   Lima
Mike   November
Oscar   Papa
Quebec   Romeo
Sierra   Tango
Uniform   Victor
Whisky   X-ray
Yankee   Zulu

Interrupter gear, 877
see also Constantinesco,
see also Fokker

Iran Air

Operating title of Iran National Airlines Corporation, formed in 1962 by amalgamating the former Iranian Airways Co and Persian Air Services, which had been in existence since 1944 and 1955 respectively. Iran Air operates internal air services and internationally to India, Pakistan, the Lebanon and Turkey, to most Western European capi-tals and to points in the Persian Gulf area. Headquarters: Teheran. Member of IATA.

Iran: Air Force

Imperial Iranian Air Force: Originated as a branch of the Army, and received its first aircraft (a Junkers F.13) in 1922; the Air Force was created officially in 1924, and received its "Imperial" prefix in 1932. In its formative years its aircraft were chiefly of British, French or German design; in the nineteen-thirties most front-line types were Hawker biplanes and Hurricane mono-plane fighters.

During World War Two

Search, rescue and utility vehicle of the Iran AF is the Kaman H-43 Huskie

the country was occupied jointly by Britain and the USSR, serving as an ideal channel for the supply of Lend-Lease aircraft to the Soviet Union. In return, Iran was given preferential terms for buying war-surplus US aircraft after the war, and by 1950 the IIAF was again a substantial and well-equipped force. Its first jets were T-33A trainers received in 1956. It is still primarily a tactical and defensive force, equipped with F-86F Sabres and Northrop F-5 fighters, with F-4D Phantoms on order in 1969.

Iraq: Air Force
Iraqi Air Force. Under its former title of Mesopotamia, Iraq was an important air base for Allied air forces during the 1914-18 War; consequently, several excellent airfields already existed when, in 1931, the Royal Iraqi Air Force was created as a branch of the Army. It maintained its close association with Britain and the RAF throughout the nineteen-thirties, although some US and Italian combat aircraft were acquired towards the end of this period. A German-inspired uprising in 1941 obliged the RAF virtually to eliminate the RIAF, but in 1946 a post-war programme of re-equipment with British aircraft was initiated. The RIAF's first jets were Vampires received in 1953, but the only British jet aircraft now in front-line service are ground-attack Hunters. After the monarchy was overthrown in a military

coup d'état late in 1958, Iraq seceded from the Baghdad Pact and has since drawn her military aircraft almost exclusively from the USSR. In 1968 these included MiG-17, MiG-19 and MiG-21 fighters and Il-28 and Tu-16 bombers, although a quantity of Mirage III and Mirage 5 tactical fighters was sought from France.

Iraqi Airways
Formed in 1945 as a branch of the State-owned railway company, Iraqi Airways became independent of this ownership in 1960, at which time former assistance from BOAC also ceased. The first international services were started in 1946 with DC-3s, but otherwise the fleet has consisted entirely of British aircraft. Although the fleet is modest in size, the airline operates international services to India, Pakistan, the Middle East, Western Europe and the UK. Headquarters: Baghdad. Member of IATA.

Irish International Air lines, see Aer Lingus
Iroquois, see Bell 204
Island Airlines
Operating title of Sky Tours Inc of Port Clinton, Ohio, originally founded in 1930. It operates a "commuter" service from Port Clinton to the nearby Bass and Peele Islands, but is remarkable chiefly for being the sole operator today of the pre-war Ford "Tin Goose" (of which it has three) and one of the very few remaining operators of the Boeing 247D, which

also dates from the 'thirties.

Israel: Air Force
Heil Avir le Isràel (Israel Defence Force/Air Force). The sheer necessity of creating an air force has seldom been more forcibly demonstrated than in Israel, which was attacked by the surrounding Arab states on the very day of its creation on 14 May 1948. A hastily-collected assortment of obsolescent aircraft beat off its numerically superior attackers then, and has been repeatedly called upon to do so ever since. The war-surplus Bf 109Gs and Spitfires and the converted Auster lightplanes which formed its initial air force equipment were later supplemented by ex-wartime US bombers and transports, together with Mustang fighters and Mosquito bombers purchased from Sweden and France. In 1953 Gloster Meteors arrived to form the first jet fighter units but thereafter Israel relied heavily upon France for her modern combat aircraft. From her she has purchased Ouragan, Mystère, Super Mystère and Mirage III fighters; Vautour jet bombers;

Dassault Ouragans still serve in Israel alongside the more recent Mirage III

Magister jet trainers; and Noratlas transports. Of these, the Mystères and Ouragans featured prominently in the Suez crisis of 1956, and the IDF/AF Mirages were the chief instruments in decimating the Arab air forces during the "six day war" in June 1967.

Italy: Air Forces
Aeronautica Militare Italiano (Italian Mili-

One of Caproni's earliest types was the Ca 9, seen here in military service

tary Aviation). One of the first nations to employ aeroplanes in a military capacity (in 1911), Italy established a Military Aviation

First aircraft to be built by Fiat was the Farman 5B

overseas by propaganda flights and competitive and display flying; while the healthy domestic industry supplied virtually all of the wide range of

The Caproni Ca 36 saw action in the 1914-18 War

first- and second-line aircraft that it employed. Numerically, Italy possessed the second most powerful air force on the European continent when it entered World War Two in 1940, and except for its fighter force its aircraft were modern and efficient types. Generally, however, its aircrew were less efficient, and the

Service in 1912 which played a major part in the Allied air campaigns of the 1914-18 War. It was run down after the war, but expanded again as the *Regia Aeronautica* after Mussolini came to power in 1923. Inspired by Air Minister General Italo Balbo (*q.v.*), the new force did much to increase Italian prestige at home and

home industry could not keep pace with the heavy losses in the Mediterranean and North Africa. To counteract this, German and captured French aircraft were acquired, but their service with the *Regia Aeronautica* was short-lived, for Marshal Badoglio surrendered to the Allies in September 1943, leaving part of Italy's aircraft and aircrew in Allied-held territory and part in German hands.

After the war the reconstituted air force, now known as the *Aeronautica Militare Italiano*, had its strength limited by the surrender terms until 4 April 1949, when Italy was admitted to NATO. From then on the AMI was built up, at first with US and British types, and later by Italian-designed aircraft. Its chief combat types today are the

The A.1 Balilla of 1918 was one of Fiat's first fighters

The C.R.32 was one of a series of Fiat fighters designed by Rosatelli

After setting up many world records in "civil" guise the S.M.79 assumed its design role as a bomber and torpedo carrier

Fiat G91R and Lockheed F-104G tactical fighters.

Marinavia (Naval Aviation). Italian naval aviation was under the aegis of the *Regia Aeronautica* and AMI until the late nineteen-fifties, but since then a small separate naval air arm has existed, chiefly to operate an anti-submarine force of Grumman Trackers and Sea King helicopters.

Italy: Chronology

1595. Fausto Veranzio of Venice published an illustrated work, *Machinae Novae*, in which the principles of the parachute were first discussed scientifically.

1652. Padre Kirker conducted experiments with balloons in Rome.

1668. The *Torre Asinelli*, containing 50 different works concerned with physics and meteorology, was published in Bologna. In the *Prodromo* there was a description by Padre Francesco de Lana-Terzi of an aerial ship, consisting of a boat-like structure supported by four copper globes from which the air had been evacuated; sails and oars were to propel the craft. The principle of the lifting power of a lighter-than-air vessel was scientifically sound, but the evacuated copper spheres

would have collapsed from the external air pressure.

1803-1812. Comte Francesco Zambeccari made numerous balloon ascents at Bologna.

1828. Vittorio Sarti of Bologna experimented in public with a steam-powered flying machine based on the principles of the kite combined with a rotating wing.

1874. Carlo Bettoni's *L'Homme Volant* was published in Venice; it contained details of a plan for a dirigible of elongated, or cigar-shaped, design.

1877-1915. At Milan in 1877, the engineer Enrico Forlanini succeeded in making a machine, based on the principles of the helicopter, rise 39 ft from the ground for a period of 20 seconds. The craft was powered by a one-fifth-hp steam engine driving two contra-rotating paddle-like rotors. He later constructed flying scale models powered by solid-propellant rockets and, in 1908, flew a model seaplane on Lake Maggiore.

Following this, he turned his attention to the dirigible and successfully flew his first semi-rigid prototype—the F.1 *Leonardo da Vinci*—near Milan in 1909. His giant airship—the F.2 *Ville de Milan*

—designed to carry 20 passengers, made its maiden flight in 1913. Other airships followed, including a semi-rigid Forlanini F.3 built in 1915 for the British government.

1909-1914. The engineer Alessandro Anzani built many successful aero engines. During the year 1909 he produced six different models, ranging from 15 to 50 hp, and Blériot selected the 25-hp model for his first crossing of the English Channel. In 1910, Anzani built engines of 50 and 100 hp, and in 1913 one of 200 hp. In 1914 he produced his first two-row engine of 70 hp.

1912. Cagliani, flying an Antoni aeroplane powered by a 50-hp Gnome rotary engine, made the first flight over the Mediterranean from Pisa to Bastia.

1922. Brack Papa established a speed record of 209 mph in a Fiat R.700 aircraft.

1925. A Savoia-Marchetti S.16*ter*, piloted by de Pinedo, flew from Italy to the Far East (Australia and Japan) in 370 hours flying time, covering a distance of 34,175 miles.

November 1926. Mario de Bernardi won the Schneider Trophy contest in a Macchi M.39, powered by an 800-hp

Fiat AS.2 engine, at an average speed of 246 mph. He later used the same aircraft to capture the world speed record at 258.8 mph.

1926. General Umberto Nobile flew with Roald Amundsen and Lincoln Ellsworth across the North Pole, from King's Bay to Teller, in the airship *Norge*.

1927. De Pinedo and Del Prete crossed the South Atlantic in a Savoia-Marchetti S.55, followed by a tour of the Southern, Central and Northern territories of the American continent, covering a total distance of 26,720 miles.

1928. General Umberto Nobile made exploratory flights over the Polar regions in the airship *Italia*. The expedition ended in tragedy when, on 25 May, the airship crashed on an ice ridge in bad weather conditions. Many of the crew perished and Amundsen lost his life in a rescue attempt.

Mario de Bernardi, flying a Macchi M.52 powered by a 1,000-hp Fiat AS.3 engine, was the first pilot to travel faster than 500 kmh, establishing a new world speed record of 512.776 kmh (318.6 mph).

Led by Italo Balbo, 61 Savoia-Marchetti S.55s made the first mass formation flight across the western Mediterranean.

Ferrario and Del Prete established world records, for a closed-circuit flight of 4,763 miles and for a straight non-stop flight of 4,466 miles, from Italy to Brazil, in a Savoia-Marchetti S.64 powered

by a 590-hp Fiat A.22 engine.

1929. Caproni flew his giant Ca 90 aircraft, powered by six 1,000-hp engines and able to carry a useful load of 15 tons. Several world altitude records were gained within the category of aircraft carrying a 10-ton payload.

Italo Balbo led his second mass formation flight with 34 Savoia-Marchetti S.55s, flying over the eastern Mediterranean and the Black Sea. This was the last formation test flight before the start of the Atlantic cruises.

Piaggio tested his P.7 seaplane, built to compete in the Schneider Trophy contest. It was powered by a 970-hp Isotta-Fraschini engine, designed by Pegna in collaboration with Giuseppe Gabrielli, then

The Piaggio P.7 was designed to take off using hydrofoil and water screw and rudder

head of the Aviation Division of the Fiat company. The P.7 was characterized by its use of hydrofoils, lighter and offering less drag than floats. The hydrofoil principle has since been adopted for the construction of high-speed surface craft.

1930. The Piaggio-

by a 100-hp Fiat A.50 engine, established several world records, for distance altitude and endurance in the rotary-wing class.

1931. Balbo led the first mass transatlantic crossing, from Rome to Brazil, with 24 Savoia-Marchetti S.55s.

1933. Balbo led a two-way formation flight across the North Atlantic, from Italy to the USA, with 24 Savoia-Marchetti S.55s.

1934. Francesco Agello broke the world speed record at 440.6 mph flying a Macchi MC.72 powered by a Fiat AS. engine of 3,300 hp.

Renato Donati broke the world altitude record in a Caproni Ca 113 AQ, at a height of 44,465 ft.

1938. Mario Pezzi, in a Caproni Ca 161*bis*

d'Ascanio helicopter powered by a 100-hp

powered by a Piaggio P.XI RC 100 engine established a new world altitude record of 56,075 ft.

1940. Mario de Bernardi made the initial test flight of the Caproni-Campini N.1, first Italian aircraft—and second in the world—to be propelled by a

In 1925 de Pinedo flew this Savoia Marchetti S.16 34,175 miles to the Far East and back

The Macchi-Castoldi MC.72 set a world speed record of 440.6 mph in 1934

jet engine.
Experimental flights of the Piaggio PD.3 helicopter.

1941. Mario de Bernardi, with Pedace as co-pilot, made the first flight between Milan and Rome in the jet-engined Caproni-Campini.

1951. The Fiat company began production of the first Italian turbo-

first turbojet aircraft designed in Italy was the Fiat G80

jet aircraft, the G80, designed by Gabrielli. Fiat also produced Italy's first transonic aircraft, the G91.

vchenko
Alexander Ivchenko (who died in June 1968) was head of an aero engine design team in the USSR, with works at Zaporojie in the Ukraine. The first engine known to be of Ivchenko design was the AI-4G piston-engine of 55 hp, which powered the Ka-10 light helicopter. Since then a variety of piston- and turbine-driven engines have emanated from the bureau, chief among which are the 3,945-4,190-ehp AI-20 turboprop and the 3,300-lb st AI-25 turbofan.

58 engine, see Pratt & Whitney
79 engine, see General Electric
85 engine, see General Electric
ack, 639, 642-644, 931

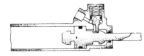

Double-acting hydraulic jack

A telescopic unit capable of transforming pneumatic or hydraulic work into mechanical work. Jacks are normally used in applications where movement is required intermittently, the main uses being for landing gear retraction and lowering, flap operation, movement of control surfaces and nose-wheel steering. Hydraulic jacks are used exclusively for control surface movement and steering.

Basically, a jack consists of a cylinder housing a piston, the application of pressure to connections either extending or retracting the assembly. Jacks are divided into two basic types,

Plunger-type jack lock

single-acting and double-acting, depending upon whether the units work in one or both directions. These two basic types are further sub-divided into jacks embodying special features. On single-acting jacks the return is effected either by an external force or by internal springs.

Some systems have to be locked mechanically, and this can be effected by internal locks embodied within the jack. The simplest type of lock consists of a spring-loaded plunger which engages with a corresponding hole in the piston rod. Unlocking is done by hydraulic pressure on the piston moving the plungers against their springs, or by withdrawing the plungers hydraulically.

For heavier loads, locks utilising collets or balls are available. In the

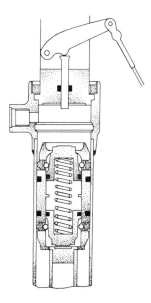

Jack ball lock, with intermediate segments and mechanical emergency release lever

latter a ring of steel balls, capable of slight radial movement, is contained within a cage attached to the piston. For locking, the ring of balls is expanded into a groove inside the cylinder near the end of the jack travel, the balls being trapped in the groove by a tapered spring-loaded "lock-bolt" Unlocking is effected hydraulically, the lock bolt being withdrawn so that the balls can retract radially, thereby freeing the piston.

Jaguar engine, see Armstrong Siddeley
Jaguar strike aircraft, see **SEPECAT**
Jan Smuts Airport, see Johannesburg
Japan: Air Forces
Japan Air and Ground Self Defence Forces. These are the successors to the former Imperial Japanese Army Air Force, which from 1911 grew out of the Provisional Balloon Corps formed in 1904. Japanese Army pilots and aircraft were involved operationally to a very limited extent in the 1914-18 War. Initial equipment was mostly of French or German design, including various Farman types, and some of these machines were built in Japan.

After the war the expansion of Army aviation was guided initially by a French aviation mission to the country in 1919, resulting in the supply, and subsequent licence manufacture, of French fighters, bombers, reconnaissance and training aircraft throughout the nineteen-twenties and early 'thirties. A Military Flying Corps was created on 1 May 1925 as a separate branch of the Army, and a few years later the first Japanese-designed combat aircraft began to enter service. The JAAF soon established air superiority over the Chinese Air Force when Japan invaded Manchuria in the Autumn of 1931. After this campaign the JAAF re-equipped its bomber and fighter squadrons with more modern aircraft, most of which, by now, were of Japanese design; and the rate of expansion increased still further after the "Shanghai Incident" early in 1932. The JAAF, however, did not emerge with conspicuous success from various clashes with Chinese forces during the later 'thirties, with the result that a further drastic re-equipment programme became necessary toward the end of the decade. This was marked by the entry into service shortly before Pearl Harbor of such excellent types as the Ki-43 fighter, Ki-21 bomber and Ki-46 reconnaissance aircraft.

Combat losses by the JAAF on the Asian mainland during World War Two were even more serious than those sustained by the Navy in the various island campaigns, and despite the introduction of several new types in the later war years the JAAF gradually lost air superiority in all theatres of the Pacific war. Finally, with the surviving Navy units, it was driven back upon

defence of its homeland against the steadily-increasing attacks by US bombers.

Home defence is still the primary object of the present-day Air Self Defence Force (*Koku Jiei Tai*), created officially on 1 July 1954 from the former National Safety Force. The JASDF has increased steadily in size, equipped chiefly with US combat types built under licence by the reconstituted aircraft industry. Its front-line squadrons today are equipped mainly with

The Fuji T1B trainer of the JASDF

F-86F Sabres and F-104 Starfighters. The JASDF is one of three air arms in present-day Japan. The Japan Ground Self Defence Force (*Kikujyo Jiei Tai*), formed at the same time, is primarily a reconnaissance/liaison/training force, most of whose equipment comprises light utility aircraft and helicopters, including the twin-turboprop Mitsubishi MU-2. The third air arm, the Maritime Self Defence Force, is described below.

Japan Maritime Self Defence Force. Successor to the former Imperial Japanese Navy Air Force, which, like that of the Army, had its origins in 1911. A Naval Air Corps was set up on 17 March 1916 with a mixture of Blériot, Farman, Short and Sopwith aircraft, some of them Japanese-built. Naval aviation after the war expanded at a slower rate than that of the Army, and was guided initially by an air mission from the UK. Gloster Sparrowhawk and Mars fighters and many ex-wartime British types were supplied from Britain, and Japanese fac-

tories built Avro 504 trainers and Short flying boats under licence. The JNAF's first carrier, the *Hosho*, was launched in November 1921, and by the time of the first take-off and landing trials in 1923 specialized shipborne combat aircraft of Japanese design (notably Mitsubishi types) had entered service. Other carriers joined the fleet in the late 'twenties, but in comparison with the Army the JNAF relied for rather longer upon foreign-designed aircraft before a generation of suitable indigenous types was evolved at the start of the 'thirties. The carriers *Hosho* and *Kaga* and their aircraft were involved in the "Shanghai Incident" of 1932, an incident that gave added impetus to the expansion of Japanese air power.

By the time of the JNAF attack on Pearl Harbor in December 1941 the JNAF was a large and powerful modern force, but losses of men and equipment during the early months of the Pacific war were never satisfactorily replaced. As the war progressed Japanese naval air power waned steadily. By the end of 1944 Japan had no carriers left in service, and her naval aircraft strength had been decimated by combat losses, destruction on the ground and the mounting scale on which machines and men were expended in suicide attacks.

A new naval air arm was created after the war, on 26 April 1952, in the form of the Japan Maritime Self Defence Force (*Kaijyoe Jiei Tai*). This is basically an anti-submarine and maritime patrol force, equipped at present with Neptune and Tracker aircraft, but shortly to receive the Japanese-designed PS-1 anti-submarine flying boat into service.

E8N1 was the Japanese designation of the Nakajima reconnaissance floatplane known by the Pacific code name "Dave"

Japan Air Lines

Successor to the privately-owned pre-war Japan Air Transport Co, the present JAL began operating in 1951. The Japanese Government has a majority holding in the company. Trans-Pacific services started in 1952, and in 1967 JAL began to operate across the Atlantic as well. It was among the first airlines to operate a Polar route, opening a Tokyo-Paris service via the North Pole in 1959 in collaboration with Air France, and is one of only four airlines to fly a complete round-the-world service. In 1968 services began to Europe and Moscow, and plans were in hand to open an Australian service in 1969.
Headquarters: Tokyo.
Member of IATA.

Japan: Chronology

27 May 1877. First flight of a balloon constructed jointly by the Army, Navy and Tokyo University. The envelope, of 14,000 cu ft capacity, was inflated with coal gas.

Ninomiya's model of 1891

29 April 1891. A rubber-powered model aeroplane designed by C. Ninomiya made a flight of 33 ft; this was the first flight of a model aircraft in Japan.
July 1904. The Provisional Balloon Corps of the Japanese Army used two Yamada-type manned balloons during the Russo-Japanese War.
8 September 1910. The 100-ft-long, 56,500-cu ft airship designed by I. Yamada flew for the first time. It was powered by an automobile engine of 14 hp.
19 December 1910. Two Army captains, Y. Tokugawa and K. Hino, flying respectively a Farman biplane and a Grade monoplane, made the first powered and sustained aircraft flights in Japan.
5 May 1911. Baron Sanji

Narahara 2, of 1911, one of the first aircraft built in Japan

Narahara made a flight of 197 ft in his Narahara No 2 biplane, powered by a 50-hp Gnome engine. This was the first powered flight in Japan of a nationally-designed and built heavier-than-air craft.
25 October 1911. First flight of the first nationally-built aeroplane for the Imperial Japanese Army. This was a Kai-1 biplane, powered by a 50-hp Gnome engine and flown by Captain Tokugawa.
26 June 1912. Naval Aeronautical Research Committee established, and the first Naval air station opened in October. Initial equipment comprised one Farman and one Curtiss hydroplane, which were flown for the first time by Lieutenants Kaneko and Kano on 2 November 1912.
November 1913. The *Wakamiya Maru* entered service as the first seaplane tender of the Imperial Japanese Navy. A converted cargo vessel, she carried four aircraft. On 1 September 1914 her aircraft were launched against German posts at Tsingtao, first Japanese naval operation of the war and the first time that Japanese combat aircraft had made operational sorties.
April 1919. A Japanese Army Aviation Depart-

First airship base in Japan was built at Tokorosawa; this was the hangar

ment was established, with a flying school at Tokorosawa.
22 June 1920. Lieutenant T. Kuwahara, flying a Sopwith Pup fighter, accomplished the first successful take-off from a Japanese ship under way. This flight was made from a special ramp built on the foredeck of the seaplane tender *Wakamiya*. Note that the suffix *Maru*—used to denote a civil vessel—had now been dropped from the tender's name.
26 June 1912. Naval
November 1921. The Imperial Japanese Navy's first aircraft carrier, the *Hosho*, was launched. With a displacement of 7,470 tons and a maximum speed of 25 knots, she carried 25 aircraft. First take-off and landing trials were carried out in the spring of 1923.
15 November 1922. The first Japanese scheduled airline service commenced, between Sakai City and Tokushima. This was operated by the Japan Air Transport and Research Institute, founded by Choichi Inouye, with Yokosho-type seaplanes and Itoh/Curtiss flying boats.
July 1923. Kawanishi Japan Air Lines Company founded, with an airport on reclaimed land at the estuary of the River Kitsu, Osaka.
23 July 1924. The Kawanishi K-6 floatplane *Harukaze*, piloted by Yukichi Goto, completed the first round-Japan flight. Sponsored by the *Mainichi Shimbun* newspaper, the 2,731-mile flight was completed in 8 days, 1 hr 29 min.

1927. The prototype Yokosho Type 1 single-seat reconnaissance seaplane was completed; it was the first Japanese aircraft designed for stowage on board submarines.
30 October 1928. Japan Air Transport Company founded. The Tozai Regular Air Transport Society and Kawanishi's Japan Air Lines Company were amalgamated with it to form the national airline.
February 1932. First air battle in Japanese history took place when Lieutenant N. Ikuta, flying a Type 3 carrier-borne biplane fighter, shot down a Boeing P-12 of Chiang Kai-Shek's air force.
1 November 1933. Japan Air Transport inaugurated Japan's first regular night air mail service between Tokyo and Osaka with a Nakajima P-1 single-seat biplane, powered by 420-hp Nakajima Jupiter 6 radial engines.

The record-breaking Mitsubishi "Kamikaze" (Divine Wind) of 1937

6-10 April 1937. The Mitsubishi two-seat monoplane *Kamikaze* (Divine Wind) of the *Asahi Shimbun* newspaper, piloted by M. Iinuma, set Japan's first international record by flying from Tokyo to London in just over 94 hours. Actual flight time was 51 hr 19 min 23 sec at an average speed of 101.2 mph.
26 August-20 October 1939. Mitsubishi Type 96 twin-engined monoplane *Nippon* made a round-the-world flight in a flight time of 194 hours. The flight was sponsored jointly by the *Osaka Mainichi* and *Tokyo Nichinichi* newspapers. Outbreak of World War Two during the flight prevented the aircraft from visiting London, Paris and Berlin.
7 December 1941. The Pacific war commenced when Japanese carrier-borne aircraft attacked the American base at Pearl Harbor, Hawaii. Taking off from the carriers *Akagi, Hiryu, Kaga, Soryu, Shokaku* and *Zuikaku*, 183 aircraft attacked in the first wave and 171 in the second, causing grave losses.
18 April 1942. First air attack on the Japanese mainland when 16 B-25 bombers led by Lieutenant-Colonel James H. Doolittle, USAAF, made a surprise raid on Tokyo, Nagoya and Kobe.
5 June 1942. Battle of Midway, turning point of the Pacific war, when Japanese naval forces lost four aircraft carriers, one cruiser and a total of 261 aircraft.

July 1945. Japan's first rocket-powered fighter aircraft, the Mitsubishi *Shusui* (J8M1), based on the German Me 163, flew for the first time.
6 August 1945. The world's first nuclear attack was carried out when a Boeing B-29 Superfortress dropped an atomic bomb on Hiroshima. A second A-bomb fell on Nagasaki on 9 August, bringing the Pacific war to a swift conclusion on 15 August.
7 August 1945. Japan's first jet-powered aircraft, the Nakajima *Kikka* naval interceptor, flew for the first time.
31 January 1951. Occupation forces gave permission for the resumption of Japanese domestic air transport.
August 1951. Japan Air Lines Company (JAL) re-established with aircraft and crews chartered from Northwest Airlines—since Japan was prohibited from owning or flying her own aircraft—to inau-

gurate services on 25 October 1951.

July 1952. Haneda Airport returned to Japanese control and re-opened as Tokyo International Airport.

28 September 1952. The Tachikawa R-52 trainer, Japan's first nationally-built post-war aircraft, was completed.

2 February 1954. JAL inaugurated Tokyo-Honolulu-San Francisco international service—first trans-Pacific airline service—with Douglas DC-6B aircraft.

14 April 1955. Tokyo University Institute of Industrial Science test-fired its first small Pencil rocket.

22 September 1960. Kappa Model 8 Type 3 rocket, developed by Tokyo University, attained an altitude of 124.3 miles.

JAT (Jugoslovenski Aero-transport)
JAT is the State-owned domestic and international airline of Yugoslavia; it was founded in 1946 and began operations in the following year to Czechoslovakia, Hungary and Rumania as well as internally. It flies aircraft of US or French design and its foreign route network is directed mainly to points in the Middle East and Western Europe, although services to Moscow and several Eastern capitals are maintained.
Headquarters:Belgrade. Member of IATA.

Jatho, Karl, 181, 185

Jerdrassik, 1283

Jet Commander, see North American Rockwell

Jet deflection
Procedure for diverting the exhaust gases of a turbine engine downward in order to give an element of upward thrust and so shorten the take-off run of an aeroplane.

Jet flap
Evolved at the National Gas Turbine Establishment in the UK, the initial concept of the jet flap was that exhaust gases from a turbojet should be ejected from a

slot running along the entire trailing-edge of the wing instead of through the usual tailpipe nozzles. This not only increases lift on its own account but, forming an effective wall between the flows above and below the wing, creates an enormous "suction bubble" over the upper surface, giving tremendous additional lift. As the aircraft gathers speed, the jet flap is swept backward, giving more and more thrust directly forward, in addition to improved lift.
From this original proposal evolved the promising "lift cylinder", consisting of a true cylinder, placed transversely in an airstream, embodying tangential slots through which air is blown at high velocity. The ejected air prevents the boundary layer (q.v.) from separating, and lift coefficients three times the best obtained with conventional aerofoils are theoretically possible.

Jet-reaction controls, 646, 647
Certain high-performance aircraft normally fly in conditions where their conventional control surfaces are partially or totally ineffective. Examples are the research aircraft which fly so high that the rarefied air is insufficient to evoke a response from the controls. The growing company of VTOL aircraft need additional control forces at the low-speed end of their speed range. For such cases a system of jet-reaction controls has been evolved. The principle is that of bleeding high-pressure air from the compressor

of the turbojet engine in the aircraft. This air is fed to jet outlets at the extremities of the aircraft: nose, tail and wing-tips. Here it is released through jet orifices to provide a reaction force moving the aircraft in the desired direction. The output of each jet nozzle is governed by a control system geared in with the normal flying controls of the aircraft—thus, in effect, taking over when ordinary aerodynamic controls are ineffective.

Jet-stream, 1535
Popular name given to a meteorological phenomenon encountered chiefly off the eastern coasts of Asia and North America and over North Africa. It is a circum-polar airstream, usually travelling in a west-to-east direction, and is identified by a narrow stream of high-velocity wind, particularly near the base of the stratosphere. Its location is not easy, as it is impossible to see or locate by instruments in an aircraft. Use has frequently been made of the phenomenon, acting as a very strong tail-wind, to reduce the scheduled west-to-east flying time of trans-oceanic airliners.
"Jetstream" was applied as a class name to the Lockheed Starliners of TWA in the late nineteen-fifties, and is currently used as a type name for the Handley Page H.P. 137 medium-range transport aircraft.

Jetstream, see Handley Page

JF-1, see Grumman

Jockey landing gear, see Messier

Johannesburg Airport (Jan Smuts), Transvaal, South Africa
One of the world's highest airports, Jan Smuts also has one of the longest runways in the world. It was opened in October 1953.
Distance from city centre: 14 miles
Height above sea level: 5,559 ft

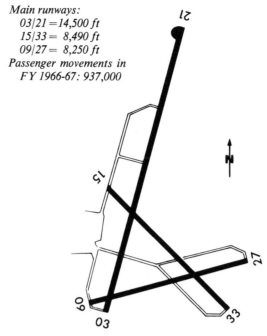

John F. Kennedy Airport, 1551, 1583, see also New York

JT3D engine, see Pratt & Whitney

JT9D engine, see Pratt & Whitney

JT11 engine, see Pratt & Whitney

JTF17 engine, see Pratt & Whitney

Jullien, Pierre, 133, 140
To this French designer goes the credit for leading airship design into a more practical stage. He designed and built two small models in 1850-51 with a sleek streamlined shape, filled with gas and with two airscrews for propulsion. These were followed by a full-sized machine, *Le Precurseur*, in 1852; but there is no record of this machine flying.

"Jumbo-jets", 745, 747
A descriptive phrase which has come into general use to describe the new generation of 350-seat-plus airliners that will enter service on air routes in the nineteen-seventies, the first of which is the Boeing 747.

Jumo engine, see Junkers

Junkers, 367, 369
Professor Hugo Junkers initiated one of the foremost German aircraft manufacturing companies, but because of his advanced ideas Junkers' designs were not put into large-scale production during the

1914-18 War. However, when eventually he developed corrugated dur-alumin and built his low-wing monoplane airliners in the early 'twenties the Junkers name began to achieve popularity and his insistence on cantilever monoplanes was beginning to be justified. The company went on to great success with a series of larger and more efficient airliners of which the three-engined Ju 52/3m achieved great prominence, both as a peacetime airliner and as a wartime troop transport. In the 'thirties Junkers forsook corrugated dural for more normal stressed-skin construction and developed the Ju 87 dive-bomber and the Ju 88 series of bombers, all of which played important parts in World War Two.

Junkers F.13, 369

Junkers J.1, J.2, J.4 etc, 367, 369

Junkers Ju 87, 909
Interest by the *Luftwaffe* in the dive-bomber rôle hardened into fact in 1935 with the creation of Henschel Hs 123 dive-bomber units; to replace these came the Ju 87, born out of the K.47 monoplane two-seat fighter which never achieved production.
The first Ju 87 flew in 1935; the aircraft en-

tered production in a refined form, joining its first unit in 1937. Three aircraft were tried out with the Condor Legion in Spain, where they were found successful. By the time World War Two broke out the Ju 87B was in service, and it was this version that blasted a way for the victorious *Wehrmacht* through Poland, the Low Countries and France, achieving a reputation as a terror weapon.
This reputation suffered a serious blow in the assault on Britain, when the Ju 87 was found to be almost useless unless the *Luftwaffe* had fighter mastery of the battle area in which it was operating. Thereafter the *Luftwaffe* endeavoured to use it either in such areas or for other rôles.

Junkers Ju 88, 397, 569, 904, 911
The Junkers Ju 88 was without doubt Germany's finest bomber of World War Two. The prototype flew for the first time on 21 December 1936, but development and initial production was protracted and the type was only just in service when World War Two began. By the time of the Battle of Britain, nearly a year later, it was in large-scale service and played a large part in the offensive. As a result of experience then gained, modifications were put in hand to improve some of the aircraft's shortcomings, and a whole "family" of Ju 88 sub-types grew up for different rôles, including torpedo-bombing, night-fighting, reconnaissance and training. The Ju 88 served wherever the *Luftwaffe* was operating, even acting as a flying bomb during the Allied landings in 1944.

Junkers Jumo engines, 909, 911, 1097, 1114-1116
The name Jumo was originally given to a series of diesel engines with which Junkers had persevered since 1929,

but which were never developed to the point of success due to the increased weight which the diesel engine necessarily brought with it. The name was subsequently applied to one of the principal motors used by the *Luftwaffe* to power its military aircraft in World War Two, the Jumo 211 being the usual variant. This engine was a supercharged twelve-cylinder inverted-Vee liquid-cooled engine. It produced 1,160 hp and powered such types as the He 111, Ju 88 and Ju 87. Junkers' first turbojet engines also bore the name Jumo.

Junkers Jumo 004 engine, 1129, 1171
Jupiter engine, see Bristol
Kai Tak Airport, see Hong Kong
Kaman H-43 Huskie, 987-988
An interesting feature of the Huskie is the configuration of the rotors, which are intermeshing and set at an angle one to the other. This aircraft first appeared in production in 1953 as a liaison helicopter for the USAF with a Pratt & Whitney piston-engine. It was followed by a radically-redesigned version with Lycoming T53 shaft-turbine, and this has become the standard USAF crash rescue and general liaison helicopter.
Kamov Ka-20 and Ka-25 ("Hormone"), 896
For anti-submarine or anti-surface vessel use the Russian Navy has in service the Kamov Ka-25 twin-turbine helicopter. The prototype was designated Ka-20 ("Harp").
Kamov Ka-22 *Vintokryl*, 1011
This large twin-engined convertiplane appeared in public for the first and only time in 1961. It has a large fuselage capable of accommodating 80 or more passengers. At the tips of the high wing are mounted two Soloviev D-25V shaft-turbine engines driving propellers

for forward thrust and at the same time believed to be linked to two four-bladed rotors. The *Vintokryl* has established helicopter records in the speed class (221.4 mph) and in the payload-to-height class (33,069 lb to 9,491 ft).
"Kangaroo" missile, 896
A large air-to-surface missile carried beneath the Russian Tu-20 ("Bear") bombers, the "Kangaroo" is about 50 ft long and resembles a swept-wing jet aircraft. It is powered by a turbojet engine and appears to have radar guidance.
Kansas City Municipal Airport, Missouri, USA

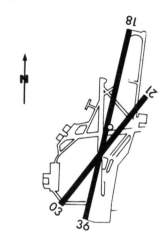

Distance from city centre: 2½ miles
Height above sea level: 758 ft
Main runways:
18/36 = 7,000 ft
03/21 = 5,052 ft
Passenger movements in 1967: 3,398,000
Karachi Airport, Mauripur, Pakistan

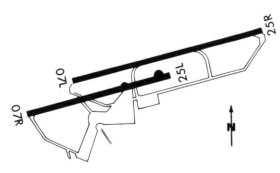

Distance from Mauripur city centre: 8 miles
Height above sea level: 81 ft
Main runways:
07L/25R = 10,000 ft

07R/25L = 7,500 ft
International passenger movements in 1966: 548,000
Kastrup Airport, see Copenhagen
Kawanishi
The Kawanishi Aircraft Co Ltd of Kobe, Japan, was established under that title in 1928, taking over the assets of a company that had produced various assorted biplanes since the early 'twenties. Early in its existence, Kawanishi acquired a licence to manufacture Short flying boats, the first of which was built for the Japanese Navy as the Type 90. The company subsequently specialized in the design and construction of flying boats, the most famous being the H6K ("Mavis") and H8K ("Emily") used during World War Two. The latter was, hydrodynamically, perhaps the best flying boat of its day, and was regarded as superior to the Short Sunderland (of which, at one time, it was thought to be a copy). Kawanishi also produced the N1K2 ("George"), rated the

best Navy fighter of the war after the Mitsubishi Zero.
Since 1949 Kawanishi has continued under the title Shin Meiwa Industry Co Ltd. It maintains its marine traditions and is currently developing the PS-1 anti-submarine flying boat for service with the Japan Maritime Self Defence Force.

Kawasaki
The Kawasaki Aircraft Engineering Co Ltd is generally associated with the production of aircraft for the Japanese Army up to the end of World War Two. However, as befitted a division of the Kawasaki Dockyard Co, it was also associated with marine aircraft and between 1923-33 engaged as chief designer Dr Richard Vogt, later of the German Blohm und Voss company. In 1924 Kawasaki began licence manufacture of the Dornier Wal flying boat and its BMW engines, and subsequently built several Dornier landplane designs as well. Kawasaki's first production type of its own design was the bomber/reconnaissance Type 88 of 1928; this was followed by several other Army warplanes during the 'thirties, including the Type 93 bomber and the Ki-3 and Ki-10 fighters. During World War Two Kawasaki was responsible particularly for the single-engined Ki-61 ("Tony") and twin-engined Ki-45 ("Nick") Army fighters, and the Ki-48 ("Lily") light bomber.
Since the war Kawasaki has continued to function under its own title. It has built the Lockheed T-33A trainer and P-2 Neptune patrol aircraft under licence and has developed an advanced jet-and-turboprop-engined version of the latter, the P-2J, for Japanese Navy service. It also manufactures in quantity the Bell 47 and Hughes 500 helicopters,

and has exclusive rights to build and sell the Boeing-Vertol 107-II to civil and non-US military orders.
Kazakov, Staff-Captain Alexander A., 857
Kennedy (John F.) Airport, see New York
Kerosene, 1159
Kestrel engine, see Rolls-Royce
Kindelberger, James ("Dutch"), 387
Kinetic heating, 445, 481, 483, 488, 501, 503-4
Basic thermodynamics rules that a molecule of a gas can trade kinetic energy (energy due to its speed) for heat energy. A high-speed aircraft flying through the atmosphere is identical, as far as the air flowing past the airframe is concerned, to a high-speed airflow rushing past a stationary aircraft. It can thus be seen that if a molecule of nitrogen, or any other atmospheric gas (or a fly, a bird or any solid matter) is slowed down by the presence of the aircraft then its temperature must be increased. The ultimate exchange of kinetic for thermal energy is seen in those molecules that actually strike the very tip of the nose or the exact foremost point of the leading-edge of a wing or fin, because such molecules are brought to rest on the metal surface and lose all their kinetic energy and thus gain a great deal of heat. In the case of subsonic aircraft the rise in temperature is not really significant, but supersonic machines encounter kinetic heating that may raise the temperature of parts of the airframe enough for the metal to be significantly weakened. The Concorde, at Mach 2.0-2.2, experiences kinetic heating sufficient to raise the temperature of the tip of the nose to about 150°C and of the leading-edges to about 130°. Faster aircraft are subject to far more intense heating, and at Mach 4

no ordinary metal structure could be used except for flights of very short duration or above the atmosphere where kinetic heating is not encountered. As the interior of most aircraft is kept cool by the presence of a large volume of cold fuel, and by the cabin air-conditioning, the heated boundary layer of air that causes kinetic heating leads to structural stresses in supersonic aircraft resulting from grossly unequal temperatures and consequent differential expansion.
Kingdom of Libya Airlines
New national airline of Libya, formed in September 1964 and entirely State-owned. First services were flown in October 1965, and international routes are now operated to Beirut, Cairo, Malta, Tunis and several European capitals.
Headquarters: Benghazi.
Member of IATA.
Kingsford Smith Airport, see Sydney
"Kipper" missile, 896
Designed for anti-shipping purposes, the "Kipper" is about 31 ft long and has a swept-wing configuration with an underslung turbojet power plant. It is normally carried by the Tu-16 ("Badger") bomber which, when so equipped, has a nose radar installation.
Kite, 145, see also Hargrave, Lawrence
Kitten (K-3), see Martin
Kitty Hawk, 609, see also Wright brothers
Kitty Hawk, in North Carolina, USA, was a little-known place until 17 December 1903 when, from its sand-dunes, the Wright brothers' biplane rose into the air to make the first powered, sustained and controlled aeroplane flight in history. A memorial has been erected there to this historic event.
Klimov
The late Vladimir Y. Klimov led the design team responsible for de-

veloping, in the USSR, the Rolls-Royce Nene turbojet engine into the 5,955-lb st VK-1 that powered the first mass-produced Soviet jet fighter, the MiG-15. Later designs have included the 6,500 lb st VK-5, a small-diameter turbojet which powers the MiG-19 and Yak-25 fighters.

KLM (Royal Dutch Airlines), 745
Fifty years of air transport is the claim of the Dutch international airline, KLM. Through many of these years it was under the guiding hand of Dr Albert Plesman, who was its first Administrator, appointed in October 1919. KLM began operations in 1920 with converted D.H.9 bombers and D.H.16s (small single-engined biplanes, one stage removed from the D.H. bombers, with accommodation for four passengers). Routes were to London, Amsterdam, Hamburg and Copenhagen.
At the end of the year KLM received its first Fokker, an F.II and this began a long association between the two firms. As early as 1924 KLM was experimenting with flights to the Dutch East Indies and by September 1930 a

regular weekly service was inaugurated. By then the airline was busily involved in European routes with an ever-increasing variety of Fokker high-wing monoplanes. In 1933 it began services to the Dutch West Indies.
KLM was among the first to see the advantages of the new generation of low-wing monoplanes in the 'thirties, and quickly ordered Douglas DC-2s; the first of these competed in the London-Australia Air Race of 1934, achieving second place. At this point KLM forsook Fokkers for American aircraft, introducing DC-3s in 1936 on its European and Far East routes and Lockheed 14s on local Western Hemisphere routes. With the fall of Holland in 1940 the greater part of KLM's equipment was lost, al-

KLM Lockheed 14, one of the fastest transports of 1937

KLM Fokker F.IX cockpit, typical flight deck of the inter-war years

though a couple of DC-3s escaped to England and were later used to maintain a wartime service to Lisbon. The East Indies service operated from Palestine until the Japanese invasion; only the West Indies division continued to flourish.
After the war came the task of rebuilding. In occupied Holland Dr Plesman had been busy planning, as best he could, the re-birth of KLM. When liberation came he put his plans into effect, and by the end of 1946 KLM had regained its pre-war route network and doubled its 1939 staff. In May 1946, KLM began transatlantic flights with Constellations, and the next decade saw a build-up of intercontinental routes. KLM has continued to select its aircraft from Douglas and Lockheed designs; the only exception was the Vickers Viscount in the late 'fifties, when no comparable turboprop airliner was available. KLM introduced jet aircraft (DC-8s) in 1960. Plesman, the presiding genius, died in 1953; his place was taken by I. A. Aler, one of KLM's early pilots.
Headquarters: The Hague.
Member of IATA.
Kloten Airport, see Zürich
Kollsman, Paul, 685
Krebs, Captain A. C., 133, 141, see also Renard
Kuwait Airways
Formed in 1954, with assistance from BOAC, as Kuwait National Airways, and retitled Kuwait Airways Corporation in 1958. It has been wholly State-owned since 1963 and today flies services within the

Middle East, to India and Pakistan, and to Geneva, Frankfurt, Paris and London.
Headquarters: Kuwait.
Member of IATA.
Kuznetsov, 843
Specializing in the production of very large and powerful turbine engines, the design bureau headed by N. D. Kuznetsov is currently responsible for what is probably the world's most powerful turbo-prop, the NK-12M. In its original form this engine developed 12,000 ehp, a figure increased in later versions to 14,795 ehp; variants of the NK-12 power the Tu-20 bomber and the An-22 and Tu-114 transports. Current products also include the NK-144 two-spool turbofan of 28,660 lb st, used in the Tu-144 supersonic airliner; this is a development of the slightly less powerful NK-8 installed in the Il-62 and Tu-154 airliners.
Kuznetsov NK-12M engine, 1309
Claimed to be the most powerful turboprop engine in the world, the NK-12M axial-flow compressor engine was developed in the early nineteen-fifties to power the Tupolev Tu-20 "Bear" long-range heavy bomber. Driving four-bladed contra-rotating airscrews, it develops 12,000 ehp at sea level. It also powers the Tu-114 airliner developed from the Tu-20.
L-1649 Starliner, see Lockheed Super Constellation
LABS (Low Altitude Bombing System)

Method of weapon delivery in which the target is approached at low level, the aircraft then pulling upward into a half-loop to release its bomb, which continues toward the target with or without radar guidance while the aircraft rolls out of the loop to fly back in the opposite direction before the weapon explodes.

Ladeco
Operating name of Linea Aera del Cobre Ldta, a privately-owned Chilean airline formed in 1958. Services are internal, with the exception of a route to Salta in the Argentine, and an air taxi service is also operated.
Headquarters: Santiago.
Member of IATA.
La France, 133, 141, 209, see also Renard, Charles
La Guardia Airport, see New York
Lana-Terzi, Francesco de, 119, 125
Lancaster, see Avro
Lancastrian, 1101, see also Avro Lancaster
LAN-Chile (Linea Aerea Nacional de Chile)
Government-sponsored and -owned airline formed in 1929, which has operated under the present title since 1932. It has an extensive network of routes inside Chile and also flies international services to points in the Argentine, Ecuador, Peru and Uruguay, and to Miami and New York in the US.
Headquarters: Santiago.
Member of IATA.
Langley, Samuel Pierpont, 181, 183, 611, 1040, 1043
Langley Aerodrome, 181, 183, 611, 1044, 1047
Latécoère 28, 257

Single-engined civil monoplane built by the French Latécoère company. It was used by the *Cie Générale Aéropostale*, in connection with whom Mermoz (*q.v.*) made his transatlantic mail flight. He used the 28-3, a seaplane version.
Launoy, 145, 161
Lebanese International Airways
Originally a charter airline, LIA was formed in 1950 and began operating scheduled passenger services in 1956. These now extend to Western Europe, India and other Middle Eastern countries. In 1967-68 the airline operated in pool with MEA (*q.v.*), with which it is expected eventually to merge.
Headquarters: Beirut.
Member of IATA.
Lebaudy, Paul and Pierre, 150
These two brothers, sugar refiners by profession, commissioned the building of a semi-rigid airship in France in 1902 under the design of Henri Julliot. It accomplished the first lengthy flight that November, a distance of 38½ miles. This was the first of a whole series of successful airships built over the ensuing nine years, one of the few such series of this period.
Le Bourget Airport, see Paris
Le Bris, Jean-Marie, 169, 173
Leduc, René, 1317
René Leduc was a proponent of the ramjet for many years, and produced his first small unit as early as 1935; however, it only pro-

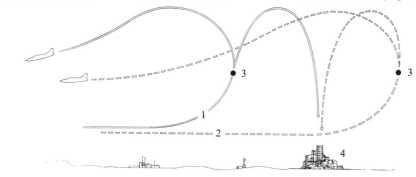

LABS technique:

1 High toss
2 "Over-the-shoulder" toss

3 Bomb release point
4 Target

duced a thrust of 8.8 lb at 985 ft/sec. It was not until after World War Two that he was able to build a full-scale ramjet aircraft. This, the 010, was first flown in unpowered form, being released from the back of an S.E.161 Languedoc in 1947. In April 1949, then fitted with a ramjet engine, it was again released from a Languedoc and flew for 12 min, reaching 450 mph on what was only half-power. The aircraft was 33 ft 8 in long and had a span of 34 ft 6 in. The fuselage was approximately the shape of an open barrel, in effect forming the powerplant. Two more prototypes were completed, the third having Turbomeca Pimene turbojets at the wingtips as auxiliary power plants. A further aircraft, the 021, was built and flown; but after that Leduc's experiments were abandoned.

Le Grand, 229, 234, see also Sikorsky, Igor

Letov

The *Vojenska Tovarna na Letadla* (Letov) company established in the Letnany suburb of Prague, Czechoslovakia, had its origins in the Military Air Arsenal in 1918. Its aircraft designations had "S" prefixes, signifying that they were the work of Letov's chief designer Alois Smolík. First Letov design was the S1 bomber and reconnaissance biplane, which was the first military aircraft type of Czechoslovak design. Best-known of the company's later products were the S16 reconnaissance bomber, built also for Latvia and Turkey; the S20 biplane fighter for the Czechoslovak and Lithuanian air forces; and the S328/528 reconnaissance-bombers of the nineteen-thirties. Such Letov types as remained in service upon the Nazi invasion in 1939 were mostly obsolete, but continued to be used

for a time by Germany and Bulgaria during World War Two.

Lewis, Colonel Isaac Newton, 863

Lewis machine-gun, 863, 871, 873, 877

Marketed in 1912 by Col Isaac Newton Lewis, the Lewis machine-gun was the first practical machine-gun for use in aircraft. It entered large-scale use with the Allied forces and continued in service until the end of World War Two.

Liberator, see Consolidated B-24

Liberty engine, 1050, 1053, 1054, 1067

Libya, see Kingdom of Libya

Lift, see Flight, principles of

Lift dumper, 615

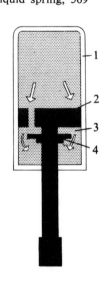

1 Lift dumper   2 Flap   3 Wing

Lift dumpers on the Hawker Siddeley Trident. These are located directly in front of the inboard flaps, and are automatically operated immediately after touchdown. The dumpers supplement the air-brakes, which also extend automatically to "spoil" the airflow over the outer sections of the wing.

A device extended to "spoil" the airflow over the wing and hence reduce or "dump" lift. Lift dumpers are used to reduce the total wing lift as soon as possible after touchdown, to increase the vertical load on the wheels, thus permitting a greater braking effort to be applied and hence shorten the landing run.

Lift jet, 1233ff

A jet engine, usually a turbojet but for some future applications possibly a turbofan, specially designed to give an upward thrust to lift a V/STOL aircraft at all

times when it must fly too slowly for wing lift to be adequate. Normally the jet lift regime occurs only at take-off and landing and lasts for only a minute or two on each flight. The lift jet may therefore be made extremely highly rated, with a relatively high turbine gas temperature and with a very compact high-intensity combustion chamber; and, since specific fuel consumption is of minor importance, the pressure ratio need not be very high and the compressor may be extremely short. Moreover, as the lift jet is required to operate only at low speeds and altitudes, it needs no elaborate control system and can be stripped of almost all its accessories. Lubrication may be by a "one shot" charge of oil and all lift jets in a battery of 8 or 12 may be started by a bleed air pipe. Thus the thrust per unit weight and per unit volume can be several times higher than the best achieved by conventional engines and it becomes practicable to instal a dozen or more lift jets, capable of lifting a 60,000-lb aircraft, without occupying such bulk or imposing such a weight penalty as to render the aircraft seriously handicapped in comparison with non-VTOL counterparts. Use of separate lift jets enables the cruise propulsion engine(s) to be tailored exactly to the task of propelling the aircraft in cruising flight and, despite the extra complication of such an arrangement, can sometimes prove more efficient than use of vectored-thrust engines.

Lightning, see BAC

Lilienthal, Otto, 169, 174, 611

Lima International Airport (Callao), Peru

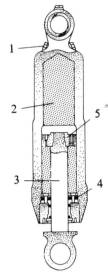

Height above sea level: 105 ft
Main runway:
15/33 = 11,506 ft

Limousine, see Westland

Lindbergh, Charles A., 253, 255, 689, 1073

Ling-Temco-Vought, see LTV-Hiller-Ryan

Linjeflyg

Linjeflyg came into being as a passenger and newspaper carrier in 1957, and also carries out ambulance and charter flying. It is owned jointly by SAS (*q.v.*) and AB Aerotransport. Headquarters: Stockholm.

Link trainer, see Flight simulators

Liquid-cooled engine, 1055, 1065, 1073

A piston-engine in which the waste (excess) heat from the cylinders is removed by liquid flowing through special galleries and passages surrounding the cylinder head and walls. As in most cars, the liquid is contained in a closed system, usually operated at a pressure higher than atmospheric and filled either with water or with a mixture of water and a fluid based upon ethylene glycol to reduce the freezing point of the coolant to below the minimum expected to be encountered. The hot liquid from the

engine is cooled in a radiator located either adjacent to the engine or else in some other part of the aircraft. For example, in the North American P-51 Mustang fighter it was positioned in an aerodynamically advantageous place under the rear fuselage, but this incurred the penalty of long pipe runs which could be vulnerable to battle damage. In the British Spitfire and German Bf 109 the radiators were again mounted well away from the engine, under the wings. The practice of cooling a piston-engine by a liquid was generally restricted to high-power in-line or Vee-type units in which the extra weight, complication and vulnerability were compensated by improved power, reduced installed drag and possibly greater efficiency.

Liquid spring, 569

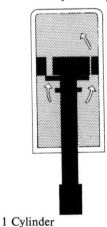

Simple principle of the liquid spring

Device used in landing gear shock-absorbers utilizing the compression of a liquid. The conception of such a spring was envisaged ever since research work in the early nineteenth century showed that liquids, normally considered incompressible, do in fact compress significantly at pressures over 10,000 lb/sq in. A patent for liquid spring shock-absorber railway buffers

Liquid spring shock absorber unit

1 Filler plugs
2 Fluid
3 Piston rod
4 High-pressure gland
5 Piston head

was taken out in 1881, and a proposal for liquid springing for motor vehicles was made in 1917. Liquid springs, however, were not practicable until the development of a

1 Cylinder
2 Piston
3 Fluid
4 High-pressure gland

special sealing gland by Dowty in England, able to withstand the very high pressures involved— over 80,000 lb/sq in. The original Dowty high pressure gland assembly had three main parts, a steel pressure plate, a resilient gland ring and a backing plate. Dowels on the pressure plate, which butted against the cylinder step, passed through the gland ring into the backing plate. A gland retaining nut supported the backing plate and a foraminated disc between the gland and pressure plate prevented a pressure lock. The outstanding feature of the design was that the hydrostatic pressure in the gland was always above the prevailing hydraulic pressure, as the cross-sectional area of the gland was less than the pressure plate, due to the dowel pin holes. A liquid spring unit consists basically of an oil-filled cylinder containing a piston operated by a rod passing through the cylinder.

Under compression, oil is compressed and forced through a valve in the piston head, thus being transferred from one side of the piston to the other. Since the cylinder is sealed and its volumetric capacity is reduced by the increased volume occupied by the entering piston rod, the fluid is compressed and energy absorption obtained.

Modern liquid spring units use a synthetic silicon-based fluid with large molecules, facilitating compression. Liquid spring units are compact, and ideal for use in lever-action legs giving a large wheel travel relative to the shock absorber deflection.

Lisbon Airport (Portela), Portugal

Distance from city centre: 6½ miles
Height above sea level: 373 ft
Main runways:
03/21 = 12,483 ft
18/36 = 7,874 ft
Passenger movements in 1967: 1,439,000

Lockheed
The original Lockheed Aircraft Co was founded in 1916 by the

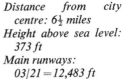

The Model G, built and flown by the Loughead brothers in 1912

brothers Allen and Malcolm Loughead with an initial design which had many advanced streamlined features. In 1926 the company was transferred to Burbank, California, where in the

Varney scheduled Lockheed Orions to link Los Angeles and San Francisco in 65 minutes

late 'twenties and early 'thirties the celebrated Vega, Orion, Altair and Sirius monoplanes were produced. During the nineteen-thirties Lockheed also produced a series of twin-engined all-metal airliners, the Models 10, 12 and 14. From these were developed the Hudson, Ventura and Harpoon of World War Two. Lockheed's other major wartime types were the excellent P-38 Lightning twin-boom fighter and the four-engined Constellation transport.

The Constellation, like the Douglas DC-4 (q.v.), became a standard post-

war airliner and was "stretched" into Super Constellation and Starliner versions with even greater capacity and range. Lockheed also built the first US turbo-prop airliner, the Electra, and its latest commercial venture is the L-1011 TriStar.

Lockheed has been a prolific producer of military aircraft since World War Two as well, beginning with the P-80 Shooting Star, America's first mass-production jet fighter. The two-seat T-33A jet trainer developed from the P-80

Final version of the Lockheed Lightning: the two-seat P-38N night fighter

has served with almost every major air force in the Western world and many remain in service

today. Other noteworthy military aircraft in the past two decades have included the U-2, military variants of the Super Constellation, the C-130 Hercules, C-141 StarLifter and C-5A Galaxy transports, and the F-104 Starfighter supersonic interceptor and strike aircraft. At the end of 1968 Lockheed had the distinction of having produced the largest aircraft in the world (the Galaxy), and also the fastest—the YF-12A, which on 1 May 1965 raised the world absolute speed record to 2,070.102 mph. The company is also in the forefront of developing compound and rigid-rotor helicopters.

Lockheed Cheyenne (AH-56A), 1313
Advanced armed helicopter, the winner of a US Army Advanced Aerial Fire Support System competition. It first flew in September 1967 and is a two-seat compound helicopter with a retractable undercarriage, powered by a General Electric T64 shaft-turbine delivering 3,435 hp. It has a nose turret, rotatable through 180° and mounting a grenade launcher or a

7.62-mm Minigun; and a central turret carrying a 30-mm cannon with a 360° arc of fire. The

Cheyenne has a maximum speed at sea level of 253 mph, a service ceiling of 26,000 ft and a range of 875 miles.

Lockheed CL-475, 971
This small experimental helicopter which first flew in November 1959 was used to pioneer the new concept of a rigid rotor. This system provides not only simplicity in the mechanics of the rotor head but also a high degree of inherent stability. This has enabled pilot conversion to helicopters to be made easier and the technique has since been used in several operational helicopters for civil and military use.

Lockheed Constellation, 307
The original Constellation was developed in 1939 as a civil airliner for TWA, but the outbreak of World War Two caused it to be modified as the C-69 military transport, the first example of which was flown on 9 January 1943.

First commercial Constellation was the Model 049, completed from C-69s cancelled at the end of the war, but the principal versions built from the outset for civilian use were the Models 649 and 749. The Model 649 first flew on 18 October 1946, and both versions were in airline service by the end of 1947.

The Constellation Model 649 had a 123-ft wing span, and could accommodate 48-64 passengers in its 95 ft 2 in fuselage. Powerplant consisted of four 2,500-hp Wright R-3350-C18 engines. Including military variants, 232 Constellations were built before, in 1951, the type was superseded in production by the "stretched" version, the Super Constellation (q.v.).

Lockheed Electra (L-188), 787

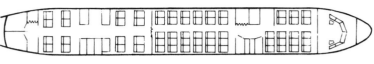

Electra

A short/medium-range turboprop airliner, the Electra entered service in January 1959 with Eastern Air Lines and American Airlines and has been supplied to twelve other operators. Powered by four Allison 501 turboprop engines of 4,050 eshp, it carries up to 98 passengers over a range of 2,770 miles at 405 mph. The Electra also formed the basis for a maritime reconnaissance aircraft, the P-3 Orion, which was developed from it and is in service with the US Navy and the air forces of Norway and New Zealand.

The name Electra was previously borne by the Lockheed Model 18 twin-engined airliner of the nineteen-thirties, a few of which are still in service today.

Lockheed Galaxy (C-5A), 573, 799-804, 1213, 1309-1311
After a design competition for the USAF's large logistics transport requirement, Lockheed was awarded the airframe contract and began building the Lockheed C-5A Galaxy in August 1966. The Galaxy is a huge, high-wing monoplane of 222 ft span, is 245 ft long and is powered by four General Electric TF39 turbofans each rated at 41,000 lb thrust, slung in pods under the swept wing. The Galaxy is designed to transport a 125,000-lb payload a distance of 8,000 miles. First flight took place on 30 June 1968.

Lockheed Hercules (C-130), 789, 887
One of the leading post-war military transports, the Lockheed Hercules was conceived in 1951 and ordered into production for the USAF the following year. It is a four-turboprop, high-

The ubiquitous C-130 "Herky bird", first of many post-war military transports from Lockheed Georgia

Reheat take-off by Lockheed F-104G of the Royal Netherlands Air Force

wing, general-purpose freighter which has proved itself capable of flying in and out of relatively unprepared airfields, needing less than 3,000 ft for landing and less than 4,000 ft for take-off. It has a large, rounded fuselage with clamshell doors at the rear making a loading ramp for vehicles, and can carry a variety of loads; in the troopcarrying rôle it can take up to 92 paratroops. The Hercules has been progressively developed and adapted to other rôles such as search and rescue, reconnaissance, as a recovery vehicle in the US space programme, and for flight refuelling; considerable improvements in power and performance have been made since it first entered service. It has been sold to many air forces outside the US, and also serves with the US Navy, Marines and Coast Guard. A civil version is in limited use with the freight airlines. There is also a ground-attack version of the C-130 in use in Vietnam, equipped with up to seven Vulcan multi-barrel cannon delivering a rate of fire of 42,000 rounds per minute.

Lockheed Hummingbird (XV-4), 1023

Lockheed L-1011, 839
While Britain, France and Germany were still discussing, in mid-1968, their respective requirements for a European airbus, the US entry in this potentially lucrative field had reached the project design stage and orders and options for the aircraft had exceeded 175. The aircraft is the Lockheed L-1011 TriStar which,

after two and a half years of discussion, has crystallized as a three-engined aircraft with two Rolls-Royce RB.211 turbo-fans in underwing pods and a third in the base of the fin. It is designed to seat 250-345 passengers in the 177-ft fuselage; the prototype is scheduled to fly late in 1970, with entry into airline service about a year later. Largest orders up to late 1968 were from Eastern Air Lines (50) and TWA (44).

Lockheed Shooting Star (P-80, later F-80), 1135, 1317
The Shooting Star was the first jet fighter to enter squadron service with the USAF. It was designed around an imported de Havilland H-1B Goblin turbojet, the XP-80 flying for the first time in January 1944. It was then redesigned for the General Electric J33 engine, which powered the first production P-80 aircraft; this had a maximum speed of 558 mph and a range, with tip-tanks, of 1,100 miles. The F-80, in its A, B and C forms, served the USAF for several years as a front-line jet fighter. In 1947 a two-seat version was produced: this became the T-33A, a tandem-seat trainer produced in large numbers both in the US and, under sub-contract, in Canada. The T-33A subsequently served with most of the non-Soviet air forces of the world and many are still in service today.

Lockheed Starfighter (F-104), 841, 852
The Starfighter was originally built in 1951 as a single-seat interceptor for defence within the

US, and went into production in January 1958. It was further developed for tactical work with the USAF, and was then released for export, the first customer being Japan. A considerable structural redesign was introduced with the F-104G tactical version in 1960, and this version was built extensively by a consortium of European companies for service with most NATO air forces. Canadair's CF-104 is basically the F-104G, built for the RCAF. Further development of the aircraft continues.
The Starfighter is characterized by its extremely small wing, spanning only 21 ft 11 in compared with the fuselage length of 54 ft 9 in. The General Electric J79 turbojet in the F-104G gives 15,800 lb st with afterburning. This confers a maximum level speed of Mach 2.2 at 36,000 ft.

Lockheed StarLifter (C-141), 697-702, 789
A turbofan-powered freighter and troop-carrier for the USAF's Military Airlift Command, the C-141 has a swept wing of 160 ft span with podded turbojet engines, and a "T" tail. Four Pratt & Whitney TF33 turbofans give it a cruising speed of 495 mph and a range of 6,140 miles. It can accommodate 154 troops; its rear-loading door and two paratroop doors permit a variety of operational loads and uses.

Lockheed Super Constellation, 1117, 1119-1122
Lockheed took advantage of the Wright Turbo-Compound en-

gine to make a considerable "stretch" of their successful Constellation airliner. The fuselage was lengthened by 18 ft, enabling 94 passengers to be carried. The Super Constellation was successively modified until the G model could fly a normal range of 4,620 miles. Super Constellations were used by many airlines in the nineteen-fifties for their long-range inter-continental operations.
Final development was the L-1649 Starliner, which matched the Super Constellation fuselage to a redesigned and more efficient wing, housing 8,000 gallons of fuel to extend the range even further. However, the Starliner was overtaken by the onset of civil jets, and relatively few entered service.

Lockheed U-2, 577
The U-2 was developed and operated under a cloak of great secrecy. The prototype flew in 1955 and 25 were built for use by Strategic Air Command as well as NASA and the Central Intelligence Agency. The aircraft had a glider-like form with an 80-ft-span wing, centrally mounted on a slender fuselage containing a Pratt & Whitney J57 (later a J75) turbojet. On its internal fuel the U-2 had a range of 2,200 miles and a cruising speed of 460 mph with a ceiling of about 70,000 ft. At least nine of these aircraft have been lost, some of them over the Soviet Union and Communist China.

Lockheed Vega, 385
Lockheed XC-35, 925

The Lockheed XC-35 was built by Lockheed Aircraft Corporation in 1936 for evaluation by the US Army Air Corps. Basically, it was a standard Lockheed Electra twin-engined airliner, with a new fuselage comprising a pressure cabin. It had a position for a pressure-compartment operator who could regulate temperature, humidity and pressure-differential. From this aircraft much useful information was obtained, leading to successful introduction of pressurized airliners, of which the Boeing 307 was the first.

Lockheed XFV-1 "Salmon", 973
Built in 1954, the Lockheed XFV-1 was one of two experimental prototypes built to examine the possibilities of an aeroplane taking off vertically from a position "sitting" on its own tail. It had a cruciform tail at the end of a short fuselage, with castoring wheels at the rear of each tailplane. The aircraft was powered by an Allison T40 turboprop driving two large contra-rotating propellers. It was nicknamed "Salmon" after "Fish" Salmon, who at the time was Lockheed's chief test pilot.

Lockheed XH-51, 971
This aircraft was a direct follow-on from the CL-475 (q.v.) rigid-rotor helicopter, and two prototypes were built to a US Navy/Army contract in 1962. With these, the military authorities made a thorough investigation of the possibilities of the rigid-rotor concept. The XH-51 provided accommodation for two, and its PT6 shaft-turbine engine gave it a maximum speed of 160 mph for a range of 345 miles.

Lockheed YF-12A, 485, 1153-1154
Most advanced airframe yet built for production is that of the Lockheed YF-12A interceptor fighter and its strategic

reconnaissance derivative, the SR-71. It was designed in 1959, originally to supersede the Lockheed U-2 high-altitude, long-range reconnaissance aircraft.
The YF-12A is a "rounded delta", with two Pratt & Whitney J58 engines installed approximately midway out on the wings and the crew compartment in a long, slender nose which itself is faired to have aerodynamic properties. Small fins are installed above the engines. The design envisaged speeds at Mach 3 at heights in excess of 70,000 ft. Only a few YF-12As were built, but approximately two dozen of the SR-71 derivative have been produced for the USAF. A YF-12A took the world absolute speed record past 2,000 mph for the first time on 1 May 1965, when it reached a speed of 2,070.102 mph.

Lod Airport, see Tel Aviv

Loftleidir
Loftleidir HF, or Icelandic Airlines, came into being in 1944 when it operated an internal service with a Stinson seaplane. Its first international service was to Copenhagen in 1947, but since early 1952 its activities have been confined to the North Atlantic route; it has made the unusual claim of being the only non-jet airline flying the Atlantic. Services today link Reykjavik with New York on the one hand and several major European cities on the other. Headquarters: Reykjavik.

Loire-Gourdou-Leseurre monoplane, 573
French single-seat, high-wing monoplane fighter of the mid-nineteen-twenties. It was powered by a 420-hp Gnome-Rhône Jupiter engine. The armament consisted of two Vickers or Darne machine-guns mounted in the fuselage firing through the propeller disc, and two in the wing. A top speed of 155 mph was claimed.

London Airport (Gatwick), Surrey, UK

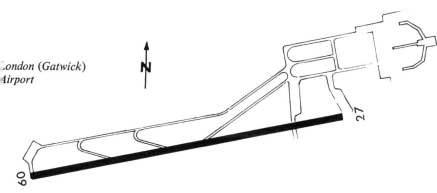

London (Gatwick) Airport

Gatwick dates from the mid-'thirties; plans to make it London's No 2 airport were formulated in the 'fifties, and it was opened for this purpose in June 1958.
Distance from London city centre: 27 miles
Height above sea level: 194 ft
Main runway:
09/27 = 8,200 ft
Passenger movements in 1967: 1,977,381

London Airport (Heathrow), Middlesex, UK, 1281, 1429, 1439

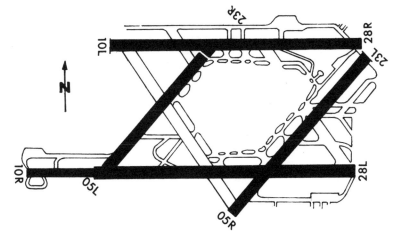

Among the world's busiest international airports, London (Heathrow) has grown up on the site of a former RAF airfield planned in 1943. It now has the highest traffic density of any airport in Europe. The location of a third airport for London is still under review.
Distance from city centre: 15 miles
Height above sea level: 80 ft
Main runways:
10R/28L = 11,000 ft
10L/28R = 9,316 ft

05R/23L = 7,735 ft
05L/23R = 6,255 ft
Passenger movements in 1967: 12,729,993
"Longhorn", see Farman
LORAN (Long Range Navigation), 1533
Los Angeles Airways
LAA was incorporated in May 1944 and became the first airline in the world to operate scheduled services by helicopter when it began an air mail service in southern California on 1 October 1947. It started passenger services in November 1954, and now has a busy regional route system served entirely by helicopters. Another landmark was passed in March 1965 when LAA became the first helicopter airline cleared by the FAA to operate scheduled services under IFR conditions.
Headquarters: Los Angeles.
Los Angeles International Airport, California, USA
First airport was established in LA in 1927, and

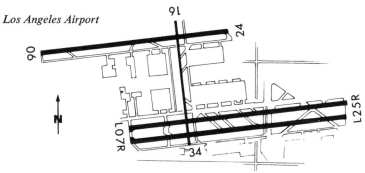

Los Angeles Airport

the National Air Races were held there in 1928 and subsequently. Outstanding feature today is the $70 million "satellite" terminal, opened in 1961-62.
Distance from city centre: 14 miles
Height above sea level: 126 ft
Main runways:
07L/25R = 12,090 ft
07R/25L = 12,000 ft
06/24 = 10,285 ft
16/34 = 6,754 ft
Passenger movements in 1967: 18,120,200
LOT (Polskie Linie Lotnicze)
State-owned national and international airline of Poland, created in 1929 to take over from

former private operators. During the late nineteen-thirties it built up a modern fleet of twin-engined airliners of US design, some of which escaped to serve with BOAC in World War Two. Post-war international services were resumed in 1946, with Soviet-designed aircraft being introduced gradually as the older types were withdrawn. LOT serves most of the major European capitals and also operates into Beirut and Cairo.
Headquarters: Warsaw.
Member of IATA.
Love Field Airport, see Dallas
Low Altitude Bombing System, see LABS
LTV F-8 Crusader, 1189
Development of the F-8 Crusader supersonic fleet fighter began in May 1953, and the first prototype flew in March 1955. Production aircraft entered service in March 1957.
Powered by a Pratt & Whitney J57 turbojet, providing 16,000 lb thrust with afterburner, the Crusader has a maximum speed of just under Mach 2 and carries four 20-mm cannon and two Sidewinder AAMs. The aircraft incorporates a variable-incidence wing and centre-section which permits a high incidence for landing and take-off without involving an exaggerated fuselage angle. The Crusader is in service with the US and French Navies; the latter's aircraft have blown flaps.
LTV-Hiller-Ryan XC-142A, 985
Lufthansa, 257
Formed in 1925 by the

amalgamation of existing airlines, Deutsche Lufthansa became the German national airline. It expanded rapidly with routes to many parts of the world, concentrating particularly on South America, where it was instrumental in fostering local airlines. With the growth of Nazi Germany the airline became a large organization up to the outbreak of World War Two.

Most of the airline became part of the Luftwaffe during World War Two, an emaciated Lufthansa flying a few services within Europe. A new DLH was formed in West Germany on 1 April 1955, which has rapidly returned to a place of importance among world airlines with routes within Europe, across the North and South Atlantic and to the Orient.
Headquarters: Cologne.
Member of IATA.
LVG, 875
Luft Verkehrs Gesellschaft (LVG) of Johannisthal began aircraft construction in Germany in 1912, building Farman-type aircraft. It acquired the services of a Swiss, Franz Schneider, whose hand had been evident in the early Nieuport aeroplanes; and his first design for LVG bore a close resemblance to the Nieuport monoplane. Schneider later left the company, and it settled into producing a long and successful series of two-seat reconnaissance and bomber biplanes. These went into quantity production during the 1914-18 War

and served in great numbers. One of the feats accomplished by LVG biplanes was an epic raid on London in 1916 by one LVG, a CII, which dropped bombs near Victoria Station but was shot down by French anti-aircraft gunners on its way home.
LVG E.VI, 875
Lycoming
In 1928 the newly-formed Aviation Division of the Lycoming Manufacturing Company, of Pennsylvania, started the design of a 200-220-hp nine-cylinder radial engine designated R-680. Deliveries began in 1931, and many thousands were made between 1941-45. In 1936 the division was taken over by Avco Manufacturing Corporation and after World War Two the firm concentrated on air-cooled "flat" (horizontally-opposed) engines for lightplanes and helicopters. These are still being made in great quantities, especially for aircraft of Piper manufacture.
After World War Two Lycoming took the bold step of entering the field of gas-turbines. This task was made much easier by the fact that they were able to build their design team around former German gas-turbine experts, many from the Junkers company. In 1953 design was completed of the T53 free-turbine engine of 700-1,200 hp, and later in the decade the T55 carried the range up to 3,500 hp. These engines have built up a reputation for robust and reliable performance in arduous military service in aeroplanes and helicopters, and have recently been supplemented by turbofan derivatives and shaft-drive versions for surface applications.
M.5K, see Fokker
McClean, Samuel Neal, 861
McDonnell Douglas Phantom II (F-4), 673-8,

852, 1181, 1381

One of the foremost military aircraft at present in use, the Phantom II first flew in May 1958 and was ordered into production for the US Navy, US Marine Corps and USAF. With a maximum level speed in excess of Mach 2, a range of about 1,000 miles and a large variety of alternative warloads, it has proved a most versatile aircraft for land-based and carrier-based operation. It is still being developed with new armament and systems, and a new variant was created by the fitting of Rolls-Royce Spey turbofans in place of the General Electric J79s. This variant is used by the British Royal Navy and Royal Air Force.

McDonnell Quail decoy missile, 896

The Quail is a small delta-wing vehicle powered by a General Electric J85 turbojet. It is launched by a B-52 and, flying at the same height and speed as the B-52 (but in a different direction), produces the same radar echo as the B-52 and so draws off the fire of radar-beamed missiles.

Macchi

Formed at Varese in Italy in 1912 as the SA Nieuport-Macchi, this company produced its first aircraft in 1913. During the 1914-18 War it built French Nieuport fighters for the Italian air service, and also developed a successful series of small flying boats following the capture of an Austrian Lohner in 1915. One of these, the M.7, won the 1921 Schneider Trophy race at 177.8 mph, and the arrival in the mid-'twenties of Ing Mario Castoldi as chief designer gave rise to a generation of successful racing aircraft. These reached their peak with the M.C.72 of 1931 which, although it failed to regain the Trophy for Italy, eventually set a new world speed record

for seaplanes of 440.7 mph in 1934 that remained unbeaten until 1961 and has never been bettered officially by a piston-engined seaplane. Experience gained in building these racing craft stood Castoldi in good stead when he designed the M.C.200 Saetta fighter for the *Regia Aeronautica* in 1937. Its only weakness was its low-powered radial engine, and when the airframe was adapted to take the German DB 601 in-line engine it became, as the M.C.202,

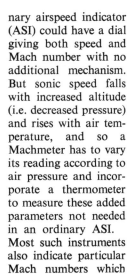
With a Daimler Benz 605 engine the Macchi C.205 fighter achieved 390 mph

the best Italian fighter in large-scale use during World War Two. Since the war Macchi has produced various piston-engined trainers and lightplanes, but its most significant product is the trainer/light attack Aermacchi M.B.326 (*q.v.*) which has been supplied to several foreign air forces and is under large-scale licence production in South Africa.

Mach, Dr Ernst,

see also Machmeter

Austrian physicist of the late nineteenth century who investigated the relationship between the speed of flow of gases and the speed of sound waves. Mach numbers, named after him, are nowadays used to express the speeds of aeroplanes in similar terms, Mach 1 being the speed of sound.

Machmeter, 677, 679, 681, 687, 696, 720

In all aircraft capable of flying in the atmosphere at speeds greater than about three-quarters that of sound it is highly desirable for the pilot to have an instrument indicating not only the airspeed but also the Mach number. If the speed of sound in air were constant an ordi-

nary airspeed indicator (ASI) could have a dial giving both speed and Mach number with no additional mechanism. But sonic speed falls with increased altitude (i.e. decreased pressure) and rises with air temperature, and so a Machmeter has to vary its reading according to air pressure and incorporate a thermometer to measure these added parameters not needed in an ordinary ASI. Most such instruments also indicate particular Mach numbers which may be specially advantageous or dangerous, and particularly the limit that should not be exceeded in any particular aircraft configuration. Frequently, high-speed aircraft are flown according to Mach number rather than according to mph, so that the Machmeter reading becomes of paramount importance. A subsonic airliner, for example, may be cruised at Mach 0.785, or some similar figure, to obtain maximum fuel economy.

MAD (Magnetic Anomaly Detector)

Anti-submarine detection device, generally housed in a "sting" fairing forming an extension of the fuselage tail-cone of aircraft employed for this function.

Mail services, see Air mail

Malév

Operating title of *Magyar Légözlekédsi Vállalat*, the State-owned Hungarian airline. Formed with Hungarian and Soviet capital in 1946, it adopted its present title in 1954 when it became wholly Hungarian-owned. Malév operates a domestic route network and also flies international services to most major terminals in Eastern and Western Europe, the USSR and the Middle East.

Headquarters: Budapest.

The Malta Airlines

Union of Malta Airways Co Ltd and Air Malta Ltd to provide air services linking Malta with Rome, Sicily, Tripoli and London. It was formed in 1948 as an associate of BEA, which still retains a financial interest and from whose fleet it draws its aircraft.

Headquarters: Sliema. Member of IATA.

Manly, Charles, 181, 1044, 1055

Map-making, see Aerial survey

Marquardt ramjets, 1319

Marquardt began development of subsonic ramjets in 1944, and its

first effective ramjet was a 20-in device which was test-flown on a Lockheed Shooting Star, giving a maximum net thrust of 1,450 lb. This was developed into the C-20, which Marquardt put into production to power US Navy pilotless target drones.

Martel missile, see Hawker Siddeley/Matra

Martin

The first Martin-built aircraft was flown at Cleveland, Ohio, by Glenn L. Martin him-

Glenn L. Martin (with XB-48 model)

self in 1909. He built his first military aeroplane, a bomber, for the US Army in 1913, and the first trainer later the same year. The Martin MB twin-engined bomber of 1918 became a standard post-war type for several years. In 1921 Martin MB-2 bombers sank the German battleship *Ostfriesland* off the US coast, demonstrating to the naval authorities that warships could be attacked and sunk by aeroplanes carrying bombs.

The MO-1 for the US Navy in 1922 was the company's first all-metal design, and in 1928 the T4M-1 became one of the largest aircraft to be launched from an aircraft carrier. Four years later, the B-10 twin-engined monoplane revolutionized bomber design in the US by proving itself 100 mph faster than the reigning US fighters. By this time the company had transferred to larger premises at Middle River, near Baltimore, where it still has its headquarters. In the late 'thirties and early 'forties Martin became known for its flying boat designs, including the 1935 China Clipper, the giant Mars —which on 1 December 1943 completed a world distance record flight of 4,375 miles from Patuxent River to Natal in Brazil—and the wartime PBM Mariner.

Martin's most notable product of World War Two, however, was undoubtedly the twin-engined B-26 Marauder

Many versions of the Martin B-26 Marauder saw action

medium bomber. This aircraft, over 5,000 of which were built, survived its early reputation of "widow-maker" and "flying coffin" to finish the war with a lower loss rate than any other American bomber

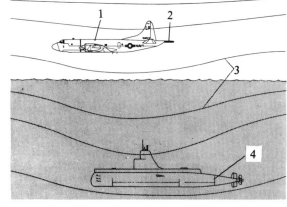

1 Orion patrol aircraft
2 MAD "sting"
3 Lines of force of earth's magnetic field
4 Submarine

Two Martin MO-1 seaplanes on the catapult of the USS Mississippi

of the war period.

Two more flying boats appeared after the war: the P5M Marlin and the jet-powered P6M SeaMaster, and the English Electric Canberra jet bomber was built under licence; but Martin's activities became increasingly directed toward missile design. The Matador, Lacrosse and Bullpup missiles were of Martin origin, and in 1955 the company withdrew largely from aeroplane manufacture. It is now known as Martin Marietta Corporation and produces the Bullpup, Blue Eye, Walleye and Pershing missiles and the Sprint antimissile system.

Martin K-3 Kitten, 551, 567

Designed by the American J. V. Martin Aeroplane Co, this US aircraft was originally planned as an altitude fighter with provision for oxygen equipment and heated clothing. It appeared, however, as a cheap "club" aircraft. It was very small, with a span of less than 18 ft, an empty weight of 350 lb and a 45-hp ABC Gnat engine. The wings were braced by unusual K-pattern struts, this feature being responsible for the K in the designation. The fuselage was constructed almost entirely of plywood. The main undercarriage was partially retractable, an unusual feature being the wheel spokes, which radiated in catherine-wheel fashion, in the manner of those of some of the wheels now being developed for vehicles intended to explore the Moon.

Martin X-24A, 338-339

Martin 2-0-2 and 4-0-4, 1117

First twin-engined airliner of completely postwar design to go into production in the United States. It flew in November 1946 and received its civil certification the following August. With accommodation for 36-40 passengers, it was powered by two 1,800-hp Double Wasps giving it a cruising speed of 286 mph and range of 635 miles. The 2-0-2 entered service with several US domestic and foreign airlines, principally in the Western hemisphere. It was followed by the 4-0-4, a pressurized version with more powerful engines.

Martin Marietta Bullpup missile, 896, 898

The Bullpup is an airto-surface tactical missile which was developed originally for the US Navy, who use it on many of their aircraft. It also serves with the USAF and Royal Navy. There are several versions of the missile, including one with a 250-lb conventional warhead and another with a nuclear warhead. It is guided by radio command signals transmitted by means of a miniature joystick operated by the pilot of the launch aircraft.

Martin Marietta Walleye missile, 896

Walleye is an unpowered air-to-surface missile, with a high-explosive warhead, guided by television and operating on the glide-bomb principle.

Martini carbine, 873

Matra R530 missile, 895

This missile is standard armament on Vautour and Mirage fighters of the French *Armée de l'Air*. It is also used on the South African Air Force's Mirages, those of Israel and Australia, and the French Navy's Crusaders. The missile has a two-stage Hotchkiss-Brandt solid-propellant rocket motor and a 60-lb high-explosive warhead, a speed of Mach 2.7 and a range of 11 miles.

Maxim, Sir Hiram, 159, 169, 873, 1047

M.B. 326, see Aermacchi

MEA (Middle East Airlines/Airliban)

Since the formation of MEA as a privately-owned airline in 1945, substantial financial holdings by Pan American (1949-55) and BOAC (1955-61) have helped the company to achieve prominent international status. It became a completely independent Lebanese enterprise in August 1961, but since 1963 Air France has held a 30 per cent interest. The company absorbed the former Air Liban in 1965, and since 1967 has operated in pool with Lebanese International Airways (*q.v.*), with which it is expected to merge. MEA has a busy and extensive route network which embraces centres throughout the Middle East, Europe, Africa and Asia.

Headquarters: Beirut. Member of IATA.

Medway engine, see Rolls-Royce

Mercedes, 367

The Mercedes-Daimler Motoren Gesellschaft produced its first aero engine in 1911 and in 1912 Mercedes engines were almost universally adopted by German constructors. A range of six- and eight-cylinder air-cooled inline engines were built during the 1914-18 War and were used in many German military aircraft, the most powerful model producing 260 hp. The company reappeared in the aero engine field in 1926, when it combined with Benz of Mannheim to build aero engines including all the engines for the civil Zeppelins. The Daimler-Benz DB600 engine appeared in 1938, a liquid-cooled inverted-Y-shaped engine of twelve cylinders; this, and its derivatives, became one of the principal series of German war-time engines. By that time the name Mercedes-Benz had been dropped in favour of Daimler-Benz (*q.v.*).

Mercury spacecraft, 330, 503-4

Project Mercury was initiated in October 1958 as America's first effort to put a man in to space. The man-carrying capsules were designed and built by McDonnell and the booster rockets were either Redstone or Atlas missiles. Redstones were used for the sub-orbital flights and the Atlas rockets for the full space flights. The first American to be projected was Commander Alan Shepard on a sub-orbital flight on 5 May 1961, followed by Captain Virgil Grissom on a flight of a similar profile two months later. The first human orbital flight took place on 20 February 1962, when Lieutenant-Colonel John Glenn completed a three-orbit flight. He was followed by M. Scott Carpenter in May 1962, Walter Schirra in October 1962 and Gordon Cooper in May 1963.

Merlin engine, see Rolls-Royce

Mermoz, Jean, 253, 257

Messerschmitt, Prof Willy, 292, 295, 299, 573, 602, 817, 902, 909, 911

Prof Willy Messerschmitt

Professor Willy Messerschmitt joined BFW (Bayerische Flugzeugwerke) in 1927 and was soon at the head of the design team. His prowess was seen in the early nineteen-thirties with the development of the Bf 108 four-seat civil monoplane, and the Bf 109 fighter which played a significant part in World War Two (and which is still flying, in modified form, today). These headed a long line of military aircraft, including the first German operational jet aircraft, the Me 262.

Messerschmitt Bf 109, 902, 909, 1339

One of the world's most widely-produced aircraft (over 34,000 were

Messerschmitt Bf 109E, used by the Luftwaffe *in 1940*

built) the Bf 109 first flew in prototype form in September 1935 with a 695-hp Rolls-Royce Kestrel engine. Early production models, with Junkers Jumo engines, saw action in the Spanish Civil War of the late nineteen-thirties, but with the Bf 109E a change was made to the DB 601A engine of 1,100 hp. Armed with two 20-mm cannon and two 7.9-mm machine-guns, the Bf 109E was the principal variant engaged in the Battle of Britain in 1940. It was succeeded by the Bf 109F (1,200-hp DB 601N) and Bf 109G (1,475-1,800-hp DB 605) and variants of the fighter served in all theatres of World War Two in which the *Luftwaffe* was engaged. Heaviest-armed version was the Bf 109G-6, with one 30-mm and two 20-mm cannon and two 13-mm machine-guns; while other models were equipped for close-support duties with underwing rocket projectiles or bombs.

After World War Two, production of the Bf 109 continued by Avia in Czechoslovakia (as the S 199) and Hispano in Spain (as the HA-1109), in the latter case with Rolls-Royce Merlin engines. It is estimated that overall production of the Bf 109 amounted to around 35,000 aircraft. A few remain in service today.

Messerschmitt Bf 110, 1135

Willy Messerschmitt first conceived the Bf 110 in 1934 as a long-range escort fighter, and the first prototype was flown on 12 May 1936. It was a twin-engined lowwing monoplane, powered by two 1,800-hp Daimler-Benz DB 600 engines; a tandem cockpit seated a crew of two in its slim fuselage, and it had twin fins and rudders.

Development continued with alternative power plants and various armament installations, and the Bf 110C was in service when Germany invaded Poland in 1939. It was successful as a strike fighter, but when used as an escort fighter against single-engined fighter opposition it was found to be useless due to its poor manœuvrability. It was used largely on the Russian Front from 1941 onwards, and in the Middle East, and could carry bombs in the strike rôle. The Bf 110 later served as a night-fighter, and for the latter part of the war the more powerful Bf 110G was the *Luftwaffe*'s principal night fighter; late models were equipped with airborne interception radar.

Messerschmitt Bf 110C twin-engined fighter, designed in 1936

Messerschmitt Me 163B Komet, 299, 583, 585, 602, 1137

Developed from a glider prototype in 1940, the Me 163 became the first rocket-propelled fighter and the fastest aircraft in operational service in World War Two. The first rocket-powered version flew in April 1941, attaining a level speed of 623.85 mph the following month. Development of the rocket motor and airframe proceeded, but it was late 1944 before the first operational unit of Me 163s was formed. Owing to their extremely short duration, these aircraft were stationed near specific targets which they were to defend, the procedure being to climb as rapidly as possible to a height a few thousand feet above the attacking formations, then glide into position and make one diving pass at the aircraft, zooming and making a further attack until height could no longer be maintained. The aircraft was not easy to fly and many accidents took place, chiefly on take-off and landing; and it had little effect on the progress of the war.

Messerschmitt Me 262, 292, 295, 817, 1135

The Messerschmitt Me 262 twin-engined jet was developed to much the same time-scale as the British Gloster Meteor (q.v.). The first prototype flew earlier, in the summer of 1941, but operational service was delayed owing to a change of operational policy which dictated that Germany's first jet fighters should be used in the fighter-bomber rôle. The aircraft was put into quantity production in several versions, the first becoming operational shortly after the Meteor, in August 1944. Further versions comprised single- and two-seat fighters and night-fighters, and a tank-busting version with a 50-mm cannon in the nose. By the end of the war, 1,294 Me 262s had been built, but comparatively few were encountered operationally.

Messerschmitt Me 323, 573

The Me 323 was a World War Two development of the Me 321, a giant glider of 181-ft span. Fitted with six Gnome-Rhône 14N radial engines, the Me 323 was used as a troop transport by the *Luftwaffe* on many fronts. It was a large, ungainly high-wing monoplane with nose-loading doors, and was capable of carrying 130 troops, two motor trucks, an 88-mm flak gun and equipment, 52 two-litre fuel drums or 60 stretcher cases. It was an easy target for enemy aircraft and many were shot down, especially during the retreat from North Africa.

Messier Jockey, 577

One of the most advanced undercarriage units developed by this French company, which has specialized in the design and building of landing gear since before World War Two and produced some of the early French retractable undercarriages. The Jockey is a tandem unit retracting into fairings on the sides of the aircraft's fuselage.

Meteor, see Gloster

Metropolitan-Vickers

Metropolitan - Vickers came into the aviation field in 1938 with no previous experience of aircraft work but with experience of turbine development. They took up war work for the RAE team, headed by Dr A. A. Griffiths, working on the then new principle of jet propulsion. At a time when most practical work on turbine engines for aircraft use concerned the centrifugal compressor, Metrovick embarked on an axial-flow design, basing their work on axial-flow air compressors in which they already had some development experience. The first engine runs took place in December 1941, and development work continued for a further eleven months. The engine was fitted first in the tail of a Lancaster bomber for flight tests and then in one of the Gloster F.9/40 twin-engined testbeds (which developed into the Gloster Meteor).

Because the axial-flow engine was an advance, technically, over other current engines, the development time was much longer; but the work was encouraging, not least in the field of fuel consumption. To improve on this further, Griffiths designed in 1944, in collaboration with Metrovick, a ducted-fan engine with an aft-fan, pre-dating the general use of such engines by about sixteen years.

Meusnier, General J.B.M., 133, 137

Miami International Airport, Florida, USA

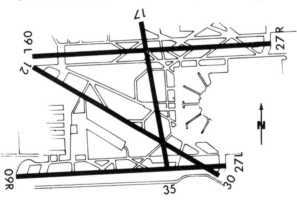

Distance from city centre: 5 miles
Height above sea level: 9 ft
Main runways:
  09L/27R = 10,500 ft
  12/30 = 9,604 ft
  09R/27L = 9,350 ft
  17/35 = 6,081 ft
Passenger movements in 1967: 8,722,300

Middle East Airlines, see MEA

Mikoyan and Gurevich, 307

The abbreviated designation "MiG" originally signified a collaborative product of the Soviet designers A. Mikoyan and N. Gurevich. It was first used in the MiG-1 and MiG-3 piston-engined fighters, the former of which flew in prototype form in March 1940. The same team was responsible after World War Two for the MiG-9, one of the USSR's first jet fighters, but it really came into world-wide prominence with the MiG-15, the prototype of which was flown on 30 December 1947. First mass-produced Soviet jet fighter, the MiG-15 was powered by a Klimov (q.v.) development of the Rolls-Royce Nene turbojet, and played a major rôle in the Korean War of 1950-53. It was superseded in production in 1953 by the generally similar MiG-17.

In the mid-nineteen-fifties appeared the MiG-19, the first supersonic fighter to go into Soviet service. By this time, Gurevich was believed to have left the partnership, although the MiG prefix has been retained for subsequent designs. These include the MiG-21 day fighter, in service with the Soviet and several other world air forces; and the latest operational fighter, the Mach 3 twin-jet MiG-23, which holds the world's closed-circuit speed record for jet aircraft.

Mil

One of the world's principal helicopter designers, Mikhail L. Mil became connected with rotary-winged development in the USSR at least as early as 1930. First product to bear his own design bureau number was the little Mi-1 of 1949, which first flew in 1950 and was subsequently built in considerable numbers, in Poland as well as in the USSR. This was followed by the larger Mi-4, also the subject of substantial home and licence production, and the Mi-6, Mi-10 and Mi-12 turbine-powered helicopters which are by far the largest in the world. The later, and much smaller, Mi-2 is built exclusively in Poland.

Mil Mi-6, 1313

The Mi-6 is the largest series-built helicopter flying today. It first appeared in Russia in 1957 and is used as a passenger or general transport helicopter. In the passenger configuration it can carry up to 65 people, whereas the freighter version has an internal payload of 26,450 lb. In 1967 a fire-fighting version appeared. The Mi-6 has one five-bladed main rotor and a directional tail rotor and may be seen with stub wings at shoulder height beneath the main rotor. It is powered by two Soloviev 5,500-shp shaft-turbine engines, giving it a maximum cruising speed of 155 mph and a range, with 9,920-lb payload, of 620 miles.

Mil Mi-10, 1313

Developed from the Mi-6, the Mi-10 is a flying crane helicopter which first appeared in 1961. It differs from the Mi-6 in having a stalky undercarriage and a shallow fuselage with provision for carrying bulky loads slung underneath the fuselage. Although dimensionally slightly smaller than the Mi-6, the Mi-10 has a higher maximum take-off weight.

Miles, 297

Miles Aircraft Ltd had its origins in the Phillips and Powis organization at Woodley Aerodrome, Reading, England. By 1932, F. G. Miles, with brother G. H. and wife, Maxine, had already designed and built two fast light biplanes at Shoreham and flew the later type, the Satyr, into Woodley. As a result of meeting Mr Powis, Miles designed a cheap, sturdy two-seat low-wing monoplane to suit the club at Reading. It sold at £400 and was named the Miles Hawk, acquiring immediate popularity. From this beginning stemmed a whole family of trainers, tourers and racing aircraft, all with the basic Miles features. They sold well at home and abroad right up to the outbreak of World War Two, when Miles Aircraft had in quantity production the Magister elementary trainer and the Master advanced trainer for the RAF.

After the war the company produced a whole new range of civil types including the Messenger, Gemini and Aerovan, but in 1947 it ran into financial difficulties and had to close down. F. G. Miles later began again at his original base of Shoreham where he has built up a flourishing business in the ancillary and electronic industry.

Minigun, 887

The multi-barrel 7.62-

Mil Mi-10 carrying large passenger coach

mm Minigun is a relatively new feature of air warfare. Based on the Gatling gun, the Minigun has a rate of fire of 6,000 rounds per minute, and it has been found that a combination of such guns fixed to fire out of the side of a large aircraft can provide a lethal swathe of

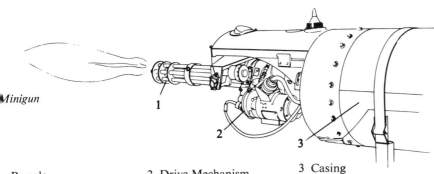

Minigun

1 Barrels

2 Drive Mechanism

3 Casing

fire. By flying a pre-determined flight path at low level, this swathe can be directed across a nominated ground target. This has become a very effective means of attack in the Vietnam War, and even old Dakota transports of World War Two vintage were modified and used in this rôle before the advent of more modern aircraft.

Mirage III, see Dassault

Mitchell, R. J., 287

Reginald Mitchell

The designer of the Spitfire, R. J. Mitchell joined Supermarine as an engineer in 1916, the first aircraft with which he was involved being an obscure flying boat for the Admiralty. Flying boats were his main concern during and after the 1914-18 War as he graduated eventually to the post of Chief Designer. He realized that, in order to win the coveted Schneider Trophy, the ultimate high-speed prize, it was necessary to design sleek monoplane seaplanes, and he commenced his series of Schneider machines with the S.4 in 1925, culminating in the S.6B of 1931 with which the Trophy was won outright. All the

experience gained with these was turned toward a high-performance fighter when Mitchell put forward a design for the Air Ministry's F.7/30 specification. This specification was too limiting for Mitchell's talents, and the resulting aircraft was unspectacular; but he followed it with his own private venture, the original Spitfire. Its subsequent history is recorded elsewhere; unhappily, Mitchell was not to see it enter service, for he died in June 1937.

Mitsubishi

Mitsubishi Heavy Industries Ltd, formed at Nagasaki towards the end of the nineteenth century, is one of the oldest industrial organizations in the Orient. It was registered as a shipbuilding company in 1917 and established an aircraft works at Nagoya in 1920 with Herbert Smith, formerly of Sopwith (*q.v.*) as its chief designer. Several fighter and torpedo-carrying biplanes, designed by Smith and powered by licence-built Hispano-Suiza or Napier engines, went into service with the Japanese Navy during the nineteen-twenties. Licence-built products during this period included Blackburn, Hanriot and Curtiss aircraft and Armstrong-Siddeley Jaguar engines. In the nineteen-thirties Mitsubishi produced its first carrier-borne fighters for the Navy, as well as Army reconnaissance-bombers and civil transports. In April 1937 the single-engined Karigane

monoplane *Kamikaze* (Divine Wind) made a flight of 9,900 miles from Tokyo to London in 94½ hours, and in the same year came the Navy specification that led to the Zero fighter, or A6M, destined to become the personification of Japanese air power during World War Two. This made its first flight on 1 April 1939, and over 10,000 Zeros were built before the war ended. Other Mitsubishi designs to see wartime service included the Ki-21 ("Sally") and G4M ("Betty") bombers, the latter also acting as carrier for the *Ohka* suicide bomb; and the Ki-46 ("Dinah") reconnaissance and night fighter aircraft. When the war ended, Mitsubishi had com-

Mitsubishi Ki 46, known to the US forces as "Dinah"

pleted a few test examples of the J8M1 *Shusui* rocket-powered interceptor based on the Messerschmitt Me 163. Up to this time Mitsubishi was also one of the foremost Japanese manufacturers of aero engines, mostly in the 1,000-2,500 hp range and including the wartime Kasei and Kinsei series.

The company was re-established on a post-war basis after the 1952 Peace Treaty, with pre-

mises at Komaki and Nagoya. After some initial overhaul work it began to assemble F-86F Sabre fighters for the Air Self Defence Force in 1955; later, with Kawasaki (*q.v.*) it built over 200 Lockheed Starfighters for the JASDF and has also built the Sikorsky S-58 helicopter under licence. Present activities include manufacture and sale of various Sikorsky helicopters, production of the Mitsubishi-designed MU-2 twin-turboprop utility transport, and development of a supersonic trainer, the XT-2A.

Mixmaster (XB-42), see Douglas

Mohawk Airlines

A major inter-state airline in the US, Mohawk was founded as Robinson Airlines in 1945 and assumed its present title in 1952. It was the first, and is still the largest, US operator of the BAC One-Eleven, and its network embraces well over 50 cities in ten neighbouring states as well as Montreal and Toronto in Canada.

Headquarters: Utica, New York.

Member of IATA.

A structure in which all the loads are taken by the outer skin and not by any internal members; an example in nature is the shell of an egg. The first monocoque used in aircraft construction was designed by Ruchonnet in France and applied by Bechereau to the fuselage of the "Monocoque Deperdussin". The fuselage was moulded in two halves, which were later joined together. In addition to being strong structurally, the smooth streamlined shape of the fuselage provided an aerodynamic bonus in the form of reduced drag.

Today most aircraft structures embody some internal stiffening members, even though most of the loads are carried by the skin. Initially the term semi-monocoque was used to describe such structures, but today the generally accepted term used is "stressed-skin".

Monocoque, 361

1 Eggshell

2 Fuselage

3 Frames

4 Stringers

Montgolfier, 121, 126

To the Montgolfier brothers goes the credit of producing the first successful lighter-than-air craft, the first ascent being at Annonay on 5 June 1783. Using the hot air formed by burning wool and straw (and believing it at the time to be a special gas), they soon had balloons ascending, later with animals and then with human beings. Pilâtre de Rozier was the first human to ascend, on 15 October 1783. This was a tethered ascent, the first free ascent taking place with a "Montgolfière," as all hot-air balloons were then described, on 21 November that year.

Montreal International Airport (Dorval), Quebec, Canada.

Original Dorval airport was completed in 1941, although Montreal's association with flying goes back much earlier (to 1910). Present terminal opened in 1960.

Distance from city centre: 14 miles.

Height above sea level: 117 ft

Main runways:
06L/24R = 11,000 ft
06R/24L = 9,600 ft
10/28 = 7,000 ft

Passenger movements in 1967: 5,050,800

Morane-Saulnier, 209, 243, 875

The Morane-Saulnier company was formed in France in 1911. It produced many aircraft before and during the 1914-18 War, its *forte* being the "parasol" monoplanes. These played an important part in the early stages of air fighting, more effectively when deflector plates had been fitted to the propellers to enable machine-guns to be fired between the blades. Later in the war the company also built biplanes, but found that the "parasol" monoplane was its most effective product, and it produced a continuous series of this type of aircraft in the nineteen-twenties and 'thirties. Many of them were used for military training and Army co-operation work. After World War Two it revived and concentrated on training and touring aircraft, producing the Rallye series and being early in the light jet field with the MS.760. In 1963 it was declared bankrupt and acquired by Potez. The Rallye series is now marketed by Socata, a subsidiary of Sud-Aviation.

Moscow Airport
(Sheremetievo), USSR

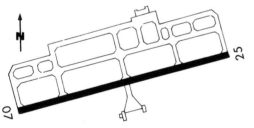

Opened in June 1950, taking over major international traffic from Vnukovo; this is now shared with Domodedovo, 25 miles from the city (opened 1963). Other airfields in Moscow area include Tushino, where the Military Aviation Day displays are held, and Bykovo, but Vnukovo remains the major centre for domestic traffic.
Distance from city centre: 15 miles
Height above sea level: 623 ft
Main runway: 07/25 = 11,483 ft
Passenger movements: (Domodedovo and Vnukovo) in 1966: 3,800,000

Mosquito (D.H. 98), see de Havilland
Moteur-canon, 883
The *moteur-canon* originally appeared as an answer to the problem of designing an interrupter gear for machine-guns firing within the arc of a propeller. The principle was to fit the gun, usually a large-calibre cannon, within the hollow crankshaft of the engine so that it would fire through the propeller boss. It was first fitted (other than experimentally) in some French fighters of the 1914-18 War, and the idea was perpetuated in peacetime French fighters and in the early Messerschmitt Bf 109 fighters in World War Two.
There were fundamental disadvantages to the layout and it eventually ceased to be used.
Mountain rescue, see Rescue services
Mozhaisky, 159, 168, 169
Musée de l'Air, 217
Museum located at Chalais-Meudon, near Paris, devoted exclusively to aeronautical exhibits. Well-known for its extensive collection of early aircraft and aero engines.
Mustang (P-51), see North American
Myasishchev Mya-4 ("Bison"), 835
First large jet bomber of the Soviet Air Force, the Mya-4 entered service in the mid-nineteen-fifties. With a wing span of 170 ft and powered by four AM-3M turbojets of 19,800 lb st each, it has been a formidable weapon although its top speed of 559 mph has now relegated it to reconnaissance and maritime rôles for which its range of 6,000 miles is particularly suitable.

Nairobi Airport
(Embakasi), Kenya

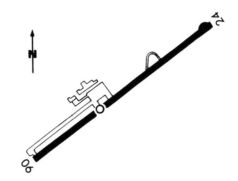

Opened in March 1958 to absorb the scheduled traffic from nearby Eastleigh and Nairobi West, which no longer had adequate handling facilities.
Distance from city centre: 11 miles
Height above sea level: 5,327 ft
Main runway: 06/24 = 13,500 ft
International passenger movements in 1967: 734,000
Nakajima
Formed in 1914 as the Japan Aircraft Works, this company was re-titled Nakajima Aircraft Co Ltd in 1924 and became second only to Mitsubishi (*q.v.*) for the quality and quantity of aircraft produced up to the end of World War Two. The founder was Captain Nakajima, a former Japanese Navy officer, and his company began building aircraft during the 1914-18 War. In 1919 the Nakajima No 4 won the first air mail contest held in Japan, and later that year the No 6 set up a new national distance record. In the early 'twenties Nakajima produced the first Japanese all-metal aircraft, the B-6 bomber and its civil counterpart the P-6. Several licences were then acquired for foreign designs, among them the Avro 504, Breguet XIX and Nieuport 29 as well as Lorraine and Bristol Jupiter aero-engines; other famous types built by Nakajima during

the 'thirties included the Fokker Universal and Douglas DC-2 airliners for Japan Airways.
Nakajima fighters for the Navy and Army during the 'thirties included the Type 91, A1N, A2N, A4N and Ki-27, but the company really came into the limelight with the attack on Pearl Harbor in 1941 when B5N ("Kate") torpedo bombers formed a major part of the attacking force. They were superseded as the war progressed by the B6N ("Jill") and C6N ("Myrt"), both excellent designs, and by the Ki-43 ("Oscar"), Ki-44 ("Tojo") and Ki-84 ("Frank") radial-engined Army fighters. Nakajima also produced large numbers of aero engines, notably the Sakae and Homare series. Among wartime products which did not become operational were the G5N1 and G8N1 four-engined bombers and the *Kikka* twin-jet fighter. When the company ceased to function at the end of World War Two it had produced altogether some 30,000 aeroplanes. Following the Peace Treaty with Japan in 1952, a successor to Nakajima emerged in 1953 as Fuji Heavy Industries Ltd, building the T-34 Mentor under

licence from Beech in the US. In January 1958 the prototype was flown of the Fuji T1F2 trainer, the first production jet aircraft of Japanese design. At the present time Fuji is producing the Bell 204 (*q.v.*) under licence and a four-seat cabin monoplane of its own design, the FA-200 Aero Subaru.
NAMC YS-11
The YS-11, produced by the Nihon Aeroplane Manufacturing Co of Japan, is the

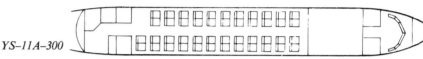
YS–11A–300

first Japanese-designed transport to appear since World War Two. It flew for the first time on 30 August 1962, and entered service with Toa Airways at the end of 1964. Later variants include the YS-11A-200 airliner, the YS-11A-300 passenger/cargo version and the YS-11A-400 freighter; these are being built for the Japanese armed forces and for several domestic and foreign airlines. The YS-11A-200 seats up to 60 passengers, and can cruise at 256 mph for up to 1,300 miles on the power of two 3,060-ehp Rolls-Royce Dart turboprop engines.
Napier Nomad engine, 1117, 1127, 1283
Napier Sabre engine, 1097
For many years the Napier Company in Great Britain were wedded to the "H" type of in-line, liquid-cooled aero engine. In this configuration, two banks of horizontally-opposed cylinders are placed side-by-side and geared to a common propeller shaft. The culmination of Napier's art was the Sabre engine; this was a 24-cylinder "H" engine, the cylinders having sleeve, not poppet, valves. The Sabre arrived on the scene just after World War Two broke out, and was selected as the

power plant for the Hawker Typhoon. Eventually it became a brilliant engine for this and the Hawker Tempest fighters, but for a long period it was dogged by development troubles, even after going into production and squadron service in the Typhoon. The first production version, the Sabre Va, produced 2,310 hp, the later Sabre VII 3,000 hp.
National Airlines
A pioneer US airline. National's first scheduled operations were flown by two Ryan monoplanes in 1934. It was the first US domestic airline to introduce jet services (in December 1958), and today operates a coast-to-coast network that includes non-stop services from Miami to New York and Los Angeles. Headquarters: Miami. Associate member of IATA.
National Air Races (USA), see Competitive flying
Natter, see Bachem Ba 349
NC-4, see Curtiss
Nene engine, see Rolls-Royce
Netherlands: Air Forces
Koninklijke Luchtmacht (Royal Netherlands Air Force). Created as an Aviation Division of the Army on 1 July 1913, equipped initially with Farman F.22s to which were added foreign combat aircraft interned or purchased during the 1914-18 War. The Fokker (*q.v.*) factory established in Holland after the war supplied most of its aircraft during the nineteen-twenties and early 'thirties, although the Army air arm in the Dutch East Indies employed several foreign types including British Avro 504s and D.H.9s and US Curtiss P-6 Hawks. Not until the late 'thirties, when the

Anthony Fokker's Spin

threat of Nazi Germany could no longer be ignored, was a major reorganization and expansion programme begun; again, Fokker was to provide the principal new aircraft, which included D.XXI and G.I fighters and T.V bombers, although Curtiss Hawk 75s and Douglas DB-8As were ordered from the US. The re-equipment was only partially complete when Germany invaded Holland in 1940, and the Dutch Air Force was soon overcome by the superior *Luftwaffe*. After the liberation of Holland, a small Dutch Army Air Force fought in the closing months of the war in Europe.

After World War Two the home and East Indies elements of the LVA (*Luchtvaart Afdeling*) began in 1947 to re-form into a single service, with aircraft and other assistance from Britain and the US. The RNethAF was created officially as an independent force on 27 March 1953, and Holland's entry into NATO was marked by bringing the RNethAF into the jet age with Thunderjet fighter-bombers supplied under MDAP. These were superseded by F-84F Thunderstreaks, which

in turn are at present being replaced by Canadian-built Northrop F-5s. The RNethAF also has in current service F-104G Starfighters and a substantial number of Alouette III helicopters.

Marine Luchtvaartdienst (Royal Netherlands Naval Air Service). Formed on 18 August 1917, initially with Martin biplanes and Friedrichshafen seaplanes. After the 1914-18 War it also built up naval aviation in the Dutch East Indies with various Fokker types, Dornier Wal flying boats and British Fairey IIIDs. Like that of the Army, Dutch naval aviation only began to expand and equip with really modern warplanes in the late nineteen-thirties, new types ordered including Fokker T. VIII-W torpedo-bombers for use at home and Dornier Do 24K flying boats and Martin 139-W bombers for the East Indies. These and several other US military aircraft were received in 1940-41, and with them Dutch aircrews fought against Japan, at first from their East Indies bases and later from Ceylon and Australia.

The naval air arm

returned to the East Indies after the war, but in 1949 the new republic of Indonesia claimed most of its existing aircraft and the Dutch personnel were withdrawn to Holland. Fairey Fireflies and Hawker Sea Furies were acquired to serve aboard the newly-introduced carrier *Karel Doorman*, and from 1953 MDAP assistance included Lockheed Harpoons and Neptunes and Grumman Avengers for anti-submarine and maritime patrol duties. Sea Hawk naval jet fighters were acquired in the late 'fifties. Some Neptunes still remain in service, but the Harpoons and Avengers have been replaced respectively by Grumman Trackers and Westland Wasp ASW helicopters. The *Karel Doorman* was withdrawn from service in 1968.

Netherlands: Chronology 29 September 1804. First successful balloon ascent by a Dutchman, at Rotterdam, when Abraham Hopman made a short flight to nearby Schiedam.
1886. A military balloon park was established in the Netherlands. Subsequently, target balloons were introduced at the Dutch Army artillery range at Oldebroeck.
19 October 1907. The Netherlands Aeronautical Association (later Royal) was established to ". . . further the interests of aeronautics in the Netherlands". Now better known as KNVvL, it plays an important part in all aspects of national aviation.
27 June 1909. Flying a Wright biplane, the Belgian Comte de Lambert made the first powered aeroplane flight in the Netherlands at Etten-Leur, near Breda.
1 July 1913. The *Luchtvaartafdeling* (LVA), or Netherlands Army Aviation Corps, was established by Royal

Dutch-built Farman, the van Meel "Brik" at Soesterberg in 1913

decree, with an airfield near Soesterberg. Initial strength was two aircraft and four pilots.
18 August 1917. The *Marine Luchtvaartdienst* (MLD), or Naval Aviation Service, was created. Initial equipment comprised new Farman F-22 training aircraft, which were based at De Kooy airfield near den Helder naval base.
5 April 1919. *Rijksstudiedienst voor de Luchtvaart*—the Dutch aviation research department—began activities from its headquarters at the Naval Establishment in Amsterdam. It is known today as the Dutch Aerospace Research Laboratory.
21 July 1919. Anthony Fokker returned from Germany and founded the Netherlands Aircraft Factories. Initial work of the company consisted of overhauling and completing some 200 aircraft and aero engines smuggled out of Germany.
1 August 1919. The first Dutch Air Transport Exhibition (ELTA) opened in Amsterdam. Some half a million visitors were able to see the progress made by world aviation during the four years of war.
7 October 1919. *Koninklijke Luchtvaart Maatschappij voor Nederland en Koloniën NV*—better known as KLM—was founded. It is the oldest airline operating under its original name.
17 May 1920. A de Havilland D.H.9 of the British Aircraft

Transport & Travel Company, chartered by KLM, landed at Schiphol airfield. This marked the beginning of scheduled air services to and from the Netherlands.
1 October 1924. Piloted by Captain Van der Hoop, a Fokker F.VII airliner took off from Schiphol Airport, bound for Batavia. This first flight to the Dutch East Indies was completed on November 24.
13 July 1926. First private flying club in the Netherlands, the Rotterdam Aeroclub was founded.
15 June 1927. A Fokker F.VIIa airliner took off on the first return flight to the Dutch East Indies. Captain G. J. Geyssendorffer and his crew landed back at Amsterdam on 24 July.
15 September 1930. Captain E. van Dijk, piloting the Fokker

Chartered by KLM, this D.H.9 flew the first services between Schiphol and London

F.VIIb *Regier* (Heron), initiated the first scheduled fortnightly air mail service between the Netherlands and the Dutch East Indies.
20 October 1934. Flying KLM's first Douglas DC-2 airliner, Captain K. D. Parmentier and his crew took first place in the handicap section of the famous McRobertson London-Melbourne air race, in a flying time of 90 hours 17 minutes.
15 December 1934. First transatlantic flight by a Dutch aircraft. The Fokker F.XVIII *Snip* (Snipe), piloted by Captain J. J. Hondong, landed at Paramaribo (Surinam) on 22 December.
10 May 1940. German forces invaded the Netherlands. Practically all military and civil aircraft were destroyed or captured within five days.
27 September 1945. KLM resumed civil airline operations with D.H.89 aircraft.
1 January 1946. The Government Civil Flying School for airline pilots was founded. First course was held at Gilze-Rijen Air Base, near Bréda. Headquarters is now established at Eelde Airport,

KLM crew who won UK-Melbourne handicap race of 1934: van Brugge, Parmentier, Moll, Prins

Republic RF-84F, standard photo-reconnaissance machine of the RNethAF for many years

Gröningen.

21 May 1946. KLM became the first airline on the European continent to initiate a scheduled service to New York. First flight was made by a Douglas DC-4 piloted by Captain E. van Dijk.

27 March 1953. The air forces of the Royal Netherlands Army became an independent service as the Royal Netherlands Air Force.

1 November 1958. KLM inaugurated its Amsterdam-Tokyo scheduled service, via the North Pole and Anchorage, Alaska. Its DC-7Cs had a flight time of 30 hours over the route. Since KLM already operated to Tokyo on a southern route, this also marked the beginning of a Dutch round-the-world air service.

10 May 1960. The Dutch National Air Museum was opened at Schiphol Airport after a quarter of a century of preparation.

Newark Airport, see New York

New York Airport (John F. Kennedy International), NY, USA

Opened to commercial flying in 1948 and known as Idlewild until renamed in 1963. Now one of the world's largest and busiest international airports, with the architecture (by Eero Saarinen) an outstanding visual feature.
Distance from city centre: 17 miles
Height above sea level: 12 ft
Main runways:
  13R/31L = 14,572 ft
  04L/22R = 11,352 ft
  13L/31R = 10,000 ft
  04R/22L = 8,400 ft
Passenger movements in 1967: 19,988,600
New York Airport (La Guardia), NY, USA

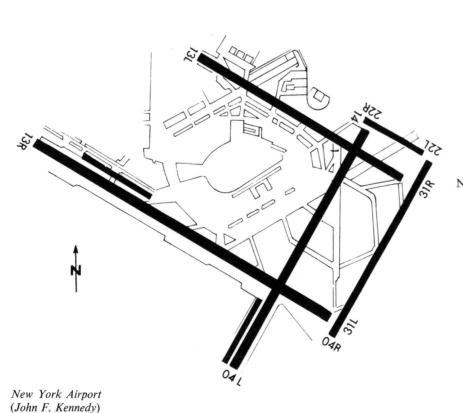

New York Airport (John F. Kennedy)

Distance from city centre: 8 miles
Height above sea level: 21 ft
Main runways:
  04/22 = 7,000 ft
  13/31 = 7,000 ft
Passenger movements in 1967: 8,136,000
New York: Newark Airport, New Jersey, USA
Built on former marshland. First flight from Newark was made in August 1928. It is now one of the biggest terminals in the New York traffic area.
Distance from Newark city centre: 3½ miles
Distance from New

New York Airport (La Guardia)

York: 13½ miles
Height above sea level: 18 ft
Main runways:
  04/22 = 7,000 ft
  11/29 = 6,796 ft
Passenger movements in 1967: 6,070,000
New York Airways
Helicopter airline, formed in 1949, which began by operating an air mail service between airports in the New York area in October 1952. This service was later extended, an inter-airport passenger service was opened in July 1953, followed by a service to Manhattan in December 1956. It now also serves the heliport on the roof of the PanAm building in New York City. Financial assistance is provided by Pan American and TWA (*qq.v.*)
Headquarters: New York (La Guardia Airport).
Associate member of IATA.
New Zealand: Air Force
Royal New Zealand Air Force. Although a Blériot monoplane was supplied to the New Zealand Government in 1913, and New Zealand aircrew served with the RFC and RNAS during the 1914-18 War, it was not until 1923 that a very modest air force was created officially. By 1930 this still comprised only about 20 aircraft and less than 50 personnel. The first signs of the creation of

a reasonable-sized air arm came in the 1935 budget, and on 1 April 1937 the Royal New Zealand Air Force was established as an independent body. Expansion up to the outbreak of World War Two was steady rather than spectacular, but during the war New Zealand flying schools made a tremendous contribution to the Empire Air Training Scheme—some 13,000 aircrew, including over 5,600 pilots, received at least a part of their flying training in New Zealand.

Operational units of the RNZAF gave distinguished wartime service, especially in the Pacific theatre once their own modest aircraft resources had been boosted by the arrival

of US combat aircraft. The RNZAF squadrons in the European theatre were equipped chiefly with British fighters and bombers. The RNZAF kept several surviving types after the war until 1950, when the acquisition of Vampire trainers and fighter-bombers marked the introduction of jet aircraft into service. These were followed later by Canberra jet bombers, which are still in service. Maritime reconnaissance work, previously

performed by Sunderland flying boats, is now carried out by land-based Lockheed Orion patrol aircraft.

New Zealand National Airways Corporation NZNAC was created in 1945 by the amalgamation of the former Air Travel (NZ), Cook Strait Airways and Union Airways. When it began operations in 1957 its acquired fleet ranged from Fox Moth biplanes to Sunderland flying boats, but a gradual re-equipment programme took place from the early 'fifties with Douglas DC-3s and from the late 'fifties with turboprop Viscounts and Friendships, which still constitute the majority of the fleet. NZNAC's present network covers some 4,000

North American Harvards of the RNZAF

route miles in and between the North and South Islands.
Headquarters: Wellington.
Associate member of IATA.
Nieuport, 629
The Nieuport monoplane became well established in France in 1911, and enhanced the company's reputation by winning the French Government's test meeting at Rheims against 28 competitors; this led to immediate purchase of the winning aircraft

Bristol 170 Freighter of 41 Squadron RNZAF

56 Nieuport "10,000" observation and patrol aircraft served with the Italian Air Force in 1915-17

and an order for ten more. The type was used to great effect by one of the leading English pilots, Claude Grahame-White, both in the UK and in the USA. A later type was used by the British Army, and in it Lieutenant B. H. Barrington-Kennett established a world's closed-circuit distance record with a flight of 249 miles 840 yards on 14 February 1912.

During the 1914-18 War Nieuport began production of a series of sesquiplane fighters, the first of which, the Nieuport 11, was one of the Allied types which finally overcame the "Fokker scourge" in 1915-16. It was followed by the Nieuport 17 and 28, both widely used during the war. A radical change of design, instigated by Gustave Delage, marked the appearance of the Nieuport-Delage 29, which became one of the most successful fighters of the nineteen-twenties. The association continued, mainly for the evolution of military aircraft, into the nineteen-thirties.

Nieuport two-seater, 861
Nigeria Airways
Previously known as WAAC (West African Airways Corporation) which was formed by Gambia, the Gold Coast, Nigeria and Sierra Leone in 1946 as an associate of BOAC. The Nigerian services of this consortium were taken over on 1 October 1958 by WAAC (Nigeria) Ltd, which now operates as Nigeria Airways. In addition to internal routes, regional services are now flown to Dakar, the Congo-

lese Republic, London and New York. The airline has been wholly owned by the Nigerian Government since 1961. Headquarters: Lagos Airport.
Member of IATA.
NK-12M engine, see Kuznetsov
Nord 262

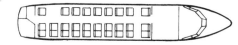

Nord 262

This French short-haul transport originated as a Max Holste design, the MH-250 Super Broussard, having Wasp piston-engines and an unpressurized fuselage. It was developed by Nord-Aviation, via the turboprop-engined MH-260, into the Nord 262 with a new, pressurized fuselage of circular cross-section. It has since been sold to several short-haul airlines around the world including Allegheny Airlines, which acts as US distributor for the type. With two 1,065-ehp Turbomeca Bastan turboprops, the Nord 262 has a maximum range of 690 miles with a 4,209-lb payload, and can cruise at up to 233 mph. A maximum of 29 passengers can be accommodated.
Nord AS.20 missile, 896
Air-to-surface missile in large-scale use by the French Navy and Air Force and with German and Italian forces. It has a dual-thrust solid-propellant rocket motor and a 66-lb conventional warhead. It is normally controlled visually from the parent aircraft, guidance being

by radio control, although it can be adapted to an optical aiming/infra-red guidance system. Its speed is Mach 1.7 and range 4.35 miles.
Nord Griffon, 301
The Griffon was developed in the late nineteen-fifties to pioneer a new approach to the fighter concept. A SNECMA Atar 101 turbojet was mounted within and at the front of a large ramjet of Nord design, the object being to use the turbojet for take-off and climb and the ramjet for boosting performance at the heights and speeds where its efficiency would be paramount. It first flew in this form on 23 January 1957, and in May 1957 exceeded Mach 1 with the ramjet at less than full power. Development continued until full exploration of the Griffon's possibilities had been carried out. This mixed-powerplant concept was not taken any further, due to the rapid advance in the potential output of the more normal turbojet.
North American
North American Aviation was incorporated in Delaware in 1928. From 1934 until the end of World War Two it was engaged exclusively on the design and production of military aircraft, and to this end a new manufacturing plant was established on the site of Los Angeles Municipal Airport at Inglewood, California, in 1935. Best-known pre-war product was the NA-16, from which developed the AT-6 Texan (or Harvard), one of the most widely-used trainers ever built; about 15,000 were produced.
Two of the most outstanding aircraft of World War Two—the

Mustang fighter and the Mitchell bomber—came from North American. Both remained in service with numerous world air forces for many years after the war, and examples of both types are still flying today. The Mustang in particular is a favourite mount of civilian racing pilots, and in 1968-9 a turboprop-engined version was being developed for present-day ground-attack duties.
Since World War Two the company has continued to specialize in military aircraft, the most successful of which has undoubtedly been the F-86 series of Sabre jet fighters. Sabres have

North American F-86H, most powerful of the original Sabre family

been supplied under MDAP to most of the world's major air forces, as well as being built under licence (and further developed) in Australia and Canada. The F-100 Super Sabre was, despite its name, a completely new design: it was the first of the "Century Series" fighters and the first supersonic US warplane. Other major types of the post-war era include the A-5 Vigilante naval bomber and the OV-10 Bronco counter-insurgency aircraft. Two other most important types are the XB-70 Valkyrie Mach 3 aerodynamic research aircraft and the rocket-powered X-15A. The latter, which was air-launched from a B-52 "mother plane", has been flown at a speed of 4,534 mph (Mach 6.72) and to an altitude of 354,200 ft, and several X-15 pilots have qualified for astronaut's "wings" by flying to

more than 50 miles above the Earth.
On 22 September 1967 North American merged with Rockwell Standard Corporation; the new organization, which is known as North American Rockwell Corporation, is now responsible also for the former Aero Commander range of light and business aircraft.
North American Hound Dog missile, 896
Standard armament on all current versions of the Boeing B-52, the Hound Dog is a stand-off bomb with a thermonuclear warhead of approximately 4-megaton yield. It is powered by a Pratt & Whitney J52 turbojet of 7,500-lb thrust, conferring a speed of Mach 2 and a range in excess of 600 miles.
North American Mustang (P-51), 1101
The Mustang was designed by North American to British requirements at the beginning of World War Two. First flight was in May 1941, and delivery to the RAF began in November 1941. It was a single-seat, single-engined fighter with laminar-flow wings and an Allison V-1710-39 engine of 1,150 hp.
It was found that this engine conferred a fine performance low down (max speed 390 mph at 8,000 ft), but this was not sustained at altitude, and the early Mustangs were therefore used for fighter-reconnaissance duties. The Mk III and Mk IV Mustangs (P-51B, C and D) were redesigned with 1,680-hp Packard-built Merlin engines and were

used extensively by the RAF and the USAAF as long-range escort fighters in the last two years of the war. The P-51D Mustang, with a maximum speed of 442 mph, range of 1,710 miles and ceiling of 42,500 ft, was one of the finest fighters of the war. It remained in extensive post-war service with the USAF for many years and was supplied to many smaller air forces all over the world. Many of these still serve today, and "civilianized" Mustangs are popular among competitive fliers in the US and elsewhere.
North American Sabreliner (T-39), 837
One of the earliest of the executive jets to appear, the Sabreliner had the advantage of winning a military contract for a jet trainer and communications aircraft, although it was first produced as a private venture. First Sabreliner flew in September 1958; 191 aircraft were subsequently produced for the USAF and US Navy, paving the way for civilian production. The aircraft has since been refined to provide increased amenities, and there is also a "stretched" version for seating up to ten (including two crew). It is a small aircraft with a wing span of 44 ft 5 in; its two rear-mounted Pratt & Whitney JT12A turbojets give it a maximum cruising speed of 563 mph at 21,500 ft.
North American Super Sabre (F-100), 1189
First of the USAF's "Century Series" fighters, the F-100 gained distinction by being the first fully supersonic fighter in service in the world.
Prototypes were ordered in November 1951, and the first aircraft flew in May 1953, going supersonic on its first flight. It set a new World Speed Record of 755.149 mph in Octo-

ber 1953. Production began with the F-100A, which entered service in September 1954, but this model encountered stability troubles and the next production version was the F-100C with a revised tail and provision for greater underwing loads. This went into large-scale service both in the US and overseas, and was followed by the F-100D variant with provision for LABS bombing, Bullpup or Sidewinder missiles.

The F-100 has been in widespread service with the Tactical Air Command of the US and has been supplied also to French, Danish, Turkish and Nationalist Chinese Air Forces. It continues to give useful service in the fighter-bomber rôle in the war in Vietnam.

North American Texan (T-6), 1093

Probably one of the finest training aeroplanes ever produced, the Texan (or Harvard, as it is known by many), has been the aircraft on which many thousands of pilots have gained their "wings".

Like many famous aircraft, it was the result of a line of development reaching back many years. For some time in the 'thirties, North American had built low-wing monoplanes with radial engines, swept-back wing leading-edges and deep fuselages with two seats in tandem under a large canopy. The first design in the genesis of the Texan/Harvard was the BC-1A for the US Army, which the US Navy used as the SNJ and which the British RAF bought as the Harvard I in 1939. Powered by a Pratt & Whitney R-1340-47 Wasp of 550 hp, this version was extensively used in the Empire Air Training Scheme, followed by its cleaned-up derivative, the Harvard II (AT-6C). This version also went into large-

scale service with the US forces, and continued in production throughout and after World War Two, over 15,000 being built.

Large numbers became available after the war, and were used both as trainers and as ground-attack aircraft by major and minor powers until the mid-'sixties, being found more suitable than many more sophisticated aircraft in the "brush-fire" type of war so prevalent since 1945.

North American Tornado (B-45), 817

First multi-jet bomber to enter service with the USAF. Powered by four Allison J35 turbojets each of 4,000-lb static thrust, it had a maximum speed of 550 mph. The turbojets were mounted, paired, in nacelles under the wings. The B-45 entered service in 1948 and served for many years, at first in the medium bomber rôle and later as a tactical bomber. It was also used for medium-range reconnaissance with the USAF.

North American Valkyrie (XB-70), 483, 485, 501-4, 645

The XB-70A began life as a Weapons System requirement of the USAF in 1954. This requirement was for a supersonic strategic bomber to replace the B-52, and North American won the contract with the XB-70 design. Before the first aircraft was built this requirement had been rendered obsolete by America's missile armoury and the XB-70 programme was curtailed, first to three and finally, in 1963, to two aircraft. It is a large, tail-first delta-winged aircraft with six General Electric YJ93 turbojets, each developing 31,000 lb st. One of the novel features of this aircraft is that the fuel is "blanketed" with a nitrogen seal around the tanks, to prevent it igniting when the temperature rises as a result of kinetic heating.

First flight was in September 1964, and soon after the aircraft was flown to a speed of Mach 2.85 and to a height of 68,000 ft. A second aircraft joined the flight test programme in 1965, but one aircraft was subsequently destroyed in an aerial collision.

North American Vigilante (A-5), 629, 644-5

This very advanced twin-jet two-seat carrier-based all-weather attack aircraft first flew in August 1958, and was the subject of large-scale production orders in 1959. It is in service now mainly in the reconnaissance rôle (as the RA-5) aboard carriers of the US Navy.

North American X-15, 297, 303, 485, 503-4, 585, 604, 646, 647, 1137

Three X-15 manned research aircraft were ordered in November 1955, and the first aircraft flew (on its B-52 "mother ship") on 10 March 1959. The X-15s were built for a joint USAF/USN/NASA advanced programme for research up to speeds of Mach 7 and heights of 264,000 ft. The first free flight was on 8 June 1959 (without using the rocket motor), and the first powered flight, made by the second aircraft, was on 17 September 1959; a speed of Mach 2.1 was attained in a climb on that first flight, the pilot being Scott Crosfield. This aircraft was badly damaged on its third flight.

All three aircraft have since been engaged in a continuous programme of research to the frontiers of manned flying, as these "highest and furthest" figures show:
Highest recorded speed to date: 4,534 mph (7,297 kmh) in the X-15A-2 on 3 October 1967—pilot Major Pete Knight of the USAF.
Highest Mach number: Mach 6.72, on the same occasion.
Highest skin temperature: 1,320 °F on 8 May

1962—pilot Lt Col Robert Rushworth, USAF.
Greatest height reached: 354,200 ft (107,960 m) on 22 August 1963—pilot Joseph Walker, chief test pilot of NASA. The second aircraft was again damaged in November 1962 and has since been rebuilt in a new form for further research; most of the modifications have been concerned with enabling it to cope with even higher skin temperatures for hypersonic flight.

North American Rockwell Jet Commander, 837

Fastest of a long line of executive aircraft from the Aero Commander stable, the first Jet Commander flew in January 1963. With accommodation for eight passengers, it has two General Electric CJ610 turbojets mounted at the rear of the fuselage, which give it a maximum cruising speed of 500 mph at 35,000 ft. In the first three years of its existence 62 aircraft were sold. In 1968, manufacturing rights were acquired by Israel Aircraft Industries Ltd.

Northeast Airlines

Major US regional trunk route operator, Northeast was founded in the nineteen-thirties as Boston-Maine Airways, taking its present title in 1940. It operates a large route network in the eastern United States, extending from Miami in the south to Montreal in the north. Headquarters: Boston, Massachusetts.

Northwest Orient Airlines

Operating title of Northwest Airlines Inc, formed in 1926 as Northwest Airways. Apart from routes to Florida and Philadelphia, main services are transcontinental ones westward of its base at Minneapolis, services across the Pacific to the south-east Asian island groups, and others to Alaska and Canada. Total route mileage is about 20,000. Headquarters: Minneapolis, St Paul Airport, Minnesota.

Member of IATA.

Norway: Air Force

Luftforsvaret (Royal Norwegian Air Force). Official support for military aviation in Norway came soon after the outbreak of the 1914-18 War, although both the Army and the Navy had possessed their own aircraft since 1912. The domestic factories built Farmans, Bristol Fighters and Brandenburg seaplanes during the war, and in the 'twenties began to turn out modest numbers of nationally-designed aircraft. Expansion during the middle and late 'thirties brought more foreign types into both the Army and Naval Air Forces, among them Curtiss Hawk fighters and Douglas DB-8 dive-bombers from the US, Gladiators from Britain and Heinkel He 115 seaplanes from Germany. These resisted the German invasion for a while in 1940, after which many of those still surviving escaped to Britain. A "little Norway" training base was

set up near Toronto in Canada, where Norwegian personnel were trained to continue their war effort in US or British aircraft. Norwegian-manned fighter squadrons joined the Allied forces in Europe after the invasion, while others performed valuable convoy escort and anti-submarine patrols over the North Sea and North Atlantic.

The two separate air arms were combined in a new Royal Norwegian Air Force in the Spring of 1944, and re-equipped after the war with British Spitfires, Mosquitos and Ansons and ex-RAF Catalinas. The first jet squadron was formed with Vampires in 1948; Republic Thunderjets from the US followed in 1951, and F-86F Sabres in 1957. The Sabres have recently been phased out in favour of Northrop F-5s, which with F-104G Starfighters form the major tactical element of the present-day Air Force. Maritime patrol units are equipped with the Lockheed P-3 Orion.

Norway: Chronology

First military aircraft to see service in Norway was a German *Taube* monoplane. Bought by five naval officers, this machine was presented to the Norwegian Navy on 1 August 1912. At about the same time, the Norwegian Army also acquired its first aircraft.

Production of aircraft in Norway started in 1914 when the Haerens Flyfabrikk (Army Aircraft Factory) and Naval Flying Boat

Northrop F-5 aircraft are the standard tactical strike fighters of the RNorAF; 65 are in service

Factory were established, and soon after this both services set up their own flying schools. The Norwegian Army also received aircraft from the Kjeller Flyfabrikk at Lillestrøm, which began production in 1915.

In May 1927 the civilian airline Det Norske Luftruter (DNL) was founded, operating services to Denmark and Germany; it was integrated into the Scandinavian Airlines System (SAS) in 1946. In the same year that DNL lost its identity, a new carrier came on the scene: Braathens South America & Far East Airtransport (currently known as Braathens SAFE). In addition to its international services, this company inaugurated domestic routes in 1952. When Norway was invaded by German forces in 1940, its Air Force was eliminated quickly, but many of its pilots and ground crews managed to escape to Britain, where they served with Norwegian squadrons within the RAF. Today, Norway and Denmark together form the Northern Command of NATO.

NOTS Sidewinder missile, 895, 897
The Sidewinder was developed by the US Naval Ordnance Test Station, first firing being in September 1953. It is extremely simple and cheap and is propelled by a solid-propellant rocket motor with a 10-lb high-explosive warhead. It uses infra-red homing and is in large-scale service aboard aircraft of the Western Powers.

Nozzle, 1171, 1181, 1187, 1189, 1212, 1216, 1235, 1243-1245, 1249, 1251, 1253, 1263, 1273ff
NY-2, see Consolidated
NYP monoplane, see Ryan
O/100 and O/400, see Handley Page
Oberursel engines, 248, 875-7
Before the 1914-18 War,

the German Oberursel company took up manufacture of the Gnome rotary engine under licence. This engine powered a number of German warplanes during that period, including the Fokker Eindecker.
O'Hare Airport, see Chicago
Oleo strut, 529, 565, 571, 583, 593
This is a unit used in undercarriages to absorb landing and taxiing loads. Although shock struts have been designed which rely entirely on air, most

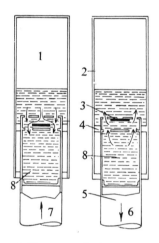

1 Compressed air
2 Cylinder
3 Restrictor valve
4 Seals
5 Piston
6 Rebound
7 Load
8 Oil

modern struts employ air and oil. In these the resilient medium (air) is trapped above or within a piston, at the top of the telescopic part of the leg, sliding in an oil-filled cylinder.
When the aircraft lands, the piston is forced up to compress the air and so absorb the energy of impact. At the same time oil passes through the piston. When the leg extends, the return flow of the oil is restricted, thus damping any rebound.
The ratio of "air" to "oil" travel can be varied for specific duties. Struts combining a long "oil" travel with a short "air" travel are often used on naval aircraft,

where the initial impact of landing on a deck may involve high velocities—and thus long travel—but where taxiing and take-off require only a short, stiff gear and one free from roll.
Olympia Aero Shows, 361 see also Air shows and exhibitions

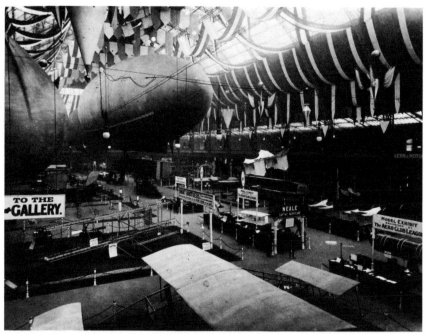
First British indoor air exhibition: the Olympia (London) Aero Show of 1910

These were the first significant aero shows in the UK in which, for the first time, real advances in the art of aircraft design were becoming apparent. At the first of them, in 1910, there were 21 complete aircraft on view, 17 of which were British-built. It was notable that, even at that early date, there was one all-steel aircraft, a tractor biplane built by the French Breguet company.
Olympic Airways
Owned by Greek shipping magnate Aristotle Onassis, Olympic Airways was formed to take over from the Government-owned national airline TAE, and began operations in April 1957. It now operates a domestic service connecting the principal Greek islands and cities, plus international routes to major countries in Europe and the Middle East and to the US. Headquarters: Athens. Member of IATA.

Olympus 593 engine, see Rolls-Royce/SNECMA
One-Eleven, see BAC
Orion engine, see Bristol
Orly Airport, see Paris
Ornithopter, 117, 607, 609, 625
A mechanically-driven aerodyne whose lift in flight is obtained by the action of "flapping" wings. The only successful ornithopters to date are natural creatures, although Man has continually sought to achieve flight by this means.
Ozark Air Lines
Since its incorporation in September 1943, Ozark has doubled its original activity and now serves nearly 60 cities in 12 mid-western states of the US with a route network of nearly 5,700 miles.
Headquarters: St Louis, Missouri.
P-40 Warhawk, see Curtiss
P-51 Mustang, see North American
P. 1040, see Hawker
P. 1127, see Hawker Siddeley
Packet (C-82 and C-119), see Fairchild
"Packplane" (XC-120), see Fairchild
Pakistan: Air Force
Pakistan Air Force: The Pakistan Air Force was formed initially with one fighter and one transport squadron from the former Royal

Indian Air Force, after the partition of India in mid-1947. Within the next few years the PAF built up three squadrons of Tempest fighter-bombers and one of Halifax heavy bombers, together with a mixed Anglo-American transport fleet. The Tempests gave way to Fury monoplanes in 1950, which were themselves superseded by Supermarine Attacker jet fighters and later by F-86F Sabres. The latter type remains in service alongside recently-acquired Chinese-built MiG-19s and French Mirage IIIs, while the tactical bomber force is still equipped with the Martin B-57 (Canberra) jet bombers acquired some years ago.
Pakistan International Airlines
The present Pakistan International Airlines Corporation was created by the Pakistan Government in 1955 as successor to the former Orient Airways, which had operated internal and overseas services since 1946. After an uneasy start, PIA's fortunes picked up rapidly after the introduction of jet services in 1960, and it now flies into London via points in the Middle East and Europe, and

via Moscow; into central Africa; eastwards to China; and, more locally, to Afghanistan and Burma. A helicopter "feeder" service was operated in East Pakistan, a task that may eventually be taken over by hydrofoils or hovercraft.
Headquarters: Karachi Airport.
Member of IATA.

Pan American World Airways, 273, 761
This famous airline was born in October 1927 with a mail and passenger service from Key West to Havana, flown by a Fokker F.VII. Although this was a landplane, PAA's prewar operations were predominantly with floatplanes and flying boats. Most of these were Sikorsky 'boats, used to extend Pan American's routes across the Caribbean, Central America and into South America, reaching as far as Buenos Aires by 1929. The next few years were involved in building up reliability and frequency on these routes, but the airline's President, Juan Trippe, was anxious to inaugurate routes across the Atlantic and Pacific as soon as suitable aircraft appeared. Two contenders appeared in 1932: the Sikorsky S-42 and the Martin 130; and PAA ordered both types. By 1934, preparations could begin for the trans-Pacific route, and on 22 November 1935 a Martin 130 flew the first scheduled air mail flight across this ocean. In 1937, survey flights were flown on two different routes across the Atlantic, but the first scheduled flight across the Atlantic, with Boeing 314s, took place in May 1939.
Up to the outbreak of World War Two PAA had been predominantly an overwater airline, but in 1940 it took delivery of Boeing 307 landplanes for operations within the Ameri-

cas. Wartime transport flying involved the airline more and more with landplanes, and when peace came again PAA set about opening up its old routes with DC-4 and Constellation landplanes. The airline soon rose to become one of the world's biggest intercontinental airlines. Always progressive, it introduced the Stratocruiser luxury airliner on its "President" de luxe transatlantic service in 1949. The airline maintained its reputation as a pace-setter by ordering Comet jet airliners in 1953, but after the disasters to the early Comets it then chose the Boeing 707, which it placed in service on transatlantic routes at the end of 1958.

Pan American has maintained its position as a topmost airline

Pan American World Airways was the first operator of any Boeing jetliner: the 707-120

ever since, with services reaching wherever Americans want to travel. Its latest innovation is the ordering of Boeing 747 "jumbo-jets" for its routes.
Headquarters: New York City.
Member of IATA.

Panther, see Parnall

Parabellum machine-gun, 877
A belt-fed machine-gun in widespread use with the German Air Force in the 1914-18 War.

Paris Airport (Le Bourget), France
French terminal for the first London—Paris services in 1919, Le Bourget now handles over 1¾ million passengers a

year, and is also celebrated for the biennial Salon de l'Aéronautique et l'Espace.
Distance from city centre: 10½ miles

Formerly a military flying school, Orly has been developed into a major international airport since World War Two, and has excellent modern terminals and easy road access to the centre of Paris,
Distance from city centre: 11¼ miles
Height above sea level: 292 ft
Main runways:
  07/25 = 11,975 ft
  08/26 = 10,892 ft
  02L/20R = 7,874 ft
  02R/20L = 6,080 ft
Passenger movements in 1967: 6,907,300

Paris Airport (Roissy), France
Alternatively known as Paris-Nord, Roissy is the new airport planned to succeed Orly as France's major international terminal when the latter reaches its traffic limit in the mid-nineteen-seventies. It is being planned to handle up to 25 million passengers and 2 million tons of freight a year. It will have,

initially, two runways each 11,800 ft long, all capable of future extension if required.

Parnall Panther, 361

Parnall Prawn, 811

Passarola, 126, see also de Gusmão

Pauly, S. J., 139

Pegasus piston-engine, see Bristol

Pegasus vectored-thrust engine, see Rolls-Royce (Bristol)

Perseus engine, see Bristol

Peru: Air Force
Fuerza Aérea Peruana (Peruvian Air Force). This small but well-equipped modern force first began to take shape in 1919, when a French Aviation Mission arrived in the country. A similar Mission from the US Navy arrived in 1920, and during the early 'twenties both Army and Naval Air Services were established in Peru with aircraft of French, American and British design. They were combined in

a single force in May 1929, and became operational in frontier incidents with Colombia four years later. In 1935 Peru entered into an association with Italy to reorganize and re-equip the Air Force, and several Caproni types were acquired; but a few years later Peru turned to the US for more modern aircraft, and the Italian influence ceased in 1940. There were further border incidents, this time with Ecuador, in 1941, but Peru took no part in World War Two and her next major re-equipment followed the Rio Mutual Defence Treaty of 1947, when substantial numbers of US wartime types replaced most of the aircraft previously in service. In 1950 the *Cuerpo de Aeronautica del Perú*, as it had hitherto been known, adopted its present title, and in 1955 the first jet aircraft (Sabre fighters and T-33A trainers) marked the start of the next major modernization programme. Sabres are still in service, together with Hunter fighters and Canberra bombers from the UK, and Peru is among the customers for the French Mirage 5 multi-mission strike fighter.

Pfalz, 361
Manufacturer of German fighter and reconnaissance aircraft in the 1914-18 War.

Phantom II, see McDonnell Douglas

Philippine Air Lines
This airline has operated under its present title since 1941, having formerly been known from its formation in 1931 as Philippine Aerial Taxi Co. Soon after World War Two it absorbed two other local carriers, and later began long-distance services to the US and Europe. Between 1954-62 it restricted itself to local and regional services only, but is now fully-international again, with routes to Australia, Japan and the US in addition to an extensive network within the Philippine Islands group.
Headquarters: Manila.
Member of IATA.

Phillips, Horatio F., 193
This Englishman took out patents in 1884 and 1891 for "sustainers" with an aerofoil cross-section having a greater curve on their upper surfaces than on the lower ones. This represented the archetype of all aerofoil-section wings used subsequently. Phillips built several aircraft employing numbers of these sustainers (which were of very narrow chord) mounted on a frame or series of frames resembling a large Venetian blind. On one of these machines he made several hop-flights in 1907, including one of just over 500 ft (150 m) which may be regarded as the first flight by a powered aeroplane in England.

The unmistakable 1904 Phillips multiplane with single "Venetian blind"

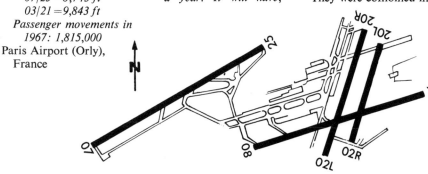

Paris Airport (Orly), France
Height above sea level: 217 ft
Main runways:
  07/25 = 8,743 ft
  03/21 = 9,843 ft
Passenger movements in 1967: 1,815,000

Phoenix, see Heston

Phoenix missile, see Hughes

Photography, see Aerial survey

Piaggio

Previously noted for its engineering and ship-building interests, SA Piaggio & Cie entered the aircraft industry at Geneva in 1916, and after the 1914-18 War also began to manufacture aero engines at Pontedera; airframe construction was then concentrated mainly at the Finale-Ligure works. The majority of its early products were licence-built versions designed by other companies, including Dornier Wal and Savoia-Marchetti S.M.55 flying boats and Bristol and Gnome-Rhône radial engines. The Piaggio P.7 of 1929 was a novel but unsuccessful seaplane built for an attempt on the Schneider Trophy. Piaggio also built the helicopter designed by Ing Corradino d'Ascanio, which flew successfully on 8 October 1930. During World War Two Piaggio designed Italy's only four-engined heavy bomber, the P.108B, and in post-war years has been noted chiefly for the P.49 basic trainer and the P.136 and P.166 light utility transports. Latest product is the Piaggio-Douglas PD-808 twin-jet rear-engined executive and communications aircraft.

Piedmont Airlines

Operating title, since 1948, of Piedmont Aviation Inc. It has built up a 9,000-mile route network in the eastern US, and its large, mixed fleet now flies to 77 cities in 11 states.
Headquarters: Winston-Salem, North Carolina.

Pilatus

The Pilatus Flugzeugwerke AG was formed late in 1939 and, as a subsidiary of the Swiss Oerlikon Company, began to manufacture aircraft in its works at Stans, near Lucerne, in 1941. First product of its own was the SB-2 Pelican single-engined light transport, which appeared toward the end of World War Two. After the war Pilatus built the P-2 and P-3 advanced trainers for the Swiss Air Force and the P-4 five-seat cabin monoplane. The company is currently enjoying considerable success in world markets with the PC-6 Porter and Turbo-Porter general-purpose STOL monoplanes and in 1968 introduced the Twin Porter, a twin-engined light transport developed from the earlier design.

Pilatus Porter, 1311

The Porter originated with the Swiss Pilatus Flugzeugwerke AG, making its first flight in May 1959. It is a high-wing utility STOL transport, able to seat up to 10 passengers or take freight items up to 16 ft long.
It was originally powered by a Lycoming engine, and achieved some measure of success. It has since been produced, as the Turbo-Porter, with Turbomeca Astazou, Pratt & Whitney PT6A and AiResearch TPE 331 turboprops, which greatly enhance its performance and usefulness. Porters are in use in many parts of the World and are licence-built by Fairchild Hiller in America.

Pilcher, Percy, 169, 175

Percy Pilcher, who had at one time assisted Sir Hiram Maxim, closely followed Otto Lilienthal in the development of gliders which were basically similar to those designed by the German. Pilcher's first glider, completed in June 1895, was relatively unsuccessful until he incorporated improvements suggested in principle during a visit to Lilienthal. Pilcher built five gliders in all, the last one being a triplane owing much to

Replica of Percy Pilcher's Hawk glider, made by Armstrong Whitworth apprentices in 1958

the influence of Hargrave's box-kites. He was about to build another glider, similar to his most successful one, the Hawk, but with an oil engine for propulsion. Before he could complete it, however, the Hawk sustained a structural failure in the air and Pilcher died from his injuries on the following day, 2 October 1899.

Piper Cherokee, 415-420

The Piper Cherokee was first introduced by Piper Aircraft Corporation in 1960 as a cheap and simple four-seat light aircraft. It was Piper's first such aircraft with a low-wing monoplane configuration and has since virtually replaced the long line of traditional Piper high-wing monoplanes stemming from the original Cub through to the Tri-Pacer family. The Cherokee has itself been developed into a "family" of aircraft, extending from the basic 140-hp aircraft, through the Cherokee Arrow with retractable undercarriage, to the Cherokee Six with accommodation for six/seven people and a choice of 260- or 300-hp Lycoming engines.

Piston, 1054, 1055, 1059, 1067, 1075, 1081-83, 1097, 1101, 1125

In a piston-engine (reciprocating engine), the plug-like unit which oscillates to and fro in the cylinder to convert the high pressure of the burning fuel/air mixture into useful mechanical work, and also to draw in the fresh charge,

compress it and expel it after combustion has been completed. The piston is usually a forging machined in a heat-resistant aluminium alloy and has the form of an inverted cup, the head (known as the crown) being solid and the bottom terminating in an open "skirt" section, The diameter of the piston is less than the bore of the cylinder, but a gas-tight fit is achieved by a series of rings of springy, high-tensile steel which fit corresponding grooves around the piston; some of these rings serve to ensure that no gas leaks down from the combustion space, while the lowest is generally an oil scraper ring to prevent lubricating oil from leaking up past the piston and being wasted by being burned in the combustion chamber. Across the centre of the piston are arranged strong bearings for the wrist pin (gudgeon pin) which pivots the piston to the connecting rod and thence to the crankshaft.

Piston ring, see Piston

Piston speed

The figure normally quoted for piston speed is based upon the actual distance a piston travels in one minute, which may be of the order of 3,000 ft at maximum crankshaft rpm. It will be appreciated that near mid-stroke the actual speed of the piston may be nearer to 5,500 ft/min because it "dwells" almost at rest at the end of each stroke.

Pitch, 1333

Pitch change bearing

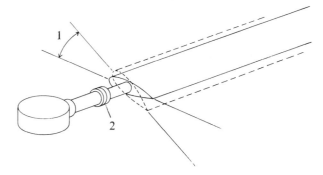

1 Pitch angle

Bearing embodied in a helicopter rotor head allowing the blade pitch angle to be varied. The angle of pitch is controlled by the swashplate (q.v.) which in turn is connected to the pilot's control levers in the cockpit.

Pitot head, 433, 939, see also Altimeter and Airspeed Indicator.
The pitot head has been installed on aircraft ever since instruments for giving flight information have been carried. The pitot head, or its later development the pressure head, comprises two tubes facing forward into the airflow. One tube is closed and maintains a standard atmospheric pressure. The other is open-ended and records the pressure of the airflow on the aircraft. The tubes feed to opposite sides of a diaphragm which moves according to the pressure differential, the diaphragm being linked to the pointers of the airspeed indicator.

Platz, Reinhold, 369

PLUNA (Primeras Lineas Uruguayas de Navegacion Aerea)
As its full title implies, PLUNA is the principal Uruguayan airline, having begun as a privately-owned company in 1936 and become entirely State-owned in 1951. It operates an internal route system as well as international flights to points in Argentina, Brazil and Paraguay. Headquarters: Montevideo.

2 Pitch change bearing

Poland: Air Force
Polskie Lotnictwo Wosjkowe (Polish Air Force). The nucleus of a permanent Polish Air Force was formed in 1918 in France, where the early pilots were trained, and the force was equipped at the outset with French Spad fighters and Breguet 14 bombers. It was involved in combat with the Ukraine in mid-1919, by which time bases and flying schools had been established in Poland and the aircraft were of an already varied assortment of nationalities. The Air Force was reorganized in 1921, and a few years later the reigning government set up a "League of Air Defence" which, by the outbreak of World War Two, had over a million members. This movement served both to provide trained fliers for the Air Force and, through its public displays of flying and gliding, to make the nation as a whole airminded. Parallel development of a healthy aviation industry ensured a steady flow of aircraft, many of which during the 'thirties were excellent types of Polish design.
Notable among these were the P-11 and P-24 fighters designed by Pulawski for the PZL (q.v.) company, and the twin-engined P-37 bomber; but the Nazi invasion in 1939 caught the Polish Air Force in the middle of a re-

Second prototype of the PZL P-1, a type which put Poland on the map in fighter design

equipment phase, and her obsolescent aircraft were no match for the might of the *Luftwaffe*. Thousands of Polish aircrew escaped, first to France and then to England, and Polish contingents with the RAF played a major rôle in the subsequent air offensive against Germany; many others escaped eastward to the USSR to continue the war from that quarter. After the war Poland came immediately under Soviet influence, and for a time most of the senior positions were occupied by Soviet Air Force officers. Post-war re-equipment was, inevitably, with Soviet aircraft. Since 1953, when it first began to build MiG-15 jet fighters under licence, many of these have been built in Poland. Principal types in current service are MiG-19 and MiG-21 fighters, Il-28 light bombers and Su-7 ground-attack aircraft.

Poland: Chronology

1650. Kazimierz Siemienowicz, Lieutenant-General of Ordnance to Ladislaus IV, King of Poland, published in Amsterdam his classic work *Artis magnae artilleriae pars prima*. In a section entitled *De Rochetis* he set forth for the first time in the world's literature the principle of multi-stage rockets. His designs included a three-stage rocket, each with its own gunpowder fuel, fuse and thrust nozzle; and a rocket with large delta-shaped stabilizing fins.

1892. Stefan Drzewiecki published in Russia and France his first work on the theory of propellers. A versatile inventor, Drzewiecki advocated fixed-wing aircraft as the most rational means of achieving mechanical flight and advanced the first practical method for propeller design and calculation. His Elemental Blade Theory, known also as the Froude-Drzewiecki Theory, formed the basis of all subsequent propeller design.

1893/1896. Czeslaw Tanski designed and built a rubber-powered

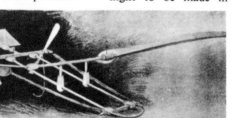

The Tanski model No 1, dating from 1893-4

model aeroplane, capable of controlled free flight, including full circles—the first successful heavier-than-air model to fly in Poland. Tanski followed this in 1896 with a man-carrying glider, named *Lotnia*, on which he made a number of short glides.

June 1910. First sus-

tained flights in Poland on a nationally-designed and built aeroplane were made by Stefan Kozlowski, in Warsaw, on a tractor biplane of his own design. The flights, all in a straight line, covered distances up to 490 ft. Further attempts ended after the aircraft had been damaged in a landing accident.

25 September 1911. The Cywinski and Zbieran-

First all-Polish aircraft: the Cywinski and Zbieranski biplane, completed in 1910

ski tractor biplane, flown by Polish pilot Michal Scipio del Campo, made a flight of 12½ miles at a height of about 170 ft. This was the first fully sustained and controlled flight to be made in

Poland on a nationally-designed aircraft—which was of advanced conception, utilizing steel-tube in its construction.

November 1918. The Polish Air Force was established.

5 November 1918. First combat mission flown by an aircraft of the

Kozlowski's Anzani-powered biplane achieved the first sustained flight by a Polish design

new national Air Force.

29 April 1919. Polish Air Force claimed its first victory when Lt Stefan Stec brought down a Ukrainian Nieuport fighter over Sokolniki.

April to October 1920. The Polish Air Force played a vital part in the Russo-Polish war, making a substantial contribution to Polish victory in the crucial

Battle of Warsaw.

27 August to 25 September 1926. Flying a Breguet XIX, Boleslaw Orlinski and Leonard Kubiak made a flight from Warsaw to Tokyo and back—a total distance of 11,955 miles— the longest intercontinental flight ever made by a Polish crew.

August 1929. The PZL P-1 all-metal Pulawski-wing monoplane fighter flew for the first time. It was the first of the long line of Pulawski-type fighters which remained in production until the outbreak of World War Two, equipping Polish, Bulgarian, Greek, Rumanian and Turkish fighter squadrons.

November 1918. The

1930-1931. Jerzy Rudlicki developed an inclined two-component tail unit—the Rudlicki Vee tail, better known as a butterfly tail. In the

A pioneer "butterfly" tail: the Hanriot built by Plage/Laskiewicz in 1931

early Summer of 1931, a Polish-built Hanriot biplane, with a Vee tail unit, began flight trials at Lublin. It was the first aircraft in the world to utilize this type of tail unit.

In 1933 the RWD 5bis single-seater became the lightest aircraft ever to cross the Atlantic

7-8 May 1933. Stanislaw Skarzynski, flying an RWD 5bis, crossed the South Atlantic from St Louis-de-Senegal, West Africa, to Maceio, Brazil, establishing a new International Distance Record in the FAI Category 2. This record of 2,224 miles is still unbeaten and the RWD 5bis, with an empty weight of 990 lb, was the lightest aircraft ever to cross the Atlantic.

14 September 1935. By winning the Gordon-Bennett Balloon Championship, Burzynski and Wysocki achieved the third consecutive Polish victory and brought Poland permanent possession of the Gordon-Bennett Trophy.

September 1939. The Polish Air Force, with a total combat strength of 400 aircraft, fought alone against the *Luftwaffe*. During the Polish campaign, it achieved

126 confirmed victories against the Germans and lost 333 of its own aircraft in the process.

August-September 1940.

Polish fighter pilots flying with the British Royal Air Force made a vital contribution to victory in the Battle of Britain. They were credited with no fewer than 11 per cent of all

enemy aircraft claimed by the RAF during the Battle.

August 1940 to May 1945. The Polish Air Force in Great Britain operating at a peak strength of 15 combat squadrons, flew a total of 102,486 combat sorties in 290,895 flying hours. Nearly 2,000 of its aircrew were killed or posted as missing in action.

23 August 1944. The Polish Air Force formed in Russia began combat activities on the eastern front.

5 February 1960. The TS-11 Iskra, designed by Tadeusz Soltyk, first nationally-designed and built jet aircraft, and powered by the first nationally-evolved turbojet engine, the SO-1, made its maiden flight near Warsaw. The Iskra entered service with the Polish Air Force as a standard basic trainer in 1964.

Police and patrol services The use of aeroplanes for policing duties— in the widest sense of that word—has existed for many years, and air patrols for one purpose or another are today comparatively common practice in many countries. They are particularly useful, both to the established police forces and to motoring organizations, for one of the most universal problems of the day— traffic control. The New York Police Department, for example, em-

One of the first "police and patrol" photographs in the files of Aerofilms Ltd: traffic on Derby Day, England, 1921

ploys a full-time fleet of helicopter "sky cops" to detect and advise on the relief of traffic congestion, or drivers who exceed the speed limits. Most other State Police and Highway Departments in the US also maintain aircraft for this purpose. Other similar patrols are maintained by the French *Gendarmerie*, and by the Automobile Association in Britain, which has used aircraft in its work for nearly 50 years. Such other activities as crowd control, life-saving or the tracking of criminals also offer useful work for aeroplanes.

Equally important is the wide use made in such countries as Canada, South America and Scandinavia for forestry patrol, for the prompt reporting of an outbreak of fire by a patrolling aircraft can often bring preventive or remedial action in time to save vast areas from devastation. In some cases special "water bomber" aircraft have been developed to deal quickly from the air with this kind of outbreak.

Poplavko, Lieutenant, 875

Poppet valves, 1097

Portela Airport, see Lisbon

Porter, see Pilatus

Portugal: Air Force

Força Aérea Portuguesa (Portuguese Air Force). Military and naval air arms were formed in Portugal in 1917, although the first official sponsorship of aviation dates from 1912 and Portuguese Army units had fought with the British Army since early in the 1914-18 War. The first aircraft were of French and British design, a pattern that continued and expanded throughout the 'twenties and 'thirties, although German types also came into service around the mid-'thirties. Reorganization and reequipment was instigated in 1937-38, and among many new German, Italian and British aircraft ordered was the Spitfire, for which Portugal was the first non-British customer. Britain and the US kept Portugal well supplied with up-to-date combat aircraft during World War Two, in return for which RAF Coastal Command was granted the use of the Azores as a base for its Atlantic patrols.

The separate Army and Navy air forces were eventually combined to form the present-day FAP in 1952, the year following Portugal's admission to NATO, and jet aircraft—F-84G Thunderjets and T-33A trainers—entered service in 1953. Some Thunderjets are still in service, although the main air defence/tactical strike rôle is now undertaken by F-86 Sabres and Fiat G91Rs. Other important operational types include Lockheed Neptunes for maritime patrol.

Postal services, see Air mail

Powered flying controls, 637, 643

Power Jets Ltd

This company was formed in 1936 to promote Frank Whittle's turbojet engine. It was largely a design agency at first, sub-contracting work to other members of industry with turbine experience. With the construction of the experimental engines, Power Jets took over the testing of them in an old foundry at Lutterworth. It continued as a design and development centre until the British Government decided it would be more advantageous for it to be nationalized and eventually it was blended into the RAE (*q.v.*).

Power/weight ratio, 1067, 1100

Pratt & Whitney, 1057-60, 1073, 1078-79, 1109, 1117, 1225-28

The Pratt & Whitney Aircraft Co was formed in 1925 as a subsidiary of the larger tool-making company. Its first product, the Wasp radial engine, was intended for fighter aircraft. It appeared in 1926 and was found to be such an advance on current engines used by the US Navy that a large number were ordered. It was followed in 1927 by the Hornet, which was designed for larger aircraft, and this too found almost immediate success. Both engines were nine-cylinder air-cooled radials and had the advantage of many common features, the Hornet having a greater capacity and thus a power output of 525 hp instead of 450 hp.

So great was the popularity of these two types that by 1930 a new factory had been built in Canada, and BMW in Germany and Nakajima in Japan acquired licences to build the engines, which were undergoing continual refinement.

In 1929 a smaller, 300-hp engine, the Wasp Junior, was put into production. This, too, had a long production run, finding popularity with the smaller, general-purpose type of aircraft. It was developed, for the military, into the Twin Wasp Junior by uniting two Wasp Juniors on a common crankshaft, thus making a fourteen-cylinder, two-row radial of 625 hp. The next logical step was to do this with the Wasp, to form the Twin Wasp; and with this too Pratt & Whitney had a winner in the early nineteen-thirties.

The whole range of engines was extended and improved continually, and by 1934 one-third of all the engines used on the world's airliners were of Pratt & Whitney design. The Twin Hornet of 1937 lifted the upper margin of the Pratt & Whitney range to 1,400 hp. Production of the Hornet and Twin Wasp Junior were dropped during World War Two to enable the company to concentrate on military and transport requirements. First new engine to be made under the spur of war was the Double Wasp, which followed the general configuration of the Twin Wasp but used eighteen cylinders instead of fourteen; the most powerful model of this range produced a maximum 2,300 hp.

In 1948 Pratt & Whitney acquired a licence to manufacture the Rolls-Royce Nene centrifugal-flow turbojet, marketing it as the JT6 Turbo Wasp. This was followed by the J48, a licence-built Rolls-Royce Tay, which was also known as the Turbo Wasp. The culmination of radial-engined development was reached with a twenty-eight cylinder, four-row engine, the Wasp Major, delivering 3,500 hp. First P & W-designed turbojet was the axial-flow J57 of 1953, the first US turbojet in the 10,000-lb static thrust class, and an engine which has been used in a large number of successful American aircraft. From this grew a prolific new "family" of turbojets and turbofans.

Pratt & Whitney J58 (JT11) engine, 1153-1154

Pratt & Whitney JT3C & JT3D engines, 1148, 1201, 1233, 1237

The JT3C two-spool axial-flow turbojet entered production in 1953 and, in various versions, is used in US military aircraft (eg, B-52, KC-135, F-100, F-101, F-102, F-8 and A-3) and civil airliners (eg, Boeing 707 and DC-8). It produces between 12,000 and 13,750 lb thrust dry and 18,000 lb with afterburning.

This well-tried and -developed engine became the subject of modifications in which the first three stages of the compressor were removed and replaced by two fan stages, turning the engine into the JT3D turbofan. This produces 50 per cent more take-off thrust, at the same time giving a 13 per cent better cruising fuel consumption. It took the air in a B-52 in 1960 and since then has been in large-scale production for Boeing 707s and Douglas DC-8s. It also powers the Lockheed C-141 StarLifter. So successful has it been that many JT3Cs have been returned to Pratt & Whitney for conversion to JT3D configuration.

Pratt & Whitney JT9D engine, 1152, 1201, 1223, 1225-1228, 1237-1238

This engine represents the next step forward towards turbofan engines to power the very large transport aircraft of the immediate future. The JT9D was developed to power the Boeing 747 "jumbo-jet", and was intended to deliver 43,500 lb thrust in its initial service form. It is a two-spool fan engine with a by-pass ratio of 5 : 1, the fan being of 8-ft diameter.

Pratt & Whitney JTF17A engine, 1187, 1251

The JTF17A is a large, two-spool turbofan engine designed for supersonic flight and has been running since 1966, three experimental engines being built. These achieved a rating of 57,000 lb thrust, and the production version was expected to give 61,000 lb. However, the engine was unsuccessful as a contender for America's supersonic transport and has been developed no further. An interesting feature of this ducted-fan engine is that the duct gases are ignited in the full-length exit duct in order to provide thrust augmentation for certain take-off conditions.

Pratt & Whitney T34 Turbo Wasp engine, 1309

Under US Navy direction, Pratt & Whitney began development of this axial-flow turboprop in 1945. In its first rating test it completed a 50-hr flight test at 5,700 hp, later being installed in the nose of a Boeing B-17 in August 1950. Later still it was evaluated under service conditions in two US Navy Super Constellations and in two Douglas C-124B transports of the USAF. It powers the Douglas C-133 Cargomaster.

Pratt & Whitney T57 engine, 1309

A remarkably powerful and promising turboprop, the T57 appeared at a time when the choice of powerful powerplants was transferring to the pure jet, and its further development was cancelled.

Pratt & Whitney TF30 engine, 1399, 1402
A high-compression two-spool turbofan produced in 1965, this is one of the first turbofans to incorporate afterburning. It was chosen as the power plant for the F-111 "swing-wing" tactical fighter and is also used in the Corsair II (without afterburner); those fitted to the Mirage III-T and III-V are built under licence by SNECMA.

Without afterburning the engine gives 11,000 lb thrust; the initial version for the F-111 gives 20,000 lb with afterburning, but the developed version will improve on this figure.

Pratt & Whitney/ SNECMA TF306 engine, 1181, 1187

Pratt & Whitney Wasp Major engine, 817, 1057, 1058, 1117
Pratt & Whitney's piston-engine range comprised air-cooled radials almost exclusively, and the Wasp Major Series represented the culmination of this long line of engines.

Coming into service in the late nineteen-forties, it was a four-row, twenty-eight-cylinder engine, each row having seven cylinders. The engine provided 3,250 hp at 2,700 rpm and, with water injection, 3,500 hp for take-off. It was installed in the very advanced and heavy American military and civil aircraft of the period, eg the Convair B-36, Boeing Stratocruiser and Douglas C-124.

Prawn, see Parnall

Précurseur, 133, 140, see also Jullien, Pierre

Pressure ratio, 1139, 1159, 1195, 1213, 1223

Pressurization, 925
Pressurization is used in most modern aircraft to enable them to operate at greater heights. The cabin or cockpit is enclosed in a metal cell which is automatically sealed when all doors and hatches are closed. As the aircraft climbs, so the pressure in the cell is maintained at that of a lower altitude than the one at which the aircraft is flying. This is accomplished by introducing more air into the cell by mechanical means. Thus, when flying at extremely high altitudes the cabin is maintained within a comfortable pressure band without the need to resort to oxygen or other assisted means of breathing.

Prier, Pierre, 209

Princess, see Saunders-Roe

Proteus engine, see Bristol Siddeley

PT-19, -23 and -26, see Fairchild Cornell

Pulitzer Trophy, 387, see also Competitive flying

Pullman, see Bristol

Pup, see Beagle

Pusher propellers, 815, 817
The "pusher" configuration is that in which the propeller is behind the engine and drives the aircraft forward from behind. It is the opposite of a "tractor" installation, in which the propeller draws the aircraft along behind it.

Puteaux gun, 883

PZL
The *Panstwowe Zaklady Lotnicze* (State Aircraft Factories) of Warsaw were first established in Poland in 1928, and by 1939 PZL was one of four major aircraft manufacturing organizations in the country. It produced a variety of cabin lightplanes and other types during this period, but probably its best-known pre-war designs were the P-11, P-23, P-24 and P-37 fighters and bombers designed by Ing Zygmunt Pulawski. After the war, the Polish aircraft industry was reorganized following the signing of the Warsaw Treaty with the USSR, Czechoslovakia and Hungary, and the initials PZL now represent any product of the present-day industry. These include the PZL-101 Gawron utility aircraft and the PZL-104 Wilga general-purpose monoplane.

Qantas
Qantas Airways Ltd (formerly Qantas Empire Airways Ltd until August 1967) was formed in 1920 as Queensland and Northern Territory Aerial Services, from the initials of which it adopted its operating name Qantas in January 1934. The first regular services, with Armstrong Whitworth F.K.8 biplanes, were started in November 1922 between Charleville and Cloncurry. In association with Imperial Airways (q.v.), Qantas operated the Brisbane-Singapore sector of the England-Australia route from late 1934. Prior to this, in 1928, the airline was responsible for establishing the first Flying Doctor service in Australia (and in the world) with a converted D.H.50 biplane.

During the nineteen-thirties de Havilland-designed aircraft figured prominently in the QEA fleet, and during the latter half of this period Short "C" (or "Empire" class) flying boats were introduced on the longer stages. These later formed the basis for a flying boat squadron of the Royal Australian Air Force in World War Two. In 1938 the Brisbane-Singapore leg was extended to Karachi, and Qantas maintained its end of the England-Australia service during the war by operating from Perth to Colombo in collaboration with BOAC. Catalina and Sandringham flying boats formed part of the immediate post-war fleet, to which were added Douglas DC-4 and Lockheed Super Constellation landplanes; the latter type inaugurated the first-ever round-the-world airline service which Qantas opened on 14 January 1958. In the following year jet services were started, Qantas being the first non-American operator to employ the Boeing 707. The large fleet is now almost entirely composed of jet-powered types, and orders have been placed for the Boeing 747 "jumbo-jet" and either the Concorde or the Boeing supersonic transport. Services are world-wide and stretch to every continent. Headquarters: Sydney, NSW.
Member of IATA.

Quail missile, see McDonnell Douglas

Quebecair
Quebecair is the title used since 1953 by the former Rimouski Airline Ltd (founded 1947); in 1965 the airline absorbed two other, smaller Canadian operators, Northern Wings Ltd and Matane Air Services. Quebecair operates regional services on either side of the St Lawrence river between Quebec and Montreal.
Headquarters: Rimouski, PQ.
Associate member of IATA.

R3Y Tradewind, see Convair

R-4B and R-5, see Sikorsky

R.100, 399, 727

Most successful British airship, the Vickers-built R.100 (shown after removal of tailcone)

One of two airships built in the UK in the late nineteen-twenties, the R.100 was built by the Airship Guarantee Co Ltd, backed by Vickers Ltd, and was completed in November 1929. It incorporated an early form of the geodetic construction (q.v.) invented by Dr Barnes Wallis and was 709 ft in length, with accommodation for 100 passengers. Four Rolls-Royce Condor engines gave it a speed of 81 mph, and it had a range of 3,600 miles at 70 mph or 5,000 miles at 50 mph. This airship made a flight to Canada and back in July/August 1930, but further flying was curtailed as a result of the disaster to R.100's sister ship, the R.101.

Death of a dream: British officials at Beauvais, October 1930

R.101, 727, 811
The last British airship R.101 was built by the Royal Airship Works at Cardington in 1929. It was slightly longer than its sister-ship R.100, which was being built by a private company at the same time. Powered by four Beardmore diesel engines of 580 hp each, R.101's speed was about 80 mph. After several successful flights it set out for India on 5 October

Qantas' four Lockheed Electras were the first turboprop-engined aircraft in Australian scheduled service; they were used on Far East routes

1930, only to be destroyed in France. This disaster marked the end of British airship development.

Radar, 707, 1377, 1379, 1381, 1521

Name derived from Radio Direction and Range. British research from 1934 showed that radio waves of suitable form and frequency are reflected by metal objects, such as aircraft, and that the reflected signal can be detected by a suitable receiver near the transmitter or elsewhere. Not only can the reflecting "target" thus be detected but the time taken by the waves to reach it and return enables the target distance to be obtained. The first radars were large ground stations which served a vital function during the Battle of Britain in providing early warning of enemy raids and enabling ground controllers to direct defending fighters to ensure successful interception. By 1941 radars had been made small enough to fit inside aircraft, and different types of equipment were used by night fighters to intercept other aircraft at night or in bad weather, by bombers to "see" the ground through dense cloud, and by coastal and naval aircraft to hunt down surface ships and submarines.

Modern radars fulfil all these functions and many others connected with navigation, traffic control, collision avoidance, missile and spacecraft guidance and such other diverse duties as providing detailed pictures of friendly or enemy territory, indicating storm-cloud or atmospheric turbulence ahead of aircraft and, using the Doppler principle (rendered aurally obvious by the falling pitch of a train whistle or hooter as it passes at speed), measuring aircraft speed over the ground.

Radar (or radio) altimeter, 1357

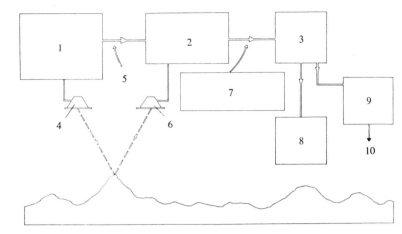

1 Transmitter
2 Receiver
3 Counter
4 Transmitter aerial
5 Sampling probe
6 Receiver aerial

7 Difference between received and transmitted frequencies gives aircraft's altitude

8 Monitoring altimeter indicator
9 Linear inter-face unit
10 To navigation and weapon aiming computer

Used as a supplementary height indicator, the radar altimeter gives an instantaneous presentation of the height of an aircraft over the land or sea directly beneath it. It does so by emitting radio waves to the ground and measuring the time taken to return to the aircraft. It thus gives an actual height-above-ground figure.

Radial engine, 385, 911, 1044, 1047, 1055, 1057, 1058, 1065, 1067, 1077, 1079

An aero engine with its cylinders arranged radially around the crankshaft, the cylinders being stationary and the crankshaft revolving.

Radio Magnetic Indicator, see RMI

Radley-England Waterplane, 229

This aircraft was a collaborative design by James Radley, whose aviation experience was predominantly on Blériots, and E. C. Gordon England, who began flying in 1910 with other people's designs which he modified. Later he flew for the Bristol Aeroplane Company. In 1913 the three-engined Waterplane was built at Shoreham, but crashed on its first flight. It was rebuilt and appeared at the *Daily Mail* trials

that year without success. It was still flying in 1914.

RAF, see UK: Air Forces

Ram air

Air supplied through a ram intake.

Ram intake

A forward-facing intake on a vehicle moving at speed through the atmosphere.

Ramjet, 301, 1315-19

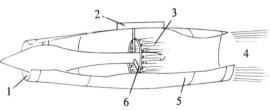

1 Intake
2 Diffuser
3 Flame
4 Supersonic jet
5 Nozzle
6 Fuel manifold

An air-breathing propulsion system in which air is compressed in a ram intake, further compressed and slowed to a velocity suitable for combustion by passing along a suitably profiled diffuser duct, heated by the combustion of fuel in a combustion chamber and finally expelled as a supersonic jet of hot gas through a propulsive nozzle. The ramjet has been used to drive the tips of helicopter rotors at peripheral speeds of the order of 400 mph or less, but is much better suited to highly-

supersonic propulsion. Its cycle efficiency depends largely upon the achieved pressure ratio, which rises rapidly at supersonic speed (since the ram compression varies with the square of the speed). At Mach numbers greater than about 3 to 3.5, the ramjet is markedly superior to the turbojet, but it

always suffers from the disadvantage that it cannot start from rest but must have air rammed into the intake.

Ram pressure

Pressure gained by converting kinetic energy of a high-speed airflow to pressure energy. A forward-facing intake may be designed to recover more than 90 per cent of the total loss of kinetic energy in air rammed into it at high speed. The pressure theoretically varies as the square of the speed of the intake through the air and thus

rises with great rapidity at high supersonic speeds.

Ranken, Lieutenant Francis, 873, see also *Flechettes*

Raymond, Arthur, 387, 737

A designer at the Douglas Aircraft Company in the early nineteen-thirties with responsibility for working with airlines in the production of the DC-1, DC-2, DC-3 series of airliners.

Raytheon Sparrow missile, 895, 897

The Sparrow, in its IIIB production form, is a large air-to-air missile tailored to the McDonnell F-4 Phantom and equips both US Navy and British versions of this aircraft. It is 12 ft long, and has a 60-lb high explosive warhead, is propelled by a Rocketdyne solid-propellant rocket motor, and utilizes a Raytheon continuous-wave semi-active radar homing system.

RB racer, see Dayton-Wright

RB.108 engine, see Rolls-Royce

RB.145 engine, see Rolls-Royce

RB.153/61 engine, see Rolls-Royce/MAN

RB.162 engine, see Rolls-Royce

RB.172/T.260 engine, see Turboméca

RB.189 engine, see Rolls-Royce/General Motors

R.E.8, 1333

One of the aircraft designed by the Royal Aircraft Factory, Farnborough, England, the R.E.8 was a two-seat Corps Reconnaissance aircraft powered by a 150-hp RAF.4a engine. It first flew in mid-1916 and was soon in production to replace the obsolete B.E. series of biplanes then being badly mauled in France. After certain modifications had been carried out to cure its proneness to spinning, the R.E.8 went into service in November 1916 and eventually re-equipped the bulk of the British Army Co-operation squadrons.

Reaction jets, 646-7

Rebecca, 705

One of the wartime airborne applications of radar, Rebecca was evolved as a radar homing device, principally for use by night bombers of the Royal Air Force. The aircraft carried a transmitter and receiver, which transmitted a signal to a "slave" ground station; this picked up the signal and returned it to the receiver in the aircraft, giving bearing and time-to-run information. It was developed after the war into the VOR/DME (*q.v.*) system.

Red Top, 895

Operational British infra-red air-to-air missile, the Hawker Siddeley Red Top has a 68-lb warhead, Mach 3 cruising speed and range of 7 miles. It is standard equipment for the later Lightnings of the RAF and Sea Vixens of the Royal Navy.

Reduction gear, 1093, 1102

Reed metal propellers, 1323, 1333

Refuelling, see Flight refuelling

Registration authorities

In all countries where civil flying is undertaken, there exists a national authority responsible for registering, or authorizing, each specific aeroplane type that flies in that country. Many of these authorities have reciprocal agreements whereby the conditions required to be satisfied by one authority are accepted by its counterpart in another country. Thus, for example, a US aircraft which receives an FAA Type Approval Certificate automatically becomes acceptable for operation in most South American states, as well as in other specified countries.

The principal national authorities include the SGAC in France; the LBA in Federal Germany; the RAI in Italy; the JCAB in Japan; the

ARB in the UK; and the FAA in the US.

Reheat, see Afterburner

Renard, Charles, 133, 141

In 1871 this Frenchman built a successful model glider consisting of ten small wings, one above the other, with stabilizers on each side of the fuselage. He later turned his attention to lighter-than-air craft with marked success and, working with Captain A. C. Krebs, produced in 1884 the first airship really capable of going where it was directed. This was *La France* (*q.v.*).

Transfer between a British lifeboat and a search/rescue Whirlwind HAR.10 of the RAF

Rescue services

Nowadays aerial rescue work is almost exclusively the province of helicopters or light aircraft, but the use of marine aircraft for air/sea rescue dates back to the 1914-18 War. In World War Two such types as the Walrus and Catalina rescued many hundreds of Allied aircrew who would otherwise have perished after coming down in the sea, and their present-day successors include such types as the Grumman Albatross amphibian. Helicopters, by their ability to hover or to land on steep mountainsides and other inaccessible places, brought tremendous advances both in rescue techniques and in making possible rescue attempts in circumstances that no fixed-wing aeroplane could overcome. Rescue, whether on sea or on land, was one of the first tasks undertaken on a large scale by helicopters, and still forms a major part of

their work in both civil and military spheres today.

Reverse pitch, 1341, see also Constant-speed propeller

Although the normal type of aircraft propeller always rotates in the same direction, its net thrust can be changed from forward to rearward by turning the blades to reverse pitch. This is a setting some 30° beyond the normal fine pitch stop; it can be achieved only in propellers designed for the purpose, and then only by the pilot selecting the "reverse" position. Even then the weight of the aircraft must compress the landing gear struts continuously for a period of some 2 seconds before the reverse position is adopted. The pilot then applies full power to impart a considerable rearward thrust from the propeller blades, which sweep air in from behind and send out their slipstream in front.

Reverse thrust, 843, 1189, 1244, 1264, 1275

A turbojet or turbofan may be fitted with various types of mechanism to deflect the exhaust gases either vertically or sideways and thus impart reverse (braking) thrust to shorten the landing run. It is in use in many large jet aircraft and is particularly useful in wet conditions as it does not incur skidding or aquaplaning.

Rhein/Main Airport, see Frankfurt

Rhodesia: Air Force

The nucleus of the

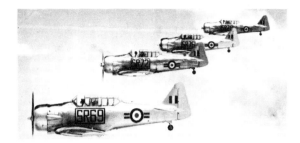

Harvard (North American T-6) trainers have served in Rhodesia for almost 30 years

then Southern Rhodesian Air Force was created in 1936, and at the outset of World War Two three Rhodesian squadrons joined forces with the RAF, subsequently fighting with distinction in the European, Mediterranean and Middle Eastern theatres of war. Southern Rhodesia also contributed greatly to the Empire Air Training Scheme, training several thousand aircrew for the Allied cause from 1940 onward. From 1945-51 the SRAF was essentially a non-combatant force, but then a squadron of Spitfires was formed, followed in 1953 by the first jet unit with Vampire fighter-bombers. The title Royal Rhodesian Air Force was granted in 1954 (and has been retained following UDI in 1967), and the RRAF currently has ground-attack Hunters and Canberra jet bombers in service, in addition to some of the original Vampires.

Richell, Professor, 133

Rio de Janeiro Airport (Galeao), Brazil

Distance from city centre: 8½ miles
Height above sea level: 20 ft
Main runway:
14/32 = 10,827 ft
International passenger movements in 1966: 606,000

RMI (Radio Magnetic Indicator)

Instrument on the flight deck of an aircraft, a link in the VOR (*q.v.*) system, which informs the pilot in which direction he is from the VOR. On large aircraft with twin VOR installations, the RMI displays two separate needles.

Robert, M.N., 129, 209

Robot 04 missile, 896

The Royal Swedish Air Force has had this air-to-surface missile in service since early 1959 aboard its Lansen aircraft. It is intended principally for use against targets at sea and is a subsonic weapon with a warhead weighing approx 660 lb and a solid-propellant rocket motor. It will also be carried by the

AJ 37 attack version of the Viggen due to enter service in the early 'seventies.

Rocker arms, 1038, 1059

Roe, Sir Alliott Verdon, 189, 193, 222, see also Avro

Sir Alliott Verdon-Roe

A. V. Roe was one of the early British pioneers in heavier-than-air flying. He began with models, winning the *Daily Mail* aeroplane competition of 1907; in the following year he built a full-sized aircraft on which he became airborne. In 1909 he built the first of his famous triplanes which he flew at Lea Marshes; developed versions took part in the first English flying meeting at Blackpool in 1910. In January 1910 he formed A. V. Roe & Co; during this year he achieved some

Avro 504K trainer, supplied by A. V. Roe to Sweden

successful flights and also turned his attention to biplanes. From these was developed the Avro 504, one of his most successful designs and an elementary trainer which played a large part in RFC,

RNAS and RAF flying training during and after the 1914-18 War.

Rohrbach, Dr Adolph, 369

Roissy Airport, see Paris

Roland, 361

"Roland" was the marketing name adopted by the Luftfahrzeug Gesellschaft mbH (LFG) of Berlin-Charlottenburg, to avoid confusion of initials with its contemporary LVG (*q.v.*). It was one of the smaller aircraft manufacturing companies in Germany during the 1914-18 War, but it kept up a steady stream of production aircraft for the Air Force. These were chiefly two-seat observation aircraft; some of them, notably the C.II, embodied very advanced design features. Monocoque construction enabled a voluminous fuselage to be built; the wing strutting was streamlined and, together with the well-faired engine, the aircraft's capability and appearance were ahead of their time.

Rolls-Royce

It is doubtful that any other company in the world is better known or more highly respected than this great British enterprise, which began in 1906 to make high-quality motor cars as a result of collaboration between the Hon Charles S. Rolls and Sir Henry Royce. What is not so well-known is that today the firm is enormous, even by US standards; it has 88,000 employees and factory floor area exceeding 16½ million square feet. It also surprises some people to find that today cars are only a small item in the Rolls budget; aero engines account for 87 per cent of the total company strength and the other 13 per cent includes rail, military and ship engines, industrial turbines, oil engines and nuclear power.

During the 1914-18 War the name became famous in the world of flight by virtue of high-quality Vee-12 engines of 200-400 hp, the Eagle being used to power the Vimy which made the first flights across the Atlantic non-stop, and from England to South Africa and Australia. During the nineteen-twenties Rolls engines powered all kinds of RAF aircraft, and a special racing unit won the Schneider Trophy outright for Britain. From this engine was developed the Merlin, which offered 1,030 hp when it entered RAF service in 1937; by 1945 it was giving 2,000 hp and was in service in greater numbers than any other British aircraft engine before or since (over 150,000).

By 1945 the company's piston-engines had an unsurpassed position, but the firm saw that the future lay in the gas-turbine and today its turbojet, turbofan, turboshaft and turboprop engines equip hundreds of types of aircraft including machines made by 29 firms outside the United Kingdom. When this book went to press during 1969, Rolls-Royce gas-turbines had flown 105 million hours and set up the following "firsts": first turboprop (1945); first jet in airline service (1952); first turboprop in airline service (1953); first jet VTOL (1953); first multi-hole air-cooled turbine blade in service (1957); first jet VTOL to make transition from jet lift to forward flight (1957); first transatlantic jet service, first airline silencer and first airline reverser (all 1958); first vectored-thrust turbofan (1960); first turbofan in airline service and first air-cooled blades in airline service (1960); first engine using glass-fibre composite material (1962); first 10-year parts cost guarantee (1963); first supersonic airline engine run (1964); first engine ever

to reach overhaul time of 10,000 hours (1966); first run of three-shaft engine (1967); and first run of engine using carbon fibre composite parts (1968). This last engine was the RB.211 turbofan, which was the first non-American engine ever to be selected to power a new type of jet aircraft by a US manufacturer.

Rolls-Royce Avon engine, 314, 843, 1181

The 300 Series Avon powers Lightnings and Drakens at ratings up to more than 17,000 lb thrust

The Avon was Rolls-Royce's first axial-flow turbojet to go into large-scale production, and was intended to replace the Nene, primarily in the military field. The first production Avons were in the 100 Series; these were soon followed by the 200 Series, and the 300 Series which, with re-heat, develops 17,000 lb thrust in the BAC Lightning and Saab Draken. For the civil field, the Avon RA.29 was developed and has found its principal application in the Comet 4 and Caravelle. Avons for both civil and military application have been built extensively under licence in various countries.

Rolls-Royce Conway engine, 1195

First turbofan in service in the world was the Conway; this one was installed in a DC-8-40

This engine was the first two-spool turbofan to enter production.

It powers the Mk 2 version of the Handley Page Victor aircraft with the RAF, is used on several versions of the Boeing 707 and Douglas DC-8 airliners, and powers the military and civil BAC VC10 and Super VC10 airliners. The latest version is rated at 21,800 lb static thrust.

Rolls-Royce Dart engine, 1211-1212, 1283, 1309

Most successful turbo-prop engine ever built, the Dart owes its inception to design work started in 1945. It has a centrifugal compressor with an axial-flow turbine. A test engine was installed in the nose of a Lancaster, commencing flight trials in October 1947. Results were so encouraging that the engine was ordered for the new Viscount airliner. From that moment the Dart be-came a much-sought-after engine, and its performance in the Viscounts of BEA brought it into world-wide notice. The original Dart gave an output of 1,400 shp plus 295 lb static thrust.

Development has been carried on continuously, until today Darts can produce up to 3,245 ehp in the most powerful variant. In their first twenty years, over 5,000 Dart engines have been built.

Rolls-Royce Derwent engine, 1283

This jet engine was derived almost directly from the original Whittle development engines. The Rover Company built a prototype W2B/26 engine in 1942, based on a Power Jets design. Rolls-Royce took over this design and developed it into the Derwent, which powered the Gloster Meteor I fighters in service with the Royal Air Force in 1944. The Mk I Derwent had a static thrust of 2,000 lb, but the engine was developed in the early post-war years, later versions giving over

The 3,030-hp Dart RDa.10 powers the Japanese YS-11 transport; a military version gives 3,245 hp

3,000 lb thrust. The Derwent had a centrifugal compressor.

Rolls-Royce Eagle engine, 253, 815, 1067, 1097

This engine, a culmination of Rolls-Royce wartime development, was a twelve-cylinder Vee-shaped engine developing 360 hp. It provided new standards in reliability and was one of the mainstays of the nineteen-twenties,

being widely used for many years in such aircraft as the Vickers Vimy in which Alcock and Brown made the first non-stop Atlantic crossing.

A second Rolls-Royce Eagle was produced towards the end of World War Two. In this engine Rolls-Royce followed Napier's lead in producing an engine of "H" configuration with twenty-four sleeve-valve cylinders and two crankshafts, producing 3,500 hp. It was fitted to the early Westland Wyverns, but was barely out of its lengthy teething troubles when it was overtaken by the jet age and was never developed further.

Rolls-Royce Falcon engine, 1067

Of very similar design and construction to the Rolls-Royce Eagle, the Falcon was a twelve-cylinder in-line water-cooled Vee engine producing 270 bhp; it was in widespread use in the nineteen-twenties.

Rolls-Royce Goshawk engine, 815

The Rolls-Royce Goshawk was not in the main stream of Rolls-Royce aero engine development, in that it was steam-cooled rather than water-cooled. The idea behind the design was that the engine should be capable of being "buried" in an aircraft and the propeller driven by a shaft, thus facilitating trim or streamlining in the aircraft's design.

A specification was issued for a four-gun fighter powered with this engine, for which many companies tendered, but the steam-cooling presented problems and the whole concept was overtaken by the rapid development of the Kestrel and racing engines for the Schneider Trophy aircraft.

Rolls-Royce Griffon engine, 286, 1283, 1339

The inception of this engine dates from the beginning of World

War Two, when a successor to the Merlin was envisaged. This engine related directly in many ways to the Rolls-Royce "R", its cylinder arrangement and dimensions being identical.

First versions developed 1,730 hp at 750 ft and 1,490 hp at 14,000 ft, and these were first installed in the Firefly naval fighter. The Mk XII version of the Spitfire was the first Griffon-engined variant of that fighter, entering service in 1943. Thereafter a whole "family" of Griffon-engined Spitfires and Sea-fires ensued. Later versions of the Griffon produced even more power, the final model being the Mk 83, designed for contra-rotating propellers, which developed 2,340 hp at 750 ft and 2,120 hp at 12,250 ft.

Rolls-Royce Kestrel engine, 909

The Rolls-Royce Kestrel began life as the Rolls-Royce "F" engine. In its design as a twelve-cylinder, Vee-shaped liquid-cooled engine, special attention was paid to the reduction of frontal area. The success of this step was seen immediately in aircraft such as the Hawker Fury biplane fighter and the Hart variants. It achieved wide use throughout the world, and several variants were produced to meet differing needs, the power output varying between 635 and 745 bhp. During the early nineteen-thirties the Kestrel was supreme, although it was later overshadowed by its successor, the Merlin, into the design of which went much that had been learnt from producing the Kestrel.

Rolls-Royce Medway engine, 1233

A medium-sized turbofan project by Rolls-Royce in the V/STOL field, fitted with jet deflectors. It was intended for use on the Hawker Siddeley 681 transport, but when this was cancelled in 1965 the Medway was cancelled with it.

Rolls-Royce Merlin engine, 282, 286, 911, 1095, 1099-1102, 1107, 1283, 1339

This engine was developed from the Kestrel, but incorporated much that was learnt from the Rolls-Royce "R" built for the Supermarine S.6 racers. The first Merlins were ready for the new generation of monoplanes entering service with the RAF from 1937 onward, the first service aircraft to use it being the Fairey Battle I bomber. Initial production Merlins were Mk IIs and Mk IIIs, which produced maximum power of 1,440 hp at 5,500 ft. These also powered the Hurricanes and Spitfires in the Battle of Britain. A vast programme of Merlin development and adaptation took place under the spur of wartime progress; the engine was used by Halifax and Lancaster bombers and for the new Mosquito light bomber/fighter.

By 1941 the Merlin 45 was giving the Spitfire V equivalent power at a greater height, and this trend went further when the Mk 61 series appeared for high-altitude operation (1,565 hp at 12,250 ft and 1,390 hp at 23,500 ft). This and earlier versions were built under licence in America by Packard and used in American P-40s and P-51s.

Women helped to make 150,000 Merlins; this picture was taken in 1942

After the war the Merlin was in demand for the first generation of civil transports: Yorks, Lancastrians, and the

Canadair North Star conversion of the DC-4; and it remained in service with various air forces into the nineteen-sixties, the Spanish Air Force having re-engined its licence-built Heinkel He 111s and Messerschmitt Bf 109s with Merlins.

Rolls-Royce Nene engines, 995, 1235

The Nene was Rolls-Royce's second production turbojet engine, succeeding the Derwent. It was a centrifugal-flow engine with a take-off thrust of 5,100 lb, and was used to power the first two Royal Naval jet fighters, the Attacker and the Sea Hawk. A modified version was developed in the USSR by Klimov (*q.v.*) to power the Soviet MiG-15 jet fighter.

Rolls-Royce "R" engine, 1087, 1107

Rolls-Royce RB.108 engine, 993, 1235

The RB.108 was produced in 1955 as a simple turbojet designed

A Nene 105 manufactured under licence by Hispano-Suiza

for lift-jet duties. It is a lightweight axial-flow turbojet of 2,340 lb static thrust. One of its features, necessary for the lift-jet rôle, is a rapid response to throttle movements. The RB.108 was first installed in the Short

Display exhibit of twin RB.108 lift-jets, first engines designed for VTOL lift

SC.1 prototypes.

Rolls-Royce RB.145 engine, 1233

This lightweight jet engine has been developed with jet-lift applications in mind. It has a nine-stage axial-flow compressor and a two-stage turbine, and is fitted with reheat. Without reheat it develops 2,750 lb static thrust. Six RB.145s powered the German VJ 101C research aircraft.

Rolls-Royce RB.162 engine, 1009, 1026, 1235

A second-generation lift-jet engine, the RB.162 is produced in a series of models for different installations. The Series 30 has the most current applications, and is used on the Do 31 and the Mirage III-V. It produces 5,500 lb static thrust and (as do all RB.162s) makes extensive use of low-

cost materials such as glass-fibre reinforced plastics.

Rolls-Royce Spey engine, 1181, 1213

The Spey is a two-spool axial-flow turbofan, designed in 1959 to a civil specification with the first engine run following at the end of December 1960. Speys are now in full service in the Trident and One-Eleven airliners, giving a take-off rating of 11,000 lb at 12,250 rpm. The engine has been adapted for military use in the Buccaneer S Mk 2, the Hawker Siddeley Nimrod MR Mk. 1 and the British McDonnell Phantoms, the thrust in these versions being uprated to 12,500 lb.

The first version of the Spey turbofan, since succeeded by many versions for transports, ocean patrol, bomber and supersonic fighter aircraft

Rolls-Royce Thrust Measuring Rig ("Flying Bedstead"), 319, 323, 991, 995-996, 1235

Rolls-Royce Trent engine, 1213, 1223, 1254, 1283

The original Trent was the first turboprop engine to fly. It was in effect a Derwent centrifugal turbojet geared to drive a five-bladed propeller, and was fitted to an early Gloster Meteor for flight-testing.

The name Trent has been revived for the first of a new generation of Rolls-Royce engines, a three-spool turbofan providing a take-off rating of 9,750 lb thrust. This engine has a specific fuel consumption (at Mach = 0.7, cruising

The RB.162 set a new standard of low weight and cost for lift-jets by using novel materials

at 25,000 ft) of 0.716. It promises to have wide applications.

Rolls-Royce Tyne engine, 450-451, 1177-1180, 1285-1286, 1335

The Tyne was developed in 1955 by Rolls-Royce as a second-generation turboprop with an immediate application to the Vickers Vanguard airliner then projected for British European Airways. It is a two-spool axial-flow turboprop engine with a six-stage LP compressor and nine-stage HP compressor; it is 9 ft in length and has a maximum diameter of 4 ft 7 in. The first version, the RTy.1, was rated at 4,785 ehp, but development brought even-

The most efficient turboprop in service: the Tyne, used by many airlines and air forces

tually the RTy.20, which has been type-tested to 6,100 ehp. The Tyne is installed in the Vanguard and Canadair CL-44 civil transports, the Short Belfast and Transall C-160 military transports and the Breguet 1150 Atlantic anti-submarine aircraft.

Rolls-Royce (Bristol) Pegasus vectored-thrust engine, 327, 997, 1006, 1009, 1021, 1233, 1273-1276
This Pegasus is the radical new engine originally developed by Bristol Siddeley as the BS 53 for use in the Hawker P.1127/Kestrel/ Harrier series of VTOL aircraft. It is a turbofan engine in which air from the low-pressure compressor is ejected through a forward pair of lateral nozzles which can be swivelled from a fully-aft position to a position just forward of vertically downward. The hot exhaust gas is also ejected from another pair of similarly-rotatable nozzles further aft. This ar-

rangement enables a single-engined aircraft to hover, move backwards, decelerate quickly or fly forwards. Air from the compressor is also bled off for stabilizing devices.
The Pegasus was originally rated at 11,500 lb thrust, but by 1968 had been developed to a rating of 19,000 lb st.

Rolls-Royce (Bristol) Viper engine, 817
The Viper was originally designed by Armstrong Siddeley Motors as a "short-life" expendable jet engine for use in Jindivik target aircraft. It held such promise that a "long-life" version was produced for the Folland Midge light fighter and subsequently developed in the middle 'fifties for service with the Gnat fighter and Gnat Trainer. Since then a wide range of Vipers has been produced for light jet trainers and executive aircraft the world over. The Viper is also used as an auxiliary power plant

in the Avro Shackleton.

Rolls-Royce/MAN RB. 153/61 engine, 1233, 1279-1281
This engine is a collaborative venture between Rolls-Royce and MAN-Turbo GmbH of Munich, and was designed for the VJ 101D supersonic VTOL aircraft. It is a two-shaft turbofan engine developing 6,850 lb static thrust, and is fitted with a reheat installation which can be used with jet deflection equipment. The reheat produces a 75 per cent increase in thrust. Its future is now obscure.

Rolls-Royce/SNECMA Olympus 593 engine, 843, 1213, 1249, 1251, 1261-1265
This is a considerably more powerful version of the subsonic Olympus turbojet which powers the Hawker Siddeley Vulcan bomber and has been developed specifically to power the Concorde supersonic transport. It was first flown in a Vulcan test-bed in 1967, and is expected to produce a dry rating of 35,080 lb st when fully developed. Modifications have been made to allow for intake temperatures in the region of 150°C.

Rome Airport (Fiumicino), Italy
Also known as Leonardo da Vinci Inter-

national, Fiumicino's opening was planned to coincide with the Rome Olympic Games in 1960, but did not actually take place until 1961. Terminal buildings are of sweeping but simple lines, while the control tower is the second tallest in the world.
Distance from city centre: $12\frac{1}{2}$ miles.
Height above sea level: 7 ft
Main runways:
$16R/34L = 12,800$ ft
$07/25 = 8,612$ ft
Passenger movements in 1967: 4,827,600

Rotary engine, 229, 877, 1043-44, 1047, 1063-64, 1067, see also Gnome
An aero engine in which the cylinders are arranged radially around the crankshaft, the latter remaining stationary and the cylinders rotating, being attached to the propeller by an extension.

Rotating weapon door
External door of an aircraft weapons bay, on the inside of which weapons can be attached. The door rotates to expose the weapons for delivery, after which it returns to the closed position. Use of a rotating door thus avoids the effects of added drag that are caused by conventional bomb-bay doors extending into the air-stream when opened.

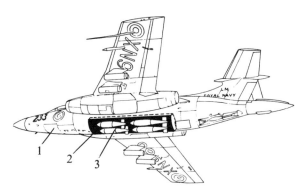

1 Buccaneer aircraft
2 Rotating weapon door
3 Weapons

Rotodyne, see Fairey
Rotol propeller, 1283
Rotor, 145, 971ff, 1203
Rotor head, 955ff, 981-984
General name given to the complex hub mechanism of the rotor of a helicopter or autogyro. The mechanism includes the drive shaft, blade hinges and swashplate controls. Subse-

quent to the development of higher-strength materials with good fatigue properties, much effort is being devoted to means of simplifying the basic design of rotor heads. (Note: the main rotating assembly of a compressor or turbine can also be called the "rotor".)

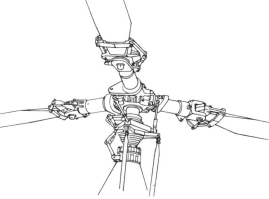

Simplified Sud-Bölkow rigid rotor on an Alouette II

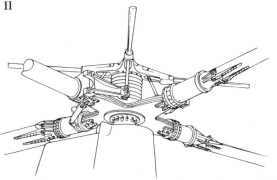

Lockheed rigid rotor, embodying three laminates for the attachment of the rotor blades

First vectored-thrust V/STOL engine to go into production was the Pegasus, for P.1127 and Harrier

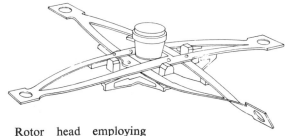

Rotor head employing flexible laminate mountings on the Hughes LOH

Rover Car Company, 1135

The Rover Car Company became involved in the development of British jet engines in 1940 when a tenuous link was established with Whittle's Power Jets company. Rover took over development of the W2B engine in August 1940, in a factory at Clitheroe, Lancashire, to "productionize" it for ease of manufacture and subsequent maintenance. Modifications included electrical starting, which soon became standard and was employed on Rover's own development of the Whittle engine, first flown in the second Gloster E.28/39 on 1 March 1943. In April 1943 Rolls-Royce took over the factory and development work, and Rover dropped out of the picture. It is interesting to note, however, that soon after the war Rover was one of the first British manufacturers to experiment with gas-turbine-powered cars.

Royal Aircraft Establishment, 661, 1095, 1129, 1223

In April 1918 the Royal Air Force was born and quickly became known as the RAF. Until then, these initials had denoted the Royal Aircraft Factory (q.v.) at Farnborough, and so to avoid confusion the latter was retitled Royal Aircraft Establishment, a title more in keeping with its ever-widening activities.

It had long been concerned more with the evaluation of aircraft than with their construction, and with the development of aircraft systems and equipment. With the coming of peace these aims were widened and many RAE departments concerned themselves more and more with basic research into fields connected with aviation, research which could be investigated experimentally on the collection of aircraft operated by the Establishment. This has remained the aim of the Establishment ever since, and its various departments have contributed an enormous amount of fundamental knowledge and practical innovations to the science and operation of flying.

During World War Two its task was multiplied and geared to developing devices for prosecuting the war more effectively and countering the enemy's aviation developments. Subsequently it returned to its more basic tasks, although few aircraft flying today have not benefited in some way or other from work done at the RAE through the years. It remains the largest aeronautical research establishment in the United Kingdom, although its researches now reach into many fields outside aeronautics.

Royal Aircraft Factory, 661

Former title of the Royal Aircraft Establishment (q.v.), which was first established in 1894 as H.M. Balloon Factory on the edge of Cove Common, near Farnborough in Hampshire, England. To keep pace with the changing nature of its activities, its name was changed to Army Aircraft Factory in April 1911 and to Royal Aircraft Factory on 11 April 1912.

The title of "Factory" became progressively more of a misnomer, for its main function was always more in matters of aeronautical research and design than in mass production. During the 1914-18 War it produced more than a score of B.E.-, R.E.- and S.E.-designated aircraft, but once a Factory design was basically proven its large-scale production was generally entrusted to outside industry.

Among the Factory's staff before and during the 1914-18 War were such designers as Geoffrey de Havilland (q.v.) and H. P. Folland, and such pilots as Frank Goodden and Edward Busk. It was Busk more than any other individual who did much to solve the mysteries of the little-known phenomenon of spinning.

Royal Air Force, see UK: Air Forces

Royal Air Force College, Cranwell, 1129

Marshal of the Royal Air Force Lord Trenchard, universally regarded as the founder of the Royal Air Force, believed it was important that this new Service should have its own cadet college, equivalent of the Army's Sandhurst and the Navy's Dartmouth.

Within two years of the formation of the RAF on 1 April 1918, this belief was put into prac-

Jet Provosts in stepped-up line to starboard flying past the RAF College headquarters building

tice when, on 5 February 1920, the RAF College at Cranwell in Lincolnshire was opened and the first course of cadets entered. The first Commandant was Air Commodore C. A. H. Longcroft. It remained as the RAF College until the outbreak of World War Two in 1939, when cadetships ceased and the station was used by various training units. It resumed its function as a College soon after hostilities ceased, and continues in this rôle today.

Royal Air Maroc

This North African airline was formed in 1953 by the merging of the former Air Maroc and Air Atlas companies, and adopted its present title in 1957. In addition to services within North Africa, it flies to several points in Western Europe and to the Canary Islands. Majority shareholder is the Moroccan Govern-

ment, and Air France (q.v.) also has a substantial interest.
Headquarters:
Casablanca Airport.

Royal Dutch Airlines, see KLM

Royal Jordanian Airlines, see Alia

Royal Nepal Airlines

Nepalese airline which, since 1958, has operated the internal routes formerly served by Indian Airlines Corporation (q.v.); it also has services to several cities in India. Fleet comprises eight Douglas DC-3s, one Friendship, two Antonov An-2s and two Mi-4 helicopters.
Headquarters:
Katmandu.

Rubber shock-absorber, 569ff

equivalents of the old-fashioned bands of "bungee" used on the legs of many early aircraft. As on the unit illustrated, the absorbers generally use rubber blocks or discs in compression, the blocks being guided by means of metal discs on a central tube. A small auxiliary block at the bottom acts as a recoil buffer.

The use of this type of absorber is restricted to relatively light aircraft. The low-temperature properties of the rubber selected are important, as on some compounds the loss of resilience at sub-zero temperatures may be serious.

Rumpler CIV, 265

The Rumpler CIV two-seat reconnaissance biplane is considered among the best of the German artillery observation aircraft produced during the 1914-18 War. Powered by a 260-hp Mercedes engine, it had a maximum speed of over 100 mph, a 4-hr endurance and, of most importance, could fly in the region of 21,000 ft, a height well above that attainable by most Allied aircraft. The high ceiling led to the development of respiratory apparatus for the crew.

Armed with a single fixed machine-gun firing forward and a movable Parabellum machine-gun for the observer in the rear cockpit, the Rumpler could hold its own against British and French single-seat scouts.

Rumpler CIs were used to operate the first post-war passenger air services in 1919. Two passengers could be carried and, on one aircraft at least, they were provided with an enclosed cockpit.

Ryan, 565

The Ryan Aircraft Corporation originally commenced as Ryan Airlines in 1922, headed by T. Claude Ryan. In 1925 the company entered the manufacturing field with the M-1 high-wing

1 Outer casing
2 Rubber blocks
3 Separator plates
4 Guide tube
5 Piston
6 Rebound rubber

Used in undercarriages to absorb landing loads, these are the modern

From the Farnborough archives: S. F. Cody (right) with one of his early kites

monoplane. It was a development of this, the NYP (*q.v.*) which hit the world's headlines when it was used by Lindbergh in 1927 to fly solo across the Atlantic. From this the Brougham six-seater was developed and sold quite extensively. Further models were initiated, but the company disappeared in the depression of 1931. Claude Ryan re-entered the manufacturing field in 1933 with a new company, building the S-T two-seat, low-wing trainer monoplane which achieved some success. In 1938 it produced the S-C 3-seat cabin monoplane but in 1939 the S-T was adopted by the US Army Air Corps as the PT-19 and quantity production followed. Variations on the S-T were produced during World War Two. Toward the end of the war the company concentrated on the development of a complex fighter for the US Navy, the Fireball, employing a Wright R-1820 radial piston-engine and a General Electric I-16 jet engine. With the return of peace the company produced the Navion cabin monoplane, designed by North American, before turning to the target drone and missile field. More recently it has participated in developing the LTV-Hiller-Ryan XC-142A (*q.v.*) experimental vertical-lift aircraft, and has also produced the XV-5A VTOL research aeroplane.

Ryan NYP monoplane, 253, 255, 689
The NYP (New York-Paris) monoplane, a development of the Ryan M-1, was built specially for Colonel Charles Lindbergh's transatlantic flight in 1927 and named *Spirit of St Louis*. From it was developed the Ryan Brougham.

Ryan XV-5A, 1012-1013, 1021

Two XV-5A's were built under US Army contract by Ryan to develop the VTOL concept of using wing fans for lift. First flight took place in May 1964, and by November the first transitions had been made. The aim of this aircraft is to enhance the performance spectrum; the estimated maximum speed of the XV-5A is 547 mph, and maximum range 1,000 miles. It is powered by two General Electric J85 turbojets rated at 6,658 lb st each.
S.5, S.6 and S.6B, see Supermarine
S-55, see Sikorsky
Saab
The initials of Saab stood originally for the Svenska Aeroplan Aktiebolaget, which was founded at Trollhättan in 1937 to develop and build military aircraft. Two years later it amalgamated with another Swedish company, ASJA, a rolling-stock manufacturer which had also been producing aircraft, both military and civil, since the mid-'twenties. Among its successful pre-war designs were the single-engined Saab-17 and twin-engined Saab-18, both of which served with the Royal Swedish Air Force as bombers and reconnaissance aircraft. Then followed the radical Saab-21, a twin-boom pusher-engined fighter/ground attack aircraft which is the only combat aircraft ever to have been produced in quantity in both piston-engined and jet-engined forms. The Saab-21R jet-powered version was among the first aircraft in the world to be fitted with an ejection seat for the pilot; this, too, was of Saab design.

Hundreds of Saab-35 Drakens are in service

After World War Two Saab produced Sweden's first aircraft designed from the outset for jet power (the Saab-29), followed by the transonic Saab-32 Lansen attack fighter and the Mach 2 Saab-35 Draken. The latter, whose original design was started some twenty years ago, is still one of the most advanced combat aircraft in production in Europe. To replace the Lansen, Saab is currently developing the Saab-37 Viggen, an advanced multi-mission warplane of "canard" layout, which will enter service in 1971.

Sabena
Operating name of Belgian World Airlines (*Société Anonyme Belge d'Exploitation de la Navigation Aérienne*), which was founded in 1923 and in addition to a domestic and European network was responsible for building up regular air services to and within the then Belgian Congo. It is now owned jointly by the Belgian Government and private interests. In 1953 Sabena opened scheduled passenger services between certain European capitals with a small fleet of helicopters, which by late 1959 had carried over a quarter of a million passengers; but the service was withdrawn in the early 'sixties. The fixed-wing fleet still maintains not only a comprehensive internal and European service, but flies regularly to the Middle East,

San Francisco International Airport, California, USA

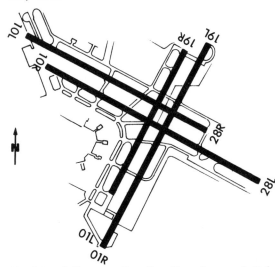

Traditional "gateway" to the Pacific for generations of sea travel, San Francisco maintains this position in terms of air travel as well, and all major trans-Pacific air routes pass through here. San Francisco is among the ten busiest international air-

San Francisco and Oakland Helicopter Airlines
First operator in the world to have a fleet consisting entirely of turbine-powered helicopters, SFO was incorporated in January 1961 and began scheduled local services in June of the same year with two S-62As leased from Sikorsky. Unlike most contemporary helicopter carriers, SFO has operated without a Government subsidy since its inception. Headquarters: Oakland International Airport, California.

Santa Cruz Airport, see Bombay

Santos-Dumont, Alberto, 133, 149, 187, 193
A Brazilian living in Paris, Santos-Dumont was largely responsible for putting the airship in the forefront of aviation development in France. He did this by building a succession of fourteen single-seat airships, most of them small and manoeuvrable. The first flew in 1898, and with the sixth

Saab System 37, the Viggen, sole European example yet flown of a modern integrated supersonic weapon system comparable in character with the F-111

Saab-105, 1253
Powered by two Turboméca Aubisque or General Electric J85 light turbojets, the Saab-105 is in service with the Swedish Air Force as a trainer and ground-attack aircraft. It is also used by the Austrian Air Force. A shoulder-wing monoplane of 31-ft span, it has a maximum speed of 478 mph at 20,000 ft with Aubisques.

Africa and North America.
Headquarters: Brussels.
Member of IATA.
Sabre engine, see Napier
Sabreliner (T-39), see North American
"Safe-life", 432, 439
Safety, see Flight safety
Salon de l'Aéronautique, see Air shows and exhibitions

*ports in the world.
Distance from city centre: 18 miles
Height above sea level: 11 ft
Main runways:
10R/28L = 10,600 ft
01R/19L = 9,500 ft
10L/28R = 8,896 ft
01L/19R = 7,005 ft
Passenger movements in 1967: 12,248,100*

Santos-Dumont, who in 1906 made the first controlled heavier-than-air flight in Europe, photographed with his "Demoiselle" of 1909

Saab-21A, later developed into the first Swedish jet design

he made the flight around the Eiffel Tower which won him the Deutsch de la Meurthe prize of 125,000 francs. His ninth was an attempt in 1903 to interest the French Army in airships, appearing at a review at Longchamps. Santos soon turned his attention to heavier-than-air craft, and in 1906 made the first recognized aeroplane flight in Europe, at Bagatelle, France, with his *14bis* canard biplane. Three years later his *Demoiselle*, a small, 18-ft-span monoplane, was a star attraction at flying meetings, and was the world's first really practical light aeroplane that the man in the street could fly without difficulty.

SAS (Scandinavian Airlines System)
Airline consortium founded in 1946 to operate the combined transatlantic services of AB Aerotransport (Swedish Air Lines), DDL (Danish Air Lines) and DNL (Norwegian Air Lines); it now operates all internal and overseas services of the three holding companies, although a proportion of its aircraft is registered in each country. In 1961 SAS was the first airline to inaugurate a service via the North Pole, from Copenhagen to Tokyo, and now has a world-wide network reaching to Europe, Africa, the Middle and Far East and North and South America, in addition to Scandinavian domestic services. SAS has a minority holding in Thai International Airways (*q.v.*).
Headquarters: Stockholm.
Member of IATA.

Caravelle III airliner of SAS

SATCO (Servicio Aereo de Transportes Commerciales)
SATCO is a commercial airline which, with a transport element of the Peruvian Air Force, is charged with pioneering new trunk routes within Peru, with Government assistance when necessary. When a new service has been established satisfactorily it is handed over to a domestic airline for continued operation. SATCO and TANS (the military transport element) were previously a single organization known as the Transportes Aereos Militar, which could trace its origins back to the Linea Aerea Nacional of 1931.
Headquarters: Lima.

Saudi Arabian Airlines
Founded as a Government-sponsored charter operator in 1946, Saudi Arabian Airlines ran its first scheduled services in March 1947. First jet services began in April 1962, and the airline was established as a Corporation in the following year. Financial and other assistance has been provided by TWA (*q.v.*) since late 1966, and the airline's present network extends to most Middle Eastern countries, India, Pakistan, North Africa, Germany, Switzerland and the UK.
Headquarters: Jeddah.
Member of IATA.

Saulnier, Raymond, 875
Chief designer of the Morane-Saulnier (*q.v.*) company for many years and designer of a rudimentary gun synchronising gear before the 1914-18 War. He is also thought to have played a part in designing the cross-Channel Blériot monoplane of 1909.

Saunders-Roe Princess, 745, 831
This large British six-engined flying boat owed its inception to a Ministry of Supply requirement of July 1945. Three prototypes were ordered, and at first BOAC were interested in using it for transatlantic operation. However, the Corporation later became committed to landplane operation across the Atlantic, and in 1952 the programme was cut back to one flying prototype and two static airframes, with the hope that RAF Transport Command would eventually adopt the type.
The Princess prototype flew in August 1952, and carried out a full flight test programme. At the end of this, its future still obscure, the Princess was cocooned at Cowes in 1954. The second and third aircraft had already been launched and laid-up in a cocooned state in 1953. Various projects for the use of these aircraft were envisaged but none came to fruition and more than a decade later they were finally scrapped.

Savoia-Marchetti, 253, 256
The Italian Savoia company, founded in 1915, became known in the 'twenties and 'thirties for the production of flying boats to the designs of Ing A. Marchetti. Especially distinctive were his twin-hulled flying boats, of which the S.55 was used in the formation flights carried out by General Italo Balbo (*q.v.*). It was also used by the *Regia Aeronautica* for other long-distance flights. These twin-hulled flying boats were developed for civil use as well. The company later turned its attention to three-engined landplanes, and in the middle 'thirties produced a highly successful series of three-engined, low-wing monoplane transports and bombers. These included the SM.79 and

the SM.81, used in the Spanish Civil War and in World War Two.
Now known as Siai-Marchetti, it concentrates chiefly on the production of light single- and twin-engined cabin monoplanes and also sponsored the Silvercraft SH-4 three-seat light helicopter.

Savoia-Marchetti S.55P, one of a series of distinctive twin-hulled flying boats

SBA (Standard Beam Approach), 1373, see also Take-off and Landing Techniques
The SBA, or Standard Beam Approach, system grew out of radio developments in the nineteen-twenties and 'thirties. It was aimed at assisting aircraft to place themselves at the threshold of a runway, prepared to land, in bad visibility, and was one of the first practical aids for bad-weather flying to come into widespread use.
The system consisted of two radio beacons: an inner marker, set in line with the runway and its threshold, and an outer marker, set in line with the runway and the inner marker, but several miles away. These transmitted a distinctive set pattern of signals vertically upward and in a 360° arc. The latter signals varied from one side of the runway to the other, i.e., if the aircraft was to the left of the runway it would receive the Morse letter "A", and if to the right the letter "N" (which in Morse is the reverse of "A"). If it was on the line of the runway the two signals would blend to form a continuous note. This enabled a pilot approaching an airfield to establish, first of all, on which side of the runway he was flying,

and then to reach the runway. Having found it, he could then, by keeping a continuous signal in his headphones, fly along the runway and establish his position on it as he flew over the marker beacons. Having done this, he could then fly a set approach pattern to bring his aircraft on to a landing approach, simply by orientating it according to the signals. The outer marker would give him the position in which to put the aircraft in landing configuration (i.e., full flaps, undercarriage down, fine pitch, etc.), while the inner marker would alert him for the final landing manoeuvre.

SBAC displays, see Air shows and exhibitions
SC-1, see Short
Scandinavian Airlines System, see SAS
Scarff, Warrant Officer F. W., 883
Schiphol Airport, see Amsterdam
Schlieren
This is one of several kinds of optical system which render visible the pattern of varying air density in an aerodynamic system, such as in the working sec-

tion of a high-speed wind tunnel. It can assume many forms, but in all of them a powerful source of monochromatic light is passed through a system of mirrors through the working section of the tunnel. The resulting interferometric effect causes the variation in density in the air around the model or other test body to be clearly visible. Usually this change in density also gives rise to contrasting colours, typical colours being a pale yellowish-green background with strong reds, whites or other hues in any shock wave pattern that may be present. Other interferometric optical systems cause a pattern of fine parallel lines to cover the field of view and these can enable direct measurements to be obtained of the density variation.

Schneider, Franz, 875
Schneider Trophy, 222, 227, 281, 387, 1087; see also Competitive flying
A bronze sculpture, the Schneider Trophy was presented by M. Jacques Schneider to the *Aéro Club de France* in 1913. Originally the trophy was intended to encourage seaworthy qualities in marine aircraft, but it turned into a speed race contest round a course over water. The first contest was held on 16 April 1913 at Monaco, and the last in the Solent on 12 Septem-

Sqn Ldr Orlebar in the Supermarine S.6 that gained a world speed record of 357.7 mph

M.67, one of a series of Schneider Trophy contenders by the Italian Macchi company

ber 1931, when the Trophy was won outright by Great Britain after her third successive win. The winning aircraft was the Supermarine S.6B at a speed of 340.08 mph.

Schwarz, David, 133

Science Museum (London), 609

London museum originally founded in 1857, when its exhibits included foods and art collections. Since 1909 those relating to science and technical industry have formed the Science Museum. The aim of the museum is to aid study of scientific and technical development. A major part of the museum is devoted to the Aeronautical Collection, and it has sponsored a number of internationally-admired publications on the development of flying.

Scout biplane, see Bristol

Scout helicopter, see Westland

S.E.4, 241

Built by the Royal Aircraft Factory (q.v.) at Farnborough, this fastest of pre-1914 scouts was designed by H. P. Folland and achieved 135 mph at ground level in 1914. Much streamlining was incorporated in its design and it was reported on favourably by those who flew it. However, its landing speed of 52 mph was considered too high, and it was not developed for operational service.

S.E.5, 1075

The S.E.5 owed its origin to the excellence of the Hispano-Suiza engine. Fifty had been bought in 1915 by the Royal Flying Corps, and licence production began in the UK in 1916. Around this engine, design staff at the Royal Aircraft Factory, headed by H. P. Folland, conceived a fighter intended for use at the Front by pilots with the limited flying skill prevalent in those days. The prototype S.E.5 flew in December 1916,

and showed great promise before crashing as a result of structural failure. This was remedied, and the aircraft went into squadron service in March 1917. It saw its first action on 22 April, and the first victory was scored the following day by Captain Albert Ball, VC. A more powerful Hispano-Suiza, delivering 200 hp, appeared later, and with this the aircraft was developed into the S.E.5A. This followed the S.E.5 into service, and many squadrons flew the type with great success, finding it very strong and a steady gun platform.

Seaboard World Airlines

Title since 1961 of the former Seaboard and Western Airlines Inc, which was formed in 1947 and began scheduled operations in April 1956. Seaboard's primary activity is transatlantic freight-carrying between New York and major cities in Western Europe, and for many years it was the only all-freight airline with scheduled flights across the North Atlantic. It also carries out charter freighting for Military Airlift Command of the USAF.
Headquarters:
New York.
Member of IATA.

Sea Fury, see Hawker

Seagull, see British Deperdussin

Sea Hawk, see Hawker

Seguin, Louis and Laurant, 229, 1043, 1065

These two French engineers were responsible, in 1909, for manufacture of the first successful mass-produced rotary engine, the seven-cylinder 50-hp Gnome.

SEPECAT Jaguar, 1243

Subject of a successful Anglo-French collaborative venture, the Jaguar has been developed jointly by Breguet in France and BAC in Great Britain to fulfil French Air Force and Royal Air Force requirements.

The Anglo-French SEPECAT Jaguar tactical strike trainer (French 2-seat strike version pictured)

The Jaguar, which first flew in September 1968, is powered by two Rolls-Royce / Turbomeca Adour turbofans. It is a shoulder-wing single- or two-seat monoplane with a radius of 370 miles and a maximum speed at altitude of Mach 1.7. It will be produced in strike and training versions for the French and British air forces, and as a French Naval carrier-based fighter.

Servo tab, 614

A servo tab is an additional control surface, fixed to a primary control surface, which always acts in the direction opposite to that in which the pilot moves the control and thus makes the actual control movement easier.

Sesquiplane, 369

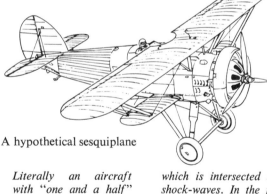

A hypothetical sesquiplane

Literally an aircraft with "one and a half" wings, a sesquiplane is an aircraft with one main lift-bearing wing together with another, much smaller wing, usually braced to the other wing by a strut or struts. The most typical form is a high-wing or "parasol" monoplane which has a stub-wing fixed to the lower longerons of the fuselage.

Shackleton, see Avro

Shaft-turbine engines, 969
see also Turboshaft

Sheremetievo Airport, see Moscow

Shock diamond, 1187

A diamond-shaped mass of luminous gas visible in the centre of any supersonic jet of hot gas

The result is that the jet is criss-crossed by a series of inclined shock-waves, and where these pass across the centre of the gas stream the enhanced luminosity gives rise to a succession of shock diamonds. Sometimes as many as 12 or 15 can be seen before the falling gas temperature makes them too faint to be visible.

Shock-waves, 297

Shooting Star (F-80), see Lockheed

Short brothers, 229, 232

Luminous shock diamonds in the supersonic jet from a rocket

which is intersected by shock-waves. In the jet from a turbojet with afterburner (reheat) in operation, or in the jet from a rocket engine, the flow velocity is invariably supersonic and the temperature high enough for the gas to appear visible as a flame of various colours. As the jet leaves the propelling nozzle it passes through an inclined shock-wave which is internally reflected from the opposite side of the jet at the boundary layer with the surrounding atmosphere.

Two of the Short brothers, Eustace and Oswald, entered aviation in 1898, building spherical balloons. In 1908 they were joined by their elder brother, Horace, and the following year began licence production of Wright biplanes in a hangar at Sheppey, England. From this they evolved their own designs, based on the Wright, and soon turned their attention to multi-engined aircraft, producing the first aircraft which could cope with engine failure in flight. These were the S.27, known as the Gnome Sandwich or Tandem Twin, and a

three-propeller aircraft known as the Triple Tractor. In 1911 the Shorts turned their attention to floatplanes and, having patented a wing-folding device, became large-scale contractors to the Royal Navy for coastal patrol seaplanes during the 1914-18 War, and for tentative work from the decks of ships. Between the wars the Short Bros company turned to flying boats, producing a long line of aircraft for civil and military use. This line culminated in the "Empire" 'boats for Imperial Airways and a military derivative, the Sunderland, which became the most famous of the RAF's maritime reconnaissance aircraft in World War Two. Short also built the Stirling, the first four-engined monoplane bomber to serve with the RAF.

Since World War Two the company has been taken over by the Government, and has built a succession of varying aircraft types, including development and experimental airframes. Most important of these was the SC.1 VTOL research aircraft. Two of these were built in 1957, the first hovering flight being made on 25 October 1958. Since then both have been used for a continuing programme of research into vertical take-off techniques. Short continues to build production aircraft, its latest designs being the Belfast freighter for the RAF and the Skyvan light STOL transport, which is still in production.

Short Belfast, 1-12, 561, 573, 789

The Belfast is a large, four-engined transport

Type 184, one of the first important Short designs for the British Admiralty

aircraft, ten of which were completed for Royal Air Force Air Support Command. It is used for the carriage of heavy freight, including guns, vehicles, missiles and helicopters, and these can be loaded through beaver-tail rear-loading doors. It is a high-wing mono-plane, powered by four Rolls-Royce Tyne tur-boprops, and has a range of 5,300 miles, a maximum cruising speed of 352 mph and a maximum payload of 78,000 lb.

Short "Empire" ("C") class) flying boats, 745
Designed to the require-ments of Imperial Air-ways for the carriage of all Empire air mail, this advanced four-en-gined monoplane flying boat was built by Short Bros in 1936. The ini-tial series of 28 was ordered from the draw-ing board, an unheard-of decision in those days. The decision was proved right, for this became one of the out-standing British aircraft of the 'thirties; it en-tered service in October 1936, regular services beginning in February 1937 when sufficient of the fleet had been de-livered. These aircraft flew through to Aus-tralia as well as to India, East Africa and South Africa.
The "Empire" 'boats established a new stan-dard of comfort both for crew and passen-gers, making a journey a sought-after pleasure. They were also used to pioneer Imperial Air-ways' attempts at trans-atlantic flying before World War Two, their range being extended by the experimental use of flight refuelling to enable them to make the journey non-stop. Their wartime service, maintaining many intra-imperial routes, is an epic in itself, and six-teen of the fleet sur-vived the war. By then the days of the com-mercial flying boat were numbered, and the last

scheduled flight by an "Empire" 'boat was on 23 December 1947.
Such was the design of these modern flying boats that they required little alteration to turn them into fine mari-time reconnaissance air-craft. Short Bros de-veloped the wartime Sunderland from their "Empire" 'boat ex-perience, a type which served admirably throughout World War Two.

Short SC.1, 993, see also Short brothers
The SC.1 is an experi-mental delta-wing re-search aircraft, two pro-totypes of which were built by Short Bros & Harland Ltd in 1957-58. It is powered by five Rolls-Royce RB.108 engines, four of which are installed vertically in the centre of the short, tubby fuselage, the fifth being placed horizon-tally at the tail to give forward thrust. The first SC.1 was flown con-ventionally in April 1957, vertically in Octo-ber 1958, and the first complete transition was made on 6 April 1960. The following year the aircraft flew from Eng-land to Paris for the Paris Air Show, becom-ing the first jet-lift air-craft to cross the English Channel. Since then both prototypes have been engaged in a con-tinuous series of trials at the Royal Aircraft Establishment, Bedford.

Short Skyvan, 1311
Built by Short Bros & Harland as a near-approach to the "flying boxcar" ideal, the Sky-van first flew in January 1963 with Continental piston-engines as a temporary measure be-fore the installation of Turboméca Astazous in October. A production

batch was laid down for this twin-turboprop STOL utility transport, but after prolonged eval-uation by several inter-ested airlines it was found that the Astazous were not able to cope with certain atmospheric conditions and the air-craft was redeveloped with the AiResearch TPE331 engine of 715 hp. With this engine, development has gone ahead, and the Skyvan is beginning to find a market in different parts of the world. It has accommodation for 19 passengers or 4,600 lb of freight in its bulky, rectangular fuselage from which the high strutted wing carries the twin turboprops. It has a cruising speed of 170 mph, service ceiling of 21,000 ft, and a range of 640 miles with a 2,910-lb payload.

Short Stirling, 569
The Stirling, conceived in 1936, was the first of the four-engined bomb-ers to see service and action with the Royal Air Force in World War Two. First prototype flew in May 1939, crash-ing on its first flight; it had been preceded by a half-scale prototype, the Short S.31, which had flown in 1938. Second prototype flew after the outbreak of war; quantity produc-tion enabled the first squadron to equip in August 1940 and to go into action in Febru-ary 1941. The Stirling's maximum bomb-load was 14,000 lb, twice that of the Whitley and over three times that of the Wellington.
Stirlings were used origi-nally by day as well as by night with a strong fighter escort on day-light raids over France. This did not continue

long, however, and after the summer of 1941 the Stirling was chiefly used on the night offensive. It was particularly suit-able for mine-laying and continued on this task for some time after it had been replaced in the main night offensive by the Halifax and Lancaster (qq.v.). It was also used on radio countermeasures and found a new lease of life in 1944 as a glider-tug and transport air-craft, being used in the various airborne land-ings in 1944-45 and in the repatriation of prisoners of war in 1945. It remained in service as a transport (in its Mks IV & V versions) until 1946.

Short Tandem Twin, 229, 232

Short Triple Tractor, 229, 232, 811

"Shorthorn", 815, see also Farman, Henry

Shuttle, see Commuter air services

Shvetsov
A.D. Shvetsov heads a design bureau in the USSR that has produced some of the most widely-used Soviet piston-engines. Best known are the 1,000-hp ASh-62 nine-cylinder, air-cooled radial that powers the Antonov An-2, and the 1,700-1,900 hp ASh-82, which was developed from the American Pratt & Whitney Twin Wasp radial (q.v.).

Sidewinder missile, see NOTS Sidewinder

Sikorsky, Igor Ivanovich, 229, 234, 317, 957, 959
Igor Sikorsky was born in Kiev, Russia, in 1889 and his early interest in aviation was marked by the construction of his first helicopter in 1909.

Neither this nor his second was a success, and he then turned his attention to fixed-wing aeroplanes; his S-2 flew briefly in June 1910. Sikorsky's first impor-tant contribution to avi-ation, however, was the building of the Le Grand, a four-engined aircraft in which the classic formula of four engines in a spanwise row was first used. Further large aircraft were built and used by the Russians for bombing during the 1914-18 War, but the Russian Revolution put an end to Sikorsky's work there.

The Sikorsky S.29A played the role of a German Gotha in the film "Hell's Angels"

Moving to the USA, he formed the Sikorsky Aero Engineering Corp-oration in March 1923, and resumed his development of large aircraft, principally fly-ing boats and amphi-bians. These were very successful in both mili-tary and civil markets between the wars, the S.42 being used by Pan American World Air-ways to pioneer their services across the At-lantic and Pacific during 1937-38. In 1939 the

ation to concentrate on helicopter development. From this stemmed the long and highly success-ful range of military and civil helicopters which has put Sikorsky in the forefront of world heli-copter development and production to this day.

Sikorsky CH-54A, 1313
The CH-54A is the mili-tary version of the Sikor-sky S-64 "flying crane" helicopter. It is powered by two 4,500-shp Pratt & Whitney JFTD12A shaft-turbines driving a six-blade main rotor and giving a maximum speed of 127 mph and a range of 253 miles. The CH-54A can carry a useful load of 22,890 lb, which is accommodated in interchangeable pods. These can be designed for many rôles—e.g., a pod with accommoda-tion for 67 troops, a cargo-transport pod, a mine-sweeping pod or a field hospital pod. The CH-54A is in action in the Vietnam War.

Sikorsky Ilya Mourometz, 723
The Ilya Mourometz was a progressive de-velopment of Sikorsky's

The Sikorsky S.42 served majestically on many of Pan American's major international routes

company amalgamated with Chance Vought Aircraft Corporation, a sister division of United Aircraft Corporation, and work on flying boats gradually tailed off. It was then that Sikorsky returned to helicopter work, design-ing the VS-300 which first flew successfully in May 1940. So promising was this aircraft that in 1943 Sikorsky Aircraft Ltd was reconstituted as a separate division of United Aircraft Corpor-

Le Grand. At the end of December 1913, this large four-engined bi-plane was completely assembled at St. Peters-burg and flew for the first time in January 1914; in February it carried 16 passengers to a height of nearly 1,000 ft, believed to be a world record at that time. A second aircraft was com-pleted in April 1914; on 18 June this stayed in the air for 6 hr 33 min, and between 29 June-11 July it flew a 1,590-

Short Skyvan STOL freight and passenger transport, here seen (unusually) at a large airport

Igor Sikorsky

mile round trip from St. Petersburg to Kiev and back, with only one stop, in 10½ hr flying time. Later in July the first aircraft flew trials on floats for the Imperial Russian Navy.

With the onset of war the *Ilya Mourometz* was developed into a bomber, a special squadron of these aircraft being assembled and making its first operational flight on 15 February 1915 from a base in Poland. Continued development of the type was forestalled by the Russian Revolution of 1917.

Sikorsky R-4B, 969

The R-4 was developed directly from Sikorsky's experience with the VS-300 and, after a development batch, went into production for the US Navy and Coast Guard and the Royal Navy. It served with these as a training aircraft, providing a growing company of helicopter-trained pilots in the British and US forces. It was powered by a 185-hp Warner R-550 radial engine which gave it a maximum speed of 75 mph and a ceiling of 8,000 ft.

Sikorsky R-5 (S-51), 969

Developed for operational duties, the Sikorsky R-5 achieved large-scale production for observation duties with the US forces. It was also licence-built by Westland, as the Dragonfly, with a British Alvis Leonides radial engine replacing the Pratt & Whitney Wasp Junior of the American version. It was the first helicopter licensed by the American CAA for commercial operation.

Sikorsky S-55, 969

The Sikorsky S-55 was the first really worthwhile helicopter to go into production from a commercial point of view. Whilst the S-51 provided a viable military operational vehicle in limited roles, the S-55 (or Whirlwind as it was called in the British services) provided sufficient payload for the

helicopter to become a battlefield transport — and the Korean War provided its proving ground.

The S-55 cabin could hold between eight and ten passengers and, with a maximum speed of 103 mph and range of 470 miles, it became a feasible vehicle for civil as well as military operations. It remained in large-scale production for many years and is still in service in many parts of the world.

Sikorsky VS-300, 317, 323, 954, 957, 959, see also Sikorsky, Igor Ivanovitch

This aircraft was the culmination of Sikorsky's early dreams, and with it he proved the practicability of the helicopter. It flew first in May 1940, and after extensive tests became the basis from which the whole line of Sikorsky helicopters has grown. Sikorsky flew it from both land and water, and broke the world's endurance record with this vehicle on 6 May 1941.

Simulators, see Flight simulators

Singapore Airport

Distance from city centre: 7 miles
Height above sea level: 59 ft
Main runway:
02/20 = 9,000 ft
International passenger movements in 1967: 958,000

Single-row radial engine, 1057ff

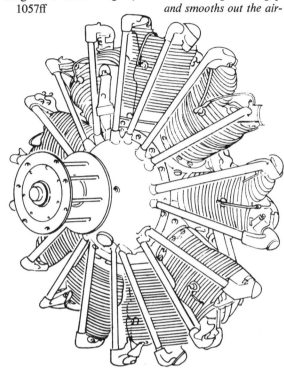

A piston-engine having cylinders arranged in a single radial row, like

Originally a military airfield, Singapore has handled commercial traffic since 1958, the new terminal building being started in the early 'sixties. Singapore is an important "staging post" for round-the-world flights and routes to the Far East.

spokes of a wheel, all the connecting rods driving a single crankpin.

Siskin, see Armstrong Whitworth

Skyvan, see Short

Slat, 641

Patented by the British

pioneer Handley Page, this device consists of an auxiliary aerofoil mounted along the leading-edge of a wing. At high angles of attack the slat lifts clear of the wing, leaving a narrow slot or gap. Air accelerates through this gap and smooths out the air-

flow over the wing, thus delaying stalling. Its development increased the speed range of aircraft and virtually eliminated the "stall-and-spin" type of accident. In 1928, the British Air Ministry ordered "slots" as they were popularly known, to be fitted to all British military aircraft, their adoption leading to a significant reduction in the accident rate.

Slats were also adopted for large transports and light aircraft; and today they are still widely used on many types of aircraft. Operation on some aircraft is automatic, the slats rising when high angles of attack are reached, but power operation is used on high-speed aircraft.

Sleeve-valve engine, 1081-1083, 1097

The sleeve-valve engine works on a different principle to the poppet-valve engine for the admission and exhaustion of the explosive mixture and burnt gases respec-

tively for each cylinder. Instead of having valves opening and closing at the cylinder head, the cylinder of the sleeve-valve engine has openings in the sides of the cylinder walls. Within the cylinder is a metal sleeve which has corresponding openings. This sleeve moves up and down and rotates under the guidance of cranks; so that at the appropriate moments in the four-stroke cycle the openings in the sleeve coincide with those in the cylinder wall, and the correct fuel and air mixtures are injected before the explosion and the exhaust gases extracted afterwards.

The two chief advantages of this type of engine are lower operating temperatures and a better-shaped combustion chamber.

Slingsby Dart, 502

An outstanding modern high-performance sailplane, the Dart first appeared in 1963 and has been used by many of the leading exponents of soaring. It has appeared high in the lists of place winners in British National and World Championships and in 1965 was awarded the OSTIV Design Prize at the

Slingsby T.51 Dart, seen on an early test flight in 15 metre form

World Championships. It is available with both 15-metre and 17-metre wings and has a retractable undercarriage as an optional extra.

Smithsonian Institution, 181

American museum located in Washington, DC, founded in 1846 for the increase and diffusion of knowledge among men. It acts as the organization for the exchange of publications with learned institutions throughout the world

and is noted for its aeronautical exhibits, which include the original Wright Flyer, first powered aeroplane to achieve sustained and controlled flight.

Snail, see Sopwith

Soloviev engines, 1011, 1313

The Soloviev design team has produced several engines which are in production and service in the USSR. The D-20P is a two-spool turbofan engine producing a maximum take-off rating of 11,905 lb static thrust. Two of these power the Tupolev Tu-124 transport of Aeroflot. The D-25V is a shaft-turbine engine for VTOL use, and powers the Mi-6, Mi-10 and Mi-10K helicopters and the Kamov Ka-22 *Vintokryl* convertiplane; it produces a take-off power of 5,500 eshp. The D-30P is a 14,990-lb two-spool turbofan engine used in the Tupolev Tu-134 airliner.

Sommer biplane, 232

Sopwith, Sir Thomas Octave Murdoch, 222

Thomas Sopwith appeared on the British aeronautical scene in 1910 with a brief debut in October, when he wrecked his aircraft at Brooklands; but in

the following month he learnt to fly and gained his brevet all in one day. In 1911 he became a test pilot, chiefly for the

Sir Thomas Sopwith

Martin and Handasyde firm, and achieved a growing reputation as a racing pilot. The following year he set up a flying school at Brooklands, but his attention was directed more and more to aeroplane manufacture. First aircraft produced at Sopwith's Kingston works was a flying boat, the Bat Boat; it was followed quickly by a three-seat tractor biplane. The Sopwith name was made with the Tabloid single-seater, which appeared in 1913 and which, slightly modified and adapted as a seaplane, won the Schneider Trophy for Great Britain at Monaco in 1914. From this basic type stemmed a whole family of aircraft in the 1914-18 War— the Baby floatplane, 1½-Strutter, Pup, Triplane, Camel, Dolphin and Snipe, all of which were put into quantity production and served in all the fighting areas. The Camel was considered to be the finest British scout of the war. After the war Sopwith's company became successively H. G. Hawker Engineering Company Ltd, and, later, Hawker Aircraft Ltd.

Sopwith Snail, 361

Sopwith Tabloid, 858, 861
The Tabloid appeared in November 1913, having been assembled in secrecy at Sopwith's factory in Kingston, Surrey. It was a small single-seat tractor biplane with a wing span of 25 ft 6 in, and its 80 hp Gnome engine almost totally enclosed in a neat cowling. It had a maximum speed of 92 mph, faster than most contemporary monoplanes.
The type went into production as a two-seater while Harry Hawker was demonstrating the prototype in Australia. One machine, fitted with a float undercarriage, won the Schneider Trophy in 1914. This aircraft was the progenitor of the Sopwith Baby of the

war years. Tabloids took part in the war in a scouting rôle with the Navy. Two of them flew a bombing raid on Düsseldorf on 8 October 1914, when Flight Lieutenant R. L. G. Marix managed to destroy Zeppelin Z.IX. The Tabloid was soon overtaken by later developments, but was a noteworthy step forward in the development of the fast, manoeuvrable biplane.

Sopwith tractor biplane, 356

South Africa: Air Force
Suid-Afrikaanse Lugmag (South African Air Force). Following recommendations made in 1912, a South African Aviation Corps was created in 1915. Later it became a squadron of the British RFC until disbanded early in 1918. A new permanent South African Air Force was created on 1 February 1920 with aircraft supplied under "Imperial Gift" from the UK. Two years later it was engaged operationally in South-West Africa, and in 1925 opened the first regular air mail service in the Union with D.H.9's. Re-equipment plans laid in the late nineteen-twenties led to the purchase and licence manufacture of several more modern British designs during the first half of the 'thirties. Then, in 1936, a more ambitious programme of expansion and modernization was begun; but by the outbreak of World War Two the only really modern warplanes in the country were six Hawker Hurricanes, one Bristol Blenheim and a Fairey Battle, and the total strength of the SAAF was only 104 aircraft and just over 2,000 officers and men, including reserves.
With the outbreak of war the commercial fleet of South African Airways (q.v.) was impressed for military service, and the airline's Ju 86s were employed on maritime patrol and

reconnaissance work. Units in South Africa were rapidly built up with British and US combat aircraft, several of them being sent to defend East African states against the threat from Italian-held Abyssinia. The SAAF was even more prominently engaged in the desert campaigns in North Africa, the invasion of Italy and the anti-U-boat campaigns over the North Atlantic. In addition, although not a participant in the Empire Air Training Scheme, South Africa provided basic training facilities which by the end of 1945 had passed out over 12,000 aircrew for the SAAF and nearly 21,000 more for the RAF.
After the war the SAAF was initially reduced to a small permanent force. It made a modest contribution to the 1948-49 Berlin Airlift, and a Mustang-equipped SAAF squadron fought with distinction in the Korean War of 1950-53. In the closing months of this conflict it received its first operational jet fighters —F-86 Sabres—although British Vampires

First jet bomber of the SAAF was the Canberra B(I).12

had been supplied to units in South Africa in 1952. The SAAF ordered Canadair Sabres in 1955, and in 1957 replaced its Sunderland flying boats with Avro Shackletons, which are still in service. The Sabres are now being replaced with French Mirage III fighters, while more recent additions include Buccaneer strike aircraft and Super Frelon transport helicopters. Large numbers of the Aermacchi M.B.326 jet trainer and light attack aircraft are being licence-built for

the SAAF by Atlas Aircraft Corporation.

South Africa: Chronology
18 December 1814. A balloon, built by Coussy, made the first balloon ascent at Cape Town, with a cat as passenger.
1875. Evidence suggests that John Goodman Household flew in a glider of his own design in the Karkloof Valley, near Howick, Natal.
1878. Paul Kruger (later State President of the South African Republic) flew as passenger in a Henri Giffard balloon while visiting Paris.
April 1885. The beginning of military aviation in South Africa. A balloon operated by British Royal Engineers was used during the Bechuanaland Expedition.
1899-1902. Four balloon units of the British Royal Engineers were active on the major fronts during the Boer War.
1907. John Weston built the first powered aircraft in South Africa. Before being tested it was taken to France; Weston returned later and played a leading part in the country's early aviation activities.
28 December 1909. A French pilot, Albert Kimmerling, made the first powered aeroplane flight in South Africa, at East London.
27 December 1911. E.F. Driver of Pietermaritzburg, Natal, flew the country's first air mail between Kenilworth and Muizenburg, near Cape Town.
1912. Lieutenant-General C. F. Beyers, Commandant General of the newly-formed Union Defence Force, visited Europe and recommended the purchase of military aircraft and the training of pilots.

1913. Ten pupil pilots sent to C. Compton Paterson's flying school at Alexanderfontein, near Kimberley.
1915. South African Aviation Corps (Zuid-Afrikaanse Vliegerskorps) raised to accompany South African forces invading German South-West Africa. First mission flown in May with a Henri Farman F-27. At the completion of the campaign, the unit went to Britain, becoming No 26 (South African) Squadron Royal Flying Corps, later serving in East Africa until disbanded early in 1918.
1914-1918. More than 3,000 South African aircrew served with the RFC, RNAS, and RAF. Distinguished officers included Captain A. W. Beauchamp Proctor, VC, DSO, MC (Bar), DFC, who destroyed 54 enemy aircraft while serving in France; and Captain I. V. J. Pyott, DSO, who destroyed the Zeppelin L-34 in November 1916. General (later Field Marshal) J. C. Smuts, was given the task of examining Britain's air defences, and his recommendations led to the amalgamation of the RFC and RNAS to form the Royal Air Force.
1919. The British government donated the "Imperial Gift" of 111 aircraft to South Africa— comprising de Havilland D.H.4's, D.H.9's, Avro 504K's and S.E.5A's— together with ground equipment to form a permanent air force in the Union.
1 February 1920. South African Air Force (Suid-Afrikaanse Lugmag) established as the first of the new Commonwealth air forces, with headquarters at Swartkop Air Station, near Pretoria.
1925. The SAAF flew the first regular air mail service in the Union, between Cape Town and Durban.
August 1926. Union Air-

ways (Pty) Ltd registered and a subsidized service inaugurated between Port Elizabeth and Cape Town.
1930. South West Africa Airways founded by German interests, using Junkers A-50 and F-13 aircraft. This company later acquired financial interest in Union Airways.
1930-1939. 42 Avro Tutors, 65 Hawker Hartbees and 27 Westland Wapitis were built for the SAAF at the Artillery and Aircraft Depot, Pretoria.
February 1934. South African Airways (q.v.) established. It took over the routes and equipment of Union Airways.
1935. South African Airways took over South West Africa Airways.
September 1939. South Africa declared war on Germany.
June 1940. South Africa declared war on Italy; several squadrons operated against the Italians in Abyssinia and East Africa.
December 1944. SAAF released six Lockheed Lodestars, complete with air and ground crews, to South African Airways to resume internal air services.
1945. Many South Africans had served with distinction in the British RAF, including Squadron Leader M. T. St J. Pattle, DFC (Bar), with 41 victories, and Group Captain A. G. ("Sailor") Malan, DSO (Bar), DFC (Bar), with 35 victories. Squadron Leader J. D. Nettleton and Captain Edwin Swales, DFC, were awarded the Victoria Cross.
1949. Air and ground crews of the SAAF seconded to the RAF to take part in the Berlin Airlift.
August 1950-July 1953. No. 2 (Fighter) Squadron, SAAF, sent to Korea for service with United Nations forces.
31 May 1961. The Union of South Africa became a republic, outside of

A flight of supersonic Mirage III-CZ fighters of the SAAF

the Commonwealth.

1961-68. With expansion of the Republic's defences, British Buccaneer and French Mirage strike aircraft entered service. Manufacture of Italian M.B.326 jet trainer undertaken by Atlas Aircraft Corporation of Johannesburg. Helicopters of the South African police sent to Rhodesia to assist in combating terrorism in 1967.

South African Airways

Successor to the former Union Airways, SAA was formed in 1934 under the aegis of the nationalized South African Railways and Harbours Administration, by which it is still controlled. A comprehensive domestic network of routes is served, while regional and international services are flown to neighbouring states, to Mozambique, to major West European capitals and to Australia.

Headquarters: Johannesburg.

Member of IATA.

Spad fighters, 883

The *Société Pour l'Aviation et ses Dérivés*, or SPAD for short, produced a series of single-seat fighters in the 1914-18 War for the French and British forces. These fighters were all very similar in layout, being twin-bay biplanes with closely-cowled Hispano-Suiza engines and one or two machine-guns mounted on the cowling. They were sturdy aircraft; the wings had neither dihedral nor stagger, consequently they were unstable and relied on their speed and strength rather than their manoeuvrability in combat. The standard versions in service with the scout squadrons were the Spad VII and XIII, while the Spad XII and XIV were fitted with early *moteur-canon* (q.v.).

Spain: Air Force

Ejercito del Aire (Spanish Air Force). Spanish military aviation was established in March 1911 with the formation of the Farman-equipped *Aeronáutica Militar Española;* additional British, French and Austro-Hungarian aircraft were purchased later in the year, some of which were engaged in military operations in Morocco in 1913. Morane-Saulnier Parasols and de Havilland D.H.4s were built under licence during the next two years, but not until after the 1914-18 War did Spain have a wide choice of modern aircraft to equip her growing air force. Several of these, by this time, were designs powered by the excellent Hispano-Suiza Vee-type engine evolved during the war years. By the middle 'twenties the expanding Spanish aircraft industry was building several excellent foreign types under licence, both for the

Army and Naval air services.

In July 1936 the Spanish Civil War was precipitated by a revolution in Morocco, and lasted until March 1939. During this period the air power both of the AME and of Franco's Nationalist forces had become a complex admixture of modern combat aircraft sent by Britain, Czechoslovakia, France, Germany, Italy and the USSR to aid one side or the other. The new Nationalist Government under General Franco established a new air force, the *Ejercito del Aire*, in November 1939. Faced with the task of replacing its motley collection of machines, this force drew on its own combat experience to select such types as the Fiat C.R.32 and Messerschmitt Bf 109 fighters and the Heinkel He 111 bomber for licence manufacture, while it purchased such second-line foreign types as it could and left the domestic industry to design and produce its training aircraft. Otherwise, during World War Two (in which Spain took no significant part), it acquired very few aircraft from outside sources. Even after the war re-equipment was difficult, and the first Hispano-built Bf 109G (with a Hispano-Suiza engine) did not fly until 1951. Only after the completion of a Defence Treaty with the US in 1953 did a real modernization become feasible. This began with jet training in 1954, followed by the supply of Sabre fighters and various piston-engined trainers, transports and air/sea rescue types in the next few years. The domestic industry is also now on a healthier footing, having in recent years produced HA-200 jet trainers and CASA-207 transports for the Air Force, but its com-

bat aircraft are still of US origin. Some Sabres remain in service; to these have been added F-104G Starfighters and Spanish-built Northrop F-5s.

Spain: Chronology

1783. First balloon in Spain, constructed by Viera Clavijo, ascended at Madrid.

3 November 1792. A manned observation balloon, built by officers of the Royal Artillery College, made an ascent at Segovia.

1883/1888. Martinez Díaz made numerous ascents in free balloons throughout Spain, and in Brazil, Colombia, Portugal and Venezuela. He made several ascents in the United States, and in 1888 worked for Edison on an airship project.

1889. At Madrid, the Regent, Queen María Cristina, was the first Royal personage to be carried as passenger in a balloon ascent.

1896. Military Balloon Service established at Guadalajara.

1900. Information supplied by military observation balloons used for the first time in Spain for correcting artillery fire.

1905. Foundation of the Royal Aero Club of Spain.

January 1906. Fernández Duro made a 21-hr balloon flight extending the length of Spain, from Pau in Southern France to Guadix, in the Sierra Nevada.

1906. Test flights at Guadalajara of the semi-rigid airship designed by Torres Quevedo. It was later exhibited in Paris, where the French Astra company acquired a licence in 1913 to build the Astra-Torres series of airships. Some were used by the British Royal Navy in the 1914-18 War.

1909. Antonio Fernández designed and exhibited a biplane at the first French *Salon de l'Aeronautique* in Paris. A licence to build the

type was acquired by Levavasseur.

1910. First powered aeroplane flight in Spain made at Barcelona, and Benito Loygorri became the first Spaniard to fly a powered aeroplane.

1910. An aerodynamics laboratory established at Cuatro-Vientos.

1911. Spanish military aviation established.

1913. First civil flying school began operating at Getafe, near Madrid.

1913. Three squadrons used on operational service in Morocco, performing visual and photographic reconnaissance as well as bombing missions.

1915. Colonel Vives, Chief of Military Aviation, gave orders for the national design of an aircraft engine. The resulting designs were sent to Hispano-Suiza of Barcelona, manufacturers of motor cars. Birkigt, the company's Swiss engineer, working in collaboration with Spanish engineers Quesada and Sousa Peco, finalized design as the 140-hp Hispano-Suiza engine. This won a contest initiated by the French Air Ministry,

thwarted production of any of these designs.

1920. Cierva's first Autogiro tested, unsuccessfully, at Getafe.

1921. Juan de la Cierva gave the Science Academy full details of his latest Autogiro design. In the same year Spain's first airline—CETA—inaugurated a service between Seville and Larache, in Morocco, across the Straits of Gibraltar.

1922. Cierva designed articulated hinges for the rotor blades of his C.4 Autogiro.

9 January 1923. First successful flight of Juan de la Cierva's C.4 Autogiro at Cuatro-Vientos, piloted by Lieutenant Gómez Spencer. It was this aircraft which pointed the way to the successful helicopter.

22 January-10 February 1926. The Dornier *Wal* flying boat *Plus Ultra*, powered by two 450-hp Napier Lion engines, crossed the South Atlantic, piloted by Commandante Ramón Franco. The flying boat took off from Palos de Moguer, Spain, arriving at Buenos Aires, via Recife, Brazil, in a

The "Plus Ultra", a Spanish Dornier Wal flying boat crossed the South Atlantic in 1926

and was subsequently mass-produced in several countries, either in Hispano-Suiza factories or under licence.

1917. An aviation section of the Spanish Navy was established.

1918/1919. A contest for nationally designed and built military aircraft was sponsored, for which a number of successful prototypes were entered, including a three-engined bomber designed by Juan de la Cierva. Availability of large numbers of surplus Allied aircraft

flight time of 62 hr 52 min.

December 1926. Three Dornier *Wal* flying boats, operated by military aviation squadrons, flew from Spanish Morocco to Spanish Guinea, returning in February 1927.

26/27 March 1929. Captains Jiménez and Iglesias completed a non-stop transatlantic flight in a CASA-built Breguet XIX GR named *Jesús del Gran Poder*, direct from Seville to Bahía, Brazil. Elapsed time for the flight was

43 hr 48 min.

April 1932. Jorge Loring, in a small aircraft of his own design, the Loring E-11, flew from Madrid to Manila. A year later he repeated the flight in a Comper Swift.

June 1933. Captain Barberán and Lieutenant Collar made a record over-water flight of 3,914 miles from Seville to Camagüay, Cuba, in 40 hr. The aircraft used was a CASA-built Breguet Super Bidon

The "Cuatro-Vientos" (Four Winds) of Barberán and Collar; CASA-built Breguet with Hispano engine

named *Cuatro-Vientos*, powered by a 650-hp Hispano-Suiza engine.

1936/1939. During the Spanish Civil War new equipment and tactics were tested operationally. Captain Haya, in a DC-2, acted as a "pathfinder", releasing flares to guide a formation of S.81 bombers; a Ju 52 unit led by Gallarza carried out a night bombing sortie during the battle of Brunete, using a pattern of ground fires for guidance; and the Condor Legion introduced the "double pair" as the basic formation for fighter units.

1945. World altitude record for two-seat sailplanes, 20,547 ft, established by Luis Vicente Juez.

1955. First flight of the first jet aircraft designed and built in Spain, the Hispano-Aviacion HA-200. It was later built in quantity both in Spain and, under licence, in Egypt.

1964. Tomás Castano won the World Aerobatic Championships at Bilbao.

Sparrow missile, see Raytheon
Specific fuel consumption, 1213
Fuel burned by an engine per hour for each horsepower or pound of thrust generated. It is an overall measure of efficiency.
Sperry, 685
Sperry turn and slip indicator, 685, see also Turn and slip indicator
Spey engine, see Rolls-Royce
Spinner, 1072, 1079, 1093
The streamlined fairing

over the hub of a propeller or a ram-air-driven windmill. Spinners began to be common on aircraft during the 1920s when speeds rose to 200 mph and beyond. Even today they are by no means universal. The Hamilton propellers used on thousands of transport aircraft have no spinner in the true sense but merely the smooth fairing provided by the propeller manufacturer over the mechan-

1 Spinner

ism inside the hub. The DC-7B and DC-7C, however, do have spinners. The spinner reduces drag considerably, improves airflow and possibly engine cooling and even aircraft control. In the drawing of the Rolls-Royce Tyne propeller on page 234 the streamlined spinner shape to improve performance of the large turboprop engine is clearly seen.

Spirit of St Louis, see Ryan NYP monoplane
Spitfire, see Supermarine
"Spookies", 887
A name given to Douglas C-47 Dakotas (DC-3s) modified with Miniguns for the ground-attack rôle in the Vietnam War in the mid-1960s.
Spruce, 359, 501, 502
Stable stall, 839
The stable stall is an aerodynamic condition at present peculiar to aircraft with high "T" tails and rear-mounted engines. With a normal

stall, an aircraft reaches a condition in which the angle of attack of the wings is so high that the airflow pattern over them breaks down and the wings are no longer able to provide the requisite lift. To disengage from such a condition it is necessary to lower the angle of attack, and this is done by lowering the elevators and thus lowering the nose.

However, in aircraft with "T" tails a condition can arise whereby the wings reach an angle of attack at which lift breaks away, and the elevators or moving tailplane become inoperative because the wings are blanketing any airflow from reaching the elevators. The condition is further aggravated by the positioning of the engines, in that the thrust from them cannot be used to give a pitching moment to lower the nose and reduce the angle of attack.
Standard Beam Approach, see SBA
Stand-off missiles, 1379
Missiles which can be carried by strike aircraft and which are released many miles from the target, being directed to the target without the need for the parent aircraft to approach the enemy defences.
Starfighter (F-104), see Lockheed
Star monoplane (1911), 629
This was a monoplane built by the Star Motor Co of Wolverhampton. It was based generally on the Antoinette (*q.v.*) design and was exhibited in its original form at the 1910 Olympia Aero Show.
StarLifter (C-141), see Lockheed
Starliner, see Lockheed Super Constellation
Stator blades, 1139
The fixed aerofoil-shaped blades of an axial compressor (*q.v.*).
Steam engines, 1045, 1047
At the time when the first designers of fixed-wing aeroplanes were coming to the stage of building practical flying machines, the internal combustion engine was so much in its infancy as to prove largely unacceptable for aircraft work. Consequently, several designers turned to what was then the standard prime mover, the steam engine. Sir George Cayley (*q.v.*) proposed steam as the propulsive power for his designs, and Sir Hiram Maxim (*q.v.*)

also built powerful steam compound engines for his biplane before suspending his activities in this direction. Langley (*q.v.*) took it up and powered his model aircraft very successfully with steam engines.
In the first decade of the 20th century, great strides in developing internal combustion engines removed the necessity for further development of steam engines as aircraft power plants.
"Stick pusher", 839
Stinson, 565
The Stinson Aircraft Corporation was formed in 1926 by Edward A. Stinson, an experienced pilot who had been flying since 1911. He marketed the Detroiter, the first American aircraft with such up-to-date features as an enclosed cabin, engine starter, brakes, heating and soundproofing. From then on the company built a series of successful high-wing, single-engined monoplanes which sold well and competitively. Two three-engined airliners were built, but these did not achieve the success of the other and later single-engined monoplanes, of which the Reliant became the most successful. In 1940 the company was acquired by Vultee Aircraft Inc, but retained its separate identity, building the Reliant as a trainer and communi-

cations aircraft and the Sentinel as an Army reconnaissance and liaison aircraft during World War Two. The company survived after the war for a short while, building the Voyager light monoplane.
Stirling, see Short
Stockholm Airports (Arlanda and Bromma)

Arlanda Airport

Distance from city centre:
26¾ *miles (Arlanda)*
6 *miles (Bromma)*
Height above sea level:
123 *ft (Arlanda)*
48 *ft (Bromma)*
Main runways (Arlanda):
01/19 = 10,827 *ft*
08/26 = 8,202 *ft*
Main runways (Bromma):
13/31 = 6,618 *ft*
05/23 = 4,002 *ft*
Passenger movements in 1967 (Arlanda and Bromma combined):
2,355,000
STOL, 1331
The generally-accepted abbreviation for aircraft in a class specially designed to take off and land in a very short space, formed from the initials of Short Take-Off and Landing. Such aircraft are usually fitted with wings incorporating high-lift devices, with auxiliary lift-jets, or with tilt-wings, or combinations of these devices.
Stormovik, see Ilyushin Il-2
Stratocruiser, see Boeing
Stratofortress, see Boeing
Stratofreighter, see Boeing
Stratojet, see Boeing

Stratoliner, see Boeing

Stressed-skin structure, 385, 387, 397, 409, 412, 430, 460, 709, 787, 975

Stringfellow, John, 157

John Stringfellow collaborated in the design of Henson's Aerial Steam Carriage (q.v.), chiefly in the steam engine for it. When Henson gave up after the failure of model experiments, Stringfellow continued on his own and designed an aircraft in 1848 on the basis of Henson's earlier design (which itself owed much to the ideas of Cayley). He built a model of his design, with a 10-ft wing span, and in June 1848 at Chard, Somerset, this model was tested without any great success. He was unable to achieve any sustained flights with his model and little interest was aroused by his experiments, so he abandoned his work on it. He reappeared on the aviation scene in 1868 at the Aeronautical Exhibition at the Crystal Palace, exhibiting a steam-powered triplane and a steam engine for model use.

Stroke

In a piston-engine, the linear distance moved by any one piston between top dead centre and bottom dead centre. It is equal to twice the radius of the crank pin. In most modern aircraft piston-engines the stroke is of the order of 4 to 6 in (10 to 15 cm).

Stromple, George, 743

Sudan Airways

State-owned airline of the Sudanese Republic, which began commercial operations in 1947 with assistance from Airwork of Brittain. It now operates domestic services throughout the Sudan and internationally to the Middle East, Ethiopia, Kenya, Athens, Rome and London. Headquarters: Khartoum. Member of IATA.

Sud-Aviation

When the French aviation industry was re-organized and nationalized in 1936 the former Lioré-et-Olivier, Potez-CAMS, Romano and SPCA companies were united in a new regional group known as the Société Nationale de Constructions Aéronautiques (SNCA) du Sud-Est. Similarly, the Bloch and Blériot companies, and additional Lioré-et-Olivier factories were amalgamated as the SNCA du Sud-Ouest.

These two groups resumed design and manufacture, after World War Two, of aircraft, with, respectively, S.E.- and S.O.-prefix designations. In March 1957, however, they themselves were amalgamated to form Sud-Aviation, which thus became one of the largest aircraft manufacturing organizations in Western Europe. Since its formation, Sud-Aviation has absorbed its subsidiary companies SFERMA and GEMS (Morane-Saulnier), creating a new subsidiary (Socata) to handle matters concerning light and business aircraft.

Sud-Aviation's present activities include continued production of the Caravelle series of jet airliners, the French share of responsibility for the Concorde supersonic transport programme, and the production of a wide range of helicopters from the small Alouette and SA.341 to the large Super Frelon anti-submarine and transport helicopter.

Sud-Aviation Alouette, 1313

The Alouette, with an Artouste engine, first flew in March 1955 and was soon found to be significantly superior to similar helicopters powered with piston-engines. On the 360 shp of the Artouste, the Alouette had a maximum speed of 115 mph, service ceiling of 7,000 ft and endurance of 4.1 hr. It had accommodation for five persons, the maximum payload being 1,200 lb.

Approximately one thousand of this version, the Alouette II, have been built since it first flew. A later version, fitted with the Astazou engine, has now replaced it in production. The Alouette III is a substantial redesign with greater power (550-hp Artouste), larger capacity (seven persons) and higher performance (max speed 131 mph).

Sud-Aviation Caravelle, 312, 317, 579, 787, 837, 852

Sud-Aviation's SO.1221 Djinn (top), SE.2130 Alouette II, SE.3160 Alouette III and SE.3200 Frelon

to the nationalized Sud-Est company to build prototypes of a new short-range twin-jet airliner; thus was the Caravelle born. Two years later the first prototype flew and the type entered service with Air France in 1959. The new concept of rear-engined layout proved an immediate success and Sud-Est soon had a full order-book. Many European airlines acquired Caravelles and a market was also found further afield, notably in America with United Air Lines. By Autumn 1969, well over 250 Caravelles had been ordered, including the turbofan Super Caravelle.

The Caravelle set the fashion for the succession of rear-engined aircraft which have followed, such as the VC10, BAC One-Eleven, Boeing 727, Trident, DC-9 and Il-62.

Suidwes Lugdiens

South African airline formed in 1946 as South West Air Transport (Pty) Ltd and changing to its present title in 1959. Since 1950 its main activity has been the provision of local "feeder" services in South-West Africa to connect at Windhoek with the services of South African Airways (q.v.). Headquarters: Windhoek. Associate member of IATA.

Supercharger, 1075, 1084-1086, 1087, 1090-1092, 1100, 1102, 1105, 1107, 1109, 1111, 1115, 1117, 1125, 1127

Super Constellation, see Lockheed

Superfortress, see Boeing B-29, also Boeing B-50

Supermarine, see Vickers

Supermarine S.5, 387, see also Mitchell, R. J.

The second of R. J. Mitchell's racing sea-planes for the Schneider Trophy race. It enabled the British team to win in 1927, paving the way for the two following wins which gained the Trophy outright for Great Britain in 1931.

Supermarine S.6, 387, see also Mitchell, R. J.

Supermarine S.6B, 281, 287, 387, see also Mitchell, R. J.

This was the final development of the racing seaplanes built during the nineteen-twenties for competition in the Schneider Trophy contests. Designed by R. J. Mitchell, the S.6B was an S.6 with larger floats, of unequal length to counteract torque, and a Rolls-Royce "R" engine with bhp raised to 2,300 by an increase in revolutions. Two were built, of which S1595 won the Schneider Trophy outright in 1931 and raised the world speed record to 407.5 mph in September 1931. The other, S1596, had previously set a record of 379.05 mph.

Supermarine Spitfire, 282, 285, 286, 287, 411, 412, 552, 567, 904, 911, 1070-2, 1095, 1099, 1107, 1339

Probably the most famous fighter of all time, the Spitfire was developed by its designer, R. J. Mitchell, from the Supermarine racing seaplanes built for the Schneider Trophy contests, via the F.7/30 Goshawk-engined monoplane. The prototype, using the new Merlin engine, first flew on 5 March 1936, and its performance was such that the first production order was given only three months later. Production began in 1937, and first deliveries were to No 19 Squadron at Duxford in June 1938. By the time World War Two broke out, nine full squadrons of Spitfire Is were serving in Fighter Command.

Their first victory was on 16 October 1939, when Spitfires destroyed two Heinkel He 111s. The Spitfire proved itself invaluable in the Battle of Britain in 1940 for, being faster than the Hurricane, was more able to out-manoeuvre the German fighters. The Mk II went into service late in 1940, followed by the cannon-armed Mks II and V in 1941, taking part in offensive sweeps and as bomber escorts over France. The design was also developed for photo reconnaissance at this time.

The Spitfire served overseas from March 1942, at first in Malta and then in the Western Desert. In 1943 it was flying in Australia and Burma, and had been converted for aircraft-carrier use with the Navy as the Seafire. The Mk IX version, with the more powerful Merlin 61, was in service from 1942 onwards; and this version, like the Mk V, was adapted as a fighter-bomber. In 1943 the Spitfire was re-engined with the Griffon and variants with this engine entered service that year. Griffon-engined versions remained in RAF service in developed forms, and as Fleet Air Arm Seafire variants, until 1950. The Seafire was the final version in action, No 800 Squadron flying it in Korea from HMS *Triumph*. In all, 22,759 Spitfires and Seafires were built.

Super Sabre (F-100), see North American

Super Skymaster, see Cessna

Super Sportster, see Gee Bee

Super VC10, see BAC

Survey, see Aerial survey

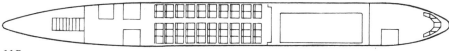

Caravelle 11R

In 1953 the French Secretariat of State for Air gave the go-ahead

Swashplate, 982

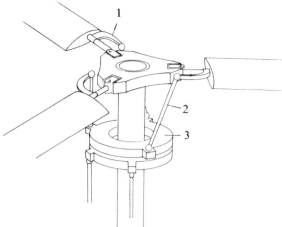

1 Blade control arm
2 Connecting rod

3 Swashplate

Device embodied in helicopter rotor head mechanisms through which cockpit control lever movements are transmitted to the rotor blades. The swashplate is in two halves, the upper part revolving with the head, while the lower part remains stationary relative to the fuselage. The swashplate is mounted on universal bearings so that it can be tilted in any direction. If the swashplate is level and then moved, the pitch of all blades alters equally. This is known as collective pitch change. If the swashplate is tilted and then moved, the pitch of the blades will alter continuously as they revolve. This is known as cyclic pitch change.

Sweden: Air Forces
Flygvapnet (Royal Swedish Air Force). The Swedish Navy and Army received their first aircraft, both donated, in 1911 and 1912 respectively, and an Army Aviation Corps was established late in 1912. By the outbreak of the 1914-18 War each service owned four aircraft, and during the ensuing four years domestic factories—notably the Army Aircraft Factory and that of Dr Enoch Thulin—produced several foreign types for the Swedish services. Among the types purchased and/or built once the war was over were the Austrian Phönix and

German Albatros single- and two-seaters, the Avro 504K and the nationally-designed O-1 Tummelisa trainer. On 1 July 1926 the *Flygvapnet* was officially created as an independent organization, a step followed by limited expansion and modernization during the late 'twenties and early 'thirties with aircraft of British, Dutch and German origin and the nationally-designed Svenska Jaktfalk. A major increase in strength did not occur until 1936-37, when the RSwAF began to receive Gladiator fighters, DB-8A and Ju 86K bombers, and Fw 44 and NA-16 basic trainers in substantial numbers. Even these, by the end of the decade, were not considered modern enough, and attempts to order more up-to-date foreign warplanes were frustrated by the gradually worsening situation in Europe culminating in the outbreak of World War Two. When orders placed with the US in 1940 for 300 front-line aircraft were cut back to 60, Sweden managed to secure replacements from Italy, but it was now obvious that she would have to rely largely on her own resources for the duration of the war. This provoked the design of the State-sponsored J 22 fighter and other new

major types from the Saab company.
In April 1945 the RSwAF acquired Mustang fighters from the US, and at the end of that year the Saab-21 fighter entered service. Vampire jet fighters arrived in 1946, followed by the jet-powered version of the Saab-21 in 1949 and the Saab-29 two years later. To

Saab J29 "flying barrel" fighters emerging from one of the RSwAF underground hangars

bridge the operational gap between the latter and the supersonic Saab-35 Draken, British Hunters and Venoms were acquired, while the multi-purpose Saab-32 Lansen entered service in 1956. The Draken (in service since 1960) and Lansen are still the major types, performing interceptor, reconnaissance and tactical strike rôles between them. The Lansens are to be replaced from mid-1970 by the multi-mission Saab-37 Viggen.
The present-day Royal Swedish Army and Navy maintain modest-sized air arms. That of the Navy (*Marinen*) is primarily an anti-submarine force equipped with Boeing-Vertol Sea Knight helicopters. That of the Army (*Armén*) is also equipped mainly with helicopters (Alouettes) which serve in the liaison/communications rôle.

Sweden: Chronology
Sweden's first historical record of a form of aircraft came in 1716, when the scientist Emanuel Swedenborg (1688-1772) published details of a flying machine which he had designed. It was not until nearly 200 years later, however, that Sweden had its first machine capable of flight, the Ask-Nyrop

No. 1 aeroplane. This was built by Hjalmar Nyrop and Oscar Ask early in 1910, and it flew for the first time during the summer of that year.
Nyrop's third aircraft was donated to the Swedish government in 1911, intended for use as a maritime reconnaissance aircraft, and it was in this aeroplane that the first military flight in Swedish history was made, during February 1912.
In 1914, Enoch Thulin (1881-1919) founded the AB Enoch Thulins Aeroplanfabrik (AETA) and this company was the first in Sweden to initiate series production of several types of aircraft for both civil and military use.

The original Svenska Aeroplan AB was founded at Trollhättan in 1936, amalgamating subsequently with Svenska Järnvägsverkstäderna in 1939, when its head office and engineering departments were established at Linköping. Since that time Saab have built some 3,000 aircraft, mainly for the Swedish Air Force, and the company is currently building the Saab-35 Draken and Saab-37 Viggen in large numbers.

Swept volume
In a piston-engine, the total volume "swept" by all the pistons during a 180° turn of the crankshaft. This is the same as the cross-section area of each cylinder (assumed to be the same as the effective area of the top of the piston), multiplied by the stroke, multiplied by the number of cylinders. In the Pratt & Whitney R-2800 Double Wasp, used in the DC-6 and Convair 340 and 440 airliners, there are 18 cylinders each of

Junkers F.13 floatplane on the first service between Stockholm and Helsinki, 1924

It was not until March 1924, however, that a civil airline was formed. This was AB Aerotransport (ABA) which operated a number of domestic services until, in 1946, it became integrated with Scandinavian Airlines System (SAS).

$5\frac{3}{4}$-in bore and 6-in piston stroke: the swept volume is thus 2,804 cu in (45.9 litres), hence the designation "R-2800". Swept volume is also known as "displacement". There is no corresponding parameter in a gas-turbine engine, the nearest equivalent being

First over the pole: SAS with the DC-7C in 1957

the air mass flow.

Swing-wing, see Variable Geometry, also General Dynamics F-111

Swissair
Operating title of *Schweizerische Luftverkehr AG*, one of the world's oldest airlines. It was formed in March 1931 from the former Ad Astra Aero AG (which ran a domestic inter-city seaplane link in 1919) and Basle Air Traffic Co (known as Balair). In the early 'thirties Swissair operated fast single-engined Lockheed Orion monoplanes, and later was among the first airlines outside the US to order the Douglas DC-2. Swissair is predominantly privately-owned, and since World War Two has built up a large international route network that now extends throughout Western Europe and Africa, to North and South America and the Middle and Far East. Headquarters: Kloten. Member of IATA.

Switzerland: Air Force. *Kommando Fliegertruppe* (Swiss Air Force). Swiss military aviation originated with the formation of one monoplane and one biplane squadron on 1 August 1914. Although Switzerland remained neutral during the 1914-18 War, several of her pilots served with the French *Aviation Militaire*, gaining experience later of use to their own air force. The *Fliegertruppe* underwent several administrative reorganizations during the nineteen-twenties, when it was manned largely by reservist personnel. Its aircraft during and after the war included several nationally designed types (by Haefeli) and others from Britain, France and Holland. Gradual expansion, and further reorganization, took place in the nineteen-thirties, with a significant change in status on 19 October 1936 when, as the *Schweizerische Flugwaffe*, it became an independent service in its own right. In the late 'thirties, in addition to imported and locally-built examples of the Potez 63, Messerschmitt Bf 109 and Morane-Saulnier M.S.406C, the Swiss Federal Aircraft Factory began to evolve the D-3801 fighter and the general-purpose C-3600 series monoplanes. During World War Two, although staunchly retaining its neutrality, Switzerland was obliged to accept some Bf 109s and other German aircraft. When the war was over, the Swiss Air Force had to choose carefully a jet fighter capable of operating in mountainous areas and from its comparatively restricted airfields. The choice fell on the de Havilland Vampire, which it began to receive in 1949. Vampires (and, later, Venoms) were subsequently licence-built in the country. The nationally-designed P-1604 jet strike fighter, first flown in 1955, has not been adopted for service, although development of it still continues. Meanwhile, the *Fliegertruppe* decided to replace its Vampires with Hunters, and these have recently been joined in service by Swiss-built French Mirage III fighters.

Britain supplied a very large number of Hunter 58s to the Swiss Air Force

Switzerland: Chronology
1807-8. Jakob Degen, a Swiss watchmaker who had emigrated to Austria, was claimed to have carried out experiments with an ornithopter during 1807-8. It would appear from contemporary reports, however, that a balloon

was used to raise the machine.
1880. Eduard Spelterini made the first of some 500 balloon ascents. Ten of these flights were made over the Alps. One flight involved carrying 180 hydrogen cylinders and a balloon to the top of a 5,000-ft-high peak.
1908. Emil Messner and Theodor Schaeck won the Gordon Bennett competition for free balloons, with a flight of 753 miles in 72½ hr.

Edmond Audemars, a 1910 Demoiselle pilot

22-26 October 1910. Switzerland's first air meeting was held at Dubendorf, and was attended by some 100,000 spectators.
September 1911. A pilot named Failloubaz was the first Swiss pilot to take part in Swiss Army manoeuvres, but his single reconnaissance flight ended in a crash landing.
18-19 August 1912. The Swiss Society of Officers

decided to organize a national collection to finance military aviation. A total of 1,734,564 francs was subscribed.
1913. Piloting a Blériot monoplane, Oscar Bider made the first flight over the Pyrenees, from Pau to Madrid.

Oscar Bider

1913. Agénor Parmèlin was the first pilot to fly over Mont Blanc, highest of the Alpine peaks, at a height of 18,373 ft.
1 August 1914. Military aviation established in Switzerland with the formation of No 1 (biplane) and No 2 (monoplane) squadrons. The force was under the command of Captain Real, who had learned to fly at a private flying school at Darmstadt in 1911.
December 1914. The centre of Swiss military aviation was established at Dubendorf.
1915. Marc Birkigt developed the famous Hispano-Suiza aeroengines which powered many Allied aircraft in the 1914-18 War.

Durafour's Caudron G-3 on Mont Blanc, 1921.

1926. At the end of the year Walter Mittelholzer, flying a Dornier Merkur floatplane, made the first flight from Zürich to Capetown.
1931. Paul Kipfer accompanied Dr Auguste Piccard on the first scientific balloon ascent into the stratosphere.

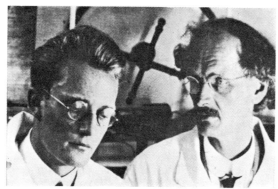

Dr Auguste Piccard (right) and Ing Paul Kipfer before their ascent on 27 May 1931

Accommodated in a pressure capsule suspended below a large balloon, they attained a height of 51,774 ft.
26 March 1931. The national airline, Swissair (*q.v.*), was established by the integration of two earlier airlines—Ad Astra Aero and Balair—with initial equipment comprising six Fokker F. VIIB/3m, two Dornier Merkur, one Fokker F. VIIA, one Messerschmitt 18D and a Comte AC/4.
1932. Jakob Ackeret, Professor of Aerodynamics at the Federal Institute of Technology, Zürich, designed the world's first supersonic wind-tunnel. It was constructed by the Swiss firm Brown Boveri.
1952-66. Hermann Geiger, in conjunction with Fredy Wissel and two Swiss military pilots, developed new techniques to permit safe landing and take-off from mountain glaciers.

Hermann Geiger, glacier pilot extraordinary, with Super Cub 125

His log book recorded some 23,000 glacier landings, in which 574 injured or stranded persons were carried to safety.
Sydney Airport (Kingsford Smith), NSW, Australia

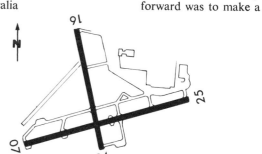

Distance from city centre: 6½ miles
Height above sea level: 10 ft
Main runways:
07/25 = 7,900 ft
16/34 = 5,500 ft
Passenger movements in 1967: 3,080,300
Synchronising gear, 875, see also Constantinesco gear
In operations with the early military types of aircraft it soon became obvious that any improvement in performance would lie only with the tractor types; the pusher aircraft at the time was capable of little further development. However, the disadvantage of the tractor aircraft was the difficulty of firing fixed guns forward.
The first attempt to alleviate the problem

was to fit metal deflector plates to the propeller and fire the gun through it so that any bullets hitting the blades would be deflected. This was obviously very inefficient, and the next step forward was to make a mechanical interrupter device which would prevent the gun from firing while the blades were in the line of fire. This was a great improvement, but still cut down the rate of fire of the gun. The final development was to link the drive of the engine to the trigger mechanism of the gun. The guns were then always synchronized so that they fired when the propeller blades were clear of the line of fire. (For development see text.)

Syrian Arab Airlines
Successor to the former Syrian Airways (founded 1946) which became a part of United Arab Airlines (*q.v.*) in 1960. Syrian Arab Airlines was created in 1961 as a State-owned carrier and now operates domestic services and international flights to other Middle Eastern countries, India and Pakistan, and selected points in Eastern and Western Europe.
Headquarters: Damascus.
Member of IATA.

T-6 Texan, see North American
T34 Turbo Wasp engine, see Pratt & Whitney
T-39 Sabreliner, see North American
T57 engine, see Pratt & Whitney
T64 engine, see General Electric
Tabloid, see Sopwith
TABSO, see Bulgarian Air Transport

Take-off and landing techniques

It is often necessary for different types of aircraft to employ special techniques for taking off or landing in restricted spaces. The most obvious example is the aircraft carrier (*q.v.*), in which the steam catapult and angled deck permit high-speed aircraft to be launched or landed on simultaneously. From above deck the steam catapult resembles a tram line, aircraft engages to apply a braking effect and so shorten the landing roll. On the largest US supercarriers, the deck area is itself sufficient for many slower-flying piston-engined aircraft to land normally without

Steam catapult

1 Ready to launch: valve open, steam builds up behind launching piston

2 Piston driven forward by steam pressure until arrested hydraulically

3 Valve closed: steam exhausts at rear, permitting return of piston

4 Valve closed, piston in position for next aircraft

5 Steam able to enter cylinder past valve

6 Projection linking aircraft to launching piston

7 Piston with braking cone

8 Hydraulic braking for piston

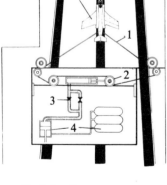

Deck landing:
1 Arrester wires
2 Pulley system
3 Controls with compressed-air cylinder
4 Compressed-air cylinder and storage bottles
5 Buccaneer aircraft

along which the aircraft is aligned automatically by a deck roller. A strop links the underside of the aircraft to the catapult mechanism, which launches the aircraft into the air over the bows after the engines have been run up to full thrust. As the strop then falls away, the pilot immediately retracts the flaps and landing gear and climbs steeply away on full power.

The most common method of deck landing entails the deployment of a series of steel cables, stretched across the landing area, in which an arrester hook lowered by the landing

recourse to this technique.

Arrester hooks and wires are now being used increasingly at land bases where high-speed military aircraft operate, to eliminate the possibility of an aircraft over-running the end of the runway. Another device fulfilling the same purpose is the nylon crash barrier, which is positioned near the end of the runway and collapsed on it, but which can be elevated quickly (rather like the starting tapes in horse racing) to "catch" an over-running aircraft and bring it to a halt.

Radio- or radar-using avionics landing systems such as Automatic Landing, GCA, ILS, and SBA appear under their respective headings in this Index.

Military aircraft, especially large aircraft or others required to operate from restricted airstrips, frequently make use of the braking parachute to reduce their landing run. This is usually stowed inside a fairing, in or near the tail of the aeroplane, and is streamed immediately after touchdown. Some very large aircraft have a cluster of small braking parachutes instead

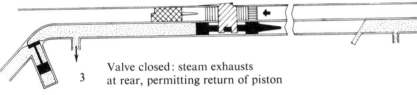

Tail-braking parachute:

1 Victor bomber
2 Air-brakes open
3 Main braking 'chute
4 Drogue 'chute

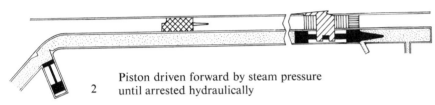

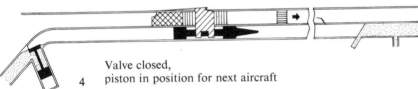

Nylon crash barrier

of the single large parachute.

Tandem Twin, see Short

TAP (Transportes Aereos Portugeses)

Portuguese airline, formed in 1944, which started its first passenger services between Lisbon and Madrid in September 1946. Originally a State-owned branch of the Civil Aviation Department, it is now mainly privately-owned and operates a comprehensive network of internal routes as well as services to Western European capitals, Johannesburg, New York, Rio de Janeiro and Buenos Aires.

Headquarters: Lisbon. Member of IATA.

Tarom

Operating title of Transporturile Aeriene Romine, the State-owned airline of Rumania, which was created in 1954 to succeed the former Soviet-assisted TARS formed in 1946. Tarom's international services are confined to Eastern and Western Europe (including Moscow), and a domestic network within Rumania is also served.

Headquarters: Bucharest.

Taurus engine, see Bristol

Taylor, Charles, 1034, 1055
Mechanic to the Wright brothers (*q.v.*) in the construction of their first aeroplanes.

TDC (Top Dead Centre)
In a piston-engine, the highest point reached by the piston, with the crankpin at the "12 o'clock" position; alternatively the term may be applied to this position of the crankpin. At TDC the velocity of the piston is instantaneously zero but its acceleration is at maximum.

ILS

1 Localizer clearance antenna
2 Localizer course antenna
3 Runway
4 "Fly down and left"
5 Localizer path width
6 "Course correct"
7 Outer marker
8 Glide path width
9 "Fly up and right"
10 Middle marker
11 Glide path

Tel Aviv Airport (Lod), Israel

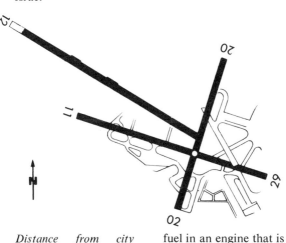

Distance from city centre: 12 miles
Height above sea level: 132 ft
Main runways:
12/30 = 8,500 ft
11/29 = 7,743 ft
02/20 = 5,807 ft
International passenger movements in 1966: 580,000

Tempest, see Hawker

Terrain-following radar, 1381, 1401, 1405

Texan (T-6), see North American

TF30 engine, see Pratt & Whitney

TF39 engine, see General Electric

TF306 engine, see Pratt & Whitney

Thai Airways International
The State-owned Thai Airways Co, formed in 1951, is responsible for all domestic air services within Thailand. This company also has the controlling interest in Thai Airways International, which was set up in 1959 in partnership with SAS (*q.v.*) and is responsible for all major international services of the country. Scheduled operations started in May 1960 and at present extend to Hong Kong, China, Japan, the Philippines, Singapore, Indonesia, Burma, India and South Vietnam. SAS still provides financial and other assistance.
Headquarters: Bangkok.

Thermal efficiency
The percentage of the total heat energy released by burning the fuel in an engine that is actually put to useful work. In a piston-engine of the conventional spark-ignition type, about 40 per cent of the heat released escapes with the exhaust and 30 per cent is dissipated in heating the coolant—air or liquid—necessary to prevent the cylinders from melting. This leaves a bare 30 per cent in the mechanical work done by the crankshaft. There are various ways in which thermal efficiency may be calculated for any given shaft-drive engine. In the case of a jet engine (turbojet or turbofan) in which there is no mechanical output, it is impossible to obtain a direct measurement, but the efficiency for any given speed may be calculated from a knowledge of the thrust generated. Empirical rules for calculating thermal efficiency are: for a shaft-drive engine it is numerically equal to 2,545 × brake horsepower divided by the number of British Thermal Units released each hour by burning the fuel. For a jet engine, one formula is: propulsive thermal efficiency equals 2,545 × aircraft speed in mph divided by 375 × engine specific fuel consumption × heat released (in BTU/lb) by burning the fuel. If the turbofan engines of the Boeing 747 have a specific fuel consumption of 0.48 when cruising at 600 mph while burning fuel giving 20,000 BTU/lb, their thermal efficiency will be about 42 per cent.

Thompson Trophy, see Competitive flying

Three-shaft engine, 1213

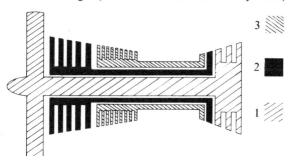

1 Low pressure
2 Intermediate pressure
3 High pressure

A gas-turbine engine in which there are three main rotating assemblies each capable of running at its own optimum speed. An example is the Rolls-Royce Trent turbofan, in which the low-pressure shaft drives a single-stage fan, the intermediate-pressure shaft a four-stage intermediate compressor, and the high-pressure shaft a five-stage high-pressure compressor. No aircraft engine has yet been built with more than three shafts, although this may become necessary as pressure ratios continue to rise. The shafts, of course, link the turbines to whatever they drive—compressor spools, fan or propeller gearbox—and are arranged concentrically one inside the other.

Three-spool engine, 1213
A gas-turbine engine having three separate axial compressors in series: low-pressure, intermediate and high. No three-spool engine has yet been built, although the Trent (described above under "Three-shaft engine") is essentially of this type, the first "spool" being a single-stage fan.

Thrust Measuring Rig, see Rolls-Royce

Thrust reversal, see Reverse thrust

Thunderbug, see British Deperdussin

THY (Turk Hava Yollari)
Formed in 1933 as *Develt Hava Yollari*, the Turkish airline was controlled successively by the Turkish Ministries of National Defence and Communications until early 1956, when it became a Corporation. It is still State-controlled, although a small financial holding is maintained by BOAC (*q.v.*). Its route network is predominantly internal, but international flights are made to Athens, Beirut, Nicosia and Rome.
Headquarters: Istanbul.
Member of IATA.

Tilt-rotor, 985

"Tin Goose", see Ford

Titanium, 399, 413, 485

Tokyo International Airport (Haneda), Japan
Built as a national airport in 1931, Haneda now occupies virtually the whole of a small island in Tokyo Bay. Most of the expansion has taken place since World War Two, especially since the opening of trans-Polar services from Europe to the Far East in 1961.
Distance from city centre: 11 miles
Height above sea level: 8 ft

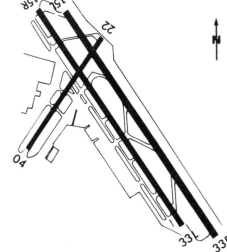

Main runways:
15L/33R = 10,335 ft
15R/33L = 9,840 ft
04/22 = 5,150 ft
Passenger movements in 1967: 5,169,300

Top Dead Centre, see TDC

Tornado (B-45), see North American

Torque-reaction, 813, 817, 957
The moment of the aerodynamic forces about the thrust line of an airscrew, which tends to turn the aeroplane in the opposite direction to that in which the airscrew is rotating. The effect is particularly pronounced when a propeller-driven aircraft is taking off, for the aircraft tends to swing off the runway or take-off path. To counter this tendency, designers may provide contra-rotating airscrews, with two sets of blades rotating in opposite directions; or, in twin-engined aircraft, "handed" airscrews rotating in opposite directions. Another remedy is to increase the fin area of the aircraft concerned.

TPE 331 engine, see AiResearch

Tractor propeller, see Pusher propeller

Tradewind (R3Y), see Convair

Transall C-160, 448-456, 781-784

Trans-Australia Airlines
Government-controlled airline administered through the Australian National Airlines Commission. TAA flew its first regular services in September 1946 and currently has a 47,000-mile route network linking about 150 towns and cities. Most of these are within the Australian sub-continent (including Tasmania) but routes are also flown to points in New Guinea and Papua. The airline also maintains and operates aircraft for the Royal Flying Doctor Service of Queensland and for the Northern Territory Aerial Medical Service.
Headquarters: Melbourne.
Associate member of IATA.

Trans-Mediterranean Airways
A Lebanese operator originally formed in 1953 to carry freight supplies to oil-drilling companies operating in the Persian Gulf, TMA inaugurated scheduled freight services in 1959. These still form the bulk of its activities, although the network has widened and now includes services as far afield as Tokyo and London.
Headquarters: Beirut.
Member of IATA.

Trans World Airlines, see TWA

Trent engine, see Rolls-Royce

Tricycle landing gear, 565, 571

Trident, see Hawker Siddeley

Trimmer, 627
A trimmer is a control which alters the setting of a trimming tab on a control surface. The tabs are small, movable auxiliary sections on the trailing-edges of control surfaces to improve the aerodynamic balance of the surface concerned. Thus an aircraft can have elevator trimmers, rudder trimmers and aileron trimmers.

Trim tab, 627
This is a small hinged flap, usually set in the trailing-edge of a control surface. Its purpose is so to offset the control surface as to trim out any out-of-balance force on the control column or rudder bar and make the flying of the aircraft easier for the pilot.

Triple Tractor, see Short

Tunis Air
Operating name of the *Société Tunisienne de l'Air*, a Tunisian airline owned jointly by the Government, Air France (*q.v.*) and private interests. It was formed in 1948 for the operation of internal and international services; the latter currently include routes to neighbouring North African states and to points in Western Europe. Technical assistance is given by Air France.
Headquarters: Tunis. Member of IATA.

Tupolev, Andrei Nikolaevich
Andrei Tupolev is the doyen of Soviet aircraft designers, having played a prominent part in aviation in the USSR since the late nineteen-twenties, when he became a designer at the TsAGI (Central Aero-Hydrodynamic Institute) in Moscow. He was responsible for the large multi-engined ANT monoplanes whose endurance flights to the Polar regions and across the Atlantic were a feature of the 'thirties (see USSR: Chronology), and several Soviet designers later to head design bureaux of their own began their careers under his guidance—men such as Petlyakov, who designed the Pe-8 or TB-7, the only four-engined Soviet bomber to see service during World War Two.
Since the war Tupolev has been associated with some of the USSR's leading bomber and commercial transport aircraft. The Tu-104, second jet airliner in service in the world after the early British Comets, was a civil adaptation of his Tu-16 medium bomber; likewise the Tu-114 (*q.v.*) four-turboprop transport is the commercial counterpart of the Tu-20 swept-wing bomber. Tupolev has since developed a whole series of civil jet transports including the Tu-144 supersonic airliner.

First supersonic transport aircraft to fly: Tupolev's Tu-144, on 31 December 1968. This design closely resembles the Concorde in many respects

On the military side, the most recent designs to enter service are the Tu-22 ("Blinder") rear-engined jet bomber and the "Fiddler" interceptor.

Tupolev Tu-16 ("Badger"), 312 ,835

Tupolev Tu-20 ("Bear"), 1309, 1311
This large four-turboprop bomber first appeared in 1955 and soon formed an important component of the Soviet bombing fleet. Its performance corresponds roughly with that of the Tu-114 airliner, which derives from it and which has a range of 6,200 miles and a maximum speed of 590 mph with a 55,000-lb payload. The Tu-20 has for some time been equipped to carry stand-off bombs.

Tupolev Tu-22 ("Blinder"), 837
The Tu-22 is a very advanced supersonic bomber in service with the Soviet Air Force, with an estimated speed of Mach 1.5 and capable of carrying stand-off missiles. It has a long slender fuselage, with a nose-probe for refuelling, a narrow cabin, and a low swept wing with trailing-edge undercarriage fairings. The most distinctive feature is the mounting of the two turbofan engines relatively high, side-by-

The USSR's first jet airliner to serve with the state airline, Aeroflot, was this adaptation of the Tu-16 bomber with a modified fuselage. It appeared in service in 1956, having first flown in 1955. It has since been built in large numbers for Aeroflot and other Soviet *bloc* airlines, and is still a standard Aeroflot type.

Tupolev Tu-114 ("Cleat"), 750-753, 787, 1309, 1311
A civil development of the Tu-20 bomber, the Tu-114 entered service with Aeroflot in April 1961. It is a long-range passenger transport powered by four Kuznetsov NK-12MV turboprops of 14,795 ehp. It has a wing span of 167 ft, a length of 177 ft, and a standard layout for 170 passengers. Its range is 5,560 miles and cruising speed 478 mph. Only a small number of these aircraft are in service.

Tupolev Tu-124 ("Cookpot"), 835
The Tu-124 is, in effect, a scaled-down Tu-104 for medium-range airline duties. It also varies from the Tu-104 in being

powered by turbofan engines, Soloviev D-20Ps rated at 11,905 lb st. It has accommodation for 44 passengers and a range of 1,305 miles and is in commercial service with Aeroflot and CSA of Czechoslovakia. It has also been supplied to the air forces of other Soviet *bloc* countries.

Tupolev Tu-134 ("Crusty"), 579, 582, 837
The Tupolev Tu-134 appeared in 1964 as a rear-engined development of the Tu-124, comprising basically the same fuselage with a new "T" tail and its two Soloviev D-30 turbofan engines of 14,990 lb st mounted one on each side of the rear fuselage. It entered service in the USSR in 1967, international service in September of that year, and has since been exported to Eastern *bloc* countries. Its normal range is 1,500 miles, cruising at 540 mph at 36,000 ft.

Tupolev Tu-144, 316, 843
First knowledge of this Russian supersonic airliner came in 1965, three years after the com-

mencement of the Concorde. In appearance it is basically similar to the Concorde and made its first flight on 31 December 1968. Its entry into service will thus bring it into direct competition with the Concorde on the world's air routes. Anticipated performance is similar, except for a slightly higher cruise speed of Mach 2.35 and slightly reduced accommodation.

Tupolev Tu-154, 837

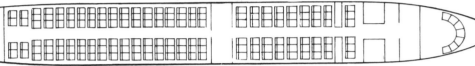

Tu-154

The Tu-154 is Tupolev's latest medium-range airliner, and is a three-engined design having one engine in the fin base and one on each side of the rear fuselage. It has three basic versions with accommodation for 128, 158 or 164 passengers. It is expected in service in 1970.

Turbine, 1161ff
A device for converting the energy of a fluid flow into mechanical power available from a rotating shaft. Various types of turbine have been built, but almost all gas-turbines used to propel aircraft incorporate an axial-flow turbine consisting of one or more discs around the edge of each of which are arranged a large number of curved blades, projecting radially, between which the flow of hot

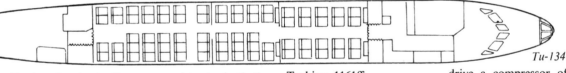

Tu-134

gas from the combustion chamber passes. In doing so the pressure and temperature of the

gas fall violently and the energy thus lost appears as a tangential force on the blades which results in a large torque being imparted to the disc. An axial-flow turbine can run stalled, prevented from rotating, without being harmed; but the flow velocity through it is always so great that the blades can be damaged by droplets of unburnt fuel or particles of carbon. Both discs and blades are made of special high-strength, heat-resistant metals and in the hottest engines the blades are cooled by an internal airflow. The number of stages of discs and blades depends upon the work which the turbine must do. In a pure turbojet a single-stage turbine often suffices to drive the compressor, but to drive a compressor of high pressure-ratio a two-stage turbine will be more efficient and possibly lighter. Three-stage turbines are often used in turbofan and turboprop engines in which much more energy is extracted from the gas flow.

Turbo-Compound engine, see Wright

Turbofan, 1009, 1026, 1137, 1181, 1201ff, see also By-pass turbojet
A gas-turbine used for propulsive purposes, in

side, at the root of the fin.

Tupolev Tu-104 ("Camel"), 309, 312, 835

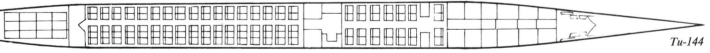

Tu-104 B

which the air delivered by the compressor is divided into two streams, one of which passes to the combustion chamber and turbine while the other is discharged directly to atmosphere through a propulsive nozzle or nozzles. Compared with a turbojet of equal thrust, a turbofan handles a much greater airflow but discharges it at a lower mean velocity. This increases propulsive efficiency at subsonic speeds, but the turbojet is a more slender engine which becomes superior at supersonic speeds.

Turbojet, 1129ff
The simplest type of gas-turbine. Air is drawn into it from the atmosphere, compressed and fed to a combustion chamber in which fuel is burned. From the combustion chamber a steady stream of very hot gas passes to the turbine, which extracts just enough energy to drive the compressor and such accessories as the fuel pump and electric generator. The gas, still with high temperature and high pressure, then escapes to the atmosphere through a nozzle in which as much as possible of the flow's energy is converted to kinetic energy in a high-velocity jet. Every other gas-turbine is based on a compressor/combustion chamber/turbine assembly which, if removed and coupled to a suitable nozzle, could form a turbojet.

Turboméca, 1243, 1311-14
This French company was originally formed in 1938 to develop blowers, compressors and turbines for aeronautical use. The war interrupted its efforts, but in 1947 it began developing the first of a long line of low-powered turbines for aircraft auxiliary power and for powering light aircraft. In the ensuing twenty years it has produced some 50 different types of engine of which about fifteen have gone into production. It has lately worked in collaboration with Rolls-Royce (q.v.) to produce a more powerful engine for the Anglo-French Jaguar. This engine is the RB.172-T.260 Adour supersonic turbofan, with a dry take-

off thrust of 4,000-5,000 lb.
Turboméca Artouste engine, 1148, 1313
Turboméca Aspin engine, 1213, 1311
The Aspin was the first application of the ducted-fan principle to light jet engines. Produced in 1951, it had a single-stage compressor fan and developed 440 lb static thrust for a specific fuel consumption of 0.63 (compared with 1.1 for a non-fan engine).
Turboméca Astazou engine, 1313
Turboméca Aubisque engine, 1253
Turboprop, 1129, 1283ff
A gas-turbine driving an air propeller. The propeller can be basically similar to those used with piston-engines of comparable power but, whereas propeller speeds are of the order of 700-2,000 rpm, gas turbine speeds are typically 8,000-50,000 rpm, so a speed-reducing gearbox is necessary. The turboprop handles a larger airflow than any other aircraft propulsion engine and thus, at low flight speeds, generates the largest thrust for a given fuel consumption. On the other hand, the thrust falls off rapidly with increasing forward speeds and at about 450 mph the propeller efficiency has begun to fall drastically as a result of the blade tips nearing or exceeding sonic velocity.
Turbopump
A pump, usually handling a flow of liquid such as fuel or hydraulic oil, driven by a turbine supplied with compressed air or hot gas. Turbopumps are used in virtually all liquid-propellant rockets, apart from certain small engines with a short burning time in which the fluids can be fed by bottled gas pressure. Another example is the high-capacity pump used for reheat fuel in an afterburner, which is usually driven by a turbine fed by compressor bleed air.

Turboshaft, 1313
A gas-turbine used to provide a mechanical power output in the form of a rotating shaft. Such an engine can be regarded either as a turbojet with an extra stage or two on the turbine, or with an added free turbine driving a mechanically independent output shaft, or else as a turboprop without its propeller and reduction gearbox. Frequently the same basic power section, or gas producer, is used as a basis for a turboshaft, turbojet, turbofan and turboprop. In aircraft propulsion a turboshaft (or shaft-turbine) engine is used to drive mechanical transmission systems of helicopters and some V/STOL machines.
Turbosupercharger, see Supercharger
Turn and slip (turn and bank) indicator, 680, 685, 690, 1666
The turn and slip indicator is a gyro instrument, introduced between the two World Wars, which gives two important indications to the pilot. The first is an indication of the rate of turn which the aircraft is making. Combined with this is an indication of whether the aircraft is in a correctly-banked turn, the other pointer indicating whether the aircraft is slipping inward or skidding outward. This information is presented either by a pointer connected to a pendulum or by a ball in a curved glass tube. By means of these two indications in the one instrument the pilot can ensure that he is making accurate turns at the desired rate.
Turnaround time 1359
With airline or military aircraft, the total time taken to perform all necessary functions between a landing and a take-off. Strictly speaking, in airline parlance the term is properly applied only at the end of a route when the aircraft literally turns round to fly back. Many func-

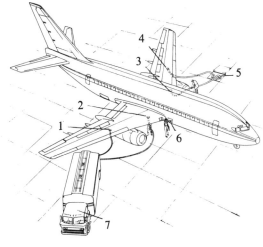

Turnaround (Multi-point refuelling):

1–4 Refuelling points
5 Hydrant

6 Refuelling control panel
7 Refuelling bowser

tions are performed which are not needed at en-route stops. A modern short-haul jet may achieve a turnaround time of 30 minutes, during which it must be cleared of passengers, baggage and cargo, swept out, have literature replaced, pantry and bar restocked, toilets purged and refilled, potable water replenished, externals visually inspected, ground air-conditioning and electrical power coupled up and later switched off, the crew briefed and their flight-plan filed, and the new load of passengers, baggage and cargo taken aboard and all documents signed and clearances obtained. The same aircraft may achieve an en-route time on the ground of but five minutes, and one engine may then not even be stopped. With a military aircraft the list of functions on a turnaround is more variable but obviously includes crew debriefing and re-briefing, fuel replenishment, rearming, inspection for battle damage and possibly realigning of an inertial platform or other system.
Tushino, see Air shows and exhibitions; see also Moscow Airports
Tutor (CL-114), see Canadair

TWA (Trans World Airlines), 737, 761
The letters TWA first appeared in 1930 when four American airlines, Western Air Express, Standard, Maddux and TAT, merged to form Trans-continental and Western Air, with a coast to coast service in a flying time of 36 hr. This was cut to 24 hr when TWA regularly commenced night flying. TWA soon introduced several "firsts" such as de-icing equipment and flight logs, and it collaborated with Douglas in the introduction of

Mainliner of TWA until the 747: the Boeing 707 Intercontinental

the DC family of airliners, ordering the DC-2 and then the DC-3 for its routes. It was one of the first airlines to introduce "over-the-weather" flying, with pressurized Boeing 307s which later served with Air Transport Command of the USAAF when war came. TWA worked with Air Transport Command on wartime routes, and it was from flying outside the US on this task that TWA was prompted, in

1946, to begin a trans-atlantic service to Paris. Lockheed Constellations, a type which TWA had helped to pioneer, were used for the service.
In 1950 it changed its name to Trans World Airlines, aiming to establish itself as a world-wide carrier—an aim which its 50,000-mile route network proclaims that it has accomplished.
Headquarters:
New York City.
Member of IATA.
Twin Otter, see de Havilland Canada
Twin Wasp engine, see Pratt & Whitney
Twiss, Peter L., 302
After serving as a pilot in the Fleet Air Arm during and after World War Two, Peter Twiss joined Fairey Aviation as a test pilot and later took a leading part in the development of the Fairey F.D.2, being the pilot who flew it on the record attempt which raised the World Speed Record to 1,132 mph in 1956.
Two-shaft engine, 1195
A gas-turbine engine in which there are two main rotating assemblies each capable of running at its own optimum speed. Some of the earliest two-shaft engines were free-turbine engines in which the high-pressure turbine drove the compressor and the low-pressure turbine drove a shaft passing down the centre to a propeller reduction gearbox at the front. An example of such an engine is the Bristol Siddeley Proteus (q.v.). It is also possible to make a two-shaft, two-spool turboprop, an example being the Rolls-Royce

Tyne (*q.v.*) in which a single-stage high-pressure turbine drives a nine-stage high-pressure compressor and a three-stage low-pressure turbine drives a six-stage low-pressure compressor and the propeller. All two-spool engines must be two-shaft engines

Two-stage supercharger, see Supercharger

Two-stroke engine, 1105
A piston-engine having only two strokes of the piston (one complete crankshaft revolution) to each complete cycle. When the piston is at top dead centre the combustion space contains the fully compressed mixture of fuel and air (or, in a diesel, air alone) plus some spent exhaust gas from the previous cycle. Firing of the mixture (or, in a diesel, injection of the metered dose of fuel) drives the piston down on the working stroke. Near the bottom of this stroke the piston uncovers the exhaust port, through which most of the exhaust escapes, while an inlet port uncovered on the opposite side of the cylinder admits fresh mixture under pressure, very often from within the crankcase. Usually the piston crown has a humped profile to deflect the fresh charge up into the cylinder head to help it expel the exhaust without itself escaping through the exhaust port. The rising piston then shuts off the two ports and compresses the fresh charge. Two-stroke engines are often diesels in which a scavenging blower (supercharger) can blow the cylinder clear and ensure that the next fuel charge is injected into fairly clean air. Two-stroke spark-ignition engines are usually small and have only one or two cylinders; they are used in some single-seat aeroplanes and rotorcraft. (And, of course, in motor cycles and small cars.)

Tyne engine, see Rolls-Royce
Typhoon, see Hawker
U-2, see Lockheed
UK: Air Forces
Royal Air Force. Prior to the formation of the Royal Air Force on 1 April 1918, British service aviation had been divided into separate military and naval forces since the establishment of the Royal Flying Corps on 13 May 1912. The RFC was formed with Naval and Military Wings, the former being generally known as the Royal Naval Air Service long before the official granting of this title in July 1914. Both Wings were served by the Central Flying School, which was created simultaneously.
For the first two years of the 1914-18 War the function of the RFC was regarded as that of providing reconnaissance and close support for the British Expeditionary Force in France, and its early aircraft were ill-equipped for the aerial fighting thrust upon them during this period. The Farnborough-designed B.E.2 in particular was an unhappy victim of the "Fokker scourge" of 1915-16, having no effective answer to the front-gunned German fighters; but by early 1916 the pusher-engined D.H.2 and F.E.2b, aided by the early French Nieuports, had begun to gain a measure of superiority. The Nieuports were among several French combat types introduced by the RFC early in the war, until adequate supplies of British aircraft became available. Included among the latter were the Bristol Scout and the Sopwith 1½-Strutter, the first front-gunned tractor biplanes to enter RFC service in quantity. In 1917 there followed the Bristol F.2B, Sopwith Camel and S.E.5A, the three best fighters in the RFC during the war period. Bombing activities improved significantly with the appearance in 1917 of the single-engined de Havilland D.H.4 and the Handley Page O/100 and O/400 "heavies". On 1 April 1918 the RFC and RNAS were amalgamated into the Royal Air Force—the world's first fully-autonomous air force; and on 5 June 1918 there was formed under General Hugh Trenchard (who had commanded the RFC since August 1914) an Independent Force within the RAF charged specifically with the strategic bombing of targets in Germany.
When the war ended the RAF had a strength of nearly 23,000 aircraft—a far cry from the 270-odd with which the RFC had entered the war—but it was run down rapidly during 1919 from 200 squadrons to 12, nine of which were stationed in the Middle East. From here, in 1921, the RAF pioneered the Cairo-Baghdad air mail route later taken over by Imperial Airways. The RAF in the early 'twenties was engaged in local conflicts in Iraq and on the North-West Frontier of India. The first post-war-designed aircraft began to enter service

from 1923, Grebe and Siskin fighters and Fawn and Virginia bombers joining the wartime Bristol Fighters, Vimy bombers and others. Naval aviation, which had come under the control of the RAF in 1918, remained so despite the new title of Fleet Air Arm granted to it in April 1924; the Royal Navy could exercise operational control only over ship-borne units while they were at sea. Understandably, perhaps, the RAF's first loyalty was to its land-based units, and although during the 'twenties and 'thirties the FAA was equipped with several excellent types of aircraft (notably those of Blackburn or Fairey design) the performance of these at any given date was generally below that of corresponding RAF types.
Among the best new RAF types of the late 'twenties was the Hawker Hart two-seat day

Hurricanes of No 111 Squadron RAF, exercising over Britain in 1937

Hawker Hart, one of the most versatile basic designs ever created

bomber, which had a performance superior to the current fighter types until Hawker's own Fury entered service in 1931. Another pointer to the future at this time could be seen in the Supermarine racing seaplanes evolved by R. J. Mitchell (*q.v.*) to compete for the Schneider Trophy, for these led ultimately to the legendary Spitfire fighter that achieved universal fame in World War Two. Before the Spitfire, however, came another now-famous name — the Hawker Hurricane. In the Hurricane, which entered ser-

vice in 1937, the RAF received not only its first monoplane fighter but its first fighter with an eight-gun armament. The Spitfire followed it into service in August 1938. Production of these two fighters followed the decision taken in 1934 to expand the strength of the RAF to counter the potential threat from the fast-growing German *Luftwaffe*. This decision resulted, in 1936, in the start of the "shadow factory" production of new combat aircraft by the motor-car industry as well as by the aviation industry. Battle, Blenheim, Wellington and Whitley bombers were among the types produced under this system, and the four-engined heavy bombers of World War Two were first projected at about this time. Concurrent with these expansion plans came a reorganization of the RAF into func-

tional instead of geographical commands, and from 1937-39 control of the Fleet Air Arm was gradually returned to the Royal Navy.
On the eve of World War Two an Advanced Air Striking Force flew to France, and Bomber Command Blenheims and Wellingtons made their first attack on an enemy target on 4 September 1939. There followed the so-called "phoney war" period until May 1940, when the Nazi offensive in Western Europe began in earnest. Thanks to the foresight of Fighter Command's chief, Air Chief Marshal Dowding, enough RAF Hurricanes and Spitfires had been retained in the UK to provide a measure of dis cover for the Dunkirk evacuation. Dowding's policy was even more firmly vindicated a few months later when those same aircraft and their pilots won the Battle of Britain over southern England. With Hitler's invasion plans shattered, the *Luftwaffe* fell back upon an all-out day and night bombing offensive against the UK, to counter which such new specialist aircraft as the Beaufighter appeared to join existing types in the defence against the *Blitzkrieg*. Further strains were placed upon the RAF's

The Camel, Sopwith's most successful wartime design and top scorer of the RFC

resources by Italy's entry into the war in June 1940, necessitating the strengthening of Middle East units, and by the opening of the war in the Pacific in December 1941. In February 1941, however, a raid by RAF Stirlings marked the entry of the first British four-engined heavy bombers into action, and Lend-Lease warplanes from the US soon began to strengthen RAF formations at home and in North Africa and the Pacific theatre. In 1942 came the highly-versatile twin-engined Mosquito, and the Avro Lancaster joined the Stirling and Halifax to complete the British trio of heavy bombers. With these the RAF joined with the US Eighth Air Force in mounting a sustained bombing offensive against German industrial targets in preparation for the Allied invasion in June 1944—in which the combined Allied air forces were commanded by an RAF officer, Air Chief Marshal Tedder. There soon

9,000 aircraft, from which a 20-squadron bomber group was preparing to join the proposed "Tiger Force" for the Far East when Japan surrendered.

After World War Two the RAF, although reduced considerably in size, remained committed to numerous overseas commands as well as to the occupation of Germany. Its post-war re-equipment began with wartime designs such as the piston-engined Brigand, Hornet and Lincoln, and the Vampire and later Meteor jet fighters. A pause in the gradual run-down came with the Berlin Airlift in 1948-49, while there were "brush-fire" engagements to be performed in the Middle East, Kenya and Malaya. American Superfortresses were purchased to bridge the operational gap between the Lincoln and the first jet V-bombers, although the Canberra light jet bomber entered service in 1951. By this time the Korean War had indicated a need to expand

Handley Page Victor B.2, strategic strike and reconnaissance bomber

forecast that the P.1B (Lightning) would be the RAF's last manned fighter, and that no replacement would be provided for the V-bomber force once it was retired. Of this trio, the Valiant and Victor bombers are no longer in service, though the Vulcan (which has a "stand-off" capability) will remain in service for several years yet. The Lightning has had to be augmented by the purchase of American Phantoms, and following the cancellation of the outstanding TSR-2 strike aircraft in 1965 there is no comparable successor to the veteran Canberra, although the Buccaneer will serve as an interim type in this rôle. More comforting are the emergence of the Nimrod as a fully-fledged

recognized officially as a separate Royal Naval Air Service in July 1914. Before the outbreak of the 1914-1918 War the Navy had several pioneering achievements to its credit. They included

Fairey Flycatcher over HMS Eagle (*the first carrier to bear the name*) *in the days when the RAF provided the naval air arm in Britain*

Furious that deck-landing techniques were pioneered by Squadron Commander E. H. Dunning, in a Sopwith Pup, in August 1917; and from *Furious*, in July 1918, seven Camel

and potent anti-submarine aircraft, and the introduction into service in 1969 of the Hawker Siddeley Harrier, the world's first V/STOL strike fighter. This streamlined RAF is thus fairly well equipped for its task now that it has been relieved of any "east of Suez" responsibility. The major question is to find a successor to the ageing V-bomber fleet, and in this connection the idea of developing a bomber version of the Concorde supersonic airliner is less far-fetched than might appear at a first glance.

Fleet Air Arm. The Fleet Air Arm of today has its origins in the RFC Naval Wing, created in May 1912 and

the first successful take-off from water by a British seaplane (November 1911); the first from the deck of a British warship (January 1912); the first from a British ship under way (May 1912); the first bomb-dropping experiments (1912); and the first successful torpedo-dropping (July 1914).

When war broke out the RNAS possessed 71 aeroplanes and 7 airships, a strength which by 1 April 1918 (when it became a part of the new RAF) had grown to nearly 3,000 aeroplanes and over 100 airships. Its wartime activities were by no means confined to action with the Fleet: it also carried out early bombing raids, over-water anti-Zeppelin and anti-submarine patrols, was responsible until Spring 1916 for the air defence of the UK, and fought alongside the RFC on the Western Front. The war period saw the conversion of several civilian ships into seaplane tenders, and the evolution of the first true aircraft carriers—HM ships *Hermes*, *Furious*, *Campania* and *Vindex* being among the earliest. It was on

fighter-bombers took off for a successful raid on the Zeppelin base at Tondern.

The title Fleet Air Arm was bestowed upon Naval aviation in April 1924, but it remained under RAF control from April 1918 until May 1939, when full control reverted to the Royal Navy. Officially it was then retitled Naval Aviation, although the old FAA title persisted in practice until it was restored officially in 1953. During the inter-war years the depleted FAA introduced many fine aeroplanes into service, and made significant advances in the evolution of satisfactory deck take-off and landing techniques, including the use of the catapult. But it remained very much the "Cinderella" air service; at the outbreak of World War Two it was a shadow of its 1918 size, with only 225 carrier-borne aircraft and 115 other types in service, of which all but a few Blackburn Skua dive-bombers were antiquated biplanes. Its first high-performance monoplane fighters were

On most criteria the Gloster Meteor was the first jet aircraft in the world to enter service. These are F.IIIs, photographed in the autumn of 1944

followed, in the Autumn, the first jet operations, by RAF Meteors against the V-1 flying bombs over England. Meanwhile, in the Far East, RAF squadrons in South-East Asia Command outnumbered those of the USAAF by more than two to one, although they were equipped predominantly with US aircraft. At the conclusion of war in Europe the RAF had a strength of just over

again, and large-scale production priority was given to the Canberra, the Hunter and Javelin fighters and the V-bomber programme. The Hunter entered service in 1954, and the Valiant, Vulcan and Victor in 1955, 1956 and 1958. Further operational engagements were undertaken in the 1956 crisis over Suez. The Defence White Paper of 1957 brought the now notoriously inaccurate

*BAC (*English Electric*) Lightning F.6, latest in a series of all-weather fighters for the RAF*

RAF Hurricanes and Spitfires, hastily converted in 1942 to operate from carriers, and not until the addition of the later Seafires and Lend-Lease Corsairs and Hellcats did it have a full-strength modern fighter force. Meanwhile, however, the ancient biplanes were acquitting themselves nobly. The most outstanding example was the Fairey Swordfish torpedo-bomber, a 1933 design which, nearly a decade later, played the major rôle in the famous attack on the Italian Fleet at Taranto. The Swordfish was to the forefront in the sinking of the German pocket battleship *Bismarck*, at the Battle of Cape Matapan, and in numerous other major engagements. Its success was such that it outlived its intended replacement, the Albacore, serving until the very end of the war in Europe.

When the war in the Far East ended in August 1945 the strength of the FAA stood at 1,300 aircraft, a total which included the Barracuda torpedo-bomber and the excellent Firefly fighter-reconnaissance monoplanes. On the verge of entering service was the Firebrand torpedo-strike fighter. On 4 December 1945, the FAA became the first naval air force in the world to acknowledge the jet age, when a modified de Havilland Vampire was landed on HMS *Ocean*. The higher speeds of the latest piston-engine fighters and the new generation of jet aircraft provoked three important innovations designed to overcome the attendant problems. These were all pioneered by the FAA, and have since been adopted for use in all other carriers in service around the world. First of these, introduced on HMS *Centaur* in 1953, was the angled deck, invented

by Captain D. R. F. Campbell and designed to increase the available deck runway area without significantly increasing the size of the vessel itself. In 1954 this device was complemented by the first steam catapult invented by Commander C. C. Mitchell, and shortly afterward by the mirror landing sight, an invention of Commander H. C. N. Goodhart. Meanwhile the FAA's first operational jet fighter squadron, equip-

Supermarine Attacker— first FAA jet fighter

ped with Supermarine Attackers, had been formed in 1951. The Attackers were followed in 1953 by turboprop Wyvern strike aircraft; in 1954 by the first anti-submarine helicopters; and in 1955 by the turboprop Gannet anti-submarine aircraft. To replace the Attacker as the principal jet fighters came first the Hawker Sea Hawk and de Havilland Sea Venom, and then the swept-wing Scimitar and Sea Vixen. The Scimitar has now given way to the US Phantom; the carrier-based strike rôle is met by the transonic Buccaneer, while the Gannet AEW Mk 3 is retained for airborne early warning patrol. The Fleet Air Arm's few remaining carriers are to be phased out during the nineteen-seventies, after which the Buccaneers will be transferred to serve with the RAF.

UK: Chronology

1250 (approx). Roger Bacon, a Franciscan friar, and one of the great pioneers in science and philosophy, was the

first man to write on the subject of flight in a true mechanical or scientific sense.

27 August 1784. First balloon ascent in Britain was made by James Tytler, at Edinburgh. Using a *Montgolfière* type of hot-air balloon, he made a flight of about half a mile.

1804. The world's first successful aeroplane, a model glider, built and tested by Sir George Cayley.

7-8 November 1836. Charles Green, first to conceive the idea of using coal gas for the inflation of free balloons, completed a 480-mile balloon flight from London to Weilburg. Also to Green goes credit for invention of the trail rope, to help control a balloon at low altitudes.

1843. William Samuel Henson produced the design for his Aerial Steam Carriage. This was the first detailed study for a complete aeroplane.

1853. Sir George Cayley designed and built a man-carrying glider which successfully carried his coachman across a small valley: the first man-carrying aeroplane flight in history.

5 September 1862. James Glaisher and Henry Coxwell exceeded a height of 20,000 ft in a balloon during a scientific flight.

12 January 1866. The Aeronautical Society of Great Britain founded, under the Presidency of the Duke of Argyll, to ensure that the whole subject of aeronautics was approached in a scientific manner. This was the first scientific organization of its kind in the world.

1906-7. Lord Northcliffe offered, through the medium of his *Daily Mail* newspaper, large money prizes to promote aviation. One of £1,000 for the first flight between France and England led to Louis Blériot's epic flight on

25 July 1909. Britain, for the first time, became conscious of the potential of the aeroplane.

1907. A. V. Roe took first prize at Alexandra Palace in a model aeroplane competition sponsored by the *Daily Mail*. His elastic-powered biplane made two successful flights, one of 60 ft and one of 100 ft.

1907. Probably the first powered aeroplane flight in Britain was made during the summer of 1907 by Horatio Phillips on one of his "Multiplanes." This embodied four parallel frames containing some 200 narrow-chord wings of "Venetian-blind" formation.

16 October 1908. S. F. Cody (then a US subject) made the first officially-recognized powered and sustained flight in Great Britain, covering a distance of 1,390 ft at a height of 30 ft in 27 sec, piloting his *British Army Aeroplane No. 1*.

2 May 1909. J. C. T. Moore-Brabazon made the first officially-recognized flight by a British pilot in the United Kingdom. Flying a Voisin biplane, at Shellbeach, he was airborne for about 1,500 ft. A. V. Roe had made several short powered hops at

The first pig to fly—with Lt-Col J. T. C. Moore-Brabazon

Brooklands in his pusher biplane during June 1908, but these failed to qualify as there was no official observer.

13 July 1909. With a

flight of some 100 ft in his No 1 Triplane, powered by a 9-hp JAP engine, A. V. Roe finally achieved acknowledgment for the first officially-recognized powered flight in Great Britain in an all-British aeroplane.

1909. The Short brothers started building the first powered aeroplane of their own design. They had been official balloon constructors to the Aero Club for a number of years and this (which was followed by an order to build six Wright biplanes under licence) enabled them to claim to be the first manufacturers of aircraft in the world.

27-28 April 1910. The London-to-Manchester Air Race, for the *Daily Mail* £10,000 prize, won by Louis Paulhan of France on a Farman biplane. In attempting to overcome his rival's lead, Claude Grahame-White made probably the first night aeroplane flight in Great Britain.

1 April 1911. Formation of the first Air Battalion of the Royal Engineers, with Headquarters and No 1 Airship Company at Farnborough and No 2 Aeroplane Company at Larkhill, marked the beginning of British military aviation.

4 July 1911. Horatio Barber, flying a Valkyrie B monoplane, carried the first air cargo—a box of Osram lamps—from Shoreham to Hove.

9 May 1912. During a Naval Review at Weymouth, Lieutenant C. R. Samson, RN, flew a

First UK ship take-off: S.27 from HMS Hibernia

Short S.27 biplane off a short staging on HMS *Hibernia* while the ship was steaming at 15 knots. This was the first flight from a ship under way.

13 May 1912. Formation of the Royal Flying Corps, with Military and Naval Wings, Central Flying School, Reserve, and Royal Aircraft Factory at Farnborough.

May 1912. Lieutenant Wilfred Parke piloted the Avro F monoplane on the first flight of a totally-enclosed cabin aeroplane.

7 May 1913. HMS *Hermes*, first seaplane carrier in the world, entered service.

3 August 1917. By landing on the deck of HMS *Furious*, Sqn Cdr E. H. Dunning, flying a Sopwith Pup, made the first deck landing on a ship under way.

1 April 1918. The Royal Air Force came into being, formed from the Royal Flying Corps and the Royal Naval Air Service. Sired by Trenchard's Independent Force, it was the first independent air force in the world.

14-15 June 1919. Flying a Vickers Vimy bomber, from St John's, Newfoundland, Captain John Alcock and Lieutenant Arthur Whitten-Brown crossed the North Atlantic to make the first non-stop Atlan tic flight. They landed at Clifden, Ireland, after being airborne for 16 hr 12 min.

2-13 July 1919. First two-way crossing of the Atlantic completed successfully by the British dirigible R.34.

25 August 1919. Aircraft Transport & Travel, founded by Holt Thomas, started the first international aeroplane passenger service, between London and Paris.

1922. Following the Cairo Conference in 1921, squadrons of the Royal Air Force took over "police" duties in Iraq. This represented the first use of air power to control a potential trouble-spot.

1922. Daimler Airway became the first airline to employ a steward to serve refreshments to passengers during flight.

22 February 1925. Maiden flight of the de Havilland Moth, the first moderately-

ally-prepared Westland P.V.3 biplane, piloted by Lord Clydesdale, and a Westland Wallace, piloted by Flight Lieutenant D. F. McIntyre, flew over the summit of Mount Everest for the first time.

19-23 October 1934. C. W. A. Scott and T. Campbell Black won the McRobertson Mildenhall-Melbourne air race. Flying a de Havilland twin-piston-engined Comet, they completed the course in just under 2 days 23 hr flying time.

6 November 1935. Maiden flight of the Hawker Hurricane (q.v.), the first heavily-armed monoplane fighter.

12 April 1937. The prototype jet-engine designed by Frank Whittle ran successfully for the first time. His improved W.1 turbojet engine of only 860-lb thrust powered Britain's first jet aircraft, the Gloster

Hurricanes and Spitfires of RAF Fighter Command—won the Battle of Britain. It was the first great and decisive aerial battle, and one in which a vital part was played by radar, the invention of Mr (later Sir) Robert Watson-Watt.

15 May 1941. The Gloster E.28/39 jet-powered research aircraft made its maiden flight. Though not the world's first jet-powered aircraft, it was far more practical than the German and Italian aircraft which had preceded it. A developed version of its Whittle turbojet engine was to form the power plant of the Gloster Meteor, the first jet-engined fighter to be used in action.

May 1945. First air tests of a Martin-Baker ejection seat, in a modified Defiant fighter. The first live ejection was made, from a Gloster Meteor jet fighter, on 24 July 1946. Martin-Baker seats are now installed in most of the world's high-performance military aircraft, and by the beginning of 1969 had been respon-

night mail service, in the eastern counties of England. This was followed in June 1950 by the first scheduled helicopter passenger service, between Liverpool and Cardiff.

2 May 1952. The de Havilland Comet 1, first jet airliner to enter production anywhere in the world, began scheduled airline service with BOAC.

April 1953. The turbo-prop-powered Vickers Viscount, Britain's most successful airliner, entered service on BEA's short- and medium-range routes.

1954. Rolls-Royce pioneered a completely new VTOL concept known as jet-lift. Their odd-looking "Flying Bedstead" test rig was raised from the ground by the efflux from two jet engines.

October 1956. Flown by test pilot Peter Twiss, the Fairey Delta 2 research aircraft made the first over-1,000-mph world speed record at an officially-recorded 1,132 mph.

6 November 1957. The Fairey Rotodyne made

jetliner, introduced the world's first transatlantic jet passenger service.

1969. The Hawker Siddeley Harrier, first VTOL strike aircraft, entered RAF service.

United Air Lines, 737

As its name implies, United began life from a merger of several airlines, the main unit being Boeing Air Transport (q.v.). This company had begun flying a San Francisco-to-Chicago route in 1927 as an off-shoot of the Boeing manufacturing company. Two years later it embraced National Air Transport and then Varney Air Lines in 1930. In all this the member airlines retained their individual identities, flying their own routes until 1933 when moves began which resulted in the new United image in 1934.

By then, United was flying the new Boeing 247 which, with the DC-2, was to revolutionize air transport in America. When the DC-3 appeared United acquired some of these also and continued

routes and revenue, acquiring four-engined equipment in the shape of DC-4s, DC-6s and, later, Stratocruisers. In 1955 United became the first domestic operator to order jet equipment (DC-8s), the first of which was delivered in June 1959. Jet operations were further extended by placing the medium-haul French Caravelle in service in June 1961.

In the same month, UAL acquired Capital Airlines, and with it a fleet of Vickers Viscounts. United's growth has continued under the guiding hand of W. A. Patterson, an executive from Boeing Air Transport days, and today United is one of the biggest US domestic airlines.

Headquarters: Chicago (O'Hare) Airport.
Member of IATA.

United Arab Airlines

Created under this title as recently as 1960, following the formation of the United Arab Republic by Egypt and Syria, UAA's origins actually go back to May 1932 and the formation

The de Havilland Moth was designed for towing on public roads behind a car in 1925

priced and reliable two-seat lightplane. This aircraft made possible the tremendous expansion of the flying club movement throughout the world.

1931. Flight Lieutenant J. N. Boothman, flying a Supermarine S.6B seaplane, won the Schneider Trophy outright for Great Britain. On 29 September 1931, Flight Lieutenant G. H. Stainforth set up the first over-400-mph world speed record on a similar aircraft, at a speed of 407.5 mph.

3 April 1933. A speci-

E.28/39, in 1941.

5-6 July 1937. First east-west airline crossing of the Atlantic by a Short "C" class flying-boat, *Caledonia*. Simultaneously, a Pan American Sikorsky S.42 made the first west-east airline crossing.

21 July 1938. The seaplane *Mercury*, upper component of the *Maia/Mercury* composite, landed at Montreal, Canada, 20 hr 20 min after being launched from *Maia* over Foynes, Ireland.

September 1940. "The Few"—the pilots of

Ejecting from a typical jet fighter

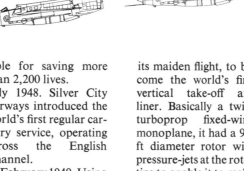

sible for saving more than 2,200 lives.

July 1948. Silver City Airways introduced the world's first regular car-ferry service, operating across the English Channel.

21 February 1949. Using Sikorsky S-51 aircraft, British European Airways initiated the world's first helicopter

its maiden flight, to become the world's first vertical take-off airliner. Basically a twin-turboprop fixed-wing monoplane, it had a 90-ft diameter rotor with pressure-jets at the rotor tips to enable it to make vertical take-offs and landings.

4 October 1958. BOAC, using the new Comet 4

steady expansion until, when war broke out, it had 69 twin-engined aircraft. Half of these were handed over to the Army Air Corps, and many United crews went with them, flying mainly trans-Pacific and Alaskan routes.

United returned to post-war airline operation by steadily building up its

of the Egyptian airline Misr Airwork, which started scheduled operations in 1933. This company became wholly Egyptian-owned in 1949, when it adopted the modified name Misrair. In 1960 Misrair merged with Syrian Airways to form UAA, although it subsequently restored the Syrian route network and equipment to the new Syrian Arab Airlines (q.v.). In addition to domestic services, UAA now has international flights to the Middle East, Africa, Eastern and Western Europe, and India; it is contemplating the resumption of a former route to the Far East. Headquarters: Cairo. Member of IATA.

Universal Airlines
New title since late 1966 of the former Zantop Air Transport, a charter and contract freight carrier whose activities include a substantial amount of "Logair" transportation of freight for the USAF and similar work for the US Navy.
Headquarters: Willow Run Airport, Michigan.

USA: Air Force
US Air Force. The present-day USAF has its origins in the US Army Signal Corps, which set up an aviation branch in 1907, although Samuel Pierpont Langley (q.v.) had been granted $50,000 by the Army in 1898 for developing his experimental *Aerodrome.* The failure of this in 1903 probably prejudiced the Army against the Wright Flyer, which it twice rejected, unseen, before finally accepting a Flyer Type A in 1909 as *Signal Corps No. 1,* the US Army's first aeroplane. Further Wright biplanes and two Curtiss "pushers" were acquired in 1910-11, and the Army's First Aero Squadron was established officially in 1913. A few weeks before the outbreak of the 1914-18 War an Aviation Section of the Signal Corps was created, followed by the procurement of Curtiss "Jenny" and Martin trainers. In 1915-16 the First Aero Squadron was engaged in the Mexican revolutionary campaigns, and in August 1916 a major expansion programme, in recognition of the seriousness of the war in Europe, was initiated. When the

US entered the war in April 1917, however, it had no combat aircraft, only trainers, and was obliged to buy French (and, later, British) warplanes to equip the American Expeditionary Force in Europe. This made an important contribution to the Allied war effort, and by the end of the war included substantial numbers of licence-built D.H.4s and some Sopwith Camels.
The Armistice brought tremendous cuts in the huge aircraft programme initiated in 1917, and in the immediate post-war period the principal US service types were Curtiss "Jenny" and Thomas-Morse trainers and Le Père LUSAC-11 and Standard SJ-1 fighters, to which were added US-built British D.H.4 and Handley Page O/400 bombers. The Army, reorganized in 1920, retained control of the Air Service, despite the efforts of General "Billy" Mitchell and others to have it recognized as a separate service; and various publicity-making flights in the mid-'twenties could not conceal that it was seriously under-equipped. On 2 July 1926 it was retitled US Army Air Corps, and began an attempt to expand and modernize despite restricted funds and economic depression. Among the new types were Boeing P-12s and the early Curtiss Hawk fighters, Curtiss and Keystone bombers, and a variety of observation/general-purpose aircraft. Really modern types,

such as the Boeing P-26 and Martin B-10, began to appear in the mid-'thirties, followed in 1935 by the formation of specialized tactical and strategic elements within the USAAC. A major result of this was the ultimate realization of a long-range heavy bomber force, heralded by the prototype Boeing 299 which flew in July 1935 and eventually developed into the B-17 Fortress.
Presidential approval, influenced by the political situation in Europe, was given in 1938 for a massive expansion programme to a total aircraft strength of 10,000, over one-third of which were to be combat types. Added to this, by the spring of 1940 Britain and France had over 8,000 US military aircraft on order—a formidable enough challenge, even to the vast production resources of such a large industrialized nation, yet the US Army Air Force (it was renamed in June 1940) alone reached an eventual peak strength of some 80,000 aircraft, quite apart from the many thousands more built on behalf of other Allied nations. This compared with a first-line aircraft strength, at the time of Pearl Harbor, of less than 3,000, of which the principal types—numerically —were P-39 and P-40 fighters, A-20 light bombers, and B-17 and B-24 heavy bombers. After initial heavy losses from air attacks on its Hawaiian and Philippine bases, the USAAF

directed its main attention to Europe and North Africa, setting up the Eighth Air Force in the UK and the Twelfth and Ninth Air Forces in the Middle East in 1942. From bases in East Anglia the Eighth Air Force mounted a steadily-increasing bomber offensive against enemy targets in Europe, which it and the Ninth Air Force maintained from French bases after June 1944. In the Pacific, USAAF strength was also increased steadily, notably the Tenth and Fourteenth Air Forces in the India-Burma-China theatre, Air Transport Command's "over the Hump" supply ferry across the Himalayas between India and China, and the 21st Bomber Command which mounted the B-29 bombing attacks on the Japanese homeland.
After World War Two the USAAF was run down to about one-third of its wartime manpower and half its aircraft strength, and huge quantities of war-surplus military aircraft were given or sold to other air forces the world over. On 18 September 1947, the former USAAF was given the new title of United States Air Force and at last became independent of Army control. Prior to this it had already adopted its new peacetime structure with Air Defense Command, Strategic Air Command and Tactical Air Command as the three main combat elements. ADC soon began to receive the new F-80, F-84 and F-86 jet fighters, while SAC expanded with a fleet of B-47 and B-52 global bombers and TAC's equipment included the Martin B-57, licence version of the British Canberra. Several of these were engaged in the Korean War of 1950-53. The first supersonic fighters (F-100 Super Sabres) and bombers (B-58

Hustlers) entered service in 1953 and 1960 respectively, and are still in use today. Other current fighters include the Convair F-102 and F-106, Lockheed F-104, Republic F-105, General Dynamics F-111 and McDonnell Douglas F-4; other major front-line types include the LTV A-7 strike aircraft, Lockheed SR-71 reconnaissance aircraft and the gigantic Lockheed C-5A transport.
US Army. US Army aviation, as distinct from the US Army Air Force described above, dates officially from 18 September 1947 but effectively from some six years earlier than this. During this period its activities have been concerned chiefly with maintaining a force of observation, reconnaissance, light transport and general utility aircraft, including helicopters.
US Navy and Marine Corps. US Naval aviation dates, effectively, from the take-off by Eugene Ely (q.v.) from the USS *Birmingham* in a Curtiss biplane in November 1910; and the Navy accepted its first aeroplane, a Curtiss A-1 floatplane, in the following year. Curtiss or Burgess-Dunne seaplanes constituted most of its early equipment, and Curtiss flying boats and their Felixstowe developments played an important rôle with the British RNAS during the 1914-18 War. When America entered the war in April 1917 the USN had nearly 140 seaplanes and over a dozen non-rigid airships on order, but only a score of seaplanes actually in service: hence, like the Army, it had to acquire combat aircraft of European design for its anti-submarine rôle. It finished the war with some 1,400 aircraft, more than 80 per cent of them seaplanes.
Eighty airships and over 600 new aircraft were ordered in the first post-

war expansion programme; and in 1922 the USN's first aircraft carrier, the USS *Langley,* was commissioned, and take-off platforms were fitted to several other capital ships. From this point the Navy procured more landplanes than seaplanes, even though in 1923 it still held nearly two-thirds of the world's seaplane records. Curtiss, Douglas, Martin and Vought provided many of the new service types of the 'twenties, although some European types were also acquired. A new expansion programme in 1926 was accompanied by two additional carriers, *Saratoga* and *Lexington,* in 1927. By 1931 the USN had a total aircraft strength of almost 1,000, with over 200 more and a fourth carrier, *Ranger,* still to come; and in 1934 further large orders were placed for aircraft and two more carriers, *Yorktown* and *Enterprise.* After two serious losses the USN abandoned its interest in rigid airships, but retained its belief in the value of non-rigid "blimps", which it continued to use for observation up to the 'sixties. By the latter half of the nineteen-thirties the USN had begun to plan the replacement of its obsolescent biplane fighters and bombers by such modern types as the Grumman Wildcat fighter, Douglas Dauntless dive-bomber and PBY (Catalina) flying boat, and had ordered yet another carrier, the *Wasp.* The proposed aircraft strength was increased five-fold, to 15,000, in mid-1940 and new orders included large quantities of the excellent Curtiss Helldiver, Chance Vought Corsair and Grumman Hellcat and Avenger. At the time of the Pearl Harbor attack the combined USN, Marine Corps and Coast Guard had over 5,000 aircraft on strength, the

The Martin 139, one of many monoplane bombers by the Baltimore firm and first to have a gun turret

Douglas SBD Dauntless, almost the only means the US Navy had of striking back in 1942

first of twelve more major carriers (*Hornet*) was in commission, and over one hundred conversions of merchantmen to escort carriers had been authorized.

In the Pearl Harbor attack all eight battleships of the Pacific fleet were put out of action, and within another six months the *Langley* and *Lexington* had been lost, but during 1942 the US Navy—and its naval air power—hit back hard against the Japanese fleet in the Battles of the Coral Sea, Guadalcanal and Midway, giving it a setback from which it was never able to recover. At Midway alone, the Japanese Navy lost four carriers and over 250 aircraft. The US Navy was also hard-hit—by the end of 1942 it had also lost the *Hornet*, *Wasp* and *Yorktown*—but it was in a better position to replace its losses. Detachments of US Naval air power saw service in South American waters, in the Atlantic, with RAF Coastal Command, and elsewhere; but its major rôle was in the Pacific theatre of war where, with the aid of superior aircrew and aircraft, it gradually swung the balance of air power in its own direction by the increasing use of carrier task forces against the Japanese-held islands. After their recapture these continued to be held chiefly by Marine Corps units. A major carrier *v.* carrier conflict came in June 1944 with the "Marianas Turkey

Shoot", more officially known as the Battle of the Philippine Sea, and later in that year the mounting wave of Japanese suicide (*Kamikaze*) attacks brought a serious but only temporary setback to Allied progress in the Pacific. But this too was overcome, and from the island bases won back by Navy and Marine Corps personnel the USAAF was able to mount the air attacks that led ultimately to the Japanese surrender in 1945. The Navy Department claimed that, for the loss of 2,700 of its own aircraft, it had destroyed 17,000 belonging to the enemy.

The aftermath of World War Two left the US clearly the most potent sea power in the world, even after the natural post-war run-down, and the Atlantic and Pacific fleets were equipped respectively with fourteen and eighteen carriers of various calibres. New post-war aircraft included the Grumman Tigercat and Bearcat piston-engined fighters and its first jet fighter, the original McDonnell Phantom I. Best of the new attack bombers was the Douglas Skyraider, an immensely successful type in the subsequent Korean War (as was the Corsair) and, even later, in Vietnam. The Korean War also marked the beginnings of the Marine Corps' marked interest in the helicopter as a multi-purpose military vehicle. The British-invented angled deck

was introduced in 1952 on the carrier *Antietam;* it was subsequently applied to all the principal USN carriers in service, as well as to the new *Forrestal* class "supercarriers" that were under construction. *Forrestal* herself was commissioned at the beginning of 1956, and by this time several swept-wing transonic aircraft were in service or on order—notably the Douglas Skyhawk and Skywarrior, Grumman Cougar, Chance Vought Cutlass and North American Fury, plus the first of several hundred Lockheed Neptune piston-engined patrol aircraft. The first "commando" carrier, USS *Thetis Bay*, was also commissioned in 1956; this was capable of carrying a large complement of marine troops and several helicopters.

Skyhawks, Skywarriors and Neptunes are still in USN/USMC service. More recent equipment includes Phantom II and Crusader fighters, Corsair II strike aircraft, Orions and Sea King helicopters for anti-submarine warfare and Tracker and Hawkeye AEW aircraft.

USA: Chronology

9 January 1793. First public balloon ascent in America made by Pierre Blanchard at Philadelphia. An earlier attempt, in 1789, failed when the balloon burned prior to launching.

2 May 1835. Ascent by John Wise at Philadelphia marked the "coming of age" of ballooning as an exhibition-scientific-sporting activity. Wise became the premier American balloonist for nearly 40 years, setting a distance record of 1,120 miles in 1859 that stood for 41 years. His career culminated in a trans-atlantic flight attempt in 1873.

23 June 1861. First US military reconnaissance from a balloon, by Professor Thaddeus Lowe at Arlington, Virginia.

17 March 1883. Professor John J. Montgomery began glider flights near San Diego, California.

6 May 1896. Samuel Pierpont Langley of the Smithsonian Institution demonstrated the proper combination of lift distribution, propulsion, and inherent stability necessary for mechanical flight, by flying a 13-ft steam-powered model, *Aerodrome No. 5*, 3,200 ft over the Potomac river.

22 June 1896. Octave Chanute began "hang-glider" experiments near Chicago. These quickly improved on the designs of the German Lilienthal (*q.v.*) and established the basic proportions and trussing of biplane structures that are still in use today.

October 1900. The Wright brothers began their gliding experiments at Kitty Hawk, North Carolina, and developed a successful three-axis control system.

17 December 1903. The Wright brothers made the first controlled and sustained powered flight, covering 120 ft in 12 sec in an enlarged version of their 1902 glider fitted with a home-made 12-hp·engine.

1 October 1907. Dr Alexander Graham Bell, inventor of the telephone, formed the Aerial Experiment Association, which organized recognized engineering talent into a task force for the purpose of developing practical flying machines.

10 February 1908. Award of first US military contract for an aeroplane to the Wright brothers. Basic price $25,000, with bonus and penalty clauses subject to measured performance.

4 July 1908. Glenn L. Curtiss won *Scientific American* Trophy for the first public flight of 1 km, in the Aerial Experiment Association's third aeroplane, the *June Bug*, at Hammondsport, NY.

18 August 1908. The Army accepted its first airship, *Signal Corps No 1*, from Captain Thomas Scott Baldwin.

17 September 1908. Lieutenant Thomas Selfridge became the first person to die in an aeroplane crash, during a demonstration of the Wright Flyer to the Army at Fort Meyer, Virginia.

14 November 1910. Eugene Ely, Curtiss test pilot, flew a Curtiss pusher from a platform on the after deck of the cruiser USS *Birmingham* anchored in Hampton Roads, Virginia, to demonstrate the feasibility of ship-borne aircraft.

18 January 1911. Ely landed a Curtiss pusher on the deck of the cruiser USS *Pennsylvania* in San Francisco Bay and took off again.

26 January 1911. Glenn Curtiss made the first successful flight in a displacement - pontoon seaplane at San Diego, California.

1 July 1911. US Navy accepted delivery of its first aeroplane, the Curtiss A-1, at Hammondsport, NY.

17 September to 10 December 1911. Calbraith P. Rogers exceeded the 30-day time limit in trying to win publisher William Randolph Hearst's cash prize for a coast-to-coast flight, covering 3,390 miles from Sheepshead Bay, NY, to Pasadena, California, in 49 days. The Pacific Ocean was not reached until 35 days later.

Spring 1912. Glenn Curtiss made the first successful flight in a flying boat at San Diego, California.

1 January 1914. Inauguration of the first scheduled airline service, across Tampa Bay from St Petersburg to Tampa, Florida, using Benoist flying boats.

25 April 1914. First US air operations against enemy forces, when Navy Curtiss flying boat AB-3, based on USS

Mississippi and piloted by Lieutenant P. N. L. Bellinger, scouted the harbour of Vera Cruz, Mexico, for mines prior to entry of the fleet.

September 1914. US Army organized its First Aero Squadron.

15 March 1916. First Aero Squadron began operations against Mexican forces of Pancho Villa at Columbus, New Mexico.

20 April 1917. First flight of US Navy's first airship, the DN-1. By World War Two the US Navy was the only user of military airships.

17 October 1917. First flight of the twelve-cylinder Liberty engine, major American contribution to aviation in the 1914-18 War. This had a detrimental effect on post-war engine and aeroplane development because of its availability in great numbers at surplus prices. The US Army did not retire its last Liberty-powered aeroplane until 1935.

15 May 1918. Inauguration of first regularly-scheduled US Air Mail; flown by Army pilots and aircraft between New York City and Washington, DC. The Post Office Department soon took over the service and operated it over the entire nation until the last Government routes were turned over to private contractors in September 1927.

8-31 May 1919. NC-4, one of three US Navy flying boats to start, completed the first-ever flight across the Atlantic, reaching Portsmouth, England, in five hops.

22 May 1919. Raymond Orteig, French-born New York hotel owner, offers cash prize of $25,000 for the first non-stop flight to be made in either direction between New York and Paris within five years. No attempts were made within the time limit, so the offer was renewed.

8 September 1919. Scheduled trans-continental air mail service inaugurated.

27 February 1920. Major Rudolph W. Schroeder flies LePère LUSAC-11 to an altitude record of 33,000 ft, in the first practical demonstration of the General Electric turbosupercharger.

1 December 1921. First flight of a helium-filled airship, the US Navy blimp C-7. Subsequent burning of hydrogen-filled airships and the US monopoly of the world's only natural helium supplies resulted in helium becoming mandatory for all US airships.

2-3 May 1923. First non-stop coast-to-coast flight by Lieutenants Macready and Kelly in Army Fokker T-2 transport, covering 2,520 miles in 36 hr 4 min.

29 June 1923. First aerial refuelling of one aeroplane by another, a hose being used for the transfer from one US Army DH-4 to another over San Diego, California.

6 April to 28 September 1924. First flight around the world completed by two of four US Army Douglas World Cruisers, starting from Seattle, Washington and covering 27,533 miles in 15 days and 11 hr flying time.

9 May 1926. Commander Richard E. Byrd and Lieutenant Floyd Bennett, US Navy, made the first flight over the North Pole in a civil Fokker F.VII trimotor from a base at Spitzbergen.

20-21 May 1927. Charles A. Lindbergh won the Orteig Prize and triggered a worldwide aviation boom by flying 3,610 miles from New York to Paris in 33 hr 30 min in the single-engined Ryan NYP monoplane *Spirit of St Louis*.

17-18 June 1928. Amelia Earhart became the first woman to cross the Atlantic by air, as a passenger in the trimotor Fokker seaplane

Friendship piloted by Wilmer Stultz and Louis Gordon.

1-7 January 1929. World endurance record of 150 hr 40 min established by Major Carl Spatz and US Army crew in Fokker C-2A trimotor *Question Mark* using hose refuelling technique of 1923. Aerial refuelling did not assume military significance in the US until 1948, when used to extend the range of World War Two aircraft to global distances.

7 July 1929. Inauguration of transcontinental plane-train service by Transcontinental Air Transport and Pennsylvania Railroad, passengers flying by day and riding Pullman cars at night.

28-29 November 1929. First flight over the South Pole, by Admiral Richard E. Byrd in a civil Ford Tri-motor piloted by Bernt Balchen.

23 June to 1 July 1931. Wiley Post and Harold Gatty flew around the world, covering 16,500 miles in 8 days and 16 hr in the Lockheed Vega *Winnie Mae*. This flight was 11,000 miles shorter than that of the 1924 Army fliers, being made at more northerly latitudes.

8 February 1933. First flight of the twin-engined Boeing 247 (*q.v.*)

Construction of the Akron, *one of two big US Navy airships by Goodyear, named for the firm's headquarters city*

The Boeing 247D, pioneer metal monoplane transport with retractable landing gear

transport, a new design concept that made existing transports obsolete and ushered in a new era of transport design and operations.

22-29 November 1935. Inauguration of transPacific passenger flights by Pan American Airways with the Martin 130 flying boat *China Clipper*.

16 February 1936. Introduction into scheduled service of the Douglas DC-3, the world's most widely-used airliner. Many of the 10,926 built (10,123 as military types in World War Two) still serve the world's airlines.

1937. Joint US Army/Lockheed Aircraft Co experiments with a pressurized cabin on the Lockheed XC-35 pointed the way to the pressurized transports and bombers of the future and won the Collier Trophy for 1937.

16 April 1941. Igor Sikorsky set a national

helicopter record of 1 hr 5 min in the Vought-Sikorsky VS-300, establishing the single main rotor with stabilizing tail rotor as the most promising helicopter configuration.

1 October 1942. First American jet-propelled aeroplane, the Bell XP-59A (*q.v.*), flown by factory test pilot Robert M. Stanley at Muroc Dry Lake, California.

14 October 1947. The first successful supersonic flight, at a speed of Mach 1.05, made by Captain Charles E. Yeager in the rocket-powered Bell XS-1 research aircraft following air-drop from a Boeing B-29.

15 July 1954. First flight of Boeing Model 367-80, first American jet transport from which the Boeing 707/720 "family" is developed.

21 September 1964. First flight of the North American XB-70A prototype aircraft, designed to cruise at Mach 3 for

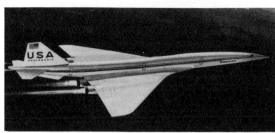
The fixed-geometry Boeing 2707-300, the future US supersonic transport

powered research aircraft attained a speed of 4,250 mph—fastest ever by a piloted aeroplane. The same aircraft has reached a record height of 354,200 ft.

prolonged periods.

21 December 1964. First flight of the General Dynamics F-111A, first variable-geometry (swing-wing) aircraft to enter production.

18 November 1966. North American X-15A-2 rocket-

and Italian aircraft were purchased for the V-VS in the early nineteen-twenties; the de Havilland D.H.9A was built under licence, and Russian-designed aircraft began to emerge

USSR: Air Forces

Sovietskaya Voenno-Vozdushnye Sily (Soviet Military Aviation Forces). An Army Aviation School was established in the Russia of Tsar Nicholas II in 1910, several French and British aircraft being purchased in 1912. By the outbreak of the 1914-18 War the Imperial Air Service had over 200 aircraft, many of them licence-built in Russia. To these were added in the next few years several indigenous designs, notably the large Sikorsky *Ilya Mourometz* (*q.v.*) four-engined bombers, as well as further quantities of French and British types. Following the November 1917 Revolution a new air force, the Red Air Fleet,

was created in May 1918, which adopted the present V-VS title in 1924. An important research establishment, the TsAGI (Central Aero-Hydrodynamic Institute) was also founded in 1918.

Initial supplies of Dutch

from the now-nationalized aviation industry. Junkers of Germany opened a factory near Moscow where, in later years, A. N. Tupolev (*q.v.*) first came into prominence as a designer; and fighter designs were produced by Grigorovich and Polikarpov. Under the first 5-Year Plan, which began in 1928, mass-produced Russian aircraft included the twin-engined Tupolev ANT-4 (TB-1) bomber, a variant of which made a Moscow-New York flight in 1929; and the I-4 and I-5 fighters and U-2 general-purpose biplane designed by Polikarpov. Further reorganization and expansion of both Soviet civil and military aviation was introduced with the second 5-Year Plan, and during this period one of the most significant types was the multi-purpose, four-engined Tupolev ANT-6, built in substantial numbers as a bomber and paratroop transport. Interest in large multi-engined aircraft waned after the loss of the Kalinin K-7 in 1933 and the ANT-20 *Maxim Gorkii* in 1935, both with heavy losses of life. In the middle 'thirties, in common with their foreign contemporaries, Soviet designers were attracted to the concept of the fast, twin-engined medium bomber monoplane, and there

First pressurized transport in service was the Boeing 307 Stratoliner of TWA

Anti-submarine hunter-killer team of carrier and Goodyear airship, 1950

Tupolev TB-3 (ANT-6), standard four-engined bomber of 1933-7 and much used for parachute training

emerged the Tupolev SB-2 and Ilyushin DB-3 which became standard V-VS equipment before and during World War Two. The SB-2 and several other Soviet types, including Polikarpov's excellent little I-15 and I-16 single-seat fighters, obtained a taste of combat service against the Nationalist forces in the Spanish Civil War of 1936-39, while in the eastern USSR other V-VS units were engaged in a bitter fight against Japan in Manchukuo and Mongolia in 1938-39. These engagements, plus a series of internal Army purges in the late 'thirties, left the V-VS seriously weakened before the outbreak of World War Two, and modern replacements such as the first MiG and Yakovlev fighters, and the Il-2 and Su-2 close-support aircraft, did not become available in quantity until much later.

During the early campaigns against Finland and Germany, therefore, the Soviet Air Force, although numerically strong, lacked up-to-date equipment. Stupendous increases in home production were made—from 500 aircraft a month at the end of 1941 to nearly 2,000 a month by the following summer—and effective new combat aircraft at last became available in quantity. Among these were the MiG-3, La-5 and Yak-1 series of fighters, Il-2 and Pe-2 attack aircraft and Pe-8 heavy bombers. To these were

added several thousand warplanes supplied under Lend-Lease: Bell P-39s and P-63s, Curtiss P-40s, Douglas A-20s, North American B-25s and many others from the US; and Hurricanes, Mosquitos and Spitfires from the UK. With their strength thus augmented, the Soviet forces were able to turn back the Nazi advance in 1942-43 and keep up a sustained pressure against rapidly-diminishing *Luftwaffe* resistance until the war ended. By then, later Soviet types in service included the La-7 and Yak-9 fighters and the Tu-2 light attack bomber.

After the war the USSR acquired the services of several leading German aircraft and aero engine designers, enabling it to lay the foundations of gas-turbine development in the country. Several experimental jet-powered aircraft were produced, but not until Klimov (*q.v.*) developed his version of the Rolls-Royce Nene turbojet was mass production undertaken of an indigenous jet fighter. This fighter, the MiG-15, was to prove one of the combat surprises of the Korean War in 1950-53. Development of jet bombers proceeded in parallel, and again, after the emergence of several interim and experimental types, a really successful and widely-built design appeared in the Ilyushin Il-28. Like the MiG-15, the Il-28 is still in diminishing service today with Soviet *bloc* air forces around the

world.

During the 'fifties and 'sixties, the MiG design bureau has been primarily responsible for the V-VS's subsequent fighter types, current examples being the MiG-21 and the MiG-23. Exceptions include the twin-jet Yakovlev Yak-25 all-weather fighter/light bomber series, and the comparatively recent re-emergence of Pavel Sukhoi

In 1961 Yakovlev revealed this much improved version of the Yak-25 known to the West as "Firebar"

hoi as designer of the fighter/ground attack Su-7 and Su-9. Medium and heavy bombers, apart from Myasishchev's four-jet "Bison", have stemmed from the Tupolev team, and include the twin-jet Tu-16, the four-turboprop Tu-20 and the supersonic Tu-22 "Blinder", all of which are currently in service.

Aviatsiya-Voenno Morkskikh Flota (Naval Air Fleet). Naval aviation in Russia also dates from 1910, and in 1912 Curtiss flying boats were acquired from the US. The Imperial Navy had nearly 100 aircraft by the outbreak of the 1914-18 War, among which were a number of Grigorovich flying boats of Russian design. After the 1917 Revolution Naval aviation became a branch of the V-VS (see above), with the title of Volga Military Flotilla. Development of the VMF through the 'twenties and early 'thirties was to a similar, but subordinate, pattern to that of Army aviation. Bomber, reconnaissance and transport variants of the Tupolev TB-1 entered service, as did the Heinkel-inspired Grigorovich MI-4 fighter floatplane. These

were joined, in the 'thirties, by the Tupolev MDR-2 and Beriev MBR-2 long- and short-range flying boats, and later the American PBY (Catalina) built under licence, and these were still the VMF's principal maritime types at the outbreak of World War Two.

Soviet Naval aviation was—and still is—primarily concerned with anti-shipping patrol in the waters around the USSR and for defensive protection of its Naval and coastal air bases. It has no carriers for fixed-wing aircraft and, apart from patrol flying boats, utilized similar front-line aircraft during 1939-45 to those of the V-VS. These were allocated to the Fleets operating in the Pacific, the Black Sea, the Baltic, and northern waters including the North Sea.

Since World War Two the VMF (now the A-VMF) has continued the practice of employing V-VS land-based types —currently typified by variants of the Tu-16 and Tu-20—but has in the past decade introduced some modern flying boat designs. First of these was the piston-engined Beriev Be-6, but this has now been superseded by the twin-turboprop Be-12 and the supersonic land-based Tu-22 for the maritime reconnaissance rôle.

USSR: Chronology

1884. First powered "hop" in Russia made by I. N. Golubev, on the twin-engined monoplane built by Alexander Mozhaisky, after gaining speed down a ramp. This was preceded only by the similar hop by du Temple's

aeroplane in France. 1909-10. Igor Sikorsky built in Russia his first two helicopters, neither of which flew.

13 May 1913. First flight of the world's first four-engined aeroplane the *Russkii Vitiaz* (Russian Knight) or *Le Grand*, designed and built by Igor Sikorsky.

8 August 1913. Endurance flight of 1 hr 54 min made by the Sikorsky *Le Grand*, carrying eight passengers and reaching an altitude of 2,723 ft.

27 August 1913. Petr Nesterov became the first pilot to loop the loop, in a Russian-built Nieuport monoplane.

January 1914. First flight of the Sikorsky *Ilya Mourometz* four-engined airliner—at that time the world's largest aeroplane.

29 June-11 July 1914. The *Ilya Mourometz* No 2 flew a round trip of 1,590 miles from St Petersburg to Kiev and back, carrying passengers, in $10\frac{1}{2}$ hr flying time, with only one stop for refuelling. During this journey a complete meal was prepared and served on board an aircraft for the first time in history.

1929 (?). First successful Soviet autogyro, the KaSkr-I, designed by N. Kamov and N. Skrzhinsky.

23 August 1929. S. Chestakov and crew of three made the first non-stop Moscow-New York flight in the ANT-4 monoplane *Strana Sovietov* (Land of the Soviets).

August 1930. Tethered tests of the first Soviet-designed and built helicopter, the TsAGI 1-EA, powered by two 120 hp M-2 rotary engines.

3 October 1933. Prokofiev and a crew of two made an ascent into the stratosphere in the balloon *U.S.S.R.*, reaching an altitude of 58,727 ft. All three were killed when attempting to better this achievement on 30 January 1934 in the balloon *Ossoaviakhim I*.

1934. S. Liapidevski and six other aircrew rescued the members of an Arctic expedition led by the explorer O. Schmidt, whose icebreaker, the *Chelyuskin*, had been trapped in the ice. All were made Heroes of the Soviet Union.

21 May 1937. Mikhail Vodopyanov landed a four-engined ANT-6 ski-plane within 20 km of the North Pole and helped to set up the first Polar scientific station.

18 June 1937. Non-stop flight from Moscow to Portland, Ohio, by Valery Chkalov and crew in an ANT-25 monoplane, over the North Pole.

12 July 1937. Mikhail Gromov and crew of two broke the world distance record by flying non-stop from Moscow to San Jacinto, California, a distance of 6,305 miles, in an ANT-25, via the North Pole.

30 December 1947. First flight of the first mass-produced Soviet jet fighter, the MiG-15,

"N-169", The Tupolev ANT-5 that flew to the Pole of Inaccessibility in the remote Arctic in 1941

More than 10,000 MiG-15 fighters were built, using a developed version of the Rolls-Royce Nene engine; they transformed the quality of Soviet first-line defence

powered by a Soviet-built version of the Rolls-Royce Nene turbojet engine.

3 July 1953. The first Soviet tandem-rotor helicopter, the Yak-24, made its first flight piloted by S. Brovtsev and Y. Milyutichev.

1955. The USSR's first level-supersonic fighter, the MiG-19, entered service.

15 September 1956. The Soviet Union's first jet airliner, the Tu-104 entered scheduled service, preceded only by the British Comet 1.

Autumn 1957. The world's largest helicopter, the Mil Mi-6, made its first flight, piloted

The impressive Mil Mi-6P in scheduled service with Aeroflot

by Rafail Kaprelian.

28 April 1961. World altitude record of 113,892 ft set up at Podmoskovnoe by Col Georgi Mossolov, in an E-66A powered by one 13,117-lb st turbojet engine and one 6,615-lb st rocket engine.

July 1961. First Soviet convertiplane, the Kamov Ka-22 *Vintokryl* ("screw-wing"), made its appearance at the Tushino air display.

UTA (Union de Transports Aériens)
This French overseas airline came into being late in 1963 as the result of a merger be-

tween the former Union Aéromaritime de Transport and Compagnie de Transports Aériens Intercontinentaux (both formed shortly after World War Two). The former routes of UAT lay across the African continent, while those of TAI, although serving parts of Africa, extended also to the Far East, Australia and New Zealand, and eastward across the Pacific to Los Angeles, where they connected with the Air France services to Paris. The present route network covers broadly the same territory, although new routes in some of these areas have been added since the formation of UTA. Headquarters: Paris. Member of IATA.

V-1, 294, see also Hawker Tempest
The first of Hitler's *Vergeltungswaffen* (reprisal weapons), this flying bomb stemmed from the Fieseler Fi 103. It was designed in 1942 and flown as a glider at the end of that year, the powered version flying soon after. It was powered by an Argus As 014 impulse duct, mounted above the rear section of a simple fuselage containing the 1,870-lb warhead and

the directional controls. Simple plank-like wings and tailplane were fitted, and a small fin and rudder between the fuselage and the jet engine.

Enormous efforts were made to prepare sufficient launching sites to swamp the south-east corner of England, but these plans were frustrated by intense and prolonged bombing by Allied bomber formations. Those that were launched, commencing in June 1944, were but a fraction of the force intended; even so they were of more than nuisance value to the civilian population of south-east England and diverted a considerable part of the fighter defence force of Great Britain. The offensive lasted for just over three months, by which time most of the launching sites had been overrun by the Allied armies. A continued offensive, on a small scale, was made possible by carrying the V-1s aloft on He 111s or Ju 88s and air-launching them over the North Sea. A few were also fitted up for piloted operation, but none were thus used operationally.

V-1a, see Vultee
Valiant, see Vickers
Valkyrie (XB-70), see North American
Valves, 1097
Any devices for controlling the flow of a fluid. In a piston-engine of conventional form the combustible mixture of fuel and air (or, in a diesel, plain air) is admitted into the cylinder through an inlet valve which in a normal four-stroke engine is

opened a few degrees before top dead centre, when the piston is almost at the top of its stroke, and closed some degrees after bottom dead centre. (The inertia of the incoming mixture enables the charge to continue to flow in even after the piston has started to rise.) The exhaust escapes through an exhaust valve which opens before bottom dead centre at the end of the firing stroke and closes after top dead centre, by which time the inlet valve is already open to admit the next charge. Most engines use poppet valves (*q.v.*) which are basically discs on the ends of rods sliding in a bearing hole in a curved intake port. The rod is kept in tension by a valve spring, usually two or more helical springs one inside the other, so that the valve is closed except when the rod is pushed down by the cam driven by the valve gear. Some engines use sleeve valves (*q.v.*), in which the piston slides in an open-ended sleeve fitted closely inside the cylinder and driven in an elliptical oscillating motion so that, at the

appropriate times in each cycle, inlet and exhaust ports in the sleeve are aligned with corresponding ports in the wall of the combustion chamber.
Vampire (D.H.100), see de Havilland
Vanguard, see Vickers
Variable geometry, 841, 1235
Term which has come into general use to cover items of structure which are able to be changed in configuration; specifically, in the aerodynamic field, structures which can change their configuration during flight. Variable geometry is prominent in two particular fields, one being that of aircraft such as the General Dynamics F-111 (p.245) whose wings have little sweepback for slow-speed flying but can be fully swept back in flight for supersonic efficiency. The principle on which this system is based was developed by Dr Barnes Wallis (*q.v.*) of Vickers after World War Two. The term is also applied to intake ducts of certain turbojet engines which have variable-geometry intakes to cater for the vastly-differing airflow problems at each end

of the speed range.
Variable-pitch propeller, 287, 1333
Variable stator engine, 1183, 1195
A gas-turbine engine in which the angle of incidence of one or more rows of the compressor stator blades may be varied while the engine is running. The stator blades, and the inlet guide vanes ahead of the first stage of rotor blades, guide the airflow so that it meets the next stage of rotor blades at the correct angle. With fixed stator blades a compressor designed to achieve a high pressure ratio may suffer from instability, surging and poor response to changed conditions. Performance and flexibility of control are greatly eased either by splitting the compressor into two or three sections, each able to choose its own rotational speed, or by using variable stators. An example of a variable stator engine is the General Electric J79, in which the inlet guide vanes and next seven stator rows have blades mounted in rotary bearings and all mechanically interlinked so that their incidence can

General Dynamics/Grumman F-111A demonstrating its ability to sweep its wings from 16° to a maximum supersonic angle of 72° while maintaining station with a KC-135 tanker

be controlled by an actuating ram fed with high-pressure engine fuel. The system is entirely automatic and responds without pilot command.

Varig

Operating name of the Brazilian airline *Empresa de Viacao Aerea Rio Grandense*, which was formed with German assistance in 1927. Originally an operator of internal services, Varig has absorbed a number of smaller Brazilian airlines during its growth, including the former REAL and Panair do Brasil during the nineteen-sixties. It now has an extensive route pattern throughout the whole of Central and South America, flies to all other continents, and in 1968 was planning to introduce a round-the-world service.
Headquarters:
Rio de Janeiro.
Member of IATA.

VASI (Visual Approach Slope Indicators), 1571

VASP (Viacão Aerea São Paulo)

Brazilian domestic airline which has operated scheduled services within the country since 1935, the year after its formation. Its present route network is over 16,000 miles long, and among its many activities is included the half-hourly "air bridge" between São Paulo and Rio de Janeiro, which it operates in collaboration with Cruzeiro and Varig (*qq.v.*).
Headquarters:
São Paulo.
Associate member of IATA.

VC10, see BAC

Vectored-thrust engine, see also Rolls-Royce (Bristol) Pegasus engine

Strictly speaking, any aero-engine having a thrust line which can be vectored in different directions falls into this category, and this definition would include all jet engines fitted with reversers. In practice the description is reserved for powerplants for V/STOL aircraft, and especially for turbojets and turbofans provided with deflectors so that their thrust vector may be rotated from rearward to downward, to provide lift to support the aircraft at times when forward speed is too low for wing lift to be adequate. Frequently the thrust vector may also be directed forward to brake the aircraft.

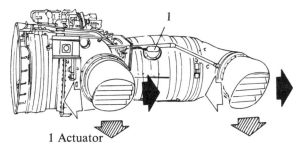

1 Actuator

Such power plants are also known as lift/cruise engines, since they serve the dual function of providing lift for the VTOL regime and forward propulsion in cruising flight. The first such engines were modified Rolls-Royce Derwent turbojets, but the first vectored-thrust unit designed as such was the Rolls-Royce (Bristol) Pegasus, in which the front fan and rear hot jet each discharge through left and right pairs of nozzles fitted with deflectors which turn the flow through approximately 90°. All four nozzles are mechanically linked and actuated by a motor and drive shafts to point in the desired direction. Arrangements in which the whole engine is pivoted, or in which an engine is fixed but drives a tilting propeller, do not properly qualify for this description.

Vega, see Lockheed

Vehicle ferries

As the main text indicates, most large military transport aircraft are capable of carrying standard Army vehicles and other wheeled equipment. On the commercial front, however, specialized aircraft for carrying motor cars or other vehicles have been comparatively little developed.
First aircraft sufficiently suited to this task to permit the introduction of regular car-ferry services was the Bristol Freighter (*q.v.*), with which Silver City Airways (now BUA, *q.v.*) introduced a regular UK-France service across the English Channel in the nineteen-fifties. Another, larger aircraft adapted for the purpose with some success is the Aviation Traders Carvair (*q.v.*) conversion of the Douglas DC-4 airliner.
In the late 'sixties, however, general indications were that the development of further specialized vehicle-carrying commercial air-

Carvair, the conversion of the DC-4 to carry cars as well as passengers without performance penalty

craft was unlikely; it seemed more probable that the car-ferrying vehicle of the immediate future would be the hovercraft.

Venezuela: Air Force

Fuerzas Aéreas Venezolanas (Venezuelan Air Force). Originated as the Military Air Service in 1920, being French-orientated as regards its initial equipment and training. A small naval air base was also established at about the same time. The original Caudron and Farman aircraft remained in service until 1936, when the MAS was expanded to the status of an Army regiment. New training aircraft were then acquired from the US, but Venezuela's first combat aircraft were Italian Fiat C.R.32 fighters and B.R.20 bombers.
During World War Two, with additional US aircraft received under Lend-Lease, Venezuelan units contributed to the air defence of the West Indies. Further US aircraft and other assistance continued after the war, and the Air Force adopted its present title. Its first jet aircraft were British Vampires and Canberras, ordered in 1950 and 1952, these being followed by American Sabres in 1955 and additional Canberras in the late 'fifties. These two types are still in service, together with the Venom fighter-bombers which replaced the original Vampires.

Vertical speed indicator, 675, 679

Originally more often called the "rate of climb" or "climb and descent" indicator, the VSI is calibrated in hundreds of feet per minute or in metres per second and is an essential instrument on all except light aircraft. The usual method of operation is for an aneroid capsule (a thin-walled convoluted drum of springy metal exactly like that found in an aneroid barometer) to be connected to a container of air at constant temperature. The latter is virtually a Thermos flask, and it is provided with a slow leak to the atmosphere.
Air enters or leaves the container through this leak, so that at steady height the pressures inside and outside the system are equal and the instrument reads zero. If the aircraft then dives, it enters air of steadily increasing pressure which acts on the outside of the instrument aneroid capsule, compresses it and causes an accurate reading of the rate of descent. At the same time air begins to leak into the main storage container, and when the aircraft has levelled off this equalizes the pressures. Climbing causes the reverse process; a reading is given according to the pressure difference inside and outside the system and the pressures equalize through the leak as the rate of climb falls to zero.

Verville-Sperry R3, 567

Two low-wing mono-

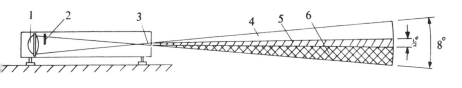

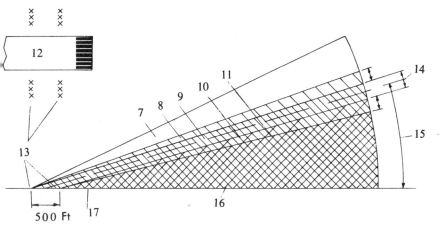

1 Light source	7 White/white*	13 Indicators
2 Red filter	8 Pink/white*	14 25 + R250 feet
3 Slot	9 Red/white*	15 Nominal glide path
4 White*	10 Red/pink*	angle
5 Pink*	11 Red/red*	16 Ground level
6 Red*	12 Plan view	17 Runway threshold

* Or lighting system based on similar colour patterns.

plane racers were built in 1922 by the Sperry company to the design of Alfred Verville. These aircraft participated in the Pulitzer Race at Detroit and finished 5th and 7th. One aircraft was redesigned in 1923 and re-engined with a 500-hp Curtiss D-12A. This aircraft was entered for the Pulitzer Race in St Louis, but withdrawn after starting through excessive instability. The R3 was the first machine with the landing gear retracting into the wings.

VHF, 705
Accepted abbreviation for Very High Frequency, referring to the radio wavebands used by most aircraft radio stations today. It ties in with the frequencies used by the VOR radio navigational aids.

VIASA (Venezolana Internacional de Aviacion SA)
Venezuelan international airline whose creation in 1961 was at Government instigation. It flies the international services formerly operated by LAV and Avensa (which are now purely domestic carriers); these two airlines are among the private interests which hold 45 per cent of the shares in VIASA, the other 55 per cent being held by the Government. Technical and other material assistance is provided by KLM (q.v.). VIASA flies long-range jet services to other South and Central American cities and to the eastern United States; its services to Western European capitals are integrated with those flown by Alitalia, Iberia and KLM (qq.v.).
Headquarters: Caracas.
Member of IATA.

Vickers
An Aviation Department of Vickers Ltd was first formed, under Captain H. F. Wood, in 1911, when manufacturing rights were acquired in the French

R.E.P. monoplane, several variants of which appeared before the 1914-18 War. To an Admiralty requirement of 1912 Vickers produced the Type 18 Destroyer pusher-engined gun-carrying fighter, from which was evolved, via successive E.F.B. (Experimental Fighting Biplane) prototypes, the F.B.5 and F.B.9 "Gunbus" fighters of the war period. A later front-gunned fighter, the tractor-engined F.B.19, was less successful, but in the Vimy bomber of 1917 Vickers produced a type that was to remain a standard RAF bomber throughout the nineteen-twenties. It was in a modified Vimy that John Alcock and Arthur Whitten-Brown (qq.v.) made the first non-stop Atlantic crossing by air in 1919. The Vimy Commercial was an airliner variant with an enlarged fuselage seating 11 passengers, and from this in turn was developed the Vernon troop transport for the RAF. The Vimy and Vernon were succeeded respectively by the Virginia and the Victoria in the middle 'twenties. In July 1928 Vickers' Aviation Department became Vickers (Aviation) Ltd, and five months later took over control of the Super-

marine Aviation Works Ltd at Southampton. Supermarine had originated as constructor of the wartime Admiralty AD flying boat in 1916 and, under its subsequent titles of Pemberton-Billing Ltd and Supermarine, had specialized in the design and manufacture of marine aircraft. Early

products after its absorption by Vickers included the Seagull and Southampton flying boats. Also, in the 'twenties, came the S.4, S.5 and S.6 series of Schneider Trophy seaplanes designed by R. J. Mitchell (q.v.).
Vickers products in the early 'thirties continued with the Vildebeest torpedo-bomber and the Vincent general-purpose biplane. In June 1935 came the first flight of the long-range Wellesley, first RAF aircraft to employ the then-revolutionary geodetic construction (q.v.) designed by Sir (then Dr) Barnes Wallis. It was followed nine months later by the prototype of Mitchell's supreme achievement, the Supermarine Spitfire (q.v.), and in June 1936 by the first Vickers Wellington bomber, another geodetic design destined to achieve great success during World War Two. In October 1938 both Vickers (Aircraft) and Supermarine became divisions of Vickers-Armstrongs Ltd, although their individual titles continued to be perpetuated in their aircraft designs.
During World War Two production of the Wellington and the multitudinous Spitfire/Seafire variants accounted for

Vickers Virginias usually had Napier Lion engines but this mark had Bristol Jupiter radials

most of the company's productive effort, continued manufacture of another successful design—the Supermarine Walrus amphibian—having been transferred to Saunders-Roe (q.v.). The Warwick was evolved to succeed the Wellington, and in August 1946 came the first flight of the VC 1

Viking, the first British post-war civil transport to enter service. Less than a fortnight earlier, at the end of July, the company's first jet aircraft, the Supermarine Attacker, had flown for the first time; it was followed in 1951 by the first swept-wing British fighter, the Swift, and in 1954 by Super-

marine's last product, the twin-jet Scimitar naval fighter.
Meanwhile, at Vickers the first British V-bomber (the Valiant) and the world's first turboprop airliner to enter service (the Viscount 700) made their maiden flights in 1950 and 1951; and the Vanguard airliner followed in 1959. In February 1960, Vickers merged with the Bristol Aeroplane Co and the English Electric Co to form the British Aircraft Corporation and is now the Weybridge Division of BAC (q.v.). It is responsible for the civil aircraft designs of BAC (except for the Concorde) and is engaged currently in production of the VC10 and BAC One-Eleven series of jet airliners.
Vickers "Gunbus", 247, 815, 873
The Vickers F.B.5, the first production "Gunbus", had its origins in the Vickers Type 18 Destroyer first exhibited at the Olympia Aero Show of 1910. This was a two-seat, front-gunned "pusher" biplane, and from it were developed a number of E.F.B. (Experimental Fighting Biplane) prototypes which entered production as the F.B.5 in

1914-15. The "pusher" configuration was adopted in this, as in the contemporary D.H.2 and F.E.2b, due to the absence at the time of a satisfactory British gun interrupter gear. The F.B.5 was followed into production by a slightly improved version, the F.B.9, but the eventual arrival of satisfac-

tory gun-synchronizing mechanism rendered large-scale production of both types unnecessary. In recent years the Weybridge Division of BAC (formerly Vickers) built a flying replica of the F.B.5, which in June 1968 was presented to

Vanguard Series 951

the RAF Museum.
Vickers Valiant, 835
The Valiant was the first of the RAF's V-bombers to enter service, in July 1955. It was built to an earlier design than the Victor and the Vulcan, providing the first effective striking force for the RAF's nuclear deterrent. It was a Valiant which was used to drop the British atomic and hydrogen bombs over the Pacific in 1956 and 1957 respectively; others were used for the development of the Blue Steel stand-off bomb carried by the Victor and Vulcan. Eight bomber squadrons of Valiants were in service at one period, although with the entry of the more advanced V-bombers the Valiant was transferred to the tanker rôle and provided the RAF's tanker force for many years. The Valiant's operational career was cut short in 1964 when wide-spread fatigue was

discovered in the wing spars.
Powered by four Rolls-Royce Avon 201 turbojets of 10,000 lb thrust each, the Valiant had a maximum speed of Mach 0.84 above 30,000 ft, a range of 3,450 miles and a service ceiling of 54,000 ft.
Vickers Vanguard, 759, 1285-1286, 1335
Conceived as a second-generation turboprop airliner to succeed the Viscount, the Vanguard appeared in 1958 and was ordered by BEA and Trans-Canada Air Lines. It was powered by four Rolls-Royce Tyne turboprops of 5,545 ehp each, a new engine which brought with it some teething troubles. These were successfully overcome and the aircraft now serves as a 139-seater for stage lengths up to 2,340 miles. In the event, no large requirement for this type of airliner materialized, as the airlines found the economics of turbojet aircraft more favourable than anticipated.
Vickers Vimy, 253, 255, 813, 815
Developed as a contemporary of the Handley Page bombers of the 1914-18 War, the Vimy, built by Vickers at Brooklands, was not ready in time for service in that war. This was due partly to an unsatisfactory original choice of engines. Many different power plants were utilized, but it was with the Rolls-Royce Eagle that the Vimy suceeded. Its niche in history is assured as a result of the non-stop transatlantic flight by Alcock and Brown in June 1919, but other record flights were flown with Vimys, such as the flight by Ross and Keith Smith from England to Australia in November 1919, and the England-Cape Town flight by van

Jet-to-jet refuelling by RAF Valiant tanker and Vulcan bomber in 1957

Ryneveld and Quintin Brand in February 1920. The Vimy saw service as a heavy bomber with the RAF until 1926 but, more important, was the forerunner of a family of large twin-engined Vickers biplanes: the Virginia, Vernon, Victoria and Valentia, which saw considerable service with the RAF up to and during World War Two.

Vickers Viscount, 284, 307, 308, 1309

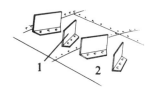

Viscount Series 800

Conceived during the later stages of World War Two as a result of the Brabazon Committee formed to postulate suitable civil types for development. It was designed to accommodate any of three turbo-prop engines being developed, the choice eventually falling upon the Rolls-Royce Dart. The first prototype flew on 16 July 1948. This was the forerunner of the Series 700 first ordered by British European Airways, who put it into service in 1953. The aircraft soon found acceptance with the airlines and orders poured in from all over the world. Trans-Canada Air Lines' order in November 1952 introduced the aircraft to the North American continent, and was followed by large orders from Capital Airlines. A longer-fuselage version, the Series 800, was developed for BEA and ushered in a new range of Viscounts for various airlines, including the heavier, more powerful Series 810. When production ceased in 1964, 444 aircraft had been built.

Vickers Wellesley, 399

Vickers Wellington, 399ff, 1244

The Vickers Wellington was the mainstay of RAF Bomber Command in the first years of World War Two. It

employed Dr Barnes Wallis's form of geodetic construction (*q.v.*) which enabled many badly-crippled Wellingtons to return successfully from raids over Europe. The type went into production at the end of 1937, though by the outbreak of war only six operational squadrons were equipped. Soon, however, the whole of No 3 Group Bomber Command had Wellingtons; they went into action the day after war broke out, with a raid on warships at Brunsbüttel. Other daylight raids followed, but the Wellington's main achievement was in the night offensive. The Wellington was the most successful of the medium bombers with the RAF at the beginning of the war and in due course was used in other Groups. It also served in the Middle East theatre, forming a long-range bombing group in Egypt in 1940.

The Mk II version was equipped with Rolls-Royce Merlin X engines, and the Mk III with Bristol Hercules, in place of the Bristol Pegasus. The Mk III took over the brunt of Wellington operations by 1942, by which time the type was also going into service with Coastal Command on anti-submarine duties, using the Leigh Light for illuminating submarines on the surface. Wellingtons also served in large numbers as operational trainers, and almost all Bomber Command's crews in the latter half of the war came together as a crew at a Wellington-equipped Operational Training Unit. In this and other specialized training rôles the Wellington survived the war

and remained in service until 1953. It remained in production throughout the war, 11,461 being built in all.

Vickers machine-gun, 787
This machine-gun was one of the infantry weapons used greatly in the 1914-18 War. In 1916 Vickers began experimenting with a synchronizing gear for this gun, and when this was perfected the gun became the standard forward-firing weapon for British aircraft. It continued to serve in most British fighter aircraft until the mid-nineteen-thirties, when its place was taken by the Browning.

Vickers pom-pom gun, 883

Vickers-Armstrongs 37-mm gun, 872

Victor, see Handley Page

Vietnam: Air Forces
North Vietnam: People's Air Force. Created after the partition of Vietnam (formerly French Indo-China) into Northern and Southern zones in 1954. Since 1955 it has been organized and equipped chiefly by the USSR, possibly with assistance on a much smaller scale from the other Communist states which recognized the Ho Chi Minh regime. Defence assistance comprises all kinds of military equipment including missiles, and the People's Air Force includes in its inventory Soviet MiG-17 and missile-armed MiG-21 fighters and Ilyushin Il-28 light bombers.
South Vietnam: Vietnamese Air Force. An air force in South Vietnam was formed under French supervision in 1951, the first combat unit being formed some two years later. After the fall of Dien Bien Phu in mid-1955 the present VAF was created on a more modern basis with advice and

assistance from the US, which began to supply American military aircraft to replace the older French types in service. The first new combat unit, with Grumman Bearcat fighter-bombers, was formed in mid-1956. These were followed in the late 'fifties and early 'sixties by T-33A jet trainers and Douglas Skyraider strike aircraft. The latter are still in service, together with Martin B-57B (Canberra) bombers and, more recently, Northrop F-5 supersonic strike fighters.

Vigilante (A-5), see North American

Vimy, see Vickers

Vinay, Louis, 573

Vintokryl, see Kamov Ka-22

Viper engine, see Rolls-Royce (Bristol)

Viscount, see Vickers

Visual Approach Slope Indicator, see VASI

VJ 101C, 1233
The VJ 101C was built by a consortium of German manufacturers as an experimental airframe to explore VTOL flight prior to the development of a supersonic VTOL interceptor. Two aircraft were built, the first beginning hovering trials in April 1963, and the first transition took place in September 1963. The aircraft itself comprises a long, slender single-seat fuselage containing two vertically-mounted RB.145 lift engines. A high wing, spanning 21 ft, carries two rotating lift pods at the extremities, each pod containing two RB.145 engines. These pods could be used for vertical lift and, rotated slowly, provide transition to horizontal flight.
Two VJ 101C's were built, the first one crashing in September 1964. It reached a maximum speed of Mach 1.08 during trials.

Voisin brothers, 188, 209, 217, 229
Gabriel Voisin, joining with another French-

man, Archdeacon, designed and built a float-plane glider in 1905. Its conception was a great advance, in that it combined ideas drawn from the Wright Flyer with the box-kite principles of Lawrence Hargrave. Gabriel was soon joined by his brother Charles in his aeronautical ventures. Together they became among the first professional aeroplane manufacturers, and many of the early pioneers, including Blériot, bought their aircraft. It was not until 1907 that the Voisin aircraft became successful. One of their customers, Henry Farman (*q.v.*), himself came to the fore as an aircraft constructor, his early aircraft improving considerably over the standard Voisins.

von Guericke, Otto, 119

von Ohain, Pabst, 295, 1135, 1191
Pabst von Ohain, a German technician, worked with Dr Ernst Heinkel to build a 1,100-lb-thrust jet engine. Known as the He S.3B, this had an axial inducer, centrifugal compressor, reverse flow combustion chambers and a single-stage turbine. The engine was fitted in the He 178 which, on 27 August 1939, became the first jet-propelled aeroplane to fly.

von Opel, Fritz, 1137

VOR (VHF Omni-Range), 696, 705, 1373, 1523
A radio navigation system developed in the United States in the late nineteen-forties and, by 1955, used in almost every country. It is based upon a network of ground transmitter stations known as VOR beacons sited at strategically useful places—usually on airways. The unmanned station is usually of circular plan and has a central aerial which, day and night, sends out a continuous radio signal in all directions. The character of the

signal varies through the complete 360° from North right round back to North again, and when it is received by an aircraft within range of the station an indicator in the cockpit responds to the variation and shows the bearing of the station as seen from the aircraft. Flying past a station abeam of the aircraft the VOR dial may read successively 124°, 125°, 126°, changing by a degree every second or so; flying straight towards a VOR, along a "VOR radial", the dial may maintain a steady 078° (or whatever the bearing is). VOR cannot give a "fix", only a position line. For a fix, either a second VOR must be brought in or DME must be used (*q.v.*).

Vortex generators, 615

1. Vortex generators
2. Surface skin

Sometimes known as "turbulators", these consist of small flat metal plates, mounted perpendicularly, to improve the flow of air across a surface. The surfaces are set at an angle to a sluggish airflow, generally in transverse rows, to disturb the drag-inducing boundary layer, and thus improve control, reduce drag or raise a speed limitation.

Vostok I, 329, 330
This historic space-ship satellite launched by the USSR on 12 April 1961 was the first vehicle to carry a human being in space. The actual satellite comprised a pilot's capsule and landing system together with a separate retro-rocket and instrumentation section, the whole weighing 4,725 kg. It was launched by a multi-stage booster with six engines totalling 20,000,000 hp.

VS-300, see Sikorsky

VTOL, 646, 647, 954-1026, 1201, 1223ff
Abbreviation for Vertical Take-Off and Landing, used to cover a class of aircraft which, beginning with the helicopter and Autogiro, is now developing apace in many fields and many shapes. If methods can be found to lift an aircraft on and off the ground vertically the economics and safety of aircraft operation will benefit greatly. Broadly speaking there are three fields in which this aim has been explored for many years and the helicopter is now a practical and useful vehicle. The second field is that of jet lift, whereby conventional aircraft are fitted with lift engines which take over for the landing and take-off, enabling the aircraft to behave normally once airborne; an alternative in this field is to have jet engines with swivelling nozzles doing both tasks. Most advanced aircraft in this field is the Hawker Siddeley Harrier in service with the Royal Air Force as a strike fighter. The third field is in the tilt-wing or tilt-rotor concept in which the whole wing (or, in the case of a helicopter, the rotor installation) is tilted from the vertical to the horizontal according to the mode of flight.

Vuia, Trajan, 529
Trajan Vuia was a Hungarian domiciled in France who was one of many caught up with the desire to build a successful flying machine during the first decade of this century. He built an aircraft employing two revolutionary ideas in 1906; one was to use a motor powered by carbonic acid, the other to use pneumatic tyres. With this aircraft he attained a limited success in that year.

Vulcan, see Hawker Siddeley

Vulcan guns, 885, 887

Vultee V-1a transport monoplane, 567, 569
The V-1a was designed by Gerald Vultee, previously Chief Engineer of the Lockheed Aircraft Corporation; it appeared in 1933 and set a new trend in fast commercial aircraft with its sleek monoplane form and retractable undercarriage. It entered service with American Airlines.

Wagner, H. A., 369

Wakefield, Captain E. W., 222

Walleye missile, see Martin Marietta

Wallis, Sir Barnes N., 399
Sir Barnes Wallis has served as a designer and inventor for forty years. Most of this time he has been associated with Vickers Ltd, to which company he went after leaving the Airship Guarantee Co. For the latter he had devised a constructional method, used in the airship R.100, developed later into the geodetic (q.v.) form of construction used in the Vickers Wellesley and Wellington bombers.
He was also responsible for design of the weapons used by the Lancasters of 617 Squadron to breach the Moehne and Eder dams in 1943 and the large "Grand Slam" (q.v.) bombs used by Bomber Command toward the

end of the war.
Since World War Two he has become best known for his original proposals for "swing-wing" or variable-geometry aircraft, in his capacity as head of research for Vickers-Armstrongs (Aircraft) and later British Aircraft Corporation.

Walter rocket motors, 1137
Walter rocket motors were first used to power

the Heinkel He 176 rocket aircraft in June 1939, this particular motor developing 1,100 lb thrust. It used hydrogen peroxide, with potassium or calcium permanganate as a catalyst.
A development of this motor, the HWK R.II, was used to power the early Me 163, but the engine was found to be too dangerous and had to be redesigned to use hydrazine hydrate and methyl alcohol as a catalyst. In this form it went into production, giving thrust values up to 3,750 lb, and was used in the Me 163. Further development took the thrust level up to 4,400 lb.

Warhawk (P-40), see Curtiss

Warping, see Wing-warping

Washington Airport (Dulles International), Washington DC, USA

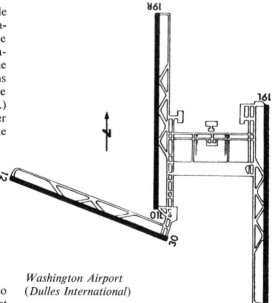

Washington Airport (Dulles International)

Dulles International was officially opened in November 1962, having been brought into being to relieve the hard-pressed National Airport which had handled most of the city's traffic since the mid-'thirties. Today, a

passenger booked to Washington is liable to be landed at any one of the undermentioned airports in the area. Dulles has the dubious distinction of being the world's most expensive, in terms of building costs, for an entirely new airport. As at John F. Kennedy Airport, New York, an outstanding feature is the terminal architecture, designed by Eero Saarinen.
Height above sea level: 313 ft
Main runways:
01L/19R = 11,500 ft
01R/19L = 11,500 ft
12/30 = 10,000 ft
Passenger movements in 1967: 1,333,000

Washington Airport (Andrews Air Force Base), Washington DC, USA
Height above sea level: 279 ft
Main runways:
01R/19L = 9,755 ft
01L/19R = 9,300 ft

10/28 = 9,450 ft
15/33 = 9,450 ft
04/22 = 6,000 ft
Passenger movements in 1967: 2,470,000

Washington National Airport, Washington DC, USA
Distance from city centre: 5 miles
Height above sea level: 15 ft
Main runways:
18/36 = 6,870 ft
15/33 = 5,212 ft
03/21 = 4,724 ft
Passenger movements in 1967: 8,497,000

Wasp engines, see Pratt & Whitney

Washington National Airport

Washington Airport (Andrews Air Force Base)

Wasp helicopter, see Westland

Weapon system
Modern combat aircraft are such complicated, close-knit creations that it is no longer appropriate to think of them as just an aeroplane into which are inserted a number of military devices. For their design to be efficient the approach has to be different. First, the overall mission to be accomplished is laid out carefully. Then the equipment needed to do the job is considered in the greatest detail. Finally, the vehicle in which to carry the equipment is developed. If it needs a crew it emerges as an aeroplane; if a crew is better left on the ground the result is a guided weapon.
Then the weapon system

Washington: Baltimore Airport (Friendship International), Maryland, USA
Distance from Baltimore city centre: 6 miles
Height above sea level: 146 ft
Main runways:

is painstakingly created by considering not only all the obvious systems, such as guns, air-launched missiles, cameras, radars, hydraulics, electrics, crew life support and emergency systems, propulsion, fuel and instruments, but also all the less obvious factors: the engineering tools, instructional manuals, even the trestles and special road vehicles needed to carry parts about, and such supplementary gear as the packing case needed for the delivery of the flight simulator. Everything must be designed with everything else in mind; there must be no afterthoughts or incompatibility. For example, it would be too late to discover, after the aircraft is built, that the side-looking radar pack

*Weapon system
(BAC Lightning)*

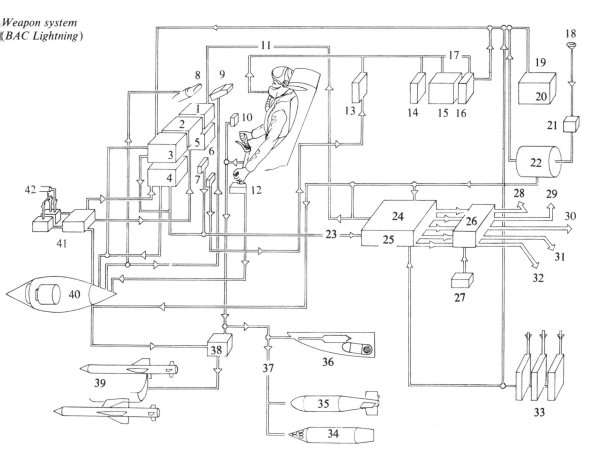

1 Attitude indicator
2 Standby instruments
3 Navigation display
4 Height display
5 Strip speed display
6 Auto-pilot control
7 IFF control
8 Radar CRT display
9 Light Fighter sight
10 Weapon selector
11 Flight director and
    MRG signals
12 Radar control

13 IFF
14 Standby
15 UHF main
16 Homer
17 Communications
    radio
18 Magnetic flux
    detector
19 Navigation radio
20 TACAN
21 Compass amplifier
22 Master reference
    gyro

23 Pilot demand
    data
24 Flight control
    computer
25 Autopilot
26 Autostabilizer
27 Three-axis rate gyro
28 Aileron
29 Rudder
30 Tailplane
31 Aileron
32 Throttle
33 ILS radio receiver

34 SNEB rockets
35 1,000-lb bombs
36 Two Aden guns in
    ventral fuel pack
37 Under- and
    over-wing stores
38 Weapon launching
    unit
39 Missiles or recce pack
40 Search and track
    radar
41 Air data computer
42 Pitot and pitot static

cannot be removed from beneath the fuselage if a refuelling pod is in place under the starboard wing. The accompanying sketch shows just the flying part of a modern weapon system. The subject is the BAC Lightning multi-purpose interceptor, strike and reconnaissance aircraft. The various devices give the pilot a wide range of options in both offensive and defensive situations. For example, he can search the sky ahead by radar, automatically lock the aircraft on to a hostile target, approach it under computer-generated steering guidance and attack with infra-red homing pursuit-course
missiles (Firestreak) or collision-course missiles (Red Top), spin-stabilized rockets or 30 mm guns. All information is fed into the airborne computers, which determine the best tactics for any situation and display the answers to keep the pilot fully informed about what is going on.

Weather conditions, see also WMO

Weather conditions, and especially visibility conditions, change so rapidly and vary so enormously from one part of the globe to another that for the purpose of air travel the ICAO (q.v.) has laid down categories of weather for aircraft

landings. Fundamentally, these categories relate to the height at which a pilot must decide whether or not he can see enough of the runway before him to attempt a safe landing. In practice, this means that commercial aircraft not equipped for automatic landing (q.v.) can operate only down to Category 2 weather conditions. The five categories in current use are as follows:

Category 1. Operation down to a decision height of 200 ft (60 m) with a visibility of more than 2,600 ft (800 m).

Category 2. Operation down to decision heights between 200 ft

and 100 ft (60 m and 30 m) with visibility between 2,600 ft and 1,200 ft (800 m and 400 m).

Category 3A. Operation to and along the surface of the runway with external visibility during the final phase of the landing down to 700 ft (200 m).

Category 3B. Operation to and along the surface of the runway and taxiways down to a visibility of 150 ft (50 m) which is only sufficient for visual taxying. Up to the end of 1968 this represented the maximum clearance yet given to aircraft with automatic landing equipment.

Category 3C. Operation

to and along the surface of the runway without external visibility.

Weir helicopter, 957

This was a twin-rotor helicopter built in Britain in 1938 which flew moderately well. To overcome the torque-reaction problem the two rotors were mounted side-by-side on outriggers from the fuselage. This entailed a complicated mechanical transmission system and as a result the machine was overweight and the lifting performance poor.

Wellesley, see Vickers

Wellington, see Vickers

Wessex helicopter, see Westland

Western Air Lines

Western makes a strong claim to be the oldest scheduled airline in the US, having originated as Western Air Express in 1925. Today it is one of the major regional trunk carriers, having nearly 10,000 miles of routes covering twelve of the western states and extending into Canada and Mexico. Services into Alaska were added after the acquisition of the former Pacific Northern Airlines in mid-1967. Headquarters: Los Angeles, California.

Westland

This famous British manufacturing company was founded in 1915 as the Westland Aircraft Works of Petters Ltd and was engaged in large-scale production of several important British combat aircraft during the

1914-18 War. During the nineteen-twenties it produced a variety of civil and military designs, among which the most notable were the Limousine commercial transports and the Widgeon lightplanes. The first type to go into really large-scale production was the Wapiti general-purpose biplane; over 500 were built for the RAF and many others for sale to Australia, China and South Africa. During the thirties its designs included the experimental Pterodactyl series, the Wessex light airliner and a number of Cierva-type Autogiros. The company was renamed Westland Aircraft Ltd in July 1935, and in the following year there appeared the prototype of one of the most celebrated Westland types, the Lysander Army Co-operation monoplane.

The Lysander was used for artillery spotting and air/sea rescue in World War Two, and its outstanding STOL qualities led to its use for many "cloak-and-dagger" flights into enemy-held territory to land or pick up Allied agents. The twin-engined Whirlwind and Welkin fighters were rather less successful, but during the late 'forties Westland built the turboprop-engined Wyvern strike fighter in quantity for the Royal Navy. This was the last fixed-wing type to be produced by the company, for the acquisi-

The STOL Westland Lysander was used in 1942-45 for landing agents in German occupied territory

tion in 1947 of a licence to build the Sikorsky S-51 helicopter for the Royal Navy marked the first step towards the eventual establishment of Westland as the major British helicopter constructor, and one of the largest in the world outside the USA and the USSR.

Westland followed the S-51 with the larger S-55, and then in 1957 undertook conversion of the Sikorsky S-58 helicopter to turbine power, the resulting aircraft becoming known as the Wessex. Westland acquired the former Saunders-Roe company in 1959, and in the following year absorbed Fairey Aviation and the helicopter division of Bristol Aircraft Ltd. Since then it has been engaged mainly in production of the Scout, Wasp, Wessex and Gnome-powered Whirlwind helicopters, most of which is now complete. It is currently engaged, in collaboration with Sud-Aviation (*q.v.*) of France, in Anglo-French development and production of the SA 330, SA 341 and WG 13 range of helicopters; and in developing a number of projects for tilting-rotor transports.

Westland F.7./30, 815
Westland Limousine, 769-771

Introduced by Westland in 1919, the idea behind the Westland Limousine was to provide a cabin with motor-car comfort. The cabin was a plywood box in

Westland Wasp ASW helicopter on board a frigate

the centre of the fuselage with room for three passengers and the pilot, the latter sitting in the left rear seat, 2½ ft higher than the passengers, with his head through a hole in the roof to see where and how he was flying. The aircraft was later converted into a Mk II with different power plant (it was a single-engined biplane) and after three such aircraft had been built a developed aircraft, the Limousine III (*q.v.*) was produced.

Westland Limousine III, 771

The Limousine III was built by Westland at Yeovil as a development of its earlier Limousines. It appeared in 1920 and had a large, enclosed cabin for five passengers and a pilot. Only two such aircraft were built, owing to the almost complete lack of demand for civil aircraft at the time.

Westland Scout, 583

Developed by Westland Aircraft from an original design by Saunders-Roe, the Scout is the standard five-seat general-purpose helicopter in

service with the British Army, and is used in a variety of rôles. It is powered by a Nimbus shaft-turbine giving it a maximum speed of 131 mph, and its skid undercarriage enables it to land on any solid surface. It has the same basic airframe as the Westland Wasp, but lacks many of the extra features of the latter aircraft.

Westland Wasp, 579, 583, 588

The Wasp is a light naval helicopter, developed by Westland from an original design by Saunders-Roe to serve on board the Royal Navy's anti-submarine frigates. With a 710-hp Nimbus engine, it can carry two homing torpedoes. It is in service on board British frigates and has also been supplied to South Africa, Brazil, New Zealand and the Netherlands for similar duties.

Westland Wessex (helicopter), 967

Under licence agreement the Westland Aircraft Co Ltd has been building Sikorsky designs since 1947 and adapting them to British use. One of the forms this has taken has been to re-engine them with British light-weight turbine engines. The Westland Wessex is the Sikorsky S-58 built under licence and given turbine power. First installation was of one Napier Gazelle shaft-turbine giving 1,450 shp. This version is in service with the Royal Navy and the Royal Australian Navy as an anti-submarine aircraft. The second installation was of two coupled 1,350-shp Bristol Siddeley Gnome shaft-turbines giving an increase in overall power and conferring the advantages of twin-engined safety and capability, including an increased range by cruising on one engine. This version serves with the Royal Air Force in the assault rôle and with

the Royal Navy as a commando assault helicopter. It is also in small-scale use with civil operators as the Series 60, and has been exported to Iraq, Ghana and Brunei.

Westland Whirlwind (helicopter), 967

Under its licence agreement with Sikorsky, Westland obtained agreement to build the Sikorsky S-55 helicopter under licence in 1953. At first the Whirlwind, as it was called, was identical to the S-55; later it was re-engined with British engines. The Series 2 version received Alvis Leonides piston engines and the Series 3 the 1,050-shp Bristol Siddeley Gnome shaft-turbine.

Almost all surviving Whirlwinds have been or are being converted to Gnome power. The type has been used by the Royal Navy for anti-submarine, commando assault and search and rescue duties, and serves with the Royal Air Force in the close-support and search and rescue rôles. In addition, many have been sold abroad. Its main cabin can hold up to eight passengers or a variety of equipment, and the Gnome-powered version has a maximum speed of 106 mph, with a range of 300 miles.

Wheels, see Aircraft wheels
Whirlwind engine, see Wright
Whirlwind helicopter, see Westland
Whitley, see Armstrong Whitworth
Whitten Brown, Lieutenant Arthur, 253, 255, 813
Whittle, Sir Frank H., 287, 289, 295, 1127, 1129ff, 1191

Frank Whittle was born at Coventry in 1907, and joined the RAF as an apprentice in 1923. In 1926 he graduated to the Royal Air Force College, Cranwell, as an officer cadet. Even while there, he was turning his thoughts to jet propulsion, and he

wrote a thesis which incorporated his ideas. His next post was as a trainee flying instructor at the Central Flying School, and here he crystallized ideas about using a turbine to produce the necessary jet for propulsion. His Commanding Officer encouraged him to put forward his ideas to the Air Ministry—which turned them down. He then approached industry, and they too lacked interest; however, in 1935 private financial sources became interested in his ideas and, while studying for a Tripos at Cambridge, he laid out plans for his first jet-propulsion turbine.

Sir Frank Whittle

The following year a company entitled Power Jets (*q.v.*) was formed, and the Air Ministry allowed Whittle to be seconded to the company for special duty. Its first engine was manufactured by the British Thomson Houston Co, which was also working on industrial turbines at the time. The first test-run of his first engine took place on 12 June 1937, and a programme of test-runs eventually so interested the Air Ministry as to evoke a contract for the W.1 engine for flight trials. This was the breakthrough which enabled Whittle to go forward and, after initial talks with the Gloster Aircraft Co, a contract for the first British jet aircraft was signed in 1940. The resultant aircraft, the Gloster E.28/39, carried Whittle's first

jet engine aloft at Cranwell, the very place where he had originally thought out the concept of jet propulsion, on 15 May 1941. Thenceforward Whittle and Power Jets embarked on a priority development programme of engine work which led eventually to the Gloster Meteor jet fighter that entered operational service with Fighter Command in 1944.

Whittle W.1 turbojet, 290, 295, 1135, 1144, 1155-1156, 1215-1216
Widerøe's Flyveselskap

Widerøe's is one of the pioneer Norwegian airlines, having been founded in 1934 and responsible for many inaugural coastal services before World War Two. Since the war it has engaged in a wide variety of non-scheduled activities including ambulance and survey work and charter operations. It also undertakes scheduled passenger services in northern Norway on behalf of SAS (*q.v.*).

Headquarters: Oslo.

Wien Air Alaska

Operating name of Wien Alaska Airlines Inc, formed in 1924 as Northern Air Transport by Sigurd Wien, who is still its president and chairman. It is the oldest scheduled-service airline in the territory, and adopted its present title in 1936.

Headquarters: Fairbanks.

Wiencziers, 567

German monoplane of 1911 embodying the first retractable undercarriage. The undercarriage was of primitive construction, consisting simply of ordinary "legs" fitted with a hinge so that the wheels could be swung rearward to lie flush beside the fuselage.

Wilkins, John, 119
Williams research turbojet, 1223
Wing

Wing lift is provided by a wing having an aerofoil section over which air flows from leading-

Westland Wessex 5 of the Royal Navy: similar twin-Gnome versions serve in Bahrain, Iraq and Ghana

edge to trailing-edge. In an aeroplane the relative motion results from forward motion of a fixed wing attached to a moving aircraft; in a helicopter it is due to rotation of a number of wings (a rotor) about a vertical axis and the "rotary wing" of the helicopter can thus give lift while the aircraft as a whole is at rest. Structurally, a wing is a cantilever beam, fixed at the root and free to deflect vertically at the tip. On the ground it sags downward; in the air it is bent upward by the lift of the air passing over it, and in turbulent air sudden variations in this lift, due to vertical air movements, cause the wing to flap up and down and possibly experience complete reversals of load. For this reason it must be designed not to suffer catastrophic failure as a result of fatigue (*q.v.*) even after 40,000 or more hours of flight. Many modern wings are filled with fuel and incorporate leading-edge slats or Krüger flaps, upper surface spoilers or air brakes and trailing-edge ailerons and flaps. The way a wing behaves aerodynamically is discussed under "Flight, principles of".

Wing-warping, 177-179, 181, 350, 611, 621-622, 625, 629, 630
A method of lateral control of an aircraft whereby stability can be achieved by increasing the angle of incidence of a dropping wing to give it more lift and raise it once more. Originally patented by Mathew Boulton, an Englishman, in 1868, it was the Wright brothers who made the first practical use of it in their gliders and the powered Flyers. It remained a standard method of lateral control for some time even after the aileron had become generally accepted.
It was normally accomplished by wires attached from the control column to the trailing-edge of the wings, so that when the control column was moved in the appropriate direction the trailing-edge of the wing was lowered, thus increasing the camber, and hence the incidence of that wing.

WMO (World Meteorological Association)
International body, sponsored by the United Nations since 1951, whose objective is to facilitate and promote world-wide co-operation in establishing a network of weather-reporting stations and a rapid and mutual exchange of meteorological information. Its headquarters is in Geneva.

World Airways
Supplemental US air transport operator, which was formed in 1948 and which since 1966 has been permanently licensed for world-wide operations. Its major activities include trooping and freight charter for Military Airlift Command and other elements of the USAF.
Headquarters: Oakland, California.

Wright
Wright Aeronautical Corporation, a division of Curtiss-Wright Corporation, has been a producer of aircraft engines for 51 years; but it is no longer the world leader it was between the wars. Early in the nineteen-twenties it produced various Vee-type eight- and twelve-cylinder engines, some of them derived from wartime Hispano-Suiza designs, but the company was put on its feet by a nine-cylinder radial, modified from the original Lawrance J-1, called the Whirlwind. This went into use at 200-220 hp in 1924 and was ultimately developed to over 300 hp. It was a Whirlwind that in 33½ hr carried Lindbergh from New York to Paris, and others flew over the Poles, across the Pacific

and to many other places where aircraft had never been before. In 1925 a 400-hp unit known as the P-1, built for the US Navy, started a line of development that led to the famous Cyclone family, and which continues in production even today as a helicopter engine rated at 1,475-1,525 hp. Early Cyclones, half as powerful, took the DC-2 to Melbourne in 1934 and set many other records; and 1,200-hp Cyclones with turbosuperchargers were the standard engines of the Boeing Fortress. In 1946 the company saw a future in compound piston-engines, gas turbines and ramjets. Wright successfully launched the 3,500-hp Turbo-Compound, the ultimate in piston-engine refinement, which is still serving in Super Constellations, DC-7s and Neptunes; but the size and quality of the firm's research was inadequate to meet the challenge of the new forms of power.

Wright Cyclone engine, 1109, 1117
In 1928, following the success of their Whirlwind radial engine, the Wright Aeronautical Corporation developed a nine-cylinder radial engine at the request of the US Navy for heavy-duty work. This engine, named Cyclone, originally developed 525 hp. Within two years it was delivering over 600 hp, and was in use on three US Navy flying boats and one twin-engined seaplane as well as a civil airliner and two amphibians.
By 1940 a whole family of Cyclones had been developed, ranging from the Cyclone 7 series of 890/900 hp up to the Duplex-Cyclone, a two-row 18-cylinder radial of 2,000 hp claimed to be the most powerful air-cooled engine of its day. With America's entry into World War Two the Cyclone series was in great demand for

military and transport aircraft and continued in production throughout the war.

A Rolls-Royce with distinguished occupants: from the left, Horace Short, The Hon Charles S. Rolls, Orville Wright, Griffith Brewer and Wilbur Wright

Development continued after the war, reaching a climax in the R-3350 Series Cyclones of 2,700 hp and the Turbo-Compound Cyclone engine.
Wright Turbo-Compound engine, 1117, 1119-22 see also Wright Cyclone
Wright Whirlwind engine, 255, 1073
The Wright Whirlwind air-cooled radial engine was produced in the nineteen-twenties by the Wright Aeronautical Corporation of Paterson, New Jersey, and was originally produced in three versions, with five, seven and nine cylinders respectively. In 1927 these engines were used to power five long-distance flights and, their reliability thus proved, they became very popular. The series was extensively refined and developed in the following years, progressing as far as a two-row fourteen-cylinder variant. Most production centred on the single-row versions, and these powered many aircraft during the 'thirties and into World War Two. The engine remained in production throughout this war, and was used principally to power basic training aircraft. Toward the end of the war, production was taken over by Continental Motors, the engine also being used in American tanks and tank-destroyers.

Wright, Orville and Wilbur, 177-181, 193, 217, 609, 611, 620-623, 1033-1038, 1055, 1063

Wright Flyer I, 181, 193, 520, 527, 620-623, 813
Built by the Wright brothers, Orville and Wilbur, it became the first aircraft to accomplish powered, controlled and sustained flight at Kitty Hawk, North Carolina, on 17 December 1903. It was the basis for the subsequent Flyer III of 1905, which was the first practical flying machine in the world and was not surpassed by any European design until several years later.
X-1, see Bell XS-1, 297, 300
X-15, see North American
X-22A, see Bell Aerosystems
X-24A, see Martin
XB-42 Mixmaster, see Douglas
XB-70 Valkyrie, see North American
XC-35, see Lockheed
XC-120 "Packplane", see Fairchild
XC-142A, see LTV-Hiller-Ryan
XFV-1, see Lockheed
XFY-1, see Convair
XF2Y-1, see Convair
XH-51, see Lockheed
XP-42, see Curtiss
XS-1, see Bell
XV-3, see Bell
XV-4 Hummingbird, see Lockheed
XV-5A, see Ryan
Yakovlev, Aleksandir Sergeivich
A. S. Yakovlev is one of the most senior, and also one of the most versatile, of the USSR's aircraft designers; for, where other design

bureaux tend generally to specialize in certain categories (eg, MiG fighters, Kamov helicopters), Yakovlev in his time has produced single-piston-engined fighters, multi-jet fighters, light sporting and training aircraft, small jet airliners and a number of helicopters. Yakovlev's name first came to prominence outside the USSR in the early part of World War Two, when the series of single-seat fighters beginning with the Yak-1 were produced in great numbers. This line was developed through to the Yak-9, which after the war was supplied to several other air forces in the Soviet *bloc*. Even more widespread in use has been the Yak-18 primary trainer: this first appeared in 1946 and has since been extensively developed and built for military flying schools and civil clubs in a score of countries. The latest versions are still in production more than twenty years later, and continue to win international championships for aerobatic flying.
The USSR's standard first-generation all-weather fighter, the twin-jet Yak-25, was also a Yakovlev design, and this too has been developed into more advanced variants, including one for specialized ultra-high altitude reconnaissance. Yakovlev produced the Soviet Air Force's only tandem-rotor helicop-

ter, the Yak-24, and his latest success is the tri-jet Yak-40 short-range transport aircraft. Yakovlev Yak-40 ("Codling"), 837

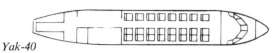

Yak-40

Yakovlev designed the Yak-40 as a short-haul jet transport capable of flying in and out of grass and inferior airfields. It is capable of carrying 24 passengers at cruising speeds around 350 mph for distances under 1,000 miles. Prototypes have been flying for a couple of years and production has now begun for Russian internal services.

Yeager, Charles H., 297, 300

Charles "Chuck" Yeager, a Major in the USAF, made the first faster-than-sound flight on 14 October 1947. The flight was made in the Bell X-1 rocket-powered research aircraft after it had been carried to a height of 30,000 ft under a Superfortress mother-plane. The speed reached on this historic occasion was Mach 1.05, which was held for a few seconds.

Yemen Airlines

Small, Government-controlled airline of the Yemen Republic, which undertakes scheduled services internally and to Aden and Eritrea. Other domestic activities include charter flights and agricultural work.

Headquarters: Taiz.

YF-12A, see Lockheed

York, see Avro

Yugoslavia: Air Force

Jugoslovensko Ratno Vazduhoplovstvo (Yugoslav Air Force). Military aviation in Serbia

(which became a part of Yugoslavia in 1918) dates back to 1912, and Serbian airmen, flying French aircraft, supported several French campaigns during the 1914-18 War. When the new state of Yugoslavia was created after the war an Army Aviation Department was established, and to its Spad VII fighters and Breguet 14 bombers were added Dewoitine D-1s and Breguet 19s purchased from France; meanwhile the newly-created Ikarus factory began to manufacture German Brandenburgs for the training rôle. In 1930 separate commands for military and naval aviation were instituted, and within the next few years most of the aircraft in Air Force or Navy service were Yugoslav-built, the principal combat types being of French or Czechoslovak design. In the mid-'thirties, fighter units were modernized with purchased French D.500s and British Hawker Furies, the latter also being locally-built with Hispano-Suiza engines. The first Yugoslav-designed fighters, the Ikarus IK-1 and IK-2, appeared at about this time, although they were not built in large numbers. Bristol Blenheim, Dornier Do 17K and Savoia-Marchetti S.M.79 bombers were acquired in the late 'thirties, and in 1938 the prototype of the nationally-designed IK-3 fighter made its first flight.

After the outbreak of World War Two, Germany supplied Bf 109 fighters to the pro-Axis Royalist government, but after this was deposed in March 1941 she and Italy attacked Yugoslavia, dividing it into Croatian and

Yugoslavia's most successful jet design, the Viper-powered Soko Galeb

Serbian zones. The Yugoslav Air Force was disbanded, being replaced by a Croatian Air Force equipped with German and Italian aircraft. To counter this action, increasing Allied support was given to the partisan forces led by Marshal Tito, and the squadrons thus formed (with US and British aircraft) provided the basis for a new Air Force after the country was liberated late in 1944.

Yugoslavia became a republic late in November 1945, since when, although a Communist state, it has remained independent of the Soviet *bloc* and has never become a satellite of the USSR. Hence, in spite of initial re-equipment after the war with Soviet aircraft, it remained free in 1951 to accept British and US aircraft under a mutual assistance treaty. With this aid, the domestic industry was also re-established and has produced several air-

craft of national design in the ensuing years. Latest of these are the Galeb and Jastreb jet trainer and light attack types, and the Kraguj piston-engined counter-insurgency aircraft. These are in current service alongside Soviet MiG-21s and American Sabres.

Zambia Airways Corporation

The original Zambia Airways was formed in 1963 as a subsidiary of Central African Airways Corporation. On 1 September 1967 this was superseded by the present State-sponsored Corporation which operates with some assistance from Alitalia (*q.v.*). Services are mainly internal.

Headquarters: Lusaka. Member of IATA.

Zeppelin, Ferdinand von, 133, 241

Count Ferdinand von Zeppelin conceived the airship as a large military vehicle, and in due course his ideas came to fruition in the 1914-18 War. The first rigid airship with a metal framework appeared in 1900, first flight being on 2 July. A decade later, five had been built, the last one (L.Z.5) making a cross-country journey of over 900 miles. By the outbreak of war twenty-five had been built, and

The Graf Zeppelin *above her bigger and newer sister, the* Hindenburg

an airline company, Delag, formed. Another 88 were built during the war and made over fifty bombing raids on England. After the war Zeppelin continued to develop airships, his most famous being the *Graf Zeppelin* which flew in airline service for nine years (1928-37), making 651 flights and carrying nearly 14,000 passengers; one trip was a round-the-world voyage, 7,000 miles of which was non-stop. The last

Zeppelin was the *Hindenburg*, which caught fire at Lakehurst, New Jersey, in 1937 and virtually ended the Zeppelin line.

Zeppelin, Graf, 147, 253, 725; see also Zeppelin, Ferdinand von

Zeppelin Z.IX, 861

Destroyed in a bombing raid on Dusseldorf on 8 October 1914 by a Royal Navy Sopwith Tabloid, the Z.IX was the first military target destroyed by bombing in the 1914-18 War.

Zero reader, 703

Zürich Airport (Kloten), Switzerland

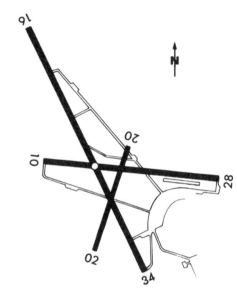

Kloten dates from before World War Two, though it is in post-war years that it has expanded to become one of the principal European traffic centres. It is also the operating base of Swissair, the national airline.
Distance from city centre: 7½ miles
Height above sea level: 1,416 ft
Main runways:
  16/34 = 12,137 ft
  10/28 = 8,202 ft
  02/20 = 5,035 ft
Passenger movements in 1967: 3,104,000